T0181525

Lecture Notes in Computer Science 13265

More information about this series at https://link.springer.com/bookseries/558

Karen Aardal · Laura Sanità (Eds.)

Integer Programming and Combinatorial Optimization

23rd International Conference, IPCO 2022
Eindhoven, The Netherlands, June 27–29, 2022
Proceedings

Editors
Karen Aardal 🄳
Delft University of Technology
Delft, The Netherlands

Laura Sanità
Eindhoven University of Technology
Eindhoven, The Netherlands

ISSN 0302-9743   ISSN 1611-3349 (electronic)
Lecture Notes in Computer Science
ISBN 978-3-031-06900-0   ISBN 978-3-031-06901-7 (eBook)
https://doi.org/10.1007/978-3-031-06901-7

This Springer imprint is published by the registered company Springer Nature Switzerland AG
The registered company address is: Gewerbestrasse 11, 6330 Cham, Switzerland

Preface

This volume collects the 33 extended abstracts presented at IPCO 2022, the 23rd Conference on Integer Programming and Combinatorial Optimization, held during June 27–29, 2022, in Eindhoven (The Netherlands).

IPCO is under the auspices of the Mathematical Optimization Society, and it is an important forum for presenting the latest results on the theory and practice of the various aspects of discrete optimization. The first IPCO conference took place at the University of Waterloo in May 1990, and the Eindhoven University of Technology organized the 23rd such event. The conference had a Program Committee consisting of 15 members. In response to the Call for Papers, we received 93 submissions. Each submission was reviewed by at least three Program Committee members. Because of the limited number of time slots for presentations, many excellent submissions could not be accepted. The page limit for contributions to this proceedings was set to 14. We expect the full versions of the extended abstracts appearing in this Lecture Notes in Computer Science volume to be submitted for publication in refereed journals, and a special issue of Mathematical Programming Series B containing such versions is in process.

For the third time, IPCO had a Best Paper Award. The IPCO 2022 Best Paper Award was given to Gennadiy Averkov and Matthias Schymura for their paper On the maximal number of columns of a Δ -modular matrix. This year, IPCO was preceded by a Summer School held during June 25–26, 2022, with lectures by Shipra Agrawal (Columbia University), Shayan Oveis Gharan (University of Washington), and Stefan Weltge (TU Munich). We thank them warmly for their contributions. We would also like to thank

- the authors who submitted their research to IPCO;
- the members of the Program Committee, who spent much time and energy reviewing the submissions;
- the expert additional reviewers whose opinions were crucial in the paper selection;
- the members of the Local Organizing Committee, who made this conference possible;
- the Mathematical Optimization Society and in particular the members of its IPCO Steering Committee, Oktay Günlük, Jochen Könemann, and Giacomo Zambelli, for their help and advice;
- EasyChair for making paper management simple and effective; and
- Springer for their efficient cooperation in producing this volume and for financial support for the Best Paper Award.

We would further like to thank the following sponsors for their financial support: NETWORKS, DIAMANT, NWO, Eurandom, Cardinal Operations, Google, Gurobi, Mosek, and the Optimization firm.

March 2022

Karen Aardal
Laura Sanità

Organization

Program Committee

Karen Aardal (Chair)	Delft University of Technology, The Netherlands
Marco Di Summa	University of Padua, Italy
Alina Ene	Boston University, USA
Samuel Fiorini	Université libre de Bruxelles, Belgium
Ricardo Fukasawa	University of Waterloo, Canada
Yusuke Kobayashi	Kyoto University, Japan
Jochen Koenemann	University of Waterloo, Canada
Nicole Megow	Universität Bremen, Germany
Gonzalo Muñoz	Universidad de O'Higgins, Chile
Neil Olver	London School of Economics and Political Science, UK
Sebastian Pokutta	Georgia Institute of Technology, USA
Domenico Salvagnin	University of Padua, Italy
Laura Sanita	Eindhoven University of Technology, The Netherlands
David Shmoys	Cornell University, USA
Vera Traub	ETH Zurich, Switzerland

Additional Reviewers

Abdi, Ahmad
Abels, Andreas
Anegg, Georg
Antoniadis, Antonios
Bhore, Sujoy
Bienkowski, Marcin
Borgwardt, Steffen
Braun, Gábor
Cao, Yixin
Caoduro, Marco
Carvalho, Margarida
Cen, Ruoxu
Chakrabarti, Amit
Chakrabarty, Deeparnab
Chalermsook, Parinya
Chandrasekaran, Karthekeyan
Chang, Yi-Jun

Chen, Chen
Chmiela, Antonia
Clautiaux, François
Coester, Christian
Cohen, Ilan
Cornuejols, Gerard
Cristi, Andrés
Dash, Sanjeeb
De Loera, Jesus
Del Pia, Alberto
Dey, Santanu
Dubey, Yatharth
Dughmi, Shaddin
Eberle, Franziska
Efstratios, Skoulakis
Eisenbrand, Friedrich
Englert, Matthias

Friggstad, Zachary
Fujii, Kaito
Fukunaga, Takuro
Gairing, Martin
Gallato, Martina
Georgiou, Konstantinos
Gijswijt, Dion
Gleixner, Ambros
Goez, Julio C.
Gomez, Andres
Goycoolea, Marcos
Grandoni, Fabrizio
Gálvez, Waldo
Halldórsson, Magnús M.
Harks, Tobias
Hertrich, Christoph
Hirai, Hiroshi
Hojny, Christopher
Hommelsheim, Felix
Huang, Zhiyi
Huchette, Joey
Huiberts, Sophie
Hunkenschröder, Christoph
Husic, Edin
Huynh, Tony
Hyatt-Denesik, Dylan
Ibrahimpur, Sharat
Im, Sungjin
Jansen, Klaus
Kafer, Sean
Kale, Sagar
Kasperski, Adam
Kazachkov, Aleksandr M.
Kell, Nathaniel
Kesselheim, Thomas
Khalil, Elias
Klein, Kim-Manuel
Klein, Nathan
Konrad, Christian
Koster, Arie
Krumke, Sven
Kurpisz, Adam
Kurtz, Jannis
Le Bodic, Pierre
Lee, Jon

Letchford, Adam
Levin, Roie
Linderoth, Jeff
Livanos, Vasilis
Ljubic, Ivana
Lodi, Andrea
Lubin, Miles
Ma, Will
Manthey, Bodo
Marandi, Ahmadreza
Martinez-Rubio, David
Michini, Carla
Mnich, Matthias
Molinaro, Marco
Moran, Diego
Murota, Kazuo
Mömke, Tobias
Nagarajan, Viswanath
Nannicini, Giacomo
Newman, Alantha
Nimbhorkar, Prajakta
Nägele, Martin
Oosterwijk, Tim
Paat, Joseph
Padmanabhan, Swati
Pap, Gyula
Papadigenopoulos, Orestis
Peis, Britta
Polak, Adam
Quanrud, Kent
Raichel, Benjamin
Ravi, R.
Rohwedder, Lars
Ryan, Chris
Röglin, Heiko
Salavatipour, Mohammad
Santiago, Richard
Schramm, Tselil
Schwiegelshohn, Chris
Serra, Thiago
Serrano, Felipe
Sharma, Kartikey
Shioura, Akiyoshi
Singla, Sahil
Sinnl, Markus

Sofranac, Boro
Soto, José A.
Spiegel, Christoph
Sumita, Hanna
Swamy, Chaitanya
Tardos, Jakab
Torrico, Alfredo
Touitou, Noam
Tsepenekas, Leonidas
Vaish, Rohit
Van Dyk, Madison
van Emde Boas, Peter
Van Vyve, Mathieu
van Zuylen, Anke

Vargas, Luis
Verdugo, Victor
Verschae, José
Vredeveld, Tjark
Wang, Guanyi
Ward, Justin
Weltge, Stefan
Wiese, Andreas
Wlodarczyk, Michal
Wu, Xiaowei
Xiao, Mingyu
Xu, Yibo
Yang, Sheng
Yuditsky, Yelena

Sponsors

Eurandom

Workshop Centre in the area of Stochastics

Contents

Total Dual Dyadicness and Dyadic Generating Sets

Ahmad Abdi[1(✉)], Gérard Cornuéjols[2], Bertrand Guenin[3], and Levent Tunçel[3]

[1] Department of Mathematics, LSE, London, UK
a.abdi1@lse.ac.uk
[2] Tepper School of Business, Carnegie Mellon University, Pittsburgh, USA
gc0v@andrew.cmu.edu
[3] Department of Combinatorics and Optimization, University of Waterloo, Waterloo, Canada
{bguenin,ltuncel}@uwaterloo.ca

Abstract. A vector is *dyadic* if each of its entries is a dyadic rational number, i.e. of the form $\frac{a}{2^k}$ for some integers a, k with $k \geq 0$. A linear system $Ax \leq b$ with integral data is *totally dual dyadic* if whenever $\min\{b^\top y : A^\top y = w, y \geq \mathbf{0}\}$ for w integral, has an optimal solution, it has a dyadic optimal solution. In this paper, we study total dual dyadicness, and give a co-NP characterization of it in terms of *dyadic generating sets for cones and subspaces*, the former being the dyadic analogue of *Hilbert bases*, and the latter a polynomial-time recognizable relaxation of the former. Along the way, we see some surprising turn of events when compared to total dual integrality, primarily led by the *density* of the dyadic rationals. Our study ultimately leads to a better understanding of total dual integrality and polyhedral integrality. We see examples from dyadic matrices, T-joins, circuits, and perfect matchings of a graph.

1 Introduction

A *dyadic rational* is a number of the form $\frac{a}{2^k}$ for some integers a, k where $k \geq 0$. The dyadic rationals are precisely the rational numbers with a finite binary representation, and are therefore relevant for (binary) floating-point arithmetic in numerical computations. Modern computers represent the rational numbers by fixed-size floating points, inevitably leading to error terms, which are compounded if serial arithmetic operations are performed such as in the case of mixed-integer linear, semidefinite, and more generally convex optimization. This has led to an effort to mitigate floating-point errors [27] as well as the need for exact solvers [6, 25].

We address a different, though natural theoretical question: *When does a linear program admit an optimal solution whose entries are dyadic rationals?* A vector is *dyadic* if every entry is a dyadic rational. Consider the following primal dual pair of linear programs for $A \in \mathbb{Z}^{m \times n}, b \in \mathbb{Z}^m$ and $w \in \mathbb{Z}^n$.

$$(P) \quad \max\{w^\top x : Ax \leq b\} \qquad (D) \quad \min\{b^\top y : A^\top y = w, y \geq \mathbf{0}\}.$$

© Springer Nature Switzerland AG 2022
K. Aardal and L. Sanitá (Eds.): IPCO 2022, LNCS 13265, pp. 1–14, 2022.
https://doi.org/10.1007/978-3-031-06901-7_1

(**0** and **1** denote respectively the all-zeros and all-ones column, or row, vectors of appropriate dimension.) When does (D) admit a dyadic optimal solution for all $w \in \mathbb{Z}^n$? How about (P)? Keeping close to the integral case, these questions lead to the notions of *totally dual dyadic* systems and *dyadic polyhedra*. In this paper, we reassure the reader that dyadic polyhedra enjoy a similar characterization as *integral polyhedra*, but in studying totally dual dyadic systems, we see an intriguing and somewhat surprising turn of events when compared to *totally dual integral (TDI)* systems [10]. As such, we shall keep the focus of the paper on total dual dyadicness and its various characterizations. The characterizations lead to *dyadic generating sets* for cones and subspaces, where the first notion is polyhedral and can be thought of as a dyadic analogue of *Hilbert bases*, while the second notion is lattice-theoretic and new. We shall see some intriguing examples of totally dual dyadic systems and dyadic generating sets from Integer Programming, Combinatorial Optimization, and Graph Theory. Our study eventually leads to a better understanding of TDI systems and integral polyhedra.

Our characterizations extend easily to the *p-adic rationals* for any prime number $p \geq 3$. For this reason, we shall prove our characterizations in the general setting. Interestingly, however, most of our examples do *not* extend to the p-adic setting for $p \geq 3$.

1.1   Totally Dual p-adic Systems and p-adic Generating Sets

Let $p \geq 2$ be a prime number. A *p-adic rational* is a number of the form $\frac{a}{p^k}$ for some integers a, k where $k \geq 0$. A vector is *p-adic* if every entry is a p-adic rational. Consider a linear system $Ax \leq b$ where $A \in \mathbb{Z}^{m \times n}, b \in \mathbb{Z}^m$. We say that $Ax \leq b$ is *totally dual p-adic* if for all $w \in \mathbb{Z}^n$ for which $\min\{b^\top y : A^\top y = w, y \geq \mathbf{0}\}$ has an optimum, it has a p-adic optimal solution. For $p = 2$, we abbreviate 'totally dual dyadic' as 'TDD'. We prove the following characterization, which relies on two key notions defined afterwards.

Theorem 1 (Proved in Sect. 4). *Let $A \in \mathbb{Z}^{m \times n}, b \in \mathbb{Z}^m$ and $P := \{x : Ax \leq b\}$. Given a nonempty face F, denote by $A_F x \leq b_F$ the subsystem of $Ax \leq b$ corresponding to the implicit equalities of F. Then the following statements are equivalent for every prime p: (1) $Ax \leq b$ is totally dual p-adic, (2) for every nonempty face F of P, the rows of A_F form a p-adic generating set for a cone, (3) for every nonempty face F of P, the rows of A_F form a p-adic generating set for a subspace.*

In fact, in (2), it suffices to consider only the minimal nonempty faces.

Let $\{a^1, \ldots, a^n\} \subseteq \mathbb{Z}^m$. The set $\{a^1, \ldots, a^n\}$ is a *p-adic generating set for a cone (p-GSC)* if every integral vector in the conic hull of the vectors can be expressed as a p-adic conic combination of the vectors (meaning that the coefficients used are p-adic). In contrast, $\{a^1, \ldots, a^n\}$ is a *p-adic generating set for a subspace (p-GSS)* if every integral vector in the linear hull of the vectors can be expressed as a p-adic linear combination of the vectors. For $p = 2$, we use the acronyms DGSC and DGSS instead of 2-GSC and 2-GSS, respectively.

The careful reader may notice that an *integral* generating set for a cone is just a *Hilbert basis* [13] (following [21], Sect. 22.3). In contrast with Hilbert bases where a satisfying characterization remains elusive, we have the following polyhedral characterization of a p-GSC:

Theorem 2 (Proved   in   Sect. 3).   *Let $\{a^1, \ldots, a^n\} \subseteq \mathbb{Z}^m$, $C := \mathrm{cone}\{a^1, \ldots, a^n\}$, and p a prime. Then $\{a^1, \ldots, a^n\}$ is a p-GSC if, and only if, for every nonempty face F of C, $\{a^i : a^i \in F\}$ is a p-GSS.*

The careful reader may notice that in contrast to total dual integrality, the characterization of totally dual p-adic systems, Theorem 1, enjoys a third equivalent condition, namely (3). This new condition, as well as the characterization of a p-GSC, Theorem 2, is made possible due to a distinguishing feature of the p-adic rationals: *density*. The p-adic rationals, as opposed to the integers, form a dense subset of $\mathbb{R}$. We shall elaborate on this in Sect. 2.

Going further, we have the following lattice-theoretic characterization of a p-GSS. We recall that the *elementary divisors* (a.k.a. *invariant factors*) of an integral matrix are the nonzero entries of the *Smith normal form* of the matrix; see Sect. 3 for more.

Theorem 3 (Proved in Sect. 3). *The following statements are equivalent for a matrix $A \in \mathbb{Z}^{m \times n}$ of rank r and every prime p: (1) the columns of A form a p-GSS, (2) the rows of A form a p-GSS, (3) whenever $y^\top A$ and Ax are integral, then $y^\top A x$ is a p-adic rational, (4) every elementary divisor of A is a power of p, (5) the GCD of the subdeterminants of A of order r is a power of p, (6) there exists a matrix B with p-adic entries such that $ABA = A$.*

Theorem 3 is used in Sect. 3 to prove that testing the p-GSS property can be done in polynomial time. Subsequently, the problem of testing total dual p-adicness belongs to co-NP by Theorem 1 (see Sect. 4), and the problem of testing the p-GSC property belongs to co-NP by Theorem 2 (see Sect. 3). Whether the two problems belong to NP, or P, remains unsolved. It should be pointed out that testing total dual *integrality*, as well as testing the Hilbert basis property, is co-NP-complete [9, 19].

1.2   Connection to Integral Polyhedra and TDI Systems

Our characterizations stated so far, as well as our characterization of p-adic polyhedra explained in Sect. 5, have the following intriguing consequence:

Theorem 4. *Let $A \in \mathbb{Z}^{m \times n}$, $b \in \mathbb{Z}^m$, and $P := \{x : Ax \leq b\}$. Then the following are equivalent: (1) $Ax \leq b$ is totally dual p-adic for all primes p, (2) $Ax \leq b$ is totally dual p- and q-adic, for distinct primes p, q, (3) for every nonempty face F of P, the GCD of the subdeterminants of A_F of order $\mathrm{rank}(A_F)$ is 1.*

Proof. **(1)** $\Rightarrow$ **(2)** is immediate. **(2)** $\Rightarrow$ **(3)** For every nonempty face F of P, the rows of A_F form both a p- and a q-GSS by Theroem 1, so the GCD of the subdeterminants of A_F of order $\mathrm{rank}(A_F)$ is both a power of p and a power of q by Theorem 3, so the GCD of the subdeterminants of A_F of order $\mathrm{rank}(A_F)$ must be 1. **(3)** $\Rightarrow$ **(1)** follows from Theorem 1 and Theorem 3 $\square$

If $Ax \leq b$ is TDI, and therefore totally dual p-adic for any prime p, then statement (3) above must hold (this is folklore, and explored in [22]. In fact, if P is pointed, then for every vertex of P, we have a stronger property known as *local strong unimodularity* [12].) It was a widely known fact that the converse is not true. Theorem 4 clarifies this further by equating (3) with (1) and (2). Going a step further, it is known that if $Ax \leq b$ is TDI, then $\{x : Ax \leq b\}$ is an integral polyhedron [10, 13]. We shall strengthen this result as follows:

Theorem 5 (Proved in Sect. 5). *If $Ax \leq b$ is totally dual p- and q-adic, for distinct primes p, q, then $\{x : Ax \leq b\}$ is an integral polyhedron.*

Fulkerson's theorem that every integral set packing system is TDI, can be seen as a (stronger) converse to Theorem 5 [11]. As for set covering systems, there is a conjecture of Paul Seymour that predicts a (stronger) converse to Theorem 5.

Conjecture 6 (The Dyadic Conjecture [20], Sect. 79.3e). Let A be a matrix with $0, 1$ entries. If $Ax \geq 1, x \geq 0$ defines an integral polyhedron, then it is TDD.

The step of the Dyadic Conjecture: If $Ax \geq 1, x \geq 0$ defines an integral polyhedron, then for every nonnegative integral w such that $\min\{w^{\top}x : Ax \geq 1, x \geq 0\}$ has optimal value two, the dual has a dyadic optimal solution [1].

1.3   Examples

Our first example comes from Integer Programming, and more precisely, from matrices with restricted subdeterminants.

Theorem 7. *Let $A \in \mathbb{Z}^{m \times n}$ be a matrix whose subdeterminants belong to $\{0\} \cup \{\pm p^k : k \in \mathbb{Z}_+\}$ for some prime p, and let $b \in \mathbb{Z}^m$. Then $Ax \leq b$ is totally dual p-adic.*

Similar, if not identical, settings have been studied previously; see for example [4, 16] (the last reference has more relevant citations); see the full version for more details and the proof [2]. The node-edge incidence matrix of a graph is known to satisfy the hypothesis for $p = 2$ (folklore), and therefore leads to a TDD system. More generally, matrices whose subdeterminants belong to $\{0\} \cup \{\pm 2^k : k \in \mathbb{Z}_+\}$ have been studied from a matroid theoretic perspective; matroids representable over the rationals by such matrices are known as *dyadic matroids* and their study was initiated by Whittle [28].

Moving on, from Combinatorial Optimization, we get examples only in the dyadic setting. Let $G = (V, E)$ be a graph, and T a nonempty subset of even cardinality. A *T-join* is a subset $J \subseteq E$ such that the odd-degree vertices of $G[J]$ is precisely T. T-joins were studied due to their connection to the *minimum weight perfect matching problem*, but also to the *Chinese postman set problem* (see [7], Chap. 5). As a consequence of a recent result [3], we obtain the following.

Theorem 8 (Proved in Sect. 6). *Let $G = (V, E)$ be a graph, and $T \subseteq V$ a nonempty subset of even cardinality. Then $x(J) \geq 1 \ \forall \ T$-joins $J; x \geq 0$ is TDD.*

The basic solutions to the dual of $\min\{\mathbf{1}^\top x : x(J) \geq 1 \ \forall\ T\text{-joins } J; x \geq \mathbf{0}\}$ may actually be non-dyadic, with many examples coming from snarks G on at least 18 vertices with $T = V(G)$ and $w = \mathbf{1}$ [18], thereby creating an interesting contrast between the proofs of Theorem 7 and Theorem 8. Also, Theorem 8 does not extend to the p-adic setting for any prime $p \geq 3$. To see this, let G be the graph with vertices $1, 2, 3, 4, 5$ and edges $\{1, 3\}, \{1, 4\}, \{1, 5\}, \{3, 2\}, \{4, 2\}, \{5, 2\}$, let $T := \{1, 2, 3, 4\}$, and let $w := \mathbf{1}$. Then the dual has a unique optimal solution, namely $y^\star = \frac{1}{2} \cdot \mathbf{1}$, which is not p-adic for any $p \geq 3$.

The system in Theorem 8 defines an integral set covering polyhedron (see [8], Chap. 2), so Theorem 8 verifies Conjecture 6 for such instances. In fact, it has been conjectured that the system in Theorem 8 is totally dual *quarter-integral* ([8], Conjecture 2.15). Observe that quarter-integrality is a stronger variant of dyadicness, and should not be confused with "4-adicness", which is not even defined in this paper.

Moving on, let $G = (V, E)$ be a graph. A *circuit* is a nonempty subset $C \subseteq E$ such that the subgraph $(V(C), C)$ is connected where every vertex has degree two. A *perfect matching* is a subset $M \subseteq E$ such that every vertex in V is incident with exactly one edge in M. Define $\mathbf{C}(G) := \{\chi_C : C \text{ a circuit of } G\}$ and $\mathbf{M}(G) := \{\chi_M : M \text{ a perfect matching of } G\}$. See [14] for an excellent survey on lattice and conic characterizations of these two sets.

Theorem 9 (Proved in Sect. 6). *Let $G = (V, E)$ be a graph such that $|V|$ is even. Then $\mathbf{M}(G)$ is a DGSC.*

This theorem does not extend to the p-adic setting for $p \geq 3$ either; this is justified in Sect. 6. If G is an r-*graph*, then the *Generalized Berge-Fulkerson Conjecture* [23] predicts that the all-ones vector can be written as a *half-integral* conic combination of $\mathbf{M}(G)$; Theorem 9 proves this can be done *dyadically*.

Theorem 10. *Let $G = (V, E)$ be a graph. Then $\mathbf{C}(G)$ is a DGSC.* □

If G is bridgeless, then the *Cycle Double Cover Conjecture* [24, 26] predicts that the all-ones vector can be written as a *half-integral* conic combination of the vectors in $\mathbf{C}(G)$; Theorem 10 implies this can be done *dyadically*. The theorem is proved in the full version [2], and uses Theorem 1 and interestingly the notion of *cuboids* [1]. There, we also note that the theorem does not extend to the p-adic setting for $p \geq 3$.

2 Density Lemma and the Theorem of the Alternative

Many of our results are made possible by an important feature of the p-adic rationals distinguishing them from the integers, namely *density*.

Remark 11. The p-adic rationals form a dense subset of $\mathbb{R}$.

Lemma 12 (Density Lemma). *Let $A \in \mathbb{Z}^{m \times n}, b \in \mathbb{Z}^m$ and p a prime. If $\{x : Ax = b\}$ contains a p-adic point, then the p-adic points in the set form a dense subset. In particular, a nonempty rational polyhedron contains a p-adic point if, and only if, its affine hull contains a p-adic point.*

Proof. It suffices to prove the first statement. Suppose $\{x : Ax = b\}$ contains a p-adic point, say $\hat{x}$. Since A has integral entries, its kernel has an integral basis, say $d^1, \ldots, d^r$. Observe that $\{x : Ax = b\}$ is the set of vectors of the form $\hat{x} + \sum_{i=1}^{r} \lambda_i d^i$ where $\lambda \in \mathbb{R}^r$. Consider the set $S := \{\hat{x} + \sum_{i=1}^{r} \lambda_i d^i : \lambda_i \text{ is } p\text{-adic for each } i\}$. By Remark 11, it can be readily checked that S is a dense subset of $\{x : Ax = b\}$. Since $\hat{x}$ is p-adic, and the d^i's are integral, the points in S are p-adic, thereby proving the lemma.   □

A natural follow-up question arises: When does a rational subspace contain a p-adic point? Addressing this question requires a familiar notion in Integer Programming. Every integral matrix of full row rank can be brought into *Hermite normal form* by means of *elementary unimodular column operations*. In particular, if A is an integral $m \times n$ matrix of full row rank, there exists an $n \times n$ *unimodular* matrix U such that $AU = (B\ \mathbf{0})$, where B is a non-singular $m \times m$ matrix, and $\mathbf{0}$ is an $m \times (n-m)$ matrix with zero entries. By a square unimodular matrix, we mean a square integral matrix whose determinant is ± 1; note that the inverse of such a matrix is also unimodular. See ([5], Sect. 1.5.2) or ([21], Chap. 4) for more details.

Lemma 13 (Theorem of the Alternative). *Let $A \in \mathbb{Z}^{m \times n}, b \in \mathbb{Z}^m$, and p a prime. Then either $Ax = b$ has a p-adic solution, or there exists a $y \in \mathbb{R}^m$ such that $y^\mathsf{T} A$ is integral and $y^\mathsf{T} b$ is non-p-adic, but not both.*

Proof. Suppose $A\hat{x} = b$ for a p-adic point $\hat{x}$, and $y^\mathsf{T} A$ is integral. Then $y^\mathsf{T} b = y^\mathsf{T}(A\hat{x}) = (y^\mathsf{T} A)\hat{x}$ is an integral linear combination of p-adic rationals, and is therefore a p-adic rational. Thus, both statements cannot hold simultaneously. Suppose $Ax = b$ has no p-adic solution. If $Ax = b$ has no solution at all, then there exists a vector y such that $y^\mathsf{T} A = \mathbf{0}$ and $y^\mathsf{T} b \neq 0$; by scaling y appropriately, we can ensure that $y^\mathsf{T} b$ is non-p-adic, as desired. Otherwise, $Ax = b$ has a solution. We may assume that A has full row rank. Then there exists a square unimodular matrix U such that $AU = (B\ \mathbf{0})$, where B is a non-singular matrix. Observe that $\{x : Ax = b\} = \{Uz : AUz = b\}$. Thus, as $Ax = b$ has no p-adic solution x, and U has integral entries, we may conclude that the system $AUz = b$ has no p-adic solution z either. Let us expand the latter system. Let I, J be the sets of column labels of $B, \mathbf{0}$ in $AU = (B\ \mathbf{0})$, respectively. Then $\{z : AUz = b\} = \{z : (B\ \mathbf{0})\left(\begin{smallmatrix} z_I \\ z_J \end{smallmatrix}\right) = b\} = \{z : Bz_I = b, z_J \text{ free}\} = \{z : z_I = B^{-1}b, z_J \text{ free}\}$. In particular, since $AUz = b$ has no p-adic solution, the vector $B^{-1}b$ is non-p-adic. Thus, there exists a row y^T of B^{-1} for which $y^\mathsf{T} b$ is non-p-adic. We claim that $y^\mathsf{T} A$ is integral, thereby showing y is the desired vector. To this end, observe that $B^{-1} AU = B^{-1}(B\ \mathbf{0}) = (I\ \mathbf{0})$, implying in turn that $B^{-1} A = (I\ \mathbf{0})U^{-1}$. As the inverse of a square unimodular matrix, U^{-1} is also unimodular and therefore has integral entries, implying in turn that $B^{-1}A$, and so $y^\mathsf{T} A$, is integral.   □

The reader may notice a similarity between Lemma 13 and its integer analogue, which characterizes when a linear system of equations admits an integral solution, commonly known as the *Integer Farkas Lemma* (see [5], Theorem 1.20). We refrain from calling Lemma 13 the "p-adic Farkas Lemma" as we reserve that title for Corollary 16 below.

Remark 14. If t is a p- and q-adic rational, for distinct primes p, q, then t is integral.

Corollary 15. *Let $A \in \mathbb{Z}^{m \times n}, b \in \mathbb{Z}^m$. If $Ax = b$ has p- and q-adic solutions, for distinct primes p and q, then the system has an integral solution.*

Proof. By the Theorem of the Alternative, whenever $y^\top A$ is integral, $y^\top b$ is both p- and q-adic, implying in turn that $y^\top b$ is integral by Remark 14. Thus, by the Integer Farkas Lemma, $Ax = b$ has an integral solution. $\qquad\square$

Finally, the Density Lemma and the Theorem of the Alternative have the following p-adic analogue of Farkas Lemma in Linear Programming.

Corollary 16 (p-Adic Farkas Lemma). *Let P be a nonempty rational polyhedron whose affine hull is $\{x : Ax = b\}$, where A, b are integral. Then for every prime p, P contains a p-adic point if, and only if, there does not exist y such that $y^\top A$ is integral and $y^\top b$ is non-p-adic.*

3 p-Adic Generating Sets for Subspaces and Cones

Recall that a set of vectors $\{a^1, \ldots, a^n\} \subseteq \mathbb{Z}^m$ forms a p-GSS if every integral vector in the linear hull of the vectors can be expressed as a p-adic linear combination of the vectors. Observe that every p-adic vector in the linear hull of a p-GSS can also be expressed as a p-adic linear combination of the vectors. We prove the following lemma in the full version [2].

Lemma 17. *Let $A \in \mathbb{Z}^{m \times n}$, and U a unimodular matrix of appropriate dimensions. Then (1) the columns of A form a p-GSS if, and only if, the columns of UA do, and (2) the columns of A form a p-GSS if, and only if, the columns of AU do.* $\qquad\square$

In order to prove Theorem 3, we need a definition. Let A be an integral matrix of rank r. It is well-known that by applying elementary row *and* column operations, we can bring A into *Smith normal form*, that is, into a matrix with a leading $r \times r$ minor D and zeros everywhere else, where D is a diagonal matrix with diagonal entries $\delta_1, \ldots, \delta_r \geq 1$ such that $\delta_1 \mid \delta_2 \mid \cdots \mid \delta_r$ (see [21], Sect. 4.4). It can be readily checked that for each $i \in [r]$, $\prod_{j=1}^{i} \delta_j$ is the GCD of the subdeterminants of A of order i. The δ_i's are referred to as the *elementary divisors*, or *invariant factors*, of A.

Proof of Theorem 3. **(1)** $\Leftrightarrow$ **(3)** Suppose (1) holds. Choose x, y such that $y^\top A$ and Ax are integral. Let $b := Ax \in \mathbb{Z}^m$. By (1), there exists a p-adic $\bar{x}$ such that $b = A\bar{x}$. Thus, $y^\top Ax = y^\top A\bar{x} = (y^\top A)\bar{x}$, which is p-adic because $y^\top A$ is integral and $\bar{x}$ p-adic, as required. Suppose conversely that (3) holds. Pick $b \in \mathbb{Z}^m$ such that $A\bar{x} = b$ for some $\bar{x}$. We need to prove that $Ax = b$ has a p-adic solution. If $y^\top A$ is integral, then $y^\top b = y^\top A\bar{x}$, which is p-adic by (3). Thus, by the Theorem of the Alternative, $Ax = b$ has a p-adic solution, as required.

(2) $\Leftrightarrow$ **(3)** holds by applying the established equivalence (1) $\Leftrightarrow$ (3) to $A^\top$.

(1)–(3) $\Leftrightarrow$ **(4)**: By Lemma 17, the equivalent conditions (1)–(3) are preserved under elementary unimodular row/column operations; these operations clearly preserve (4) as well. Thus, it suffices to prove the equivalence between (1)–(3) and (4) for integral matrices in Smith normal form. That is, we may assume that A has a leading $r \times r$ minor D and zeros everywhere else, where D is a diagonal matrix with diagonal entries $\delta_1, \ldots, \delta_r \geq 1$ such that $\delta_1 \mid \delta_2 \mid \cdots \mid \delta_r$. Suppose (1)–(3) hold. We need to show that each δ_i is a power of p. Consider the feasible system $Ax = e_i$; every solution x to this system satisfies $x_j = 0, j \in [r] - \{i\}$ and $x_i = \frac{1}{\delta_i}$. Since the columns of A form a p-GSS, $\frac{1}{\delta_i}$ must be p-adic, so δ_i is a power of p, as required. Suppose conversely that (4) holds. We need to show that whenever $Ax = b, b \in \mathbb{Z}^m$ has a solution, then it has a p-adic solution. Clearly, it suffices to prove this for $b = e_i, i \in [r]$, which holds because each $\delta_i, i \in [r]$ is a power of p.

(4) $\Leftrightarrow$ **(5)** is rather immediate; the only additional remark is that every divisor of a power of p is also a power of p.

(6) $\Rightarrow$ **(3)** If $y^\top A$ and Ax are integral, then $y^\top Ax = y^\top(ABA)x = (y^\top A)B(Ax)$, which is p-adic since $y^\top A, Ax$ are integral and B has p-adic entries, as required.

(4) $\Rightarrow$ **(6)** Choose unimodular matrices U, W such that UAW is in Smith normal form with elementary divisors $\delta_1, \ldots, \delta_r$. Let B' be the $n \times m$ matrix with a leading diagonal matrix $D^{-1} = \mathrm{Diag}(\frac{1}{\delta_1}, \ldots, \frac{1}{\delta_r})$, and zeros everywhere else. Let $B := WB'U$, which is a matrix with p-adic entries since each δ_i is a power of p. We claim that $ABA = A$, thereby proving (6). This equality holds if, and only if, $UABAW = UAW$. To this end, we have $UABAW = UA(WB'U)AW = (UAW)B'(UAW) = UAW$, where the last equality holds due to the definition of B' and the Smith normal form of UAW. $\qquad\square$

In light of the previous proposition we may say that an integral *matrix* forms a p-GSS if its rows, respectively its columns, form a p-GSS. Consider the following complexity problem: (A) *Given an integral matrix, does it form a p-GSS?* The Smith normal form of an integral matrix, and therefore its elementary divisors, can be computed in polynomial time [15]. Thus, Theorem 3 has the following consequence.

Corollary 18. (A) *belongs to* P.

Recall that a set of vectors $\{a^1, \ldots, a^n\} \subseteq \mathbb{Z}^m$ forms a p-GSC if every integral vector in the conic hull of the vectors can be expressed as a p-adic conic combination of the vectors. Observe that every p-adic vector in the conic hull of a p-GSC can also be expressed as a p-adic conic combination of the vectors.

Proposition 19. *If* $\{a^1, \ldots, a^n\} \subseteq \mathbb{Z}^m$ *is a p-GSC, then it is a p-GSS.*

Proof. Let $A \in \mathbb{Z}^{m \times n}$ be the matrix whose columns are $a^1, \ldots, a^n$. Take $b \in \mathbb{Z}^m$ such that $A\bar{x} = b$ for some $\bar{x}$. We need to show that the system $Ax = b$ has a p-adic solution. To this end, let $\bar{x}' := \bar{x} - \lfloor \bar{x} \rfloor \geq \mathbf{0}$ and $b' := A\bar{x}' = b - A\lfloor \bar{x} \rfloor \in \mathbb{Z}^m$. Thus, $Ax = b', x \geq \mathbf{0}$ has a solution, namely $\bar{x}'$, so it has a p-adic solution, say $\bar{z}'$, as the columns of A form a p-GSC. Let $\bar{z} := \bar{z}' + \lfloor \bar{x} \rfloor$, which is also p-adic. Then $A\bar{z} = A\bar{z}' + A\lfloor \bar{x} \rfloor = b' + A\lfloor \bar{x} \rfloor = b$, so $\bar{z}$ is a p-adic solution to $Ax = b$, as required. $\qquad\square$

The converse of this result, however, does not hold. For example, let $k \geq 3$ be an integer, $n := p^k + 1$, and m an integer in $\{4, \ldots, p^k\}$ such that $m - 1$ is not a power of p. Consider the matrix

$$A := \left(E_n - I_n \left| \frac{E_m - I_m}{0} \right. \right)$$

where E_d, I_d denote the all-ones square and identity matrices of dimension d, respectively. We claim that the columns of A form a p-GSS but not a p-GSC. To see the former, note that A has rank n, and since $\det(E_n - I_n) = n - 1 = p^k$, the GCD of the subdeterminants of A of order n is a power of p, so the columns of A form a p-GSS by Theorem 3. To see the latter, consider the vector $b \in \{0, 1\}^n$ whose first m entries are equal to 1, and whose last $n - m$ entries are equal to 0. Then $Ay = b, y \geq 0$ has a unique solution, namely $\bar{y}$ defined as $\bar{y}_i = 0$ for $1 \leq i \leq n$, and $\bar{y}_i = \frac{1}{m-1}$ for $n + 1 \leq i \leq n + m$. In particular, as $m - 1$ is not a power of p, b is an integral vector in the conic hull of the columns of A, but it cannot be expressed as a p-adic conic combination of the columns. Thus, the columns of A do not form a p-GSC.

However, we do have the following sort of converse.

Remark 20. If $\{a^1, \ldots, a^n\} \subseteq \mathbb{Z}^m$ is a p-GSS, then $\{\pm a^1, \ldots, \pm a^n\}$ is a p-GSC.

Proposition 21. *Let $\{a^1, \ldots, a^n\} \subseteq \mathbb{Z}^m$ be a p-adic generating set for a cone, and F a nonempty face of the cone. Then $\{a^i : a^i \in F\}$ is a p-adic generating set for the cone F.*

Proof. Let b be an integral vector in the face F. Since $b \in C$, we can write b as a p-adic conic combination of the vectors in $\{a^1, \ldots, a^n\}$. However, since b is contained in the face F, the conic combination can only assign nonzero coefficients to the vectors in F, implying in turn that b is a p-adic conic combination of the vectors in $\{a^i : a^i \in F\}$. As this holds for every b, $\{a^i : a^i \in F\}$ forms a p-GSC. $\square$

Proof of Theorem 2. ($\Rightarrow$) follows from Proposition 21 and Proposition 19. ($\Leftarrow$) Let b be an integral vector in C, and F the minimal face of C containing b. Let B be the matrix whose columns are the vectors $\{a^i : a^i \in F\}$. We need to show that $Q := \{y : By = b, y \geq 0\}$, which is nonempty, contains a p-adic point. By the Density Lemma, it suffices to show that $\mathrm{aff}(Q)$, the affine hull of Q, contains a p-adic point. Our minimal choice of F implies that Q contains a point $\mathring{y}$ such that $\mathring{y} > 0$, implying in turn that $\mathrm{aff}(Q) = \{y : By = b\}$. As the columns of B form a p-GSS, and b is integral, it follows that $\mathrm{aff}(Q)$ contains a p-adic point, as required. $\square$

Consider the following complexity problem: (B) *Given a set of vectors, does it form a p-GSC?* Theorem 2 and Theorem 18 have the following consequence.

Corollary 22. *(B) belongs to co-NP.*

4   Totally Dual p-adic Systems

Given integral A, b, recall that $Ax \le b$ is totally dual p-adic if for every integral w for which $\min\{b^\top y : A^\top y = w, y \ge 0\}$ has an optimal solution, it has a p-adic optimal solution. It can be readily checked that the rows of A form a p-GSS if, and only if, $Ax = 0$ is totally dual p-adic; and the rows of A form a p-GSC if, and only if, $Ax \le 0$ is totally dual p-adic.

Proof of Theorem 1. Consider the following pair of dual linear programs, for w later specified.

$$(P) \quad \max\{w^\top x : Ax \le b\} \qquad (D) \quad \min\{b^\top y : A^\top y = w, y \ge 0\}$$

For every nonempty face F of P, denote by $A_{\bar{F}}$ the row submatrix of A corresponding to the rows not in A_F. For every vector y, denote by $y_F, y_{\bar{F}}$ the variables corresponding to the rows in $A_F, A_{\bar{F}}$, respectively.

(1) $\Rightarrow$ (2) Consider a nonempty face F of P. We need to show that the rows of A_F form a p-GSC. Let w be an integral vector in the conic hull of the rows of A_F. It suffices to express w as a p-adic conic combination of the rows of A_F. To this end, observe that every point in F is an optimal solution to (P). As $Ax \le b$ is TDD, (D) has a p-adic optimal solution, say $\bar{y} \ge 0$. As Complementary Slackness holds for all pairs $(\bar{x}, \bar{y}), \bar{x} \in F$, it follows that $\bar{y}_{\bar{F}} = 0$. Subsequently, we have $w = A^\top \bar{y} = A_F^\top \bar{y}_F$, thereby achieving our objective. (2) $\Rightarrow$ (3) follows from Proposition 19. (3) $\Rightarrow$ (1) Choose an integral w for which (D) has an optimal solution; we need to show now that it has a p-adic optimal solution. Denote by F the face of the optimal solutions to the primal linear program (P). By Complementary Slackness, the set of optimal solutions to the dual (D) is $Q := \{y : A^\top y = w, y \ge 0, y_{\bar{F}} = 0\}$. We need to show that Q contains a p-adic point. In fact, by the Density Lemma, it suffices to find a p-adic point in $\mathrm{aff}(Q)$, the affine hull of Q. By Strict Complementarity, Q contains a point $\mathring{y}$ such that $\mathring{y}_F > 0$, implying in turn that $\mathrm{aff}(Q) = \{y : A_F^\top y_F = w, y_{\bar{F}} = 0\}$. Since the rows of A_F form a p-GSS, and w is integral, we get that $\mathrm{aff}(Q)$ contains a p-adic point, as required. □

The careful reader may notice that by applying polarity to Theorem 1 with $b = 0$, we obtain another proof of Theorem 2. Moving on, consider the following complexity problem: (C) *Given a system $Ax \le b$ where A, b are integral, is the system totally dual p-adic?* Theorem 1 (3) and Corollary 18 have the following consequence.

Corollary 23. (C) *belongs to co-NP.*

5 p-Adic Polyhedra

A nonempty rational polyhedron is *p-adic* if every nonempty face contains a p-adic point. In this section we provide a characterization of p-adic polyhedra.

Remark 24. Let $A \in \mathbb{Z}^{m \times n}, b \in \mathbb{Z}^m, y \in \mathbb{R}^m$ and $y' := y - \lfloor y \rfloor \geq 0$. Then $A^\top y \in \mathbb{Z}^n$ if and only if $A^\top y' \in \mathbb{Z}^n$, y is p-adic if and only if y' is p-adic, and $b^\top y$ is p-adic if and only if $b^\top y'$ is p-adic.

Theorem 25. *Let $A \in \mathbb{Z}^{m \times n}, b \in \mathbb{Z}^m$ and $P := \{x : Ax \leq b\}$. Then the following are equivalent for every prime p: (1) P is a p-adic polyhedron, (2) for every nonempty face F of P, $\mathrm{aff}(F)$ contains a p-adic point, (3) for every nonempty face F of P, and z, if $A_F^\top z$ is integral then $b_F^\top z$ is p-adic, (4) for all $w \in \mathbb{R}^n$ for which $\max\{w^\top x : x \in P\}$ has an optimum, it has a p-adic optimal solution, (5) for all $w \in \mathbb{Z}^n$ for which $\max\{w^\top x : x \in P\}$ has an optimum, it has a p-adic optimal value.* □

There is an intriguing contrast between this characterization and that of integral polyhedra (see [5], Theorem 4.1), namely the novelty of statements (2) and (3), which are ultimately due to Strict Complementarity and the Density Lemma.

Proof. **(1)** $\Rightarrow$ **(2)** follows immediately from definition. **(2)** $\Rightarrow$ **(1)** By the Density Lemma, every nonempty face contains a p-adic point, so P is a p-adic polyhedron. **(2)** $\Leftrightarrow$ **(3)** follows from the Theorem of the Alternative. **(1)** $\Rightarrow$ **(4)** Suppose $\max\{w^\top x : x \in P\}$ has an optimum. Let F be the set of optimal solutions. As F is in fact a face of P, and P is p-adic, it follows that F contains a p-adic point. **(4)** $\Rightarrow$ **(5)** If x is a p-adic vector, and w an integral vector, then $w^\top x$ is a p-adic rational.

(5) $\Rightarrow$ **(3)** We prove the contrapositive. Suppose (3) does not hold, that is, there exist a nonempty face F and z such that $w := A_F^\top z \in \mathbb{Z}^n$ and $b_F^\top z$ is not p-adic. By Remark 24, we may assume that $z \geq 0$. Consider the following pair of dual linear programs:

$$(P) \quad \max\{w^\top x : Ax \leq b\} \qquad (D) \quad \min\{b^\top y : A^\top y = w, y \geq 0\}$$

Denote by $A_{\bar{F}}$ the row submatrix of A corresponding to rows not in A_F. Denote by $y_F, y_{\bar{F}}$ the variables of (D) corresponding to rows A_F and $A_{\bar{F}}$ of A, respectively. Define $\bar{y} \geq 0$ where $\bar{y}_F = z$ and $\bar{y}_{\bar{F}} = 0$. Then $A^\top \bar{y} = A_F^\top z = w$, so $\bar{y}$ is feasible for (D). Moreover, Complementary Slackness holds for every pair $(x, \bar{y}), x \in F$. Subsequently, $\bar{y}$ is an optimal solution to (D), and $b^\top \bar{y} = b_F^\top z$ is the common optimal value of the two linear programs. Since w is integral and $b_F^\top z$ is not p-adic, (5) does not hold, as required. □

Corollary 26. *Let $A \in \mathbb{Z}^{m \times n}, b \in \mathbb{Z}^m$, and p a prime. If $Ax \leq b$ is totally dual p-adic, then $\{x : Ax \leq b\}$ is a p-adic polyhedron.*

Proof. This follows immediately from Theorem 25 (5) $\Rightarrow$ (1). □

Proof of Theorem 5. By Corollary 26, $P := \{x : Ax \leq b\}$ is a p- and q-adic polyhedron, that is, every minimal nonempty face of P contains a p-adic point and a q-adic point. Each minimal nonempty face of P is an affine subspace, so by Corollary 15, it contains an integral point. Thus, every minimal nonempty face of P contains an integral point, so P is an integral polyhedron. □

6 T-joins and Perfect Matchings

Let $G = (V, E)$ be a graph, and T a nonempty subset of even cardinality. A T-*cut* is a cut of the form $\delta(U)$ where $|U \cap T|$ is odd. Recall that a T-join is a subset $J \subseteq E$ such that the set of odd-degree vertices of $G[J]$ is precisely T. It can be readily checked that every T-cut and T-join intersect (see [8], Chap. 2). The following result was recently proved:

Theorem 27 ([3]). *Let* $G = (V, E)$ *be a graph, and* T *a nonempty subset of even cardinality. Let* τ *be the minimum cardinality of a* T-*cut. Then there exists a dyadic assignment* $y_J \geq 0$ *to every* T-*join* J *such that* $\mathbf{1}^\top y = \tau$ *and* $\sum (y_J : J \text{ a } T\text{-join containing } e) \leq 1 \ \forall e \in E.$

The proof of Theorem 27 uses the Density Lemma, the Theorem of the Alternative, and a result of Lovász on the *matching lattice* [17].

Proof of Theorem 8. Let A be the matrix whose columns are labeled by E, and whose rows are the incidence vectors of the T-joins. We need to show that $\min\{w^\top x \ : \ Ax \geq \mathbf{1}, x \geq \mathbf{0}\}$ yields a TDD system. Choose an integral w such that the dual $\max\{\mathbf{1}^\top y : A^\top y \leq w, y \geq \mathbf{0}\}$ has an optimal solution, that is, $w \geq \mathbf{0}$. Let G' be obtained from G after replacing every edge e with w_e parallel edges (if $w_e = 0$, then e is deleted). Let τ_w be the minimum cardinality of a T-cut of G', which is also the minimum weight of a T-cut of G. By Theorem 27, there exists a dyadic assignment $\bar{y}_J \geq 0$ to every T-join of G' such that $\mathbf{1}^\top \bar{y} = \tau_w$ and $\sum (\bar{y}_J : J \text{ a } T\text{-join of } G' \text{ containing } e) \leq 1 \ \forall e \in E(G')$. This naturally gives a dyadic assignment $y_J^\star \geq 0$ to every T-join of G such that $\mathbf{1}^\top y^\star = \tau_w$ and $A^\top y^\star \leq w$. Now let $\delta(U)$ be a minimum weight T-cut of G. Then $\chi_{\delta(U)}$ is a feasible solution to the primal which has value τ_w. As a result, $\chi_{\delta(U)}$ is optimal for the primal, and $y^\star$ is optimal for the dual. Thus, the dual has a dyadic optimal solution, as required.   □

Moving on, let $G = (V, E)$ be a graph such that $|V|$ is even. Let us prove that $\mathbf{M}(G)$, which is equal to the set $\{\chi_M : M \text{ a perfect matching of } G\}$, is a DGSC.

Proof of Theorem 9. We may assume that G contains a perfect matching. Let $T := V$. Note that every T-join has cardinality at least $\frac{|V|}{2}$, with equality holding precisely for the perfect matchings. By Theorem 8, the linear system $x(J) \geq 1 \ \forall \ T$-joins $J; x \geq \mathbf{0}$ is TDD. Let P be the corresponding polyhedron, and F the minimal face containing the point $\frac{2}{|V|} \cdot \mathbf{1}$. The tight constraints of F are precisely $x(M) \geq 1$ for perfect matchings M, so by Theorem 1 for $p = 2$, the rows of the corresponding coefficient matrix form a DGSC, implying in turn that $\mathbf{M}(G)$ is a DGSC.   □

Let P_{10} be the Petersen graph. Then P_{10} has six perfect matchings. Let M be the matrix whose columns are labeled by $E(P_{10})$, and whose rows are the incidence vectors of the perfect matchings. It can be checked that the elementary divisors of M are $(1, 1, 1, 1, 1, 2)$. Thus, for any prime $p \geq 3$, the rows of M which are the vectors in $\mathbf{M}(P_{10})$ do not form a p-GSS by Theorem 3, and so they do not form a p-GSC by Proposition 19. Thus, Theorem 9 does not extend to the p-adic setting for $p \geq 3$.

Acknowledgement. We would like to thank the referees whose comments on an earlier draft improved the final presentation. Bertrand Guenin was supported by NSERC grant 238811. Levent Tunçel was supported by Discovery Grants from NSERC and ONR grant N00014-18-1-2078.

References

1. Abdi, A., Cornuéjols, G., Guenin, B., Tunçel, L.: Clean clutters and dyadic fractional packings. SIAM J. Discret. Math. **36**, 1012–1037 (2022)
2. Abdi, A., Cornuéjols, G., Guenin, B., Tunçel, L.: Total dual dyadicness and dyadic generating sets, November 2021. full version, available online
3. Abdi, A., Cornuéjols, G., Palion, Z.: On dyadic fractional packings of T-joins (2021). submitted
4. Appa, G., Kotnyek, B.: Rational and integral k-regular matrices. Discrete Math. **275**, 1–15 (2004)
5. Conforti, M., Cornuéjols, G., Zambelli, G.: Integer Programming. GTM, vol. 271. Springer, Cham (2014). https://doi.org/10.1007/978-3-319-11008-0
6. Cook, W., Koch, T., Steffy, D.E., Wolter, K.: An exact rational mixed-integer programming solver. In: Günlük, O., Woeginger, G.J. (eds.) IPCO 2011. LNCS, vol. 6655, pp. 104–116. Springer, Heidelberg (2011). https://doi.org/10.1007/978-3-642-20807-2_9
7. Cook, W., Cunningham, W., Pulleyblank, W.R., Schrijver, A.: Combinatorial Optimization. Wiley, Hoboken (1998)
8. Cornuéjols, G.: Combinatorial Optimization: Packing and Covering, vol. 74. SIAM (2001)
9. Ding, G., Feng, L., Zang, W.: The complexity of recognizing linear systems with certain integrality properties. Math. Program. **114**, 321–334 (2008). https://doi.org/10.1007/s10107-007-0103-y
10. Edmonds, J., Giles, R.: A min-max relation for submodular functions on graphs. In: Studies in Integer Programming (Proceedings of the Workshop, Bonn, 1975), vol. 1, pp. 185–204 (1977). Ann. of Discrete Math
11. Fulkerson, D.: Blocking and anti-blocking pairs of polyhedra. Math. Program. **1**, 168–194 (1971). https://doi.org/10.1007/BF01584085
12. Gerards, B., Sebő, A.: Total dual integrality implies local strong unimodularity. Math. Program. **38**, 69–73 (1987)
13. Giles, F., Pulleyblank, W.: Total dual integrality and integer polyhedra. Linear Algebra Appl. **25**, 191–196 (1979). https://doi.org/10.1016/0024-3795(79)90018-1
14. Goddyn, L.: Cones, lattices, and hilbert bases of circuits and perfect matchings. In: Contemporary Mathematics. vol. 147, pp. 419–439. Amer. Math. Soc. (1993)
15. Kannan, R., Bachem, A.: Polynomial algorithms for computing the Smith and Hermite normal forms of an integer matrix. SIAM J. Comput. **8**, 499–507 (1979)
16. Lee, J.: Subspaces with well-scaled frames. Linear Algebra Appl. **114**(115), 21–56 (1989)
17. Lovász, L.: Matching structure and the matching lattice. J. Comb. Theor. Ser. B **43**(2), 187–222 (1987). https://doi.org/10.1016/0095-8956(87)90021-9
18. Palion, Z.: On dyadic fractional packings of T-joins. Undergraduate thesis, London School of Economics and Political Science, April 2021
19. Pap, J.: Recognizing conic TDI systems is hard. Math. Program. Ser. A **128**, 43–48 (2011). https://doi.org/10.1007/s10107-009-0294-5

20. Schrijver, A.: Combinatorial Optimization: Polyhedra and Efficiency. Springer, Heidelberg (2003)
21. Schrijver, A.: Theory of Linear and Integer Programming. Wiley, Hoboken (1998)
22. Sebő, A.: Hilbert bases, Carathéodory's theorem, and combinatorial optimization. In: Kannan, R., Pulleyblank, W. (eds.) Integer Programming and Combinatorial Optimization (IPCO 1990), Lecture Notes in Computer Science, pp. 431–456 (1990)
23. Seymour, P.D.: On multi-colourings of cubic graphs, and conjectures of Fulkerson and Tutte. Proc. London Math. Soc. **38**(3), 423–460 (1979). https://doi.org/10.1112/plms/s3-38.3.423
24. Seymour, P.D.: Matroids and multicommodity flows. Europ. J. Combinatorics **2**, 257–290 (1981)
25. Steffy, D.E.: Topics in Exact Precision Mathematical Programming. Ph.D. thesis, Georgia Institute of Technology (2011)
26. Szekeres, G.: Polyhedral decomposition of cubic graphs. Bull. Austral. Math. Soc. **8**, 367–387 (1973)
27. Wei, H.: Numerical Stability in Linear Programming and Semidefinite Programming. Ph.D. thesis, University of Waterloo (2006)
28. Whittle, G.: A characterization of the matroids representable over GF(3) and the rationals. J. Comb. Theor. Ser. B **65**, 222–261 (1995)

Faster Goal-Oriented Shortest Path Search for Bulk and Incremental Detailed Routing

Markus Ahrens[1], Dorothee Henke[2], Stefan Rabenstein[3(✉)], and Jens Vygen[3]

[1] IBM Deutschland R & D GmbH, Böblingen, Germany
markus.johannes.ahrens@ibm.com
[2] TU Dortmund University, Dortmund, Germany
dorothee.henke@math.tu-dortmund.de
[3] University of Bonn, Bonn, Germany
{rabenstein,vygen}@dm.uni-bonn.de

Abstract. We develop new algorithmic techniques for VLSI detailed routing. First, we improve the goal-oriented version of Dijkstra's algorithm to find shortest paths in huge incomplete grid graphs with edge costs depending on the direction and the layer, and possibly on rectangular regions. We devise estimates of the distance to the targets that offer better trade-offs between running time and quality than previously known methods, leading to an overall speed-up. Second, we combine the advantages of the two classical detailed routing approaches—global shortest path search and track assignment with local corrections—by treating input wires (such as the output of track assignment) as reservations that can be used at a discount by the respective net. We show how to implement this new approach efficiently.

1 Introduction

The task of VLSI routing [4,22] is to connect the set of pins of every net on a chip by wires so that wires of different nets are sufficiently far apart and various other constraints are met. See Fig. 1 (left) for an example. Typically, one first computes a *global routing*, a rough packing of wires that ignores all local constraints but guarantees that the wires in certain areas do not require more space than available. This allows for globally optimizing objectives such as power consumption and timing constraints [10,16].

The output of global routing then restricts the search space for every net in detailed routing, where many complicated rules need to be obeyed and one essentially routes one net at a time. While the detailed routing graph formed by routing tracks on an entire chip can contain about 10^{13} vertices on 10–20 layers, the restricted area corresponding to the global routing solution for a net results in a much smaller detailed routing graph, with rarely more than 10^8 vertices. Nevertheless, these subgraphs are still huge, and there are millions of nets to connect. Two general strategies have been proposed (cf. [22]).

K. Aardal and L. Sanitá (Eds.): IPCO 2022, LNCS 13265, pp. 15–28, 2022.
https://doi.org/10.1007/978-3-031-06901-7_2

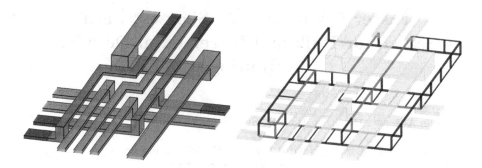

Fig. 1. Left: tiny part of a routed chip. The blue wires connect the two dark blue pins and the red wires connect the three dark red pins. Pins of the same color belong to the same net. The gray wires are part of the connections of multiple other nets with pins outside of the visible region. Right: the relevant part of the detailed routing graph before routing the blue and the red net. (Color figure online)

The first approach is based on a fast subroutine to find a shortest path that connects two metal components, each of which can consist of a pin or a set of previously computed wires connecting a subset of the pins of that net. The subgraph is given by the global routing solution, excluding vertices and edges that would result in a conflict to previously routed wires. For an example of the resulting graph, see Fig. 1 (right). To allow for an efficient packing of wires and to model various aspects such as signal delays, one uses different costs for horizontal and for vertical edges on each layer as well as for vias connecting two adjacent layers.

The second approach first considers the layers one after the other and assigns wires to routing tracks so that the most important detailed routing rules are satisfied, at least for most wires. This is often called *track assignment* [5,21]. Then detailed routing tries to correct violations locally. A very similar problem occurs when a detailed routing has already been computed, but a few changes to the input have been made (e.g., corrections of the logical behavior or to speed up signals that arrived too late). In both cases, one asks for an *incremental* detailed routing, largely following the input but deviating where necessary. However, local corrections are often not possible if the routing is very dense.

One classical speed-up technique of Dijkstra's shortest path algorithm [6] (sometimes called A^*) is to use reduced costs, based on a feasible potential that estimates the distance to the targets [9,14,19]. Instead of the undirected graph with the original edge cost $c(e)$, we orient each edge in both ways and run Dijkstra's algorithm with the *reduced cost* $c_\pi(e) := c(e) - \pi(v) + \pi(w)$ for every edge e directed from v to w, where the vertex potentials π are chosen so that c_π is nonnegative and $\pi(t) = 0$ for every target t. These conditions imply that $\pi(v)$ is a lower bound on the distance between v and the closest target. The better this lower bound is, the fewer vertices this goal-oriented version of Dijkstra's algorithm must label before it knows a shortest path to a target.

Hence, there is a trade-off between a possible preprocessing time, the query time to compute the potential of a vertex, and the quality of the lower bound.

For example, in subgraphs of unweighted grid graphs, the ℓ_1-distance to the nearest target can be a reasonable choice for π [11]. A better estimate, which however requires substantial preprocessing, was suggested by [17]. In this paper, we propose new methods with better trade-offs than previously known.

Moreover, we combine the advantages of the two classical detailed routing approaches mentioned above. Our new, more global approach treats given input wires as so-called *reservations* and encourages, but not forces, the detailed router to follow the reservations where feasible. This is achieved by finding a shortest path where reservations can be used at a discount.

However, this does not work well together with the classical goal-oriented techniques. For example, if there are some reservations (edges) that can be used at a 50 % discount, the ℓ_1-distance would have to be divided by 2 in order to induce a feasible potential. This would often be a very inaccurate estimate, leading to an increased number of labels in Dijkstra's algorithm and hence larger running time. We show that our better potentials make goal-oriented Dijkstra not only as fast as without reservations, but in fact faster. Overall, this yields a new efficient incremental detailed routing algorithm.

1.1 Problem Statement

Our core problem will consist of computing distances in a weighted grid graph with a simple structure. To define the grid graph, we number the layers $1, \ldots, l$ and let $V = \mathbb{Z} \times \mathbb{Z} \times \{1, \ldots, l\}$ and

$$E = \left\{ \{(x, y, z), (x', y', z')\} \in \binom{V}{2} \mid |x - x'| + |y - y'| + |z - z'| = 1 \right\}$$

be the vertex set and edge set of an infinite grid with l layers. Edges connecting adjacent layers are called *vias*, edges in x-direction are *horizontal* and edges in y-direction *vertical*. We will consider finite subgraphs of $G = (V, E)$. These subgraphs correspond to the area defined by the global routing solution and to the restriction to the routing tracks that can be used for the current net. Often, many vertices of these subgraphs will have degree 2 and will not be considered explicitly, but we ignore this here for the sake of a simpler exposition.

Every layer has a preference direction ($\leftrightarrow$ or $\updownarrow$); edges in the other direction are more expensive. Horizontal and vertical layers alternate. Moreover, the layers have very different electrical properties which is reflected by appropriate edge costs. In the simplest model, the cost of an edge depends only on its direction and the layer: let $c_z^{\leftrightarrow}, c_z^{\updownarrow} > 0$ for $z \in \{1, \ldots, l\}$ and $c_{z,z+1} > 0$ for $z \in \{1, \ldots, l-1\}$; then

$$c(\{(x, y, z), (x', y', z')\}) = \begin{cases} c_z^{\leftrightarrow} & \text{if } x' = x + 1 \\ c_z^{\updownarrow} & \text{if } y' = y + 1 \\ c_{z,z'} & \text{if } z' = z + 1 \end{cases} \cdot$$

In a more general model, a rectilinear grid induces rectangular regions, called *tiles*, and the cost also depends on the tile. Let

$$-\infty = \xi^0 < \xi^1 \le \ldots \le \xi^p < \xi^{p+1} = \infty, \quad -\infty = v^0 < v^1 \le \ldots \le v^q < v^{q+1} = \infty$$

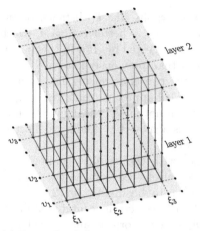 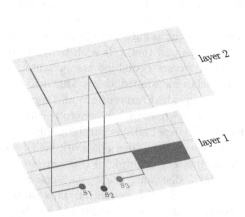

(a) Example of the general cost model. The red edges form the set E_1^{12}. The blue horizontal edges have cost $c_1^{21\leftrightarrow}$, the blue vertical edges have cost $c_1^{21\updownarrow}$, and the blue via edges have cost $c_{1,2}^{21}$. The cost of the blue vertical edges on ξ_2 is given by $\min\{c_1^{11\updownarrow}, c_1^{21\updownarrow}\}$, which is $c_1^{21\updownarrow}$ in this example. The edges that are not drawn have cost infinity in this example.

(b) A target T consisting of three rectangles and the coarsest partition consistent with T. For three possible query locations, the shortest s_1-T-path, s_2-T-path, and s_3-T-path for a cost function depending only on direction and layer are shown. The preference directions of layer 1 and 2 are $\leftrightarrow$ and $\updownarrow$, respectively. Nevertheless it is cheapest for the s_3-T-path to stay on layer 1 since its vertical segment is very short.

Fig. 2. Examples for grids, edge costs, target sets, and shortest paths (Color figure online)

be integer coordinates that define the rectangular tiles

$$V_z^{ij} = \left\{ (x, y, z) \in V \mid \xi^i \le x \le \xi^{i+1},\ v^j \le y \le v^{j+1} \right\},$$

and set

$$E_z^{ij} = \big\{ \{(x, y, z), (x', y', z')\} \in E \mid \xi^i \le x \le x' \le \xi^{i+1},$$
$$v^j \le y \le y' \le v^{j+1},\ z \le z' \big\}.$$

Now we have costs $c_z^{ij\leftrightarrow}, c_z^{ij\updownarrow}, c_{z,z+1}^{ij} > 0$ that also depend on the tile and define the edge costs accordingly. If an edge belongs to more than one tile, the minimum cost applies. See Fig. 2a for an example. We allow that two (but not three) consecutive coordinates are identical, i.e., $\xi^i = \xi^{i+1}$ or $v^j = v^{j+1}$, in order to model a cheap cost at one x- or y-coordinate only.

With this more general model, one can, for example, punish wires on low layers near the electrical source of a net (which would lead to poor delays) or implement a discount on reservations as we will describe in detail in Sect. 5.2. Moreover, we can set edge costs to infinity outside the area corresponding to the global routing solution so that the distances in (G, c) reflect necessary detours that are implied by routing in this subgraph.

Given a finite subgraph $G' = (V', E')$ of G and sets $S, T \subseteq V'$, we look for a shortest path from S to T in G' with respect to the length function c. For a goal-oriented path search, we define a potential $\pi(v)$ for every vertex $v \in V'$ by the distance to T in G (instead of G'):

$$\pi(v) := \mathrm{dist}_{(G,c)}(v, T).$$

The idea is that distances in G are much easier to compute than in the subgraph G' (we will see how fast), but often still give a good lower bound. The reason is that (G, c) has a simple structure, given by the tiles, while G' can be very complicated as it does not contain vertices or edges whose use would result in a conflict to nets routed previously. This allows us to use Dijkstra's algorithm with the reduced costs c_π in the digraph resulting from G' by orienting every edge in both ways, since the reduced costs are nonnegative.

Some of our algorithms work best for simple targets. We often assume that T is represented as the union of t rectangles, where a rectangle is a vertex set of the form $\{(x, y, z) \in V \mid \xi^- \leq x \leq \xi^+, v^- \leq y \leq v^+, z = \zeta\}$ for some $\xi^-, \xi^+, v^-, v^+ \in \mathbb{Z}$ and $\zeta \in \{1, \ldots, l\}$. Often t is small in practice. Sometimes it will be useful to assume that this representation is *consistent* with the partition of V into tiles in the following sense: each of the t rectangles representing T fits into the grid, i.e., is of the form $\{(x, y, z) \in V \mid \xi^{i^-} \leq x \leq \xi^{i^+}, v^{j^-} \leq y \leq v^{j^+}, z = \zeta\}$ for some indices i^-, i^+, j^- and j^+. This can be achieved by adding at most $2t$ new x-coordinates ξ^i and $2t$ new y-coordinates v^i. We call this procedure *refining the grid with respect to the targets*. See Fig. 2b for an example of the empty grid refined with respect to several target rectangles.

1.2 Previous Work and Our Results

In the simple model without regions (i.e., for $p = q = 0$), one can query π in $O(tl^2)$ time without preprocessing. We improve this to $O(tl)$ in Theorem 2. With a preprocessing time polynomial in t and l, we obtain a query time of $O(\log(t + l))$; see Theorem 3. We present these results in Sects. 2 and 3.

For the more general model, which is the subject of Sect. 4, Peyer et al. [17] refined the grid with respect to the targets and showed that then the restriction of $\pi : V \to \mathbb{R}_{\geq 0}$ to V_z^{ij} is the minimum of k^2 affine functions for any i, j, z, where k is the number of different horizontal and vertical edge costs, i.e.,

$$k := \left| \left\{ c_z^{ijd} \mid i \in \{0, \ldots, p\}, j \in \{0, \ldots, q\}, d \in \{\leftrightarrow, \updownarrow\}, z \in \{1, \ldots, l\} \right\} \right|. \tag{1}$$

They also showed that all these functions can be computed in $O((p + t)(q + t)lk^4 \log(p + q + l + t))$ time, allowing $O(k^2)$ time queries after this preprocessing (plus $O(\log(p + q + t))$) to find the region containing the given vertex by binary search; here and henceforth p and q refer to the original number of rows and columns, before refining the grid).

We make multiple improvements over the approach of Peyer et al. [17]. By considering domination between affine functions with different slopes, we reduce the number of affine functions needed to describe the minimum. By first computing the distances from the edges on the boundaries of the tiles to the targets, we can compute these affine functions faster. Finally, we use a regional query data structure to reduce query time. For any $0 < \varepsilon \leq 1$, we can obtain an algorithm with preprocessing time $O((p+t)(q+t)\min\{k,(p+q+1)l\}l^{1+\varepsilon}\frac{1}{\varepsilon}\log(p+q+l+t))$ and query time $O(\log(p+q+t)+\frac{1}{\varepsilon}\log(k+l))$. See Table 1 for an overview.

Our second contribution is a new approach to incremental routing. Rather than trying to correct a given infeasible input routing with local transformations only, we compute a new routing from scratch, at least for all nets for which the input routing does not obey all rules. However, to compute a solution similar to the input where reasonable, we reserve the space occupied by legal input wires for the respective net and allow to use edges corresponding to input wires at a discount. By letting each input wire be a separate tile V_z^{ij}, we can model the discount in the cost function c and work with reduced costs efficiently. When most of the input routing can be used, we can find a shortest path much faster than without a discount.

This makes this new approach not only useful for incremental routing, but also for bulk routing. Treating the output of a track assignment as reservations (wherever it is legal) and then pursuing our new incremental routing approach can combine the advantages of the two classical bulk routing approaches, successive shortest paths and track assignment with local corrections. We explain our new approach in detail in Sect. 5, where we also show experimental results.

Full proofs and more detailed experimental results can be found in [1].

2   Distances Without Preprocessing in the Simple Model

In the simple model, there is always a shortest path with a very simple structure:

Lemma 1. *Let $c : E \to \mathbb{R}_{>0}$ depend only on direction and layer, and let $r, s \in V$. Then there is a shortest path P between r and s in (G, c) that consists of at most one sequence of horizontal edges, at most one sequence of vertical edges, and hence at most three sequences of vias.*

Table 1. Various methods to compute $\pi(v)$, possibly after preprocessing. The running times depend on the number t of target rectangles, the number l of layers, and in the general model on the numbers p and q of coordinates that define the $(p+1)(q+1)$ regions, and on the number k of different horizontal and vertical edge costs (cf. (1)). Note that $k \leq 2(p+1)(q+1)l$. For simplicity, we assume T to be consistent with the grid in this table, which may increase p and q by up to $2t$.

Model	Preprocessing time	Query time	Reference
Simple	–	$O(tl)$	Theorem 2
Simple	$O(t^2 l^3 \log l)$	$O(\log(t+l))$	Theorem 3
General	$O(pqlk^4 \log(p+q+l))$	$O(\log(p+q)+k^2)$	[17]
General	$O(pql^{1+\varepsilon}k\frac{1}{\varepsilon}\log(p+q+l))$	$O(\log(p+q)+\frac{1}{\varepsilon}\log(k+l))$	Theorem 7

Proof. Let P be a shortest path, and $P_{[v,w]}$ and $P_{[v',w']}$ two maximal subpaths of P in the same direction (all-horizontal or all-vertical), say from v to w and from v' to w', respectively, and let $P_{[w,v']}$ be the subpath in between. Suppose, w.l.o.g., that $P_{[v,w]}$ and $P_{[v',w']}$ are horizontal paths and that the cost of an edge of $P_{[v,w]}$ is not more expensive than the cost of an edge of $P_{[v',w']}$ (note that these paths may be on different layers). Then translating $P_{[v',w']}$ by adding $w - v'$ to all its vertices, translating $P_{[w,v']}$ by adding $w' - v'$ to all its vertices, and swapping these two paths yields a walk from r to s with one maximal horizontal subpath less and at most the same number of maximal vertical subpaths, while the cost does not increase. If the walk is not a path, we can shortcut it to a path. By induction, the assertion follows. □

Hence, in order to compute a shortest path, we can enumerate the targets and then the layers on which the horizontal sequence and the vertical sequence are, and which of the two comes first. This has running time $O(tl^2)$. We show how to improve on this, obtaining a linear dependence on the number of layers:

Theorem 2. *Let $c : E \to \mathbb{R}_{>0}$ depend only on direction and layer. Then, without preprocessing, one can compute $\mathrm{dist}_{(G,c)}(s, T)$ for any given $s \in V$ and given $T \subseteq V$ consisting of t rectangles in $O(tl)$ time.*

Proof. We enumerate over all t rectangles of T. For each such rectangle R, we determine the vertex $r \in R$ that is closest to s (geometrically) in constant time. First compute the total cost c_{z_1, z_2} of a path of vias between layer z_1 and layer z_2 for all $z_1, z_2 \in \{1, \ldots, l\}$ with $\{z_1, z_2\} \cap \{z_r, z_s\} \neq \emptyset$, where z_r and z_s denote the layers of r and s, respectively. This can easily be done in $O(l)$ time.

Now we compute, in $O(l)$ time, the minimum length of a path from r to s that (when traversed from r to s) consists of a path of vias, then a horizontal path, then a path of vias, then a vertical path, then a path of vias. We will then do the same with exchanging the roles of r and s, and we are done by Lemma 1.

For each layer $z \in \{1, \ldots, l\}$, consider the vertex v_z on layer z whose y-coordinate is the one of r and whose x-coordinate is the one of s. We first compute for $z = l, l - 1, \ldots, 1$ the distance $\bar{d}_z$ between r and v_z in the subgraph of G that contains no horizontal edges on the layers $1, \ldots, z-1$: set $\bar{d}_{l+1} = \infty$ and $\bar{d}_z = \min\{\bar{d}_{z+1} + c_{z,z+1}, c_{z_r,z} + c_z^{\leftrightarrow} \cdot |x_r - x_s|\}$. Then we compute for $z = 1, \ldots, l$ the distance d_z from r to v_z by setting $d_0 = \infty$ and $d_z = \min\{\bar{d}_z, d_{z-1} + c_{z-1,z}\}$. Finally, the shortest path from r to s that goes first horizontal and then vertical has length $\min\{d_z + c_z^{\updownarrow} \cdot |y_r - y_s| + c_{z,z_s} \mid z \in \{1, \ldots, l\}\}$. □

3   Logarithmic Query Time in the Simple Model

We now achieve $O(\log(t + l))$ query time with polynomial preprocessing time:

Theorem 3. *Let $c : E \to \mathbb{R}_{>0}$ depend only on direction and layer, and let $T \subseteq V$ consist of t rectangles. Then there is a data structure that requires $O(t^2 l^3 \log l)$ preprocessing time and, for any given $s \in V$, can then determine $\mathrm{dist}_{(G,c)}(s, T)$ in $O(\log(t + l))$ query time.*

Proof (sketch). We first interpret our instance as an instance of the general model by choosing $p = q = 0$ and then refine the grid with respect to the targets, which yields $O(t^2 l)$ tiles V_z^{ij}. By Lemma 1, and since the number of possible slopes in either direction is at most $2l + 1$, the distance from any tile to T can be expressed as a minimum of at most $4l^2 + 4l + 1$ affine functions. A dynamic program can compute all required functions for every tile in $O(t^2 l^3)$ time. Since a tile containing a query location can be determined in $O(\log t)$ time using binary search, it now suffices to evaluate the minimum of at most $4l^2 + 4l + 1$ affine functions in $O(\log l)$ query time after $O(l^2 \log l)$ preprocessing time for each tile independently using the following lemma. □

Lemma 4. *Let F be a set of affine functions $f \colon \mathbb{R}^2 \to \mathbb{R}$ and $R := [x^-, x^+] \times [y^-, y^+]$ a rectangle. Then there is a data structure that requires $O(|F| \log|F|)$ preprocessing time and, given any query point $p \in R$, can then determine the value $\min_{f \in F} f(p)$ in $O(\log|F|)$ query time.*

Proof. First intersect the half spaces $\{(x, y, \phi) \in \mathbb{R}^3 \mid \varphi \leq f(x, y)\}$ for $f \in F$ in $O(|F| \log |F|)$ time [18]. By projecting the lower faces of the resulting polyhedron into the plane, we obtain a subdivision of R into at most $|F|$ convex polygons and the minimizing function $f \in F$ for each polgyon. Using [15], we can build a data structure in $O(|F| \log |F|)$ preprocessing time and can then determine a polygon containing any given query point $p \in R$ in $O(\log |F|)$ time. □

Point location algorithms which attain the same theoretic guarantees as [15], but successively improve practical performance and ease of implementation have been described in [7,12,20].

4   The General Model

In this section, we develop an algorithm to compute the potential $\mathrm{dist}_{(G,c)}(v, T)$ for any v in the general model efficiently after preprocessing. We will assume T to be consistent with the grid, i.e., we have already refined the grid if it was not. Our preprocessing will work on the horizontal and vertical line segments of the grid, i.e., the sets $\mathrm{Hor}_z^{ij} := \{(x, y, z) \in V_z^{ij} \mid y = v^j\}$ and $\mathrm{Ver}_z^{ij} := \{(x, y, z) \in V_z^{ij} \mid x = \xi^i\}$. The exposition will focus on the horizontal line segments; vertical segments can be handled analogously. Our algorithm consists of two preprocessing steps and a query step. The first preprocessing step is a variant of Dijkstra's algorithm. For its correctness, the following observation about the structure of shortest paths, which can be shown similarly to Lemma 1 for the simple model, is essential:

Lemma 5. *Let $c \colon E \to \mathbb{R}_{>0}$ depend on tile and direction, let $T \subseteq V$ be consistent with the grid, and $s \in V \setminus T$. Then there is a shortest path P from s to T in (G, c) that uses only one type of edges (either horizontal, vertical or via) before entering some tile in which s does not lie.*

Our algorithm will first compute $\text{dist}_{(G,c)}(s, T)$ for all s lying in horizontal segments of the grid. More precisely, for each horizontal segment Hor_z^{ij}, the algorithm maintains a set F_z^{ij} of affine functions $f \colon [\xi^i, \xi^{i+1}] \to \mathbb{R}_{\geq 0}$ such that each value $f(x)$ corresponds to the length of a path between (x, v^j, z) and T. At any point during the algorithm, for every vertex (x, v^j, z), the value $\min\{f(x) \mid f \in F_z^{ij}\}$ can be considered to be its current label. If $\xi^i = \xi^{i+1}$, we simply store that value. Otherwise we use a binary search tree to keep only those functions that are not dominated, i.e., attain the pointwise minimum in more than one point.

In addition, we maintain a binary heap representing all functions in $\bigcup\{F_z^{ij} \mid i \in \{0, \ldots, p\}, j \in \{1, \ldots, q\}, z \in \{1, \ldots, l\}\}$ that have not been processed yet, where the key of a function is the minimum of the labels it implicitly represents. The functions that are added to or removed from some F_z^{ij} must be added to or removed from the heap at the same time, and whenever a key changes, it must be updated also in the heap.

The algorithm starts by initializing $F_z^{ij} := \emptyset$ for all $i \in \{0, \ldots, p\}, j \in \{1, \ldots, q\}$, and $z \in \{1, \ldots, l\}$. If $\text{Hor}_z^{ij} \subseteq T$, we add the constant function $x \mapsto 0$ to the corresponding set F_z^{ij}. If not the whole segment but one (or both) of its endpoints is in T, we add the affine function describing the distance to this point, i.e., $x \mapsto \min\{c_z^{ij\leftrightarrow}, c_z^{i(j-1)\leftrightarrow}\} \cdot (x - \xi^i)$ or $x \mapsto \min\{c_z^{ij\leftrightarrow}, c_z^{i(j-1)\leftrightarrow}\} \cdot (\xi^{i+1} - x)$.

In every iteration, a function f with minimum key is chosen and removed from the heap. The function f describes the labels of a subset of some horizontal segment Hor_z^{ij}, corresponding to an interval $[x_f^-, x_f^+]$. We now propagate the labels from these vertices to the neighboring horizontal segments by computing at most six new affine functions:

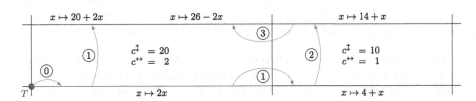

Fig. 3. Example run of the algorithm computing the distance from all horizontal line segments to T. The instance consists of two horizontally adjacent tiles with coordinates $\xi^1 = 0$, $\xi^2 = 4$, $\xi^3 = 7$, $v^1 = 0$, and $v^2 = 1$. The target $T = \{(0, 0, 1)\}$ consists of the single point in the bottom left corner. We disregard the outside tiles (by setting their costs to infinity). All other costs are as written in the centers of the respective tiles. During the algorithm, five affine functions are added to the four horizontal segments, which are colored by the function attaining the minimum in the end of the algorithm. The incoming arrow depicts the propagation by which that function was added and is numbered by the iteration of the algorithm (where 0 stands for initialization). (Color figure online)

(down) If $z > 1$, add the function $x \mapsto f(x) + \min\{c^{ij}_{z-1,z}, c^{i(j-1)}_{z-1,z}\}$ to F^{ij}_{z-1}.

(up)   If $z < l$, add the function $x \mapsto f(x) + \min\{c^{ij}_{z,z+1}, c^{i(j-1)}_{z,z+1}\}$ to F^{ij}_{z+1}.

(south) If $j > 1$, add the function $x \mapsto f(x) + c^{i(j-1)\updownarrow}_z \cdot (v^j - v^{j-1})$ to $F^{i(j-1)}_z$.

(north) If $j < q$, add the function $x \mapsto f(x) + c^{ij\updownarrow}_z \cdot (v^{j+1} - v^j)$ to $F^{i(j+1)}_z$.

(west) If $i > 0$, add the function

$$x \mapsto f(\xi^i) + \min\{c^{(i-1)j\leftrightarrow}_z, c^{(i-1)(j-1)\leftrightarrow}_z\} \cdot (\xi^i - x) \text{ to } F^{(i-1)j}_z.$$

(east) If $i < p$, add the function

$$x \mapsto f(\xi^{i+1}) + \min\{c^{(i+1)j\leftrightarrow}_z, c^{(i+1)(j-1)\leftrightarrow}_z\} \cdot (x - \xi^{i+1}) \text{ to } F^{(i+1)j}_z.$$

The algorithm stops when the heap is empty. For an example run of the algorithm, see Fig. 3.

It is easy to show that each function we add gives an upper bound on the distance from the corresponding horizontal segment to the target. To show that the algorithm works correctly, we prove that there will be a function attaining the distance from any point to the target once the minimum key in the heap exceeded that distance. The proof uses induction on the distance to T. By Lemma 5, we may assume that the corresponding shortest path never uses a horizontal edge or a via in the interior of a tile. We exploit this structure to find an appropriate predecessor function in the dynamic program.

To ensure that the algorithm terminates and has the desired running time, we first observe that the functions in F^{ij}_z have at most $2k' + 1$ different slopes, where $k' := \min\{k, (q + 1)l\}$. This allows us to bound the number of iterations by $(p + 1)q(2k' + 1)l$, since two functions propagated from the same horizontal segment in different iterations can be shown to have different slopes. This yields:

Theorem 6. *There is an algorithm that computes for each horizontal segment Hor^{ij}_z a set F^{ij}_z of at most $2k' + 1$ affine functions such that $\min\{f(x) \mid f \in F^{ij}_z\} = \mathrm{dist}_{(G,c)}((x, v^j, z), T)$ for all $(x, v^j, z) \in \mathrm{Hor}^{ij}_z$. The algorithm can be implemented to run in $O(pqk'l \log(p + q + l))$ time.*

The second preprocessing step builds regional query data structures using Lemma 4. Each affine function in each horizontal and vertical segment induces an affine function on each query layer. To avoid a quadratic dependence on the number of layers in the preprocessing running time, we limit the number of query data structures to which any affine function contributes: any subset of query layers $1, \dots, z$ can share a data structure for any subset $z, \dots, l$ of layers containing the segments and vice versa. We choose a trade-off factor $0 < \varepsilon \leq 1$ between the number of data structures which need to be considered during each query and the number of layers sharing the same data structures. By carefully choosing the correct subranges of layers, we obtain the following result:

Theorem 7. *Let $0 < \varepsilon \leq 1$, let $c : E \rightarrow \mathbb{R}_{>0}$ depend on tile and direction, and let $T \subseteq V$, not necessarily consistent with the grid. Then there is a data structure that requires $O((p+t)(q+t) \min\{k, (p+q+1)l\} l^{1+\varepsilon} \frac{1}{\varepsilon} \log(p+q+l+t))$ preprocessing time and, for any given $s \in V$, can then determine $\mathrm{dist}_{(G,c)}(s, T)$ in $O(\log(p + q + t) + \frac{1}{\varepsilon} \log(k + l))$ query time.*

5   Practical Aspects

5.1   Implementation

With some improvements to practical performance that we describe in the full version [1], we implemented the algorithms presented in the previous sections as part of BonnRoute [2,3,8,13], a detailed router developed at the University of Bonn in joint work with IBM. BonnRoute is the main detailed routing tool used by IBM for the design of its processor chips.

Up to parallelization and conflict resolution, BonnRoute routes one net after the other. Each net is routed by iteratively connecting two of its components by a path until the net is fully connected. The path search is the algorithmic core of BonnRoute and requires approximately 80–90 % of the total runtime.

All experiments were performed on the same AMD EPYC 7601 machine with 64 CPUs and 1024 GB main memory using 64 threads. Our testbed consists of nine real-world instances from three recent IBM processor chips in 7 nm and 5 nm technology nodes. We started all experiments on the same instance from the same snapshot, which was taken right before the detailed routing. At this point, a (three-dimensional) global routing was already computed for each net.

Table 2. Performance of the following four different feasible potentials on our testbed. In the rows **without** potential, each query returns 0 in constant time. When using ℓ_1**-distance**, the $O(t)$ query computes the minimum required cost in each of the three directions separately. This requires an $O(l)$ preprocessing. In the **simple** and **general** rows, the shortest distance to T in the respective models is returned. Here the difference between the two is that the general model restricts to the area corresponding to the global routing solution (outside of it, the costs are infinite). Runtimes are summed over all 64 threads except for the last column, which shows the BonnRoute wall time.

| Potential | Preprocessing | All Dijkstra calls | | Standard Dijkstra calls | | Total BonnRoute |
| | | Runtime | Labels | Runtime | Labels | Wall time |
	h:mm	h:mm	10^9	h:mm	10^9	h:mm
Without	0:00	7933:04	5246.5	4619:12	3633.9	142:52
ℓ_1-distance	0:02	5457:19	2867.0	2535:45	1618.6	104:12
Simple	0:39	4191:51	2232.0	1517:52	1082.4	83:54
General	46:40	3608:00	1916.2	1094:36	848.9	75:44

Table 2 compares the performance of path searches using different feasible potentials. Each potential is the distance to T in the same supergraph G of G', but with respect to different edge costs c.

The results show that the general potential performs much better than the simple potential, which already performs much better than the ℓ_1-distance potential. Both the number of labels and the runtime improve significantly, even when considering the additional preprocessing time. Certain instances benefit less from these potentials. If there is no path to be found, all vertices are labeled regardless of which potential is used. After a path search failed, BonnRoute may perform a backup path search which allows routing through existing wires at high cost (and then would rip-up such wires and try to re-route them). Since these rip-up costs are not modeled in any of our potentials, a large portion of the graph may be labeled regardless of which potential is used. The column *Standard Dijkstra calls* in Table 2 excludes these situations and hence shows an even larger gain than the column *All Dijkstra calls*. The question how to model rip-up costs efficiently when computing potentials remains for future research.

5.2   Reservations and Discounts for Incremental Routing

In chip design practice, there are two main scenarios where a detailed routing is not computed from scratch, using just a global routing as input, but in an incremental way, using an almost feasible detailed routing as input. The first scenario is when a detailed routing has already been computed, but now a few changes have been made, for example in order to correct the logical function of the chip or to improve its timing behavior. The second scenario is when a step in between global and detailed routing is used, typically called track assignment, that maps the global wires to routing tracks in a way that obeys most—but not all—design rules. In both scenarios, the task is to compute a completely feasible detailed routing by doing only few changes. While it is not exactly specified what "few" means, the motivation is that the input routing has already been optimized, for example with respect to the timing behavior of the chip; moreover, one aims at saving runtime.

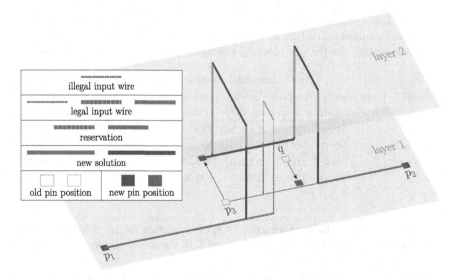

Fig. 4. Example of a net (consisting of the pins p_1, p_2, and p_3) that is re-routed using reservations after changes to the input have been made. Even though all of the green input wires are legal, we may choose to create reservations only for the thick green wires, e.g., if we expect the harm of blocking other nets to outweigh the benefit of keeping them usable for this net. (Color figure online)

We suggest to repair violations globally but with a preference of using the initial solution. To this end, we convert any detailed wire in the input to a global wire and possibly a reservation. A reservation reserves that space for the particular net. When other nets are routed earlier, this space is blocked. Therefore, reservations are created only for (parts of) detailed wires that do not conflict with other detailed wires in the input. For an example, see Fig. 4.

Once a net is routed for which we have created reservations, we would like to encourage, but not force, the net to use the reserved space. We do this by defining a discount factor $0 < \delta < 1$ and multiplying all edge costs on a reservation of that net by δ. In our experiments, we have chosen $\delta = \frac{3}{4}$. With the traditional goal-oriented search techniques, reservations would lead to a slow-down, since many or even all of the edge costs in the supergraph G would need to be multiplied by δ in order for the potential to stay feasible. Our generalized framework, however, allows us to refine the grid not only with respect to the targets, but also with respect to the reservations, and define individual (discounted) costs on the edges corresponding to reservations.

On six instances from real design practice, in which a detailed routing is no longer completely legal after some changes to the input, we compare two runs, each computing a new solution for every path containing a violation. The run that creates reservations based on the initial solution attains a 33% speed-up in the sum of Dijkstra run times over the run which creates no reservations, from 50 h and 51 min to 34 h and 9 min. If we instead start from an almost legal solution—the output of BonnRoute in one of the runs in Table 2—and repeat all path searches, this speed-up increases to 55%. Naturally, the overall BonnRoute wall time improves less, namely by 11% and 42%, respectively. See the full version [1] for details.

Acknowledgements. We thank the many other contributors to BonnRoute, in particular Niko Klewinghaus, Christian Roth, and Niklas Schlomberg. Thanks also to Lukas Kühne, who started the initial implementation of the reservations concept. We also thank Niklas Schlomberg for carefully reading a preliminary version of our manuscript. Dorothee Henke has partially been supported by Deutsche Forschungsgemeinschaft (DFG) under grant no. BU 2313/6, and the other authors under grants EXC 59 and EXC-2047 (Hausdorff Center for Mathematics).

References

1. Ahrens, M., Henke, D., Rabenstein, S., Vygen, J.: Faster goal-oriented shortest path search for bulk and incremental detailed routing. arXiv:2111.06169 (2021)
2. Ahrens, M.: Efficient algorithms for routing a net subject to VLSI design rules. Ph.D. thesis, University of Bonn (2020)
3. Ahrens, M., et al.: Detailed routing algorithms for advanced technology nodes. IEEE Trans. Comput. Aided Des. Integr. Circuits Syst. **34**(4), 563–576 (2015)
4. Alpert, C.J., Mehta, D.P., Sapatnekar, S.S.: Handbook of Algorithms for Physical Design Automation. CRC Press, Boca Raton (2008)

5. Batterywala, S., Shenoy, N., Nicholls, W., Zhou, H.: Track assignment: a desirable intermediate step between global routing and detailed routing. In: Proceedings of the 2002 IEEE/ACM International Conference on Computer-Aided Design, pp. 59–66 (2002)
6. Dijkstra, E.W.: A note on two problems in connexion with graphs. Numer. Math. **1**(1), 269–271 (1959)
7. Edelsbrunner, H., Guibas, L., Stolfi, J.: Optimal point location in a monotone subdivision. SIAM J. Comput. **15**(2), 317–340 (1986)
8. Gester, M., Müller, D., Nieberg, T., Panten, C., Schulte, C., Vygen, J.: BonnRoute: algorithms and data structures for fast and good VLSI routing. ACM Trans. Des. Autom. Electron. Syst. **18**(2), 1–24 (2013)
9. Hart, P., Nilsson, N., Raphael, B.: A formal basis for the heuristic determination of minimum cost paths. IEEE Trans. Syst. Sci. Cybern. **4**, 100–107 (1968)
10. Held, S., Müller, D., Rotter, D., Scheifele, R., Traub, V., Vygen, J.: Global routing with timing constraints. IEEE Trans. Comput. Aided Des. Integr. Circuits Syst. **37**(2), 406–419 (2018)
11. Hetzel, A.: A sequential detailed router for huge grid graphs. In: Proceedings of Design, Automation and Test in Europe, pp. 332–338. IEEE (1998)
12. Kirkpatrick, D.: Optimal search in planar subdivisions. SIAM J. Comput. **12**(1), 28–35 (1983)
13. Klewinghaus, N.: Efficient detailed routing on optimized tracks. Ph.D. thesis, University of Bonn (2022)
14. Lawler, E., Luby, M., Parker, B.: Finding shortest paths in very large networks. In: Nagl, M., Perl, J. (eds.) Proceedings of Graph-Theoretic Concepts in Computer Science. Trauner, Linz (1983)
15. Lipton, H.J., Tarjan, R.E.: Applications of a planar separator theorem. In: 18th Annual IEEE Symposium on Foundations of Computer Science, pp. 162–170 (1977)
16. Müller, D., Radke, K., Vygen, J.: Faster min-max resource sharing in theory and practice. Math. Program. Comput. **3**(1), 1–35 (2011). https://doi.org/10.1007/s12532-011-0023-y
17. Peyer, S., Rautenbach, D., Vygen, J.: A generalization of Dijkstra's shortest path algorithm with applications to VLSI routing. J. Discrete Algorithms **7**(4), 377–390 (2009)
18. Preparata, F.P., Müller, D.E.: Finding the intersection of n half-spaces in time $O(n \log n)$. Theoret. Comput. Sci. **8**(1), 45–55 (1979)
19. Rubin, F.: The Lee path connection algorithm. IEEE Trans. Comput. **23**, 907–914 (1974)
20. Sarnak, N., Tarjan, R.: Planar point location using persistent search trees. Commun. ACM **29**(7), 669–679 (1986)
21. Sarrafzadeh, M., Lee, D.T.: Restricted track assignment with applications. Int. J. Comput. Geom. Appl. **4**(1), 53–68 (1994)
22. Tellez, G., Hu, J., Wei, Y.: Routing. In: Lavagno, L., Markov, I.L., Martin, G., Scheffer, L.K. (eds.) Electronic Design Automation for IC Implementation, Circuit Design, and Process Technology. CRC Press (2016)

On the Maximal Number of Columns
of a Δ-modular Matrix

Gennadiy Averkov and Matthias Schymura[✉]

BTU Cottbus-Senftenberg, Platz der Deutschen Einheit 1, 03046 Cottbus, Germany
{averkov,schymura}@b-tu.de

Abstract. We study the maximal number of pairwise distinct columns in a Δ-modular integer matrix with m rows. Recent results by Lee et al. provide an asymptotically tight upper bound of $\mathcal{O}(m^2)$ for fixed Δ. We complement this and obtain an upper bound of the form $\mathcal{O}(\Delta)$ for fixed m, and with the implied constant depending polynomially on m.

1 Introduction

Full row rank integer matrices with minors bounded by a given constant Δ in the absolute value have been extensively studied in integer linear programming as well as matroid theory: The interest for optimization was coined by the paper of Artmann, Weismantel & Zenklusen [1] who showed that integer linear programs with a *bimodular* constraint matrix, meaning that all its maximal size minors are bounded by two in absolute value, can be solved in strongly polynomial time. With the goal of generalizing the results of Artmann et al. beyond the bimodular case, Nägele, Santiago & Zenklusen [8] studied feasibility and proximity questions of a subclass of integer programs with bounded subdeterminants. Fiorini et al. [3] obtained a strongly polynomial-time algorithm for integer linear programs whose defining coefficient matrix has the property that all its subdeterminants are bounded by a constant and all of its rows contain at most two nonzero entries. For more information on the development regarding this topic, we refer to the three cited contributions above and the references therein.

For a matrix $A \in \mathbb{R}^{m \times n}$ and for $1 \leq k \leq \min\{m,n\}$, we write

$$\Delta_k(A) := \max\{|\det(B)| : B \text{ is a } k \times k \text{ submatrix of } A\}$$

for the maximal absolute value of a $k \times k$ minor of A. Given an integer $\Delta \in \mathbb{Z}_{>0}$, a matrix $A \in \mathbb{R}^{m \times n}$ of rank m is said to be Δ-*modular* and Δ-*submodular*, if $\Delta_m(A) = \Delta$ and $\Delta_m(A) \leq \Delta$, respectively.[1] Moreover, a matrix $A \in \mathbb{R}^{m \times n}$ is said to be *totally* Δ-*modular* and *totally* Δ-*submodular*, if $\max_{k \in [m]} \Delta_k(A) = \Delta$ and $\max_{k \in [m]} \Delta_k(A) \leq \Delta$, respectively, where $[m] := \{1, 2, \ldots, m\}$.

---

[1] The authors of [4,7] use the term Δ-modular for what we call Δ-submodular.

© Springer Nature Switzerland AG 2022
K. Aardal and L. Sanitá (Eds.): IPCO 2022, LNCS 13265, pp. 29–42, 2022.
https://doi.org/10.1007/978-3-031-06901-7_3

Our object of studies is the *generalized Heller constant*, which we define as

$$\mathrm{h}(\Delta, m) := \max\{n \in \mathbb{Z}_{>0} : A \in \mathbb{Z}^{m \times n} \text{ has pairwise distinct columns}$$
$$\text{and } \Delta_m(A) = \Delta\}.$$

If $\Delta = 1$, we are concerned with the classical notion of unimodular integer matrices. Lee [6, Sect. 10] initiated the study of the maximal number of columns beyond unimodular matrices in 1989 and proved a bound of order $\mathcal{O}(r^{2\Delta})$, for totally Δ-submodular integer matrices of row-rank r. Glanzer, Weismantel & Zenklusen [4] revived the story by extending the investigation to Δ-submodular integer matrices and obtaining a polynomial bound in the parameter m.

The value $\mathrm{h}(\Delta, m)$ is directly related to the value $\mathfrak{c}(\Delta, m)$ studied in [4,7] and defined as the maximum number n of columns in a Δ-submodular integer matrix A with m rows with the properties that A has no zero columns and for any two distinct columns A_i and A_j with $1 \leq i < j \leq n$ one has $A_i \neq A_j$ and $A_i \neq -A_j$. It is clear that

$$\mathfrak{c}(\Delta, m) = \frac{1}{2}\big(\max\{\mathrm{h}(1, m), \dots, \mathrm{h}(\Delta, m)\} - 1\big)$$

holds, showing that $\mathfrak{c}(\Delta, m)$ and $\mathrm{h}(\Delta, m)$ are "equivalent" in many respects. However, our proofs are more naturally phrased in terms of $\mathrm{h}(\Delta, m)$ rather than $\mathfrak{c}(\Delta, m)$, as we prefer to prescribe $\Delta_m(A)$ rather than providing an upper bound on $\Delta_m(A)$ and we do not want to eliminate the potential symmetries within A coming from taking columns A_i and A_j that satisfy $A_i = -A_j$.

Upper bounds on the number of columns in Δ-(sub)modular integer matrices with m rows have been gradually improved over time as described in the introduction of [4]. Glanzer et al. [4] showed that for each fixed $\Delta \geq 2$, $\mathrm{h}(\Delta, m)$ is of order at most $\mathcal{O}(\Delta^{2+\log_2 \log_2 \Delta} \cdot m^2)$, a result that has been recently improved by Lee, Paat, Stallknecht & Xu [7] to the following estimate:

Theorem 1 ([7, Thm. 2 & Prop. 1 & Prop. 2]). *Let $\Delta, m \in \mathbb{Z}_{>0}$. Then,*

$$\mathrm{h}(\Delta, m) = m^2 + m + 1 + 2m(\Delta - 1) \qquad if \qquad \Delta \leq 2 \ \ or \ \ m \leq 2,$$

and, for all other cases (Δ, m), one has the bounds[2]

$$m^2 + m + 1 + 2m(\Delta - 1) \leq \mathrm{h}(\Delta, m) \leq (m^2 + m)\Delta^2 + 1. \tag{1}$$

Note that the case $\Delta = 1$ is the classical result of Heller [5] stating that the maximal number of pairwise distinct columns in a unimodular integer matrix with m rows is $\mathrm{h}(1, m) = m^2 + m + 1$. As a conjecture Lee et al. [7] formulate that the lower bound in (1) is actually the correct value of $\mathrm{h}(\Delta, m)$, for any choice of $\Delta, m \in \mathbb{Z}_{>0}$. A Δ-modular integer matrix with m rows and that many columns has the difference set of

$$\{0, e_1, e_2, \dots, e_m\} \cup \{2e_1, 3e_1, \dots, \Delta e_1\}$$

---

[2] Lee et al. [7, p. 23] remark that their techniques provide $\mathrm{h}(\Delta, m) \leq \mathcal{O}(m^2 \cdot \Delta^{1.95})$.

as its columns, where e_i denotes the ith coordinate unit vector.

The upper bound in (1) is quadratic both in m and in Δ. However, it is known that, for each fixed m, $h(\Delta, m)$ is linear in Δ, see the comment on page 24 in [7]. Still, so far there has not been any bound that is polynomial in m and linear in Δ. The authors of [7, p. 24] ask if there exists a bound of the form $\mathcal{O}(m^d)\Delta$ for some constant $d \in \mathbb{Z}_{>0}$. As our main result, we answer this question in the affirmative by showing that a bound of order $\mathcal{O}(m^4)\Delta$ exists.

Theorem 2. Let $\Delta \in \mathbb{Z}_{>0}$ and $m \geq 5$. Then,

$$h(\Delta, m) \leq m(m+1) + 1 + 2(\Delta - 1) \cdot \sum_{i=0}^{4} \binom{m}{i}.$$

It remains an open question whether our bound can be improved to a bound of order $\mathcal{O}(m^d)\Delta$ for some exponent $d < 4$.

2   Counting by Residue Classes

Our main idea is to count the columns of a Δ-modular integer matrix by residue classes of a certain lattice. This is the geometric explanation for the linearity in Δ of our upper bound in Theorem 2.

To be able to count in the non-trivial residue classes, we need to extend the Heller constant $h(1, m)$ to a *shifted* setting. Given a translation vector $t \in \mathbb{R}^m$ and a matrix $A \in \mathbb{R}^{m \times n}$, the shifted matrix $t + A := t\mathbf{1}^\intercal + A$ has columns $t + A_i$, where $A_1, \ldots, A_n$ are the columns of A, and $\mathbf{1}$ denotes the all-one vector.

Definition 1. *For any $m \in \mathbb{Z}_{>0}$, we define the* shifted Heller constant $h_s(m)$ *as the maximal number n such that there exists a translation vector $t \in [0, 1)^m \setminus \{\mathbf{0}\}$ and a matrix $A \in \{-1, 0, 1\}^{m \times n}$ with pairwise distinct columns such that $t + A$ is totally 1-submodular, that is, $\max_{k \in [m]} \Delta_k(t + A) \leq 1$.*

Note that, in contrast to the generalized Heller constant $h(\Delta, m)$, we do not necessarily require $t + A$ to have full rank, but we restrict A to have entries in $\{-1, 0, 1\}$ only. Moreover, the reason for restricting the non-zero translation vectors to the half-open unit cube $[0, 1)^m$ becomes apparent in the proof of the following crucial estimate.

Lemma 1. *For every $\Delta, m \in \mathbb{Z}_{>0}$, we have*

$$h(\Delta, m) \leq h(1, m) + (\Delta - 1) \cdot h_s(m).$$

Proof. Let $A \in \mathbb{Z}^{m \times n}$ be a matrix with $\Delta_m(A) = \Delta$ and pairwise distinct columns and let $X_A \subseteq \mathbb{Z}^m$ be the set of columns of A. Further, let $b_1, \ldots, b_m \in X_A$ be such that $|\det(b_1, \ldots, b_m)| = \Delta$ and consider the parallelepiped

$$P_A := [-b_1, b_1] + \ldots + [-b_m, b_m] = \left\{ \sum_{i=1}^{m} \alpha_i b_i : -1 \leq \alpha_i \leq 1, \forall i \in [m] \right\}.$$

Observe that $X_A \subseteq P_A$. Indeed, assume to the contrary that there is an $x = \sum_{i=1}^{m} \alpha_i b_i \in X_A$, with, say $|\alpha_j| > 1$. Then, $|\det(b_1, \ldots, b_{j-1}, x, b_{j+1}, \ldots, b_m)| = |\alpha_j| \Delta > \Delta$, which contradicts that A was chosen to be Δ-modular.

Now, consider the sublattice $\Lambda := \mathbb{Z} b_1 + \ldots + \mathbb{Z} b_m$ of $\mathbb{Z}^m$, whose index in $\mathbb{Z}^m$ equals Δ. We seek to bound the number of elements of X_A that fall into a fixed residue class of $\mathbb{Z}^m$ modulo Λ. To this end, let $x \in \mathbb{Z}^m$ and consider the residue class $x + \Lambda$. Every element $z \in (x + \Lambda) \cap P_A$ is of the form $z = \sum_{i=1}^{m} \alpha_i b_i$, for some uniquely determined $\alpha_1, \ldots, \alpha_m \in [-1, 1]$, and can be written as

$$z = \sum_{i=1}^{m} \lfloor \alpha_i \rfloor b_i + \sum_{i=1}^{m} \{\alpha_i\} b_i, \tag{2}$$

where $\{\alpha_i\} = \alpha_i - \lfloor \alpha_i \rfloor \in [0, 1)$ is the fractional part of α_i, and where $\bar{x} := \sum_{i=1}^{m} \{\alpha_i\} b_i$ is the unique representative of $x + \Lambda$ in the half-open parallelepiped $[0, b_1) + \ldots + [0, b_m)$, and in particular, is independent of z. We use the notation $\lfloor z \rfloor := (\lfloor \alpha_1 \rfloor, \ldots, \lfloor \alpha_m \rfloor) \in \{-1, 0, 1\}^m$ and $\{z\} := (\{\alpha_1\}, \ldots, \{\alpha_m\}) \in [0, 1)^m$ and thus have $z = B(\lfloor z \rfloor + \{z\})$, where $B = (b_1, \ldots, b_m) \in \mathbb{Z}^{m \times m}$.

Because the vectors $(x + \Lambda) \cap X_A$ constitute a Δ-submodular system and since $|\det(b_1, \ldots, b_m)| = \Delta$, the set of vectors $\{\lfloor z \rfloor + \{z\} : z \in (x + \Lambda) \cap X_A\}$ are a 1-submodular system. For the residue class Λ, this system is given by $\{\lfloor z \rfloor : z \in \Lambda \cap X_A\} \subseteq \{-1, 0, 1\}^m$ and moreover has full rank as it contains $e_1, \ldots, e_m$, and we are thus in the setting of the classical Heller constant $h(1, m)$.

For the $\Delta - 1$ non-trivial residue classes $x + \Lambda$, $x \notin \Lambda$, we are in the setting of the shifted Heller constant $h_s(m)$. Indeed, as the matrix with columns $\{b_1, \ldots, b_m\} \cup ((x + \Lambda) \cap X_A) \subseteq X_A$ is Δ-submodular, the matrix with columns

$$\{e_1, \ldots, e_m\} \cup \{\lfloor z \rfloor + \{z\} : z \in (x + \Lambda) \cap X_A\}$$

has all its minors, of any size, bounded by 1 in absolute value. By the definition of $h_s(m)$, the second set in this union has at most $h_s(m)$ elements.

As a consequence, we get $n = |X_A| \leq h(1, m) + (\Delta - 1) \cdot h_s(m)$, as desired. $\square$

Remark 1. The proof above shows that we actually want to bound the number of columns n of a matrix $A \in \{-1, 0, 1\}^{m \times n}$ such that the system

$$\{e_1, \ldots, e_m\} \cup \{t + A_1, \ldots, t + A_n\}$$

is 1-submodular, for some $t \in [0, 1)^m \setminus \{0\}$. However, $t + A$ is totally 1-submodular if and only if $\{e_1, \ldots, e_m\} \cup (t + A)$ is 1-submodular.

Remark 2. As any matrix $A \in \{-1, 0, 1\}^{m \times n}$ with pairwise distinct columns can have at most 3^m columns, one trivially gets the bound $h_s(m) \leq 3^m$. Thus, Lemma 1 directly implies the estimate $h(\Delta, m) \leq 3^m \cdot \Delta$.

2.1   Small Dimensions and Lower Bounds in the Shifted Setting

Recall that the original Heller constant is given by $h(1, m) = m^2 + m + 1$. The following exact results for dimensions two and three show the difference between this original (unshifted) and the shifted setting grasped by $h_s(m)$. Note that as in the shifted setting we require $t \neq 0$, the Heller constant $h(1, m)$ is not a lower bound on the shifted Heller constant $h_s(m)$.

Proposition 1. *We have* $h_s(2) = 6$ *and* $h_s(3) = 12$.

Proof. First, we show that $h_s(2) = 6$. Let $A \in \{-1, 0, 1\}^{2 \times n}$ have distinct columns and let $t \in [0, 1)^2 \setminus \{0\}$ be such that $t + A$ is totally 1-submodular. Since $t \neq 0$, it has a non-zero coordinate, say $t_1 > 0$. As the 1×1 minors of $t + A$, that is, the entries of $t + A$, are bounded in absolute value by 1, we get that the first row of A can only have entries in $\{-1, 0\}$. This shows already that $n \leq 6$, as there are simply only 6 options for the columns of A respecting this condition.

An example attaining this bound is given by

$$A = \begin{bmatrix} -1 & -1 & -1 & 0 & 0 & 0 \\ -1 & 0 & 1 & -1 & 0 & 1 \end{bmatrix} \quad \text{and} \quad t = \begin{bmatrix} 1/2 \\ 0 \end{bmatrix}.$$

One can check that (up to permutations of rows and columns) this is actually the unique example (A, t) with 6 columns in A.

Now, we turn our attention to proving $h_s(3) = 12$. The lower bound follows by the existence of the following matrix and translation vector

$$A = \begin{bmatrix} -1 & -1 & -1 & -1 & -1 & -1 & 0 & 0 & 0 & 0 & 0 & 0 \\ -1 & -1 & -1 & 0 & 0 & 0 & -1 & -1 & -1 & 0 & 0 & 0 \\ -1 & 0 & 1 & -1 & 0 & 1 & -1 & 0 & 1 & -1 & 0 & 1 \end{bmatrix} \quad \text{and} \quad t = \begin{bmatrix} 1/2 \\ 1/2 \\ 0 \end{bmatrix}.$$

Checking that $t + A$ is indeed totally 1-submodular is a routine task that we leave to the reader.

For the upper bound, let $A \in \{-1, 0, 1\}^{3 \times n}$ and $t \in [0, 1)^3 \setminus \{0\}$ be such that $t + A$ is totally 1-submodular. Let s be the number of non-zero entries of $t \neq 0$. Just as we observed for $h_s(2)$, we get that there are $s \geq 1$ rows of A only containing elements from $\{-1, 0\}$. Thus, if $s = 3$ there are only $2^3 = 8$ possible columns and if $s = 2$, there are only $2^2 \cdot 3 = 12$ possible columns, showing that $n \leq 12$ in both cases.

We are left with the case that $s = 1$, and we may assume that A has no entry equal to 1 in the first row and that $t_1 > 0$. Assume for contradiction that $n \geq 13$. There must be $\ell \geq 7$ columns of A with the same first coordinate, which we subsume into the submatrix A'. By the identity $h(1, 2) = 7$ applied to the last two rows, and $t_2 = t_3 = 0$, we must have $\ell = 7$ and up to permutations and multiplication of any of the last two rows by -1, $A' = \begin{bmatrix} a & a & a & a & a & a & a \\ 0 & 1 & 0 & -1 & 0 & 1 & -1 \\ 0 & 0 & 1 & 0 & -1 & -1 & 1 \end{bmatrix}$, for some $a \in \{-1, 0\}$. Since the absolute values of the 2×2 minors of $t + A$ are bounded by 1, the remaining $n - \ell \geq 6$ columns of A are different from $(b, 1, 1)^\mathsf{T}$ and $(b, -1, -1)^\mathsf{T}$, where b is such that $\{a, b\} = \{-1, 0\}$. Under these conditions, we find that A contains either $B = \begin{bmatrix} -1 & -1 & 0 \\ 1 & 0 & 1 \\ 0 & -1 & -1 \end{bmatrix}$, $B' = \begin{bmatrix} -1 & -1 & 0 \\ -1 & 0 & -1 \\ 0 & 1 & 1 \end{bmatrix}$, $C = \begin{bmatrix} 0 & 0 & -1 \\ 1 & 0 & 1 \\ 0 & -1 & -1 \end{bmatrix}$ or $C' = \begin{bmatrix} 0 & 0 & -1 \\ -1 & 0 & -1 \\ 0 & 1 & 1 \end{bmatrix}$ as a submatrix. However, both the conditions $|\det(t + B)| \leq 1$ and $|\det(t + B')| \leq 1$ give $t_1 \geq 1$, and both $|\det(t + C)| \leq 1$ and $|\det(t + C')| \leq 1$

give $t_1 \le 0$. Hence, in either case we get a contradiction to the assumption that $0 < t_1 < 1$. □

Combining Lemma 1, the identity $h(1, m) = m^2 + m + 1$, and Proposition 1 yields the bounds $h(\Delta, 2) \le 6\Delta + 1$ and $h(\Delta, 3) \le 12\Delta + 1$. The latter bound improves upon Theorem 1. However, as $h(\Delta, 2) = 4\Delta + 3$ by Theorem 1, we see that the approach via the shifted Heller constant $h_s(m)$ cannot give optimal results for all m.

A quadratic lower bound on $h_s(m)$ can be obtained as follows:

Proposition 2. *For every $m \in \mathbb{Z}_{>0}$, we have*

$$h_s(m) \ge h(1, m - 1) = m(m - 1) + 1.$$

Proof. Let $A' \in \{-1, 0, 1\}^{(m-1) \times n}$ be a totally unimodular matrix with $n = h(1, m - 1)$ columns, and let $A \in \{-1, 0, 1\}^{m \times n}$ be obtained from A' by simply adding a zero-row as the first row. Then, for the translation vector $t = (\frac{1}{m}, 0, \ldots, 0)^\mathsf{T}$ the matrix $t + A$ is totally 1-submodular.

Indeed, we only need to look at its $k \times k$ minors, for $k \le m$, that involve the first row, as A' is totally unimodular by choice. But then, the triangle inequality combined with developing the given minor by the first row, shows that its absolute value is bounded by 1. □

3   A Polynomial Upper Bound on $h_s(m)$

An elegant and alternative proof for Heller's result that $h(1, m) = m^2 + m + 1$ has been suggested by Bixby & Cunningham [2] and carried out in detail in Schrijver's book [10, § 21.3]. They first reduce the problem to consider only the supports of the columns of a given (totally) unimodular matrix and then apply Sauer's Lemma from extremal set theory that guarantees the existence of a large cardinality set that is shattered by a large enough family of subsets of $[m]$.

We show that this approach can in fact be adapted for the shifted Heller constant $h_s(m)$. The additional freedom in the problem that is introduced by the translation vectors $t \in [0, 1)^m \setminus \{\mathbf{0}\}$ makes the argument a bit more involved, but still gives a low degree polynomial bound. To this end, we write $\mathrm{supp}(y) := \{j \in [m] : y_j \ne 0\}$ for the *support* of a vector $y \in \mathbb{R}^m$ and

$$\mathcal{E}_A := \{\mathrm{supp}(A_i) : i \in [n]\} \subseteq 2^{[m]}$$

for the family of supports in a matrix $A \in \mathbb{R}^{m \times n}$ with columns $A_1, \ldots, A_n$. We use the notation 2^Y for the power set of a finite set Y.

Just as in the unshifted Heller setting, each support can be realized by at most two columns of A, if there exists a translation vector $t \in [0, 1)^m$ such that $t + A$ is totally 1-submodular.

Proposition 3. *Let $A \in \{-1, 0, 1\}^{m \times n}$ and $t \in [0, 1)^m$ be such that $\Delta_k(t + A) \le 1$, for $k \in \{1, 2\}$. Then, each $E \in \mathcal{E}_A$ is the support of at most two columns of A.*

Proof. Observe that in view of the condition $\Delta_1(t + A) \leq 1$ and the assumption that $t_i \geq 0$, for every $i \in [m]$, we must have $t_r = 0$, as soon as there is an entry equal to 1 in the rth row of A.

Now, assume to the contrary that there are three columns A_i, A_j, A_k of A having the same support $E \in \mathcal{E}_A$. Then, clearly $|E| \geq 2$ and the restriction of the matrix $(A_i, A_j, A_k) \in \{-1, 0, 1\}^{m \times 3}$ to the rows indexed by E is a ± 1-matrix. Also observe that there must be two rows $r, s \in E$ so that (A_i, A_j, A_k) contains an entry equal to 1 in both of these rows. Indeed, if there is at most one such row, then the columns A_i, A_j, A_k cannot be pairwise distinct. Therefore, we necessarily have $t_r = t_s = 0$. Now, there are two options. Either two of the columns A_i, A_j, A_k are such that their restriction to the rows r, s give linearly independent ± 1-vectors. This however would yield a 2×2 submatrix of $t + A$ with minor ± 2, contradicting that $\Delta_2(t + A) \leq 1$. In the other case, the restriction of the three columns to the rows r, s has the form $\pm \begin{bmatrix} 1 & 1 & 1 \\ 1 & 1 & 1 \end{bmatrix}$ or $\pm \begin{bmatrix} 1 & 1 & -1 \\ 1 & 1 & -1 \end{bmatrix}$, up to permutation of the indices i, j, k. If $|E| = 2$, then this cannot happen as A is assumed to have pairwise distinct columns. So, $|E| \geq 3$, and considering the columns, say A_i, A_j, which agree in the rows r, s, there must be another index $\ell \in E \setminus \{r, s\}$ such that $(A_i)_\ell = 1$ and $(A_j)_\ell = -1$, or vice versa. In any case this means that also $t_\ell = 0$ and that there is a 2×2 submatrix of $t + A$ in the rows r, ℓ consisting of linearly independent ± 1-vectors. Again this contradicts that $\Delta_2(t + A) \leq 1$, and thus proves the claim. $\square$

As mentioned above, this observation on the supports allows to use *Sauer's Lemma* from extremal set theory which we state for the reader's convenience. It was independently published by Sauer [9] and Shelah [11] (who also credits M. Perles) in 1972, and again independently by Vapnik & Chervonenkis [13] a few years earlier.

Lemma 2. *Let $m, k \in \mathbb{Z}_{>0}$ be such that $m > k$. If $\mathcal{E} \subseteq 2^{[m]}$ is such that $|\mathcal{E}| > \binom{m}{0} + \binom{m}{1} + \ldots + \binom{m}{k}$, then there is a subset $Y \subseteq [m]$ with $k + 1$ elements that is shattered by $\mathcal{E}$, meaning that $\{E \cap Y : E \in \mathcal{E}\} = 2^Y$.*

Now, the strategy to bounding the number of columns in a matrix $A \in \{-1, 0, 1\}^{m \times n}$ such that $t + A$ is totally 1-submodular for some $t \in [0, 1)^m$ is to use the inequality $|\mathcal{E}_A| \geq \frac{1}{2}n$, which holds by Proposition 3, and then to argue by contradiction. Indeed, if $n > 2 \sum_{i=0}^{k-1} \binom{m}{i}$, then by Sauer's Lemma there would be a k-element subset $Y \subseteq [m]$ that is shattered by $\mathcal{E}_A$. In terms of the matrix A, this means that (possibly after permuting rows or columns) it contains a submatrix of size $k \times 2^k$ which has exactly one column for each of the 2^k possible supports and where in each column the non-zero entries are chosen arbitrarily from $\{-1, 1\}$. For convenience we call any such matrix a *Sauer Matrix* of size k. For concreteness, a Sauer Matrix of size 3 is of the form

$$\begin{bmatrix} 0 & \pm 1 & 0 & 0 & \pm 1 & \pm 1 & 0 & \pm 1 \\ 0 & 0 & \pm 1 & 0 & \pm 1 & 0 & \pm 1 & \pm 1 \\ 0 & 0 & 0 & \pm 1 & 0 & \pm 1 & \pm 1 & \pm 1 \end{bmatrix},$$

for any choice of signs.

The combinatorial proof of $h(1, m) = m^2 + m + 1$ is based on the fact that no Sauer Matrix of size 3 is totally 1-submodular. This is discussed in Schrijver [10, §21.3], Bixby & Cunningham [2], and Tutte [12], and also implicitly in the analysis of the first equation on page 1361 of Heller's paper [5]. In order to extend this kind of argument to the shifted setting, we need some more notation.

Definition 2. *Let S be a Sauer Matrix of size k. We say that a vector $r \in [0, 1)^k$ is* feasible *for S if $r + S$ is totally 1-submodular. Further, we say that S is* feasible for translations *if there exists a vector $r \in [0, 1)^k$ that is feasible for S, and otherwise we say that S is* infeasible for translations.

Moreover, the Sauer Matrix S is said to be of type $(s, k - s)$*, if there are exactly s rows in S that contain at least one entry equal to 1.*

Note that there is (up to permuting rows or columns) only one Sauer Matrix of type $(0, k)$. As feasibility of a Sauer Matrix of type $(s, k - s)$ is invariant under permuting rows, we usually assume that each of its first s rows contains an entry equal to 1.

Proposition 4. *Let $m, k \in \mathbb{Z}_{>0}$ be such that $m > k$ and assume that no Sauer Matrix of size k is feasible for translations. Then,*

$$h_s(m) \leq 2 \cdot \sum_{i=0}^{k-1} \binom{m}{i} \in \mathcal{O}(m^{k-1}).$$

Proof. Assume for contradiction that there is a matrix $A \in \{-1, 0, 1\}^{m \times n}$ and a translation vector $t \in [0, 1)^m$ such that $t + A$ is totally 1-submodular and $n > 2 \sum_{i=0}^{k-1} \binom{m}{i}$. By Proposition 3, we have $|\mathcal{E}_A| \geq \frac{1}{2}n > \sum_{i=0}^{k-1} \binom{m}{i}$ and thus by Sauer's Lemma (up to permuting rows or columns) the matrix A has a Sauer Matrix S of size k as a submatrix. Writing $r \in [0, 1)^k$ for the restriction of t to the k rows of A in which we find the Sauer Matrix S, we get that by the total 1-submodularity of $t + A$, the matrix $r + S$ necessarily must be totally 1-submodular as well. This however contradicts the assumption. □

In contrast to the unshifted setting, for the sizes 3 and 4, there are Sauer Matrices S and vectors r, such that $r + S$ is totally 1-submodular. For instance,

$$S = \begin{bmatrix} 0 & -1 & 0 & 0 & -1 & -1 & 0 & -1 \\ 0 & 0 & -1 & 0 & -1 & 0 & -1 & -1 \\ 0 & 0 & 0 & -1 & 0 & -1 & -1 & -1 \end{bmatrix}, \quad r = \begin{bmatrix} 1/2 \\ 1/2 \\ 1/2 \end{bmatrix},$$

and

$$S = \begin{bmatrix} 0 & -1 & 0 & 0 & 0 & -1 & -1 & -1 & 0 & 0 & 0 & -1 & -1 & -1 & 0 & -1 \\ 0 & 0 & -1 & 0 & 0 & -1 & 0 & 0 & -1 & -1 & 0 & -1 & -1 & 0 & -1 & -1 \\ 0 & 0 & 0 & -1 & 0 & 0 & -1 & 0 & -1 & 0 & -1 & -1 & 0 & -1 & -1 & -1 \\ 0 & 0 & 0 & 0 & -1 & 0 & 0 & -1 & 0 & -1 & -1 & 0 & -1 & -1 & -1 & -1 \end{bmatrix}, \quad r = \begin{bmatrix} 1/2 \\ 1/2 \\ 1/2 \\ 1/2 \end{bmatrix}.$$

In both cases, $2(r + S)$ is a matrix all of whose entries are either 1 or -1. By Hadamard's inequality, the determinant of any ± 1-matrix of size $k \leq 4$ is at most 2^k, and thus $\Delta_k(r + S) \leq 1$ for all $k \leq 4$, in the two examples above.

Our aim is to show that this pattern does not extend to higher dimensions, and that no Sauer Matrix of size 5 is feasible for translations. The proof requires a more detailed study of Sauer Matrices of special types and sizes 4 and 5.

Proposition 5.

(i) *The vector $r \in [0,1)^4$ is feasible for the Sauer Matrix of type $(0,4)$ if and only if $r = (\frac{1}{2}, \frac{1}{2}, \frac{1}{2}, \frac{1}{2})^\mathsf{T}$.*

(ii) *The Sauer Matrix of type $(0,5)$ is infeasible for translations.*

(iii) *No Sauer Matrix of type $(1,4)$ is feasible for translations.*

(iv) *If $r \in [0,1)^4$ is feasible for a Sauer Matrix of type $(1,3)$, then $r = (0, \frac{1}{2}, \frac{1}{2}, \frac{1}{2})^\mathsf{T}$.*

(v) *No Sauer Matrix of type $(2,3)$ is feasible for translations.*

The proof of these statements is based on identifying certain full-rank submatrices of the respective Sauer Matrix for which the minor condition provides a strong obstruction for feasibility. The details are given in Sect. 4.

Lemma 3. *There does not exist a Sauer Matrix S of size 5 and a translation vector $r \in [0,1)^5$ such that $r + S$ is totally 1-submodular.*

Proof. Assume that there is a Sauer Matrix S of size 5 and a vector $r \in [0,1)^5$ such that $\Delta_k(r + S) \leq 1$, for all $k \leq 5$. Note that if in the ith row of S there is an entry equal to 1, then $r_i = 0$, because of $\Delta_1(r + S) \leq 1$. So, if there are three rows in S containing an entry equal to 1, then they contain a Sauer Matrix of size 3 that is itself totally 1-submodular. However, we already noted that no such Sauer Matrix exists.

Thus, we may assume that S is a Sauer Matrix whose type is either $(0,5)$, $(1,4)$, or $(2,3)$. We have proven in Proposition 5 (ii), (iii), and (v), however, that all such Sauer Matrices are infeasible for translations. $\square$

With these preparations we are now able to prove our main result.

Proof (Theorem 2). In view of Lemma 1, we have $h(\Delta, m) \leq h(1, m) + (\Delta - 1) \cdot h_s(m)$. The claimed bound now follows by Heller's identity $h(1, m) = m^2 + m + 1$ and the fact that $h_s(m) \leq 2 \sum_{i=0}^4 \binom{m}{i}$, which holds by combining Proposition 4 and Lemma 3. $\square$

4   Feasibility of Sauer Matrices in Low Dimensions

Here, we complete the discussion from the previous section and give the proof of Proposition 5. Parts of the argument are based on the observation that the condition $|\det(r + M)| \leq 1$, for any $M \in \mathbb{R}^{k \times k}$, is equivalent to a pair of linear inequalities in the coordinates of $r \in \mathbb{R}^k$. This turns the question on whether a given Sauer Matrix is feasible for translations into the question of whether an associated polyhedron is non-empty.

Proof (Proposition 5). **(i)**: Assume that $r \in [0, 1)^4$ is such that $r + S$ is totally 1-submodular, and consider the following two 4×4 submatrices of S:

$$M = \begin{bmatrix} 0 & 0 & 0 & 0 \\ 0 & 0 & -1 & -1 \\ 0 & -1 & 0 & -1 \\ 0 & -1 & -1 & 0 \end{bmatrix} \quad \text{and} \quad N = \begin{bmatrix} -1 & -1 & -1 & -1 \\ -1 & -1 & 0 & 0 \\ -1 & 0 & -1 & 0 \\ -1 & 0 & 0 & -1 \end{bmatrix}.$$

By the 4×4 minor condition on $r + S$, we have

$$|\det(r + M)| = r_1 \cdot \det \begin{bmatrix} 0 & 1 & 1 \\ 1 & 0 & 1 \\ 1 & 1 & 0 \end{bmatrix} = 2r_1 \leq 1,$$

and hence $r_1 \leq \frac{1}{2}$. Likewise, we have

$$|\det(r + N)| = (1 - r_1) \cdot \det \begin{bmatrix} 0 & 1 & 1 \\ 1 & 0 & 1 \\ 1 & 1 & 0 \end{bmatrix} = 2(1 - r_1) \leq 1,$$

and hence $r_1 \geq \frac{1}{2}$, so that actually $r_1 = \frac{1}{2}$. Analogous arguments for the other coordinates of r, show that $r = (\frac{1}{2}, \frac{1}{2}, \frac{1}{2}, \frac{1}{2})^\mathsf{T}$ as claimed. The fact that $r + S$ is totally 1-submodular has been already discussed above.

(ii): The argument is similar to the one for the first part. Assume for contradiction, that there is a vector $r \in [0, 1)^5$ such that $\Delta_5(r + S) \leq 1$. Consider the following two 5×5 submatrices of S:

$$X = \begin{bmatrix} 0 & 0 & 0 & 0 & 0 \\ 0 & 0 & -1 & -1 & -1 \\ 0 & -1 & 0 & -1 & -1 \\ 0 & -1 & -1 & 0 & -1 \\ 0 & -1 & -1 & -1 & 0 \end{bmatrix} \quad \text{and} \quad Y = \begin{bmatrix} -1 & -1 & -1 & -1 & -1 \\ -1 & -1 & 0 & 0 & 0 \\ -1 & 0 & -1 & 0 & 0 \\ -1 & 0 & 0 & -1 & 0 \\ -1 & 0 & 0 & 0 & -1 \end{bmatrix}.$$

By the 5×5 minor condition on $r + S$, we have

$$|\det(r + X)| = r_1 \cdot \det \begin{bmatrix} 0 & 1 & 1 & 1 \\ 1 & 0 & 1 & 1 \\ 1 & 1 & 0 & 1 \\ 1 & 1 & 1 & 0 \end{bmatrix} = 3r_1 \leq 1,$$

and hence $r_1 \leq \frac{1}{3}$. Likewise, we have

$$|\det(r + Y)| = (1 - r_1) \cdot \det \begin{bmatrix} 0 & 1 & 1 & 1 \\ 1 & 0 & 1 & 1 \\ 1 & 1 & 0 & 1 \\ 1 & 1 & 1 & 0 \end{bmatrix} = 3(1 - r_1) \leq 1.$$

Therefore, we get $r_1 \geq \frac{2}{3}$, a contradiction.

(iii): Without loss of generality, we may assume that the first row of S contains an entry equal to 1, and we assume for contradiction that there is some $r \in [0, 1)^5$ such that $r + S$ is totally 1-submodular. As the entries of $r + S$ are contained in $[-1, 1]$, we get that $r_1 = 0$. Moreover, the last four rows of S contain a Sauer Matrix of type $(0, 4)$. By part (i), this means that $r_2 = r_3 = r_4 = r_5 = \frac{1}{2}$, so that in summary there is only one possibility for the translation vector r.

Now, as $r_1 = 0$, we may multiply the first row of S with -1 if needed, and can assume that the vector $(-1, -1, -1, -1, -1)^\mathsf{T}$ is a column of S. If M denotes any of the four matrices

$$\begin{bmatrix} -1 & -1 & 0 & 0 & 0 \\ -1 & -1 & 0 & 0 & 0 \\ -1 & 0 & -1 & 0 & 0 \\ -1 & 0 & 0 & -1 & 0 \\ -1 & 0 & 0 & 0 & -1 \end{bmatrix}, \begin{bmatrix} -1 & 0 & -1 & 0 & 0 \\ -1 & -1 & 0 & 0 & 0 \\ -1 & 0 & -1 & 0 & 0 \\ -1 & 0 & 0 & -1 & 0 \\ -1 & 0 & 0 & 0 & -1 \end{bmatrix}, \begin{bmatrix} -1 & 0 & 0 & -1 & 0 \\ -1 & -1 & 0 & 0 & 0 \\ -1 & 0 & -1 & 0 & 0 \\ -1 & 0 & 0 & -1 & 0 \\ -1 & 0 & 0 & 0 & -1 \end{bmatrix}, \begin{bmatrix} -1 & 0 & 0 & 0 & -1 \\ -1 & -1 & 0 & 0 & 0 \\ -1 & 0 & -1 & 0 & 0 \\ -1 & 0 & 0 & -1 & 0 \\ -1 & 0 & 0 & 0 & -1 \end{bmatrix}$$

then the absolute value of the determinant of $r + M$ equals $3/2$. Thus, if indeed $\Delta_5(r + S) \leq 1$, then these matrices cannot be submatrices of S. In particular, this implies that

$$M' = \left[\begin{array}{c|cccc} 0 & 1 & 1 & 1 & 1 \\ \hline -1 & -1 & 0 & 0 & 0 \\ -1 & 0 & -1 & 0 & 0 \\ -1 & 0 & 0 & -1 & 0 \\ -1 & 0 & 0 & 0 & -1 \end{array}\right]$$

must be a submatrix of S. However, the determinant of $r + M'$ equals -2, in contradiction to $r + S$ being totally 1-submodular.

(iv): We assume that the first row of each considered Sauer Matrix S of type $(1, 3)$ contains an entry equal to 1, so that $r_1 = 0$. As in (iii) we can moreover assume that $(-1, -1, -1, -1)^\intercal$ is a column of S (by possibly multiplying the first row by -1). We now employ a case distinction based on the signs of the entries in the first row of the columns $a = (\pm 1, -1, 0, 0)^\intercal$, $b = (\pm 1, 0, -1, 0)^\intercal$, and $c = (\pm 1, 0, 0, -1)^\intercal$ of S.

Case 1: $a_1 = b_1 = c_1 = -1$.

Under this assumption, S contains the matrix N from part (i) as a submatrix and thus $r_1 \geq \frac{1}{2}$, contradicting that $r_1 = 0$.

Case 2: $a_1 = b_1 = c_1 = 1$.

In this case, S contains the submatrices

$$A = \left[\begin{array}{c|ccc} 0 & 1 & 1 & 1 \\ \hline 0 & -1 & 0 & 0 \\ 0 & 0 & -1 & 0 \\ 0 & 0 & 0 & -1 \end{array}\right] \quad \text{and} \quad B = \left[\begin{array}{c|ccc} 0 & 1 & 1 & 1 \\ \hline -1 & -1 & 0 & 0 \\ -1 & 0 & -1 & 0 \\ -1 & 0 & 0 & -1 \end{array}\right].$$

The conditions $|\det(r + A)| \leq 1$ and $|\det(r + B)| \leq 1$ translate into the contradicting inequalities $r_2 + r_3 + r_4 \leq 1$ and $r_2 + r_3 + r_4 \geq 2$, respectively.

Case 3: Exactly two of the entries a_1, b_1, c_1 equal -1.

Without loss of generality, we may permute the last three rows of S, and assume that $a_1 = b_1 = -1$. We find that S now contains the submatrices

$$C = \left[\begin{array}{c|ccc} 0 & -1 & -1 & 0 \\ \hline 0 & -1 & 0 & 0 \\ 0 & 0 & -1 & 0 \\ 0 & 0 & 0 & -1 \end{array}\right], \quad D = \left[\begin{array}{c|ccc} 0 & -1 & -1 & 0 \\ \hline -1 & -1 & 0 & 0 \\ -1 & 0 & -1 & 0 \\ -1 & 0 & 0 & -1 \end{array}\right] \quad \text{and} \quad E = \left[\begin{array}{c|ccc} -1 & -1 & -1 & 0 \\ \hline -1 & -1 & 0 & -1 \\ -1 & 0 & -1 & -1 \\ -1 & 0 & 0 & 0 \end{array}\right].$$

The conditions $|\det(r + C)| \leq 1$, $|\det(r + D)| \leq 1$ and $|\det(r + E)| \leq 1$ translate into the contradicting inequalities $r_2 + r_3 \leq 1$, $r_4 \geq \frac{1}{2}$, and $r_4 + 1 \leq r_2 + r_3$, respectively.

Case 4: Exactly two of the entries a_1, b_1, c_1 equal 1.

As in Case 3, we may assume that $a_1 = b_1 = 1$. Here, the following six matrices can be found as submatrices in S:

$$\left[\begin{array}{c|ccc} -1 & 0 & 0 & -1 \\ \hline -1 & -1 & 0 & 0 \\ -1 & 0 & -1 & 0 \\ -1 & 0 & 0 & -1 \end{array}\right], \quad \left[\begin{array}{c|ccc} -1 & 0 & 0 & -1 \\ \hline -1 & -1 & -1 & 0 \\ -1 & 0 & 0 & 0 \\ -1 & 0 & -1 & -1 \end{array}\right], \quad \left[\begin{array}{c|ccc} -1 & 0 & 0 & -1 \\ \hline -1 & 0 & 0 & 0 \\ -1 & -1 & -1 & 0 \\ -1 & -1 & 0 & -1 \end{array}\right],$$

$$\left[\begin{array}{c|ccc} 0 & 1 & 1 & 0 \\ \hline -1 & -1 & 0 & 0 \\ -1 & 0 & -1 & 0 \\ -1 & 0 & 0 & -1 \end{array}\right], \quad \left[\begin{array}{c|ccc} 0 & 1 & 1 & 0 \\ \hline -1 & -1 & 0 & -1 \\ -1 & 0 & -1 & -1 \\ -1 & 0 & 0 & 0 \end{array}\right], \quad \left[\begin{array}{c|ccc} 0 & 1 & 1 & 0 \\ \hline 0 & -1 & 0 & 0 \\ 0 & 0 & -1 & 0 \\ 0 & 0 & 0 & -1 \end{array}\right].$$

The minor conditions for these matrices translate into the inequality system

$$r_4 \leq \tfrac{1}{2} \qquad\qquad r_3 \leq r_2 \qquad\qquad r_2 \leq r_3$$
$$r_4 \geq \tfrac{1}{2} \qquad\qquad r_2 + r_3 \geq 1 \qquad\qquad r_2 + r_3 \leq 1$$

in the same order as the matrices were given above. Solving this system of inequalities shows that necessarily $r_2 = r_3 = r_4 = \frac{1}{2}$, and the proof is complete.

(v): Assume that there is a Sauer Matrix S of type $(2, 3)$ and a vector $r \in [0, 1)^5$ that is feasible for S. Observe that S contains feasible Sauer Matrices of types $(1, 3)$ in its rows indexed by $\{1, 3, 4, 5\}$ and by $\{2, 3, 4, 5\}$. By part (iv) this means that necessarily we have $r = (0, 0, \frac{1}{2}, \frac{1}{2}, \frac{1}{2})^\mathsf{T}$, and we can now argue similarly as we did in part (iii).

First of all, as $r_1 = r_2 = 0$, we may multiply the first or second row of S with -1 if needed, and can assume that the vectors $(-1, 0, -1, -1, -1)^\mathsf{T}$ and $(0, -1, 0, 0, 0)^\mathsf{T}$ are columns of S. We distinguish cases based on the signs of the entries in the first or second row of the columns $a = (\pm 1, 0, -1, 0, 0)^\mathsf{T}$, $b = (\pm 1, 0, 0, -1, 0)^\mathsf{T}$, $c = (\pm 1, 0, 0, 0, -1)^\mathsf{T}$, and $a' = (0, \pm 1, -1, 0, 0)^\mathsf{T}$, $b' = (0, \pm 1, 0, -1, 0)^\mathsf{T}$, $c' = (0, \pm 1, 0, 0, -1)^\mathsf{T}$ of S.

Case 1: $a_1 = b_1 = c_1 = 1$ or $a'_2 = b'_2 = c'_2 = -1$.

Here, one of the matrices

$$
C_1 = \left[\begin{array}{cc|ccc}
0 & 0 & 1 & 1 & 1 \\
0 & -1 & 0 & 0 & 0 \\
\hline
-1 & 0 & -1 & 0 & 0 \\
-1 & 0 & 0 & -1 & 0 \\
-1 & 0 & 0 & 0 & -1
\end{array}\right]
\quad \text{or} \quad
C_2 = \left[\begin{array}{cc|ccc}
0 & -1 & 0 & 0 & 0 \\
0 & 0 & -1 & -1 & -1 \\
\hline
0 & -1 & -1 & 0 & 0 \\
0 & -1 & 0 & -1 & 0 \\
0 & -1 & 0 & 0 & -1
\end{array}\right]
$$

must be a submatrix of S, but the absolute value of the determinant of both $r + C_1$ and $r + C_2$ equals $3/2$.

Case 2: Two of the entries a_1, b_1, c_1 equal -1 or two of the entries a'_2, b'_2, c'_2 equal 1.

Without loss of generality, we may permute the last three rows of S, and assume that either $a_1 = b_1 = -1$ or $a'_2 = b'_2 = 1$. Now, one of the matrices

$$
C_3 = \left[\begin{array}{cc|ccc}
-1 & 0 & -1 & -1 & 0 \\
0 & -1 & 0 & 0 & 0 \\
\hline
-1 & 0 & -1 & 0 & 0 \\
-1 & 0 & 0 & -1 & 0 \\
-1 & 0 & 0 & 0 & -1
\end{array}\right]
\quad \text{or} \quad
C_4 = \left[\begin{array}{cc|ccc}
-1 & 0 & 0 & 0 & 0 \\
0 & -1 & 1 & 1 & 0 \\
\hline
-1 & 0 & -1 & 0 & 0 \\
-1 & 0 & 0 & -1 & 0 \\
-1 & 0 & 0 & 0 & -1
\end{array}\right]
$$

must be a submatrix of S, but again the absolute value of the determinant of both $r + C_3$ and $r + C_4$ equals $3/2$.

Case 3: Up to permuting the last three rows of S we have $\left[\begin{smallmatrix} a_1 & b_1 & c_1 \\ a_2 & b_2 & c_2 \end{smallmatrix}\right] = \left[\begin{smallmatrix} -1 & 1 & 1 \\ 1 & -1 & -1 \end{smallmatrix}\right]$.

With this assumption, one of the matrices

$$
\left[\begin{array}{cc|ccc}
-1 & 0 & 0 & 1 & 1 \\
-1 & -1 & 0 & 0 & 0 \\
\hline
-1 & 0 & -1 & 0 & 0 \\
-1 & 0 & 0 & -1 & 0 \\
-1 & 0 & 0 & 0 & -1
\end{array}\right], \;
\left[\begin{array}{cc|ccc}
-1 & -1 & 0 & 0 & 0 \\
1 & 0 & 1 & 0 & 0 \\
\hline
-1 & -1 & -1 & 0 & 0 \\
-1 & 0 & 0 & -1 & 0 \\
-1 & 0 & 0 & 0 & -1
\end{array}\right], \;
\left[\begin{array}{cc|ccc}
1 & 0 & -1 & 1 & 1 \\
-1 & -1 & 0 & 0 & 0 \\
\hline
-1 & 0 & -1 & 0 & 0 \\
-1 & 0 & 0 & -1 & 0 \\
-1 & 0 & 0 & 0 & -1
\end{array}\right], \;
\left[\begin{array}{cc|ccc}
1 & -1 & 0 & 0 & 0 \\
1 & 0 & 0 & -1 & -1 \\
\hline
-1 & -1 & -1 & 0 & 0 \\
-1 & 1 & 0 & -1 & 0 \\
-1 & -1 & 0 & 0 & -1
\end{array}\right]
$$

must be a submatrix of S, because one of the four vectors $(\pm 1, \pm 1, -1, -1, -1)^\mathsf{T}$ must be a column of S. As before, if F denotes any of these four matrices, then the absolute value of the determinant of $r + F$ equals $3/2$.

Case 4: Up to permuting the last three rows of S we have $\left[\begin{smallmatrix} a_1 & b_1 & c_1 \\ a_2 & b_2 & c_2 \end{smallmatrix}\right] = \left[\begin{smallmatrix} -1 & 1 & 1 \\ -1 & -1 & 1 \end{smallmatrix}\right]$.

In this case, one of the matrices

$$
C_7 = \left[\begin{array}{cc|ccc}
-1 & 0 & 1 & 0 & 0 \\
0 & -1 & 0 & -1 & 0 \\
\hline
0 & -1 & 0 & 0 & 0 \\
0 & 0 & -1 & -1 & 0 \\
0 & 0 & 0 & 0 & -1
\end{array}\right]
\quad \text{or} \quad
C_8 = \left[\begin{array}{cc|ccc}
1 & -1 & 0 & 0 & 0 \\
0 & 0 & -1 & -1 & 0 \\
\hline
0 & -1 & -1 & 0 & 0 \\
0 & 0 & 0 & -1 & 0 \\
0 & 0 & 0 & 0 & -1
\end{array}\right]
$$

must be a submatrix of S, because one of the vectors $(\pm 1, 0, 0, 0, 0)^{\mathsf{T}}$ must be a column of S. As before, the absolute value of the determinant of both $r + C_7$ and $r + C_8$ equals $3/2$.

In conclusion, in all cases we found a 5×5 minor of $r + S$ whose absolute value is greater than 1, and thus no feasible Sauer Matrix of type $(2, 3)$ can exist. $\qquad\square$

5 Discussion and Open Problems

The determination of the exact value of $\mathrm{h}(\Delta, m)$ remains an open problem. Note that the bounds from other sources and the bound we prove here are incomparable when both m and Δ vary. In order to understand the limits of our method for upper bounding $\mathrm{h}(\Delta, m)$, it is necessary to determine the exact asymptotic behavior of $\mathrm{h_s}(m)$. Finally, for (partial) verification of the conjecture by Lee et al. one could try checking this conjecture in the cases where m and/or Δ are fixed to small values. The smallest choice of Δ, for which the conjecture is open is $\Delta = 3$. As for the case of fixed m, we suspect that our upper bounds on $\mathrm{h}(\Delta, m)$, for $m = 3$ and $m = 4$, are not tight.

References

1. Artmann, S., Weismantel, R., Zenklusen, R.: A strongly polynomial algorithm for bimodular integer linear programming. In: Proceedings of the 49th Annual ACM SIGACT Symposium on Theory of Computing, pp. 1206–1219 (2017)
2. Bixby, R.E., Cunningham, W.H.: Short cocircuits in binary matroids. Eur. J. Comb. **8**, 213–225 (1987)
3. Fiorini, S., Joret, G., Weltge, S., Yuditsky, Y.: Integer programs with bounded sub-determinants and two nonzeros per row. https://arxiv.org/abs/2106.05947 (2021)
4. Glanzer, C., Weismantel, R., Zenklusen, R.: On the number of distinct rows of a matrix with bounded subdeterminants. SIAM J. Discrete Math. **32**(3), 1706–1720 (2018)
5. Heller, I.: On linear systems with integral valued solutions. Pac. J. Math. **7**, 1351–1364 (1957)
6. Lee, J.: Subspaces with well-scaled frames. Linear Algebra Appl. **114–115**, 21–56 (1989)
7. Lee, J., Paat, J., Stallknecht, I., Xu, L.: Polynomial upper bounds on the number of differing columns of Δ-modular integer programs. https://arxiv.org/abs/2105.08160 (2021)
8. Nägele, M., Santiago, R., Zenklusen, R.: Congruency-Constrained TU Problems Beyond the Bimodular Case. https://arxiv.org/abs/2109.03148 (2021)
9. Sauer, N.: On the density of families of sets. J. Combin. Theory Ser. A **13**, 145–147 (1972)
10. Schrijver, A.: Theory of linear and integer programming. Wiley-Interscience Series in Discrete Mathematics, John Wiley & Sons Ltd, Chichester (1986), a Wiley-Interscience Publication

11. Shelah, S.: A combinatorial problem; stability and order for models and theories in infinitary languages. Pac. J. Math. **41**, 247–261 (1972)
12. Tutte, W.T.: A homotopy theorem for matroids. II. Trans. Am. Math. Soc. **88**, 161–174 (1958)
13. Vapnik, V.N., Chervonenkis, A.Y.: On the uniform convergence of relative frequencies of events to their probabilities. Theory Probab. Appl. **16**, 264–280 (1971)

The Simultaneous Semi-random Model for TSP

Eric Balkanski$^{(\boxtimes)}$, Yuri Faenza, and Mathieu Kubik

IEOR Department, Columbia University, New York, USA
eb3224@columbia.edu

Abstract. Worst-case analysis is a performance measure that is often too pessimistic to indicate which algorithms we should use in practice. A classical example is in the context of the Euclidean Traveling Salesman Problem (TSP) in the plane, where local search performs extremely well in practice even though it only achieves an $\Omega(\frac{\log n}{\log \log n})$ worst-case approximation ratio. In such cases, a natural alternative approach to worst-case analysis is to analyze the performance of algorithms in semi-random models.

In this paper, we propose and investigate a novel semi-random model for the Euclidean TSP. In this model, called the simultaneous semi-random model, an instance over n points consists of the union of an adversarial instance over $(1 - \alpha)n$ points and a random instance over αn points, for some $\alpha \in [0, 1]$. As with smoothed analysis, the semi-random model interpolates between distributional (random) analysis when $\alpha = 1$ and worst-case analysis when $\alpha = 0$. In contrast to smoothed analysis, this model trades off allowing some completely random points in order to have other points that exhibit a fully arbitrary structure.

We show that with only an $\alpha = \frac{1}{\log n}$ fraction of the points being random, local search achieves an $\mathcal{O}(\log \log n)$ approximation in the simultaneous semi-random model for Euclidean TSP in fixed dimensions. On the other hand, we show that at least a polynomial number of random points are required to obtain an asymptotic improvement in the approximation ratio of local search compared to its worst-case approximation, even in two dimensions.

Keywords: Traveling Salesman Problem · Semi-random Models · Local Search

1 Introduction

The Traveling Salesman Problem (TSP) is a cornerstone of integer programming and combinatorial optimization, having been investigated for more than 60 years. Since Dantzig, Fulkerson, and Johnson [10] developed the cutting plane method to solve a (then astonishing) 42 cities instance, the TSP has been at the forefront of research in optimization, pushing the limits of computation in practice (see, e.g., [2,5,9,25]), while at the same time being the test bed for many new ideas in the theory of algorithms (see, e.g., [16,18,24,31,32]).

© Springer Nature Switzerland AG 2022
K. Aardal and L. Sanitá (Eds.): IPCO 2022, LNCS 13265, pp. 43–56, 2022.
https://doi.org/10.1007/978-3-031-06901-7_4

The simplest non-trivial TSP instances are arguably those that go under the name of *Euclidean*: given a set of points in the d-dimensional unit cube, find a cycle of minimum total length containing all those points (a *tour*), where the length of an edge between any two points is given by their Euclidean distance. Euclidean TSP is NP-hard [15,26], but in fixed dimension a PTAS can be obtained using approximate dynamic programming ideas [3]. However, in practice, even simple algorithms perform very well on Euclidean instances. Take for instance the 2-*opt* local search algorithm: given the current tour T, orient it arbitrarily and let (a,b) and (c,d) be two edges of T, traversed in this order. Consider the tour T' obtained from T by replacing (a,b), (c,d) with (a,c), (b,d) (i.e., performing a *swap*). If T' has a strictly smaller length than T', let $T = T'$ and iterate; else, attempt to swap two different pairs of edges from T. The algorithm halts when all pairs of edges from the current tour T have been tested for a swap, with none leading to an improved tour.

2-opt is known to perform extremely well on many Euclidean instances, such as those from the TSPLIB library, both in terms of convergence time and quality of the output [12,17,27]. However, classical worst-case analysis does not seem adequate to match these empirical findings with theorems on the performance of 2-opt. For instance, it is known that 2-opt only gives an $\Omega(\frac{\log n}{\log \log n})$-approximation for Euclidean TSP in the plane and may terminate after a number of steps exponential in n [7,8], where n is the number of points. A fundamental quest(ion) is thus to find a theoretical explanation for the empirical performance of 2-opt:

Why does local search perform well on TSP in practice?

A first, natural alternative model assumes that the n points are distributed independently and uniformly at random, instead of being given adversarially. Following [28], we call this model *Distributional*. In the distributional model, the performance of 2-opt – and, more generally, optimal solutions to the Euclidean TSP – are well-understood for fixed dimensions d. The expected number of iterations of 2-opt is polynomial in n [8], while its output obtains, with high probability, a constant factor approximation to the optimal tour. This latter fact holds since the value of the solution found by 2-opt on *any* set of n points in the d-dimensional unit cube is $\mathcal{O}(n^{1-1/d})$ [8] and the length of the optimal tour in the distributional model is, with high probability, $\Omega(n^{1-1/d})$ [30]. However, a main limitation of the distributional model is that random instances have a very particular structure. For example, for a random instance of size n in the unit square, any region of constant size $c \in [0,1]$ contains, with high probability, $cn \pm \varepsilon$ points.

In order to interpolate between worst-case scenarios and distributional models, much research in optimization has been devoted to define and study *semi-random* models. Such models contain both an adversarial and a random component. A classical example is *smoothed analysis*, where all the input data is perturbed by some noise, and the performance of the algorithm is then studied on the perturbed instance. In the Euclidean TSP case, this perturbation is usually achieved by adding to the positions of each point a value sampled from the same Gaussian distribution $N(0, \sigma)$. It is known that, in this model, the expected running time and approximation ratio of 2-opt are polynomial in σ and logarithmic in $1/\sigma$, respectively [12,22]. Other common semi-random models for discrete optimization problems first generate a random instance, e.g., a graph, and then adversarialy perturbs it, e.g., by adding/removing edges of the graph, or vice versa (see e.g., [28]).

1.1 Our Contributions

A New Semi-random Model. As a step towards answering the motivating question of this paper, we define and study a new semi-random model for Euclidean TSP instances, that we dub *Simultaneous Semi-Random*. In this model, a $1 - \alpha$ fraction of the points are chosen by an adversary and an α fraction of the points are uniformly random, for some parameter $\alpha \in [0, 1]$. This semi-random model provides an explanation for the approximation performance of algorithms that complements the explanation provided by smoothed analysis. In order to appreciate this complementarity, we distinguish two different levels of the structure of a point set instance in the unit square. Given a parameter $c < 1$, consider a $c^{-1} \times c^{-1}$ grid that partitions the unit square into squares of size $c \times c$ called *local regions*. The *global structure* of an instance is the number of points inside of each local region. The *local structure* of a local region is the positions of the points in that region.

Informally, smoothed instances exhibit an arbitrary global structure and random local structures. In contrast, simultaneous semi-random instances have arbitrary local structures, except for a small random fraction of the local regions. Thus, smoothed analysis explains the performance of local search on instances with specific global structures, e.g., instances where all the points are only in a constant number of local regions. In contrast, our simultaneous semi-random model explains the performance of local search on instances with specific local structures, e.g., points that form perfectly straight lines. In other words, this semi-random model tradeoffs allowing some completely random points in order to capture instances where there is a subset of the points that exhibit a fully arbitrary structure.

Bounds. We show that an $\alpha = 1/\log n$ fraction of random points are sufficient for local search to obtain an $\mathcal{O}(\log \log n)$ approximation ratio in the simultaneous semi-random model in constant dimensions, which improves over the lower bound $\Omega(\log n / \log \log n)$ from worst-case analysis, which holds even in two dimensions [8].

Theorem 1. *For Euclidean TSP in $[0, 1]^d$ where d is constant, 2-opt obtains, with probability $1 - o(1)$, an $\mathcal{O}(\log \log n)$-approximation ratio in the simultaneous semi-random model, with $\alpha = \frac{1}{\log n}$.*

Theorem 1 is proved in Sect. 3. This result implies that the hard instances of Euclidean TSP are not robust to the addition of a small number of random points. Combined with smoothed analysis, we get that either a small amount of random noise to all points or a small fraction of completely random points improves the performance of local search. Interestingly, even though the analyses are completely different, the "amount of noise" σ needed for smoothness, and the fraction of points α needed for the simultaneous model, to improve the approximation to $\mathcal{O}(\log \log n)$ is $1/\log n$ in both cases.

We note that we actually prove a result that is stronger than Theorem 1 in many ways. For instance, one can take $\alpha = \frac{1}{\log^\delta n}$ for any constant $\delta > 0$ without changing the approximation ratio. We refer the reader to Sect. 3 for details. From our proof, it is also easy to see that we obtain the same result in the more challenging model where the adversary may first observe the random points before placing the adversarial points.

Our second main result is that if $\alpha \leq n^{-3/5-\varepsilon}$, then the approximation ratio of local search cannot be improved in the simultaneous semi-random model compared to its worst-case approximation.

Theorem 2. *For Euclidean TSP in $[0,1]^2$, 2-opt achieves, with probability $1 - o(1)$, an $\Omega\left(\frac{\log n}{\log \log n}\right)$ approximation ratio in the simultaneous semi-random model with $\alpha = n^{-3/5-\varepsilon}$, for any constant $\varepsilon > 0$.*

Theorem 2, which is proved in Sect. 4, implies that polynomially many random points are required to obtain an approximation that asymptotically improves over the worst-case approximation. We believe that closing the gap between Theorem 1 and Theorem 2 is an intriguing open problem, and in particular resolving whether there is some $\alpha = o(1)$ such that local search obtains a constant approximation. Answering this question could shed further light on the relationship between the simultaneous semi-random model and "real-world" behavior of the local search algorithm. Obtaining bounds on the running time of local search in this model and investigating it in the context of other optimization problems are also interesting paths forward.

1.2   Technical Overview

The Upper Bound. The upper bound consists of two main steps. We first show a new upper bound on the worst-case length of a 2-optimal set of edges over an instance V that gives an $\mathcal{O}(\log \frac{n^{1-1/d}}{\mathrm{OPT}_V})$ approximation (here and throughout the paper, an instance V is given by a set of point in the Euclidean space). This bound is useful because it separates adversarial instances V into two regimes. In the first regime, the optimal length of a tour is large ($\mathrm{OPT}_V = \Omega(\frac{n^{1-1/d}}{\log n})$) and the approximation of local search on V, even without random points, is $\mathcal{O}(\log \log n)$. In the second regime, OPT_V is small and we get that the optimal tour length OPT_R over the random points R, with $\alpha = 1/\log n$, is such that $\mathrm{OPT}_R \geq \mathrm{OPT}_V$. We then use our newly proved worst-case bound to analyze the lengths of 2-optimal tours and optimal tours in the simultaneous semi-random model by combining bounds from both worst-case and distributional analysis.

The Lower Bound. We first present a framework that reduces proving lower bounds in the simultaneous semi-random model to constructing an adversarial instance V and a Hamiltonian path P over V from a point $s \in V$ to a point $t \in V$ that satisfy three parametrized properties: the length $\ell(P)$ of P is such that $\ell(P) \geq \gamma \text{OPT}_V$ (called the γ-*bad property*), $\text{OPT}_V \geq L$ (called L-*long*), and finally, $P \cup \{(s,x)\}$, $P \cup \{(x,y)\}$ and $P \cup \{(t,y)\}$ are 2-optimal for some $K \subseteq [0,1]^2$ and any $x, y \in K$ (called K-*resistant*). Existing hard instances of course satisfy the γ-bad property with $\gamma = \Theta(\log n / \log \log n)$, but they do not satisfy the L-long and the K-resistant properties for desirable parameters L and K. These latter two properties cause significant additional challenges.

Our construction starts with the construction from [8] for the $\Omega(\frac{\log n}{\log \log n})$ worst-case lower bound and then consists of three steps that modify it. We first make the construction thinner, so that it fits in $[0,1] \times [0,\varepsilon]$ for some small ε. Then, we stack multiple copies of the thin instance, without incurring any loss in the γ-bad parameter and while keeping the construction relatively thin, to increase the L-long parameter. Finally, the most challenging step is to satisfy the K-resistance property. For that, we carefully add a small number of additional points to the construction so that, with high probability, there is a Hamiltonian path P on the adversarial instance V that can be connected to a 2-optimal Hamiltonian path on the random vertices R to obtain a 2-optimal tour on $V \cup R$.

1.3   Additional Related Work

Approximation Algorithms for Euclidean TSP. For TSP in the plane, Karp [19] showed that a partitioning algorithm that subdivides the points into groups of size t obtains an $\mathcal{O}(\sqrt{n/t})$ approximation, which improves to $\mathcal{O}(t^{-1/2})$ if the points are uniformly random. A seminal result by Arora [3] obtained a PTAS for d-dimensional Euclidean TSP, for any constant d. The approximation ratio of the 2-opt algorithm was recently improved from $\mathcal{O}(\log n)$ [8] to $\mathcal{O}(\frac{\log n}{\log \log n})$ [7] for Euclidean TSP in the plane, which is the best approximation achievable [8]. For additional approximation algorithms results on Euclidean TSP, see e.g. [1,20].

We next discuss three different families of semi-random models.

Semi-random Models with a Monotone Adversary. Seminal work by Blum and Spencer [6] proposed semi-random models for k-coloring. In the colorgame model, edges are first placed at random between pairs of vertices and then an adversary places additional edges. Similar semi-random models where an adversary manipulates randomly generated instances were considered for problems such as minimum bisection and maximum independent set [14].

Smoothed analysis is a semi-random model where random perturbations are applied to an adversarial instance. It was first studied by Spielman and Teng [29] to explain the fast running time of the simplex method in practice. Smoothed analysis of both the running time and approximation of local search (2-opt) for TSP was first studied by [12] who obtained an $O(1/\sigma)$ approximation when Gaussian random variables with mean 0 and standard deviation σ are added to each point. They also obtained more general bounds for any distributions with bounded densities. This approximation was then improved to $O(\log 1/\sigma)$ by [22]. Smoothed analysis of local search has also been studied for general, non-Euclidean, graphs [13] and in the context of clustering [4]. We are not aware of other semi-random models that have been studied for TSP.

Multi-stage Semi-random Models. More complex semi-random models that generate instances in three or more steps, where some steps are adversarial and the others are made randomly, have also been studied for many problems, including unique games [21], partitioning [23], 3-coloring [11], and clustering mixtures of Gaussians [33].

In contrast to all previous semi-random models, where the randomized and adversarial steps occur sequentially, in the simultaneous semi-random model that we introduce and study in this paper these steps occur simultaneously and independently of each other. We are not aware of previously studied semi-random models that have this property.

2   Preliminaries

In the following, given $n, d \in \mathbb{N}$, an instance of size n and dimension d, or n-instance, is a set of n points in $[0,1]^d$. When the dimension is not mentioned, it is assumed to be 2. For $m \in \mathbb{N}$, the *random instance* $R(m)$ is a set of m points drawn uniformly and independently from $[0,1]^d$. For an instance V, we indifferently call $v \in V$ a *point* or a *vertex*. Given $x \in \mathbb{R}^d$, we let $\|x\|$ be its Euclidean norm. For an edge $e = (v_1, v_2)$, we often write $\|e\| = \|v_1 - v_2\|$. The angle between two edges $e = (v_1, v_2)$ and $e' = (v_1', v_2')$ is the angle between the vectors $u = v_2 - v_1$ and $u' = v_2' - v_1'$, which is equal to $\arccos \frac{u \cdot u'}{\|u\| \|u'\|} \in [0, \pi]$. We let $\sqcup$ denote the disjoint union operator of sets.

Given an instance V and a set of m edges $T = \{(v_{i_1}, v_{i_2}) \mid i \in [m]\}$, the *length* of T is $\ell(T) = \sum_{i=1}^m \|v_{i_1} - v_{i_2}\|$. A *tour* on an instance V is a set of $|V|$ edges T that form a cycle. Given an instance V, OPT_V is the length of a tour on V of minimum length. Assume T is an arbitrary collection of edges, a *2-swap* replaces (v_{i_1}, v_{i_2}) and (v_{j_1}, v_{j_2}) in T with (v_{i_1}, v_{j_1}) and (v_{i_2}, v_{j_2}). We say that T is *2-optimal* if there is no set T' of strictly smaller length obtained from T via a 2-swap. In particular, when T defines a tour, the concept of 2-optimality coincides with the stopping criterion for 2-opt. We now present some general facts about optimal and 2-optimal TSP tours on Euclidean instances. The first is a bound on the length of any 2-optimal set of edges.

Lemma 1. ([8]). *Let T be a 2-optimal set of n edges on an instance $V \in [0,1]^d$, and assume d to be a constant. Then $\ell(T) = \mathcal{O}\left(n^{1-1/d}\right)$.*

This in particular implies that the optimal tour, up to a constant factor, always has length at most $n^{1-1/d}$. We also know the behavior of OPT on random instances.

Lemma 2. ([30]). *With probability $1 - o(1)$, we have $\mathrm{OPT}_{R(n)} = \Theta\left(n^{1-1/d}\right)$.*

From Lemma 1 and Lemma 2, we immediately deduce the following corollary.

Corollary 1. *On a random instance, the approximation ratio of 2-opt is constant with probability* $1 - o(1)$.

Last, we recall the following best-known upper bounds on the performance of 2-opt on general instances.

Lemma 3. ([7,8]). *Let V be an arbitrary d-dimensional instance of size n with d constant. Let* $T \subset V^2$ *be a 2-optimal set of edges. Then* $\ell(T) = \mathcal{O}(\log n)\mathrm{OPT_V}$. *Moreover, if* $d = 2$, *then* $\ell(T) = \mathcal{O}\left(\frac{\log n}{\log \log n}\right) \mathrm{OPT_V}$.

3   An Improved Approximation for Local Search in the Simultaneous Semi-random Model

In this section, we show that an $\alpha = 1/\log n$ fraction of random points is sufficient to improve the approximation achieved by local search to $\mathcal{O}(\log \log n)$ in the simultaneous semi-random model.

3.1   An Improved Worst-Case Approximation for Local Search

We first show a new upper bound on the worst case approximation of 2-opt.

Lemma 4. *Let V be an arbitrary d-dimensional instance of size n with d constant. Let* $T \subset V^2$ *be a 2-optimal set of edges. Then* $\ell(T) = \mathcal{O}\left(\mathrm{OPT_V} \log \frac{n^{1-1/d}}{\mathrm{OPT_V}}\right)$.

This new bound is helpful because it separates instances into two regimes. The first is when the length of the optimal tour is large, when $\mathrm{OPT_V} = \Omega(\frac{n^{1-1/d}}{\log n})$. In this regime, we immediately get from Lemma 4 that $\ell(T) = \mathcal{O}(\log \log n \cdot \mathrm{OPT_V})$ for any 2-optimal set of edges T, so a locally optimal tour performs well on the adversarial instance, without even needing random points.

In the second regime, when $\mathrm{OPT_V} = o(\frac{n^{1-1/d}}{\log n})$, we have that for $\alpha = 1/\log n$, the length of the optimal tour on the random points R dominates the length of the optimal tour on the adversarial instance V: $\mathrm{OPT_R} = \Theta(|R|^{1-1/d}) = \Theta((\frac{n}{\log n})^{1-1/d}) \geq \mathrm{OPT_V}$ where the first equality is by Lemma 2. We later combine the constant approximation obtained by 2-opt on random instances and the fact that $\mathrm{OPT_R} \geq \mathrm{OPT_V}$ to get that 2-opt obtains a constant approximation on $V \sqcup R$ in that regime. In summary, Lemma 4 is helpful because it shows that it is only when the length of the optimal tour is small that 2-opt performs poorly on adversarial instances. We will show that, in this regime, adding random points to an adversarial instance improves the approximation obtained by 2-opt.

The remainder of Sect. 3.1 is devoted to the proof of Lemma 4. We first introduce the concepts of similarly-oriented edges and similar-length edges that will be used in the proof.

Similarly-Oriented Edges. We use the notion of similarly-oriented edges from [8].

Definition 1. ([8]). *Edges e and e' are* similarly-oriented *if the angle between e and e' is at most* $\arctan \frac{1}{4}$.

Edges can be partitioned into a constant number of families of edges such that every pair of edges in a same family are similarly-oriented. For a vector $u \in \mathbb{R}^d$, we denote by T^u the collection of all vectors $u' \in T$ such that the angle between u and u' is at most $\frac{1}{2} \arctan \frac{1}{4}$ and by $\mathbb{S}^{d-1}$ the unit sphere in $\mathbb{R}^d$. Thus, for any $u \in \mathbb{R}^d$, every pair of edges in T^u are similarly-oriented. Using the topological definition of compactness, we know that there exists a constant I and $u_1, \ldots u_I \in \mathbb{S}^{d-1}$ such that $T = \cup_{i=1}^I T^{u_i}$. Hence, up to a constant loss, it is sufficient to bound the total length of all edges in T^{u_i} for an arbitrary i. For the remainder of this section, we abuse notation and write T^i instead of T^{u_i}. Similarly oriented edges have the following useful property.

Lemma 5. ([8]). *Let $e = (v_1, v_2)$ and $e' = (v_1', v_2')$ be two similarly-oriented edges which form a 2-optimal set. Then $\|v_1' - v_1\| \geq \frac{1}{2} \min (\|e\|, \|e'\|)$.*

Similar-Length Edges. In addition to being partitioned into families of similarly-oriented edges, edges are also partitioned into families of similar length edges. In particular, let $1 > \eta > \varepsilon > 0$, we define $T_< = \{e \in T \mid \|e\| < \varepsilon\}$, $T_> = \{e \in T \mid \|e\| \geq \eta\}$, and, for any $j \geq 0$ such that $2^j \varepsilon \leq \eta$, $T_j := \{e \in T \mid 2^j \varepsilon \leq \|e\| < 2^{j+1}\varepsilon\}$. Thus writing $J = \lfloor \log_2 \frac{\eta}{\varepsilon} \rfloor$, we have $T = T_< \sqcup T_> \sqcup \bigsqcup_{j=0}^J T_j$. The following result is known for long edges.

Lemma 6. ([8]). *For any $\eta > 0$ and constant dimension d, $\ell(T_>) = \mathcal{O}(\eta^{1-d})$.*

We denote families of edges that are both similarly-oriented and of similar-length by $T_j^i = T^i \cap T_j$. We similarly denote $T_<^i = T^i \cap T_<$ and $T_>^i = T^i \cap T_>$. Now we are ready to prove Lemma 4.

Proof. (of Lemma 4). Let $i \in [I]$ and consider the family $T^i \subset T$ of similarly-oriented edges. First, we have $\ell(T_<^i) \leq n\varepsilon$. Second, by Lemma 6, we have $\ell(T_>^i) \leq \mathcal{O}(\eta^{1-d})$. To bound $\ell(T^i) = \ell(T_<^i) + \ell(T_>^i) + \sum_{j=0}^J \ell(T_j^i)$, it remains to bound the length of the family of similarly-oriented and similar length edges T_j^i for an arbitrary $j \in [J]$.

Let T^* be an optimal tour on V: if we fix any point to be the first one, T^* defines an order on V, and we can use it to order T_j^i by saying that $(v_1, v_2) < (v_1', v_2')$ if $v_1 < v_1'$ in T^*. Hence, for fixed i and j, we can enumerate $T_j^i = \{e^l = (v_1^l, v_2^l) \mid 1 \leq l \leq N\}$ (where $N = |T_j^i|$), such that v_1^l appears before v_1^{l+1} in T^*.

Let $\widetilde{T}^* = \{(v_1^l, v_1^{l+1}) \mid 1 \leq l \leq N\}$ (where we let $v_1^{N+1} = v_1^1$). By the triangular inequality, we have $\ell(\widetilde{T}^*) \leq \ell(T^*)$. Moreover, for any l, since $T_j^i = \{e^l = (v_1^l, v_2^l) \mid 1 \leq l \leq N\} \subseteq T$ and is therefore 2-optimal, we have by Lemma 5 $\|v_1^l - v_1^{l+1}\| \geq \frac{1}{2} \min (\|e^l\|, \|e^{l+1}\|)$. As $e^l, e^{l+1} \in T_j^i$, and the length of every vector in T_j^i is between $2^j \varepsilon$ and $2^{j+1}\varepsilon$, the longest of $\|e^l\|, \|e^{l+1}\|$ is at most two

times larger than the shortest. Thus, we get $||v_1^l - v_1^{l+1}|| \geq \frac{1}{4}||e^l||$. Summing for $l \in [N]$, we obtain $\ell(T_j^i) = \sum_{l=1}^{N} ||e^l|| \leq 4 \sum_{l=1}^{N} ||v_1^l - v_1^{l+1}|| = 4\ell(\widetilde{T}^*) \leq 4\ell(T^*)$.
Hence, putting the bounds on $\ell(T_<^i)$, $\ell(T_>^i)$, $\sum_{j=0}^{J} \ell(T_j^i)$ together, we have

$$\ell(T^i) \leq n\varepsilon + \mathcal{O}(\eta^{1-d}) + \lfloor \log_2 \frac{\eta}{\varepsilon} \rfloor 4\mathrm{OPT} = \mathcal{O}\left(n\varepsilon + \eta^{1-d} + \mathrm{OPT} \log \frac{\eta}{\varepsilon} \right).$$

Summing over all families of similarly-oriented edges and letting $\varepsilon = \frac{\mathrm{OPT}}{n}$ and $\eta = \mathrm{OPT}^{\frac{1}{1-d}}$, we get

$$\ell(T) \leq \sum_{i=1}^{I} \ell(T^i) = \mathcal{O}\left(n\varepsilon + \eta^{1-d} + \mathrm{OPT} \log \frac{\eta}{\varepsilon} \right) = \mathcal{O}\left(\mathrm{OPT} \log \frac{n^{1-\frac{1}{d}}}{\mathrm{OPT}} \right).$$

$\square$

3.2   Proof of Theorem 1

By combining the new worst-case bound from Sect. 3.1 together with the bound in Lemma 1, one obtains the following upper bound on the length of a 2-optimal tour on the union of two instances.

Lemma 7. *Let V and U be disjoint instances of sizes n and m. Then, if d is constant, for any 2-optimal tour T on $V \sqcup U$, $\ell(T) = \mathcal{O}(\mathrm{OPT}_V \log \frac{n^{1-\frac{1}{d}}}{\mathrm{OPT}_V} + m^{1-\frac{1}{d}})$.*

The last lemma needed is a bound on the optimal length of a tour on an instance of the simultaneous semi-random model.

Lemma 8. *Let $n, m \in \mathbb{N}$ and V be a d-dimensional n-instance with d constant. With probability $1 - o(1)$, $\mathrm{OPT}_{V \sqcup R(m)} = \Omega\left(\max\left(\mathrm{OPT}_V, m^{1-\frac{1}{d}} \right) \right)$.*

We are now ready to prove the main result for this section, from which Theorem 1 follows.

Theorem 3. *Let V be any d-dimensional n-instance, with d a fixed constant. With probability $1 - o(1)$, the approximation ratio of 2-opt on $V \sqcup R(m)$ is $\mathcal{O}(1)$ if $m^{1-\frac{1}{d}} > \mathrm{OPT}_V \log \frac{n^{1-\frac{1}{d}}}{\mathrm{OPT}_V}$ and $\mathcal{O}(\log \frac{n^{1-\frac{1}{d}}}{\mathrm{OPT}_V})$ otherwise. In particular, for $m = \frac{n}{\log^c n}$, for any constant $c > 0$, the approximation ratio is $\mathcal{O}(\log \log n)$.*

Proof. Let T be any 2-optimal tour on $V \sqcup R(m)$. If $m^{1-\frac{1}{d}} > \mathrm{OPT}_V \log \frac{n^{1-\frac{1}{d}}}{\mathrm{OPT}_V}$, by Lemma 7 we obtain that $\ell(T) = \mathcal{O}(m^{1-\frac{1}{d}})$ while by Lemma 8 we have $\mathrm{OPT}_{V \sqcup R(m)} = \Omega(m^{1-\frac{1}{d}})$. Hence, the approximation ratio of 2-opt on this instance is $\mathcal{O}(1)$. However, when $m^{1-\frac{1}{d}} \leq \mathrm{OPT}_V \log \frac{n^{1-\frac{1}{d}}}{\mathrm{OPT}_V}$ Lemma 7 tells us that $\ell(T) = \mathcal{O}(\mathrm{OPT}_V \log \frac{n^{1-\frac{1}{d}}}{\mathrm{OPT}_V}) = \mathcal{O}(\mathrm{OPT}_{V \sqcup R(m)} \log \frac{n^{1-\frac{1}{d}}}{\mathrm{OPT}_V})$, hence we can deduce that the approximation ratio is $\mathcal{O}(\log \frac{n^{1-\frac{1}{d}}}{\mathrm{OPT}_V})$.

Now take $m = \frac{n}{\log^c n}$, for any $c > 0$. For $\mathrm{OPT}_V = \mathcal{O}((\frac{n}{\log^c n})^{1-\frac{1}{d}} / \log n)$, the first regime applies and 2-opt gives a constant approximation. Else, using the second regime, we obtain a $\mathcal{O}(\log \log n)$-approximation. $\square$

4   Improved Approximations Require poly(n) Random Points

In this section, we complement the upper bound from the previous section by showing that, with $\alpha = n^{-3/5+\varepsilon}$ for any constant $\varepsilon > 0$, local search obtains an $\Omega(\frac{\log n}{\log \log n})$ approximation ratio in the simultaneous semi-random model when $d = 2$. This lower bound implies that more than $n^{2/5-\varepsilon}$ random points are required to obtain an asymptotic improvement over the $\mathcal{O}(\frac{\log n}{\log \log n})$ worst-case approximation in the simultaneous semi-random model.

To show this lower bound, we construct a 2-optimal tour T over an adversarial instance V that is far from optimal and is such that, with high probability, T can be augmented to obtain a 2-optimal tour $T_{V \sqcup R}$ over $V \sqcup R$ that contains T, where R consists of m random points. The length of the optimal tour on $V \sqcup R$ is upper bounded by combining the lengths of the optimal tours on both V and R. We first develop a framework for proving lower bounds in the simultaneous semi-random model, see Sect. 4.1. We then sketch the construction of the bad instance, which builds upon the construction from [8], in Sect. 4.2, and then we analyze it using the framework from Sect. 4.1. Complete details can be found in the full version of the paper.

4.1   A Framework for Simultaneous Semi-random Lower Bounds

In this section, we define parametrized properties of an instance that, if satisfied, guarantee a lower bound. In other words, this section reduces the problem of showing a lower bound to constructing an instance that satisfies the following properties. A path of an instance V is called *Hamiltonian* if it passes exactly once through each vertex of V.

Definition 2. *An instance V and a 2-optimal Hamiltonian path P over V from $s \in V$ to $t \in V$ are*

- *L-**long** if $\mathrm{OPT}_V \geq L$;*
- *γ-**bad** if $\ell(P) \geq \gamma \mathrm{OPT}_V$;*
- *K-**resistant**, $K \subseteq [0,1]^2$, if for any $x, y \in K$, $P \cup \{(s,x)\}$, $P \cup \{(x,y)\}$ and $P \cup \{(t,y)\}$ are 2-optimal.*

We now give some intuition on why we care about the above properties. If K is sufficiently big and α sufficiently small, then, with high probability, the αn random points R all lie in K. Combined with the K-resistance condition, this implies that a 2-optimal Hamiltonian path P over V can be extended to obtain a 2-optimal tour over $V \sqcup R$. More precisely, we have the following lemma.

Lemma 9. *Let V be an instance and $K \subseteq [0,1]^2$ be any region. If V has a Hamiltonian path P that is K-resistant, then for any instance $U \subseteq K$ disjoint from V, there exists a 2-optimal tour T on $U \sqcup V$ which extends P, i.e., such that P is a subpath of T.*

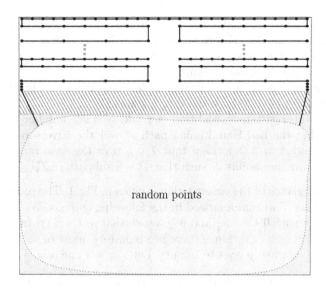

Fig. 1. High-level illustration of the construction from Sect. 4 (image not to scale). The gray area K accounts for most of the area of the unit square, hence w.h.p. all random points v are such that $v \in K$. The adversarial construction is in the top part of the square. A 2-optimal tour is given by the bold path P (which is a 2-optimal Hamiltonian path on the adversarial instance and far from optimal), plus an optimal Hamiltonian path covering the random points. The striped area is w.h.p. empty, and serves as a buffer between the deterministic and the random points, in order to ensure K-resistance of P in the adversarial instance.

The γ-bad condition guarantees that there is a bad 2-opt tour on V, which we care about because the tour we want to expand must have a bad approximation ratio for the lower bound to be effective. The L-long condition guarantees that the length of the part of the optimal tour on $V \sqcup R$ that connects vertices in V dominates the length of the part that connects the random vertices R, which is important since a 2-opt tour on random vertices R performs well compared to OPT_R. The intuition discussed above is summarized in the next lemma.

Lemma 10. *Let $\alpha \in (0,1)$ be some parameter that can depend on n. If there exists a $\sqrt{\alpha n}$-long n-instance V with a γ-bad, K-resistant Hamiltonian path P, for some region K with area $1 - o(1/(\alpha n))$, then with probability $1 - o(1)$ the approximation ratio of 2-opt on the α-semi-random instance $R(\alpha n) \sqcup V$ is $\Omega(\gamma)$.*

4.2   The Construction

In this section, we give a sketch of the construction of an instance that satisfies Lemma 10. The full construction can be found in the full version of the paper. Our starting point is the construction of [8], which does not satisfy the γ-bad and K-resistant properties for desirable parameters γ and K. We gradually modify and extend it so that it acquires the desired properties:

1. first, we modify it by making it fit in a small region and having a "bad" Hamiltonian path that only consists of "short" edges. Hence, this ensures a property useful for the separation condition, i.e., the adversarial instance is w.h.p. separated from the random points.
2. then, we modify it again to ensure the longness condition by stacking and connecting multiple copies of the thin instance;
3. finally, for the resistance condition, we add a small number of additional points so that the bad Hamiltonian path P over the adversarial instance V can be extended to a 2-optimal tour $T_{V \sqcup R}$ over the semi-random instance $V \sqcup R$ with random points R such that P is a subpath of $T_{V \sqcup R}$.

A high-level sketch of the construction is given in Fig. 1. The properties of the resulting instance V are summarized in the following lemmas. We let $\varepsilon > 0$ be a rational number such $0 < \varepsilon < \frac{1}{4}$ and $p \geq 3$ such that $p/4$ and εp are integers. We also assume that εp is odd. Since there are infinitely many of such p, all limits are understood as when p goes to infinity. Let $z = p/4$ and $s = (1 - \varepsilon)p$.

Lemma 11. *V is included in $S = [0,1] \times [1 - p^{2(z-s)} - 2p^{-s}, 1]$ and has $n = \Theta(p^{2(z+p)}) = \Theta(p^{\frac{5}{2}p})$ points. In particular, we have $p = \Theta\left(\frac{\log n}{\log \log n}\right)$.*

Lemma 12. *The optimal tour on V has length $\Theta(p^{2z}) = \Theta(n^{1/5})$.*

Lemma 13. *There exists a 2-optimal Hamiltonian path P on the adversarial instance V of length $\ell(P) \geq \frac{2}{3}\varepsilon p^{2z+1} = \Theta\left(\varepsilon \frac{\log n}{\log \log n} n^{1/5}\right)$.*

Let $K = [0,1] \times [0, 1 - p^{2(z-s)} - 4p^{-s}]$. The following is the main technical lemma of this section. Its proof requires a careful geometric analysis of V.

Lemma 14. *The instance V and the path P are K-resistant.*

Combining the previous lemmas, we get the following.

Lemma 15. *The n-instance V is $n^{1/5}$-long, and the Hamiltonian path P is $\Omega\left(\varepsilon \frac{\log n}{\log \log n}\right)$-bad and K-resistant, for some region K of area $1 - o(n^{-(2/5-\varepsilon)})$.*

Proof. By Lemma 11, Lemma 12, and Lemma 13, V with P are $n^{1/5}$-long and $\Omega\left(\varepsilon \frac{\log n}{\log \log n}\right)$-bad. Moreover, for $K = [0,1] \times [0, 1 - p^{2(z-s)} - 4p^{-s}]$, V with P are K-resistant by Lemma 14. Note that $p^{2(z-s)} = p^{p/2-(1-\varepsilon)p} p^{-s}$, and since $\varepsilon < 1/4$, $p^{2(z-s)} = o(p^{-s})$. Finally, $p^{-s} = \Theta(n^{-\frac{2}{5}(1-\varepsilon)}) = o(n^{-(\frac{2}{5}-\varepsilon)})$, thus K has area $1 - o(n^{-(\frac{2}{5}-\varepsilon)})$. □

The previous lemma combined with Lemma 10 immediately give us the proof of Theorem 2. Indeed, let $1/4 > \varepsilon > 0$ be constant. Let n be a number and V an instance as above (recall that there are infinitely many of them). By Lemma 15, the instance V verifies the hypothesis of Lemma 10 with $\alpha = n^{-3/5-\varepsilon}$ and $\gamma = \Omega\left(\frac{\log n}{\log \log n}\right)$; thus by Lemma 10 the approximation ratio of 2-opt on V is $\Omega\left(\frac{\log n}{\log \log n}\right)$ with probability $1 - o(1)$.

References

1. Antoniadis, A., Fleszar, K., Hoeksma, R., Schewior, K.: A PTAS for Euclidean TSP with hyperplane neighborhoods. In: SODA, pp. 1089–1105. SIAM (2019)
2. Applegate, D.L., Bixby, R.E., Chvátal, V., Cook, W.J.: The Traveling Salesman Problem. Princeton University Press (2011)
3. Arora, S.: Polynomial time approximation schemes for Euclidean traveling salesman and other geometric problems. J. ACM (JACM) **45**(5), 753–782 (1998)
4. Arthur, D., Vassilvitskii, S.: Worst-case and smoothed analysis of the ICP algorithm, with an application to the k-means method. In: FOCS, pp. 153–164. IEEE (2006)
5. Asadpour, A., Goemans, M.X., Madry, A., Gharan, S.O., Saberi, A.: An $O(\log n/ \log \log n)$-approximation algorithm for the asymmetric traveling salesman problem. Oper. Res. **65**(4), 1043–1061 (2017)
6. Blum, A., Spencer, J.: Coloring random and semi-random k-colorable graphs. J. Algorithms **19**(2), 204–234 (1995)
7. Brodowsky, U.A., Hougardy, S.: The approximation ratio of the 2-opt heuristic for the Euclidean traveling salesman problem. arXiv preprint arXiv:2010.02583 (2020)
8. Chandra, B., Karloff, H., Tovey, C.: New results on the old k-opt algorithm for the traveling salesman problem. SIAM J. Comput. **28**(6), 1998–2029 (1999)
9. Christofides, N.: Worst-case analysis of a new heuristic for the travelling salesman problem. Technical report, Carnegie-Mellon Univ Pittsburgh Pa Management Sciences Research Group (1976)
10. Dantzig, G., Fulkerson, R., Johnson, S.: Solution of a large-scale traveling-salesman problem. J. Oper. Res. Soc. Am. **2**(4), 393–410 (1954)
11. David, R., Feige, U.: On the effect of randomness on planted 3-coloring models. In: Proceedings of the Forty-Eighth Annual ACM Symposium on Theory of Computing, pp. 77–90 (2016)
12. Englert, M., Röglin, H., Vöcking, B.: Worst case and probabilistic analysis of the 2-opt algorithm for the TSP. Algorithmica **68**(1), 190–264 (2014)
13. Englert, M., Röglin, H., Vöcking, B.: Smoothed analysis of the 2-opt algorithm for the general TSP. TALG **13**(1), 1–15 (2016)
14. Feige, U., Kilian, J.: Heuristics for semirandom graph problems. J. Comput. Syst. Sci. **63**(4), 639–671 (2001)
15. Garey, M.R., Graham, R.L., Johnson, D.S.: Some NP-complete geometric problems. In: Proceedings of the Eighth Annual ACM Symposium on Theory of Computing, pp. 10–22 (1976)
16. Held, M., Karp, R.M.: A dynamic programming approach to sequencing problems. J. Soc. Ind. Appl. Math. **10**(1), 196–210 (1962)
17. Johnson, D.S., McGeoch, L.A.: 8. The traveling salesman problem: a case study. In: Local Search in Combinatorial Optimization, pp. 215–310. Princeton University Press (2018)
18. Karlin, A.R., Klein, N., Gharan, S.O.: A (slightly) improved approximation algorithm for metric TSP. In: Proceedings of the 53rd Annual ACM SIGACT Symposium on Theory of Computing, pp. 32–45 (2021)
19. Karp, R.M.: Probabilistic analysis of partitioning algorithms for the traveling-salesman problem in the plane. Math. Oper. Res. **2**(3), 209–224 (1977)
20. Klein, P.N.: A linear-time approximation scheme for tsp in undirected planar graphs with edge-weights. SIAM J. Comput. **37**(6), 1926–1952 (2008)

21. Kolla, A., Makarychev, K., Makarychev, Y.: How to play unique games against a semi-random adversary: study of semi-random models of unique games. In: 2011 IEEE 52nd Annual Symposium on Foundations of Computer Science, pp. 443–452. IEEE (2011)
22. Künnemann, M., Manthey, B.: Towards understanding the smoothed approximation ratio of the 2-opt heuristic. In: Halldórsson, M.M., Iwama, K., Kobayashi, N., Speckmann, B. (eds.) ICALP 2015. LNCS, vol. 9134, pp. 859–871. Springer, Heidelberg (2015). https://doi.org/10.1007/978-3-662-47672-7_70
23. Makarychev, K., Makarychev, Y., Vijayaraghavan, A.: Approximation algorithms for semi-random partitioning problems. In: STOC, pp. 367–384 (2012)
24. Mömke, T., Svensson, O.: Removing and adding edges for the traveling salesman problem. J. ACM (JACM) 63(1), 1–28 (2016)
25. Padberg, M., Rinaldi, G.: A branch-and-cut algorithm for the resolution of large-scale symmetric traveling salesman problems. SIAM Rev. 33(1), 60–100 (1991)
26. Papadimitriou, C.H.: The Euclidean travelling salesman problem is NP-complete. Theor. Comput. Sci. 4(3), 237–244 (1977)
27. Reinelt, G.: TSPLIB-a traveling salesman problem library. ORSA J. Comput. 3(4), 376–384 (1991)
28. Roughgarden, T.: Beyond the Worst-Case Analysis of Algorithms. Cambridge University Press (2021)
29. Spielman, D.A., Teng, S.H.: Smoothed analysis of algorithms: why the simplex algorithm usually takes polynomial time. J. ACM (JACM) 51(3), 385–463 (2004)
30. Steele, J.M.: Probability Theory and Combinatorial Optimization. SIAM (1997)
31. Svensson, O., Tarnawski, J., Végh, L.A.: A constant-factor approximation algorithm for the asymmetric traveling salesman problem. J. ACM (JACM) 67(6), 1–53 (2020)
32. Traub, V., Vygen, J.: Approaching 3/2 for the s-t-path TSP. J. ACM (JACM) 66(2), 1–17 (2019)
33. Vijayaraghavan, A., Awasthi, P.: Clustering semi-random mixtures of Gaussians. In: ICML, pp. 5055–5064. PMLR (2018)

A Simple LP-Based Approximation Algorithm for the Matching Augmentation Problem

Étienne Bamas[(✉)], Marina Drygala, and Ola Svensson

EPFL, Lausanne, Switzerland
etienne.bamas@epfl.ch

Abstract. The Matching Augmentation Problem (MAP) has recently received significant attention as an important step towards better approximation algorithms for finding cheap 2-edge connected subgraphs. This has culminated in a $\frac{5}{3}$-approximation algorithm. However, the algorithm and its analysis are fairly involved and do not compare against the problem's well-known LP relaxation called the cut LP.

In this paper, we propose a simple algorithm that, guided by an optimal solution to the cut LP, first selects a DFS tree and then finds a solution to MAP by computing an optimum augmentation of this tree. Using properties of extreme point solutions, we show that our algorithm always returns (in polynomial time) a better than 2-approximation when compared to the cut LP. We thereby also obtain an improved upper bound on the integrality gap of this natural relaxation.

1 Introduction

Designing cheap networks that are robust to edge failures is a basic and important problem in the field of approximation algorithms. The area containing these problems is often referred to as *survivable network design*. Generally, one has to compute the cheapest network that satisfies some connectivity requirements in-between some prespecified set of vertices. Classic examples are for instance the Minimum Spanning Tree problem in which one has to augment the connectivity of a graph from 0 to 1 or related questions such as the Steiner Tree/Forest problem. Another type of network design problem is to build 2-edge connected spanning subgraph (2-ECSS) or multisubgraph (2-ECSM), where one has to augment the connectivity of a graph from 0 to 2. The latter problems are closely related to the famous Traveling Salesman Problem (TSP). Unfortunately, most of the problems in this area are NP-hard (or even APX-hard), and what one can hope for is generally to compute an approximate solution in polynomial time. Powerful and versatile techniques such as *primal-dual* [15,31] or *iterative rounding* [19,24] guarantee an approximation within factor 2 for many of these problems but improving on this bound for any connectivity problem is often quite challenging. In the case of 2-ECSS, a 4/3-approximation is known if the underlying graph G is unweighted [18,28]. However, a similar result for the weighted case has remained elusive, and the best approximation algorithm only

© Springer Nature Switzerland AG 2022
K. Aardal and L. Sanitá (Eds.): IPCO 2022, LNCS 13265, pp. 57–69, 2022.
https://doi.org/10.1007/978-3-031-06901-7_5

guarantee a factor 2 approximation. A prominent special case of the weighted 2-ECSS problem is the so-called Forest Augmentation Problem (FAP). In such instances of 2-ECSS all edge weights are either 0 or 1 (we will refer to edges of cost 0 as *light edges* and edges of cost 1 as *heavy edges*). The name stems from the fact that one can assume that the light edges form a forest F, and the goal is to find the smallest set of heavy edges E' such that $F \cup E'$ is 2-edge connected.

A famous special case of FAP is the Tree Augmentation Problem (TAP) which has been extensively studied for decades. In this problem, the forest F is a single spanning tree, and one has to find the smallest set of edges to make the tree 2-edge connected. For this problem, several better-than-2 approximations were designed in a long line of research [1,3,6–9,11,12,14,17,20,22,23,26,27,29,30]. One can see TAP as an extreme case of FAP where the forest is a single component. Another interesting special case is the Matching Augmentation Problem (MAP), in which the forest of light edges forms a matching M and one has to find the smallest set of heavy edges E' such that $M \cup E'$ is 2-edge connected. It can be seen as the other extreme case in which the forest forms as many components as possible. We also remark that MAP generalizes the unweighted 2-ECSS problem, which can be viewed as an instance of MAP with an empty matching. For MAP, only recently a better-than-2 approximation was given by Cheriyan et al. [4,5]. These two works culminate in a 5/3-approximation, obtained via a fairly involved algorithm and analysis.

For many of these network design problems, there is a simple linear programming relaxation called the cut LP. In the case of FAP, for a given graph $G = (V, E)$, forest $F \subseteq E$ the cut LP is written as follows, with a variable x_e to decide to take each edge e or not. Recall that $\delta(S)$ denotes the edges with exactly one endpoint in S.

$$LP(G,F): \quad \min \sum_{e \in E \setminus F} x_e$$

$$\sum_{e \in \delta(S)} x_e \geq 2, \quad \text{for all } S, \emptyset \subsetneq S \subsetneq V$$

$$0 \leq x_e \leq 1, \qquad \forall e \in E.$$

The integrality gap of this linear program is an interesting question by itself. Recently, in the case of TAP (i.e. F is a spanning tree), Nutov [27] showed that the integrality gap is at most $2 - 2/15 \approx 1.87$. Cheriyan et al. [8] showed that the integrality gap is at least $3/2$ in the case of TAP. In the case of MAP, the best upper bound on the integrality gap is 2, and the best lower bound is 9/8 [2,28]. We note that the recent works [4,5] do not seem to compare against the cut LP, and therefore do not show an integrality gap better than 2 for MAP.

1.1   Our Results

In this paper, we give an algorithm that guarantees an approximation ratio $2 - c$ (for some absolute constant $c > 0$) with respect to the best fractional solution of the cut LP. The algorithm is the following. We note that some of our techniques are reminiscent of the algorithm of Mömke and Svensson [25] for the travelling salesman problem.

The LP-based algorithm:

1. Compute an optimal extreme point solution x^* to $LP(G, M)$.
2. Let $E' = \{e \in E, x_e^* > 0\}$ be the *support* of x^*, and run a DFS on the support graph $G' = (V, E')$ which always give priority first to an available light edge and second to the available heavy edge e maximizing x_e^*.
3. Compute an optimum augmentation A to the TAP problem with respect to the DFS tree T computed in the previous step and return $H = T \cup A$.

We note that the LP-based algorithm indeed runs in polynomial time. Step 2 computes a DFS in which some edges are explored in priority (if possible). Step 3 can also be completed in polynomial time because the tree T is a DFS tree. This implies that all non-tree edges are back-edges (i.e. one endpoint is an ancestor of the other). In the language of TAP, these edges are often referred to as "uplinks", and it is well-known that TAP instances in which the edges are only "uplinks" are solvable in polynomial time [9,13].

Finally, the solution given by the algorithm is feasible since Step 2 increases connectivity from 0 to 1 and Step 3 from 1 to 2. One can check that no edge is taken twice in the process since A and T are disjoint.

In this paper, our main result shows that this simple algorithm guarantees an approximation within factor strictly better than 2 with respect to the cut LP relaxation.

Theorem 1. *The LP-based algorithm returns a feasible solution to any MAP instance of cost at most $2 - c$ times the cost of the fractional solution x^*, for some absolute constant $c > 0$.*

For the sake of exposition, we did not try to optimize the constant c but we believe that improving the ratio of 5/3 in [5] (that holds with respect to the optimum *integral* solution) would require new techniques in the analysis. Since Nutov [27] proved the integrality gap of the cut LP to be strictly better than 2 for TAP, the cut LP seems a promising relaxation for the general FAP. Additionally, we prove the following simple theorem.

Theorem 2. *The integrality gap of the cut LP for MAP is at least 4/3.*

Proof. Consider the example given in Fig. 1a, which is a simple adaptation of a classic example for the related TSP problem. One can check that the fractional solution that gives 1/2 fractional value to all heavy edges and value 1 to all light edges is feasible for a total cost of $6/2 = 3$. However, any integral solution costs at least 4.

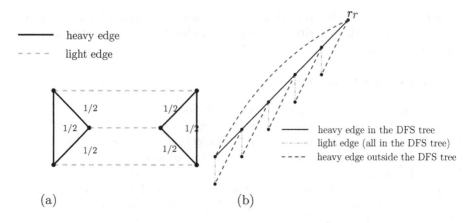

Fig. 1. (a) An integrality gap example. (b) An example of a bad DFS tree.

1.2 Our Techniques

The proof of Theorem 1 relies on several crucial observations that we sketch here. The first observation is that the total cost of the DFS tree T is always at most the cost of x^* (denoted $c(x^*)$). This follows because T must contain all the light edges since they are given priority over any other edge (note that since we assume that M is a matching, it cannot happen that two distinct light edges want priority at the same time). Therefore, the total cost of the tree T is exactly equal to $n - 1 - |M|$, while it is easy to show that $c(x^*) \geq (n - |M|)$.

Another interesting fact is that if one considers the LP solution x^* restricted to edges *not* in the tree T (denote this solution by $x^*_{E \setminus T}$), then this is a feasible solution to the cut LP of the TAP instance with respect to the tree T (i.e. $x^*_{E \setminus T}$ is a feasible solution to $LP(G, T)$). Hence, if we denote by y^* the optimum fractional solution to $LP(G, T)$, we have that $c(y^*) \leq c(x^*_{E \setminus T})$.

Because T is a DFS tree, the TAP instance with respect to the tree T contains only "uplinks" and therefore $LP(G, T)$ is known to be integral [1]. We note that this already gives a simple proof that the integrality gap of $LP(G, M)$ is at most 2. To get better than 2, we only need to show that

$$(n - |M| - 1) + c(y^*) \leq (2 - c)c(x^*).$$

Conceptually, we distinguish between two cases. If $c(x^*) > (1 + c)(n - |M|)$ (i.e. the LP solution is expensive), then the DFS tree is significantly cheaper than $c(x^*)$ and it is easy to conclude that the cost of our solution $T \cup A$ is better than $2c(x^*)$. Otherwise, assume that the LP value is close to the trivial lower bound of $(n - |M|)$. In this case, we show that $c(y^*) \leq (1 - c)c(x^*)$.

To show this, we consider two possibilities. We can prove that either we can scale down a significant portion of $x^*_{E \setminus T}$ to obtain a cheaper feasible solution to $LP(G, T)$, or that $c(x^*_{E \setminus T})$ itself is already significantly smaller than $c(x^*)$. When a lot of the tree cuts in T (i.e. the cuts defined by removing an edge from

T to obtain two trees and taking the edges with one endpoint in each tree) have some slack in the TAP solution $x^*_{E \setminus T}$ (that is when a lot of tree cuts S satisfy $x^*_{E \setminus T}(\delta(S)) > 1 + c$), the first case is realized. Otherwise, when almost all of the tree cuts are nearly tight (i.e. satisfy $x^*_{E \setminus T}(\delta(S)) \leq 1 + c$), we can show that the DFS must have captured a good fraction of the value of $c(x^*)$ *inside* the tree T. This step uses some crucial properties of extreme point solutions as well as our choice of DFS. Therefore the cost of $x^*_{E \setminus T}$ is significantly smaller than the cost of x^* completing the argument.

Before proceeding to the proof, it is worthwhile to mention that we are not aware of any example on which our algorithm has a ratio worse than $4/3$ times the cost of x^*. It remains open to give a tighter analysis of this algorithm. We also note that [21] also makes use of DFS for the related problem of unweighted 2-ECSS. They obtain a ratio of $3/2$ for the unweighted 2-ECSS problem. However, their DFS is not LP-based and we remark that if we do not guide the DFS with the LP solution, the approximation ratio can be arbitrarily close to 2. We give an example in Fig. 1b. One can see that the DFS tree (rooted at r) contains all the matching edges, and the tree augmentation problem requires us to take all but one of the back-edges. However, the optimum solution to the MAP instance is to take a Hamiltonian tour containing all the light edges. Generalizing the same example by simply increasing the depth of the tree leads to an approximation arbitrarily close to 2.

2 The Analysis of the LP-Based Algorithm

In this section, we prove Theorem 1. It is organized as follows. In Subsect. 2.1, we introduce some basic definitions. In the subsequent subsection, we proceed via a case distinction to prove the theorem.

2.1 Preliminaries

We will use T to refer to the DFS tree computed by the algorithm, and we will list edges in G as uv, where u is an ancestor of v in T. Since T is a DFS tree, all edges in G must have the property that one endpoint is an ancestor of the other in T. We will let $B = E \setminus T$ denote the set of *back-edges* of G. As in the introduction, we will call an edge of weight 1 a *heavy edge* and an edge of weight 0 a *light edge*. For every edge e in the DFS tree T computed, we let $T(e)$ denote the tree cut corresponding to the edge e in the tree T. Formally, $T(e) = \delta(T_v)$, where $e = uv$ and T_v is the sub-tree rooted at v. We call an edge $e \in T$ α-*tight* if we have

$$x^*(T(e)) - x^*_e < 1 + \alpha.$$

Implicitly, if we call an edge e α-tight, this will mean that e belongs to the tree T. In addition, we denote by $N_t^{(\alpha)}$ the number of α-tight edges in the tree T. For a tree T, we denote by x^*_T the restriction of x^* to the edges in the tree T. We note that for any instance of the MAP, it must be that $c(x^*) \geq (n - |M|)$. This follows by a simple double counting argument on the fractional degree of each component (precisely we have $n - |M|$ components that must have fractional degree 2 each). It is also clear that the DFS tree T must contain all the light edges in M since they are given priority. Hence the cost of T is at most $n - |M| - 1 \le c(x^*)$. In the following, we will fix two parameters $\epsilon = 10^{-1}, \gamma = 10^{-3}$.

2.2   The Analysis of the Algorithm

We note that if $c(x^*) \ge (1+\gamma)(n - |M|)$, it is easy to show that the cost of the returned solution $T \cup A$ is at most

$$(n - |M| - 1) + c(x^*) \le \frac{c(x^*)}{1+\gamma} + c(x^*) = c(x^*)\left(2 - \frac{\gamma}{1+\gamma}\right). \tag{1}$$

However, if $c(x^*) < (1+\gamma)(n-|M|)$ and $N_t^{(\gamma)} \le (1-\gamma)(n-|M|)$ (i.e. there are few γ-tight tree cuts), then we proceed as follows. We partition the set of back edges in our graph B into $B_t^{(\gamma)} \cup B_s^{(\gamma)}$, where $B_t^{(\gamma)}$ contains all edges $e \in B$ that are contained in $T(e')$ for some γ-tight edge $e' \in T$. Then x', defined by

$$x'(e) = \begin{cases} x^*(e) & e \in B_t^{(\gamma)} \\ \frac{x^*(e)}{1+\gamma} & e \in B_s^{(\gamma)} \\ 1 & \text{otherwise} \end{cases}$$

is also a feasible solution to $LP(G, T)$. The total fractional value represented by edges in $B_t^{(\gamma)}$ is at most $(1+\gamma)N_t^{(\gamma)}$. Hence, $c(x')$ can be upper bounded as follows.

$$c(x') \le \frac{c(x^*) - (1+\gamma)N_t^{(\gamma)}}{1+\gamma} + (1+\gamma)N_t^{(\gamma)} = \frac{c(x^*)}{1+\gamma} + \gamma N_t^{(\gamma)}.$$

Since the cost of $T \cup A$ is at most $c(x^*) + c(x')$ and we assume that $N_t^{(\gamma)} \le (1-\gamma)(n-|M|)$, it is easy to get the upper-bound of

$$c(x^*)\left(1 + \frac{1}{1+\gamma}\right) + \gamma(1-\gamma)(n-|M|) \le c(x^*)\left(2 - \frac{\gamma^3}{1+\gamma}\right), \tag{2}$$

where the last inequality follows because $n - |M| \le c(x^*)$. Since these two cases clearly give a better than 2 approximation, we assume in the rest of the analysis that

$$(n - |M|) \le c(x^*) < (1+\gamma)(n - |M|), \tag{3}$$

and

$$N_t^{(\gamma)} > (1-\gamma)(n - |M|). \tag{4}$$

We will show that $c(x_T^*)$ is at least a constant fraction times $c(x^*)$. Since the cost of the returned solution $T \cup A$ is at most $2c(x^*) - c(x_T^*)$, this will conclude the proof. First, we partition the γ-tight tree cuts into two sets of cuts $\mathcal{S}_0$ and

S_1 containing the tight tree cuts associated with light edges and heavy edges, respectively. We can then distinguish between two sub-cases. For each edge $e = uv \in T$, we say that e is a *leaf* edge if v is a leaf in the tree T (recall that we always write an edge e as $e = uv$ such that v is a descendant of u in T). We denote S_0^+ the non-leaf edges in S_0 and S_0^- the leaf edges in S_0. We have two main cases.

Suppose that $|S_1| \geq \gamma(n - |M|)$ or that $|S_0^+| \geq \gamma(n - |M|)$. By feasibility of x^* at least 2 units of x^* must cross any tree cut. Hence $x^*(\delta(T_v)) \geq 2$, for any $v \in V$. By definition of γ-tightness we know that for any γ-tight edge $e = uv$ we have $x^*(e) \geq x^*(\delta(T_v)) - (1 + \gamma) \geq 1 - \gamma$.

Hence if $|S_1| \geq \gamma(n - |M|)$, we have that

$$c(x_T^*) \geq \gamma(1 - \gamma)(n - |M|) \geq \frac{\gamma(1 - \gamma)}{1 + \gamma} c(x^*),$$

which concludes the case when $|S_1|$ is large. In the following we use some properties of extreme point solutions. We say that an edge e is *fractional* (with respect to the fractional solution x^*) if $0 < x_e^* < 1$. A vertex v is said to be α-*fractional* if it has more than $1/\alpha$ incident fractional edges in the support of x^* (for any $\alpha > 0$). We claim the following lemma, the proof of which relies on standard techniques and can be found in Appendix A. We note that a similar result was used in [25].

Lemma 1. *If x^* is an extreme point solution of the cut LP, then there are at most $2n - 1$ fractional edges in G. Moreover, for any $\alpha > 0$, there are at most $4\alpha n$ α-fractional vertices with respect to x^*.*

Using Lemma 1 with $\alpha = \gamma/16$, we get that if $|S_0^+| \geq \gamma(n - |M|)$, then (recall that $n - |M| \geq n/2$) there are at least

$$\gamma(n - |M|) - (\gamma/4)n \geq (\gamma/2)(n - |M|)$$

edges $uv \in S_0^+$ such that v is not $\gamma/16$-fractional. We then claim the following simple lemma.

Lemma 2. *Fix any $\alpha, \alpha' > 0$. Suppose that $e = uv$ is an α-tight light edge, such that v is not a leaf in T. Then if v is not α'-fractional there exists some edge $e' = vw$ in T such that $x^*(e') \geq (1 - \alpha)\alpha'$.*

Proof. By feasibility of x^* we know that $x^*(\delta(T_v \setminus v)) \geq 2$. Since e is α-tight and T is a DFS tree we have that $x^*(\delta(T_v)) - x^*(e) \leq 1 + \alpha$. We know that $E(T_v \setminus v, v) = \delta(T_v \setminus v) \setminus \delta(T_v)$, and as a result $x^*(E(T_v \setminus v, v)) \geq 1 - \alpha$. Since v is not α'-fractional there must be an edge $e' \in E(T_v \setminus v, v)$ with value at least $x^*(E(T_v \setminus v, v))\alpha' \geq (1 - \alpha)\alpha'$. Since our DFS selects always the highest possible fractional value if there is no light edge to explore, the first edge selected after exploring v must be of fractional value at least $(1 - \alpha)\alpha'$. $\square$

Combining Lemma 2 with the previous observation, if $|\mathcal{S}_0^+| \geq \gamma(n - |M|)$ we get that

$$c(x_T^*) \geq (\gamma/2)(n - |M|)(1 - \gamma)(\gamma/16) \geq c(x^*)\frac{\gamma^2(1 - \gamma)}{32(1 + \gamma)}.$$

Combining these two cases we get that if $|\mathcal{S}_1| \geq \gamma(n - |M|)$ or $|\mathcal{S}_0^+| \geq \gamma(n - |M|)$ then

$$c(x_T^*) \geq c(x^*) \cdot \min\left(\frac{\gamma^2(1 - \gamma)}{32(1 + \gamma)}, \frac{\gamma(1 - \gamma)}{1 + \gamma}\right),$$

hence the cost of $T \cup A$ is upper bounded by

$$2c(x^*) - c(x_T^*) \leq c(x^*)\left(2 - \frac{\gamma^2(1 - \gamma)}{32(1 + \gamma)}\right), \tag{5}$$

which is clearly better than 2. Hence we are left with the last case, in which

$$|\mathcal{S}_0^-| > (1 - \gamma)(n - |M|) - |\mathcal{S}_1| - |\mathcal{S}_0^+| > (1 - 3\gamma)(n - |M|).$$

Suppose $|\mathcal{S}_0^-| > (1 - 3\gamma)(n - |M|)$. This is the most interesting case. Note that for each edge $e = uv \in \mathcal{S}_0^-$, the fractional degree of v restricted to heavy edges must be at least 1, and all of this fractional degree is carried by backedges in T. Denote by B' this subset of backedges. Next we define $B'' \subseteq B'$ to be the subset of B' containing only edges with fractional value at least $\epsilon = 10^{-1}$. We claim that

$$|B''| \geq n/10. \tag{6}$$

Assume the contrary, since the fractional value of any edge is at most 1 then the total value carried by edges in $B' \setminus B''$ must be at least

$$|\mathcal{S}_0^-| - (n/10) > (1 - 3\gamma)(n - |M|) - n/10 > 3n/8 - n/10.$$

(Recall that $(n - |M|) \geq n/2$ and $(1 - 3\gamma) > 3/4$). Since all the edges in $B' \setminus B''$ have fractional value at most ϵ, there must be at least $(3n/8 - n/10)/\epsilon = 30n/8 - n > 2n - 1$ such edges, contradicting Lemma 1. Hence $|B''| \geq n/10$.

For completeness we consider the case when E contains heavy edges that are parallel to light edges. Partition B'' into $B_1'' \cup B_2''$, where B_2'' is the set of edges in B'' parallel to an edge in $\mathcal{S}_0^-$. We define B_1'' to be the remaining edges in B''.

We claim that $|B_2''| \leq n/100$, and thus loosely $|B_1''| \geq n/20$.

To see this note that,

$$c(x^*) - (n - |M|) \geq |B_2''|\epsilon. \tag{7}$$

Equation (7) holds as the lower bound of $(n - |M|)$ on $c(x^*)$ is obtained only by counting the fractional degree of each component in M. Since those parallel edges are not counted in this bound (they are only *within* a single component), they directly count in the value of $c(x^*) - (n - |M|)$, which counts the surplus of $c(x^*)$ above $(n - |M|)$.

Then as $c(x^*) - (n - |M|) \leq \gamma(n - |M|)$, by choice of ϵ and γ we obtain that $|B_2''| \leq n/100$.

Consider the set of vertices X that contains the ancestor vertices of the edges in B_1''. We claim that

$$|X| \geq n/500. \tag{8}$$

To prove this, we first claim that

$$c(x^*) - (n - |M|) \geq |B_1''|\epsilon - 2|X|. \tag{9}$$

To see this, note again that the value $c(x^*) - (n - |M|)$ represents the surplus value of $c(x^*)$ above the lower bound that gives fractional degree 2 to every vertex. This trivial lower bound gives a fractional value—which is the fractional degree restricted to heavy edges—of at most 2 to every vertex, hence a fractional value of at most $2|X|$ to the set of vertices X. Since every edge in B_1'' has fractional value of at least ϵ and is adjacent to a single vertex in X, we get that the surplus value of $c(x^*)$ above the trivial lower bound is at least $|B_1''|\epsilon - 2|X|$ which proves Eq. (9).

Since by assumption we have $c(x^*) < (1 + \gamma)(n - |M|)$ we conclude with Eq. (9) that

$$\gamma(n - |M|) > c(x^*) - (n - |M|) \geq |B_1''|\epsilon - 2|X|$$

which implies, by our lower bound on $|B_1''|$ and our choice of γ and ϵ,

$$|X| \geq \frac{|B_1''|\epsilon - \gamma(n - |M|)}{2} \geq \frac{n/200 - n/10^3}{2} = n/500. \tag{10}$$

For each vertex $u \in X$, denote by e_u the first edge selected by the DFS after reaching u. Denote $X' \subseteq X$ the subset of X containing only vertices $u \in X$ such that e_u does *not* belong to S_0^+. Then, we have by assumption,

$$|X'| \geq |X| - |S_0^+| \geq n/500 - \gamma(n - |M|) \geq n/500 - n/10^3 = n/10^3.$$

We finally claim the following, which crucially uses how the DFS selects the edges to explore in priority.

Claim.

$$c(x_T^*) \geq \epsilon|X'|.$$

Proof. There are two cases to consider (depicted in Fig. 2).

If $u \in X'$ is such that $e_u = uv$ is a heavy edge, by definition of X' there must be an edge $e' = uf$ coming from a leaf f in the tree T to u of fractional value $x_{e'} \geq \epsilon$. At the first time the DFS visits the vertex u, the leaf f was not explored yet hence the edge e' was a valid choice of edge to explore. Since our DFS always takes the highest fractional value, it must be that

$$x_{e_u} \geq x_{e'} \geq \epsilon.$$

If $u \in X'$ is such that $e_u = uv$ is a light edge, recall that by definition of X', v must be a leaf in T. Then when the DFS arrived at v, it must be that

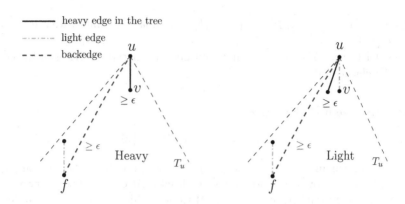

Fig. 2. On the left side, the case when the first edge selected out of u is heavy. On the right the case when the first edge selected out of u is light.

all reachable vertices from v were already visited. Hence the DFS must have backtracked to u. Now note that by our construction of B_1'', we know that there must be another leaf f such that $e' = uf$ is a back-edge in the tree of fractional value $x_{uf} \geq \epsilon$ (recall that uf is not parallel to the edge e_u). Since f is a leaf, u must have been explored before f therefore, after backtracking from v to u the edge uf was a valid edge to take. Therefore the DFS must have selected a second edge e'' in the tree from u such that

$$x_{e'} \geq x_{uf} \geq \epsilon.$$

Hence we proved that all vertices u in X must be adjacent to at least one heavy edge of fractional value ϵ that belongs to the tree T and goes to a child of u. Hence the proof of the claim. □

By the previous claim, we have $c(x_T^*) \geq \epsilon|X'|$ hence the cost of the returned solution $T \cup A$ is at most

$$2c(x^*) - c(x_T^*) \leq 2c(x^*) - n/10^4 \leq c(x^*)\left(2 - 10^{-4}\right), \tag{11}$$

which ends the proof of Theorem 1.

3   Conclusion

In this paper, we gave a simple $2 - c$ approximation algorithm for MAP with respect to the standard cut LP. Our algorithm computes a DFS tree using an optimal extreme point solution to the above-mentioned LP solution as a guide when selecting edges and then augments the resulting tree optimally. We leave it as an open problem to see if the analysis can be refined to get an improved guarantee for the algorithm. We remark that it is not difficult to see that if the LP solution is f-fractional, then our algorithm produces a solution of at most

$2 - f$ times its value. In particular, this gives an upper bound of $3/2$ for half-integral solutions. We wonder if a better understanding of the algorithm will lead to a $\frac{4}{3}$-approximation for the half-integral case.

Another interesting connection of our work to related works is by its relevance to the Path Augmentation Problem (PAP). An instance of PAP is an instance of FAP where the forest F contains only paths. We note that our techniques generalize to instances of PAP, if the cut LP returns a solution of cost equal to the number of components in F. This follows because the support of the optimal extreme point solution of the cut LP for such instances has no fractional value incident to internal nodes of the paths in F. Some independent work [16] shows that the general FAP reduces to special instances of PAP where the cost of the LP solution is almost equal to the number of components in F. However these techniques do not preserve the integrality gap. Determining whether we can bound the integrality gap of the cut LP for the FAP strictly below 2 remains an interesting open problem.

A   Deferred Proofs

Suppose that x^* is an extreme point solution of $LP(G, M)$. We know that x^* can be defined as the unique solution to the following system of $|E|$ equations, for some $\mathcal{S} \subseteq 2^V$ and $E_0 \cup E_1 \subseteq E$.

$$\sum_{e \in \delta(S)} x_e = 2, \quad \text{for all } S \in \mathcal{S}$$
$$x_e = 0 \qquad \forall e \in E_0$$
$$x_e = 1 \qquad \forall e \in E_1$$

Lemma 3 shows that we can select $\mathcal{S}$ not too large. The proof of this lemma is the same as Theorem 4.9 from [10].

Lemma 3 (Theorem 4.9 in [10]). *Let x^* be an extreme point of the MAP cut LP then the family of equations $\mathcal{S}$ can be chosen to be a laminar family.*

It is well known that any laminar family has size at most $2n - 1$. Therefore the number of fractional edges is at most $|E| - |E_0| - |E_1| = |\mathcal{S}| \leq 2n - 1$.

References

1. Adjiashvili, D.: Beating approximation factor two for weighted tree augmentation with bounded costs. ACM Trans. Algorithms (TALG) **15**(2), 1–26 (2018)
2. Alexander, A., Boyd, S., Elliott-Magwood, P.: On the integrality gap of the 2-edge connected subgraph problem. Technical report. Citeseer (2006)

3. Cecchetto, F., Traub, V., Zenklusen, R.: Bridging the gap between tree and connectivity augmentation: unified and stronger approaches. In: Proceedings of the 53rd Annual ACM SIGACT Symposium on Theory of Computing, pp. 370–383 (2021)
4. Cheriyan, J., Dippel, J., Grandoni, F., Khan, A., Narayan, V.V.: The matching augmentation problem: a 7/4-approximation algorithm. Math. Program. **182**(1), 315–354 (2020)
5. Cheriyan, J., Cummings, R., Dippel, J., Zhu, J.: An improved approximation algorithm for the matching augmentation problem. arXiv preprint arXiv:2007.11559 (2020)
6. Cheriyan, J., Gao, Z.: Approximating (unweighted) tree augmentation via lift-and-project, part i: stemless tap. Algorithmica **80**(2), 530–559 (2018)
7. Cheriyan, J., Gao, Z.: Approximating (unweighted) tree augmentation via lift-and-project, part ii. Algorithmica **80**(2), 608–651 (2018)
8. Cheriyan, J., Karloff, H., Khandekar, R., Könemann, J.: On the integrality ratio for tree augmentation. Oper. Res. Lett. **36**(4), 399–401 (2008)
9. Cohen, N., Nutov, Z.: A (1+ ln2)-approximation algorithm for minimum-cost 2-edge-connectivity augmentation of trees with constant radius. Theor. Comput. Sci. **489**, 67–74 (2013)
10. Cornuéjols, G., Fonlupt, J., Naddef, D.: The traveling salesman problem on a graph and some related integer polyhedra. Math. Program. **33**(1), 1–27 (1985)
11. Even, G., Feldman, J., Kortsarz, G., Nutov, Z.: A 1.8 approximation algorithm for augmenting edge-connectivity of a graph from 1 to 2. ACM Trans. Algorithms (TALG) **5**(2), 1–17 (2009)
12. Fiorini, S., Groß, M., Könemann, J., Sanità, L.: Approximating weighted tree augmentation via Chvátal-Gomory cuts. In: Proceedings of the Twenty-Ninth Annual ACM-SIAM Symposium on Discrete Algorithms, pp. 817–831. SIAM (2018)
13. Frederickson, G.N., Ja'ja, J.: On the relationship between the biconnectivity augmentation and travelling salesman problems. Theor. Comput. Sci. **19**(2), 189–201 (1982)
14. Frederickson, G.N., Ja'Ja', J.: Approximation algorithms for several graph augmentation problems. SIAM J. Comput. **10**(2), 270–283 (1981)
15. Goemans, M.X., Williamson, D.P.: A general approximation technique for constrained forest problems. SIAM J. Comput. **24**(2), 296–317 (1995)
16. Grandoni, F., Ameli, A.J., Traub, V.: Breaching the 2-approximation barrier for the forest augmentation problem. arXiv preprint arXiv:2112.11799 (2021)
17. Grandoni, F., Kalaitzis, C., Zenklusen, R.: Improved approximation for tree augmentation: saving by rewiring. In: Proceedings of the 50th Annual ACM SIGACT Symposium on Theory of Computing, pp. 632–645 (2018)
18. Hunkenschröder, C., Vempala, S., Vetta, A.: A 4/3-approximation algorithm for the minimum 2-edge connected subgraph problem. ACM Trans. Algorithms **15**(4) (2019). https://doi.org/10.1145/3341599
19. Jain, K.: A factor 2 approximation algorithm for the generalized Steiner network problem. Combinatorica **21**(1), 39–60 (2001)
20. Khuller, S., Thurimella, R.: Approximation algorithms for graph augmentation. J. Algorithms **14**(2), 214–225 (1993)
21. Khuller, S., Vishkin, U.: Biconnectivity approximations and graph carvings. J. ACM **41**(2), 214–235 (1994). https://doi.org/10.1145/174652.174654
22. Kortsarz, G., Nutov, Z.: A simplified 1.5-approximation algorithm for augmenting edge-connectivity of a graph from 1 to 2. ACM Trans. Algorithms (TALG) **12**(2), 1–20 (2015)

23. Kortsarz, G., Nutov, Z.: LP-relaxations for tree augmentation. Discret. Appl. Math. **239**, 94–105 (2018)
24. Lau, L.C., Ravi, R., Singh, M.: Iterative Methods in Combinatorial Optimization, vol. 46. Cambridge University Press (2011)
25. Mömke, T., Svensson, O.: Removing and adding edges for the traveling salesman problem. J. ACM (JACM) **63**(1), 1–28 (2016)
26. Nagamochi, H.: An approximation for finding a smallest 2-edge-connected subgraph containing a specified spanning tree. Discret. Appl. Math. **126**(1), 83–113 (2003)
27. Nutov, Z.: On the tree augmentation problem. Algorithmica **83**(2), 553–575 (2021)
28. Sebö, A., Vygen, J.: Shorter tours by nicer ears: 7/5-approximation for the graph-TSP, 3/2 for the path version, and 4/3 for two-edge-connected subgraphs. Combinatorica **34**(5), 597–629 (2014). https://doi.org/10.1007/s00493-014-2960-3
29. Traub, V., Zenklusen, R.: A better-than-2 approximation for weighted tree augmentation. Corr abs/2104.07114 (2021). Proceedings of the 62nd IEEE FOCS (2021)
30. Traub, V., Zenklusen, R.: Local search for weighted tree augmentation and Steiner tree. In: Proceedings of the 2022 Annual ACM-SIAM Symposium on Discrete Algorithms (SODA), pp. 3253–3272. SIAM (2022)
31. Williamson, D.P., Shmoys, D.B.: The Design of Approximation Algorithms. Cambridge University Press (2011)

Faster Connectivity in Low-Rank Hypergraphs via Expander Decomposition

Calvin Beideman[1]([✉]), Karthekeyan Chandrasekaran[1], Sagnik Mukhopadhyay[2], and Danupon Nanongkai[3]

[1] University of Illinois, Urbana-Champaign, Champaign, IL, USA
{calvinb2,karthe}@illinois.edu
[2] University of Sheffield, Sheffield, UK
s.mukhopadhyay@sheffield.ac.uk
[3] University of Copenhagen, Copenhagen, Denmark

Abstract. The connectivity of a hypergraph is the minimum number of hyperedges whose deletion disconnects the hypergraph. We design an $\hat{O}_r(p + \min\{\lambda^{\frac{r-3}{r-1}}n^2, n^r/\lambda^{\frac{r}{r-1}}, \lambda^{\frac{5r-7}{4r-4}}n^{\frac{7}{4}}\})$ (The $\hat{O}_r(\cdot)$ notation hides terms that are subpolynomial in the main parameter and terms that depend only on r) time algorithm for computing hypergraph connectivity, where $p := \sum_{e \in E} |e|$ is the input size of the hypergraph, n is the number of vertices, r is the rank (size of the largest hyperedge), and λ is the connectivity of the input hypergraph. Our algorithm also finds a minimum cut in the hypergraph. Our algorithm is faster than existing algorithms if $r = O(1)$ and $\lambda = n^{\Omega(1)}$. The heart of our algorithm is a structural result showing a trade-off between the number of hyperedges taking part in all minimum cuts and the size of the smaller side of any minimum cut. This structural result can be viewed as a generalization of an acclaimed structural theorem for simple graphs [Kawarabayashi-Thorup, JACM 19 (Fulkerson Prize 2021)]. We extend the framework of expander decomposition to hypergraphs to prove this structural result. In addition to the expander decomposition framework, our faster algorithm also relies on a new near-linear time procedure to compute connectivity when one of the sides in a minimum cut is small.

Keywords: Hypergraphs · Connectivity · Expander decomposition

1 Introduction

A hypergraph $G = (V, E)$ is specified by a vertex set V and a collection E of hyperedges, where each hyperedge $e \in E$ is a subset of vertices. In this work, we address the problem of computing connectivity/global min-cut in hypergraphs with low rank (e.g., constant rank). The *rank* of a hypergraph, denoted r, is the size of the largest hyperedge—in particular, if the rank of a hypergraph is 2, then the hypergraph is a graph. In the global min-cut problem, the input

© Springer Nature Switzerland AG 2022
K. Aardal and L. Sanitá (Eds.): IPCO 2022, LNCS 13265, pp. 70–83, 2022.
https://doi.org/10.1007/978-3-031-06901-7_6

is a hypergraph with hyperedge weights $w : E \to \mathbb{R}_+$, and the goal is to find a minimum weight subset of hyperedges whose removal disconnects the hypergraph. Equivalently, the goal is to find a partition of the vertex set V into two non-empty parts $(C, V \setminus C)$ so as to minimize the weight of the set of hyperedges intersecting both parts. For a subset $C \subseteq V$, we will denote the weight of the set of hyperedges intersecting both C and $V \setminus C$ by $d(C)$, the resulting function $d : V \to \mathbb{R}_+$ as the cut function of the hypergraph, and the weight of a min-cut by $\lambda(G)$ (we will use λ when the graph G is clear from context).

If the input hypergraph is *simple*—i.e., each hyperedge has unit weight and no parallel copies—then the weight of a min-cut is also known as the *connectivity* of the hypergraph. We focus on finding connectivity in hypergraphs. We emphasize that, in contrast to graphs whose representation size is the number of edges, the representation size of a hypergraph $G = (V, E)$ is $p := \sum_{e \in E} |e|$. We note that $p \le rm$, where r is the rank and m is the number of hyperedges in the hypergraph, and moreover, $r \le n$, where n is the number of vertices. We emphasize that the number of hyperedges m in a hypergraph could be exponential in the number of vertices.

Previous Work. Since the focus of our work is on simple unweighted hypergraphs, we discuss previous work for computing global min-cut in simple unweighted hypergraphs/graphs (i.e., computing connectivity) here. Although global min-cut in weighted graphs has a rich literature, fast computation of global min-cut in simple unweighted graphs was initiated more recently in a seminal work by Kawarabayashi and Thorup (Fulkerson Prize 2021) [20]. The current fastest algorithms to compute graph connectivity (i.e., when $r = 2$) are randomized and run in time $\tilde{O}(m)$ [11,13,15,18,20,25]. In contrast, algorithms to compute hypergraph connectivity are much slower. Furthermore, for hypergraph connectivity/global min-cut, the known randomized approaches are not always faster than the known deterministic approaches. There are two broad algorithmic approaches for global min-cut in hypergraphs: vertex-ordering and random contraction. We discuss these approaches now.

Nagamochi and Ibaraki [26] introduced a groundbreaking vertex-ordering approach to solve global min-cut in graphs in time $O(mn)$. In independent works, Klimmek and Wagner [21] as well as Mak and Wong [24] gave two different generalizations of the vertex-ordering approach to compute hypergraph connectivity in $O(pn)$ time. Queyranne [29] generalized the vertex-ordering approach further to solve *non-trivial symmetric submodular minimization*.[1] Queyranne's algorithm can be implemented to compute hypergraph connectivity in $O(pn)$ time. Thus, all three vertex-ordering based approaches to compute hypergraph connectivity have a run-time of $O(pn)$. This run-time was improved to $O(p + \lambda n^2)$

---

[1] The input here is a symmetric submodular function $f : 2^V \to \mathbb{R}$ via an evaluation oracle and the goal is to find a partition of V into two non-empty parts $(C, V \setminus C)$ to minimize $f(C)$. We recall that a function $f : 2^V \to \mathbb{R}$ is symmetric if $f(A) = f(V \setminus A)$ for all $A \subseteq V$ and is submodular if $f(A) + f(B) \ge f(A \cap B) + f(A \cup B)$ for all $A, B \subseteq V$. The cut function of a hypergraph $d : V \to \mathbb{R}_+$ is symmetric and submodular.

by Chekuri and Xu [7]: They designed an $O(p)$-time algorithm to construct a *min-cut-sparsifier*, namely a subhypergraph G' of the given hypergraph with size $p' = O(\lambda n)$ such that $\lambda(G') = \lambda(G)$. Applying the vertex-ordering based algorithm to G' gives the connectivity of G within a run-time of $O(p + \lambda n^2)$.

We emphasize that all algorithms discussed in the preceding paragraph are deterministic. Karger [16] introduced the influential random contraction approach to solve global min-cut in graphs which was adapted by Karger and Stein [17] to design an $\tilde{O}(n^2)$ time algorithm[2]. Kogan and Krauthgamer [22] extended the random contraction approach to solve global min-cut in r-rank hypergraphs in time $\tilde{O}_r(mn^2)$. Ghaffari, Karger, and Panigrahi [12] suggested a non-uniform distribution for random contraction in hypergraphs and used it to design an algorithm to compute hypergraph connectivity in $\tilde{O}((m + \lambda n)n^2)$ time. Chandrasekaran, Xu, and Yu [4] refined their non-uniform distribution to obtain an $O(pn^3 \log n)$ time algorithm for global min-cut in hypergraphs. Fox, Panigrahi, and Zhang [10] proposed a branching approach to exploit the refined distribution leading to an $O(p + n^r \log^2 n)$ time algorithm for hypergraph global min-cut, where r is the rank of the input hypergraph. Chekuri and Quanrud [5] designed an algorithm based on isolating cuts which achieves a runtime of $\tilde{O}(\sqrt{pn}(m + n)^{1.5})$ for global min-cut in hypergraphs.

Thus, the current fastest known algorithm to compute hypergraph connectivity is a combination of the algorithms of Chekuri and Xu [7], Fox, Panigrahi, and Zhang [10], and Chekuri and Quanrud [5] with a run-time of

$$\tilde{O}\left(p + \min\left\{\lambda n^2, n^r, \sqrt{pn}(m + n)^{1.5}\right\}\right).$$

1.1   Our Results

In this work, we improve the run-time to compute hypergraph connectivity in low rank simple hypergraphs.

Theorem 1. *[Algorithm] Let G be an r-rank n-vertex simple hypergraph of size p. Then, there exists a randomized algorithm that takes G as input and runs in time*

$$\hat{O}_r\left(p + \min\left\{\lambda^{\frac{r-3}{r-1}}n^2, \frac{n^r}{\lambda^{\frac{r}{r-1}}}, \lambda^{\frac{5r-7}{4r-4}}n^{\frac{7}{4}}\right\}\right)$$

to return the connectivity λ of G with high probability. Moreover, the algorithm returns a min-cut in G with high probability.

Our techniques can also be used to obtain a deterministic algorithm that runs in time

$$\hat{O}_r\left(p + \min\left\{\lambda n^2, \lambda^{\frac{r-3}{r-1}}n^2 + \frac{n^r}{\lambda}\right\}\right).$$

---

[2] For functions $f(n)$ and $g(n)$ of n, we say that $f(n) = \tilde{O}(g(n))$ if $f(n) = O(g(n)\text{polylog}(n))$ and $f(n) = \hat{O}(g(n))$ if $f(n) = O(g(n)^{1+o(1)})$, where the $o(1)$ is with respect to n. We say that $f(n) = O_r(g(n))$ if $f(n) = O(g(n)h(r))$ for some function h. We define $\tilde{O}_r(f(n))$ and $\hat{O}_r(f(n))$ analogously.

Our deterministic algorithm is faster than Chekuri and Xu's algorithm when r is a constant and $\lambda = \Omega(n^{(r-2)/2})$, while our randomized algorithm is faster than known algorithms if r is a constant and $\lambda = n^{\Omega(1)}$. We summarize the previous fastest algorithms and our results in Table 1.

Table 1. Comparison of results to compute hypergraph connectivity (simple unweighted r-rank n-vertex m-hyperedge p-size hypergraphs with connectivity λ).

	Deterministic	Randomized
Previous run-time	$O(p + \lambda n^2)$ [7]	$\tilde{O}(p + \min\{\lambda n^2, n^r, \sqrt{pn}(m+n)^{1.5}\})$
		[5,7,10]
Our run-time	$\hat{O}_r\left(p + \min\left\{\lambda n^2, \lambda^{\frac{r-3}{r-1}}n^2 + \frac{n^r}{\lambda}\right\}\right)$	$\hat{O}_r\left(p + \min\left\{\lambda^{\frac{r-3}{r-1}}n^2, \frac{n^r}{\lambda^{\frac{r}{r-1}}}, \lambda^{\frac{5r-7}{4r-4}}n^{\frac{7}{4}}\right\}\right)$

Our algorithm for Theorem 1 proceeds by considering two cases: either (i) the hypergraph has a min-cut where one of the sides is small or (ii) both sides of every min-cut in the hypergraph are large. To account for case (i), we design a near-linear time algorithm to compute a min-cut; to account for case (ii), we perform contractions to reduce the size of the hypergraph without destroying a min-cut and then run known algorithms on the smaller-sized hypergraph leading to savings in run-time. Our contributions in this work are twofold: (1) On the algorithmic front, we design a near-linear time algorithm to find a min-cut where one of the sides is small (if it exists); (2) On the structural front, we show a trade-off between the number of hyperedges taking part in all minimum cuts and the size of the smaller side of any minimum cut (see Theorem 2). This structural result is a generalization of the acclaimed Kawarabayashi-Thorup graph structural theorem [19, 20] (Fulkerson prize 2021). We use the structural result to reduce the size of the hypergraph in case (ii). We elaborate on this structural result now.

Theorem 2. *[Structure] Let $G = (V, E)$ be an r-rank n-vertex simple hypergraph with m hyperedges and connectivity λ. Suppose $\lambda \geq r(4r^2)^r$. Then, at least one of the following holds:*

1. *There exists a min-cut $(C, V \setminus C)$ such that*

$$\min\{|C|, |V \setminus C|\} \leq r - \frac{\log\left(\frac{\lambda}{4r}\right)}{\log n},$$

2. *The number of hyperedges in the union of all min-cuts is*

$$O\left(r^{9r^2+2}\left(\frac{6r^2}{\lambda}\right)^{\frac{1}{r-1}} m \log n\right) = \tilde{O}_r\left(\frac{m}{\lambda^{\frac{1}{r-1}}}\right).$$

The Kawarabayashi-Thorup structural theorem for graphs [19, 20] states that if every min-cut is non-trivial, then the number of edges in the union of all min-cuts is $O(m/\lambda)$, where a cut is defined to be *non-trivial* if it has at least two vertices on each side. Substituting $r = 2$ in our structural theorem recovers this known Kawarabayashi-Thorup structural theorem for graphs. We emphasize that the Kawarabayashi-Thorup structural theorem for graphs is the backbone of the current fastest algorithms for computing connectivity in graphs and has been proved in the literature via several different techniques [13,15,20,30,31]. Part of the motivation behind our work was to understand whether the Kawarabayashi-Thorup structural theorem for graphs could hold for constant rank hypergraphs and if not, then what would be an appropriate generalization. We discovered that the Kawarabayashi-Thorup graph structural theorem *does not* hold for constant rank hypergraphs: There exist hypergraphs in which (i) the min-cut capacity λ is $\Omega(n)$, (ii) there are no trivial min-cuts, and (iii) the number of hyperedges in the union of all min-cuts is a constant fraction of the number of hyperedges— see the full version of this work [1] for such an example. The existence of such examples suggests that we need an alternative definition of *trivial min-cuts* if we hope to extend the Kawarabayashi-Thorup structural theorem for graphs to r-rank hypergraphs. Conclusion 1 of Theorem 2 can be viewed as a way to redefine the notion of *trivial min-cuts*. We denote the *size* of a cut $(C, V \setminus C)$ to be $\min\{|C|, |V \setminus C|\}$—we emphasize that the size of a cut refers to the size of the smaller side of the cut as opposed to the capacity of the cut. A min-cut is *small-sized* if the smaller side of the cut has at most $r - \log(\lambda/4r)/\log n$ many vertices. With this definition, Conclusion 2 of Theorem 2 can be viewed as a generalization of the Kawarabayashi-Thorup structural theorem to hypergraphs which have no small-sized min-cuts: it says that if there is no small-sized min-cut, then the number of hyperedges in the union of all min-cuts is $\tilde{O}_r(m/\lambda^{\frac{1}{r-1}})$.

We mention that the factor $\lambda^{-1/(r-1)}$ in Conclusion 2 of Theorem 2 cannot be improved: There exist hypergraphs in which every min-cut has at least $\sqrt{n}$ vertices on both sides and the number of hyperedges in the union of all min-cuts is $\Theta(m \cdot \lambda^{-1/(r-1)})$—see the full version of this work [1]. We also note that the structural theorem holds only for *simple* hypergraphs/graphs and is known to fail for weighted graphs. As a consequence, our algorithmic techniques are applicable only in simple hypergraphs and not in weighted hypergraphs.

1.2   Technical Overview

Concepts used in the proof strategy of Theorem 2 will be used in the algorithm of Theorem 1 as well, so it will be helpful to discuss the proof strategy of Theorem 2 before the algorithm. We discuss this now. We define a cut $(C, V \setminus C)$ to be *moderate-sized* if $\min\{|C|, |V \setminus C|\} \in (r - \log(\lambda/4r)/\log n, 4r^2)$ and to be *large-sized* if $\min\{|C|, |V \setminus C|\} \geq 4r^2$; we recall that the cut $(C, V \setminus C)$ is small-sized if $\min\{|C|, |V \setminus C|\} \leq r - \log(\lambda/4r)/\log n$.

Proof Strategy for the Structural Theorem (Theorem 2). We assume that $\lambda > r(4r^2)^r$ as in the statement of Theorem 2. The first step of our proof is to show that every min-cut in a hypergraph is either large-sized or small-sized but not moderate-sized—in particular, we prove that if $(C, V \setminus C)$ is a min-cut with $\min\{|C|, |V \setminus C|\} < 4r^2$, then it is in fact a small-sized min-cut (see

Lemma 2 with the additional assumption that $\lambda > r(4r^2)^r$). Here is the informal argument: For simplicity, we will show that if $(C, V \setminus C)$ is a min-cut with $\min\{|C|, |V \setminus C|\} < 4r^2$, then $\min\{|C|, |V \setminus C|\} \leq r$. For the sake of contradiction, suppose that $\min\{|C|, |V \setminus C|\} > r$. The crucial observation is that since the hypergraph has rank r, no hyperedge can contain the smaller side of the min-cut entirely. The absence of such hyperedges means that even if we pack hyperedges in G as densely as possible while keeping $(C, V \setminus C)$ as a min-cut, we cannot pack sufficiently large number of hyperedges to ensure that the degree of each vertex is at least λ. A more careful counting argument extends this approach to show that $\min\{|C|, |V \setminus C|\} \leq r - \log \lambda / \log n$.

Now, in order to prove Theorem 2, it suffices to prove Conclusion 2 under the assumption that all min-cuts are large-sized, i.e., $\min\{|C|, |V \setminus C|\} \geq 4r^2$ for every min-cut $(C, V \setminus C)$. Our strategy to prove Conclusion 2 is to find a partition of the vertex set V such that (i) every hyperedge that is completely contained in one of the parts does not cross any min-cut, and (ii) the number of hyperedges that intersect multiple parts (and therefore, possibly cross some min-cut) is small, i.e., $\tilde{O}_r(m \cdot \lambda^{-1/(r-1)})$. To this end, we start by partitioning the vertex set of the hypergraph G into $X_1, \ldots, X_k$ such that the total number of hyperedges intersecting more than one part of the partition is $\tilde{O}_r(m \cdot \lambda^{-1/(r-1)})$ and the subhypergraph induced by each X_i has conductance $\Omega_r(\lambda^{-1/(r-1)})$ (see Sect. 1.3 for the definition of conductance)—such a decomposition is known as an *expander decomposition*. An expander decomposition immediately satisfies (ii) since the number of hyperedges intersecting more than one part is small. Unfortunately, it may not satisfy (i); yet, it is very close to satisfying (i)—we can guarantee that for every min-cut $(C, V \setminus C)$ and every X_i, either C includes very few vertices from X_i, or C includes almost all the vertices of X_i i.e., $\min\{|X_i \cap C|, |X_i \setminus C|\} = O_r(\lambda^{1/(r-1)})$. We note that if $\min\{|X_i \cap C|, |X_i \setminus C|\} = 0$ for every min-cut $(C, V \setminus C)$ and every part X_i then (i) would be satisfied; moreover, if a part X_j is a singleton vertex part (i.e., $|X_j| = 1$), then $\min\{|X_j \cap C|, |X_j \setminus C|\} = 0$ holds. So, our strategy, at this point, is to remove some of the vertices from X_i to form their own singleton vertex parts in the partition in order to achieve $\min\{|X_i \cap C|, |X_i \setminus C|\} = 0$ while controlling the increase in the number of hyperedges that cross the parts. This is achieved by a TRIM operation and a series of SHAVE operations.

The crucial parameter underlying TRIM and SHAVE operations is the notion of degree within a subset: We will denote the degree of a vertex v as $d(v)$ and define the degree contribution of a vertex v inside a vertex set X, denoted by $d_X(v)$, to be the number of hyperedges containing v that are completely contained in X. The TRIM operation on a part X_i repeatedly removes from X_i vertices with small degree contribution inside X_i, i.e., $d_{X_i}(v) < d(v)/2r$ until no such vertex can be found. Let X_i' denote the set obtained from X_i after the TRIM operation. We note that our partition now consists of $X_1', \ldots, X_k'$ as well as singleton vertex parts for each vertex that we removed with the TRIM operation. This operation alone makes a lot of progress towards our goal—we show that $\min\{|X_i' \cap C|, |X_i' \setminus C|\} = O(r^2)$, while the number of hyperedges crossing the partition blows up only by an $O(r)$ factor (see Claims 3 and 4). The little progress that is left to our final goal is achieved by a series of ($O(r^2)$ many) SHAVE operations. The SHAVE operation finds the set of vertices in each X_i' whose degree contribution inside X_i' is not very large, i.e., $d_{X_i'}(v) \leq (1 - r^{-2})d(v)$ and removes this set of vertices from X_i' in one shot—such vertices are again declared as singleton vertex parts in the partition. We show that the SHAVE operation strictly reduces $\min\{|X_i' \cap C|, |X_i' \setminus C|\}$ without adding too many hyperedges across the parts (see Claims 3 and 5)—this argument crucially uses the assumption that all min-cuts are large-sized (i.e., $\min\{|C|, |V \setminus C|\} \geq 4r^2$). Because of our guarantee from the TRIM operation regarding $\min\{|X_i' \cap C|, |X_i' \setminus C|\}$, we need to perform the SHAVE operation $O(r^2)$ times to obtain a partition that satisfies conditions (i) and (ii) stated in the preceding paragraph.

Algorithm from Structural Theorem (Theorem 1). We now briefly describe our algorithm: Given an r-rank hypergraph G, we estimate the connectivity λ to within a constant factor in $O(p)$ time using an algorithm of Chekuri and Xu [7]. Next, we use the estimated connectivity value $k = \Theta(\lambda)$ to obtain a subhypergraph G' with size $p' = O_r(\lambda n)$ such that all min-cuts are preserved in time $O(p)$. The rest of the steps are run on this subhypergraph G'. We have two possibilities as stated in Theorem 2. We account for these two possibilities by running two different algorithms: (i) Assuming that some min-cut has size less than $r - \log(\lambda/4r)/\log n$, we design a near-linear time algorithm to find a min-cut. This algorithm is inspired by recent vertex connectivity algorithms, in particular the local vertex connectivity algorithm of [9,28] and the sublinear-time kernelization technique of [23]. This algorithm runs in $\tilde{O}_r(p)$ time. (ii) Assuming that every min-cut is large-sized, we design a fast algorithm to find a min-cut. For this, we find an expander decomposition $\mathcal{X}$ of G', perform a TRIM operation followed by a series of $O(r^2)$ SHAVE operations, and then contract each part of the trimmed and shaved expander decomposition to obtain a hypergraph G''. This reduces the number of vertices in G'' to $O_r(n/\lambda^{1/(r-1)})$ and consequently, running the global min-cut algorithm of either [10] or [6] or [5] (whichever is faster) on G'' leads to an overall run-time of $\hat{O}_r(p + \min\{\lambda^{(r-3)/(r-1)}n^2, n^r/\lambda^{r/(r-1)}, \lambda^{(5r-7)/(4r-4)}n^{7/4}\})$ for step (ii). We return the cheaper of the two cuts found in steps (i) and (ii). The correctness of the algorithm follows by the structural theorem and the total run-time is $\hat{O}_r(p + \min\{n^r/\lambda^{r/(r-1)}, \lambda^{(r-3)/(r-1)}n^2, \lambda^{(5r-7)/(4r-4)}n^{7/4}\})$.

We note here that the expander decomposition framework for graphs was developed in a series of works for the dynamic connectivity problem [8,27,32,33]. Very recently, it has found applications for other problems [2,3,14]. Closer to our application, Saranurak [31] used expander decomposition to give an algorithm to compute edge connectivity in graphs via the use of TRIM and SHAVE operations. The TRIM and SHAVE operations were introduced by Kawarabayashi and Thorup [20] to compute graph connectivity in deterministic $O(m \log^{12} n)$ time. Our line of attack is an adaptation of Saranurak's approach. Since our structural theorem is meant for hypergraph connectivity (and is hence, more complicated than what is used by [31]), we have to work more.

Organization. We prove the structural theorem in Sect. 2. We defer the proof of the algorithmic result and all missing proofs to the full version of the work [1] due to space limitations. We also elaborate on relevant previous work in the full version.

1.3 Preliminaries

Let $G = (V, E)$ be a hypergraph. Let $S, T \subseteq V$ be subsets of vertices. We define $E[S]$ to be the set of hyperedges completely contained in S, $E(S, T)$ to be the set of hyperedges contained in $S \cup T$ and intersecting both S and T, and $E^o(S, T)$ to be the set of hyperedges intersecting both S and T. With this notation, if S and T are disjoint, then $E(S, T) = E[S \cup T] - E[S] - E[T]$ and moreover, if the hypergraph is a graph, then $E(S, T) = E^o(S, T)$. A cut is a partition $(S, V \setminus S)$ where both S and $V \setminus S$ are non-empty. Let $\delta(S) := E(S, V \setminus S)$. For a vertex $v \in V$, we let $\delta(v)$ represent $\delta(\{v\})$. We define the capacity of $(S, V \setminus S)$ as $|\delta(S)|$, and call a cut as a min-cut if it has minimum capacity among all cuts in G. The connectivity of a simple hypergraph G is the capacity of a min-cut in G.

We recall that the *size* of a cut $(S, V \setminus S)$ is $\min\{|S|, |V \setminus S|\}$. We emphasize the distinction between the size of a cut and the capacity of a cut: size is the cardinality of the smaller side of the cut while capacity is the number of hyperedges crossing the cut.

For a vertex $v \in V$ and a subset $S \subseteq V$, we define the degree of v by $d(v) := |\delta(v)|$ and its degree inside S by $d_S(v) := |e \in \delta(v) : e \subseteq S|$. We define $\delta := \min_{v \in V} d(v)$ to be the minimum degree in G. We define $\text{vol}(S) := \sum_{v \in S} d(v)$ and for $T \subseteq V$, $\text{vol}_S(T) := \sum_{v \in T} d_S(v)$. We define the *conductance* of a set $X \subseteq V$ as $\min_{\emptyset \neq S \subsetneq X} \{ \frac{|E^o(S, X \setminus S)|}{\min\{\text{vol}(S), \text{vol}(X \setminus S)\}} \}$. For positive integers, $i < j$, we let $[i, j]$ represent the set $\{i, i+1, \ldots, j-1, j\}$. The following proposition will be useful while counting hyperedges within nested sets.

Proposition 1. *Let $G = (V, E)$ be an r-rank n-vertex hypergraph and let $T \subseteq S \subseteq V$. Then,*

$$|E(T, S \setminus T)| \geq \left(\frac{1}{r-1} \right) (\text{vol}_S(T) - r |E[T]|).$$

2 Structural Theorem

We prove Theorem 2 in this section. We call a min-cut $(C, V \setminus C)$ *moderate-sized* if its size $\min\{|C|, |V \setminus C|\}$ is in the range $(r - \log(\lambda/4r)/\log n, (\lambda/2)^{1/r})$. In Sect. 2.1, we show that a hypergraph has no moderate-sized min-cuts. In Sect. 2.2, we define TRIM and SHAVE operations and prove properties about these operations. We prove Theorem 2 in Sect. 2.3. We begin with the following lemma showing the existence of an expander decomposition for low-rank hypergraphs (which follows from the existence of an expander decomposition for graphs).

Lemma 1 (Existential hypergraph expander decomposition). *For every r-rank n-vertex hypergraph $G = (V, E)$ with $p := \sum_{e \in E} |e|$ and every positive real value $\phi \leq 1/(r-1)$, there exists a partition $\{X_1, \ldots, X_k\}$ of the vertex set V such that the following hold:*

1. *$\sum_{i=1}^{k} |\delta(X_i)| = O(r\phi p \log n)$, and*
2. *For every $i \in [k]$ and every non-empty set $S \subset X_i$, we have that*

$$|E^o(S, X_i \setminus S)| \geq \phi \cdot \min\{\mathrm{vol}(S), \mathrm{vol}(X_i \setminus S)\}.$$

2.1   No Moderate-Sized Min-Cuts

The following lemma is the main result of this section. It shows that there are no moderate-sized min-cuts.

Lemma 2. *Let $G = (V, E)$ be an r-rank n-vertex hypergraph with connectivity λ such that $\lambda \geq r2^{r+1}$. Let $(C, V \setminus C)$ be an arbitrary min-cut. If $\min\{|C|, |V \setminus C|\} > r - \log(\lambda/4r)/\log n$, then $\min\{|C|, |V \setminus C|\} \geq (\lambda/2)^{1/r}$.*

Proof. Without loss of generality, let $|C| = \min\{|C|, |V \setminus C|\}$. Let $t := |C|$ and $s := r - \log(\lambda/4r)/\log n$. We know that $s < t$. Suppose for contradiction that $t < (\lambda/2)^{1/r}$. We will show that there exists a vertex v with $|\delta(v)| < \lambda$, thus contradicting the fact that λ is the min-cut capacity. We classify the hyperedges of G which intersect C into three types as follows: $E_1 := \{e \in E : e \subseteq C\}, E_2 := \{e \in E : C \subsetneq e\}$, and $E_3 := \{e \in E : \emptyset \neq e \cap C \neq C$ and $e \cap (V \setminus C) \neq \emptyset\}$. We distinguish two cases:

Case 1: Suppose $t < r$. Then, the number of hyperedges that can be fully contained in C is at most 2^r, so $|E_1| \leq 2^r$. Since $(C, V \setminus C)$ is a min-cut, we have that $\lambda = |\delta(C)| = |E_2| + |E_3|$. We note that the number of hyperedges of size i that contain all of C is at most $\binom{n-t}{i-t}$. Hence,

$$|E_2| \leq \sum_{i=t+1}^{r} \binom{n-t}{i-t} = \sum_{i=1}^{r-t} \binom{n-t}{i} \leq \sum_{i=1}^{r-t} n^i \leq 2n^{r-t}.$$

Since each hyperedge in E_3 contains at most $t-1$ vertices of C, a uniform random vertex of C is in such a hyperedge with probability at most $(t-1)/t$. Therefore, if we pick a uniform random vertex from C, the expected number of hyperedges from E_3 incident to it is at most $(\frac{t-1}{t})|E_3|$. Hence, there exists a vertex $v \in C$ such that

$$|\delta(v) \cap E_3| \leq \left(\frac{t-1}{t}\right)|E_3| \leq \left(\frac{t-1}{t}\right)|\delta(C)| \leq \left(\frac{r-1}{r}\right)\lambda.$$

Combining the bounds for E_1, E_2, and E_3, we have that

$$|\delta(v)| = |\delta(v) \cap E_1| + |E_2| + |\delta(v) \cap E_3| \leq |E_1| + |E_2| + |\delta(v) \cap E_3|$$

$$\leq 2^r + 2n^{r-t} + \left(\frac{r-1}{r}\right)\lambda < 2^r + 2n^{r-s} + \left(\frac{r-1}{r}\right)\lambda$$

$$= 2^r + \frac{\lambda}{2r} + \left(\frac{r-1}{r}\right)\lambda = \lambda + \frac{r2^{r+1} - \lambda}{2r} \leq \lambda.$$

Case 2: Suppose $t \geq r$. Then, no hyperedge can contain C as a proper subset, so $|E_2| = 0$. For each $v \in C$, the number of hyperedges e of size i such that $v \in e \subseteq C$ is at most $\binom{t-1}{i-1}$. Hence,

$$|\delta(v) \cap E_1| \leq \sum_{i=2}^{r} \binom{t-1}{i-1} = \sum_{i=1}^{r-1} \binom{t-1}{i} \leq \sum_{i=1}^{r-1} t^i \leq 2t^{r-1}.$$

Since each hyperedge in E_3 contains at most $r-1$ vertices of C, a random vertex of C is in such a hyperedge with probability at most $(r-1)/t$. Therefore, if we pick a random vertex from C, the expected number of hyperedges from E_3 incident to it is at most $(\frac{r-1}{t})|E_3|$. Hence, there exists a vertex $v \in C$ such that

$$|\delta(v) \cap E_3| \leq \left(\frac{r-1}{t}\right)|E_3| \leq \left(\frac{r-1}{t}\right)\lambda.$$

Since $t < (\lambda/2)^{1/r}$ and $t \geq r$, we have that $2t^r/\lambda < t - r + 1$. Combining this with our bounds on $|\delta(v) \cap E_1|$ and $|\delta(v) \cap E_3|$, we have that

$$|\delta(v)| = |\delta(v) \cap E_1| + |\delta(v) \cap E_3| \leq 2t^{r-1} + \left(\frac{r-1}{t}\right)\lambda = \left(r - 1 + \frac{2t^r}{\lambda}\right)\frac{\lambda}{t} < \lambda.$$

2.2 Trim and Shave Operations

In this section, we define the trim and shave operations and prove certain useful properties about them. Throughout this section, let $G = (V, E)$ be an r-rank, n-vertex hypergraph with minimum degree δ and min-cut capacity λ. For $X \subseteq V$, let $\text{TRIM}(X)$ be the set obtained by repeatedly removing from X a vertex v with $d_X(v) < d(v)/2r$ until no such vertices remain, $\text{SHAVE}(X) := \{v \in X : d_X(v) > (1 - 1/r^2)d(v)\}$, and $\text{SHAVE}_k(X) := \text{SHAVE}(\text{SHAVE} \cdots (\text{SHAVE}(X)))$ be the result of applying k consecutive shave operations to X. We emphasize that TRIM is an adaptive operation while SHAVE is a non-adaptive operation and $\text{SHAVE}_k(X)$ is a sequence of shave operations. The next claim shows that TRIM and SHAVE operations could increase the cut value only by a small factor.

Claim 3. *Let X be a subset of V, $X' := \text{TRIM}(X)$, and $X'' := \text{SHAVE}(X)$. Then*

1. *$|E[X] - E[X']| \leq |\delta(X)|$, $|E[X] - E[X'']| \leq r^2(r-1)|\delta(X)|$, and*
2. *$|\delta(X')| \leq 2|\delta(X)|$, and $|\delta(X'')| \leq r^3|\delta(X)|$.*

The following claim shows that the TRIM operation on a set X that has small intersection with a min-cut further reduces the intersection.

Claim 4. *Let $(C, V \setminus C)$ be a min-cut. Let X be a subset of V and $X' := \text{TRIM}(X)$. If $\min\{|X \cap C|, |X \cap (V \setminus C)|\} \leq (\delta/6r^2)^{1/(r-1)}$, then*

$$\min\{|X' \cap C|, |X' \cap (V \setminus C)|\} \leq 3r^2.$$

The following claim shows that the SHAVE operation on a set X which has small intersection with a large-sized min-cut further reduces the intersection.

Claim 5. *Suppose* $\lambda \geq r(4r^2)^r$. *Let* $(C, V \setminus C)$ *be a min-cut with* $\min\{|C|, |V \setminus C|\} \geq 4r^2$. *Let* X' *be a subset of* V *and* $X'' := \text{SHAVE}(X')$. *If* $0 < \min\{|X' \cap C|, |X' \cap (V \setminus C)|\} \leq 3r^2$, *then*

$$\min\{|X'' \cap C|, |X'' \cap (V \setminus C)|\} \leq \min\{|X' \cap C|, |X' \cap (V \setminus C)|\} - 1.$$

Proof. Without loss of generality, we assume that $|X' \cap C| = \min\{|X' \cap C|, |X' \cap (V \setminus C)|\}$. Since $X'' \subseteq X'$, we have that $|X'' \cap C| \leq |X' \cap C|$. Thus, we only need to show that this inequality is strict. Suppose for contradiction that $|X'' \cap C| = |X' \cap C|$. We note that $0 < |X'' \cap C| \leq 3r^2$.

Let $Z := X' \cap C = X'' \cap C$, and let $C' := C - X'$. Since $|C| \geq \min\{|C|, |V \setminus C|\} \geq 4r^2$ and $|Z| \leq 3r^2$, we know that C' is nonempty.

We note that $Z \subseteq X''$. By definition of SHAVE, we have that $\text{vol}_{X'}(Z) = \sum_{v \in Z} d_{X'}(v) > \sum_{v \in Z} \left(1 - \frac{1}{r^2}\right) d(v) = \left(1 - \frac{1}{r^2}\right) \text{vol}(Z)$.

We note that $|E(Z, V \setminus C)| \geq |E(Z, X' \setminus C)| = |E(Z, X' \setminus Z)|$, so by Proposition 1, we have that $|E(Z, V \setminus C)| \geq |E(Z, X' \setminus Z)| \geq \left(\frac{1}{r-1}\right) (\text{vol}_{X'}(Z) - r|E[Z]|) > \left(\frac{1}{r-1}\right) \left(\left(1 - \frac{1}{r^2}\right) \text{vol}(Z) - r|Z|^r\right)$. We also know from the definition of SHAVE that $|E(Z, C \setminus Z)| \leq \sum_{v \in Z} |E(\{v\}, C \setminus Z)| \leq \sum_{v \in Z} \frac{1}{r^2} d(v) = \frac{\text{vol}(Z)}{r^2}$.

Thus, using our assumption that $\lambda \geq r(4r^2)^r$, we have that $|E(Z, (V \setminus C))| >$

$$\left(\frac{1}{r-1}\right) \left(\left(1 - \frac{1}{r^2}\right) \text{vol}(Z) - r|Z|^r\right) = \left(\frac{r^2 - 1}{r^2(r-1)}\right) \text{vol}(Z) - \left(\frac{r}{r-1}\right) |Z|^r$$

$$= \frac{\text{vol}(Z)}{r^2} + \frac{\text{vol}(Z)}{r} - \left(\frac{r}{r-1}\right) |Z|^r \geq \frac{\text{vol}(Z)}{r^2} + \frac{\sum_{v \in Z} d(v)}{r} - r|Z|^r$$

$$\geq \frac{\text{vol}(Z)}{r^2} + \frac{|Z|\lambda}{r} - r|Z|^r \geq \frac{\text{vol}(Z)}{r^2} + (4r^2)^r|Z| - r|Z|^r$$

$$\geq \frac{\text{vol}(Z)}{r^2} + (4r^2)|Z|^r - r|Z|^r \geq \frac{\text{vol}(Z)}{r^2} \geq |E(Z, C \setminus Z)|.$$

We note that $E(Z, (V \setminus C))$ is the set of hyperedges which are cut by C but not C', while $E(Z, C \setminus Z)$ is the set of hyperedges which are cut by C' but not C. Since we have shown that $|E(Z, V \setminus C)| > |E(Z, C \setminus Z)|$, we conclude that $|\delta(C)| > |\delta(C')|$. Since $(C, V \setminus C)$ is a min-cut and $\emptyset \neq C' \subseteq C \subsetneq V$, this is a contradiction.

2.3   Proof of Theorem 2

Proof (Proof of Theorem 2). Suppose the first conclusion does not hold. Then, by Lemma 2, the smaller side of every min-cut has size at least $(\lambda/2)^{1/r} \geq 4r^2$. Let $(C, V \setminus C)$ be an arbitrary min-cut. We use Lemma 1 with $\phi = (6r^2/\lambda)^{1/(r-1)}$ to get an expander decomposition $\mathcal{X} = \{X_1, \ldots, X_k\}$. We note that $\phi \leq 1/(r-1)$ holds by the assumption that $\lambda \geq r(4r^2)^r$. For $i \in [k]$, let $X_i' := \text{TRIM}(X_i)$ and $X_i'' := \text{SHAVE}_{3r^2}(X_i')$.

Let $i \in [k]$. By the definition of the expander decomposition and our choice of $\phi = (6r^2/\lambda)^{1/(r-1)}$, we have that $\lambda \geq |E^o(X_i \cap C, X_i \cap (V \setminus C))| \geq \left(\frac{6r^2}{\lambda}\right)^{\frac{1}{r-1}} \min\{\text{vol}(X_i \cap C), \text{vol}(X_i \setminus C)\} \geq \left(\frac{6r^2}{\lambda}\right)^{\frac{1}{r-1}} \delta \min\{|X_i \cap C|, |X_i \setminus C|\}$.

Thus, $\min\{|X_i \cap C|, |X_i \setminus C|\} \leq (\lambda/\delta)(\lambda/6r^2)^{1/(r-1)} \leq (\lambda/6r^2)^{1/(r-1)} \leq (\delta/6r^2)^{1/(r-1)}$. Therefore, by Claim 4, we have that $\min\{|X_i' \cap C|, |X_i' \cap (V \setminus C)|\} \leq 3r^2$. We recall that $\lambda \geq r(4r^2)^r$ and every min-cut has size at least $4r^2$. By $3r^2$ repeated applications of Claim 5, we have that $\min\{|X_i'' \cap C|, |X_i'' \cap (V \setminus C)|\} = 0$.

Let $\mathcal{X}'' := \{X_1'', \ldots, X_k''\}$. Since $\min\{|X_i'' \cap C|, |X_i'' \cap (V \setminus C)|\} = 0$ for every min-cut $(C, V \setminus C)$ and every $X_i'' \in \mathcal{X}''$, it follows that no hyperedge crossing a min-cut is fully contained within a single part of $\mathcal{X}''$. Thus, it suffices to show that $|E - \bigcup_{i=1}^k E[X_i'']|$ is small—i.e., the number of hyperedges not contained in any of the parts of $\mathcal{X}''$ is $\tilde{O}_r(m/\lambda^{\frac{1}{r-1}})$.

By the first part of Claim 3, we have that $|E[X_i] - E[X_i']| \leq 2|\delta(X_i)|$ and $|\delta(X_i')| \leq 2|\delta(X_i)|$ for each $i \in [k]$. By the second part of Claim 3, we have that $|\delta(\text{SHAVE}_{j+1}(X_i'))| \leq r^3|\delta(\text{SHAVE}_j(X_i'))|$ for every non-negative integer j. Therefore, by repeated application of the second part of Claim 3, for every $j \in [3r^2]$, we have that $|\delta(\text{SHAVE}_j(X_i'))| \leq 2r^{3j}|\delta(X_i)|$. By the first part of Claim 3, for every $j \in [3r^2]$, we have that $|E[\text{SHAVE}_{j-1}(X_i')] - E[\text{SHAVE}_j(X_i')]| \leq r^3|\delta(\text{SHAVE}_{j-1}(X_i'))| \leq 2r^{3j}|\delta(X_i)|$.

Therefore, $\left| E - \bigcup_{i=1}^k E[X_i''] \right| - \left| E - \bigcup_{i=1}^k E[X_i] \right|$

$$\leq \sum_{i=1}^k |E[X_i] - E[X_i'']|$$

$$= \sum_{i=1}^k \left(|E[X_i] - E[X_i']| + \sum_{j=1}^{3r^2} |E[\text{SHAVE}_{j-1}(X_i')] - E[\text{SHAVE}_j(X_i')]| \right)$$

$$\leq \sum_{i=1}^k \left(2|\delta(X_i)| + \sum_{j=1}^{3r^2} 2r^{3j}|\delta(X_i)| \right) = \sum_{i=1}^k |\delta(X_i)| \left(2 + \sum_{j=1}^{3r^2} 2r^{3j} \right)$$

$$\leq \sum_{i=1}^k 3r^{9r^2}|\delta(X_i)| \leq 3r^{9r^2} \sum_{i=1}^k |E(X_i, V \setminus X_i)|.$$

Hence, $\left| E - \bigcup_{i=1}^k E[X_i''] \right| \leq 4r^{9r^2} \sum_{i=1}^k |E(X_i, V \setminus X_i)|$. By Lemma 1, since $\mathcal{X}$ is an expander decomposition for $\phi = (6r^2/\lambda)^{1/(r-1)}$ and since $p = \sum_{e \in E} |e| \leq mr$, we have that

$$\sum_{i=1}^k |E(X_i, V \setminus X_i)| = O(r\phi p \log n) = O(r) \left(\frac{6r^2}{\lambda}\right)^{\frac{1}{r-1}} p \log n = O(r^2) \left(\frac{6r^2}{\lambda}\right)^{\frac{1}{r-1}} m \log n.$$

Thus, $|E - \bigcup_{i=1}^k E[X_i'']| = O(r^{9r^2+2}(6r^2/\lambda)^{1/(r-1)} m \log n)$, thus proving the second conclusion.

Acknowledgement. This project has received funding from the European Research Council (ERC) under the European Union's Horizon 2020 research and innovation programme under grant agreement No 715672. The last two authors are also supported by the Swedish Research Council (Reg. No. 2015-04659 and 2019-05622). Karthekeyan and Calvin are supported in part by NSF grants CCF-1814613 and CCF-1907937.

References

1. Beideman, C., Chandrasekaran, K., Mukhopadhyay, S., Nanongkai, D.: Faster connectivity in low-rank hypergraphs via expander decomposition. CoRR abs/2011.08097 (2021)
2. Bernstein, A., et al.: Fully-dynamic graph sparsifiers against an adaptive adversary. CoRR abs/2004.08432 (2020)
3. Bernstein, A., Gutenberg, M.P., Saranurak, T.: Deterministic decremental reachability, SCC, and shortest paths via directed expanders and congestion balancing. In: FOCS. IEEE Computer Society (2020)
4. Chandrasekaran, K., Xu, C., Yu, X.: Hypergraph k-cut in randomized polynomial time. Mathematical Programming (Preliminary version in SODA 2018), November 2019
5. Chekuri, C., Quanrud, K.: Isolating cuts, (Bi-)submodularity, and faster algorithms for connectivity. In: ICALP, pp. 50:1–50:20 (2021)
6. Chekuri, C., Xu, C.: Computing minimum cuts in hypergraphs. In: SODA, pp. 1085–1100. SIAM (2017)
7. Chekuri, C., Xu, C.: Minimum cuts and sparsification in hypergraphs. SIAM J. Comput. **47**(6), 2118–2156 (2018)
8. Chuzhoy, J., Gao, Y., Li, J., Nanongkai, D., Peng, R., Saranurak, T.: A deterministic algorithm for balanced cut with applications to dynamic connectivity, flows, and beyond. In: FOCS. IEEE Computer Society (2020)
9. Forster, S., Nanongkai, D., Yang, L., Saranurak, T., Yingchareonthawornchai, S.: Computing and testing small connectivity in near-linear time and queries via fast local cut algorithms. In: SODA, pp. 2046–2065. ACM/SIAM (2020)
10. Fox, K., Panigrahi, D., Zhang, F.: Minimum cut and minimum k-cut in hypergraphs via branching contractions. In: SODA, pp. 881–896. SIAM (2019)
11. Gawrychowski, P., Mozes, S., Weimann, O.: Minimum cut in $O(m \log^2 n)$ time. In: ICALP, pp. 57:1–57:15 (2020)
12. Ghaffari, M., Karger, D., Panigrahi, D.: Random contractions and sampling for hypergraph and hedge connectivity. In: SODA, pp. 1101–1114. ACM/SIAM (2017)
13. Ghaffari, M., Nowicki, K., Thorup, M.: Faster algorithms for edge connectivity via random 2-out contractions. In: SODA. ACM/SIAM (2020)
14. Goranci, G., Räcke, H., Saranurak, T., Tan, Z.: The expander hierarchy and its applications to dynamic graph algorithms. In: SODA, pp. 2212–2228 (2021)
15. Henzinger, M., Rao, S., Wang, D.: Local flow partitioning for faster edge connectivity. In: SODA, pp. 1919–1938. ACM/SIAM (2017)
16. Karger, D.: Global min-cuts in RNC, and other ramifications of a simple min-cut algorithm. In: SODA, pp. 21–30. ACM/SIAM (1993)
17. Karger, D., Stein, C.: A new approach to the minimum cut problem. J. ACM **43**(4), 601–640 (1996)
18. Karger, D.R.: Minimum cuts in near-linear time. J. ACM **47**(1), 46–76 (2000). Announced at STOC 1996

19. Kawarabayashi, K., Thorup, M.: Deterministic edge connectivity in near-linear time. In: STOC, pp. 665–674. ACM (2015)
20. Kawarabayashi, K., Thorup, M.: Deterministic edge connectivity in near-linear time. J. ACM **66**(1), 4:1–4:50 (2019)
21. Klimmek, R., Wagner, F.: A simple hypergraph min cut algorithm. Technical report B 96–02, Institute of Computer Science, Freie Universitat (1996)
22. Kogan, D., Krauthgamer, R.: Sketching cuts in graphs and hypergraphs. In: ITCS, pp. 367–376 (2015)
23. Li, J., Nanongkai, D., Panigrahi, D., Saranurak, T., Yingchareonthawornchai, S.: Vertex connectivity in poly-logarithmic max-flows (2021, unpublished)
24. Mak, W.K., Wong, M.D.F.: A fast hypergraph min-cut algorithm for circuit partitioning. Integr.: VLSI J. **30**(1), 1–11 (2000)
25. Mukhopadhyay, S., Nanongkai, D.: Weighted min-cut: sequential, cut-query, and streaming algorithms. In: STOC, pp. 496–509. ACM (2020)
26. Nagamochi, H., Ibaraki, T.: Computing edge-connectivity in multigraphs and capacitated graphs. SIAM J. Discret. Math. **5**(1), 54–66 (1992)
27. Nanongkai, D., Saranurak, T.: Dynamic spanning forest with worst-case update time: adaptive, Las Vegas, and $O(n^{1/2-\epsilon})$-time. In: STOC, pp. 1122–1129. ACM (2017)
28. Nanongkai, D., Saranurak, T., Yingchareonthawornchai, S.: Breaking quadratic time for small vertex connectivity and an approximation scheme. In: STOC, pp. 241–252. ACM (2019)
29. Queyranne, M.: Minimizing symmetric submodular functions. Math. Program. **82**(1–2), 3–12 (1998)
30. Rubinstein, A., Schramm, T., Weinberg, S.M.: Computing exact minimum cuts without knowing the graph. In: ITCS, pp. 39:1–39:16 (2018)
31. Saranurak, T.: A simple deterministic algorithm for edge connectivity. In: SOSA. SIAM (2021)
32. Saranurak, T., Wang, D.: Expander decomposition and pruning: faster, stronger, and simpler. In: SODA, pp. 2616–2635. SIAM (2019)
33. Wulff-Nilsen, C.: Fully-dynamic minimum spanning forest with improved worst-case update time. In: STOC, pp. 1130–1143. ACM (2017)

Improving the Cook et al. Proximity Bound Given Integral Valued Constraints

Marcel Celaya[1]([✉])[ID], Stefan Kuhlmann[2], Joseph Paat[3], and Robert Weismantel[1]

[1] Department of Mathematics, Institute for Operations Research,
ETH Zürich, Zürich, Switzerland
{`marcel.celaya,robert.weismantel`}`@ifor.math.ethz.ch`
[2] Institut für Mathematik, Technische Universität Berlin, Berlin, Germany
`kuhlmann@math.tu-berlin.de`
[3] Sauder School of Business, University of British Columbia, Vancouver, BC, Canada
`joseph.paat@sauder.ubc.ca`

Abstract. Consider a linear program of the form $\max\{c^\top x : Ax \le b\}$, where A is an $m \times n$ integral matrix. In 1986 Cook, Gerards, Schrijver, and Tardos proved that, given an optimal solution x^*, if an optimal integral solution z^* exists, then it may be chosen such that $\|x^* - z^*\|_\infty < n\Delta$, where Δ is the largest magnitude of any subdeterminant of A. Since then an open question has been to improve this bound, assuming that b is integral valued too. In this manuscript we show that $n\Delta$ can be replaced with $n/2 \cdot \Delta$ whenever $n \ge 2$ and x^* is a vertex. We also show that, in certain circumstances, the factor n can be removed entirely.

1 Introduction

Suppose A is an integral full-column-rank $m \times n$ matrix. The polyhedron corresponding to a right hand side $b \in \mathbb{Q}^m$ is

$$\mathcal{P}(A, b) := \{x \in \mathbb{R}^n : Ax \le b\}.$$

The linear program corresponding to $\mathcal{P}(A, b)$ and an objective vector $c \in \mathbb{Q}^n$ is

$$\mathrm{LP}(A, b, c) := \max\{c^\top x : x \in \mathcal{P}(A, b)\},$$

and the corresponding integer linear program is

$$\mathrm{IP}(A, b, c) := \max\{c^\top x : x \in \mathcal{P}(A, b) \cap \mathbb{Z}^n\}.$$

The *proximity* question in integer linear programming can be stated as follows: Given an optimal vertex solution x^* of $\mathrm{LP}(A, b, c)$, how far away is the nearest optimal solution z^* to $\mathrm{IP}(A, b, c)$ (if one exists)? Proximity has a wide array of applications in integer linear programming. Perhaps not too surprisingly, upper bounds on proximity can help identify integer vectors in $\mathcal{P}(A, b)$ from vertices; this is relevant in search techniques such as the feasibility pump [6] and dynamic

© Springer Nature Switzerland AG 2022
K. Aardal and L. Sanitá (Eds.): IPCO 2022, LNCS 13265, pp. 84–97, 2022.
https://doi.org/10.1007/978-3-031-06901-7_7

programming [5,10]. An upper bound of π on proximity also leads to a trivial enumeration algorithm to optimize $\mathrm{IP}(\boldsymbol{A}, \boldsymbol{b}, \boldsymbol{c})$: Solve $\mathrm{LP}(\boldsymbol{A}, \boldsymbol{b}, \boldsymbol{c})$ to identify an optimal vertex $\boldsymbol{x}^*$ and then enumerate up to $(2\pi + 1)^n$ many integer points $\boldsymbol{z}^*$ satisfying $\|\boldsymbol{x}^* - \boldsymbol{z}^*\|_\infty \leq \pi$.

Proximity has been studied for decades with perhaps the most foundational result due to Cook, Gerards, Schrijver, and Tardos. To state their result, we denote the largest absolute $k \times k$ minor of $\boldsymbol{A}$ by

$$\Delta_k(\boldsymbol{A}) := \max\left\{|\det \boldsymbol{M}| : \boldsymbol{M} \text{ is a } k \times k \text{ submatrix of } \boldsymbol{A}\right\}.$$

Theorem 1 (Theorem 1 in [3]). *Let $\boldsymbol{b} \in \mathbb{Q}^m$ and $\boldsymbol{c} \in \mathbb{Q}^n$. Let $\boldsymbol{x}^*$ be an optimal vertex of $\mathrm{LP}(\boldsymbol{A}, \boldsymbol{b}, \boldsymbol{c})$. If $\mathrm{IP}(\boldsymbol{A}, \boldsymbol{b}, \boldsymbol{c})$ is feasible, then there exists an optimal solution $\boldsymbol{z}^*$ such that[1]*

$$\|\boldsymbol{x}^* - \boldsymbol{z}^*\|_\infty \leq n \cdot \Delta_{n-1}(\boldsymbol{A}).$$

Cook et al.'s result is truly a cornerstone result. Their proof technique has been used to establish proximity bounds involving other data parameters [21] and different norms [11,12]. Furthermore, their result has been extended to derive proximity results for convex separable programs [7,9,20] (where the bound in Theorem 1 remains valid), for mixed integer programs [16], and for random integer programs [15].

Lovász [18, Section 17.2] and Del Pia and Ma [4, Section 4] identified tuples $(\boldsymbol{A}, \boldsymbol{b}, \boldsymbol{c})$ such that proximity is arbitrarily close to the upper bound in Theorem 1. However, their examples crucially rely on the fact that $\boldsymbol{b}$ can take arbitrary rational values. In fact, Lovász's example uses a totally unimodular matrix $\boldsymbol{A}$ while Del Pia and Ma use a unimodular matrix. Therefore, if the right hand sides $\boldsymbol{b}$ in their examples were to be replaced by the integral rounded down vector $\lfloor \boldsymbol{b} \rfloor$, then the polyhedron $\mathcal{P}(\boldsymbol{A}, \lfloor \boldsymbol{b} \rfloor)$ would only have integral vertices. From an integer programming perspective, replacing $\boldsymbol{b}$ with $\lfloor \boldsymbol{b} \rfloor$ is natural as it strengthens the linear relaxation without cutting off any feasible integer solutions.

It remains an open question whether Cook et al.'s bound is tight when $\boldsymbol{b} \in \mathbb{Z}^m$. Under this assumption, Paat et al. [16] conjecture that the true bound is independent of n. This conjecture is supported by various results: Aliev et al. [2] prove that proximity is upper bounded by the largest entry of $\boldsymbol{A}$ for knapsack polytopes, Veselov and Chirkov's result [19] implies a proximity bound of 2 when $\Delta_n(\boldsymbol{A}) \leq 2$, and Aliev et al. [1] prove a bound of $\Delta_n(\boldsymbol{A})$ for corner polyhedra.

Our main result is the first improvement on Cook et al.'s result. Furthermore, our proof technique generalizes theirs, and we believe that it can be applied in the multiple settings where their technique is used.

Theorem 2. *Let $n \geq 2$, $\boldsymbol{b} \in \mathbb{Z}^m$, and $\boldsymbol{c} \in \mathbb{Q}^n$. Let $\boldsymbol{x}^*$ be an optimal vertex of $\mathrm{LP}(\boldsymbol{A}, \boldsymbol{b}, \boldsymbol{c})$. If $\mathrm{IP}(\boldsymbol{A}, \boldsymbol{b}, \boldsymbol{c})$ is feasible, then there exists an optimal solution $\boldsymbol{z}^*$*

---

[1] Their upper bound is stated as $n \cdot \max\left\{\Delta_k(\boldsymbol{A}) : k = 1, \ldots, n\right\}$, but their argument actually yields an upper bound of $n \cdot \Delta_{n-1}(\boldsymbol{A})$. Furthermore, their result holds for any (not necessarily vertex) optimal LP solution $\boldsymbol{x}^*$.

such that

$$\|\boldsymbol{x}^* - \boldsymbol{z}^*\|_\infty < \frac{n}{2} \cdot \Delta_{n-1}(\boldsymbol{A}).$$

Our proof of Theorem 2 consists of three parts. First, we relate proximity to the volume of a certain polytope associated with the matrix $\boldsymbol{A}$. Second, we establish lower bounds on the volume of this polytope when $n = 2$ and $n = 3$; see Sect. 2. Third, we show how these proximity bounds in lower dimensions can be used to derive proximity bounds in higher dimensions; see Sect. 3. We can improve Theorem 2 in certain settings. For instance, if n is a multiple of 3 then we can replace the factor $n/2$ with $(\sqrt{2}/3) \cdot n$, and the factor $(\sqrt{2}/3) \cdot n + 1$ can be used for every n; see Remark 2.

We also consider the case when $\boldsymbol{A}$ is *strictly Δ-modular*, that is, $\boldsymbol{A} = \boldsymbol{TB}$ for a totally unimodular matrix $\boldsymbol{T}$ and a square integer matrix $\boldsymbol{B}$ with determinant Δ; see Sect. 5. Here we show the factor $n/2$ can be removed entirely, generalizing a recent result of Nägele, Santiago, and Zenklusen [14, Theorem 5]. We also give essentially matching lower bounds; see Sect. 6. It is more or less straightforward to find a 1-dimensional polytope $\mathcal{P}(\boldsymbol{A}, \boldsymbol{b}) \subseteq \mathbb{R}^n$ with a matching lower bound on proximity when $n \geq 2$. Thus, our contribution with this lower bound is a *full-dimensional* polytope $\mathcal{P}(\boldsymbol{A}, \boldsymbol{b}) \subseteq \mathbb{R}^n$ with a unique integral point $\boldsymbol{z}^*$, and a vertex $\boldsymbol{x}^*$ *sharing no common facet with* $\boldsymbol{z}^*$, such that the proximity is, up to a constant additive factor, equal to $\Delta_{n-1}(\boldsymbol{A})$.

Theorem 3. *Let $\Delta \geq 1$ and $n \geq 2$.*

1. *For all feasible instances* IP $(\boldsymbol{A}, \boldsymbol{b}, \boldsymbol{c})$ *with $\boldsymbol{A}$ strictly Δ-modular and $\boldsymbol{b}$ integral, and for all optimal vertices $\boldsymbol{x}^*$ of* LP$(\boldsymbol{A}, \boldsymbol{b}, \boldsymbol{c})$, *there exists an optimal solution $\boldsymbol{z}^*$ of* IP$(\boldsymbol{A}, \boldsymbol{b}, \boldsymbol{c})$ *such that*

$$\|\boldsymbol{x}^* - \boldsymbol{z}^*\|_\infty \leq \max\{\Delta_{n-1}(\boldsymbol{A}), \Delta_n(\boldsymbol{A})\} - 1.$$

2. *Let $\Delta \geq 3$. There exists a feasible instance* IP $(\boldsymbol{A}, \boldsymbol{b}, \boldsymbol{c})$ *with $\boldsymbol{A}$ strictly Δ-modular and $\boldsymbol{b}$ integral, and an optimal vertex $\boldsymbol{x}^*$ of* LP$(\boldsymbol{A}, \boldsymbol{b}, \boldsymbol{c})$, *such that every feasible integral solution $\boldsymbol{z}^*$ of* IP$(\boldsymbol{A}, \boldsymbol{b}, \boldsymbol{c})$ *satisfies*

$$\|\boldsymbol{x}^* - \boldsymbol{z}^*\|_\infty = \max\{\Delta_{n-1}(\boldsymbol{A}), \Delta_n(\boldsymbol{A})\} - 2.$$

Moreover, $\mathcal{P}(\boldsymbol{A}, \boldsymbol{b})$ is full-dimensional, and $\boldsymbol{x}^*$ and $\boldsymbol{z}^*$ do not lie on a common facet of $\mathcal{P}(\boldsymbol{A}, \boldsymbol{b})$.

1.1 Preliminaries and Notation

Here we outline the key objects and parameters used in the paper.

Let $\boldsymbol{A} \in \mathbb{Z}^{m \times n}$ be a full-column-rank matrix, and $\boldsymbol{b} \in \mathbb{Z}^m$ be such that $\mathcal{P}(\boldsymbol{A}, \boldsymbol{b}) \cap \mathbb{Z}^n \neq \emptyset$. For $I \subseteq [m] := \{1, \ldots, m\}$, we use $\boldsymbol{A}_I$ and $\boldsymbol{b}_I$ to denote the rows of $\boldsymbol{A}$ and $\boldsymbol{b}$ indexed by I. If $I = \{i\}$, then we write $\boldsymbol{a}_i^\top := \boldsymbol{A}_I$. We use $\boldsymbol{0}$ and $\boldsymbol{1}$ to denote the all zero and all one vector (in appropriate dimension). For a polyhedron $\mathcal{Q} \subseteq \mathbb{R}^n$, the dimension of $\mathcal{Q}$ is the dimension of the linear span of $\mathcal{Q}$ and is denoted by $\dim \mathcal{Q}$. We also define, for $I \subseteq [m]$,

$$\gcd \boldsymbol{A}_I := \gcd \left\{ |\det \boldsymbol{M}| : \ \boldsymbol{M} \text{ is a } \operatorname{rank}(\boldsymbol{A}_I) \times \operatorname{rank}(\boldsymbol{A}_I) \text{ submatrix of } \boldsymbol{A}_I \right\},$$

with $\gcd \boldsymbol{A}_{\emptyset} = 1$. In the case when $\mathcal{P} \cap \mathbb{Z}^n = \{\boldsymbol{0}\}$, bounding proximity is equivalent to bounding

$$\max_{\boldsymbol{x} \in \mathcal{P}(\boldsymbol{A}, \boldsymbol{b})} \|\boldsymbol{x}\|_{\infty} = \max_{\boldsymbol{\alpha} \in \{\pm \mathbf{e}_1, \dots, \pm \mathbf{e}_n\}} \max \left\{ \boldsymbol{\alpha}^{\top} \boldsymbol{x} : \ \boldsymbol{x} \in \mathcal{P}(\boldsymbol{A}, \boldsymbol{b}) \right\}, \tag{1}$$

where $\mathbf{e}_1, \dots, \mathbf{e}_n \in \mathbb{Z}^n$ are the standard unit vectors. As we shall see in the proof of Theorem 2, the general case then follows from this case. In light of this, we analyze the maximum of an arbitrary linear form $\boldsymbol{\alpha}^{\top} \boldsymbol{x}$ over $\mathcal{P}(\boldsymbol{A}, \boldsymbol{b})$ for $\boldsymbol{\alpha} \in \mathbb{Z}^n$.

We provide non-trivial bounds on the maximum of these linear forms for small values of n; see Sect. 2. In order to lift low dimensional proximity results to higher dimensions (see Sect. 3), we consider slices of $\mathcal{P}(\boldsymbol{A}, \boldsymbol{b})$ through the origin induced by rows of $\boldsymbol{A}$. Given $I \subseteq [m]$ such that $|I| \leq n - 1$ and $\operatorname{rank} \boldsymbol{A}_I = |I|$, define

$$\mathcal{P}_I(\boldsymbol{A}, \boldsymbol{b}) := \mathcal{P}(\boldsymbol{A}, \boldsymbol{b}) \cap \ker \boldsymbol{A}_I.$$

We specify $\ker \boldsymbol{A}_{\emptyset} = \mathbb{R}^n$, so that $\mathcal{P}_{\emptyset}(\boldsymbol{A}, \boldsymbol{b}) = \mathcal{P}(\boldsymbol{A}, \boldsymbol{b})$. The bounds that we provide on $\boldsymbol{\alpha}^{\top} \boldsymbol{x}$ are given in terms of the parameter

$$\Delta_I(\boldsymbol{A}, \boldsymbol{\alpha}) := \frac{1}{\gcd \boldsymbol{A}_I} \cdot \max \left\{ \left| \det \begin{pmatrix} \boldsymbol{\alpha}^{\top} \\ \boldsymbol{A}_K \end{pmatrix} \right| : I \subseteq K \subseteq [m], \ |K| = n - 1 \right\},$$

and we write $\Delta(\boldsymbol{A}, \boldsymbol{\alpha}) := \Delta_{\emptyset}(\boldsymbol{A}, \boldsymbol{\alpha})$. In particular, we define $\kappa_I(\boldsymbol{A}, \boldsymbol{b}, \boldsymbol{\alpha})$ to be the number satisfying

$$\max_{\boldsymbol{x} \in \mathcal{P}_I(\boldsymbol{A}, \boldsymbol{b})} \boldsymbol{\alpha}^{\top} \boldsymbol{x} = \kappa_I(\boldsymbol{A}, \boldsymbol{b}, \boldsymbol{\alpha}) \, \Delta_I(\boldsymbol{A}, \boldsymbol{\alpha}). \tag{2}$$

Maximizing over all $I \subseteq [m]$ such that $\mathcal{P}_I(\boldsymbol{A}, \boldsymbol{b})$ has a fixed dimension d, define

$$\kappa_d(\boldsymbol{A}, \boldsymbol{b}, \boldsymbol{\alpha}) := \max_{I : \dim \mathcal{P}_I(\boldsymbol{A}, \boldsymbol{b}) = d} \kappa_I(\boldsymbol{A}, \boldsymbol{b}, \boldsymbol{\alpha}).$$

Equation (2) looks similar to the proximity bound we seek. However, $\Delta_I(\boldsymbol{A}, \boldsymbol{\alpha})$ depends on $\boldsymbol{\alpha}$, whereas our main result (Theorem 2) only depends on $\Delta_{n-1}(\boldsymbol{A})$. Later (see Sect. 4), we will substitute $\pm \mathbf{e}_1, \dots, \pm \mathbf{e}_n$ in for $\boldsymbol{\alpha}$ as in (1). We also want to consider $I = \emptyset$ because $\mathcal{P}_{\emptyset}(\boldsymbol{A}, \boldsymbol{b}) = \mathcal{P}(\boldsymbol{A}, \boldsymbol{b})$ by definition. These substitutions will convert $\Delta_I(\boldsymbol{A}, \boldsymbol{\alpha})$ to proving proximity of $\Delta_{n-1}(\boldsymbol{A})$ over $\mathcal{P}(\boldsymbol{A}, \boldsymbol{b})$ as desired in our proximity theorem.

Another important object for us is the following cone. For $\boldsymbol{x}^* \in \mathbb{R}^n$, define

$$\mathcal{C}(\boldsymbol{A}, \boldsymbol{x}^*) := \left\{ \boldsymbol{x} \in \mathbb{R}^n : \begin{array}{l} \operatorname{sign}(\boldsymbol{a}_i^{\top} \boldsymbol{x}^*) \cdot \boldsymbol{a}_i^{\top} \boldsymbol{x} \geq 0 \ \forall \, i \in [m] \text{ such that } \boldsymbol{a}_i^{\top} \boldsymbol{x}^* \neq 0 \\ \boldsymbol{a}_i^{\top} \boldsymbol{x} = 0 \ \forall \, i \in [m] \text{ such that } \boldsymbol{a}_i^{\top} \boldsymbol{x}^* = 0 \end{array} \right\}.$$

The cone $\mathcal{C}\left(\boldsymbol{A}, \boldsymbol{x}^*\right)$ serves as a key ingredient in the proof of Theorem 1 in [3]. We also define the polytope

$$\mathcal{S}\left(\boldsymbol{A}, \boldsymbol{x}^*\right) := \mathcal{C}\left(\boldsymbol{A}, \boldsymbol{x}^*\right) \cap \left(\boldsymbol{x}^* - \mathcal{C}\left(\boldsymbol{A}, \boldsymbol{x}^*\right)\right).$$

One checks that if $\boldsymbol{x}^* \in \mathcal{P}\left(\boldsymbol{A}, \boldsymbol{b}\right)$, then $\mathcal{S}\left(\boldsymbol{A}, \boldsymbol{x}^*\right) \subseteq \mathcal{P}\left(\boldsymbol{A}, \boldsymbol{b}\right)$. Moreover, if $\boldsymbol{y}^* \in \mathcal{S}\left(\boldsymbol{A}, \boldsymbol{x}^*\right)$ then $\mathcal{S}\left(\boldsymbol{A}, \boldsymbol{y}^*\right) \subseteq \mathcal{S}\left(\boldsymbol{A}, \boldsymbol{x}^*\right)$. Polytopes of this form, namely, ones in which every facet is incident to one of two distinguished vertices, known as *spindles*, were used in [17] to construct counterexamples to the Hirsch conjecture.

In Sects. 2 and 3, we fix $\boldsymbol{A} \in \mathbb{Z}^{m \times n}$ and $\boldsymbol{b} \in \mathbb{Z}^m$. Thus, in our notation we drop the dependence on $\boldsymbol{A}$ and $\boldsymbol{b}$, and write $\mathcal{P}_I$ for $\mathcal{P}_I(\boldsymbol{A}, \boldsymbol{b})$, $\Delta_I(\boldsymbol{\alpha})$ for $\Delta_I(\boldsymbol{A}, \boldsymbol{\alpha})$, $\kappa(\boldsymbol{\alpha})$ for $\kappa(\boldsymbol{A}, \boldsymbol{b}, \boldsymbol{\alpha})$, $\mathcal{S}(\boldsymbol{x}^*)$ for $\mathcal{S}(\boldsymbol{A}, \boldsymbol{x}^*)$ and so on.

1.2   Dimension Reduction

When analyzing proximity, one must consider those polyhedra $\dim \mathcal{P} < n$. A useful fact for us is that we need only consider the case when $\dim \mathcal{P} = n$, by replacing a not-necessarily full-dimensional instance with an equivalent full-dimensional instance in a lower-dimensional space.

Lemma 1. *Let* $\boldsymbol{\alpha} \in \mathbb{Z}^n$ *such that the maximum of* $\max\{\boldsymbol{\alpha}^\top \boldsymbol{x} : \boldsymbol{x} \in \mathcal{P}\}$ *is attained and is finite. Assume* $I \subseteq [m]$ *determines a linearly independent subset of the rows of* $\boldsymbol{A}$ *such that the linear span of* $\mathcal{P}_I$ *is* $\ker \boldsymbol{A}_I$, *which has dimension* d. *Then there exists a linear isomorphism* $\ker \boldsymbol{A}_I \to \mathbb{R}^d$ *given by* $\boldsymbol{x} \mapsto \boldsymbol{P}\boldsymbol{x}$ *where* $\boldsymbol{P} \in \mathbb{Z}^{d \times n}$, *which maps* $\ker \boldsymbol{A}_I \cap \mathbb{Z}^n$ *onto* $\mathbb{Z}^d$ *and maps* $\mathcal{P}_I(\boldsymbol{A}, \boldsymbol{b})$ *onto* $\mathcal{P}(\hat{\boldsymbol{A}}, \hat{\boldsymbol{b}})$ *for some* $\hat{\boldsymbol{A}} \in \mathbb{Z}^{(m-n+d) \times d}$, $\hat{\boldsymbol{b}} \in \mathbb{Z}^{m-n+d}$, *and satisfies*

$$\kappa_I\left(\boldsymbol{A}, \boldsymbol{b}, \boldsymbol{\alpha}\right) = \kappa_d\left(\hat{\boldsymbol{A}}, \hat{\boldsymbol{b}}, \hat{\boldsymbol{\alpha}}\right)$$

where $\hat{\boldsymbol{\alpha}} \in \mathbb{Z}^d$ *is the unique vector satisfying* $\hat{\boldsymbol{\alpha}}^\top \boldsymbol{P} = \boldsymbol{\alpha}^\top$.

Proof. Without loss of generality, suppose $I = [n - d]$. Set $J := [n - d]$, $\bar{J} := \{n - d + 1, \ldots, n\}$, and $\bar{I} := \{n - d + 1, \ldots, m\}$. Choose a unimodular matrix $\boldsymbol{U} \in \mathbb{Z}^{n \times n}$ (e.g., via the Hermite Normal Form of $\boldsymbol{A}_{[n]}$) such that

$$\boldsymbol{A}\boldsymbol{U} = \begin{pmatrix} (\boldsymbol{A}\boldsymbol{U})_{I,J} & \boldsymbol{0} \\ (\boldsymbol{A}\boldsymbol{U})_{\bar{I},J} & (\boldsymbol{A}\boldsymbol{U})_{\bar{I},\bar{J}} \end{pmatrix}$$

with $(\boldsymbol{A}\boldsymbol{U})_{I,J}$ square and invertible.

Set $\hat{\boldsymbol{A}} := (\boldsymbol{A}\boldsymbol{U})_{\bar{I},\bar{J}}$, $\hat{\boldsymbol{b}} := \boldsymbol{b}_{\bar{I}}$, and $\hat{\boldsymbol{\alpha}}^\top := \left(\boldsymbol{\alpha}^\top \boldsymbol{U}\right)_{\bar{J}}$. For $\boldsymbol{x} \in \ker \boldsymbol{A}_I$, we have

$$\boldsymbol{0} = \boldsymbol{A}_I \boldsymbol{x} = \boldsymbol{A}_I \boldsymbol{U} \boldsymbol{U}^{-1} \boldsymbol{x} = \left[(\boldsymbol{A}\boldsymbol{U})_{I,J}\ \boldsymbol{0}\right] \boldsymbol{U}^{-1} \boldsymbol{x} = (\boldsymbol{A}\boldsymbol{U})_{I,J}(\boldsymbol{U}^{-1}\boldsymbol{x})_J.$$

Thus, $(\boldsymbol{U}^{-1}\boldsymbol{x})_J = \boldsymbol{0}$. Hence, the map $\boldsymbol{x} \mapsto \left(\boldsymbol{U}^{-1}\boldsymbol{x}\right)_{\bar{J}}$ is a linear isomorphism from $\ker \boldsymbol{A}_I$ to $\mathbb{R}^{|\bar{J}|} = \mathbb{R}^d$, which restricts to a lattice isomorphism from $\ker \boldsymbol{A}_I \cap \mathbb{Z}^n$

to $\mathbb{Z}^d$ and maps $\mathcal{P}_I(\boldsymbol{A}, \boldsymbol{b})$ to $\mathcal{P}(\hat{\boldsymbol{A}}, \hat{\boldsymbol{b}})$. It follows that $\mathcal{P}(\hat{\boldsymbol{A}}, \hat{\boldsymbol{b}}) \cap \mathbb{Z}^d = \{\boldsymbol{0}\}$. For $\boldsymbol{x} \in \ker \boldsymbol{A}_I$, the equation $\left(\boldsymbol{U}^{-1}\boldsymbol{x}\right)_J = \boldsymbol{0}$ implies that

$$\boldsymbol{\alpha}^\top \boldsymbol{x} = \boldsymbol{\alpha}^\top \boldsymbol{U}\boldsymbol{U}^{-1}\boldsymbol{x} = \hat{\boldsymbol{\alpha}}^\top \left(\boldsymbol{U}^{-1}\boldsymbol{x}\right)_{\bar{J}}. \tag{3}$$

Moreover, if $K \subseteq \bar{I}$ with $|K| = d - 1$, then

$$
\left|\det\begin{pmatrix} \hat{\boldsymbol{\alpha}}^\top \\ \hat{\boldsymbol{A}}_K \end{pmatrix}\right| = \left|\det\begin{pmatrix} (\boldsymbol{\alpha}^\top \boldsymbol{U})_{\bar{J}} \\ (\ \boldsymbol{AU})_{K,\bar{J}} \end{pmatrix}\right|
$$
$$
= \frac{1}{\left|\det\left(\boldsymbol{AU}\right)_{I,J}\right|} \cdot \left|\det\begin{pmatrix} (\ \boldsymbol{AU})_{I,J} & \boldsymbol{0} \\ (\boldsymbol{\alpha}^\top \boldsymbol{U})_J & (\boldsymbol{\alpha}^\top \boldsymbol{U})_{\bar{J}} \\ (\ \boldsymbol{AU})_{K,J} & (\ \boldsymbol{AU})_{K,\bar{J}} \end{pmatrix}\right|
$$
$$
= \frac{1}{\gcd \boldsymbol{A}_I} \cdot \left|\det\begin{pmatrix} \boldsymbol{\alpha}^\top \\ \boldsymbol{A}_{I\cup K} \end{pmatrix}\right|,
$$

where we have used $\left|\det\left(\boldsymbol{AU}\right)_{I,J}\right| = \gcd(\boldsymbol{AU})_I = \gcd \boldsymbol{A}_I \boldsymbol{U} = \gcd \boldsymbol{A}_I$. Taking the maximum over all such K, we get

$$\Delta(\hat{\boldsymbol{A}}, \hat{\boldsymbol{\alpha}}) = \Delta_I(\boldsymbol{A}, \boldsymbol{\alpha}). \tag{4}$$

Putting (3) and (4) together, we get

$$\kappa_d(\hat{\boldsymbol{A}}, \hat{\boldsymbol{b}}, \hat{\boldsymbol{\alpha}}) = \max_{\boldsymbol{y} \in \mathcal{P}(\hat{\boldsymbol{A}}, \hat{\boldsymbol{b}})} \frac{\hat{\boldsymbol{\alpha}}^\top \boldsymbol{y}}{\Delta(\hat{\boldsymbol{A}}, \hat{\boldsymbol{\alpha}})} = \max_{\boldsymbol{x} \in \mathcal{P}_I(\boldsymbol{A}, \boldsymbol{b})} \frac{\boldsymbol{\alpha}^\top \boldsymbol{x}}{\Delta_I(\boldsymbol{A}, \boldsymbol{\alpha})} = \kappa_I(\boldsymbol{A}, \boldsymbol{b}, \boldsymbol{\alpha}). \qquad \square$$

2   Proximity for 1, 2, and 3-Dimensional Polyhedra

The Cook et al. bound roughly translates to the statement $\kappa_n < n$ for $n = 1, 2, 3$ whenever $\mathcal{P} \cap \mathbb{Z}^n = \{\boldsymbol{0}\}$. In this section, in particular in Lemma 4, we improve upon these bounds. Define the polyhedron

$$\mathcal{P}_{\boldsymbol{\alpha}} := \left\{\boldsymbol{x} \in \mathbb{R}^n : |\boldsymbol{A}\boldsymbol{x}| \leq 1, \ \boldsymbol{\alpha}^\top \boldsymbol{x} = 0\right\}.$$

This is an $(n-1)$-dimensional polyhedron, which is bounded since $\boldsymbol{A}$ has full column rank by assumption. We use $\mathrm{vol}_i(\cdot)$ to denote the i-dimensional Lebesgue measure.

Lemma 2. *Let $\boldsymbol{\alpha} \in \mathbb{Z}^n$ be non-zero. Assume $\dim \mathcal{P} = n$ and $\mathcal{P} \cap \mathbb{Z}^n = \{\boldsymbol{0}\}$. Then*

$$\kappa_n(\boldsymbol{\alpha}) < \frac{2^{n-1}\|\boldsymbol{\alpha}\|_2}{\mathrm{vol}_{n-1}(\mathcal{P}_{\boldsymbol{\alpha}}) \Delta(\boldsymbol{\alpha})}.$$

Proof. Recall $\mathcal{P} = \mathcal{P}(\boldsymbol{A}, \boldsymbol{b})$. Let $\boldsymbol{x}^* \in \mathcal{P}$ attain the maximum of

$$\kappa_n(\boldsymbol{\alpha}) = \max_{\boldsymbol{x} \in \mathcal{P}} \frac{\boldsymbol{\alpha}^\top \boldsymbol{x}}{\Delta(\boldsymbol{\alpha})},$$

which we assume is positive without loss of generality. Define the polytope

$$\mathcal{Q}\left(\boldsymbol{x}^*\right) := \mathcal{P}_{\boldsymbol{\alpha}} + \left[-\boldsymbol{x}^*, \boldsymbol{x}^*\right],$$

which is $\mathbf{0}$-symmetric and full-dimensional in $\mathbb{R}^n$. Observe that

$$\mathrm{vol}_n\left(\mathcal{Q}\left(\boldsymbol{x}^*\right)\right) = \frac{2\kappa_n(\boldsymbol{\alpha})\Delta(\boldsymbol{\alpha})}{\|\boldsymbol{\alpha}\|_2} \cdot \mathrm{vol}_{n-1}\left(\mathcal{P}_{\boldsymbol{\alpha}}\right).$$

All integer points not in $\mathcal{Q}(\boldsymbol{x}^*)$ are a positive distance away from $\mathcal{Q}(\boldsymbol{x}^*)$, hence there exists $\delta > 0$ such that $\mathcal{Q}((1+\delta)\boldsymbol{x}^*)$ and $\mathcal{Q}(\boldsymbol{x}^*)$ contain precisely the same set of integer points. This choice of δ uniquely determines $\varepsilon > 0$ for which

$$\mathcal{Q}'\left(\boldsymbol{x}^*\right) := (1-\varepsilon)\,\mathcal{Q}\left((1+\delta)\,\boldsymbol{x}^*\right)$$

has the same n-dimensional volume as $\mathcal{Q}(\boldsymbol{x}^*)$, and furthermore

$$\mathcal{Q}'\left(\boldsymbol{x}^*\right) \cap \mathbb{Z}^n \subseteq \mathcal{Q}\left(\boldsymbol{x}^*\right) \cap \mathbb{Z}^n.$$

Assume to the contrary that $\mathrm{vol}_n\left(\mathcal{Q}\left(\boldsymbol{x}^*\right)\right) \geq 2^n$. By Minkowski's convex body theorem, there exists $\boldsymbol{z}^* \in \mathcal{Q}(\boldsymbol{x}^*) \cap \mathcal{Q}'(\boldsymbol{x}^*) \cap \mathbb{Z}^n \setminus \{\mathbf{0}\}$ by the above inclusion. Therefore, with respect to the vector space decomposition of $\mathbb{R}^n$ into the line $\mathbb{R} \cdot \boldsymbol{x}^*$ and the hyperplane $\boldsymbol{\alpha}^\top \boldsymbol{x} = 0$, the vector $\boldsymbol{z}^*$ decomposes uniquely as $\boldsymbol{z}^* = \lambda \boldsymbol{x}^* + (\boldsymbol{z}^* - \lambda \boldsymbol{x}^*)$ with $\lambda \in [0,1]$ and $\boldsymbol{z}^* - \lambda \boldsymbol{x}^* \in (1-\varepsilon)\,\mathcal{P}_{\boldsymbol{\alpha}}$. Hence,

$$\left|\boldsymbol{A}\left(\boldsymbol{z}^* - \lambda \boldsymbol{x}^*\right)\right| \leq (1-\varepsilon)\,\mathbf{1}.$$

As $\mathcal{P} \cap \mathbb{Z}^n = \{\mathbf{0}\}$ and $\boldsymbol{z}^* \neq \mathbf{0}$, there exists some row $\boldsymbol{a}_j^\top$ of $\boldsymbol{A}$ such that $\boldsymbol{a}_j^\top \boldsymbol{z}^* \geq b_j + 1$. Since $\boldsymbol{x}^* \in \mathcal{P}(\boldsymbol{A}, \boldsymbol{b})$, we also have $\boldsymbol{a}_j^\top \boldsymbol{x}^* \leq b_j$. Thus, we get

$$b_j + 1 \leq \boldsymbol{a}_j^\top \boldsymbol{z}^* = \boldsymbol{a}_j^\top\left(\lambda \boldsymbol{x}^*\right) + \boldsymbol{a}_j^\top\left(\boldsymbol{z}^* - \lambda \boldsymbol{x}^*\right) \leq \lambda b_j + (1-\varepsilon) < b_j + 1.$$

This is a contradiction. Hence,

$$\frac{2\kappa_n(\boldsymbol{\alpha})\Delta(\boldsymbol{\alpha})}{\|\boldsymbol{\alpha}\|_2} \cdot \mathrm{vol}_{n-1}\left(\mathcal{P}_{\boldsymbol{\alpha}}\right) = \mathrm{vol}_n\left(\mathcal{Q}\left(\boldsymbol{x}^*\right)\right) < 2^n.$$

Rearranging yields the desired inequality. $\square$

Remark 1. Integrality of $\boldsymbol{b}$, which is the key assumption of this paper, is used above in the assertion $\boldsymbol{a}_j^\top \boldsymbol{z}^* \geq b_j + 1$. If $\boldsymbol{b}$ were not integral, then we would only be able to assert that $\boldsymbol{a}_j^\top \boldsymbol{z}^* \geq \lceil b_j \rceil$, which is not sufficient to complete the proof.

For the proof of Lemma 4 we apply the following classical result of Mahler [13] (see also [8, Page 177]) on the relationship between the area of a nonempty compact convex set $K \subseteq \mathbb{R}^2$ and the area of its polar

$$K^\circ := \left\{\boldsymbol{x} \in \mathbb{R}^2 : \boldsymbol{y}^\top \boldsymbol{x} \leq 1 \text{ for all } \boldsymbol{y} \in K\right\}.$$

Lemma 3 (Mahler's Inequality when $n = 2$ [13]). *Let $K \subseteq \mathbb{R}^2$ be nonempty compact and convex whose interior contains $\mathbf{0}$. Then* $\mathrm{vol}_2(K) \, \mathrm{vol}_2(K^\circ) \geq 8$.

Lemma 4. *Let $\boldsymbol{\alpha} \in \mathbb{Z}^n$ be non-zero. Suppose $\mathcal{P} \cap \mathbb{Z}^n = \{\mathbf{0}\}$. Then $\kappa_1(\boldsymbol{\alpha}) < 1$, $\kappa_2(\boldsymbol{\alpha}) < 1$ and $\kappa_3(\boldsymbol{\alpha}) < \sqrt{2}$.*

Proof. By Lemma 1 we may assume $\mathcal{P}$ is full-dimensional. If $n = 1$, then $\mathcal{P}(\boldsymbol{A}, \boldsymbol{b})$ is contained in the open interval $(-1, 1)$, which immediately implies $\kappa_1(\boldsymbol{\alpha}) < 1$. If $n = 2$, then the polytope $\mathcal{P}_{\boldsymbol{\alpha}}$ is an origin-symmetric line segment $[-\boldsymbol{y}^*, \boldsymbol{y}^*]$, where $\boldsymbol{y}^* \in \mathbb{R}^2$ satisfies $\boldsymbol{\alpha}^\top \boldsymbol{y}^* = 0$ and $\boldsymbol{a}_j^\top \boldsymbol{y}^* = 1$ for some $j \in [m]$. Hence

$$\mathrm{vol}_1(\mathcal{P}_{\boldsymbol{\alpha}}) = 2 \, \|\boldsymbol{y}^*\|_2 = \frac{2 \, \|\boldsymbol{\alpha}\|_2}{|\det(\boldsymbol{\alpha} \; \boldsymbol{a}_j)|}.$$

Applying Lemma 2, we get

$$\kappa_2(\boldsymbol{\alpha}) < \frac{2 \, \|\boldsymbol{\alpha}\|_2}{\mathrm{vol}_1(\mathcal{P}_{\boldsymbol{\alpha}}) \, \Delta(\boldsymbol{\alpha})} = \frac{|\det(\boldsymbol{\alpha} \; \boldsymbol{a}_j)|}{\Delta(\boldsymbol{\alpha})} \leq 1.$$

If $n = 3$, then choose $I \subseteq [m]$ with $|I| = 2$ such that

$$\boldsymbol{B} := \begin{pmatrix} \boldsymbol{\alpha}^\top \\ \boldsymbol{A}_I \end{pmatrix}$$

satisfies $|\det \boldsymbol{B}| = \Delta$. Let $\boldsymbol{A}'$ denote the last two columns of $\boldsymbol{A}\boldsymbol{B}^{-1}$. Then

$$\boldsymbol{B} \cdot \mathcal{P}_{\boldsymbol{\alpha}} = \{0\} \times \mathcal{Q},$$

where

$$\mathcal{Q} := \left\{ \boldsymbol{x} \in \mathbb{R}^2 : |\boldsymbol{A}'\boldsymbol{x}| \leq 1 \right\}.$$

We enumerate the rows of $\boldsymbol{A}'$ as $\boldsymbol{a}_1', \ldots, \boldsymbol{a}_m'$. Since $\mathcal{P}_{\boldsymbol{\alpha}}$ is a polytope, so is $\mathcal{Q}$. The polar of $\mathcal{Q}$ is the convex hull of the rows of $\boldsymbol{A}'$:

$$\mathcal{Q}^\circ := \mathrm{conv} \left\{ \boldsymbol{a}_i' : i \in [m] \right\}.$$

Let $\tau : \mathbb{R}^2 \to \mathbb{R}^2$ denote the $90°$ counterclockwise rotation in $\mathbb{R}^2$. Observe that $\tau(\mathcal{Q}^\circ) \subseteq \mathcal{Q}$. Indeed, for each pair $\{i, j\} \subseteq [m]$, we have

$$\left| \tau(\boldsymbol{a}_i')^\top \boldsymbol{a}_j' \right| = |\det(\boldsymbol{a}_i' \; \boldsymbol{a}_j')| = \frac{|\det(\boldsymbol{\alpha} \; \boldsymbol{a}_i \; \boldsymbol{a}_j)|}{|\det \boldsymbol{B}|} \leq 1.$$

Hence, by Mahler's Inequality,

$$\mathrm{vol}_2(\mathcal{Q}) \geq \sqrt{\mathrm{vol}_2(\mathcal{Q}) \, \mathrm{vol}_2(\tau(\mathcal{Q}^\circ))} = \sqrt{\mathrm{vol}_2(\mathcal{Q}) \, \mathrm{vol}_2(\mathcal{Q}^\circ)} \geq 2\sqrt{2}.$$

We have

$$\mathrm{vol}_2(\mathcal{Q}) = \frac{|\det \boldsymbol{B}|}{\|\boldsymbol{\alpha}\|_2} \cdot \mathrm{vol}_2(\mathcal{P}_{\boldsymbol{\alpha}}).$$

By Lemma 2, we get

$$\kappa_3(\boldsymbol{\alpha}) < \frac{4 \, \|\boldsymbol{\alpha}\|_2}{\mathrm{vol}_2(\mathcal{P}_{\boldsymbol{\alpha}}) \, \Delta(\boldsymbol{\alpha})} \leq \frac{4}{2\sqrt{2}} = \sqrt{2}.$$

$\square$

3   Lifting Proximity Results to Higher Dimensions

The next step is to prove Theorem 2 by showing how proximity results for low dimensional polytopes can be used to derive proximity results for higher dimensional polytopes.

Lemma 5. *Let $x^* \in \mathbb{R}^n$, and let $d = \dim(\mathcal{S}(x^*))$. Let $y^* \in \mathcal{S}(x^*)$, let $k := \dim \mathcal{S}(y^*)$, and fix $d \in \{1, \ldots, k\}$. There exists a d-face of $\mathcal{S}(y^*)$ incident to y^* that intersects some $(k-d)$-face of $\mathcal{S}(y^*)$ incident to $\mathbf{0}$.*

Proof. Let $I \subseteq [m]$ index the components i such that $a_i^\top y^* \neq 0$. For $i \in I$ let $\hat{a}_i = \mathrm{sign}\left(a_i^\top y^*\right) \cdot a_i$. The spindle $\mathcal{S}(y^*)$ can be written as

$$\mathcal{S}(y^*) = \left\{ x \in \mathbb{R}^n : \ 0 \leq \hat{a}_i^\top x \leq \hat{a}_i^\top y^* \ \forall \ i \in I \text{ and } a_i^\top x = 0 \ \forall \ i \notin I \right\}.$$

The constraints are indexed by the disjoint union $I_0 \cup I_{y^*} \cup \bar{I}$, where I_0 and I_{y^*} denote the two copies of I indexing constraints tight at $\mathbf{0}$ and at y^*, respectively. Let $J_0, J_1, \ldots, J_r$ be a sequence of feasible bases of this system, with corresponding basic feasible solutions $\mathbf{0} = y^{(0)}, y^{(1)}, \ldots, y^{(r)} = y^*$ such that for each $i < r$, the symmetric difference of J_{i+1} and J_i is a 2-element subset of $I_0 \cup I_{y^*}$. We have $|J_0 \cap I_{y^*}| = 0$ and $|J_r \cap I_{y^*}| = k$, and $|J_{i+1} \backslash J_i| = 1$ for each $i < r$. It follows that there must exist some ℓ such that $|J_\ell \cap I_{y^*}| = k - d$. Since we always have $|J_i \cap (I_0 \cup I_{y^*})| = k$ for every choice of i, we also get $|J_\ell \cap I_0| = d$.

The basic feasible solution $y^{(\ell)}$ associated to J_ℓ is a vertex of the face of $\mathcal{S}(y^*)$ obtained by making the constraints of $J_\ell \cap I_{y^*}$ tight. It is also a vertex of the face of $\mathcal{S}(y^*)$ obtained by making the constraints of $J_\ell \cap I_0$ tight. These faces are contained in a d-face and a $(k-d)$-face, respectively. $\square$

Lemma 5 will be used to create a path from one vertex of a spindle to another by traveling over d dimensional faces. In the next result, we apply proximity results to each d dimensional face that we travel over. This generalizes the proof of Cook et al., which can be interpreted as walking along edges of a spindle.

Lemma 6. *Let $\alpha \in \mathbb{Z}^n$ be non-zero. Let $\dim \mathcal{P} =: d = \sum_{i=0}^k d_i$ where each d_i is a positive integer. Then*

$$\kappa_d(\alpha) \leq \sum_{i=0}^k \kappa_{d_i}(\alpha).$$

Proof. In this proof, we supress in our notation dependence on α. Let x^* maximize $\alpha^\top x$ over $\mathcal{P}$. Build a sequence $x^* =: x_0^*, x_1^*, \ldots, x_t^* := \mathbf{0}$ of points inductively as follows. Assume $i \geq 0$ and $x_0^*, \ldots, x_i^*$ have been determined already. If both

$$i \leq k \quad \text{and} \quad d_i < \dim \mathcal{S}(x_i^*), \tag{5}$$

then we use Lemma 5 to choose a vertex x_{i+1}^* of $\mathcal{S}(x_i^*)$ that is incident to both a d_i-dimensional face F_i of $\mathcal{S}(x_i^*)$ containing x_i^*, as well as a $(\dim \mathcal{S}(x_i^*) - d_i)$-dimensional face G_i of $\mathcal{S}(x_i^*)$ containing $\mathbf{0}$. Otherwise, if (5) fails, then we set $F_i = \mathcal{S}(x_i^*)$ and $x_{i+1}^* = \mathbf{0}$, and we terminate the sequence by setting $t = i + 1$.

Let $i \in \{0, \ldots, t-2\}$. We show $\boldsymbol{x}_{i+1}^* \neq \boldsymbol{0}$. If not, then F_i contains both $\boldsymbol{0}$ and $\boldsymbol{x}_i^*$. But the only face of $\mathcal{S}(\boldsymbol{x}_i^*)$ containing $\boldsymbol{0}$ and $\boldsymbol{x}_i^*$ is $\mathcal{S}(\boldsymbol{x}_i^*)$ itself. One can see this by observing that the centre of symmetry of the centrally symmetric spindle $\mathcal{S}(\boldsymbol{x}_i^*)$ is $1/2 \cdot \boldsymbol{x}_i^*$. But this contradicts the fact that G_i has positive dimension by (5). Thus, $\boldsymbol{x}_{i+1}^*$ is non-zero, which implies

$$\dim \mathcal{S}(\boldsymbol{x}_{i+1}^*) \geq 1. \tag{6}$$

Moreover, as both G_i and $\mathcal{S}(\boldsymbol{x}_{i+1}^*)$ are contained in the affine (equivalently, linear) span of G_i, we must have

$$\dim \mathcal{S}(\boldsymbol{x}_{i+1}^*) \leq \dim G_i = \dim \mathcal{S}(\boldsymbol{x}_i^*) - d_i. \tag{7}$$

Applying (6) and then (7) sequentially with $s \in \{t-2, t-3, \ldots, 0\}$, we have

$$1 \leq \dim \mathcal{S}(\boldsymbol{x}_{t-1}^*) \leq \dim \mathcal{S}(\boldsymbol{x}_0^*) - \sum_{s=0}^{t-2} d_s \leq d - \sum_{s=0}^{t-2} d_s,$$

which is to say $d = \sum_{s=0}^{k} d_s > \sum_{s=0}^{t-2} d_s$. It follows that $t-1 \leq k$.

Suppose $I \subseteq [m]$ indexes linearly independent rows of $\boldsymbol{A}$ such that $\kappa_d = \kappa_I$, so that in particular $\ker \boldsymbol{A}_I$ is the linear span of $\mathcal{P}$. Let $i \in \{0, \ldots, t-1\}$. We have that $\boldsymbol{x}_i^* - F_i$ is a face of $\mathcal{S}(\boldsymbol{x}_i^*)$ containing $\boldsymbol{0}$. Choose an index set I_i, where $I \subseteq I_i \subseteq [m]$, such that the rows of $\boldsymbol{A}_{I_i}$ are linearly independent and $\ker \boldsymbol{A}_{I_i}$ is the linear span of $\boldsymbol{x}_i^* - F_i$. We have

$$\boldsymbol{\alpha}^\top (\boldsymbol{x}_i^* - \boldsymbol{x}_{i+1}^*) \leq \max_{\boldsymbol{x} \in \boldsymbol{x}_i^* - F_i} \boldsymbol{\alpha}^\top \boldsymbol{x} \leq \max_{\boldsymbol{x} \in \mathcal{P}_{I_i}} \boldsymbol{\alpha}^\top \boldsymbol{x} \leq \kappa_{I_i} \Delta_{I_i}.$$

If $i < t-1$, then since F_i is a d_i-dimensional face, we have $\kappa_{I_i} \Delta_{I_i} \leq \kappa_{d_i} \Delta_I$ for $i \in \{0, \ldots, t-2\}$. Otherwise $i = t-1$, in which case one of the inequalities in (5) fails. We have established that $t-1 \leq k$, thus

$$d_{t-1} \geq \dim \mathcal{S}(\boldsymbol{x}_{t-1}^*) = \dim F_{t-1}. \tag{8}$$

and hence $\kappa_{I_{t-1}} \Delta_{I_{t-1}} \leq \kappa_{d_{t-1}} \Delta_I$. Putting these all together we get

$$\Delta_I \cdot \kappa_d = \boldsymbol{\alpha}^\top \boldsymbol{x}^* = \sum_{i=0}^{t-1} \boldsymbol{\alpha}^\top (\boldsymbol{x}_i^* - \boldsymbol{x}_{i+1}^*) \leq \sum_{i=0}^{t-1} \kappa_{I_i} \Delta_{I_i} \leq \Delta_I \cdot \sum_{i=0}^{k} \kappa_{d_i}. \qquad \square$$

4   Proof of the Main Theorem

Proof (of Theorem 2). Suppose $\boldsymbol{x}^*$ is an optimal vertex of LP$(\boldsymbol{A}, \boldsymbol{b}, \boldsymbol{c})$. Let $\boldsymbol{z}^*$ be any optimal solution to IP$(\boldsymbol{A}, \boldsymbol{b}, \boldsymbol{c})$. By LP duality, there exists an optimal LP basis $I^* \subseteq [m]$, i.e., $\boldsymbol{x}^* = \boldsymbol{A}_{I^*}^{-1} \boldsymbol{b}_{I^*}$, and a vector $\boldsymbol{y} \in \mathbb{R}_{\geq 0}^{I^*}$ that satisfies $\boldsymbol{c}^\top = \boldsymbol{y}^\top \boldsymbol{A}_{I^*}$. The polytope $\mathcal{P}(\overline{\boldsymbol{A}}, \overline{\boldsymbol{b}}) := \{\boldsymbol{x} \in \mathcal{P}(\boldsymbol{A}, \boldsymbol{b}) : \boldsymbol{A}_{I^*} \boldsymbol{x} \geq \boldsymbol{A}_{I^*} \boldsymbol{z}^*\}$ contains $\boldsymbol{x}^*$ and $\boldsymbol{z}^*$ and $\Delta_k(\overline{\boldsymbol{A}}) = \Delta_k(\boldsymbol{A})$ for all $k \in [n]$. Any integer vector $\boldsymbol{w}^* \in \mathcal{P}(\overline{\boldsymbol{A}}, \overline{\boldsymbol{b}}) \setminus \{\boldsymbol{z}^*\}$ is also an optimal solution to IP$(\boldsymbol{A}, \boldsymbol{b}, \boldsymbol{c})$ because $\boldsymbol{c}^\top \boldsymbol{w}^* = \boldsymbol{y}^\top (\boldsymbol{A}_{I^*} \boldsymbol{w}^*) \geq \boldsymbol{y}^\top (\boldsymbol{A}_{I^*} \boldsymbol{z}^*) = \boldsymbol{c}^\top \boldsymbol{z}^*$, and $\boldsymbol{A}_{I^*} \boldsymbol{w}^* \geq \boldsymbol{A}_{I^*} \boldsymbol{z}^*$ with at least one of the n inequalities satisfied strictly because I^* is a basis. Thus, by replacing z^* by an integer vector in $\mathcal{P}(\overline{A}, \overline{b}) \setminus \{z^*\}$ finitely many times, we may assume that $\mathcal{P}(\overline{A}, \overline{b}) \cap \mathbb{Z}^n = \{z^*\}$. Translating the instance, we may further assume that $z^* = 0$, so that our objective is now to show $\|x^*\|_\infty < \frac{n}{2} \cdot \Delta_{n-1}(A)$.

Now let $s \in \{-1, 1\}$, let $i \in \{1, 2, \ldots, n\}$, and let $\alpha = s e_i$. We have

$$s x_i^* \leq \max_{x \in \mathcal{P}} \alpha^\top x = \kappa_d(\alpha) \cdot \Delta(A, \alpha) \leq \kappa_d(\alpha) \cdot \Delta_{n-1}(A).$$

By Lemma 4, $\kappa_1(\alpha) < 1$, so we may assume $d \geq 2$. We write $d = 3a + 2b$, where a, b are nonnegative integers, and we further specify $a = \lfloor d/3 \rfloor$. Applying Lemma 6, then Lemma 4, then the fact $d \leq n$, we get

$$\kappa_d(\alpha) \leq \kappa_3(\alpha) \cdot a + \kappa_2(\alpha) \cdot b < \sqrt{2} \lfloor d/3 \rfloor + \frac{d - 3\lfloor d/3 \rfloor}{2} \leq \frac{d}{2} \leq \frac{n}{2}. \qquad \square$$

Remark 2. The right hand side above could also be replaced with $\frac{\sqrt{2}}{3} \cdot n + 1$.

5   Proximity in the Strictly Δ-Modular Case

In this section we assume $A = TB$, where T is totally unimodular and B is an invertible square matrix with $|\det B| = \Delta_n(A)$. The following lemma is similar to [14, Lemma 29 and Lemma 30] after linear transformation with B. We say a nonzero vector x in a lattice Λ is *primitive* if $kx \notin \Lambda$ for all $k \in (0, 1)$.

Lemma 7. *Set $\Lambda := B^{-1}\mathbb{Z}^n$ and let $x^* \in \Lambda$. Then each ray of $\mathcal{C}(x^*)$ contains a primitive vector in Λ, and x^* can be written as a non-negative integral combination of those vectors.*

Proof. Let $I \subseteq [m]$ index a one-dimensional subspace $\ker A_I$ which contains a ray of $\mathcal{C}(x^*)$. Further, let $j \in [m] \setminus I$ index another row of A such that $A_{I \cup \{j\}}$ is invertible with last row A_j. We choose the following scaled vector

$$r := A_{I \cup \{j\}}^{-1} e_n = B^{-1} T_{I \cup \{j\}}^{-1} e_n \in \mathcal{C}(x^*) \cap \ker A_I. \qquad (9)$$

We have $T_{I \cup \{j\}}^{-1} e_n \in \mathbb{Z}^n$, so $r \in \Lambda$, because T is totally unimodular. The first claim follows, as the existence of a nonzero lattice vector on a ray implies the existence of a primitive lattice vector.

For the latter statement we study $\mathcal{S}(x^*)$. If $\mathcal{S}(x^*)$ is zero-dimensional, then $x^* = 0$ and we are done. If $\mathcal{S}(x^*)$ is one-dimensional, then x^* is by construction an integer multiple of some primitive lattice ray. Hence, we assume that $\mathcal{S}(x^*)$ is at least two-dimensional. Choose a vertex v adjacent to 0 which is not x^*. As the constraint matrix defining $\mathcal{S}(x^*)$ is strictly $\Delta_n(A)$-modular, every vertex of $\mathcal{S}(x^*)$ is in Λ, in particular $v \in \Lambda$. Thus, v is an integer multiple of some primitive vector in Λ. Furthermore, the symmetry of $\mathcal{S}(x^*)$ implies $x^* - v \in \Lambda$ and is a vertex of $\mathcal{S}(x^*)$ adjacent to x^*. It follows there exists a constraint of $\mathcal{S}(x^*)$ tight at $x^* - v$ and 0 but not x^*, and this implies that the dimension of $\mathcal{S}(x^* - v)$ is strictly smaller than $\mathcal{S}(x^*)$. We may therefore repeat the procedure with $\mathcal{S}(x^* - v)$ and so on, and termination is guaranteed when we reach the origin. $\qquad \square$

Proof (of Theorem 3 Part 1). Recall $\mathcal{P} = \mathcal{P}(\boldsymbol{A}, \boldsymbol{b})$. Set $\Lambda := \boldsymbol{B}^{-1}\mathbb{Z}^n$, and choose a vertex $\boldsymbol{x}^*$ of $\mathcal{P}$ such that $\|\boldsymbol{x}^*\|_\infty$ is as large as possible. Every vertex of $\mathcal{P}$ is in Λ, so $\boldsymbol{x}^* \in \Lambda$. Lemma 7 yields

$$\boldsymbol{x}^* = \sum_{s=1}^t \lambda_s \boldsymbol{r}_s \tag{10}$$

where $\boldsymbol{r}_1, \dots, \boldsymbol{r}_t \in \mathcal{C}(\boldsymbol{x}^*)$ denote the primitive vectors in Λ and $\lambda_1, \dots, \lambda_t \in \mathbb{Z}_{\geq 0}$.

Observe that each subsum of the right side in (10) is an element in $\mathcal{S}(\boldsymbol{x}^*)$. Further, recall that $\mathcal{S}(\boldsymbol{x}^*) \subseteq \mathcal{P}$. Let $N := \lambda_1 + \cdots + \lambda_t$. We choose a sequence $\boldsymbol{0} = \boldsymbol{x}^{(0)}, \boldsymbol{x}^{(1)}, \dots, \boldsymbol{x}^{(N)} = \boldsymbol{x}^*$ such that

$$\boldsymbol{x}^{(i)} - \boldsymbol{x}^{(i-1)} \in \{\boldsymbol{r}_1, \dots, \boldsymbol{r}_t\}$$

for all $i \in [N]$. Thus, $\boldsymbol{x}^{(i)} \in \Lambda$ for $i \in [N]$ and all these vectors are pairwise distinct elements in $\mathcal{S}(\boldsymbol{x}^*)$. If $N \geq \Delta_n(\boldsymbol{A})$, then by $\mathcal{S}(\boldsymbol{x}^*) \cap \mathbb{Z}^n = \{\boldsymbol{0}\}$ and the pigeonhole principle there are $\boldsymbol{x}^{(i)}$ and $\boldsymbol{x}^{(j)}$ for $i < j$ that lie in the same residue class of Λ modulo $\mathbb{Z}^n$. Hence, we have the contradiction

$$\boldsymbol{0} \neq \boldsymbol{x}^{(j)} - \boldsymbol{x}^{(i)} \in \mathcal{S}(\boldsymbol{x}^*) \cap \mathbb{Z}^n \subseteq \mathcal{P} \cap \mathbb{Z}^n.$$

We proceed with $N \leq \Delta_n(\boldsymbol{A}) - 1$. We have $\|\boldsymbol{r}_s\|_\infty \leq \frac{\Delta_{n-1}(\boldsymbol{A})}{\Delta_n(\boldsymbol{A})}$ for all $s \in [t]$ by Cramer's rule applied to (9). Altogether, this yields

$$\|\boldsymbol{x}^*\|_\infty \leq \sum_{s=1}^t \lambda_s \|\boldsymbol{r}_s\|_\infty \leq \frac{\Delta_n(\boldsymbol{A}) - 1}{\Delta_n(\boldsymbol{A})} \Delta_{n-1}(\boldsymbol{A}) \leq \max\{\Delta_{n-1}(\boldsymbol{A}), \Delta_n(\boldsymbol{A})\} - 1.$$

$\square$

6 A Lower Bound Example

The following construction proves Theorem 3, Part 2. Let $\delta \geq 3$ be an integer. Fix the matrix

$$\boldsymbol{B}^{(k)} := \begin{pmatrix} \boldsymbol{I}_{n-1} & \boldsymbol{0} \\ \beta^k & \delta \end{pmatrix} \in \mathbb{Z}^{n \times n},$$

where $\boldsymbol{I}_{n-1}$ denotes the $(n-1) \times (n-1)$ unit matrix and, for $0 \leq k \leq n-1$,

$$\beta^k := (\underbrace{0, \dots, 0}_{k \text{ zeros}}, \delta - 1, \dots, \delta - 1).$$

As a first step, we define the parallelepiped

$$\mathcal{P}(\boldsymbol{B}^{(k)}) := \left\{ \boldsymbol{x} \in \mathbb{R}^n : \boldsymbol{0} \leq \boldsymbol{B}^{(k)} \boldsymbol{x} \leq \begin{pmatrix} \boldsymbol{1}_k \\ \delta - n + k \\ \boldsymbol{1}_{n-k-1} \end{pmatrix} \right\}$$

for $\delta - n + k \geq 1$. Using the fact that the first k columns of $\boldsymbol{B}^{(k)-1}$ are integral, one can show $\left|\mathcal{P}(\boldsymbol{B}^{(k)}) \cap \mathbb{Z}^n\right| = 2^k$. In order to cut off all non-zero integer points in $\mathcal{P}(\boldsymbol{B}^{(k)})$, we define for $k \geq 1$ the row vectors

$$\boldsymbol{a}_{i,(k)}^\top := (0, \ldots, \underbrace{1}_{i\text{-th column}}, \ldots, 0, \underbrace{-\delta}_{(k+1)\text{-st column}}, \ldots, -\delta)$$

for each $i \in [k]$. The resulting polytope is

$$\mathcal{P}_{\delta,n,k} := \mathcal{P}(\boldsymbol{B}^{(k)}) \cap \left\{ \boldsymbol{x} \in \mathbb{R}^n : \boldsymbol{a}_{i,(k)}^\top \boldsymbol{x} \leq 0 \text{ for all } i \in [k] \right\}.$$

If $k = 0$, then $\mathcal{P}_{\delta,n,0} = \mathcal{P}(\boldsymbol{B}^{(0)})$. Let $\boldsymbol{x}^*$ denote the only vertex in $\mathcal{P}(\boldsymbol{B}^{(k)})$ that does not share a facet with $\boldsymbol{0}$. Further, let $\boldsymbol{A}$ and $\boldsymbol{b}$ be such that $\mathcal{P}(\boldsymbol{A}, \boldsymbol{b}) = \mathcal{P}_{\delta,n,k}$. Then one can show that

1. $\boldsymbol{x}^* \in \mathcal{P}_{\delta,n,k}$ and $\boldsymbol{x}^*$ does not share a facet with $\boldsymbol{0}$,
2. $\mathcal{P}_{\delta,n,k} \cap \mathbb{Z}^n = \{\boldsymbol{0}\}$,
3. $\boldsymbol{A}$ is strictly δ-modular.

We select $\mathcal{P}_{\delta,n,n-2}$ and get

$$\|\boldsymbol{x}^* - \boldsymbol{0}\|_\infty = \left|\boldsymbol{x}_{n-1}^*\right| = \delta - 2.$$

Observe that $\Delta_{n-1}(\boldsymbol{A}) = \delta$ for $\mathcal{P}_{\delta,n,n-2}$ which proves Part 2 of Theorem 3.

Remark 3. The polytope $\mathcal{P}_{\delta,n,n-1}$ does not work as an example since in this instance, the greatest common divisor of the n-th row of $\boldsymbol{B}^{(n-1)}$ is δ. As a result, we only obtain the weak proximity bound $\|\boldsymbol{x}^*\|_\infty = 1$. However, a question related to the proximity question is to bound $\|\boldsymbol{b}\|_\infty$ given that $\mathcal{P} \cap \mathbb{Z}^n = \{\boldsymbol{0}\}$ and all constraints of $\mathcal{P}$ are tight. The polytope $\mathcal{P}_{\delta,n,n-1}$ yields an example with $\|\boldsymbol{b}\|_\infty = \delta - 1$.

Acknowledgements. The third author was supported by a Natural Sciences and Engineering Research Council of Canada (NSERC) Discovery Grant [RGPIN-2021-02475]. The fourth author was supported by the Einstein Foundation Berlin. The authors are grateful to the referees for their valuable suggestions and comments.

References

1. Aliev, I., Celaya, M., Henk, M., Williams, A.: Distance-sparsity transference for vertices of corner polyhedra. SIAM J. Optim. **31**, 200–126 (2021). https://doi.org/10.1137/20M1353228
2. Aliev, I., Henk, M., Oertel, T.: Distances to lattice points in knapsack polyhedra. Math. Program. 175–198 (2019). https://doi.org/10.1007/s10107-019-01392-1
3. Cook, W., Gerards, A., Schrijver, A., Tardos, E.: Sensitivity theorems in integer linear programming. Math. Program. **34**, 251–264 (1986). https://doi.org/10.1007/BF01582230

4. Del Pia, A., Ma, M.: Proximity in concave integer quadratic programming. arXiv:2006.01718 (2021)
5. Eisenbrand, F., Weismantel, R.: Proximity results and faster algorithms for integer programming using the Steinitz lemma. ACM Trans. Algorithms **16**, 1–14 (2020). https://doi.org/10.1145/3340322
6. Fischetti, M., Glover, F., Lodi, A.: The feasibility pump. Math. Program. **104**, 91–104 (2005). https://doi.org/10.1007/s10107-004-0570-3
7. Granot, F., Skorin-Kapov, J.: Some proximity and sensitivity results in quadratic integer programming. Math. Program. **47**, 259–268 (1990). https://doi.org/10.1007/BF01580862
8. Gruber, P.: Convex and Discrete Geometry. Springer, Heidelberg (2007)
9. Hochbaum, D.S., Shanthikumar, J.G.: Convex separable optimization is not much harder than linear optimization. J. ACM **37**, 843–862 (1990). https://doi.org/10.1145/96559.96597
10. Jansen, K., Rohwedder, L.: On integer programming and convolution. In: 10th Innovations in Theoretical Computer Science, vol. 43, pp. 43:1–43:17 (2019). https://doi.org/10.4230/LIPIcs.ITCS.2019.43
11. Lee, J., Paat, J., Stallknecht, I., Xu, L.: Improving proximity bounds using sparsity. In: Baïou, M., Gendron, B., Günlük, O., Mahjoub, A.R. (eds.) ISCO 2020. LNCS, vol. 12176, pp. 115–127. Springer, Cham (2020). https://doi.org/10.1007/978-3-030-53262-8_10
12. Lee, J., Paat, J., Stallknecht, I., Xu, L.: Polynomial upper bounds on the number of differing columns of Δ-modular integer programs. arXiv:2105.08160 (2021)
13. Mahler, K.: Ein übertragungsprinzip für konvexe Körper. Časopis Pešt. Mat. Fyz. **68**, 93–102 (1939). (in German)
14. Nägele, M., Santiago, R., Zenklusen, R.: Congruency-constrained TU problems beyond the bimodular case. In: Proceedings of SODA 2022. arXiv:2109.03148 (2022)
15. Oertel, T., Paat, J., Weismantel, R.: The distributions of functions related to parametric integer optimization. SIAM J. Appl. Algebra Geom. 422–440 (2020). https://doi.org/10.1137/19M1275954
16. Paat, J., Weismantel, R., Weltge, S.: Distances between optimal solutions of mixed-integer programs. Math. Program. 455–468 (2018). https://doi.org/10.1007/s10107-018-1323-z
17. Santos, F.: A counterexample to the Hirsch Conjecture. Ann. Math. **176**, 383–412 (2012). https://doi.org/10.4007/annals.2012.176.1.7
18. Schrijver, A.: Theory of Linear and Integer Programming. Wiley, New York (1986)
19. Veselov, S., Chirkov, A.: Integer programming with bimodular matrix. Discret. Optim. **6**, 220–222 (2009). https://doi.org/10.1016/j.disopt.2008.12.002
20. Werman, M., Magagnosc, D.: The relationship between integer and real solutions of constrained convex programming. Math. Program. **51**, 133–135 (1991). https://doi.org/10.1007/BF01586929
21. Xu, L., Lee, J.: On proximity for k-regular mixed-integer linear optimization. In: Le Thi, H.A., Le, H.M., Pham Dinh, T. (eds.) WCGO 2019. AISC, vol. 991, pp. 438–447. Springer, Cham (2020). https://doi.org/10.1007/978-3-030-21803-4_44

Sparse Multi-term Disjunctive Cuts for the Epigraph of a Function of Binary Variables

Rui Chen[1($\boxtimes$)] and James Luedtke[2]

[1] Cornell Tech, New York, NY, USA
rui.chen@cornell.edu
[2] University of Wisconsin-Madison, Madison, WI, USA
jim.luedtke@wisc.edu

Abstract. We propose a new method for generating cuts valid for the epigraph of a function of binary variables. The proposed cuts are disjunctive cuts defined by many disjunctive terms obtained by enumerating a subset I of the binary variables. We show that by restricting the support of the cut to the same set of variables I, a cut can be obtained by solving a linear program with $2^{|I|}$ constraints. While this limits the size of the set I used to define the multi-term disjunction, the procedure enables generation of multi-term disjunctive cuts using far more terms than existing approaches. Experience on three MILP problems with block diagonal structure using $|I|$ up to size 10 indicates the sparse cuts can often close nearly as much gap as the multi-term disjunctive cuts without this restriction and in a fraction of the time.

Keywords: Disjunctive cuts · Epigraph · Sparsity · Valid inequalities

1 Introduction

We explore techniques for generating valid inequalities (cuts) for the epigraph E of a function $Q : X \to \mathbb{R}$ over binary variables:

$$E = \{(\theta, x) \in \mathbb{R} \times X : \theta \geq Q(x)\}, \tag{1}$$

where $X \subseteq \{0,1\}^n$. An important application motivating this study is stochastic mixed-integer programming (SMIP) [1], or more generally mixed-integer linear programs (MILPs) with block diagonal structures. Such MILPs take the following form

$$
\begin{aligned}
\min \quad & c^T x + \sum_{k=1}^{N} (d^k)^T y^k \\
\text{s.t.} \quad & T^k x + W^k y^k = h^k, \ y^k \geq 0, \quad k \in [N], \\
& x \in X \subseteq \{0,1\}^n.
\end{aligned}
\tag{2}
$$

This research is supported by the Office of Naval Research under grant N00014-21-1-2574.

In the case of two-stage SMIPs, the binary variables x represent first-stage decisions, N is the number of scenarios representing the possible outcomes, and for each $k \in [N] := \{1, \ldots, N\}$, the continuous decision variables y^k represent recourse actions taken in response to observing the data (d^k, T^k, W^k, h^k) in scenario k. A common approach to solving such problems is Benders decomposition, which works with a reformulation of the form

$$\min_{x,\theta} \left\{ c^T x + \sum_{k=1}^{N} \theta_k : \theta_k \geq Q_k(x) \text{ for } k \in [N], \ x \in X \right\}, \qquad (3)$$

where $Q_k(x) = \min_y \{ (d^k)^T y : T^k x + W^k y = h^k, \ y \geq 0 \}$ for $k \in [N]$, $x \in X$. The epigraph of Q_k of the form (1) shows up as a substructure in (3). In Benders decomposition, valid inequalities (Benders cuts) for this epigraph are derived via linear programming (LP) duality, but these are not generally sufficient to define the convex hull of the epigraph, thus motivating the need to derive stronger valid inequalities for sets of this form. This topic has been extensively studied both theoretically and computationally; see [15, 23, 25, 30, 34, 37–39, 41] as just a sample of the literature. Aside from SMIPs, this epigraph substructure appears in a variety of other optimization problems (e.g., [10, 32, 40]).

We study a technique for generating inequalities for E based on a disjunctive relaxation having many terms, specifically obtained by enumerating all $2^{|I|}$ feasible values for a subset I of the binary variables. Disjunctive programming has been a central tool in MILP since its origin in 1970s [4, 5]. A disjunction is a union of sets, and if the feasible region of an MILP is contained within such a union, inequalities valid for the disjunction are valid for the MILP, and are referred to as disjunctive cuts. Most disjunctive cuts used in practice are based on two disjunctive terms, e.g., split cuts [16] and lift-and-project cuts [6–8, 13]. While there has been significant work on classes of cuts that are derived from multiple-term disjunctions [2, 9, 18, 21, 31], the current methods remain focused on disjunctions with a relatively small number of terms. Perregaard and Balas [36] considered an iterative scheme for generating disjunctive cuts from many terms (see Sect. 2), but the approach remains computationally demanding.

Our proposal for generating multi-term disjunctive cuts more efficiently is based on restricting the support of the generated cut to the index set I, the same set used to define the disjunctive terms. We refer to such cuts as I-sparse cuts. Thus, our approach aligns with the spirit of generation of sparse cuts, which is motivated by the benefit of sparse constraints in terms of solution time of the LP relaxations and recent studies that have investigated the theoretical strength of sparse cuts [20, 22, 23]. Our use of sparsity is with respect to the generated cut, which differentiates it from Fukasawa et al. [24] who empirically show that split cuts derived from (two-term) split disjunctions defined by a sparse integer vector can close the majority of the split closure gap. In Sect. 2 we show that our proposed sparsity restriction enables the generation of a multi-term disjunctive cut by solving a single subproblem per term, and then solving a single cut-generating LP. Thus, while this remains a computationally demanding cut generation process, we find empirically that it is feasible to use many more disjunctive terms than have previously been considered. In Sect. 3, we propose two rules for selecting the support I to generate I-sparse inequalities. In Sect. 4, we present results of a computational study using the I-sparse inequalities based on up to 2^{10} disjunctive terms on three test problems. We find that in many cases the I-sparse cuts close nearly as much gap as multi-term disjunctive cuts without the sparsity restriction, and can be generated orders of magnitude faster.

2   Sparse Multi-term 0-1 Disjunctive Cuts

We study the problem of generating valid inequalities for the epigraph E defined in (1). Without loss of generality, we assume the domain X of the function Q is full-dimensional. (Otherwise, we can project out certain variables to make the set full-dimensional after projection.) Let $R(X)$ be a (continuous) relaxation of X with $R(X) \cap \{0,1\}^n = X$. We assume Q has a real-valued extended definition on $R(X)$. We require that minimizing Q over $R(X)$ be a problem that can be efficiently solved. E.g., this would be the case if $R(X)$ and Q over $R(X)$ are closed and convex. If this is not the case, one can replace Q with a relaxation $\hat{Q}$ satisfying $\hat{Q}(x) \leq Q(x)$ for all $x \in X$. For example, in the case of an SMIP having integer second-stage decisions, the recourse function $Q(x)$ is nonconvex and expensive to evaluate, in which case one may use instead $\hat{Q}$ defined using an LP relaxation of the recourse problem. The strength of the resulting cuts will naturally depend on the quality of the relaxation $\hat{Q}$, which could for example be improved using standard MILP valid inequalities.

We let $E^R := \{(\theta, x) \in \mathbb{R} \times R(X) : \theta \geq Q(x)\}$ denote the epigraph of Q over $R(X)$ and let I be a nonempty subset of $[n]$. We derive valid inequalities for E by finding valid inequalities for the following multi-term disjunctive relaxation of E:

$$E_I^R := \bigcup_{\chi \in \{0,1\}^I} E_I^R(\chi), \tag{4}$$

where $E_I^R(\chi) := \{(\theta, x) \in E^R : x_I = \chi\}$, x_I refers to the subvector of x with indices I, and $\{0,1\}^I := \{x_I : x_i \in \{0,1\}, i \in I\}$. We call the relaxation E_I^R of E a multi-term 0-1 disjunction, and any cut valid for E_I^R a multi-term 0-1 disjunctive cut.

2.1   Generating Multi-term 0-1 Disjunctive Cuts

By (4), an inequality of the form $\pi_0 \theta + \pi^T x \geq \eta$ is valid for E_I^R if and only if

$$\min_{\theta, x} \left\{ \pi_0 \theta + \pi^T x : (\theta, x) \in E_I^R(\chi) \right\} \geq \eta \text{ for all } \chi \in \{0,1\}^I. \tag{5}$$

Therefore, to separate a point $(\hat{\theta}, \hat{x})$ from E_I^R, in principle one can solve the following problem:

$$\min_{\pi_0, \pi, \eta} \quad \pi_0 \hat{\theta} + \pi^T \hat{x} - \eta \tag{6a}$$

$$\text{s.t. } \pi_0 \theta + \pi^T x \geq \eta, \ \forall (x, \theta) \in E_I^R(\chi), \chi \in \{0,1\}^I, \tag{6b}$$

$$\pi_0 \geq 0, \|(\pi_0, \pi)\|_1 \leq 1, \tag{6c}$$

where (6c) is just one example of a normalization constraint that can be used to ensure the separation has an optimal solution. The iterative row generating algorithm of Perregaard and Balas [36] can be used for solving (6) by starting with a relaxation defined by a small subset of the constraints (6b) and adding missing constraints (corresponding to extreme points of $E_I^R(\chi)$) that are violated by the current relaxation solution. Given coefficients of candidate cut $(\hat{\pi}_0, \hat{\pi}, \hat{\eta})$ obtained from this relaxation, determining if there is a violated constraint in (6b) is accomplished by solving the problem

$$\min\{\hat{\pi}_0\theta + \hat{\pi}^T x : (x, \theta) \in E_I^R(\chi)\} = \min\{\hat{\pi}_0 Q(x) + \hat{\pi}^T x : x \in R(X), x_I = \chi\} \quad (7)$$

for each $\chi \in \{0, 1\}^I$. While this approach is guaranteed to yield a valid inequality for E_I^R that cuts off $(\hat{\theta}, \hat{x})$ whenever one exists, it is computationally demanding when the number of terms is larger than just a few. In particular, the scalability of the algorithm is limited by the multiplied effect of (a) the size of $\{0, 1\}^I$, and (b) the potential need to solve (7) multiple times for each $\chi \in \{0, 1\}^I$. Numerical experiments in [36] generate valid inequalities for MILPs using only up to 16 disjunctive terms. In this work, we propose to restrict attention to cuts supported on I, which we find eliminates the effect of (b).

2.2 I-Sparse Inequalities

We next explore how restricting the support of the generated cut can be used to accelerate the generation of multi-term 0-1 disjunction cuts for E_I^R for a fixed I.

Definition 1. *Let $I \subseteq [n]$. We say an inequality $\theta \geq \mu^T x + \eta$ is an I-sparse inequality(/cut) for E if the following two conditions hold:*

1. *$\theta \geq \mu^T x + \eta$ is valid for E_I^R;*
2. *$\mu_i = 0$ for all $i \notin I$.*

The following proposition characterizes I-sparse inequalities.

Proposition 2. *An inequality $\theta \geq \mu^T x + \eta$ with $\mu_i = 0$ for all $i \notin I$ is an I-sparse inequality for E if and only if*

$$\sum_{i \in I} \mu_i \chi_i + \eta \leq \nu_I^R(\chi), \ \forall \chi \in \{0, 1\}^I, \quad (8)$$

where for each $\chi \in \{0, 1\}^I$,

$$\nu_I^R(\chi) := \min\{Q(x) : x \in R(X), x_I = \chi\}. \quad (9)$$

Observe that the problem (9) has a similar form as (7) which is used when applying the Perregaard and Balas algorithm [36] to solve (6).

The following result follows from Proposition 2 and provides a condition under which every nontrivial valid inequality for E with coefficients supported on the index set I is an I-sparse inequality.

Corollary 3. *If $X = \{0,1\}^n$, $R(X) = [0,1]^n$ and Q is componentwise monotonically increasing or decreasing on $R(X)$, then an inequality $\theta \geq \mu^T x + \eta$ with $\mu_i = 0$ for all $i \notin I$ is valid for E if and only if it is an I-sparse inequality.*

Based on Proposition 2, for a fixed I, finding an I-sparse inequality that is violated by a point $(\hat{\theta}, \hat{x})$ if one exists can be done by solving the LP

$$g_{\hat{x}}(I) = \max\left\{\sum_{i \in I} \mu_i \hat{x}_i + \eta : \sum_{i \in I} \mu_i \chi_i + \eta \leq \nu_I^R(\chi), \ \chi \in \{0,1\}^I\right\}. \qquad (10)$$

Specifically, the optimal solution of (10) defines an inequality that cuts off $(\hat{\theta}, \hat{x})$ if and only if $g_{\hat{x}}(I) > \hat{\theta}$. When it is easy to determine whether or not a vector is in $\text{proj}_I(X)$, we can replace $\chi \in \{0,1\}^I$ in (10) with $\chi \in \text{proj}_I(X)$ since $\nu_I^R(\chi) = +\infty$ if $\chi \notin \text{proj}_I(X)$. Since Q is finite valued in $R(X)$, $\nu_I^R(\chi) \in \mathbb{R}$ for $\chi \in \text{proj}_I(X)$. When $(\hat{x}_i)_{i \in I} \in \text{conv}(\text{proj}_I(X))$, the LP (10) is guaranteed to have an optimal solution since X being full-dimensional implies that $\text{proj}_I(X)$ is full-dimensional. When $(\hat{x}_i)_{i \in I} \notin \text{conv}(\text{proj}_I(X))$, $(\hat{\theta}, \hat{x})$ can be cut off by an inequality separating $(\hat{x}_i)_{i \in I}$ from $\text{proj}_I(X)$.

The main work to generate an I-sparse inequality is evaluating $\nu_I^R(\chi)$ by solving (9) for each $\chi \in \{0,1\}^I$, and then solving the LP (10) once. Note that (10) has $|I| + 1$ variables in contrast to $n + 2$ variables in the problem (6) used in the Perregaard and Balas (PB) [36] algorithm, and requires solving at most $2^{|I|}$ subproblems of the form (9), in contrast to the PB algorithm which solves $2^{|I|}$ subproblems of this form in multiple iterations until convergence.

2.3   Accelerating the Evaluation of $\nu_I^R(\cdot)$

Evaluating $\nu_I^R(\chi)$ for all $\chi \in \{0,1\}^I$ is the most significant computational component of generating an I-sparse inequality. We thus discuss techniques to accelerate this evaluation, focusing on our motivating example of MILPs with block diagonal structures (2). In this context, assume $R(X) = \{x : Ax \leq b\}$ is a polyhedral relaxation of X and for a fixed $k \in [N]$ let $Q_k(x) = \min_y\{(d^k)^T y : T^k x + W^k y = h^k, \ y \geq 0\}$ and assume it is finite valued for all $x \in R(X)$. In this case, when generating an I-sparse inequality for the set $E_k = \{(\theta_k, x) \in \mathbb{R} \times X : \theta_k \geq Q_k(x)\}$ the evaluation of $\nu_I^R(\chi)$ for $\chi \in \{0,1\}^I$ can be formulated as the following LP

$$\nu_I^R(\chi) = \min_{x,y}\{(d^k)^T y^k : T^k x + W^k y = h^k, \ y \geq 0, \ Ax \leq b, \ x_I = \chi\}. \qquad (11)$$

A first simple idea for accelerating the solution of (11) for all $\chi \in \{0,1\}^I$ is to exploit the possibility to warm-start these LPs (see, e.g., [11] for background). LP solvers like Gurobi [29] and CPLEX [19] automatically implement a simplex warm start when only variable bounds are changed in a LP. Thus, solving the sequence of problems (11) for $\chi \in \{0,1\}^I$ by making changes to variable bounds implied by the constraints $x_I = \chi$ will naturally benefit from these warm-start capabilities. This motivates a careful selection of the sequence these problems are solved in. For example, by following the sequence defined by a Gray code [26], at most one variable bound will change from one subproblem to the next.

We do not explore this in our computational experiments, but another possibility for reducing the time required for evaluating $\nu_I^R(\chi)$ is to use a simpler to evaluate lower bound on Q. E.g., in the context of a Benders decomposition approach for solving MILPs with block diagonal structure, one could obtain a lower bound on $\nu_I^R(\chi)$ by solving a problem of the form:

$$\hat{\nu}_I^R(\chi) = \min\{\hat{Q}_k(x) : x \in R(X), x_I = \chi\}$$

where $\hat{Q}_k$ is the current piecewise-linear convex lower bound of Q_k defined by Benders cuts. These lower bounds could then be used in (10) which would yield a valid but potentially weaker inequality. This inequality could then be improved by exactly evaluating $\nu_I^R(\chi)$ for the χ that correspond to binding constraints in the solution of (10), and then re-solving (10) with these improved values.

3   Two Selection Rules for the Support I

We now discuss techniques for choosing the set I when generating I-sparse cuts. Given a point $(\hat{\theta}, \hat{x})$, the goal is to select I in order to maximize the cut violation $g_{\hat{x}}(I)$ (defined in (10)). Since the complexity of generating these cuts grows exponentially with $|I|$ we investigate techniques that choose I satisfying $|I| \leq K$ for some fixed (small) integer K. We describe two selection rules that are derived from two different approximations of Q.

3.1   A Greedy Rule Based on a Monotone Submodular Approximation

The problem of choosing I that maximizes $g_{\hat{x}}(I)$ is a set function optimization problem. For notational convenience, we do not distinguish between a set function and a function with binary variables, i.e., we interchangeably use $f(A)$ for $f(\chi_A)$ for all $A \subseteq [n]$ where $\chi_A \in \{0,1\}^n$ is the indicator vector of A. One particular class of set functions satisfying good theoretical properties is monotone submodular functions [27]. Given $\hat{x} \in [0,1]^n$, we can show that the cut violation function $g_{\hat{x}}(I)$ is monotone submodular in I if Q is monotone submodular.

Proposition 4. *Assume Q is monotone submodular, $X = \{0,1\}^n$ and $R(X) = [0,1]^n$. Then the cut violation function $g_{\hat{x}}(I)$ is also monotone submodular in I.*

Although maximizing a monotone submodular function subject to a cardinality constraint is NP-hard [17] in general, the well-known greedy algorithm of Nemhauser et al. [33] attains a good approximation ratio to this problem. However, directly applying a greedy algorithm for choosing I may not be a good choice because (i) Q is not necessarily monotone submodular, and (ii) the greedy algorithm requires evaluating $g_{\hat{x}}(\cdot)$ many times, which is computationally expensive. Therefore, we seek alternatives to this approach by applying the greedy algorithm to a different cut violation function $\tilde{g}_{\hat{x}}$ associated with an approximation $\tilde{Q} : \{0,1\}^n \to \mathbb{R}$ of the function Q. We choose $\tilde{Q}$ (and hence $\tilde{g}_{\hat{x}}$) such that $\tilde{Q}$ is monotone and submodular and the cut violation $\tilde{g}_{\hat{x}}(\cdot)$ can be evaluated much more efficiently than $g_{\hat{x}}(\cdot)$.

To construct such an approximation, we first state a result for convexifying a special set, studied in [3,28].

Theorem 5 ([3,28]). *Let*

$$F = \{(\theta, x) \in \mathbb{R} \times \{0,1\}^n : \theta \geq a_i x_i + b, \ i = 1, \ldots, n\} \tag{12}$$

with $0 \leq a_1 \leq \ldots \leq a_n$. Then

$$conv(F) = \{(\theta, x) \in \mathbb{R} \times [0,1]^n : \theta \geq a_{i_1} x_{i_1} + \sum_{k=2}^{m} (a_{i_k} - a_{i_{k-1}}) x_{i_k} + b,$$

for all subsequences $(i_k)_{k=1}^m$ of $[n]$ such that $1 \leq i_1 \leq \ldots \leq i_m = n\}$.

The inequalities defining F have only one x_i variable on the right-hand side and share the same constant term b. The set F defines the epigraph of a monotone submodular function $\max_{i \in [n]} \{a_i x_i + b\}$. Although this characterization of the convex hull consists of exponentially many inequalities, it has been shown that the corresponding separation can be solved in polynomial time [3,28]. Such separation results can be easily extended for separating from valid inequalities for F that are supported on I.

We next describe how to construct an approximation $\tilde{Q}$ of Q of the form $\max_{i \in [n]} \{a_i x_i + b\}$. The first step is to construct an underestimate of Q by deriving the I-sparse inequalities with $I = \{i\}$ for each $i \in [n]$. The polyhedron defined by (8) has a unique extreme point $\left(\nu_{\{i\}}^R(1) - \nu_{\{i\}}^R(0), \nu_{\{i\}}^R(0)\right)$ when $I = \{i\}$, which corresponds to a valid inequality of E:

$$\theta \geq \left(\nu_{\{i\}}^R(1) - \nu_{\{i\}}^R(0)\right) x_i + \nu_{\{i\}}^R(0). \tag{13}$$

For $i \in [n]$, $\mathrm{LB}_i := \min\{\nu_{\{i\}}^R(0), \nu_{\{i\}}^R(1)\}$ is a lower bound of Q on $\{0,1\}^n$, and so is $\mathrm{LB}^* := \max_{i \in [n]} \mathrm{LB}_i$. Therefore, we can strengthen (13) to be $\theta \geq \left(\tilde{\nu}_{\{i\}}^R(1) - \tilde{\nu}_{\{i\}}^R(0)\right) x_i + \tilde{\nu}_{\{i\}}^R(0)$, where $\tilde{\nu}_{\{i\}}^R(k) = \max\{\nu_{\{i\}}^R(k), \mathrm{LB}^*\}$ for $k \in \{0,1\}$. After complementing the variables $x_i \leftarrow 1 - x_i$ for $i \in [n]$ with $\tilde{\nu}_{\{i\}}^R(1) - \tilde{\nu}_{\{i\}}^R(0) < 0$, we obtain inequalities of the form $\theta \geq a_i x_i + b$ for $i \in [n]$ with $a \geq 0$, which are valid for (reflected) E. Assume without loss of generality that $0 \leq a_1 \leq \ldots \leq a_n$. We can then apply the greedy algorithm to generate I using $\tilde{g}_{\hat{x}}$, where the evaluation of $\tilde{g}_{\hat{x}}(\cdot)$ is similar to the separation algorithm proposed in [28] (see [14] for details).

3.2   A Cutting-Plane Approximation Rule

We next describe an alternative selection rule for I that is based on a single cutting-plane approximation of Q. By Corollary 3, the following result characterizes the most violated I-sparse inequality for a function on $\{0,1\}^n$ defined by a single cutting plane.

Proposition 6. *Let*

$$F_{(a,b)} = \{(\theta, x) \in \mathbb{R} \times \{0,1\}^n : \theta \geq a^T x + b\}.$$

Given $(\hat{\theta}, \hat{x}) \in \mathbb{R} \times [0,1]^n$, *the maximum violation of a valid inequality of* $F_{(a,b)}$ *of the form* $\theta \geq \sum_{i \in I} \mu_i x_i + \eta$ *by* $(\hat{\theta}, \hat{x})$ *is* $-\sum_{i=1}^{n} a_i^- + \sum_{i \in I}(a_i \hat{x}_i + a_i^-) + b - \hat{\theta}$, *where* $a_i^- = \max\{-a_i, 0\}$.

By Proposition 6, the value $a_i \hat{x}_i + a_i^-$ in some sense measures the importance of variable x_i for the cutting plane $\theta \geq a^T x + b$ at $\hat{x}$. We use this intuition to construct a selection rule. We first pick a cutting plane $\theta \geq a^T x + b$ that approximates the epigraph of Q at $\hat{x}$. Then indices $i \in [n]$ are added to the set I in decreasing order of the value $a_i \hat{x}_i + a_i^-$ until $|I| = K$. Note that $a_i \hat{x}_i + a_i^- \geq 0$ for any $a_i \in \mathbb{R}$ and $\hat{x}_i \in [0,1]$. If the cutting plane approximation $\theta \geq a^T x + b$ is sparse (i.e., $|\{i \in [n] : a_i \neq 0\}|$ is small), it is possible that $|\{i \in [n] : a_i \hat{x}_i + a_i^- > 0\}| < K$. In such cases, we first add those indices with positive $a_i \hat{x}_i + a_i^-$ values into I, then pick another cutting plane and repeat the procedure until $|I| = K$. A potential advantage of this selection rule is that it does not require any evaluation of $g_{\hat{x}}$ and therefore can be implemented efficiently. And unlike the selection rule in Sect. 3.1, this selection rule can take advantage of the availability of dense cutting plane approximations. The potential limitation, of course, is the reliance on the single cutting-plane approximation.

The final detail we need to specify for this approach is how to choose the cutting-plane approximation(s). Assume a collection $\mathcal{A}$ of cutting planes of the form $\theta \geq a^T x + b$ is available. A natural choice for $\mathcal{A}$ is the set of cutting planes (e.g., Benders cuts) that have been added in the algorithm so far for approximating E. A natural ordering for choosing which cutting plane in $\mathcal{A}$ to use first is based on the tightness of the cutting plane at the point $\hat{x}$. The inequality in $\mathcal{A}$ with coefficients (a, b) that yield the highest $a^T \hat{x} + b$ value is chosen first, etc.

4   Computational Results

To provide insight into the potential of our method, we conduct numerical experiments on three MILP problems with block diagonal structures (2):

- The stochastic network interdiction (SNIP) problem [35]: $n = 320$ for these instances.
- The latent-class logit assortment (LLA) problem [32]: $n = 500$ for these instances.
- A stochastic version of the capacitated facility location (CAP) problem [12]: n ranges between 25 and 50 for these instances.

For the first two test problems, each block of their MILP formulations is sparse in variables x, but in distinct ways. For the SNIP problem, we observe that when applying Benders decomposition to solve its LP relaxation the Benders cuts are mostly very sparse in x. In the LLA problem each block of the MILP formulation only uses a small portion (between 12 and 20) of the x variables, making the use of sparse cuts very natural for this problem. Neither of these two sparsity properties holds for the CAP problem. See [14] for details of the test instances.

The constraints $x \in X$ in all our test problems consist of x being binary and either a lower-bounding or upper-bounding cardinality constraint on the number of nonzero x_i variables. Therefore, we use $R(X) = \text{conv}(X)$ for all our tests instances. We use the direct LP relaxation as Q_k for each block of the MILP as described in Sect. 2.3. For testing the effectiveness of the generated cuts, we add I-sparse cuts on top of the LP relaxation that is first solved by Benders decomposition. The cut generating process is described in Algorithm 1. We consider the following variants of Algorithm 1:

- Greedy-K: Use the greedy rule described in Sect. 3.1 for generating the support I of size up to K;
- Cutpl-K: Use the cutting plane approximation rule described in Sect. 3.2 for generating the support I of size up to K.

For Cutpl-K, we use the collection of all the Benders cuts added for block k in line 1 of Algorithm 1 as $\mathcal{A}$ for Q_k. To improve the efficiency of the algorithm, when applying Greedy-K, we only select I from indices for which the corresponding variables have a nonzero coefficient in at least one of the Benders cuts for block k. This restriction is also implicitly implemented when using Cutpl-K since indices i with $a_i = 0$ for all $(a, b) \in \mathcal{A}$ can never be selected by Cutpl-K. It significantly improves the efficiency of Greedy-K on SNIP instances (by skipping the generation of $\{i\}$-sparse cuts for most $i \in [n]$).

To visually compare the performance of I-sparse cuts across multiple test instances, we present results in the form of an integrality-gap-closed profile. Each curve in such a profile corresponds to a particular selection rule for I and size limit K, and its value at time t represents the average (over the set of instances for that problem class) integrality gap closed by time t, where the integrality gap closed at time t is calculated as $(z_R(t) - z_{LP})/(z^* - z_{LP}) \times 100\%$, where $z_R(t)$ is the bound obtained by the algorithm at time t, z_{LP} is the basic LP relaxation bound, and z^* is the optimal value.

The results for the SNIP, LLA, and CAP test problems are given in Figs. 1, 2, and 3, respectively, where in each case we vary $K \in \{4, 7, 10\}$ and compare the Greedy-K and Cutpl-K selection rules. In each case we find that the two different selection rules have similar trends in gap closed over time. The Cutpl-K rule has better performance on the SNIP test instances, whereas Greedy-K has significantly better performance on the LLA and CAP instances when $K = 4$ or $K = 7$. In terms of the effect of K, as expected smaller values of K yield quicker initial gap improvement, whereas larger values of K require more time to close the gap but eventually lead to more gap closed. For the SNIP instances we find that using $K = 4$ already closes most of the gap, and does so much more quickly than with $K = 7$ or $K = 10$. For the LLA instances we find that increasing K leads to more gap closed, although significant gap is already closed with $K = 4$, and the additional gap closed using $K = 10$ is marginal, while

---

Algorithm 1: Generating I-sparse cuts

---

1 Initialize a master LP using Benders decomposition;
2 **repeat**
3 | Solve the master LP to obtain solution $(\hat{\theta}, \hat{x})$;
4 | **for** *block $k \in [N]$* **do**
5 | | Choose a support I;
6 | | Generate an I-sparse cut by solving (10);
7 | | Add the I-sparse cut to the master LP if it is violated by $(\hat{\theta}_k, \hat{x})$;
8 | **end**
9 **until** *No violated cut can be generated or time limit is reached;*

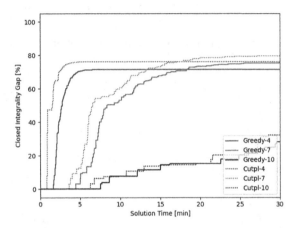

Fig. 1. Integrality-gap-closed profiles for SNIP instances obtained by Greedy-K (solid) and Cutpl-K (dashed) for $K \in \{4, 7, 10\}$

requiring significantly more time. For the CAP instances, we find that the I-sparse cuts close significantly less gap than the other test problems, although the gap closed is still significant. Large values of K yield significantly more gap closed on the CAP instances, but also requires considerably longer running time.

We observe that the number of I-sparse cuts added by the algorithm does not increase when K increases. Thus, the improvement in the bound is attributable to stronger cuts rather than an increase in the number of cuts added.

As a final experiment, we compare the I-sparse cuts with the multi-term 0-1 disjunctive cuts without the sparsity restriction, but generated from the same sets I, where the cuts are generated using the Perregaard and Balas (PB) [36] approach. Our interest in this comparison is to demonstrate the potential time reductions from using the I-sparse cuts and to estimate the extent to which the sparsity restriction degrades the quality of the relaxation. We conduct this experiment only on the CAP test instances, since we have already seen that the I-sparse cuts are sufficient to close most of the gap in the SNIP and LLA instances, and thus there is little potential to close more gap when eliminating

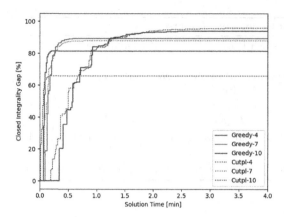

Fig. 2. Integrality-gap-closed profiles for LLA instances obtained by Greedy-K (solid) and Cutpl-K (dashed) for $K \in \{4, 7, 10\}$

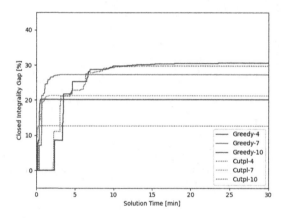

Fig. 3. Integrality-gap-closed profiles for CAP instances obtained by Greedy-K (solid) and Cutpl-K (dashed) for $K \in \{4, 7, 10\}$

the sparsity restriction. We set a 24-h time limit for the PB algorithm. For both the I-sparse and PB cuts, we use Greedy-K as the rule for selecting the set I to define the multi-term disjunction.

Figure 4 displays the integrality gap closed over time for two specific CAP instances, one for which I-sparse cuts were able to close a significant portion of the gap (CAP101), and one for which they were not (CAP111). The figures on the left display results for both the I-sparse cuts (solid lines) and PB cuts (dashed lines), with the time-scale (x-axis) determined by the time required to generate all I-sparse cuts for the largest value of K. From these figures we observe that for any value of K, within this time frame the I-sparse cuts close significantly more gap than the PB cuts. To estimate the potential for PB cuts to eventually close more gap, we show the gap closed by the PB cuts over the

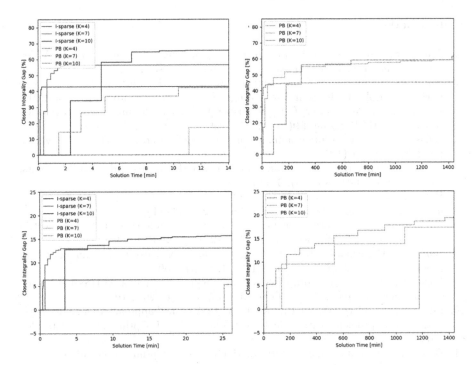

Fig. 4. Integrality gap closed by I-sparse cuts and cuts generated by the PB algorithm on instances CAP101 (top) and CAP111 (bottom)

full 24-h time limit in the figures on the right. For CAP101 we find that the PB cuts do not close more gap than the I-sparse cuts, suggesting that the sparsity restriction is not significantly degrading the strength of the cuts in this case. On the other hand, for CAP111, we find that when given enough time the PB cuts can close significantly more gap, as seen particularly for the $K = 4$ results, although requiring far more time to do so. For both CAP instances, we observe that most of the generated PB cuts are as sparse as the I-sparse cuts in the first few iterations but become significantly denser (e.g., with non-zeros on more than half the variables) in later iterations.

5   Future Directions

A natural idea inspired by our computational results is to adaptively choose the size K of the set I when generating I-sparse cuts. While not presented here, the proposed approach for generating I-sparse inequalities can also be used to improve the coefficients on a sparse subset of variables for a given valid inequality for E. It would be interesting to explore this idea further.

References

1. Ahmed, S.: Two-stage stochastic integer programming: a brief introduction. Wiley encyclopedia of operations research and management science (2010)
2. Andersen, K., Louveaux, Q., Weismantel, R., Wolsey, L.: Cutting planes from two rows of a simplex tableau. In: Proceedings of IPCO XII, vol. 4513, pp. 1–15 (2007)
3. Atamtürk, A., Nemhauser, G.L., Savelsbergh, M.W.: The mixed vertex packing problem. Math. Program. **89**(1), 35–53 (2000)
4. Balas, E.: Disjunctive programming. In: Annals of Discrete Mathematics, vol. 5, pp. 3–51. Elsevier (1979)
5. Balas, E.: Disjunctive programming: properties of the convex hull of feasible points. Discret. Appl. Math. **89**(1–3), 3–44 (1998)
6. Balas, E., Ceria, S., Cornuéjols, G.: A lift-and-project cutting plane algorithm for mixed 0-1 programs. Math. Program. **58**(1), 295–324 (1993)
7. Balas, E., Ceria, S., Cornuéjols, G.: Mixed 0-1 programming by lift-and-project in a branch-and-cut framework. Manag. Sci. **42**(9), 1229–1246 (1996)
8. Balas, E., Perregaard, M.: Lift-and-project for mixed 0-1 programming: recent progress. Discret. Appl. Math. **123**(1–3), 129–154 (2002)
9. Basu, A., Conforti, M., Cornuéjols, G., Zambelli, G.: Maximal lattice-free convex sets in linear subspaces. Math. Oper. Res. **35**(3), 704–720 (2010)
10. Bertsimas, D., Cory-Wright, R., Pauphilet, J.: A unified approach to mixed-integer optimization problems with logical constraints. SIAM J. Optim. **31**(3), 2340–2367 (2021)
11. Bertsimas, D., Tsitsiklis, J.N.: Introduction to Linear Optimization, vol. 6. Athena Scientific, Belmont (1997)
12. Bodur, M., Dash, S., Günlük, O., Luedtke, J.: Strengthened Benders cuts for stochastic integer programs with continuous recourse. INFORMS J. Comput. **29**(1), 77–91 (2017)
13. Ceria, S., Pataki, G.: Solving integer and disjunctive programs by lift and project. In: Bixby, R.E., Boyd, E.A., Ríos-Mercado, R.Z. (eds.) IPCO 1998. LNCS, vol. 1412, pp. 271–283. Springer, Heidelberg (1998). https://doi.org/10.1007/3-540-69346-7_21
14. Chen, R.: Valid inequalities and integer programming formulations for stochastic integer programming and some machine learning problems. Ph.D. thesis, The University of Wisconsin-Madison (2021)
15. Chen, R., Luedtke, J.: On generating Lagrangian cuts for two-stage stochastic integer programs. arXiv preprint arXiv:2106.04023 (2021)
16. Cook, W., Kannan, R., Schrijver, A.: Chvátal closures for mixed integer programming problems. Math. Program. **47**(1), 155–174 (1990)
17. Cornuejols, G., Fisher, M.L., Nemhauser, G.L.: Exceptional paper-location of bank accounts to optimize float: an analytic study of exact and approximate algorithms. Manag. Sci. **23**(8), 789–810 (1977)
18. Cornuéjols, G., Margot, F.: On the facets of mixed integer programs with two integer variables and two constraints. Math. Program. **120**(2), 429–456 (2009)
19. Cplex, IBM ILOG: V12. 1: User's Manual for CPLEX (2009)
20. Dey, S.S., Iroume, A., Molinaro, M.: Some lower bounds on sparse outer approximations of polytopes. Oper. Res. Lett. **43**(3), 323–328 (2015)
21. Dey, S.S., Lodi, A., Tramontani, A., Wolsey, L.A.: Experiments with two row tableau cuts. In: Eisenbrand, F., Shepherd, F.B. (eds.) IPCO 2010. LNCS, vol. 6080, pp. 424–437. Springer, Heidelberg (2010). https://doi.org/10.1007/978-3-642-13036-6_32

22. Dey, S.S., Molinaro, M., Wang, Q.: Approximating polyhedra with sparse inequalities. Math. Program. **154**(1), 329–352 (2015)
23. Dey, S.S., Molinaro, M., Wang, Q.: Analysis of sparse cutting planes for sparse MILPs with applications to stochastic MILPs. Math. Oper. Res. **43**(1), 304–332 (2018)
24. Fukasawa, R., Poirrier, L., Yang, S.: Split cuts from sparse disjunctions. Math. Program. Comput. **12**(2), 295–335 (2020). https://doi.org/10.1007/s12532-020-00180-9
25. Gade, D., Küçükyavuz, S., Sen, S.: Decomposition algorithms with parametric Gomory cuts for two-stage stochastic integer programs. Math. Program. **144**(1–2), 39–64 (2014)
26. Gray, F.: Pulse code communication. United States Patent Number 2632058 (1953)
27. Grötschel, M., Lovász, L., Schrijver, A.: Geometric Algorithms and Combinatorial Optimization, vol. 2. Springer, Heidelberg (2012)
28. Günlük, O., Pochet, Y.: Mixing mixed-integer inequalities. Math. Program. **90**(3), 429–457 (2001)
29. Gurobi Optimization, LLC: Gurobi Optimizer Reference Manual (2021). https://www.gurobi.com
30. Laporte, G., Louveaux, F.V.: The integer L-shaped method for stochastic integer programs with complete recourse. Oper. Res. Lett. **13**(3), 133–142 (1993)
31. Li, Y., Richard, J.P.P.: Cook, Kannan and Schrijver's example revisited. Discret. Optim. **5**(4), 724–734 (2008)
32. Méndez-Díaz, I., Miranda-Bront, J.J., Vulcano, G., Zabala, P.: A branch-and-cut algorithm for the latent-class logit assortment problem. Discret. Appl. Math. **164**, 246–263 (2014)
33. Nemhauser, G.L., Wolsey, L.A., Fisher, M.L.: An analysis of approximations for maximizing submodular set functions-I. Math. Program. **14**(1), 265–294 (1978)
34. Ntaimo, L.: Fenchel decomposition for stochastic mixed-integer programming. J. Global Optim. **55**(1), 141–163 (2013)
35. Pan, F., Morton, D.P.: Minimizing a stochastic maximum-reliability path. Networks **52**(3), 111–119 (2008)
36. Perregaard, M., Balas, E.: Generating cuts from multiple-term disjunctions. In: Aardal, K., Gerards, B. (eds.) IPCO 2001. LNCS, vol. 2081, pp. 348–360. Springer, Heidelberg (2001). https://doi.org/10.1007/3-540-45535-3_27
37. Rahmaniani, R., Ahmed, S., Crainic, T.G., Gendreau, M., Rei, W.: The Benders dual decomposition method. Oper. Res. **68**, 878–895 (2020)
38. Sen, S., Higle, J.L.: The C3 theorem and a D2 algorithm for large scale stochastic mixed-integer programming: set convexification. Math. Program. **104**(1), 1–20 (2005)
39. Sen, S., Sherali, H.D.: Decomposition with branch-and-cut approaches for two-stage stochastic mixed-integer programming. Math. Program. **106**(2), 203–223 (2006)
40. Wiegele, A.: Biq Mac Library-a collection of max-cut and quadratic 0-1 programming instances of medium size. Preprint 51 (2007)
41. Zhang, M., Küçükyavuz, S.: Finitely convergent decomposition algorithms for two-stage stochastic pure integer programs. SIAM J. Optim. **24**(4), 1933–1951 (2014)

A 2-Approximation for the Bounded Treewidth Sparsest Cut Problem in **FPT** Time

Vincent Cohen-Addad[1], Tobias Mömke[2], and Victor Verdugo[3]([✉])

[1] Google Research Zurich, Zürich, Switzerland
[2] Department of Computer Science, University of Augsburg, Augsburg, Germany
moemke@informatik.uni-augsburg.de
[3] Institute of Engineering Sciences, Universidad de O'Higgins, Rancagua, Chile
victor.verdugo@uoh.cl

Abstract. In the non-uniform sparsest cut problem, we are given a supply graph G and a demand graph D, both with the same set of nodes V. The goal is to find a cut of V that minimizes the ratio of the total capacity on the edges of G crossing the cut over the total demand of the crossing edges of D. In this work, we study the non-uniform sparsest cut problem for supply graphs with bounded treewidth k. For this case, Gupta, Talwar and Witmer [STOC 2013] obtained a 2-approximation with polynomial running time for fixed k, and the question of whether there exists a c-approximation algorithm for a constant c independent of k, that runs in **FPT** time, remained open. We answer this question in the affirmative. We design a 2-approximation algorithm for the non-uniform sparsest cut with bounded treewidth supply graphs that runs in **FPT** time, when parameterized by the treewidth. Our algorithm is based on rounding the optimal solution of a linear programming relaxation inspired by the Sherali-Adams hierarchy. In contrast to the classic Sherali-Adams approach, we construct a relaxation driven by a tree decomposition of the supply graph by including a carefully chosen set of lifting variables and constraints to encode information of subsets of nodes with super-constant size, and at the same time we have a sufficiently small linear program that can be solved in **FPT** time.

Keywords: Sparsest Cut · Linear Programming · Approximation Algorithms

1 Introduction

In the non-uniform sparsest cut problem, we are given two weighted graphs G and D on the same set of nodes V, such that $G = (V, E_G)$ is the so-called supply graph, and $D = (V, E_D)$ is the so-called demand graph. For every edge

Partially supported by DFG Grant 439522729 (Heisenberg-Grant), DFG Grant 439637648 (Sachbeihilfe), and ANID Grants ACT210005 and FONDECYT 11190789.

K. Aardal and L. Sanitá (Eds.): IPCO 2022, LNCS 13265, pp. 112–125, 2022.
https://doi.org/10.1007/978-3-031-06901-7_9

$e \in E_G$ we have a positive integer weight $\mathsf{cap}(e)$ called capacity, and for every edge $e \in E_D$ we have a positive integer weight $\mathsf{dem}(e)$ called the demand. An instance $\mathcal{I}$ is given by a tuple $(G, D, \mathsf{cap}, \mathsf{dem})$ and we denote by $|\mathcal{I}|$ the encoding length of an instance $\mathcal{I}$. The goal is to compute a non-empty subset of nodes $S \subseteq V$ that minimizes

$$\phi(S) = \frac{\sum_{e \in \delta_G(S)} \mathsf{cap}(e)}{\sum_{e \in \delta_D(S)} \mathsf{dem}(e)},$$

where $\delta_G(S) = \{e \in E_G : |e \cap S| = 1\}$ and $\delta_D(S) = \{e \in E_D : |e \cap S| = 1\}$. Since this problem is NP-hard [25], the focus has been on the design of approximation algorithms. In this line of work, Agrawal, Klein, Rao and Ravi [19, 20] took the first major step by describing a $O(\log D \log C)$-approximation algorithm, where D is the total sum of the demands and C is the total sum of the capacities. Currently, the best approximation factor is $O(\sqrt{\log n} \log \log n)$ due to Arora, Lee and Naor [2]. The *uniform* version of the problem, where the demand graph is unweighted and complete, has received a lot of attention through the years. The best bound for this problem is slightly better: $O(\sqrt{\log n})$ [3].

The non-uniform sparsest cut problem is hard to approximate within a constant factor, for any constant, under the unique games conjecture [9, 17, 18]. Therefore, the problem has also been studied under the assumption that the supply graph belongs to a specific family of graphs. Most notable examples include planar graphs, graph excluding a fixed minor, and bounded treewidth graphs. In this paper, we focus on the latter (see Sect. 1.1 for further related work on minor-closed families).

For inputs to the problem where the supply graph has treewidth at most k, Chlamtac, Krauthgamer and Raghavendra [11] designed a $C(k)$-approximation algorithm that runs in time $2^{O(k)}|\mathcal{I}|^{O(1)}$, where C is a double exponential function of k. Later, Gupta, Talwar and Witmer designed a 2-approximation algorithm that runs in time $|\mathcal{I}|^{O(k)}$ [17]. However, these two results are only complementary: Chlamtac, Krauthgamer and Raghavendra's algorithm is Fixed-Parameter Tractable (FPT) in the treewidth k of the supply graph, (that is, $f(k)|\mathcal{I}|^{O(1)}$ time for some computable function f), while the algorithm of Gupta, Talwar and Witmer is not. The approximation factor achieved by Gupta, Talwar and Witmer is independent of k, and furthermore, they show that assuming the unique games conjecture, there is no $(2 - \varepsilon)$-approximation algorithm for any $\varepsilon > 0$ on graphs with constant treewidth and that there is no $1.138 - \varepsilon$ approximation algorithm for treewidth 2 graphs unless $\mathsf{P} = \mathsf{NP}$. This left open the question of whether there exists a 2-approximation algorithm that runs in FPT time, when parameterized by the treewidth. We answer this question in the affirmative and show the following result.

Theorem 1. *There is an algorithm that computes a 2-approximation for every instance $\mathcal{I} = (G, D, \mathsf{cap}, \mathsf{dem})$ of the non-uniform sparsest cut problem in time*

$$2^{2^{O(k)}}|\mathcal{I}|^{O(1)},$$

where k is the treewidth of the supply graph G.

As a corollary, and following the argumentation of Gupta et al. [17], for treewidth k graphs our result implies the existence of a 2-approximation for the minimum-distortion ℓ_1 embedding problem, with $2^{2^{O(k)}}|\mathcal{I}|^{O(1)}$ running time.

The results obtained in the predecessor papers [11, 17] were based on rounding certain linear programs obtained through the Sherali-Adams *lift & project* hierarchy [27]. Our approximation algorithm is also based on rounding a linear program with a fractional objective given by the non-uniform sparsest cut value, but we construct this linear program in a different way, with the goal of obtaining a linear program of smaller size, but sufficiently strong in terms of integrality gap. If we followed the classic Sherali-Adams approach, the relaxation of *level* ℓ would be constructed by using a variable encoding the value of any subset of the original variables up to size ℓ, and it would take $n^{O(\ell)}$ time to solve this relaxation, where n is the number of nodes in the graph. In particular, solving a relaxation of level $\Theta(k)$ would take $n^{O(k)}$ time, which in principle rules out the possibility of achieving FPT running time by applying directly this approach. In order to overcome this problem, we construct a linear programming relaxation driven by a tree decomposition of the supply graph G, where the variables are carefully chosen with the goal of encoding information of subsets of nodes with super-constant size, and at the same time the number of variables and constraints is sufficiently small so we can solve the relaxation in FPT time. We show that the relaxation is strong enough to get a 2-approximation by rounding the optimal fractional solution. The construction of our relaxation and the analysis of our algorithm can be found in Sect. 3.[1]

1.1 Related Work

Despite the difficulties in approximating the non-uniform sparsest cut problem in general graphs, there are several other results for restricted families of graphs. The case in which G is planar has received a lot of attention. Quite recently, Cohen-Addad, Gupta, Klein and Li [12] showed the existence of a quasi-polynomial time $(2+\varepsilon)$-approximation for the non-uniform sparsest cut problem in the planar case. To get this result they combine a patching lemma approach with linear programming techniques. We remark that for the planar case there is no polynomial time $1/(0.878 + \varepsilon) \approx (1.139 - \varepsilon)$-approximation algorithm under the unique games conjecture [17].

Other families with constant factor approximation algorithms are outerplanar graphs [26], series-parallel [10, 16, 21], k-outerplanar graphs [9], graphs obtained by 2-sums of K_4 [7] and graphs with constant pathwidth [22]. The impact of the treewidth parameter has also been studied in the context of polynomial optimization [4]. Finally, we mention that the Sherali-Adams hierarchy has been useful to design algorithms in other minor-free and bounded treewidth graph problems, including independent set and vertex cover [5, 23], and also in several recent results on scheduling and clustering [1, 14, 15, 24, 28].

---

[1] A full version of this article is available in Arxiv and can be found in [13].

Very recently and independent of our work, Chalermsook et al. [8] obtained a $O(k^2)$-approximation algorithm for sparsest cut in treewidth k graphs, with running time $2^{O(k)} \cdot \text{poly}(n)$ and, for arbitrary $\varepsilon > 0$, an $O(1/\varepsilon^2)$-approximation algorithm with running time $2^{O(k^{1+\varepsilon}/\varepsilon)} \cdot \text{poly}(n)$. Observe that these results are incomparable with our result: they obtain an asymptotically lower running time, whereas the obtained (constant) approximation ratio is considerably larger than 2. Similar to our result, they build on the techniques from [11,17]. However, their approach is based on a new measure for tree decompositions which they call the combinatorial diameter.

2 Preliminaries: Tree Decompositions

A tree decomposition of a graph $G = (V, E)$ is a pair $(\mathcal{X}, \mathcal{T})$ where $\mathcal{T} = (\mathcal{X}, E_{\mathcal{T}})$ is a tree and $\mathcal{X}$ is a collection of subsets of nodes in V called *bags*. Each bag is a node in the tree $\mathcal{T}$. Furthermore, the pair $(\mathcal{X}, \mathcal{T})$ satisfies the following conditions.

(1) Every node in V is in at least one bag, that is, $\cup_{X \in \mathcal{X}} X = V$.
(2) For every edge $\{u, v\} \in E$ there exists a bag $X \in \mathcal{X}$ such that $\{u, v\} \subseteq X$.
(3) For every node $u \in V$ the bags containing u induce a subtree of $\mathcal{T}$.

The *width* of the tree decomposition $(\mathcal{X}, \mathcal{T})$ corresponds to the size of the largest bag in the tree decomposition, minus one. The treewidth of G is the minimum possible width of a tree decomposition for G. We typically consider the tree $\mathcal{T}$ to be rooted, and we denote its root by $\mathcal{R}$. We denote by $\text{depth}(\mathcal{T})$ the depth of the tree $\mathcal{T}$ and we say that a bag X is at level ℓ if the distance from the root $\mathcal{R}$ to X in the tree $\mathcal{T}$ is equal to ℓ. We denote by $\mu(X)$ the parent of X in the tree $\mathcal{T}$. The intersection between a non-root bag X and the parent bag, $\mu(X) \cap X$, is the called the *adhesion* of the bag X. We say that a bag Y is a *descendant* of X if $X \neq Y$ and the bag X belongs to the unique path in $\mathcal{T}$ from Y to the root, and in this case we say that X is an *ancestor* of Y.

3 The LP Relaxation and the Rounding Algorithm

Our algorithm is based on rounding the optimal solution of a linear programming relaxation for the non-uniform sparsest cut problem. In Sect. 3.1 we provide the construction of our linear programming relaxation and in Sect. 3.2 we provide the rounding algorithm and the proof of Theorem 1. In the following lemma we show the existence of a tree structure that we use to construct the linear program. The proof of this lemma can be found in the full version of the article [13].

Lemma 1. *Let G be a graph with treewidth k and let ℓ be a positive integer. Then, there exists a tree decomposition $(\mathcal{Y}, \mathcal{E})$ of G such that the following holds:*

(a) The width of $(\mathcal{Y}, \mathcal{E})$ is $O(2^{\ell} k)$ and $\text{depth}(\mathcal{E}) \in O(\log(n)/\ell)$.

(b) For every non-root bag $Y \in \mathcal{Y}$, the size of the adhesion of Y is $O(k)$.

The decomposition $(\mathcal{Y}, \mathcal{E})$ can be found in $2^{O(k^3)} n$ time.

Definition 1. *Given a graph G, we say that a tree decomposition $\Theta = (\mathcal{Y}, \mathcal{E})$ satisfying properties (a)-(b) is a (k, ℓ)-decomposition of G.*

Given a bag $Y \in \mathcal{Y}$, we denote by $\mathcal{P}_\Theta^Y$ the subset of bags that belong to the path from Y to the root $\mathcal{R}$ in the tree $\mathcal{E}$. We denote by $\mathcal{J}_Y$ the adhesion of Y. Furthermore, let $\mathcal{V}_\Theta^Y = \bigcup_{Z \in \mathcal{P}_\Theta^Y} \mathcal{J}_Z$, and for every pair of non-root bags $Y, Z \in \mathcal{Y}$ let $\mathcal{S}_\Theta(Y, Z)$ be the power set of $(Y \cup \mathcal{V}_\Theta^Y) \cup (Z \cup \mathcal{V}_\Theta^Z)$. Finally, let $\mathcal{S}_\Theta = \bigcup_{Y, Z \in \mathcal{Y}} \mathcal{S}_\Theta(Y, Z)$. Observe that for every bag $Y \in \mathcal{Y}$, the size of $Y \cup \mathcal{V}_\Theta^Y$ is $O(2^\ell k + k \log(n)/\ell)$.

3.1   The LP Relaxation

Consider a positive integer ℓ and an instance $(G, D, \mathsf{cap}, \mathsf{dem})$ where G has treewidth k. Let $\Theta = (\mathcal{Y}, \mathcal{E})$ be a (k, ℓ)-decomposition of the supply graph G. In what follows we describe our LP relaxation, inspired by the Sherali-Adams hierarchy [27] and the predecessor works [11,17]. In this linear program there are two types of variables. The variable $x(S, T)$, with $S \in \mathcal{S}_\Theta$ and $T \subseteq S$, indicates that the cut solution C satisfies that $C \cap S = T$. The variable $y(\{u, v\})$ for $u, v \in V$ with $u \neq v$, indicates whether the nodes u and v fall in different sides of the cut. For notation simplicity, we sometimes denote the union between a set A and a singleton $\{a\}$ by $A + a$. Consider the following linear fractional program:

$$\min \quad \frac{\sum_{e \in E_G} \mathsf{cap}(e) y(e)}{\sum_{e \in E_D} \mathsf{dem}(e) y(e)} \tag{1}$$

$$\text{st} \quad x(\{u, v\}, u) + x(\{u, v\}, v)) = y(\{u, v\}) \quad \forall u, v \in V \text{ with } u \neq v, \tag{2}$$

$$\sum_{A \subseteq S} x(S, A) = 1 \quad \forall S \in \mathcal{S}_\Theta, \tag{3}$$

$$x(S, A) \geq 0 \quad \forall S \in \mathcal{S}_\Theta \text{ and } A \subseteq S, \tag{4}$$

$$x(S + u, A) + x(S + u, A + u) = x(S, A) \quad \begin{matrix} \forall S \subseteq V, u \notin S \text{ such that} \\ S + u \in \mathcal{S}_\Theta \text{ and } A \subseteq S. \end{matrix} \tag{5}$$

The feasible region of this linear program is a polytope encoding the cuts in V. Indeed, given any cut C, define $U_j = 1$ if $j \in C$ and zero otherwise. For every $S \in \mathcal{S}_\Theta$ and $A \subseteq S$, define $x(S, A) = \prod_{j \in A} U_j \prod_{j \in S \setminus A}(1 - U_j)$ and $y(\{u, v\}) = U_u(1 - U_v) + U_v(1 - U_u)$. The solution (x, y) satisfies conditions (2)–(5). We remark that (5) is valid for every cut since given a subset S and a node $u \notin S$, the intersection between $S + u$ and a cut C is either $C \cap S$ or $(C \cap S) + u$, which are the two possibilities in the left hand side of (5). Since for every bag $Y \in \mathcal{Y}$ the size of $Y \cup \mathcal{V}_\Theta^Y$ is $O(2^\ell k + k \log(n)/\ell)$, we get

$$|\mathcal{S}_\Theta(Y,Z)| = 2^{O(k(2^\ell + \log(n)/\ell))} \text{ for any pair of bags } Y, Z \in \mathcal{Y},$$

$$|\mathcal{S}_\Theta| \le \sum_{Y,Z \in \mathcal{Y}} |\mathcal{S}_\Theta(Y,Z)| = n^2 2^{O(k(2^\ell + \log(n)/\ell))},$$

and therefore the number of variables and constraints in the linear fractional program is

$$O\left(|\mathcal{S}_\Theta| \cdot 2^{\max\{|S|:S \in \mathcal{S}_\Theta\}}\right) = n^2 2^{O(k(2^\ell + \log(n)/\ell))}.$$

By using a standard reformulation, the linear fractional program (1)–(5) can be solved by a linear program with one additional variable and constraint [6].

3.2   The Rounding Algorithm

In this section we describe our algorithm for the non-uniform sparsest cut problem. Before stating the algorithm, we introduce an object that will be used in the analysis. Recall that G is of treewidth k and $\Theta = (\mathcal{Y}, \mathcal{E})$ is a (k, ℓ)-decomposition of G.

Definition 2. *Given a feasible solution (x, y) satisfying (2)–(5), we define the function given by $f^{\mathcal{R}}_{x,\Theta}(A) = x(\mathcal{R}, A)$ for every $A \subseteq \mathcal{R}$, where $\mathcal{R}$ is the root bag of $\mathcal{E}$. Furthermore, given any non-root bag $Y \in \mathcal{Y}$ and a subset $T \subseteq V^Y_\Theta$ such that $x(V^Y_\Theta, T) > 0$, we define the function given by*

$$f^{T,Y}_{x,\Theta}(A) = \frac{x(V^Y_\Theta \cup Y, T \cup A)}{x(V^Y_\Theta, T)} \tag{6}$$

for every $A \subseteq Y \setminus \mu(Y)$, where $\mu(Y)$ is the parent of Y (see Fig. 1).

The functions introduced in Definition 2 have a probabilistic interpretation that will be at the basis of our rounding algorithm. The structure provided by constraints (3)–(5) induces probability distributions over subsets of a bag in the decomposition Θ. For a bag Y, the value (6) can be interpreted as a conditional probability given the choice of $T \subseteq V^Y_\Theta$. The following proposition summarizes these properties.

Proposition 1. *Consider an instance $(G, D, \mathsf{cap}, \mathsf{dem})$ with G of treewidth k and let ℓ be a positive integer. Let $\Theta = (\mathcal{Y}, \mathcal{E})$ be a (k, ℓ)-decomposition of the graph G and let (x, y) be a solution satisfying (2)–(5). Then, the following holds:*

(a) Let $L, I \in \mathcal{S}_\Theta$ such that $L \subseteq I$. Then, for every $C \subseteq L$, we have $x(L, C) = \sum_{I' \subseteq I \setminus L} x(I, C \cup I')$.

(b) $\sum_{A \subseteq \mathcal{R}} f^{\mathcal{R}}_{x,\Theta}(A) = 1$.

(c) For every non-root bag $Y \in \mathcal{Y}$ and $T \subseteq V^Y_\Theta$ such that $x(V^Y_\Theta, T) > 0$, we have $\sum_{A \subseteq Y \setminus \mu(Y)} f^{T,Y}_{x,\Theta}(A) = 1$.

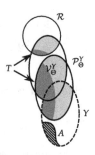

Fig. 1. Sets used in Definition 2. The set Y is the ellipse with dashed boundary. $\mathcal{P}_\Theta^Y$ is the set of bags in the path from Y to the root $\mathcal{R}$. The set $\mathcal{V}_\Theta^Y$ contains all areas depicted in gray (both light and dark gray). The dark gray part of $\mathcal{V}_\Theta^Y$ is T. The hatched subset of Y is A.

The proof of Proposition 1 can be found in the full version of the article [13]. We first design a randomized algorithm to show the existence of 2-approximation by rounding an optimal solution of the linear fractional program (1)–(5) defined by a (k,ℓ)-decomposition Θ. We start by constructing a solution at the root level, and then by conditioning on this assignment we construct a solution for the children, and we continue this propagation process until we recover an integral solution. Theorem 1 is finally obtained by optimizing the running time of our algorithm as a function of ℓ, and by performing a derandomization to get a deterministic 2-approximation algorithm. We provide the detailed randomized algorithm below.

Algorithm 1. Randomized Rounding

Input: $(G, D, \mathsf{cap}, \mathsf{dem})$ with G of treewidth k and a positive integer number ℓ.
Output: A cut in the nodes V.
1: Compute a (k,ℓ)-decomposition $\Theta = (\mathcal{Y}, \mathcal{E})$ of G.
2: Let (x, y) be an optimal solution of (1)-(5).
3: Sample a subset $B_\mathcal{R} \subseteq \mathcal{R}$ according to the probability distribution $f_{x,\Theta}^\mathcal{R}$ and let $H_\mathcal{R} = \emptyset$.
4: **for** $\ell = 1$ to depth($\mathcal{E}$) **do**
5:    For every bag Y of level ℓ in the tree $\mathcal{E}$, let $H_Y = H_{\mu(Y)} \cup (B_{\mu(Y)} \cap \mathcal{J}_Y)$.
6:    Sample a subset of nodes $B_Y \subseteq Y \setminus \mu(Y)$ according to the probability distribution $f_{x,\Theta}^{H_Y, Y}$.
7: Return $\mathcal{B} = \bigcup_{Y \in \mathcal{Y}} B_Y$.

For a bag $Y \in \mathcal{Y}$, the set $B_{\mu(Y)} \cap \mathcal{J}_Y$ is a subset of the adhesion of Y, and the set H_Y collects the union of these subsets in the path of Θ that goes from the root to Y. Then, the set $B_Y \subseteq Y \setminus \mu(Y)$ is sampled according to a conditional probability that depends on H_Y. The output of Algorithm 1 is a random subset of nodes in V and we denote by $\mathsf{P}_{x,\Theta}$ the probability measure induced by this random set-valued variable. The following lemmas summarize some properties of the algorithm. The proof of Lemma 2 can be found in the full version of the article [13].

Lemma 2. *Consider* $(G, D, \mathsf{cap}, \mathsf{dem})$ *with* G *of treewidth* k *and let* ℓ *be a positive integer. Let* $\Theta = (\mathcal{Y}, \mathcal{E})$ *be a* (k, ℓ)-decomposition *of* G *and let* (x, y) *be a solution satisfying* (2)–(5). *Then, the following holds:*

(a) *For every* $Y \in \mathcal{Y}$ *and every* $S \subseteq Y \cup \mathcal{V}_\Theta^Y$, *we have* $\mathsf{P}_{x,\Theta}(\mathcal{B} \cap S = T) = x(S, T)$ *for every* $T \subseteq S$.
(b) *For every edge* $e \in E_G$ *in the supply graph, we have* $\mathsf{P}_{x,\Theta}(|e \cap \mathcal{B}| = 1) = y(e)$.

Lemma 3. *Consider* $(G, D, \mathsf{cap}, \mathsf{dem})$ *with* G *of treewidth* k *and let* ℓ *be a positive integer. Let* $\Theta = (\mathcal{Y}, \mathcal{E})$ *be a* (k, ℓ)-decomposition *of* G *and let* (x, y) *be a solution satisfying* (2)–(5). *Then, for every edge* $e \in E_D$ *in the demand graph we have* $\mathsf{P}_{x,\Theta}(|e \cap \mathcal{B}| = 1) \geq y(e)/2$.

Proof. Let $e = \{s, t\} \in E_D$ be a demand edge. When $e \in E_G$ we are done since $\mathsf{P}_{x,\Theta}(|e \cap \mathcal{B}| = 1) = y(e)$ by Lemma 2 (b). Suppose in what follows that $e \notin E_G$, and let Y_s and Y_t be the least depth bags in the tree $\mathcal{E}$ such that $s \in Y_s$ and $t \in Y_t$. Furthermore, let Y be the lowest common ancestor of the bags Y_s, Y_t in the tree $\mathcal{Y}$. Let $\mathcal{C}_e = (Y_s \cup \mathcal{V}_\Theta^{Y_s}) \cup (Y_t \cup \mathcal{V}_\Theta^{Y_t})$. For every $T \subseteq \mathcal{C}_e$ consider the value $g_e(T) = x(\mathcal{C}_e, T)$. Since x satisfies (3), we have that $\sum_{T \subseteq \mathcal{C}_e} g_e(T) = 1$ and therefore g_e defines a probability mass function over $\mathcal{S}_\Theta(Y_s, Y_t)$, which is the power set of $\mathcal{C}_e$. Consider the set-valued random variable W distributed according to g_e and let Q_e the probability measure induced by this random variable. Then, we have

$$\mathsf{Q}_e(|e \cap W| = 1) = \mathsf{Q}_e(e \cap W = \{s\}) + \mathsf{Q}_e(e \cap W = \{t\})$$
$$= \sum_{C' \subseteq \mathcal{C}_e \setminus \{e\}} x(\mathcal{C}_e, s + C') + \sum_{C' \subseteq \mathcal{C}_e \setminus \{e\}} x(\mathcal{C}_e, t + C')$$
$$= x(e, s) + x(e, t) = y(e),$$

where the third equality holds by Proposition 1 (a) and the last equality holds since x satisfies condition (2). Let Z_s and Z_t be the children bags of Y such that Z_s belongs to unique path from Y_s to the root $\mathcal{R}$ and Z_t belongs to unique path from Y_t to the root $\mathcal{R}$, in the tree $\mathcal{E}$. Define the set $\Lambda = \mathcal{V}_\Theta^Y \cup \mathcal{J}_{Z_s} \cup \mathcal{J}_{Z_t}$. Observe that

$$\mathsf{P}_{x,\Theta}(|e \cap \mathcal{B}| = 1) = \sum_{T \subseteq \Lambda} \mathsf{P}_{x,\Theta}(|e \cap \mathcal{B}| = 1 \mid \mathcal{B} \cap \Lambda = T) \cdot \mathsf{P}_{x,\Theta}(\mathcal{B} \cap \Lambda = T)$$
$$= \sum_{T \subseteq \Lambda} \mathsf{P}_{x,\Theta}(|e \cap \mathcal{B}| = 1 \mid \mathcal{B} \cap \Lambda = T) \cdot x(\Lambda, T),$$

where the last equality holds by Lemma 2 (a) and the fact that $\Lambda \subseteq \mathcal{V}_\Theta^Y \cup Y$. On the other hand, for any $L \subseteq W$ and every $I \subseteq L$ we have

$$\mathsf{Q}_e(W \cap L = I) = \sum_{C' \subseteq \mathcal{C}_e \setminus L} x(\mathcal{C}_e, I \cup C') = x(L, I), \tag{7}$$

where the last equality holds by Proposition 1 (a). Therefore, we have

$$y(e) = Q_e(|e \cap W| = 1) = \sum_{T \subseteq \Lambda} Q_e(|e \cap W| = 1 \mid W \cap \Lambda = T) \cdot Q_e(W \cap \Lambda = T)$$

$$= \sum_{T \subseteq \Lambda} Q_e(|e \cap W| = 1 \mid W \cap \Lambda = T) \cdot x(\Lambda, T),$$

where the last equality holds by applying (7) with $L = \Lambda$. Then, in order to conclude the lemma it is sufficient to show that $Q_e(|e \cap W| = 1 \mid W \cap \Lambda = T) \leq 2 \cdot P_{x,\Theta}(|e \cap B| = 1 \mid B \cap \Lambda = T)$. Given $T \subseteq \Lambda$, consider the random variable $\omega_{s,T} \in \{0,1\}$ that indicates whether $s \in W$ given $W \cap \Lambda = T$, and let $\beta_{s,T} \in \{0,1\}$ be the random variable that indicates whether $s \in B$ given $B \cap \Lambda = T$. We define analogously the random variables $\omega_{t,T}$ and $\beta_{t,T}$. Since $s, t \notin \Lambda$, we observe that for any $T \subseteq \Lambda$ and $v \in \{s,t\}$ it holds that $v \in W$ and $W \cap \Lambda = T$ if and only if $W \cap (\Lambda + v) = T + v$. Therefore, for every $T \subseteq \Lambda$, we have that

$$Q_e(\omega_{s,T} = 1) = \frac{Q_e(s \in W, W \cap \Lambda = T)}{Q_e(W \cap \Lambda = T)} = \frac{x(\Lambda + s, T + s)}{x(\Lambda, T)} = P_{x,\Theta}(\beta_{s,T} = 1),$$

$$Q_e(\omega_{t,T} = 1) = \frac{Q_e(t \in W, W \cap \Lambda = T)}{Q_e(W \cap \Lambda = T)} = \frac{x(\Lambda + t, T + t)}{x(\Lambda, T)} = P_{x,\Theta}(\beta_{t,T} = 1),$$

where, in both cases, the first equality comes from the above observation and (7) and the second equality is a consequence of the above observation and Proposition 1 (a). We conclude that for every $T \subseteq \Lambda$ the random variables $\omega_{v,T}$ and $\beta_{v,T}$ are identically distributed, for $v \in \{s,t\}$.

Claim. Suppose we have two random variables G and K, not necessarily independent, and taking values in $\{0,1\}$. Then, we have $\Pr(G \neq K) \leq 2(\Pr(G = 1) \Pr(K = 0) + \Pr(G = 0) \Pr(K = 1))$.

The proof of the claim can be found in the full version of the article [13]. We show how to conclude the lemma using the claim. Taking $G = \omega_{s,T}$ and $K = \omega_{t,T}$, we have

$$Q_e(|e \cap W| = 1 \mid W \cap \Lambda = T)$$
$$= Q_e(\omega_{s,T} \neq \omega_{t,T})$$
$$\leq 2(Q_e(\omega_{s,T} = 1)Q_e(\omega_{t,T} = 0) + Q_e(\omega_{s,T} = 0)Q_e(\omega_{t,T} = 1))$$
$$= 2(P_{x,\Theta}(\beta_{s,T} = 1)P_{x,\Theta}(\beta_{t,T} = 0) + P_{x,\Theta}(\beta_{s,T} = 0)P_{x,\Theta}(\beta_{t,T} = 1))$$
$$= 2 \cdot P_{x,\Theta}(\beta_{s,T} \neq \beta_{t,T}) = 2 \cdot P_{x,\Theta}(|e \cap B| = 1 \mid B \cap \Lambda = T),$$

which concludes the lemma. □

Definition 3. *Let G be a graph of treewidth k, let $\Theta = (\mathcal{Y}, \mathcal{E})$ be a (k, ℓ)-decomposition of G, and consider a node $u \in V$. Let X be the least depth bag in the tree containing the node u. Given a bag $Z \in \mathcal{P}_\Theta^X$ and $H \subseteq \mathcal{V}_\Theta^Z \cup Z$, we say that a pair (M, N) is an H-extension for the node u if the following holds:*

(i) $N \subseteq X \setminus V_\Theta^X$ and $u \in N$,
(ii) $M = (H \cap V_\Theta^X) \cup L$ where $L \subseteq V_\Theta^X \setminus (V_\Theta^Z \cup Z)$.

We denote by $\Delta_\Theta(H, u)$ the set of H-extensions for u.

Observe that for any node u and X being the least depth bag containing u, for any bag $Z \in \mathcal{P}_\Theta^X$ and any $H \subseteq V_\Theta^Z \cup Z$, the set $\Delta_\Theta(H, u)$ has cardinality at most

$$2^{O(k(2^\ell + \log(n)/\ell))} \tag{8}$$

when Θ is a (k, ℓ)-decomposition. This holds since, by Lemma 1, we have $|X \setminus V_\Theta^X| \in O(2^\ell k)$ and $|V_\Theta^X \setminus (V_\Theta^Z \cup Z)| \in O(k \log(n)/\ell)$. We need one more lemma before proving Theorem 1. The proof of Lemma 4 can be found in the full version of the article [13].

Lemma 4. *For every positive real value $x \geq 4$, there exists a unique value α^* such that alpha $\alpha^* 2^{\alpha^*} = x$, and it satisfies the inequality $2^{\lceil \alpha^* \rceil} + x/\lceil \alpha^* \rceil \leq 12x/\log(x)$.*

Proof (Proof of Theorem 1). Let $\mathcal{I} = (G, D, \mathsf{cap}, \mathsf{dem})$ be an instance of the non-uniform sparsest cut problem. Recall that we denote by n the number of nodes in the instance. Let α_n^* be the unique positive real solution of the equation $\alpha 2^\alpha = \log(n)$ and let $\ell^* = \lceil \alpha_n^* \rceil$. We run Algorithm 1 over the instance $\mathcal{I}$, using the value ℓ^*, and let Θ be the (k, ℓ^*)-decomposition computed in step 1 of the algorithm. Let (x, y) be an optimal solution of the optimization problem (1)–(5) solved in step 2 of the algorithm, and we denote by $\mathsf{opt}_{\mathsf{LP}}$ the optimal value $\sum_{e \in E_G} \mathsf{cap}(e)y(e) / \sum_{e \in E_D} \mathsf{dem}(e)y(e)$. Let $\mathcal{B}$ be the solution computed by the randomized algorithm. For every pair of nodes $e = \{u, v\} \subseteq V$, with $u \neq v$, let $\xi(e)$ be equal to one if $|e \cap \mathcal{B}| = 1$ and zero otherwise. This random variable indicates when a pair of nodes is cut by the algorithm solution. Consider $\mathcal{C} = \sum_{e \in E_G} \mathsf{cap}(e)\xi(e)$ and $\mathcal{D} = \sum_{e \in E_D} \mathsf{dem}(e)\xi(e)$. By Lemmas 2 (b) and 3 we have that

$$\mathsf{E}_{x,\Theta}(\mathcal{C}) = \sum_{e \in E_G} \mathsf{cap}(e) \cdot \mathsf{P}_{x,\Theta}(|e \cap \mathcal{B}| = 1) = \sum_{e \in E_G} \mathsf{cap}(e)y(e),$$

$$\mathsf{E}_{x,\Theta}(\mathcal{D}) = \sum_{e \in E_D} \mathsf{dem}(e) \cdot \mathsf{P}_{x,\Theta}(|e \cap \mathcal{B}| = 1) \geq \frac{1}{2} \sum_{e \in E_D} \mathsf{dem}(e)y(e),$$

and therefore we get $\mathsf{E}_{x,\Theta}(\mathcal{C})/\mathsf{E}_{x,\Theta}(\mathcal{D}) \leq 2 \cdot \mathsf{opt}_{\mathsf{LP}} \leq 2 \cdot \min_{S \subseteq V} \phi(S)$, where the last inequality holds since the sparsest cut of value $\min_{S \subseteq V} \phi(S)$ defines a feasible solution for (1)–(5). We now show how to derandomize the solution $\mathcal{B}$ to get a deterministic 2-approximation. We use the method of conditional expectations. Define the random variable $\Gamma = \mathcal{C} - 2\mathcal{D} \cdot \mathsf{opt}_{\mathsf{LP}}$. Then, we have that $0 \geq \mathsf{E}_{x,\Theta}(\Gamma) = \mathsf{E}(\mathsf{E}_{x,\Theta}(\Gamma | B_\mathcal{R}))$ and therefore there exists $R' \subseteq \mathcal{R}$ such that $\mathsf{E}_{x,\Theta}(\Gamma | B_\mathcal{R} = R') \leq 0$. Fix any subset $Y_1' \subseteq \mathcal{R}$ with $\mathsf{E}_{x,\Theta}(\Gamma | B_\mathcal{R} = Y_1') \leq 0$ and let $\mathcal{R} = Y_1, Y_2, \ldots, Y_{|\mathcal{Y}|}$ be the bags visited according to some BFS ordering. Suppose we have computed for some $t \in \{1, \ldots, |\mathcal{Y}| - 1\}$ the set $A_t = \cup_{\ell=1}^t Y_\ell' \subseteq \cup_{\ell=1}^t Y_\ell$, with $Y_\ell' \subseteq Y_\ell \setminus \mu(Y_\ell)$

for each $\ell \in \{1, \ldots, t\}$, and such that $\mathsf{E}_{x,\Theta}(\Gamma | \mathcal{B} \cap (\cup_{\ell=1}^{t} Y_\ell) = A_t) \leq 0$. Then, we have

$$0 \geq \mathsf{E}_{x,\Theta}(\Gamma | \mathcal{B} \cap (\cup_{\ell=1}^{t} Y_\ell) = A_t)$$
$$= \sum_{Y' \subseteq Y_{t+1} \setminus \mu(Y_{t+1})} \mathsf{E}_{x,\Theta}(\Gamma | \mathcal{B} \cap (\cup_{\ell=1}^{t} Y_\ell) = A_t, B_{Y_{t+1}} = Y') \cdot \mathsf{P}_{x,\Theta}(B_{Y_{t+1}} = Y')$$
$$= \sum_{Y' \subseteq Y_{t+1} \setminus \mu(Y_{t+1})} \mathsf{E}_{x,\Theta}(\Gamma | \mathcal{B} \cap (\cup_{\ell=1}^{t+1} Y_\ell) = A_t \cup Y') \cdot \mathsf{P}_{x,\Theta}(B_{Y_{t+1}} = Y'),$$

and therefore there exists $Y' \subseteq Y_{t+1} \setminus \mu(Y_{t+1})$ such that $\mathsf{E}_{x,\Theta}(\Gamma | \mathcal{B} \cap (\cup_{\ell=1}^{t+1} Y_\ell) = A_t \cup Y') \leq 0$. Fix any of these subsets and we denote it by Y'_{t+1}. By the end of this process, let $\mathcal{A}$ be the union of $Y'_1, \ldots, Y'_{|\mathcal{Y}|}$. By construction, we have recovered a solution such that $\mathsf{E}_{x,\Theta}(\Gamma | \mathcal{B} = \mathcal{A}) \leq 0$ and therefore $\mathcal{A}$ is a 2-approximation.

We now study the running time of the derandomization, and more specifically, the running time that we need to compute the conditional expectations. Let $t \in \{1, \ldots, |\mathcal{Y}|\}$ and let $T \subseteq \cup_{\ell=1}^{t} Y_\ell$. To compute the value of the expectation $\mathsf{E}_{x,\Theta}(\Gamma | \mathcal{B} \cap (\cup_{\ell=1}^{t} Y_\ell) = T)$, it is sufficient to compute the probability value $\mathsf{P}_{x,\Theta}(|e \cap \mathcal{B}| = 1 | \mathcal{B} \cap (\cup_{\ell=1}^{t} Y_\ell) = T)$ for any $e \in E_G$ or $e \in E_D$. Furthermore, when $e \subseteq \cup_{\ell=1}^{t} Y_\ell$ the value of the probability is determined and equal to one or zero. Then, we suppose that $e = \{u, v\}$ is not contained in $\cup_{\ell=1}^{t} Y_\ell$. For every node $a \in V \setminus (Y_1 \cup \cdots \cup Y_t)$ let X_a be the least depth bag in $\mathcal{Y}$ that contains a. In particular, we have that $X_a \notin \{Y_1, \ldots, Y_t\}$ and let Z_a be the lowest bag in $\{Y_1, \ldots, Y_t\}$ such that Z_a belongs to the path from X_a to the root. For every $a \in V \setminus (Y_1 \cup \cdots \cup Y_t)$ consider the quantity

$$g_a = \mathsf{P}_{x,\Theta}\left(a \in \mathcal{B} \ \middle| \ \mathcal{B} \cap (\mathcal{V}_\Theta^{Z_a} \cup Z_a) = T \cap (\mathcal{V}_\Theta^{Z_a} \cup Z_a)\right).$$

Case 1. Suppose that $u \notin \cup_{\ell=1}^{t} Y_\ell$ and $v \notin \cup_{\ell=1}^{t} Y_\ell$ and that $Z_u \neq Z_v$. Then, $\mathcal{P}_\Theta^{Z_u}$ and $\mathcal{P}_\Theta^{Z_v}$ are contained in the subtree induced by the bags $Y_1, \ldots, Y_t$. By construction in Algorithm 1 we have that $\mathsf{P}_{x,\Theta}(|e \cap \mathcal{B}| = 1 | \mathcal{B} \cap (\cup_{\ell=1}^{t} Y_\ell) = T) = g_u(1 - g_v) + g_v(1 - g_u)$. Furthermore, by denoting $T_a = T \cap (\mathcal{V}_\Theta^{Z_a} \cup Z_a)$, we have

$$g_a = \sum_{(M,N) \in \Delta_\Theta(T_a, u)} f_{x,\Theta}^{M, X_a}(N) = \sum_{(M,N) \in \Delta_\Theta(T_a, u)} \frac{x(\mathcal{V}_\Theta^{X_a} \cup X_a, M \cup N)}{x(\mathcal{V}_\Theta^{X_a}, M)}$$

for each $a \in \{u, v\}$. By the observation in (8), g_u and g_v can be computed in time $2^{O(k(2^{\ell^*} + \log(n)/\ell^*))}$.

Case 2. Suppose that $u \notin \cup_{\ell=1}^{t} Y_\ell$ and $v \notin \cup_{\ell=1}^{t} Y_\ell$, and that $Z_u = Z_v = Z$. Let W be the lowest common ancestor of Y_u and Y_v. In particular, Z is an ancestor of W and $W \notin \{Y_1, \ldots, Y_t\}$. For every $H \subseteq \mathcal{V}_\Theta^{W} \setminus (\mathcal{V}_\Theta^{Z} \cup Z)$ and $K \subseteq W \setminus \mu(W)$ consider the quantity

$$\beta(H, K) = \frac{x(\mathcal{V}_\Theta^{W} \cup W, (T \cap \mathcal{V}_\Theta^{W}) \cup H \cup K)}{x(\mathcal{V}_\Theta^{W}, (T \cap \mathcal{V}_\Theta^{W}) \cup H)}.$$

Furthermore, for every $H \subseteq \mathcal{V}_\Theta^W \setminus (\mathcal{V}_\Theta^Z \cup Z)$, $K \subseteq W \setminus \mu(W)$ and $a \in \{u, v\}$ let

$$\gamma_a(H, K) = \sum_{(M,N) \in \Delta_\Theta((T \cap \mathcal{V}_\Theta^W) \cup H \cup K, u)} \frac{x(\mathcal{V}_\Theta^{Y_a} \cup Y_a, M \cup N)}{x(\mathcal{V}_\Theta^{Y_a}, M)}$$

Then, we have that $\mathsf{P}_{x,\Theta}(|e \cap \mathcal{B}| = 1 | \mathcal{B} \cap (\cup_{\ell=1}^t Y_\ell) = T)$ is equal to

$$\sum_{H \subseteq \mathcal{V}_\Theta^W \setminus (\mathcal{V}_\Theta^Z \cup Z)} \sum_{K \subseteq W \setminus \mu(W)} \beta(H, K)\Big(\gamma_u(H, K)(1 - \gamma_v(H, K)) + \gamma_v(H, K)(1 - \gamma_u(H, K))\Big).$$

As before, the above summation can be computed in time $2^{O(k(2^{\ell^*} + \log(n)/\ell^*))}$.

Case 3. Suppose that $u \in \cup_{\ell=1}^t Y_\ell$ and $v \notin \cup_{\ell=1}^t Y_\ell$ (the other case is symmetric). In this case, we have that $\mathsf{P}_{x,\Theta}(|e \cap \mathcal{B}| = 1 | \mathcal{B} \cap (\cup_{\ell=1}^t Y_\ell) = T)$ is equal to $1 - g_v$, and therefore we can compute it in time

$$2^{O(k(2^{\ell^*} + \log(n)/\ell^*))}.$$

As we observe at the end of Sect. 3.1, the optimization problem (1)–(5) can be solved in time $2^{O(k(2^{\ell^*} + \log(n)/\ell^*))} |\mathcal{I}|^{O(1)}$. On the other hand, for every $n \geq 16$, by Lemma 4 we have

$$k2^{\ell^*} + \frac{k \log(n)}{\ell^*} = k2^{\lceil \alpha_n^* \rceil} + \frac{k \log(n)}{\lceil \alpha_n^* \rceil} \leq \frac{12k \log(n)}{\log \log(n)},$$

and therefore, the randomized algorithm and the derandomization can be all performed in time

$$2^{O\left(k \frac{\log(n)}{\log \log(n)}\right)} |\mathcal{I}|^{O(1)} = 2^{2^{O(k)}} |\mathcal{I}|^{O(1)}.$$

To finish the proof, we verify the above equality by considering two cases. If $k < \log \log(n)$, we have $k \log(n) / \log \log(n) < \log(n)$ and the equality holds. Otherwise, if $k \geq \log \log(n)$ and $n \geq 4$ we have $k \log(n) / \log \log(n) \leq k \log(n) = 2^{\log(k) + \log \log(n)} \leq 2^{\log(k) + k} = 2^{O(k)}$. $\qquad\square$

References

1. Aprile, M., Drescher, M., Fiorini, S., Huynh, T.: A tight approximation algorithm for the cluster vertex deletion problem. In: Integer Programming and Combinatorial Optimization (IPCO) (2021)
2. Arora, S., Lee, J., Naor, A.: Euclidean distortion and the sparsest cut. J. Am. Math. Soc. **21**(1), 1–21 (2008)
3. Arora, S., Rao, S., Vazirani, U.V.: Expander flows, geometric embeddings and graph partitioning. J. ACM **56**(2), 5:1–5:37 (2009)

4. Bienstock, D., Munoz, G.: LP formulations for polynomial optimization problems. SIAM J. Optim. **28**(2), 1121–1150 (2018)
5. Bienstock, D., Ozbay, N.: Tree-width and the Sherali-Adams operator. Discret. Optim. **1**(1), 13–21 (2004)
6. Boyd, S., Vandenberghe, L.: Convex Optimization. Cambridge University Press, Cambridge (2004)
7. Chakrabarti, A., Jaffe, A., Lee, J.R., Vincent, J.: Embeddings of topological graphs: lossy invariants, linearization, and 2-sums. In: IEEE Symposium on Foundations of Computer Science (FOCS) (2008)
8. Chalermsook, P., Kaul, M., Mnich, M., Spoerhase, J., Uniyal, S., Vaz, D.: Approximating sparsest cut in low-treewidth graphs via combinatorial diameter. CoRR 2111.06299 (2021)
9. Chawla, S., Krauthgamer, R., Kumar, R., Rabani, Y., Sivakumar, D.: On the hardness of approximating multicut and sparsest-cut. Comput. Complex. **15**(2), 94–114 (2006)
10. Chekuri, C., Shepherd, F.B., Weibel, C.: Flow-cut gaps for integer and fractional multiflows. J. Comb. Theory Ser. B **103**(2), 248–273 (2013)
11. Chlamtac, E., Krauthgamer, R., Raghavendra, P.: Approximating sparsest cut in graphs of bounded treewidth. In: Serna, M., Shaltiel, R., Jansen, K., Rolim, J. (eds.) APPROX/RANDOM 2010. LNCS, vol. 6302, pp. 124–137. Springer, Heidelberg (2010). https://doi.org/10.1007/978-3-642-15369-3_10
12. Cohen-Addad, V., Gupta, A., Klein, P.N., Li, J.: A quasipolynomial $(2 + \varepsilon)$-approximation for planar sparsest cut. In: ACM Symposium on Theory of Computing (STOC) (2021)
13. Cohen-Addad, V., Mömke, T., Verdugo, V.: A 2-approximation for the bounded treewidth sparsest cut problem in FPT time. CoRR 2111.06163 (2021)
14. Davies, S., Kulkarni, J., Rothvoss, T., Tarnawski, J., Zhang, Y.: Scheduling with communication delays via LP hierarchies and clustering ii: weighted completion times on related machines. In: ACM-SIAM Symposium on Discrete Algorithms (SODA) (2021)
15. Garg, S.: Quasi-PTAS for scheduling with precedences using LP hierarchies. In: International Colloquium on Automata, Languages, and Programming (ICALP) (2018)
16. Gupta, A., Newman, I., Rabinovich, Y., Sinclair, A.: Cuts, trees and $\ell1$-embeddings of graphs. Combinatorica **24**(2), 233–269 (2004)
17. Gupta, A., Talwar, K., Witmer, D.: Sparsest cut on bounded treewidth graphs: algorithms and hardness results. In: ACM Symposium on Theory of Computing (STOC) (2013)
18. Khot, S.A., Vishnoi, N.K.: The unique games conjecture, integrality gap for cut problems and embeddability of negative-type metrics into $\ell1$. J. ACM **62**(1), 1–39 (2015)
19. Klein, P., Agrawal, A., Ravi, R., Rao, S.: Approximation through multicommodity flow. In: IEEE Symposium on Foundations of Computer Science (FOCS) (1990)
20. Klein, P., Rao, S., Agrawal, A., Ravi, R.: An approximate max-flow min-cut relation for undirected multicommodity flow, with applications. Combinatorica **15**(2), 187–202 (1995)
21. Lee, J.R., Raghavendra, P.: Coarse differentiation and multi-flows in planar graphs. Discret. Comput. Geom. **43**(2), 346–362 (2010)
22. Lee, J.R., Sidiropoulos, A.: Pathwidth, trees, and random embeddings. Combinatorica **33**(3), 349–374 (2013). https://doi.org/10.1007/s00493-013-2685-8

23. Magen, A., Moharrami, M.: Robust algorithms for on minor-free graphs based on the Sherali-Adams hierarchy. In: Dinur, I., Jansen, K., Naor, J., Rolim, J. (eds.) APPROX/RANDOM 2009. LNCS, vol. 5687, pp. 258–271. Springer, Heidelberg (2009). https://doi.org/10.1007/978-3-642-03685-9_20

24. Maiti, B., Rajaraman, R., Stalfa, D., Svitkina, Z., Vijayaraghavan, A.: Scheduling precedence-constrained jobs on related machines with communication delay. In: IEEE Symposium on Foundations of Computer Science (FOCS) (2020)

25. Matula, D.W., Shahrokhi, F.: Sparsest cuts and bottlenecks in graphs. Discret. Appl. Math. **27**(1–2), 113–123 (1990)

26. Okamura, H., Seymour, P.D.: Multicommodity flows in planar graphs. J. Comb. Theory Seri. B **31**(1), 75–81 (1981)

27. Sherali, H.D., Adams, W.P.: A hierarchy of relaxations between the continuous and convex hull representations for zero-one programming problems. SIAM J. Discret. Math. **3**(3), 411–430 (1990)

28. Verdugo, V., Verschae, J., Wiese, A.: Breaking symmetries to rescue sum of squares in the case of makespan scheduling. Math. Program. (3), 583–618 (2020). https://doi.org/10.1007/s10107-020-01511-3

Optimal Item Pricing in Online Combinatorial Auctions

José Correa[1], Andrés Cristi[1(✉)], Andrés Fielbaum[2], Tristan Pollner[3], and S. Matthew Weinberg[4]

[1] Universidad de Chile, Santiago, Chile
andres.cristi@ing.uchile.cl
[2] TU Delft, Delft, The Netherlands
[3] Stanford University, Stanford, USA
[4] Princeton University, Princeton, USA

Abstract. We consider a fundamental pricing problem in combinatorial auctions. We are given a set of indivisible items and a set of buyers with randomly drawn monotone valuations over subsets of items. A decision maker sets item prices and then the buyers make sequential purchasing decisions, taking their favorite set among the remaining items. We parametrize an instance by d, the size of the largest set a buyer may want. Our main result asserts that there exist prices such that the expected (over the random valuations) welfare of the allocation they induce is at least a factor $1/(d+1)$ times the expected optimal welfare in hindsight. Moreover we prove that this bound is tight. Thus, our result not only improves upon the $1/(4d-2)$ bound of Dütting et al., but also settles the approximation that can be achieved by using item prices. We further show how to compute our prices in polynomial time. We provide additional results for the special case when buyers' valuations are known (but a posted-price mechanism is still desired).

Keywords: Combinatorial Auctions · Online allocations

1 Introduction

In combinatorial auctions, a set of valuable items is to be allocated among a set of interested agents. Who should get which items in order to maximize the social welfare? This is a fundamental economic question, and a ubiquitous allocation mechanism is to simply set a price for each item and let the agents buy their preferred subset of items under those prices. The study of these mechanisms dates back to the investigations of Leon Walras over a century ago, and is closely related to the notion of Walrasrian equilibrium. Understanding the existence and approximation of Walrasrian equilibrium and related notions under pricing mechanisms has been an active area of research in recent years [3,4,13,14,20].

In this paper, we follow the approach of online combinatorial auctions and study the welfare achieved by posted-price mechanisms in a very general setup.

© Springer Nature Switzerland AG 2022
K. Aardal and L. Sanitá (Eds.): IPCO 2022, LNCS 13265, pp. 126–139, 2022.
https://doi.org/10.1007/978-3-031-06901-7_10

Specifically, our mechanisms post a price p_i on each item i. Then, buyers with randomly-drawn monotone valuations over the subsets of items arrive in arbitrary order, and upon arrival pick their preferred subset among those items that are left (at the posted prices). Of course, in this generality little can be said about the social welfare induced by posted-price mechanisms, so it is common to parametrize the instances by d, the largest size of a set a buyer might be interested in.[1] This parametrization is interesting from a combinatorial perspective: finding a socially optimal allocation is NP-hard already when $d \geq 3$.[2] Moreover, if we restrict the buyers' valuations to be deterministic and single-minded,[3] we recover the classic hypergraph matching problem.

Our main result in this paper is to determine the tight approximation guarantee of item pricing as a function of d. Specifically, we prove that there always exist a posted-price mechanism such that the expected welfare of the resulting allocation, when buyers arrive in adversarial order and iteratively purchase their preferred set, is at least a $1/(d+1)$ fraction of the expected welfare of an optimal allocation (Theorem 1). Furthermore, we prove this bound is tight (Proposition 1).

Interestingly, our result generalizes and/or improves upon several results in the literature, which we now provide context for.

1.1   Context and Related Work

Posted-Price Mechanisms. Posted-price mechanisms are ubiquitous within the economics and computation literature due to their simplicity. They are commonly used as subroutines in truthful mechanisms that approximately maximize welfare [1,2,8,9,19]. They are also used as subroutines in simple mechanisms to approximately maximize revenue in Bayesian settings [5–7,18]. Our work considers the same model (welfare maximization in Bayesian settings) initiated by Feldman et al. [13]. Other works consider restrictions on the valuations such as subadditive [11], while others consider the unrestricted case [10]. Our paper contributes to this line of work by nailing the tight approximation guarantee of posted-price mechanisms in this model for unrestricted valuations over sets of size at most d. In particular, our results improve the bound of $1/(4d-2)$ of Dütting et al. [10], to $1/(d+1)$, which is tight.

Prophet Inequalities. When there is a single item (and thus $d = 1$) our problem is equivalent to the single-item prophet inequality and thus our result takes the same form as the classic result of Samuel-Cahn [21], who proved that the optimal prophet inequality (whose factor is $1/2$) can be achieved with a single threshold. A special case of our problem when buyers are single-minded corresponds to various multiple-choice prophet inequality settings, and our results improve upon the state-of-the-art. In particular, all prophet inequalities deduced from our main result are *non-adaptive*: for each element e, a threshold T_e is set

---

[1] That is, for all sets A with $|A| > d$, $v(A) := \max_{B \subset A, |B| \leq d}\{v(B)\}$.
[2] And quite hard to approximate [22].
[3] That is, each buyer has a fixed set T, and values all sets S at $v(S) := I(T \subseteq S) \cdot v(T)$.

at the beginning of the algorithm. Element e is accepted if and only if $w_e \geq T_e$ (and it is feasible to accept e).

When $d = 2$ and buyers are single-minded, our problem translates into the matching prophet inequality problem. Our results when $d = 2$ therefore extend the 1/3-approximation of Gravin and Wang [16] from bipartite to general graphs. Note that recent work of Ezra et al. [12] provides a .337-approximation in this case, although it sets thresholds adaptively. In the full version, we further contribute to the $d = 2$ case by proving that no prophet inequality (adaptive or not) can guarantee better than a 3/7-approximation for the bipartite graph prophet inequality.

For arbitrary d when buyers are single-minded, our problem translates into the d-dimensional hypergraph prophet inequality, which generalizes the prophet inequality problem over the intersection of d partition matroids. Here, a $1/(4d - 2)$-approximation was first given by Kleinberg and Weinberg [18], and improved to $1/(e(d + 1))$ by Feldman et al. [15]. A corollary of our main result improves this to $1/(d+1)$, and with non-adaptive thresholds. A lower bound of Kleinberg and Weinberg [18] proves that it is not possible to achieve an $\omega(1/\sqrt{d})$ approximation even for this special case, but it remains an open problem to determine the tight ratio for prophet inequalities for the intersection of d partition matroids (and for the d-dimensional hypergraph prophet inequality).

1.2   A Technical Highlight and Additional Results

The proof of our main result breaks down the expected welfare into the "revenue" and "utility" achieved by setting prices, and searches for properly "balanced thresholds" as in [10,13,16,18]. In particular, we target prices that are "low enough" so that a buyer with high value for some set will choose to purchase it, yet also "high enough" so that the revenue gained when a bidder purchases items they should not receive in the optimal allocation compensates for the lost welfare. In comparison to prior work using a similar approach, the conditions that guarantee such prices are more involved, and we prove their existence using Brouwer's fixed point theorem.

As our proof makes use of Brouwer's fixed point theorem, it is inherently non-constructive; however, in the full version, we show that the prices can be efficiently computed by making use of an LP relaxation to cope with the APX-hardness of optimizing welfare, and we further provide a convex optimization formulation to find our fixed point.

In Sect. 4, we consider the special case that arises when valuations are deterministic and buyers are single-minded. In this situation the welfare optimization problem corresponds to matching in a hypergraph with edges of size at most d. So the problem of finding item prices boils down to finding a set of thresholds, one for each vertex, such that the value of the solution in which hyperedges arrive sequentially (for any order) and greedily included in the solution so long as their weight is higher than the sum of the corresponding vertex thresholds, is as close as possible to the optimal solution. For the case of standard matching ($d = 2$) we prove that there exist prices guaranteeing a factor of 1/2 of the optimal solution and that there do not exist prices guaranteeing a factor better than $2/3$. The tight factor is left as an open problem. More generally, we prove that there are prices obtaining a fraction $1/d$ of the optimal solution (thus slightly improving our general $1/(d+1)$), and that it is not possible to do better than $\Omega(1/\sqrt{d})$.

2   Model

In our basic model, we have a (multi)set of items M in which there are $k_j \geq 1$ copies of each item $j \in M$.[4] The set of buyers, denoted by N, arrive sequentially (in arbitrary order) and buy some of those items. Each buyer $i \in N$ has a valuation function $v_i : 2^M \to \mathbb{R}_+$, which is randomly chosen according to a given distribution $\mathcal{F}_i$ (defined over a set of possible valuation functions). As is standard, we assume that each possible realization of each v_i is monotone (i.e., $A \subseteq B \Rightarrow v_i(A) \leq v_i(B)$). We parametrize an instance of the problem by d, the size of the largest set a buyer might be interested in. Thus we assume that if $A \subseteq M$ is such that $|A| > d$, then

$$v(A) = \max_{B \subseteq A, |B| = d} v(B). \tag{1}$$

Note that while there are $k_i \geq 1$ copies of each item $i \in M$, no single buyer can purchase more than one copy of an item.

In this paper, we are interested in exploring the limits of using item prices as the mechanism to assign items to buyers. In a pricing mechanism, we set item prices $p \in \mathbb{R}_+^M$ and then consider an arbitrary arrival order of the buyers.[5] Thus, buyer i buys the set of remaining items that optimizes

$$\max_{A \subseteq R_i} v_i(A) - \sum_{j \in A} p_j, \tag{2}$$

where R_i stands for the *remaining items* for which there exists at least one unsold copy when i arrives. Note that (2) might be solved by $A = \emptyset$, i.e., buyer i might opt not to buy anything. When there is a tie between different sets, the buyer can choose arbitrarily, meaning that our results need to be valid even for the worst case.[6]

More precisely, if σ is the arrival order of the buyers, so that buyer i comes at time $\sigma(i)$, then buyer i gets the set $B_i(\sigma) = \arg\max_{A \subseteq R_i(\sigma)} v_i(A) - \sum_{j \in A} p_j$, where $R_i(\sigma) = \{j \in M : k_j > |\{\ell \in N : \sigma(\ell) < \sigma(i) \text{ and } j \in B_\ell(\sigma)\}|\}$. With this, given an instance of the problem (determined by M, k_j for all $j \in M$, N, and $\mathcal{F}_i$ for all $i \in N$), the quality measure of a price vector $p \in \mathbb{R}_+^M$ is the worst case (over the arrival orders) expected (over the valuations) welfare of the allocation it induces. Denoting this quantity as $ALG(p)$ we have that

---

[4] Throughout the paper M is actually a set and refers to the set of different items.
[5] Note that different copies of the same item need to get the same price.
[6] In some of the constructions in Sect. 4 we break ties conveniently but all the results hold by slightly tweaking the instances.

$$ALG(p) := \min_{\sigma} \mathbb{E}\left(\sum_{i \in N} v_i(B_i(\sigma))\right).$$

On the other hand, the benchmark we compare to throughout the paper is the welfare maximizing allocation, OPT, which is defined as

$$OPT := \mathbb{E}\left(\max_{A_i,\, i \in N}\left\{\sum_{i \in N} v_i(A_i) : \text{ s.t. } |\{i \in N : j \in A_i\}| \le k_j, \text{ for all } j \in M\right\}\right).$$

We denote by OPT_i the random set that buyer i gets in an optimal allocation.

In Sect. 4 we consider the special case of our problem in which

(i) valuations are deterministic,
(ii) there is a single copy of each item ($k_j = 1$ for all $j \in M$), and
(iii) buyers are *single-minded*, i.e., each buyer i has a set A_i, with $|A_i| \le d$, such that $A_i \nsubseteq B \Rightarrow v_i(B) = 0, A_i \subseteq B \Rightarrow v_i(B) = v_i(A_i)$.

Interestingly, already in this particular setup, the problem of maximizing the welfare of an allocation corresponds to the classic combinatorial optimization problem of hypergraph matching with hyperedges of size at most d. Indeed, in an optimal allocation buyer i either gets A_i or $\emptyset$, implying that maximizing the (now deterministic) welfare of the allocation is equivalent to finding a subset of pairwise disjoint A_i's of maximum total valuation.

3   Random Valuations

In this section we prove there exists a vector of item prices such that the resulting allocation yields in expectation at least a $1/(d+1)$ fraction of the optimal social welfare. Additionally, we show that this bound is tight.

Theorem 1. *There exists a vector of prices $p \in \mathbb{R}_+^M$ such that*

$$(d+1) \cdot ALG(p) \ge OPT.$$

To prove the theorem we will make use of the following function. For each $A \subseteq M$ and $i \in N$, we define

$$z_{i,A}(p) = \left[\mathbb{E}(\mathbb{1}_{OPT_i=A} \cdot v_i(A)) - \mathbb{P}(OPT_i = A)\sum_{j \in A} p_j\right]_+,$$

where $[\cdot]_+$ denotes the positive part. We assume without loss of generality that $|OPT_i| \le d$ for all $i \in N$, so $z_{i,A}(p) = 0$ if $|A| > d$. We start by showing a lower bound for $ALG(p)$ in terms of the functions $z_{i,A}(p)$.

Lemma 1. *For any vector of prices $p \in \mathbb{R}_+^M$,*

$$ALG(p) \ge \min_{C \subseteq M}\left\{\sum_{j \notin C} k_j \cdot p_j + \sum_{i \in N}\sum_{A \subseteq C} z_{i,A}(p)\right\}.$$

Proof. In this proof we assume the arrival order σ is arbitrary, and for simplicity we drop the dependency of $B_i(\sigma)$ and $R_i(\sigma)$ on σ and simply denote them by B_i and R_i. We separate the welfare of the resulting allocation into *revenue* and *utility*, i.e., we separate $\sum_{i \in N} v_i(B_i)$ into

$$\text{Revenue} = \sum_{i \in N} \sum_{j \in B_i} p_j \quad \text{and} \quad \text{Utility} = \sum_{i \in N} \left(v_i(B_i) - \sum_{j \in B_i} p_j \right).$$

Recall that R_i is the set of items such that there are remaining copies when i arrives. Similarly, denote by R the set of items that have remaining copies by the end of the process. We have that

$$\mathbb{E}(\text{Revenue}) \geq \mathbb{E} \left(\sum_{j \notin R} k_j \cdot p_j \right).$$

For the utility, for any $i \in N$, by the definition of B_i, it holds that

$$v_i(B_i) - \sum_{j \in B_i} p_j = \max_{A \subseteq R_i} \left(v_i(A) - \sum_{j \in A} p_j \right)$$

Note now that v_i and R_i are independent. Thus, let $(\tilde{v}_i)_{i \in N}$ be independent realizations of the valuations, and $\widetilde{OPT}_i$ the corresponding optimal solution. With this we can rewrite the expected utility of agent i as

$$\mathbb{E} \left(\max_{A \subseteq R_i} v_i(A) - \sum_{j \in A} p_j \right) = \mathbb{E} \left(\max_{A \subseteq R_i} \tilde{v}_i(A) - \sum_{j \in A} p_j \right) \geq \mathbb{E} \left(\max_{A \subseteq R} \tilde{v}_i(A) - \sum_{j \in A} p_j \right).$$

We replace the maximization over subsets of R with a particular choice, $\widetilde{OPT}_i$, whenever it is contained by R and gives positive utility (otherwise we take $\emptyset$), to obtain the following lower bound.

$$\mathbb{E} \left(\mathbb{1}_{\{\widetilde{OPT}_i \subseteq R\}} \cdot \left[\tilde{v}_i(\widetilde{OPT}_i) - \sum_{j \in \widetilde{OPT}_i} p_j \right]_+ \right)$$

$$= \mathbb{E} \left(\sum_{A \subseteq R} \mathbb{1}_{\{\widetilde{OPT}_i = A\}} \cdot \left[\tilde{v}_i(A) - \sum_{j \in A} p_j \right]_+ \right)$$

$$= \mathbb{E} \left(\sum_{A \subseteq R} \mathbb{E} \left(\left[\mathbb{1}_{\{\widetilde{OPT}_i = A\}} \left(\tilde{v}_i(A) - \sum_{j \in A} p_j \right) \right]_+ \right) \right) \geq \mathbb{E} \left(\sum_{A \subseteq R} z_{i,A}(p) \right).$$

The last inequality comes from Jensen's inequality, noting that $[\cdot]_+$ is a convex function. Summing over all agents, we get that

$$\mathbb{E}(\text{Utility}) \geq \mathbb{E} \left(\sum_{i \in N} \sum_{A \subseteq R} z_{i,A}(p) \right).$$

Therefore, adding the revenue and the utility we get that

$$ALG(p) \geq \mathbb{E} \left(\sum_{j \notin R} k_j \cdot p_j + \sum_{i \in N} \sum_{A \subseteq R} z_{i,A}(p) \right).$$

Replacing the expectation over R with a minimization over subsets of M we obtain the bound of the lemma. □

Lemma 2. *For any vector of prices $p \in \mathbb{R}_+^M$,*

$$OPT \leq \sum_{j \in M} k_j \cdot p_j + \sum_{i \in N} \sum_{A \subseteq M} z_{i,A}(p).$$

Proof. We have that OPT equals

$$\sum_{i \in N} \mathbb{E}(v_i(OPT_i)) = \mathbb{E} \left(\sum_{i \in N} \sum_{j \in OPT_i} p_j \right) + \sum_{i \in N} \mathbb{E} \left(v_i(OPT_i) - \sum_{j \in OPT_i} p_j \right).$$

Now we upper bound these two terms separately. Note that in the first term each item $j \in M$ appears at most k_j times, so

$$\mathbb{E} \left(\sum_{i \in N} \sum_{j \in OPT_i} p_j \right) \leq \sum_{j \in M} k_j \cdot p_j.$$

For the second part we upper bound with the positive part of the difference, and sum over all possible values of OPT_i.

$$\sum_{i \in N} \mathbb{E} \left(v_i(OPT_i) - \sum_{j \in OPT_i} p_j \right)$$

$$\leq \sum_{i \in N} \sum_{A \subseteq M} \mathbb{E} \left(\mathbb{1}_{\{OPT_i = A\}} \left(v_i(OPT_i) - \sum_{j \in OPT_i} p_j \right) \right)$$

$$\leq \sum_{i \in N} \sum_{A \subseteq M} z_{i,A}(p).$$

Putting together the two upper bounds we obtain the bound on OPT. □

Lemma 3. *There exists a vector of prices $p \in \mathbb{R}_+^M$ that satisfies the equation*

$$p_j = \frac{1}{k_j} \sum_{i \in N} \sum_{A \subseteq M : j \in A} z_{i,A}(p).$$

Proof. Denote the set $K = [0, OPT]^M \subseteq \mathbb{R}_+^M$. We define the function $\psi : K \to K$ as follows. For an element $p \in K$ and $j \in M$, the j-th coordinate of ψ is

$$\psi_j(p) = \frac{1}{k_j} \sum_{i \in N} \sum_{A \subseteq M : j \in A} z_{i,A}(p).$$

We prove now that ψ is a well defined continuous function, from the compact set K into itself, and therefore it has a fixed point by Brouwer's theorem. Note that a fixed point of ψ is exactly the vector of prices we are looking for.

Recall that $z_{i,A}(p) = [\mathbb{E}(\mathbb{1}_{OPT_i=A} \cdot v_i(A)) - \mathbb{P}(OPT_i = A) \sum_{j \in A} p_j]_+$, which is a decreasing function of p_j, for all $j \in M$. Moreover, note that since $[\cdot]_+$ is a convex function, $z_{i,A}$ is also a convex function of p_j for all $j \in M$. The monotonicity of $z_{i,A}$ implies that for all $p \in K$ and $j \in M$, $\psi_j(p) \leq \psi_j(0) \leq \frac{1}{k_j}\mathbb{E}(OPT)$, and therefore $\psi(p) \in K$ for all $p \in K$. The convexity of $z_{i,A}$ implies it is also continuous, so ψ is a continuous function. $\qquad\square$

Proof (of Theorem 1). Using the vector of prices from Lemma 3 in the bound of Lemma 1 results in

$$ALG(p) \geq \sum_{i \in N} \sum_{A \subseteq M} z_{i,A}(p).$$

To compare to OPT, we use the upper bound of Lemma 2, which shows

$$OPT \leq \sum_{j \in M} \sum_{i \in N} \sum_{A \subseteq M : j \in A} z_{i,A}(p) + \sum_{i \in N} \sum_{A \subseteq M} z_{i,A}(p)$$

$$= \sum_{i \in N} \sum_{A \subseteq M} (|A| + 1) \cdot z_{i,A}(p)$$

$$\leq (d+1) \sum_{i \in N} \sum_{A \subseteq M} z_{i,A}(p).$$

Comparing the two bounds we get that $(d+1)ALG(p) \geq OPT$. $\qquad\square$

To wrap up the section, we establish that the bound of Theorem 1 is best possible, by modifying a simple example of Dütting et al. [10].

Proposition 1. *For all d, and all $\delta > 0$, there exists an instance on $|N| = 2$ bidders and $|M| = d$ items such that for all p, $ALG(p) = 1$, yet $OPT(p) = d + 1 - \delta$.*

Proof. Consider a set M of exactly d items with a single copy of each, and a very small $\varepsilon > 0$. There are two buyers. The first buyer values any nonempty subset of the items at 1. The second buyer only assigns value to getting all d items, and this value is $d - \varepsilon$ with probability $1 - \varepsilon$ and it is $1/\varepsilon$ with probability ε. Now we consider setting prices p_j for all $j \in M$. If we set the prices so that $\sum_{j \in M} p_j \leq d - \varepsilon$ then there exists an item with price at most $1 - \varepsilon/d$. Therefore, the first buyer will get this item and thus the total welfare will be 1. If, on the contrary, buyer one does not purchase an item, then we must have $\sum_{j \in M} p_j \geq d$, and the second buyer will only purchase items with probability ε. In this case, the expected total welfare is also 1. This establishes that $ALG(p) = 1$ for all p. Finally, it is clear that in this instance the optimal welfare is achieved by always assigning all items to the second buyer, which results in an expected welfare of $(d - \varepsilon) \cdot (1 - \varepsilon) + \varepsilon \cdot (1/\varepsilon) \geq d + 1 - (d+1)\varepsilon$. Setting $\varepsilon = \delta/(d+1)$ completes the proof. □

Efficient Computation. Our above proof is nonconstructive as it requires a fixed point computation. However, in the full version of the paper, we show that despite this challenge and others, there exists a polynomial-time algorithm to compute the prices using only *demand queries*.

4   Single-Minded Valuations

In this section, we consider the special case where there is a single copy of each item ($k_i = 1$ for all $i \in M$), buyers' valuations are deterministic, and buyers are *single-minded*. The latter means each buyer i has a set A_i, with $|A_i| \leq d$, such that $A_i \nsubseteq B \Rightarrow v_i(B) = 0$ and $A_i \subseteq B \Rightarrow v_i(B) = v_i(A_i)$. The problem of maximizing the welfare of an allocation in this context can be seen as the classic combinatorial problem of hypergraph matching with hyperedges of size at most d, where the buyers correspond to the hyperedges and the items are the vertices. Indeed, in an optimal allocation for this setting buyer i either gets A_i or $\emptyset$, implying that maximizing the welfare of the allocation is equivalent to finding a subset of pairwise disjoint A_i's of maximum total valuation. As this is a traditional problem, in the rest of this section we will refer to hypergraphs, hyperedges and vertices, rather than buyers and items, using the usual notation $G = (V, E)$ and denoting by $w(e)$ the valuation (or weight) of the hyperedge e.

4.1   Matching in Graphs: $d = 2$

We first focus on the traditional matching problem, showing that using prices has limits even for this scenario. As argued in Lemmas 4 and 6, there are instances in which no pricing scheme can guarantee recovering more than 2/3 of the optimal solution. This is true even if the graph is bipartite or if there is a unique optimal matching; on the other hand, if both conditions are fulfilled—i.e., the graph is bipartite and there is a unique optimal matching, we show that using the dual prices leads precisely to such optimal solution.

Lemma 4. *Prices cannot guarantee obtaining more than 2/3 of the optimal matching, even if the graph is bipartite.*

Proof. Consider the graph depicted in Fig. 1, in which all edges have unit weight. There are two optimal solutions, given by the black and the red perfect matchings. Assume we have prices that are able to build an optimal solution (i.e., include three edges) regardless of the order in which the edges arrive. This implies

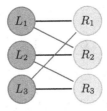

Fig. 1. Example of a bipartite graph in which, when all edges have the same weight, no pricing scheme can guarantee obtaining more than 2/3 of the optimal solution. (Color figure online)

that for at least one of the optimal solutions, all the edges will be included if their vertices are available when they arrive. Without loss assume this is the case for the black matching, i.e. for $i = 1, 2, 3$, we have $p_{L_i} + p_{R_i} \leq 1$.

On the other hand, we need to prevent the red edges to be included if they appear: to see why this is necessary, consider for instance the case in which the edge (L_1, R_2) is not discarded when appearing first; then, if the edge (L_3, R_3) appears second, no more edges could be added. To preclude this, we need to impose that for $i = 1, 2, 3$, $p_{L_i} + p_{R_{(i \bmod 3)+1}} > 1$. A contradiction follows by adding these as well as the previous three inequalities.   □

In the case of bipartite graphs, it is natural to consider the usual linear programming formulation, since it has integral optimal solutions. The following lemma shows that when we require the additional hypothesis that there is a unique optimal matching, the prices given by the optimal solution of the dual problem lead to the optimal assignment.

Lemma 5. *If the graph $G = (V, E)$ is bipartite and has a unique optimal matching, then such a matching is obtained using the dual prices.*

Proof. Because the graph is bipartite, the problem reduces to solving the linear program $\max\{\sum_{e \in E} x_e w(e) : \sum_{e \in \delta(v)} x_e \leq 1 \text{ for all } v \in V, x \geq 0\}$, which has an integral optimal solution. Because there is only one optimal matching, the LP has a unique optimal solution $(x_e^*)_{e \in E}$. Consider the prices $(p_u^*)_{u \in V}$ corresponding to an optimal dual solution, satisfying strict complementary slackness.

Consider an edge $e = (u, v)$ that is not part of the optimal matching. Hence, the corresponding primal variable takes the value $x_e^* = 0$. By complementary slackness, the corresponding dual constraint is not tight, i.e. $p_u^* + p_v^* > w(e)$. This last condition implies that buyer e will not buy the edge upon arrival. On the other hand, if e is part of the optimal solution, the corresponding dual constraint must be tight (again due to strict complementary slackness), so that those buyers will choose to buy.   □

The assumption of a unique solution is crucial for the dual prices to be useful. Indeed, when there is more than one solution, using the dual prices can be arbitrarily inefficient. Indeed, consider the same example depicted in Fig. 1,

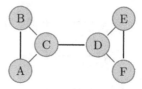

Fig. 2. Example of a graph in which, when all edges have the same weight, there is a unique optimal matching but no pricing scheme can guarantee obtaining more than 2/3 of its weight.

but modify the weight of the edges $f = (L_1, R_1)$ and $g = (L_2, R_3)$ to be ε, so that that the optimal solution has value $2 + \varepsilon$. On the other hand, consider an edge $e = (u, v)$ and the resulting dual prices p_u, p_v: complementary slackness now states that we have $p_u + p_v = w(e)$ iff e is part of any optimal solution. Edge f is part of the black optimal solution, and edge g is part of the red, hence those edges will be bought if the corresponding vertices are available when they appear. In particular, if they are the first two edges to appear, then they will both be in the final solution, and no other edge can be added, leading to a final weight of 2ε.

However, in general graphs, even the uniqueness assumption is not enough. Indeed we have the following result.

Lemma 6. *Prices cannot guarantee obtaining more than 2/3 of the optimal matching in a general graph, even if there is only one optimal matching.*

Proof. Consider the graph depicted in Fig. 2, where every edge has unit weight. The optimal matching is given by the three black edges with total value of 3. On the other hand, if any red edge enters the solution, the resulting total weight will be at most 2. We now show that any pricing scheme in which every black edge is willing to buy will also include at least one red edge if it comes first. Let $(p_i)_{i=A,\dots,F}$ prices such that for every black edge, the sum of the involved vertices is lower than 1. In particular, we have that $p_C + p_D \leq 1$, so without loss of generality we assume that $p_C \leq 1/2$. If $p_B \leq 1/2$ as well, then the red edge (B, C) will want to buy and the proof is complete. Otherwise, i.e. if $p_B > 1/2$, it implies that $p_A \leq 1/2$ because the black edge (A, B) wants to buy. But this implies that the red edge (A, C) will buy if appearing first.

Finally, if all vertex prices are 1/2, then it is straightforward to see that at least two edges will be added regardless of the order in which they appear. $\square$

In general, there are item prices that guarantee obtaining at least half of the optimal welfare. This is achieved by splitting the weight of the edges of an optimal matching uniformly between the two corresponding vertices. We present this result in Lemma 8 for general d.

4.2 Hypergraph Matching: $d > 2$

We begin this section by proving two negative results. First we show an upper bound of $\sim \sqrt{1/d}$ on the fraction of the optimal solution that can be guaranteed with prices. We then show a specific bound for the case $d = 3$, in which we cannot guarantee obtaining more than $1/2$ of the optimal welfare. Finally, we provide a pricing scheme that always obtains at least $1/d$ of the optimal welfare.

Lemma 7. *Prices cannot guarantee welfare more than an $\sim \sqrt{1/d}$ fraction of the optimal welfare, even if the arrival order is known.*

Proof. Our example is based on constructions for finite projective planes; namely, we will use the fact that if $q - 1$ is a prime power there exists a hypergraph on $q^2 - q + 1$ vertices with $q^2 - q + 1$ hyperedges that are q-regular, q-uniform and intersecting, i.e. every pair of hyperedges has at least one shared vertex (see, e.g., [17, Chapter 12] for a reference).

To build our example, we will assume that for each hyperedge there exists a corresponding buyer interested in exclusively that subset of items with a total valuation of q. We will also add one buyer whose only subset of interest is the entire set of items, with a valuation of $d = q^2 - q + 1$. Note that clearly the optimal welfare attainable is $q^2 - q + 1$.

It hence suffices to show that prices cannot achieve welfare greater than q. Assume the buyer interested in the entire set of items arrives last. Note that if there is any edge e such that the sum of the prices of the vertices in e is at most than q, we are guaranteed welfare at most q. However, if every the sum of the prices of the vertices in every hyperedge is more than q, because our graph is q-uniform that means the sum of the prices of *all* vertices is more than $q^2 - q + 1$, meaning the final buyer would not select anything and the welfare attained is zero. Hence, the total welfare attainable by prices is at most a

$$\frac{q}{q^2 - q + 1} \sim \frac{1}{\sqrt{q^2 - q + 1}}$$

fraction of the optimum.

Finally, if d cannot be written as $q^2 - q + 1$, we replicate the same construction for the largest $d' < d$ that can, and the result holds.   □

When $d = 3$ the upper bound given by Lemma 7 is $2/3$. We briefly note that this bound can be tightened to $1/2$. Our instance consists of a hypergraph $G = (V, E)$ with $V = \{1, 2, 3, 4, 5, 6\}$ and hyperedges $\{1, 2, 3\}$, $\{4, 5, 6\}$, $\{1, 2, 4\}$, $\{1, 3, 5\}$, $\{2, 5, 6\}$, $\{3, 4, 6\}$ all with unit weight; the short proof that this attains an upper bound of $1/2$ is deferred to the full version.

We conclude with our positive result. Consider a hypergraph $G = (V, E)$, with weights $(w(e))_{e \in E}$. To define the prices, take an optimal matching given by the hyperedges $OPT_1, \ldots, OPT_\ell$. For each $a \in OPT_j$, define $p_a = w(OPT_j)/d$. The prices of the items not covered by the optimal solution are set to ∞. The following simple result shows that these prices obtain at least a fraction $1/d$ of the optimal welfare.

Lemma 8. *Consider prices defined as above, and hyperedges arriving in an arbitrary order. Let Q denote the set of edges that are bought. Then*

$$\sum_{e \in Q} w(e) \geq \frac{1}{d} \sum_{j=1}^{\ell} w(OPT_j)$$

Proof. First note that for each $e \in Q$, it must hold that

$$w(e) \geq \sum_{i \in e} p_i \tag{3}$$

as otherwise the buyer associated to e would have decided not to buy. Therefore

$$\sum_{e \in Q} w(e) \geq \sum_{e \in Q} \sum_{i \in e} p_i \tag{4}$$

On the other hand, for each OPT_j in the optimal solution, there must be at least one vertex, with its corresponding price $w(OPT_j)/d$ that is covered by the edges in Q. To see this, note that there are two possible cases: either $OPT_j \in Q$ and all its vertices are covered, or $OPT_j \notin Q$, meaning that when OPT_j arrived, at least one of its vertices was not available, i.e., it was covered by an edge previously bought. The result follows directly, noting that in the RHS of (4), we are summing at least once $w(OPT_j)/d$ for each $j = 1 \ldots, \ell$.   □

References

1. Assadi, S., Kesselheim, T., Singla, S.: Improved truthful mechanisms for subadditive combinatorial auctions: breaking the logarithmic barrier. In: Proceedings of the 2021 ACM-SIAM Symposium on Discrete Algorithms (SODA), pp. 653–661. SIAM (2021)
2. Assadi, S., Singla, S.: Improved truthful mechanisms for combinatorial auctions with submodular bidders. In: 2019 IEEE 60th Annual Symposium on Foundations of Computer Science (FOCS), pp. 233–248. IEEE (2019)
3. Babaioff, M., Lucier, B., Nisan, N., Paes Leme, R.: On the efficiency of the Walrasian mechanism. In: Proceedings of the Fifteenth ACM Conference on Economics and Computation, pp. 783–800 (2014)
4. Baldwin, E., Klemperer, P.: Understanding preferences: "demand types", and the existence of equilibrium with indivisibilities. Econometrica **87**(3), 867–932 (2019)
5. Cai, Y., Zhao, M.: Simple mechanisms for subadditive buyers via duality. In: Hatami, H., McKenzie, P., King, V. (eds.) Proceedings of the 49th Annual ACM SIGACT Symposium on Theory of Computing, STOC 2017, Montreal, QC, Canada, 19–23 June 2017, pp. 170–183. ACM (2017)
6. Chawla, S., Hartline, J.D., Malec, D.L., Sivan, B.: Multi-parameter mechanism design and sequential posted pricing. In: Proceedings of the Forty-second ACM Symposium on Theory of Computing, pp. 311–320 (2010)
7. Chawla, S., Miller, J.B.: Mechanism design for subadditive agents via an ex ante relaxation. In: Conitzer, V., Bergemann, D., Chen, Y. (eds.) Proceedings of the 2016 ACM Conference on Economics and Computation, EC 2016, Maastricht, The Netherlands, 24–28 July 2016, pp. 579–596. ACM (2016)

8. Dobzinski, S.: Breaking the logarithmic barrier for truthful combinatorial auctions with submodular bidders. In: Wichs, D., Mansour, Y. (eds.) Proceedings of the 48th Annual ACM SIGACT Symposium on Theory of Computing, STOC 2016, Cambridge, MA, USA, 18–21 June 2016, pp. 940–948. ACM (2016)

9. Dobzinski, S., Nisan, N., Schapira, M.: Approximation algorithms for combinatorial auctions with complement-free bidders. In: Gabow, H.N., Fagin, R. (eds.) Proceedings of the 37th Annual ACM Symposium on Theory of Computing, Baltimore, MD, USA, 22–24 May 2005, pp. 610–618. ACM (2005)

10. Dutting, P., Feldman, M., Kesselheim, T., Lucier, B.: Prophet inequalities made easy: stochastic optimization by pricing nonstochastic inputs. SIAM J. Comput. **49**(3), 540–582 (2020)

11. Dütting, P., Kesselheim, T., Lucier, B.: An o(log log m) prophet inequality for subadditive combinatorial auctions. In: Irani, S. (ed.) 61st IEEE Annual Symposium on Foundations of Computer Science, FOCS 2020, Durham, NC, USA, 16–19 November 2020, pp. 306–317. IEEE (2020)

12. Ezra, T., Feldman, M., Gravin, N., Tang, Z.G.: Online stochastic max-weight matching: prophet inequality for vertex and edge arrival models. In: Proceedings of the 21st ACM Conference on Economics and Computation, pp. 769–787 (2020)

13. Feldman, M., Gravin, N., Lucier, B.: Combinatorial auctions via posted prices. In: Proceedings of the Twenty-Sixth Annual ACM-SIAM Symposium on Discrete Algorithms, pp. 123–135. SIAM (2014)

14. Feldman, M., Gravin, N., Lucier, B.: Combinatorial walrasian equilibrium. SIAM J. Comput. **45**(1), 29–48 (2016)

15. Feldman, M., Svensson, O., Zenklusen, R.: Online contention resolution schemes. In: Proceedings of the Twenty-Seventh Annual ACM-SIAM Symposium on Discrete Algorithms, pp. 1014–1033. SIAM (2016)

16. Gravin, N., Wang, H.: Prophet inequality for bipartite matching: merits of being simple and non adaptive. In: Proceedings of the 2019 ACM Conference on Economics and Computation, pp. 93–109 (2019)

17. Hall, M.: Combinatorial Theory, vol. 71. Wiley, Hoboken (1998)

18. Kleinberg, R., Weinberg, S.M.: Matroid prophet inequalities. In: Proceedings of the Forty-Fourth Annual ACM Symposium on Theory of Computing, pp. 123–136 (2012)

19. Krysta, P., Vöcking, B.: Online mechanism design (randomized rounding on the fly). In: Czumaj, A., Mehlhorn, K., Pitts, A., Wattenhofer, R. (eds.) ICALP 2012. LNCS, vol. 7392, pp. 636–647. Springer, Heidelberg (2012). https://doi.org/10.1007/978-3-642-31585-5_56

20. Leme, R.P., Wong, S.C.W.: Computing walrasian equilibria: fast algorithms and structural properties. Math. Program. **179**(1), 343–384 (2020)

21. Samuel-Cahn, E.: Comparison of threshold stop rules and maximum for independent nonnegative random variables. Ann. Probab. **12**(4), 1213–1216 (1984)

22. Trevisan, L.: Non-approximability results for optimization problems on bounded degree instances. In: Proceedings of the Thirty-Third Annual ACM Symposium on Theory of Computing, pp. 453–461 (2001)

On Circuit Diameter Bounds via Circuit Imbalances

Daniel Dadush[1] , Zhuan Khye Koh[2(✉)] , Bento Natura[2] ,
and László A. Végh[2]

[1] Centrum Wiskunde & Informatica, Amsterdam, The Netherlands
dadush@cwi.nl
[2] London School of Economics and Political Science, London, UK
{z.koh3,b.natura,l.vegh}@lse.ac.uk

Abstract. We study the circuit diameter of polyhedra, introduced by
Borgwardt, Finhold, and Hemmecke (SIDMA 2015) as a relaxation of
the combinatorial diameter. We show that the circuit diameter of a sys-
tem $\{x \in \mathbb{R}^n : Ax = b, \mathbb{0} \leq x \leq u\}$ for $A \in \mathbb{R}^{m \times n}$ is bounded by
$O(m^2 \log(m + \kappa_A) + n \log n)$, where κ_A is the circuit imbalance measure
of the constraint matrix. This yields a strongly polynomial circuit diameter bound if e.g., all entries of A have polynomially bounded encoding
length in n. Further, we present circuit augmentation algorithms for LPs
using the minimum-ratio circuit cancelling rule. Even though the standard minimum-ratio circuit cancelling algorithm is not finite in general,
our variant can solve an LP in $O(n^3 \log(n + \kappa_A))$ augmentation steps.

1 Introduction

The *combinatorial diameter* of a polyhedron P is the diameter of the vertex-edge
graph associated with P. Hirsch's famous conjecture from 1957 asserted that
the combinatorial diameter of a d-dimensional polytope (bounded polyhedron)
with f facets is at most $f - d$. This was disproved by Santos in 2012 [24]. The
polynomial Hirsch conjecture, i.e., finding a poly(f) bound on the combinatorial
diameter remains a central question in the theory of linear programming.

The first quasipolynomial bound was given by Kalai and Kleitman [20,21],
see [27] for the best current bound and an overview of the literature. Dyer
and Frieze [14] proved the polynomial Hirsch conjecture for totally unimodular (TU) matrices. For a system $\{x \in \mathbb{R}^d : Mx \leq b\}$ with integer constraint
matrix M, polynomial diameter bounds were given in terms of the maximum
subdeterminant Δ_M [2,7,9,15]. These arguments can be strengthened to using

**This is an extended abstract. The full version of the paper with all proofs is
available on** arXiv:2111.07913. This project has received funding from the European
Research Council (ERC) under the European Union's Horizon 2020 research and innovation programme (grant agreements ScaleOpt–757481 and QIP–805241). This work
was done while the authors participated in the Discrete Optimization Trimester Program at the Hausdorff Institute for Mathematics in Bonn in 2021.

K. Aardal and L. Sanitá (Eds.): IPCO 2022, LNCS 13265, pp. 140–153, 2022.
https://doi.org/10.1007/978-3-031-06901-7_11

a parametrization by a 'discrete curvature measure' $\delta_M \geq 1/(d\Delta_M^2)$. The best such bound was given by Dadush and Hähnle [9] as $O(d^3 \log(d/\delta_M)/\delta_M)$, using a shadow vertex simplex algorithm.

As a natural relaxation of the combinatorial diameter, Borgwardt, Finhold, and Hemmecke [4] initiated the study of *circuit diameters*. Consider a polyhedron in standard equality form

$$P = \{x \in \mathbb{R}^n : Ax = b, x \geq 0\} \tag{P}$$

for $A \in \mathbb{R}^{m \times n}$, $b \in \mathbb{R}^m$; we assume $\mathrm{rk}(A) = m$. For the linear space $W = \ker(A) \subseteq \mathbb{R}^n$, $g \in W$ is an *elementary vector* if g is a support-minimal nonzero vector in W, that is, no $h \in W\backslash\{0\}$ exists such that $\mathrm{supp}(h) \subsetneq \mathrm{supp}(g)$. A *circuit* in W is the support of some elementary vector; these are precisely the circuits of the associated linear matroid $\mathcal{M}(W)$. We let $\mathcal{F}(W) = \mathcal{F}(A) \subseteq W$ and $\mathcal{C}(W) = \mathcal{C}(A) \subseteq 2^n$ denote the set of elementary vectors and circuits in the space $W = \ker(A)$, respectively. All edge directions of P are elementary vectors, and the set of elementary vectors $\mathcal{F}(A)$ equals the set of all possible edge directions of P in the form (P) for varying $b \in \mathbb{R}^m$ [26].

A *circuit walk* is a sequence of points $x^{(1)}, x^{(2)}, \ldots, x^{(k+1)}$ in P such that for each $i = 1, \ldots, k$, $x^{(i+1)} = x^{(i)} + g^{(i)}$ for some $g^{(i)} \in \mathcal{F}(A)$, and further, $x^{(i)} + (1+\varepsilon)g^{(i)} \notin P$ for any $\varepsilon > 0$, i.e., each consecutive circuit step is *maximal*. The *circuit diameter* of P is the minimum length (number of steps) of a circuit walk between any two vertices $x, y \in P$. Note that, in contrast to walks in the vertex-edge graph, circuit walks are non-reversible and the minimum length from x to y may be different from the one from y to x; this is due to the maximality requirement. The circuit-analogue of Hirsch conjecture, formulated in [4], asserts that the circuit diameter of a d-dimensional polyhedron with f facets is at most $f - d$; this may be true even for unbounded polyhedra, see [5]. For P in the form (P), $d = n - m$ and the number of facets is at most n. Hence, the conjectured bound is m.

Circuit diameter bounds have been shown for some combinatorial polytopes such as dual transportation polyhedra [4], matching, travelling salesman, and fractional stable set polytopes [19]. The paper [3] introduced several other variants of circuit diameter, and explored the relation between them.

Circuit Augmentation Algorithms. Circuit diameter bounds are inherently related to *circuit augmentation algorithms*. This is a general algorithmic scheme to solve an LP

$$\min \langle c, x \rangle \quad \text{s.t.} \quad Ax = b, \, x \geq 0. \tag{LP}$$

The algorithm proceeds through a sequence of feasible solutions $x^{(t)}$. An initial feasible $x^{(0)}$ is required in the input. For $t = 0, 1, \ldots$, the current $x^{(t)}$ is updated to $x^{(t+1)} = x^{(t)} + \alpha g$ for some $g \in \mathcal{F}(A)$ such that $\langle c, g \rangle \leq 0$, and $\alpha > 0$ such that $x^{(t)} + \alpha g$ is feasible. The elementary vector g is an *augmenting direction* if $\langle c, g \rangle < 0$ and such an $\alpha > 0$ exists; by LP duality, $x^{(t)}$ is optimal if and only if no augmenting direction exists. The augmentation is *maximal* if $x^{(t)} + \alpha' g$ is infeasible for any $\alpha' > \alpha$; α is called the maximal stepsize for $x^{(t)}$ and g. Clearly, an upper bound on the number of steps of a circuit augmentation algorithm with maximal augmentations for arbitrary cost c and starting point $x^{(0)}$ yields an upper bound on the circuit diameter.

Simplex is a circuit augmentation algorithm that is restricted to using special elementary vectors corresponding to edges of the polyhedron. Many network optimization algorithms can be seen as special circuit augmentation algorithms. Bland [1] introduced a circuit augmentation algorithm for LP, that generalizes the Edmonds–Karp–Dinic algorithm and its analysis, see also [22, Proposition 3.1]. Circuit augmentation algorithms were revisited by De Loera, Hemmecke, and Lee in 2015 [12], analyzing different augmentation rules and also extending them to integer programming. De Loera, Kafer, and Sanità [13] studied the convergence of these rules on 0/1-polytopes, as well as the computational complexity of performing them. We refer the reader to [12] and [13] for a more detailed overview of the background and history of circuit augmentations.

The Circuit Imbalance Measure. For a linear space $W = \ker(A) \subseteq \mathbb{R}^n$, the *circuit imbalance* $\kappa_W = \kappa_A$ is defined as the maximum of $|g_j/g_i|$ over all elementary vectors $g \in \mathcal{F}(W)$, $i, j \in \mathrm{supp}(g)$. It can be shown that $\kappa_W = 1$ if and only if W is a unimodular space, i.e., the kernel of a totally unimodular matrix. This parameter and related variants have been used implicitly or explicitly in many areas of linear programming and discrete optimization, see [16] for a recent survey. It is closely related to the Dikin–Stewart–Todd condition number $\bar{\chi}_W$ that plays a key role in layered-least-squares interior point methods introduced by Vavasis and Ye [31]. An LP of the form (LP) for $A \in \mathbb{R}^{m \times n}$ can be solved in time $\mathrm{poly}(n, m, \log \kappa_A)$, which is strongly polynomial if $\kappa_A \leq 2^{\mathrm{poly}(n)}$; see [10,11] for recent developments and references.

Imbalance and Diameter. The combinatorial diameter bound $O(d^3 \log (d/\delta_M)/\delta_M)$ from [9] mentioned above translates to a bound $O((n - m)^3 m \kappa_A \log(\kappa_A + n))$ for the system in the form (P), see [16]. For circuit diameters, the Goldberg-Tarjan minimum-mean cycle cancelling algorithm for minimum-cost flows [18] naturally extends to a circuit augmentation algorithm for general LPs using the *steepest-descent* rule. This yields a circuit diameter bound $O(n^2 m \kappa_A \log(\kappa_A + n))$ [16], see also [17]. However, note that these bounds may be exponential in the bit-complexity of the input.

1.1 Our Contributions

Our first main contribution improves the κ_A dependence to a $\log \kappa_A$ dependence for circuit diameter bounds.

Theorem 1. *The circuit diameter of a system in the form (P) with constraint matrix $A \in \mathbb{R}^{m \times n}$ is $O(m^2 \log(m + \kappa_A))$.*

The proof in Sect. 3 is via a simple 'shoot towards the optimum' scheme. We need the well-known concept of *conformal circuit decompositions*. We say that $x, y \in \mathbb{R}^n$ are *sign-compatible* if $x_i y_i \geq 0$ for all $i \in [n]$, and write $x \sqsubseteq y$ if they are sign-compatible and further $|x_i| \leq |y_i|$ for all $i \in [n]$. It follows from Carathéodory's theorem and Minkowski–Weyl theorem that for any linear space $W \subseteq \mathbb{R}^n$ and $x \in W$, there exists a decomposition $x = \sum_{j=1}^k h^{(j)}$ such that $h^{(j)} \in \mathcal{F}(W)$, $h^{(j)} \sqsubseteq x$ for all $j \in [k]$ and $k \leq n$. This is called a *conformal circuit decomposition* of x.

Let $B \subseteq [n]$ be a feasible basis and $N = [n] \backslash B$, i.e., $x^* = (A_B^{-1} b, \mathbb{0}_N) \geq \mathbb{0}$ is a basic feasible solution. This is the unique optimal solution to (LP) with cost function $c = (\mathbb{0}_B, \mathbb{1}_N)$. Let $x^{(0)} \in P$ be an arbitrary starting vertex. We may assume that $n \leq 2m$, by restricting to the union of the support of x^* and $x^{(0)}$, and setting all other variables to 0. For the current iterate $x^{(t)}$, let us consider a conformal circuit decomposition $x^* - x^{(t)} = \sum_{j=1}^k h^{(j)}$. Note that the existence of such a decomposition *does not* yield a circuit diameter bound of n, due to the maximality requirement in the definition of circuit walks. For each $j \in [k]$, $x^{(t)} + h^{(j)} \in P$, but there might be a larger augmentation $x^{(t)} + \alpha h^{(j)} \in P$ for some $\alpha > 1$.

Still, one can use this decomposition to construct a circuit walk. Let us pick the most improving circuit from the decomposition, i.e., the one maximizing $-\langle c, h^{(j)} \rangle = \|h_N^{(j)}\|_1$, and obtain $x^{(t+1)} = x^{(t)} + \alpha h^{(j)}$ for the maximum stepsize $\alpha \geq 1$. The proof of Theorem 1 is based on analyzing this procedure. The first key observation is that $\langle c, x^{(t)} \rangle = \|x_N^{(t)}\|_1$ decreases geometrically. Then, we look at the sets $L_t = \{i \in [n] : x_i^* > n\kappa_A \|x_N^{(t)}\|_1\}$ and $R_t = \{i \in [n] : x_i^{(t)} \leq n x_i^*\}$, and show that indices may never leave these sets once they enter. Moreover, a new index is added to either set every $O(m \log(m + \kappa_A))$ iterations. In Sect. 4, we extend this bound to the setting with upper bounds on the variables.

Theorem 2. *The circuit diameter of a system in the form $Ax = b$, $\mathbb{0} \leq x \leq u$ with constraint matrix $A \in \mathbb{R}^{m \times n}$ is $O(m^2 \log(m + \kappa_A) + n \log n)$.*

There is a straightforward reduction from the capacitated form to (P) by adding n slack variables; however, this would give an $O(n^2 \log(n + \kappa_A))$ bound. For the stronger bound, we use a preprocessing that involves cancelling circuits in the support of the current solution; this eliminates all but $O(m)$ of the capacity bounds in $O(n \log n)$ iterations, independently from κ_A.

For rational input, $\log(\kappa_A) = O(L_A)$ where L_A denotes the total encoding length of A [10]. Hence, our result yields an $O(m^2 L_A + n \log n)$ diameter bound on $Ax = b$, $\mathbb{0} \leq x \leq u$. This can be compared with the bounds $O(n L_{A,b})$ using deepest descent augmentation steps in [12,13], where $L_{A,b}$ is the encoding length of (A, b). (Such a bound holds for every augmentation rule that decreases the optimality gap geometrically, including the minimum-ratio circuit rule discussed below). Thus, our bound is independent of b. Furthermore, it is also applicable to systems given by irrational inputs, in which case arguments based on subdeterminants and bit-complexity cannot be used.

In light of these results, the next important step towards the polynomial Hirsch conjecture might be to show a $\text{poly}(n, \log \kappa_A)$ bound on the combinatorial diameter of (P). Note that—in contrast with the circuit diameter—not even a $\text{poly}(n, L_{A,b})$ bound is known. In this context, the best known general bound is $O((n - m)^3 m \kappa_A \log(\kappa_A + n))$ implied by [9].

Circuit Augmentation Algorithms. The diameter bounds in Theorems 1 and 2 rely on knowing the optimal solution x^*; thus, they do not provide efficient LP algorithms. We next present circuit augmentation algorithms with poly$(n, m, \log \kappa_A)$ bounds on the number of iterations. Such algorithms require subroutines for finding augmenting circuits. In many cases, such subroutines are LPs themselves. However, they may be of a simpler form, and might be easier to solve in practice. Borgwardt and Viss [6] exhibit an implementation of a steepest-descent circuit augmentation algorithm with encouraging computational results.

Our main subroutine assumption RATIO-CIRCUIT(A, c, w) is the well-known minimum-ratio circuit rule. It takes as input a matrix $A \in \mathbb{R}^{m \times n}$, $c \in \mathbb{R}^n$, $w \in (\mathbb{R}_+ \cup \{\infty\})^n$, and returns a basic optimal solution to the system

$$\min \langle c, z \rangle \quad \text{s.t.} \quad Az = \mathbb{0}, \langle w, z^- \rangle \leq 1, \tag{1}$$

where $(z^-)_i := \max\{0, -z_i\}$ for $i \in [n]$. Note that $w_i z_i = 0$ if $w_i = \infty$ and $z_i = 0$. This system can be equivalently written as an LP using auxiliary variables. If bounded, a basic optimal solution is an elementary vector $z \in \mathcal{F}(A)$ that minimizes $\langle c, z \rangle / \langle w, z^- \rangle$.

Given $x \in P$, we use weights $w_i = 1/x_i$ (and let $w_i = \infty$ if $x_i = 0$). For minimum-cost flow problems, this rule was proposed by Wallacher [32]; such a cycle can be found in strongly polynomial time for flows. The main advantage of this rule is that the optimality gap decreases by a factor $1 - 1/n$ in every iteration. This rule, along with the same convergence property, can be naturally extended to linear programming [23], and has found several combinatorial applications, e.g., [33,34], and has also been used in the context of integer programming [25].

On the negative side, Wallacher's algorithm is *not* strongly polynomial: it does not terminate finitely for minimum-cost flows, as shown in [23]. In contrast, our algorithms achieve a strongly polynomial running time whenever $\kappa_A \leq 2^{\text{poly}(n)}$. An important modification is the occasional use of a second type of circuit augmentation step SUPPORT-CIRCUIT that removes circuits in the support of the current (non-basic) iterate $x^{(t)}$ (see Subroutine 2.1); this can be implemented using simple linear algebra. Our first result addresses the feasibility setting:

Theorem 3. *Consider an LP of the form* (LP) *with cost function* $c = (\mathbb{0}_{[n]\setminus N}, \mathbb{1}_N)$ *for some* $N \subseteq [n]$. *There exists a circuit augmentation algorithm that either finds a solution* x *such that* $x_N = \mathbb{0}$ *or a dual certificate that no such solution exists, using* $O(n^2 \log(n + \kappa_A))$ RATIO-CIRCUIT *and* n^2 SUPPORT-CIRCUIT *augmentation steps.*

Such problems typically arise in Phase I of the Simplex method when we add auxiliary variables in order to find a feasible solution. The algorithm is presented in Sect. 5. The analysis extends that of Theorem 1, tracking large coordinates $x_i^{(t)}$. Our second result considers general optimization:

Theorem 4. *Consider an LP of the form* (LP). *There exists a circuit augmentation algorithm that finds an optimal solution or concludes unboundedness using* $O(n^3 \log(n + \kappa_A))$ RATIO-CIRCUIT *and* n^3 SUPPORT-CIRCUIT *augmentation steps.*

The proof is given in Sect. 6. The main subroutine identifies a new index $i \in [n]$ such that $x_i^{(t)} = 0$ in the current iteration and $x_i^* = 0$ in an optimal solution; we henceforth fix this variable to 0. To derive this conclusion, at the end of each phase the current iterate $x^{(t)}$ will be optimal to (LP) with a slightly modified cost function $\tilde{c}$; the conclusion follows using a proximity argument. The overall algorithm repeats this subroutine n times. The subroutine is reminiscent of the feasibility algorithm (Theorem 3) with the following main difference: whenever we identify a new 'large' coordinate, we slightly perturb the cost function.

Comparison to Black-Box LP Approaches. An important milestone towards strongly polynomial linear programming was Tardos's 1986 paper [28] on solving (LP) in time $\text{poly}(n, m, \log \Delta_A)$, where Δ_A is the maximum subdeterminant of A. Her algorithm makes $O(nm)$ calls to a weakly polynomial LP solver for instances with small integer constraints and costs, and uses proximity arguments to gradually learn the support of an optimal solution. This approach was extended to the real model of computation for a $\text{poly}(n, m, \log \kappa_A)$ bound [11]. The latter result uses proximity arguments with circuit imbalances κ_A, and eliminates all dependence on bit-complexity.

Our circuit augmentation algorithms are inspired by the feasibility and optimization algorithms in [11]. However, using a circuit augmentation oracle instead of an approximate LP oracle changes the setup. Our arguments become simpler since we proceed through a sequence of feasible solutions, whereas much effort in [11] is needed to deal with infeasibility of the solutions returned by the approximate solver. On the other hand, we need to be more careful as all steps must be implemented using circuit augmentations in the original system, in contrast to the higher degree of freedom in [11] where we can make approximate solver calls to arbitrary projections and modifications of the input LP.

2   Preliminaries

Circuit Oracles. In Sects. 4, 5, 6, we use a simple circuit finding subroutine SUPPORT-CIRCUIT(A, c, x, S) that will be used to identify circuits in the support of a solution x. This can be implemented easily using Gaussian elimination. Note that the constraint $\langle c, z \rangle \leq 0$ is superficial as $-z$ is also an elementary vector.

Subroutine 2.1. SUPPORT-CIRCUIT(A, c, x, S)

For a matrix $A \in \mathbb{R}^{m \times n}$, vectors $c, x \in \mathbb{R}^n$ and $S \subseteq [n]$, the output is an elementary vector $z \in \mathcal{F}(A)$ with $\text{supp}(z) \subseteq \text{supp}(x)$, $\text{supp}(z) \cap S \neq \emptyset$ with $\langle c, z \rangle \leq 0$, or concludes that no such elementary vector exists.

The circuit augmentation algorithms in Sects. 5 and 6 will use the subroutine RATIO-CIRCUIT(A, c, w). For $\alpha \in \mathbb{R}$ we define $\alpha^+ = \max\{0, \alpha\}$ and $\alpha^- = \max\{0, -\alpha\}$. For a vector $z \in \mathbb{R}^n$ we define $z^+, z^- \in \mathbb{R}^n$ with $(z^+)_i = (z_i)^+$, $(z^-)_i = (z_i)^-$ for $i \in [n]$.

Subroutine 2.2. RATIO-CIRCUIT(A, c, w)

The input is a matrix $A \in \mathbb{R}^{m \times n}$, $c \in \mathbb{R}^n$, $w \in (\mathbb{R}_+ \cup \{\infty\})^n$, and returns a basic optimal solution to the system

$$\min \langle c, z \rangle \quad \text{s.t.} \quad Az = 0, \langle w, z^- \rangle \leq 1, \tag{2}$$

and a basic optimal solution (y, s) to the following dual program:

$$\max -\lambda \quad \text{s.t.} \quad s = c + A^\top y \quad 0 \leq s \leq \lambda w \tag{3}$$

We use the convention that $w_i z_i^- = 0$ whenever $w_i = \infty$ and $z_i^- = 0$ in (2). Note that (2) can be reformulated as an LP using additional variables, and its dual LP can be equivalently written as (3). If (2) is bounded, then a basic optimal solution is an elementary vector $z \in \mathcal{F}(A)$ that minimizes $\langle c, z \rangle / \langle w, z^- \rangle$. Moreover, observe that every feasible solution to (3) is also feasible to the dual of (LP).

For P as in (P), $x \in P$ and an elementary vector $g \in \mathcal{F}(A)$, we let $\mathrm{aug}_P(x, g) := x + \alpha g$ where $\alpha = \arg\max\{\bar{\alpha} : x + \bar{\alpha} g \in P\}$. For $z \in \mathbb{R}^n$ we let $1/z \in (\mathbb{R} \cup \{\infty\})^n$ denote the vector $(1/z_i)_{i \in [n]}$, with the convention $1/0 = \infty$. The following lemma is well-known, see e.g., [23, Lemma 2.2].

Lemma 1. *Let* OPT *denote the optimum value of* (LP). *Given a feasible solution* x *to* (LP), *let* g *be the elementary vector returned by* RATIO-CIRCUIT($A, c, 1/x$), *and* $x' = \mathrm{aug}_P(x, g)$. *Then,*

$$\langle c, x' \rangle - \mathrm{OPT} \leq (1 - 1/n)(\langle c, x \rangle - \mathrm{OPT}).$$

Furthermore, $\alpha \geq 1$ *for the augmentation step.*

Proof. Let x^* be an optimal solution to (LP), and let $z = (x^* - x)/n$. Then, z is feasible to (2) for $w = 1/x$. The claim easily follows by noting that $\langle c, g \rangle \leq \langle c, z \rangle = (\mathrm{OPT} - \langle c, x \rangle)/n$, and noting that $x + g \in P$ is implied by $\langle 1/x, g^- \rangle \leq 1$. $\qquad \square$

Proximity Results. The imbalance measure κ_A is mainly used for proving norm bounds that can be interpreted as special forms of Hoffman-proximity results. See [11,16] for such general results and background, in particular, similar proximity bounds via Δ_A in e.g., [28] and [8].

Lemma 2. *For $A \in \mathbb{R}^{m \times n}$, let $z \in \ker(A)$, and let $N \subseteq [n]$ such that $A_{[n] \setminus N}$ has full column rank. Then, $\|z\|_\infty \leq \kappa_A \|z_N\|_1$.*

Proof. Let $h^{(1)}, \ldots, h^{(k)}$ be a conformal circuit decomposition of z. Conformality implies that $\|z\|_\infty \leq \sum_{t=1}^k \|h^{(t)}\|_\infty$. For each $h^{(t)}$, we have $\mathrm{supp}(h^{(t)}) \cap N \neq \emptyset$ because $A_{[n] \setminus N}$ has full column rank. Hence, $\|h^{(t)}\|_\infty \leq \kappa_A |h_j^{(t)}|$ for some $j \in N$. By conformality again, we obtain $\sum_{t=1}^k \|h^{(t)}\|_\infty \leq \kappa_A \|z_N\|_1$ as desired. $\qquad \square$

The following proximity theorem will be key to derive $x_i^* = 0$ for certain variables in our optimization algorithm; see [11] and [16, Theorem 6.5]. For $\tilde{c} \in \mathbb{R}^n$, we use $\mathrm{LP}(\tilde{c})$ to denote (LP) with cost vector $\tilde{c}$, and $\mathrm{OPT}(\tilde{c})$ as the optimal value of $\mathrm{LP}(\tilde{c})$.

Theorem 5. *Let $c, c' \in \mathbb{R}^n$ be two cost vectors, such that both $\mathrm{LP}(c)$ and $\mathrm{LP}(c')$ have finite optimum values. Let s' be a dual optimal solution to $\mathrm{LP}(c')$. If there exists an index $j \in [n]$ such that*

$$s_j' > (m+1)\kappa_A \|c - c'\|_\infty,$$

then $x_j^* = 0$ for every optimal solution x^* to $\mathrm{LP}(c)$.

Estimating Circuit Imbalances. The circuit augmentation algorithms in Sects. 5 and 6 explicitly use the circuit imbalance measure κ_A. However, this is NP-hard to approximate within a factor $2^{O(n)}$, see [10,29]. We circumvent this problem using a standard guessing procedure, see e.g., [10,31]. Instead of κ_A, we use an estimate $\hat{\kappa}$, initialized as $\hat{\kappa} = n$. Running the algorithm with this estimate either finds the desired feasible or optimal solution (which one can verify), or fails. In case of failure, we conclude that $\hat{\kappa} < \kappa_A$, and replace $\hat{\kappa}$ by $\hat{\kappa}^2$. Since the running time of the algorithms is linear in $\log(n + \hat{\kappa})$, the running time of all runs will be dominated by the last run, giving the desired bound. For simplicity, the algorithm descriptions use the explicit value κ_A. Throughout, we use the shorthand $\kappa = \kappa_A$ whenever A is clear from the context.

3   The Circuit Diameter Bound

In this section, we show Theorem 1, namely the bound $O(m^2 \log(m + \kappa))$ on the circuit diameter of a polyhedron in standard form (P). As outlined in the Introduction, let $B \subseteq [n]$ be a feasible basis and $N = [n] \setminus B$ such that $x^* = (A_B^{-1}b, \mathbb{0}_N)$ is a basic solution to (LP). We can assume $n \leq 2m$: the union of the supports of the starting vertex $x^{(0)}$ and the target vertex x^* is at most $2m$; we can fix all other variables to 0. The simple 'shoot towards the optimum' procedure is shown in Algorithm 3.1.

Algorithm 3.1. *Diameter Bound*

- Start from $t = 0$ and $x^{(0)}$.
- At each iteration t, let $h^{(1)}, \ldots, h^{(k)}$ be a conformal circuit decomposition of $x^* - x^{(t)}$. Let $g^{(t)}$ be the elementary vector in the decomposition that maximises $\|h_N^{(i)}\|_1$ for $i \in [k]$ and update $x^{(t+1)} = \mathrm{aug}_P(x^{(t)}, g^{(t)})$.
- Terminate once $x^{(t+1)} = x^*$.

A priori, even finite termination is not clear. The first key lemma shows that $\|x_N^{(t)}\|_1$ decreases geometrically, and bounds the relative error to x^*.

Lemma 3. *For every iteration $t \geq 0$ in Algorithm 3.1, we have $\|x_N^{(t+1)}\|_1 \leq (1 - \frac{1}{n})\|x_N^{(t)}\|_1$ and for all $i \in [n]$ we have $|x_i^{(t+1)} - x_i^{(t)}| \leq n|x_i^* - x_i^{(t)}|$.*

Proof. Let $h^{(1)}, \ldots, h^{(k)}$ with $k \leq n$ be the conformal circuit decomposition of $x^* - x^{(t)}$ used in Algorithm 3.1. Note that $h_N^{(i)} \leq \mathbb{0}_N$ for $i \in [k]$ as $x_N^* = \mathbb{0}_N$ and $x^{(t)} \geq \mathbb{0}$. Then

$$\|g_N^{(t)}\|_1 = \max_{i \in [k]} \|h_N^{(i)}\|_1 \geq \frac{1}{k} \sum_{i \in [k]} \|h_N^{(i)}\|_1 = \frac{1}{k}\|x_N^{(t)}\|_1, \quad \text{and so}$$

$$\|x_N^{(t+1)}\|_1 = \| \operatorname{aug}_P (x^{(t)}, g^{(t)})_N \|_1 \leq \|x_N^{(t)} + g_N^{(t)}\|_1 \leq \left(1 - \frac{1}{k}\right)\|x_N^{(t)}\|_1. \tag{4}$$

Let $\alpha^{(t)}$ be such that $x^{(t+1)} = x^{(t)} + \alpha^{(t)}g^{(t)}$. Then, by conformality and (4),

$$\alpha^{(t)} = \frac{\|x_N^{(t+1)} - x_N^{(t)}\|_1}{\|g_N^{(t)}\|_1} \leq \frac{\|x_N^{(t)}\|_1}{\|g_N^{(t)}\|_1} \leq k,$$

and so for all i we have $|x_i^{(t+1)} - x_i^{(t)}| = \alpha^{(t)}|g_i^{(t)}| \leq k|g_i^{(t)}| \leq k|x_i^* - x_i^{(t)}|$.

We analyze the sets

$$L_t = \{i \in [n] : x_i^* > n\kappa\|x_N^{(t)}\|_1\}, \quad T_t = [n] \backslash L_t, \quad R_t = \{i \in [n] : x_i^{(t)} \leq nx_i^*\}. \tag{5}$$

Lemma 4. *For every iteration $t \geq 0$, we have $L_t \subseteq L_{t+1} \subseteq B$ and $R_t \subseteq R_{t+1}$.*

Proof. Clearly, $L_t \subseteq L_{t+1}$ as $\|x_N^{(t)}\|_1$ is monotonically decreasing by Lemma 3, and $L_t \subseteq B$ as $x_N^* = \mathbb{0}_N$. Next, let $j \in R_t$. If $x_j^{(t)} \geq x_j^*$, then $x_j^{(t+1)} \leq x_j^{(t)}$ by conformality. Otherwise, if $x_j^{(t)} < x_j^*$, then $x_j^{(t+1)} \leq x_j^{(t)} + n|x_j^* - x_j^{(t)}| \leq nx_j^*$ by Lemma 3. In both cases, we conclude that $j \in R_{t+1}$.

Lemma 5. *If $\|x_{T_t}^{(t)} - x_{T_t}^*\|_\infty > 2mn^2\kappa^2\|x_{T_t}^*\|_\infty$, then $R_t \subsetneq R_{t+1}$.*

Proof. Let $i \in \operatorname{supp}(x^{(t)})\backslash\operatorname{supp}(x^{(t+1)})$; such a variable exists by the maximality of the augmentation. Lemma 2 for $x^{(t+1)} - x^* \in \ker(A)$ implies that

$$x_i^* \leq \|x^{(t+1)} - x^*\|_\infty \leq \kappa\|x_N^{(t+1)} - x_N^*\|_1 = \kappa\|x_N^{(t+1)}\|_1 < \kappa\|x_N^{(t)}\|_1, \tag{6}$$

and so $i \notin L_t$. Noting that $x^{(t+1)} - x^{(t)}$ is an elementary vector and $x_i^{(t+1)} = 0$, it follows that

$$\|x_N^{(t)} - x_N^{(t+1)}\|_1 \leq (m\kappa + 1)x_i^{(t)} \leq 2m\kappa x_i^{(t)}. \tag{7}$$

On the other hand, let $h^{(1)}, \ldots, h^{(k)}$ be the conformal circuit decomposition of $x^* - x^{(t)}$ used in iteration t in Algorithm 3.1. Let $j \in T_t$ such that $|x_j^{(t)} - x_j^*| =$

$\|x_{T_t}^{(t)} - x_{T_t}^*\|_\infty$. There exists $\widetilde{h}$ in this decomposition such that $|\widetilde{h}_j| \geq \frac{1}{n}|x_j^{(t)} - x_j^*|$. Since A_B has full column rank, we have $\mathrm{supp}(\widetilde{h}) \cap N \neq \emptyset$ and so

$$\|\widetilde{h}_N\|_1 \geq \frac{|\widetilde{h}_j|}{\kappa} \geq \frac{|x_j^{(t)} - x_j^*|}{n\kappa}. \tag{8}$$

From (7), (8) and noting that $\|\widetilde{h}_N\|_1 \leq \|g_N^{(t)}\|_1 \leq \|x_N^{(t)} - x_N^{(t+1)}\|_1$ we get

$$x_i^{(t)} \geq \frac{\|x_N^{(t)} - x_N^{(t+1)}\|_1}{2m\kappa} \geq \frac{\|\widetilde{h}_N\|_1}{2m\kappa} \geq \frac{\|x_{T_t}^{(t)} - x_{T_t}^*\|_\infty}{2mn\kappa^2}. \tag{9}$$

In particular, if as in the assumption of the lemma $\|x_{T_t}^{(t)} - x_{T_t}^*\|_\infty > 2mn^2\kappa^2\|x_{T_t}^*\|_\infty$, then $x_i^{(t)} > n\|x_{T_t}^*\|_\infty \geq nx_i^*$. We conclude that $i \notin R_t$ and $i \in R_{t+1}$ as $x_i^{(t+1)} = 0$.

We are ready to give the convergence bound.

Proof (Proof of Theorem 1). In light of Lemma 4, it suffices to show that either L_t or R_t is extended in every $O(n\log(n+\kappa))$ iterations; recall the assumption $n \leq 2m$. By Lemma 5, if $\|x_{T_t}^{(t)} - x_{T_t}^*\|_\infty > 2mn^2\kappa^2\|x_{T_t}^*\|_\infty$, then $R_t \subsetneq R_{t+1}$ is extended.

Otherwise, $\|x_{T_t}^{(t)} - x_{T_t}^*\|_\infty \leq 2mn^2\kappa^2\|x_{T_t}^*\|_\infty$. Assuming $\|x_N^{(t)}\|_1 > 0$, by Lemma 3, there is an iteration $r = t + O(n\log(n+\kappa))$ such that $n^2\kappa(2mn^2\kappa^2 + 1)\|x_N^{(r)}\|_1 < \|x_N^{(t)}\|_1$. In particular,

$$(2mn^2\kappa^2 + 1)\|x_{T_t}^*\|_\infty \geq \|x_{T_t}^{(t)}\|_\infty \geq \|x_N^{(t)}\|_\infty \geq \frac{1}{n}\|x_N^{(t)}\|_1 > n\kappa(2mn^2\kappa^2 + 1)\|x_N^{(r)}\|_1. \tag{10}$$

Therefore $\|x_{T_t}^*\|_\infty > n\kappa\|x_N^{(r)}\|_1$ and so $L_t \subsetneq L_r$.

4 Diameter Bounds for the Capacitated Case

In this section we consider diameter bounds for systems of the form

$$P_u = \{x \in \mathbb{R}^n : Ax = b, 0 \leq x \leq u\}. \tag{Cap-P}$$

We prove Theorem 2 via the following new procedure. A basic feasible point $x^* \in P_u$ is characterised by a partition $B \cup L \cup H = [n]$ where A_B is a basis (has full column rank), $x_L^* = 0_L$ and $x_H^* = u_H$. In $O(n\log n)$ iterations, we fix all but $2m$ variables to the same bound as in x^*; for the remaining system with $2m$ variables, we can use the standard reformulation. The diameter of the polytope obtained in the reformulation equals the diameter of the original polytope, and it is easy to verify that κ is also preserved.

The analysis starts by showing that after $O(n\log n)$ circuit cancellations from the conformal decompositions in Algorithm 4.1, we get to $\langle c, x^{(t)}\rangle < -|H| + 1$; this is maintained after subsequent support circuit cancellations. For such a solution, $x_i^{(t)} < u_i$ for $i \in L$ and $x_i^{(t)} > 0$ for $i \in H$. Every support circuit cancellation sets $x_i^{(t)} \in \{0, u_i\}$ for some $i \in L \cup H$, and by the above property, it always sets the 'correct' bound, i.e., 0 if $i \in L$ and u_i if $i \in H$.

Algorithm 4.1. *Capacitated Diameter Bound*

Let $B \cup L \cup H = [n]$ be the partition for x^*, i.e., $\mathbb{0}_B \le A_B^{-1}b \le u_B$, $x_L^* = \mathbb{0}_L$ and $x_H^* = u_H$. Set the cost $c \in \mathbb{R}_+^n$ as $c_i = 0$ if $i \in B$, $c_i = 1/u_i$ if $i \in L$, and $c_i = -1/u_i$ if $i \in H$.

 – Start from $t = 0$ and some $x^{(0)} \in P_u$.
 – At each iteration t:
   • If $\langle c, x^{(t)} \rangle \ge -|H| + 1$, let $h^{(1)}, \ldots, h^{(k)}$ be a conformal circuit decomposition of $x^* - x^{(t)}$. Let $g^{(t)} \in \arg\min_{i \in [k]} \langle c, h^{(i)} \rangle$.
   • Else, let $g^{(t)}$ be the circuit returned by SUPPORT-CIRCUIT$(A_{S_t}, c_{S_t}, x_{S_t}^{(t)}, S_t)$, where $S_t := \{i \in L \cup H : x_i^{(t)} \ne x_i^*\}$.
   Update $x^{(t+1)} = \text{aug}_P(x^{(t)}, g^{(t)})$.
 – Repeat until $|S_t| \le m$. Then, run Algorithm 3.1 on $\tilde{A} := \begin{bmatrix} A_{B \cup S_t} & 0 \\ I & I \end{bmatrix}$
   and $\tilde{b} = \begin{bmatrix} b \\ u \end{bmatrix}$.

---

5 A Circuit-Augmentation Algorithm for Feasibility

In this section we prove Theorem 3: given a system (LP) with cost $c = (\mathbb{0}_{[n]\setminus N}, \mathbb{1}_N)$ for some $N \subseteq [n]$, find a solution x with $x_N = \mathbb{0}$, or show that no such solution exists. Our algorithm is presented in Algorithm 1. We maintain a set $\mathcal{L}_t \subseteq [n]\setminus N$, initialized as $\emptyset$. Whenever $x_i^{(t)} \ge 8n^3\kappa^2\|x_N^{(t)}\|_1$ for the current iterate $x^{(t)}$, we add i to $\mathcal{L}_t$. The key part of the analysis is to show that $\mathcal{L}_t$ is extended in every $O(n \log(n + \kappa))$ iterations.

We let $T_t = [n]\setminus\mathcal{L}_t$ denote the complement set. At each iteration when $\mathcal{L}_t$ is extended, we run a sequence of at most n SUPPORT-CIRCUIT$(A, c, x^{(t)}, T_t)$ steps. These are repeated as long as $\|x_{T_t}^{(t)}\|_\infty < 4n\kappa\|x_N^{(t)}\|_1$ and there are circuits in $\text{supp}(x^{(t)})$ intersecting T_t. Afterwards, we run a sequence of RATIO-CIRCUIT iterations until a new index is added to $\mathcal{L}_t$.

The crux of the proof is showing that $\mathcal{L}_t$ is extended after a sequence of $O(n \log(n + \kappa))$ RATIO-CIRCUIT iterations. Similarly to Lemma 1, it follows that either $\|x_N^{(t+1)}\|_1 \le \left(1 - \frac{1}{n}\right)\|x_N^{(t)}\|_1$, or the algorithm terminates with a dual certificate. This is used to derive that if $x_j^{(t)}/\|x_N^{(t)}\|_1$ is sufficiently large at some point of the algorithm, it remains large throughout. There are two possible terminations of a sequence of support circuit cancellations: either $\|x_{T_t}^{(t)}\|_\infty$ is above the threshold, or there are no more circuits to cancel. In both cases, $x_j^{(r)}/\|x_N^{(r)}\|_1$ will grow above the threshold after $\|x_N^{(r)}\|_1$ has sufficently decreased.

Algorithm 1: FEASIBILITY-ALGORITHM

Input   : Linear program in standard form (LP) with cost $c = (\mathbb{0}_{[n]\setminus N}, \mathbb{1}_N)$ for some $N \subseteq [n]$, and initial feasible solution $x^{(0)}$.

Output: A solution x with $x_N = 0$, or a dual solution y with $\langle b, y \rangle > 0$.

1   $t \leftarrow 0; \mathcal{L}_{-1} \leftarrow \emptyset$;

2   **while** $x_N^{(t)} > 0$ **do**

3   $\quad$ $\mathcal{L}_t \leftarrow \mathcal{L}_{t-1} \cup \{i \in [n] : x_i^{(t)} \geq 8n^3\kappa^2\|x_N^{(t)}\|_1\}; T_t \leftarrow [n]\setminus\mathcal{L}_t$;

4   $\quad$ **if** $t = 0$ or $\mathcal{L}_t\setminus\mathcal{L}_{t-1} \neq \emptyset$ **then**

5   $\quad\quad$ **while** $\|x_{T_t}^{(t)}\|_\infty < 4n\kappa\|x_N^{(t)}\|_1$ *and there is a circuit in* $\mathrm{supp}(x^{(t)})$ *intersecting* T **do**

6   $\quad\quad\quad$ $g^{(t)} \leftarrow$ SUPPORT-CIRCUIT$(A, c, x^{(t)}, T_t)$ such that $g_k^{(t)} < 0$ for some $k \in T_t$;

7   $\quad\quad\quad$ $x^{(t+1)} \leftarrow \mathrm{aug}_P(x^{(t)}, g^{(t)}); t \leftarrow t+1$;

8   $\quad$ $(g^{(t)}, y^{(t)}, s^{(t)}) \leftarrow$ RATIO-CIRCUIT$(A, c, 1/x^{(t)})$;

9   $\quad$ **if** $\langle b, y^{(t)} \rangle > 0$ **then**

10 $\quad\quad$ Terminate with infeasibility certificate

11 $\quad$ $x^{(t+1)} \leftarrow \mathrm{aug}_P(x^{(t)}, g^{(t)}); t \leftarrow t+1$;

12 **return** x;

---

6   A Circuit-Augmentation Algorithm for Optimization

In this section, we give a circuit-augmentation algorithm for solving (LP), assuming an initial feasible solution $x^{(0)}$ is provided. At all times, the algorithm maintains a feasible primal solution $x^{(t)}$ to (LP), initialized with $x^{(0)}$. The goal is to augment $x^{(t)}$ using the subroutines SUPPORT-CIRCUIT and RATIO-CIRCUIT until the emergence of a set $\emptyset \neq N \subseteq [n]$ which satisfies $x_N^{(t)} = x_N^* = 0$ for every optimal solution x^* to (LP). This conclusion can be derived using the proximity result Theorem 5. When this happens, we have reached a lower dimensional face of the feasible region (P) that contains the optimal face. Then, the same procedure is repeated on a smaller LP; these circuit walks can be concatenated to obtain the overall circuit walk in the original instance.

In what follows, we focus on the aforementioned VARIABLE-FIXING procedure (see pseudocode in the full version). We start by orthogonally projecting the original cost vector c to $\ker(A)$. This does not change the optimal face of (LP). If $c = 0$, then we terminate and return the current feasible solution $x^{(0)}$ as it is optimal. Otherwise, we scale the cost to $\|c\|_2 = 1$, and use RATIO-CIRCUIT to obtain a basic dual feasible solution $s^{(-1)}$ to LP(c).

The majority of VARIABLE-FIXING consists of repeated *phases*, ending when $\langle x^{(t)}, s^{(t-1)} \rangle = 0$. At the start of a phase, the set S of coordinates with large dual slack $s_i^{(t-1)} \geq \delta$ is identified for some $\delta < 1/(n^{3/2}(m+2)\kappa)$. Based on this, a modified cost function $\tilde{c} \geq 0$ is derived from $s^{(t-1)}$ by truncating the entries not in S to zero. This modified cost $\tilde{c}$ will be used until the end of the phase. Next, we augment our current primal solution $x^{(t)}$ by calling SUPPORT-CIRCUIT$(A, \tilde{c}, x^{(t)}, S)$ to eliminate circuits in supp$(x^{(t)})$ intersecting S with nonpositive $\tilde{c}$-cost. Note that there are at most n such calls because each call sets a primal variable $x_i^{(t)}$ to zero.

In the remaining part of the phase, we augment $x^{(t)}$ using RATIO-CIRCUIT$(A, \tilde{c}, 1/x^{(t)})$ for $T = O(n \log(n + \kappa))$ iterations. In every iteration, RATIO-CIRCUIT$(A, \tilde{c}, 1/x^{(t)})$ returns a minimum cost-to-weight ratio circuit $g^{(t)}$, where the choice of weights $1/x^{(t)}$ follows Wallacher [32]. Recall that the oracle also gives a basic dual feasible solution $s^{(t)}$ to LP$(\tilde{c})$. If $g^{(t)}$ does not improve the current solution $x^{(t)}$, i.e., $\langle \tilde{c}, g^{(t)} \rangle = 0$, then we terminate the phase early as $x^{(t)}$ is already optimal to LP$(\tilde{c})$. In this case, $s^{(t)}$ is an optimal dual solution to LP$(\tilde{c})$ because $\langle x^{(t)}, s^{(t)} \rangle = 0$. This finishes the description of a phase. We show that there are at most n phases. Then, applying Theorem 5 for c' with $\|c' - c\|_\infty \le n\delta$ allows us to conclude that a variable can be fixed to zero.

References

1. Bland, R.G.: On the generality of network flow theory. Presented at the ORSA/TIMS Joint National Meeting, Miami, FL (1976)
2. Bonifas, N., Di Summa, M., Eisenbrand, F., Hähnle, N., Niemeier, M.: On sub-determinants and the diameter of polyhedra. Discrete Comput. Geom. **52**(1), 102–115 (2014)
3. Borgwardt, S., De Loera, J.A., Finhold, E.: Edges versus circuits: a hierarchy of diameters in polyhedra. Adv. Geom. **16**(4), 511–530 (2016)
4. Borgwardt, S., Finhold, E., Hemmecke, R.: On the circuit diameter of dual trans-portation polyhedra. SIAM J. Discret. Math. **29**(1), 113–121 (2015)
5. Borgwardt, S., Stephen, T., Yusun, T.: On the circuit diameter conjecture. Discrete Comput. Geom. **60**(3), 558–587 (2018)
6. Borgwardt, S., Viss, C.: An implementation of steepest-descent augmentation for linear programs. Oper. Res. Lett. **48**(3), 323–328 (2020)
7. Brunsch, T., Röglin, H.: Finding short paths on polytopes by the shadow vertex algorithm. In: Fomin, F.V., Freivalds, R., Kwiatkowska, M., Peleg, D. (eds.) ICALP 2013. LNCS, vol. 7965, pp. 279–290. Springer, Heidelberg (2013). https://doi.org/10.1007/978-3-642-39206-1_24
8. Cook, W., Gerards, A.M., Schrijver, A., Tardos, É.: Sensitivity theorems in integer linear programming. Math. Program. **34**(3), 251–264 (1986)
9. Dadush, D., Hähnle, N.: On the shadow simplex method for curved polyhedra. Discrete Comput. Geom. **56**(4), 882–909 (2016)
10. Dadush, D., Huiberts, S., Natura, B., Végh, L.A.: A scaling-invariant algorithm for linear programming whose running time depends only on the constraint matrix. In: Proceedings of the 52nd Annual ACM Symposium on Theory of Computing (STOC), pp. 761–774 (2020)
11. Dadush, D., Natura, B., Végh, L.A.: Revisiting Tardos's framework for linear pro-gramming: faster exact solutions using approximate solvers. In: Proceedings of the 61st Annual IEEE Symposium on Foundations of Computer Science (FOCS), pp. 931–942 (2020)

12. De Loera, J.A., Hemmecke, R., Lee, J.: On augmentation algorithms for linear and integer-linear programming: from Edmonds-Karp to Bland and beyond. SIAM J. Optim. **25**(4), 2494–2511 (2015)
13. De Loera, J.A., Kafer, S., Sanità, L.: Pivot rules for circuit-augmentation algorithms in linear optimization. arXiv preprint arXiv:1909.12863 (2019)
14. Dyer, M., Frieze, A.: Random walks, totally unimodular matrices, and a randomised dual simplex algorithm. Math. Program. **64**(1), 1–16 (1994)
15. Eisenbrand, F., Vempala, S.: Geometric random edge. Math. Program. **164**(1–2), 325–339 (2017)
16. Ekbatani, F., Natura, B., Végh, L.A.: Circuit imbalance measures and linear programming. arXiv preprint arXiv:2108.03616 (2021)
17. Gauthier, J.B., Desrosiers, J.: The minimum mean cycle-canceling algorithm for linear programs. Eur. J. Oper. Res. **298**, 36–44 (2021)
18. Goldberg, A.V., Tarjan, R.E.: Finding minimum-cost circulations by canceling negative cycles. J. ACM (JACM) **36**(4), 873–886 (1989)
19. Kafer, S., Pashkovich, K., Sanità, L.: On the circuit diameter of some combinatorial polytopes. SIAM J. Discret. Math. **33**(1), 1–25 (2019)
20. Kalai, G.: A subexponential randomized simplex algorithm. In: Proceedings of the 24th annual ACM Symposium on Theory of Computing, pp. 475–482 (1992)
21. Kalai, G., Kleitman, D.J.: A quasi-polynomial bound for the diameter of graphs of polyhedra. Bull. Am. Math. Soc. **26**(2), 315–316 (1992)
22. Lee, J.: Subspaces with well-scaled frames. Linear Algebra Appl. **114**, 21–56 (1989)
23. McCormick, S.T., Shioura, A.: Minimum ratio canceling is oracle polynomial for linear programming, but not strongly polynomial, even for networks. Oper. Res. Lett. **27**(5), 199–207 (2000)
24. Santos, F.: A counterexample to the Hirsch conjecture. Ann. Math. **176**(1), 383–412 (2012)
25. Schulz, A.S., Weismantel, R.: An oracle-polynomial time augmentation algorithm for integer programming. In: Proceedings of the 10th annual ACM-SIAM Symposium on Discrete Algorithms (SODA), pp. 967–968 (1999)
26. Sturmfels, B., Thomas, R.R.: Variation of cost functions in integer programming. Math. Program. **77**(2), 357–387 (1997)
27. Sukegawa, N.: Improving bounds on the diameter of a polyhedron in high dimensions. Discret. Math. **340**(9), 2134–2142 (2017)
28. Tardos, É.: A strongly polynomial algorithm to solve combinatorial linear programs. Oper. Res. **34**, 250–256 (1986)
29. Tunçel, L.: Approximating the complexity measure of Vavasis-Ye algorithm is NP-hard. Math. Program. **86**(1), 219–223 (1999)
30. Vavasis, S.A.: Stable numerical algorithms for equilibrium systems. SIAM J. Matrix Anal. Appl. **15**(4), 1108–1131 (1994)
31. Vavasis, S.A., Ye, Y.: A primal-dual interior point method whose running time depends only on the constraint matrix. Math. Program. **74**(1), 79–120 (1996)
32. Wallacher, C.: A generalization of the minimum-mean cycle selection rule in cycle canceling algorithms (1989). Unpublished manuscript, Institute für Angewandte Mathematik, Technische Universität Braunschweig
33. Wallacher, C., Zimmermann, U.T.: A polynomial cycle canceling algorithm for submodular flows. Math. Program. **86**(1), 1–15 (1999)
34. Wayne, K.D.: A polynomial combinatorial algorithm for generalized minimum cost flow. Math. Oper. Res. **27**, 445–459 (2002)

A Simple Method for Convex Optimization in the Oracle Model

Daniel Dadush[1], Christopher Hojny[2], Sophie Huiberts[1], and Stefan Weltge[3(⊠)]

[1] Centrum Wiskunde and Informatica, Amsterdam, The Netherlands
{dadush,s.huiberts}@cwi.nl
[2] Eindhoven University of Technology, Eindhoven, The Netherlands
c.hojny@tue.nl
[3] Technical University of Munich, Munich, Germany
weltge@tum.de

Abstract. We give a simple and natural method for computing approximately optimal solutions for minimizing a convex function f over a convex set K given by a separation oracle. Our method utilizes the Frank–Wolfe algorithm over the cone of valid inequalities of K and subgradients of f. Under the assumption that f is L-Lipschitz and that K contains a ball of radius r and is contained inside the origin centered ball of radius R, using $O(\frac{(RL)^2}{\varepsilon^2} \cdot \frac{R^2}{r^2})$ iterations and calls to the oracle, our main method outputs a point $x \in K$ satisfying $f(x) \le \varepsilon + \min_{z \in K} f(z)$.

Our algorithm is easy to implement, and we believe it can serve as a useful alternative to existing cutting plane methods. As evidence towards this, we show that it compares favorably in terms of iteration counts to the standard LP based cutting plane method and the analytic center cutting plane method, on a testbed of combinatorial, semidefinite and machine learning instances.

Keywords: convex optimization · separation oracle · cutting plane method

1 Introduction

We consider the problem of minimizing a convex function $f \colon \mathbb{R}^n \to \mathbb{R}$ over a compact convex set $K \subseteq \mathbb{R}^n$. We assume that K contains an (unknown) Euclidean ball of radius $r > 0$ and is contained inside the origin centered ball of radius $R > 0$, and that f is L-Lipschitz. We have first-order access to f that yields $f(x)$ and a subgradient of f at x for any given x. Moreover, we only have access to K through a separation oracle (SO), which, given a point $x \in \mathbb{R}^n$, either asserts that $x \in K$ or returns a linear constraint valid for K but violated by x.

This project has received funding from the European Research Council (ERC) under the European Union's Horizon 2020 research and innovation programme (grant agreement QIP–805241).

K. Aardal and L. Sanitá (Eds.): IPCO 2022, LNCS 13265, pp. 154–167, 2022.
https://doi.org/10.1007/978-3-031-06901-7_12

Convex optimization in the SO model is one of the fundamental settings in optimization. The model is relevant for a wide variety of implicit optimization problems, where an explicit description of the defining inequalities for K is either too large to store or not fully known. The SO model was first introduced in [29] where it was shown that an additive ε-approximate solution can be obtained using $O(n \log(LR/(\varepsilon r)))$ queries via the center of gravity method and $O(n^2 \log(LR/(\varepsilon r)))$ queries via the ellipsoid method. This latter result was used by Khachiyan [27] to give the first polynomial time method for linear programming. The study of oracle-type models was greatly extended in the classic book of Grötschel, Lovász, and Schrijver [23], where many applications to combinatorial optimization were provided. Further progress on the SO model was given by Vaidya [36], who showed that the $O(n \log(LR/(\varepsilon r)))$ oracle complexity can be efficiently achieved using the so-called volumetric barrier as a potential function, where the best current running time for such methods was given very recently [25, 28].

From the practical perspective, two of the most popular methods in the SO model are the standard linear programming (LP) based cutting plane method, independently discovered by Kelley [26], Goldstein-Cheney [9] as well as Gomory [22] (in the integer programming context), and the analytic center cutting plane method [34] (ACCPM).

The LP based cutting plane method, which we henceforth dub the standard cut loop, proceeds as follows: starting with finitely many linear underestimators of f and linear constraints valid for K, in each iteration it solves a linear program that minimizes the lower envelope of f subject to the current linear relaxation of K. The resulting point x is then used to query f and the SO to obtain a new underestimator for f and a new constraint valid for K. Note that if f is a linear function, it repeatedly minimizes f over linear relaxations of K. While it is typically fast in practice, it can be unstable, and no general quantitative convergence guarantees are known for the standard cut loop.

To link to integer programming, in that context K is the convex hull of integer points of some polytope P and the objective is often linear, and the method is initialized with a linear description of P. A crucial difference there is that the separator SO is generally only efficient when queried at vertices of the current relaxation.

ACCPM is a barrier based method, in which the next query point is the minimizer of the barrier for the current inequalities in the system. ACCPM is in general a more stable method with provable complexity guarantees. Interestingly, while variants of ACCPM with $O(n \log(RL/(r\varepsilon))^2)$ convergence exist, achieved by judiciously dropping constraints [1], the more practical variants have worse guarantees. For instance, if K is the ball of radius R, the standard variant of ACCPM is only shown to achieve $O(n(RL/\varepsilon)^2 \log(RL/\varepsilon))$ convergence [30].

In this paper, we describe a new method for convex optimization in the SO model that computes an additive ε-approximate solution within $O(R^4 L^2/r^2 \varepsilon^2)$ iterations. Our algorithm is easy to implement, and we believe it can serve as a useful alternative to existing methods. In our experimental results, we show that it compares favorably in terms of iteration counts to the standard cut loop and the analytic center cutting plane method, on a testbed of combinatorial, semidefinite and machine learning instances.

Before explaining our approach, we review the relevant work in related models. To begin, there has been a tremendous amount of work in the context of first-order methods [3,5], where the goal is to minimize a possibly complicated function, given by a gradient oracle, over a *simple domain* K (e.g., the simplex, cube, ℓ_2 ball). These methods tend to have cheap iterations and to achieve poly$(1/\varepsilon)$ convergence rates. They are often superior in practice when the requisite accuracy is low or moderate, e.g., within 1% of optimal. For these methods, often variants of (sub-)gradient descent, it is generally assumed that computing (Euclidean) projections onto K as well as linear optimization over K are easy. If one only assumes access to a linear optimization (LO) oracle on K, K can become more interesting (e.g., the shortest-path or spanning-tree polytope). In this context, one of the most popular methods is the so-called *Frank–Wolfe* algorithm [19] (see [24] for a modern treatment), which iteratively computes a convex combination of vertices of K to obtain an approximate minimizer of a smooth convex function.

In the context of combinatorial optimization, there has been a considerable line of work on solving (implicit) packing and covering problems using the so-called multiplicative weights update (MWU) framework [20,31,33]. In this framework, one must be able to implement an MWU oracle, which in essence computes optimal solutions for the target problem after the "difficult" constraints have been aggregated according to the current weights. This framework has been applied for getting fast $(1 \pm \varepsilon)$-approximate solutions to multicommodity flow [20,33], packing spanning trees [8], the Held–Karp approximation for TSP [7], and more, where the MWU oracle computes shortest paths, minimum cost spanning trees, minimum cuts respectively in a sequence of weighted graphs. The MWU oracle is in general just a special type of LO oracle, which can often be interpreted as a SO that returns a maximally violated constraint. While certainly related to the SO model, it is not entirely clear how to adapt MWU to work with a general SO, in particular in settings unrelated to packing and covering.

A final line of work, which directly inspires our work, has examined simple iterative methods for computing a point in the interior of a cone Σ that directly apply in the SO model. The application of simple iterative methods for solving conic feasibility problems can be traced to Von Neumann in 1948 (see [15]), and a variant of this method, the perceptron algorithm [32] is still very popular today. Von Neumann's algorithm computes a convex combination of the defining inequalities of the cone, scaled to be of unit length, of nearly minimal Euclidean norm. The separation oracle is called to find an inequality violated by the current convex combination, and this inequality is then used to make the current convex combination shorter, in an analogous way to Frank–Wolfe. This method is guaranteed to find a point in the cone in $O(1/\rho^2)$ iterations, where ρ is the so-called width of Σ (the radius of the largest ball contained in Σ centered at a point of norm 1). Starting in 2004, polynomial time variants of this and related methods (i.e., achieving $\log(1/\rho)$ dependence) have been found [6,10,17], which iteratively "rescale" the norm to speed up the convergence. These rescaled variants can also be applied in the oracle setting [4,11,14] with appropriate adaptations. The main shortcoming of existing conic approaches is that they are currently not well-adapted for solving optimization problems rather than feasibility problems.

Our Approach. In this work, we build upon von Neumann's approach and utilize the Frank–Wolfe algorithm over the cone of valid inequalities of K as well as the subgradients of f in a way that yields a clean, simple, and flexible framework for solving general convex optimization problems in the SO model. For simpler explanation, let us assume that $f(x) = \langle c, x \rangle$ is a linear function and that we know an upper bound UB on the minimum of f over K. Given some linear inequalities $\langle a_i, x \rangle \leq b_i$ that are valid for all $x \in K$, our goal is to find convex combinations p of the *homogenized* points (c, UB) and (a_i, b_i) that are "close" to the origin. Note that if $p = 0$, the fact that K is full-dimensional implies that (c, UB) appears with a nonzero coefficient and hence $(-c, -\mathrm{UB})$ is a nonnegative combination of the points (a_i, b_i), which in turn shows that UB is equal to the minimum of f over K. In view of this, we will consider a potential $\Phi \colon \mathbb{R}^{n+1} \to \mathbb{R}_+$ with the property that if $\Phi(p)$ is sufficiently small, then the convex combination will yield an explicit certificate that UB is close to the minimum of f over K.

Given a certain convex combination p, note that the gradient of Φ at p provides information about whether moving towards one of the known points will (significantly) decrease $\Phi(p)$. However, if no such known point exists, it turns out that the "dehomogenization" of the gradient (a scaling of its projection onto the first n coordinates) is a natural point $x \in \mathbb{R}^n$ to query the SO with. In fact, if $x \in K$, it will have improved objective value with respect to f. Otherwise, the SO will provide a linear inequality such that moving towards its homogenization decreases $\Phi(p)$.

In this work, we will show that the above paradigm immediately yields a rigorous algorithm for various natural choices of Φ and scalings of inequalities. We will also see that general convex functions can be directly handled in the same manner by simply replacing (c, UB) with all subgradient cuts of f learned throughout the iterations. The same applies to pure feasibility problems for which we set $f = 0$. The convergence analysis of our algorithm is simple and based on standard estimates for the Frank–Wolfe algorithm.

Besides its conceptual simplicity and distinction to existing methods for convex optimization in the SO model, we also regard it as a practical alternative. In fact, in terms of iterations, our vanilla implementation in Julia[1] performs similarly and often even better than the standard cut loop and the analytic center cutting plane method evaluated on a testbed of oracle-based linear optimization problems for matching problems, semidefinite relaxations of the maximum cut

---

[1] https://github.com/christopherhojny/supplement_simple-iterative-methods-linopt-convex-sets.

problem, and LPBoost. Moreover, the flexibility of our framework leaves several degrees of freedom to obtain optimized implementations that outperform our naive implementation.

2   Algorithm

Recall that we are given first-order access to a convex function $f\colon \mathbb{R}^n \to \mathbb{R}$ that we want to minimize over a convex body $K \subseteq \mathbb{R}^n$. In the case where f is not differentiable, with a slight abuse of notation we interpret $\nabla f(x)$ to be *any* subgradient of f at x. We can access K by a separation oracle that, given a point $x \in \mathbb{R}^n$, either asserts that $x \in K$ or returns a point $(a, b) \in \mathcal{A} \subseteq \mathbb{R}^{n+1}$ with $\langle a, x \rangle > b$ such that $\langle a, y \rangle \le b$ holds for all $y \in K$. Here, $\langle \cdot, \cdot \rangle$ denotes the standard scalar product and we assume that all points in $\mathcal{A}$ correspond to linear constraints valid for K. To state our algorithm, let $\| \cdot \|$ denote any norm on $\mathbb{R}^{n+1}$ and $\| \cdot \|_*$ its dual norm. Moreover, let $\Phi\colon \mathbb{R}^{n+1} \to \mathbb{R}_+$ be any strictly convex and differentiable function with $\min_{x \in \mathbb{R}^{n+1}} \Phi(x) = \Phi(0) = 0$. Our method is given in Algorithm 1, in which we denote the number of iterations by T for later reference. However, T does not need to be specified in advance, and the algorithm may be stopped at any time, e.g., when a solution or bound of desired accuracy has been found.

Algorithm 1

1: UB $\leftarrow \infty$, $A_1 \leftarrow \{(\mathbf{0}, 1)/\|(\mathbf{0}, 1)\|_*\}$, $G_1 \leftarrow \emptyset$
2: **for** $t = 1, 2, \dots, T$ **do**
3:     $p_t \leftarrow \arg\min\{\Phi(p) : p \in \mathrm{conv}(A_t \cup G_t)\}$
4:     **if** $p_t = \mathbf{0}$ **then return** UB.
5:     $x_t \leftarrow -\nabla\Phi(p_t)[1 : n]/\nabla\Phi(p_t)[n + 1]$
6:     **if** $x_t \in K$ **then**
7:         UB $\leftarrow \min\{\text{UB}, f(x_t)\}$
8:         $A_{t+1} \leftarrow A_t$.
9:         $G_{t+1} \leftarrow G_t \cup \{(\nabla f(x_t), \langle \nabla f(x_t), x_t \rangle)\}$
10:    **else**
11:        get $(a, b) \in \mathcal{A}$, with $\langle a, x_t \rangle > b$ and $\|(a, b)\|_* = 1$
12:        $A_{t+1} \leftarrow A_t \cup \{(a, b)\}$.
13:        $G_{t+1} \leftarrow G_t$.
14: **return** UB.

In Line 5, $\nabla\Phi(p_t)[1 : n]$ denotes the first n components of $\nabla\Phi(p_t)$, and $\nabla\Phi(p_t)[n+1]$ denotes the last component of $\nabla\Phi(p_t)$. The sets A_t and G_t denote the already known/separated inequalities and objective gradients during iteration t.

Lemma 1. *When $x_t \in \mathbb{R}^n$ is computed in iteration t of Algorithm 1, it is well-defined and we have $\langle c, x_t \rangle \le d$ for every $(c, d) \in A_t \cup G_t$.*

Proof. Since p_t minimizes Φ over $\text{conv}(A_t \cup G_t)$, for every $q \in \text{conv}(A_t \cup G_t)$ we have $\langle \nabla\Phi(p_t), q - p_t \rangle \geq 0$. If $p_t \neq \mathbf{0}$ then from strict convexity of Φ and $\min_{x \in \mathbb{R}^{n+1}} \Phi(x) = \Phi(\mathbf{0}) = 0$ we get

$$\langle \nabla\Phi(p_t), q \rangle \geq \langle \nabla\Phi(p_t), p_t \rangle > 0. \tag{1}$$

First, apply (1) to $q = (\mathbf{0}, 1)/\|(\mathbf{0}, 1)\|_* \in A_t$ and conclude $\nabla\Phi(p_t)[n+1] > 0$. This makes sure that x_t can be computed. Second, we apply Inequality (1) to $q = (c, d) \in A_t \cup G_t$ and find that $d - \langle c, x_t \rangle = \frac{1}{\nabla\Phi(p_t)[n+1]} \langle \nabla\Phi(p_t), (c, d) \rangle > 0$, thus x_t satisfies $\langle c, x_t \rangle \leq d$ for all $(c, d) \in A_t \cup G_t$. $\qquad\square$

Note that, for the sake of presentation, in Line 3 we require p_t to be the convex combination of minimum Φ-value. However, it is usually not necessary to compute such a minimum. The same convergence rates can be obtained if, in every iteration, p_t is a suitable convex combination of p_{t-1} and some $(c, d) \in A_t \cup G_t$ with $\langle \nabla\Phi(p_{t-1}), (c, d) \rangle < 0$. If the last coordinate of p_{t-1}, as discussed in the above proof, is not positive, then such an update can be made towards $(\mathbf{0}, 1)/\|(\mathbf{0}, 1)\|_* \in A_t$. Any such update will significantly decrease $\Phi(p_t)$, and the computation in Line 3 is guaranteed to make at least that much progress. This shows that simple updates of p_t, which may be more preferable in practice, still suffice to achieve the claimed convergence rates.

Lemma 2. *Suppose that Φ is 1-smooth with respect to $\|\cdot\|_*$ and that*

$$\|(\nabla f(x), \langle \nabla f(x), x \rangle)\|_* \leq 1$$

for every $x \in K$. Then for every $t = 1, \ldots, T$, Algorithm 1 satisfies $\Phi(p_t) \leq \frac{8}{t+2}$.

Proof Idea. In every iteration, we add a point $q_t \in A_{t+1} \cup G_{t+1}$ with $\|q_t\|_* \leq 1$ such that $\langle \nabla\Phi(p_t), q_t \rangle \leq 0$ holds. The line segment between p_t and q_t contains a point p' with $\Phi(p') \leq \Phi(p_t) - \frac{1}{8}\Phi(p_t)^2$, which is enough to prove the claim. A complete proof can be found in the arXiv [13] version. $\qquad\square$

The following lemma yields conditions under which a small value of $\Phi(p_t)$ implies that UB is close to the minimum of f over K. Note in particular that it proves that if $\|p_t\| = 0$ then UB = OPT.

Lemma 3. *Assume that $\|(x, -1)\| \leq 2$ holds for every $x \in K$, and there exist $z \in K$ and $\alpha \in (0, 1]$ such that $\langle (a, b), (-z, 1) \rangle \geq \alpha \|(-z, 1)\| \|(a, b)\|_*$ holds for every $(a, b) \in A \cup \{(\mathbf{0}, 1)\}$. Moreover, assume that $\|(\nabla f(x), \langle \nabla f(x), x \rangle)\|_* \leq 1$ holds for every $x \in K$. If $\|p_T\| \leq \alpha/2$ in Algorithm 1, then the returned value satisfies $\text{UB} \geq \text{OPT} \geq \text{UB} - \frac{4\|p_T\|_*(1+\alpha)}{\alpha}$.*

Proof. Let $x^* \in K$ minimize $f(x)$ over $x \in K$ and let $F \subset [T-1]$ be the set of iterations (except the last one) in which $x_t \in K$. Now write the point p_T as a convex combination

$$p_T = \sum_{(a,b) \in A_T} \lambda_{(a,b)}(a, b) + \sum_{t \in F} \gamma_t (\nabla f(x_t), \langle \nabla f(x_t), x_t \rangle)$$

where $\lambda \geq 0, \gamma \geq 0$ and $\|(\lambda, \gamma)\|_1 = 1$. Then we have

$$
\sum_{t \in F} \gamma_t (f(x_t) - f(x^*)) \leq \sum_{t \in F} \gamma_t \langle \nabla f(x_t), x_t - x^* \rangle
$$

$$
= \Big\langle \sum_{t \in F} \gamma_t (\nabla f(x_t), \langle \nabla f(x_t), x_t \rangle), \; (-x^*, 1) \Big\rangle
$$

$$
\leq \Big\langle \sum_{t \in F} \gamma_t (\nabla f(x_t), \langle \nabla f(x_t), x_t \rangle) + \sum_{(a,b) \in A_T} \lambda_{(a,b)} (a, b), \; (-x^*, 1) \Big\rangle
$$

$$
= \langle p_T, (-x^*, 1) \rangle \leq \|p_T\|_* \cdot \|(-x^*, 1)\| \leq 2 \|p_T\|_*.
$$

Here, the inequalities respectively arise from convexity of f, that $x^* \in K$ satisfies $\langle (a, b), (-x^*, 1) \rangle \geq 0$ for every $(a, b) \in A_T$, and the Cauchy–Schwarz inequality. In particular, we find that $\min_{t \in F} f(x_t) - f(x^*) \leq \frac{2 \|p_T\|_*}{\sum_{t \in F} \gamma_t}$ whenever $\sum_{t \in F} \gamma_t > 0$. To lower bound this latter quantity, we use the assumptions on z to derive the inequalities

$$
\alpha \Big(1 - \sum_{t \in F} \gamma_t \Big) \|(-z, 1)\| = \alpha \|(-z, 1)\| \sum_{(a,b) \in A_T} \lambda_{(a,b)}
$$

$$
\leq \langle \sum_{(a,b) \in A_T} \lambda_{(a,b)} (a, b), (-z, 1) \rangle \quad (\text{since} \|(a, b)\|_* = 1)
$$

$$
= \langle p_T, (-z, 1) \rangle - \sum_{t \in F} \gamma_t \langle (\nabla f(x_t), \langle \nabla f(x_t), x_t \rangle), (-z, 1) \rangle
$$

$$
\leq \|p_T\|_* \cdot \|(-z, 1)\| + \sum_{t \in F} \gamma_t \|(\nabla f(x_t), \langle \nabla f(x_t), x_t \rangle)\|_* \cdot \|(-z, 1)\|.
$$

Now observe that $\|(\nabla f(x_t), \langle \nabla f(x_t), x_t \rangle)\|_* \leq 1$ for every $t \in F$ and divide through by $\|(-z, 1)\|$ to find $\alpha(1 - \sum_{t \in F} \gamma_t) \leq \|p_T\|_* + \sum_{t \in F} \gamma_t$. Hence, if $\|p_T\|_* \leq \frac{\alpha}{2}$ then $\alpha/2 \leq (\alpha + 1) \sum_{t \in F} \gamma_t$. This lower bound on $\sum_{t \in F} \gamma_t$ suffices to prove the lemma. □

Combining the previous two lemmas, we obtain the following convergence rate of our algorithm:

Theorem 1. *Assume that $\beta > 0$ is such that $\Phi(x) \geq \beta \|x\|_*^2$ for all $x \in \mathbb{R}^{n+1}$. Under the assumptions of Lemmas 2 and 3, Algorithm 1 computes, for every $T \geq \frac{32}{\beta \alpha^2}$, a value $\mathrm{UB} < \infty$ satisfying $\mathrm{UB} \geq \min_{x \in K} f(x) \geq \mathrm{UB} - \frac{16}{\sqrt{\beta(T+2)}} \cdot \frac{1+\alpha}{\alpha}$.*

Proof. After T iterations, we have $\beta \|p_T\|_*^2 \leq \Phi(p_T) \leq \frac{8}{T+2} \leq \beta \alpha^2 / 4$ per Lemma 2. Since then $\|p_T\|_* \leq \frac{\sqrt{8}}{\sqrt{\beta(T+2)}} \leq \alpha/2$, Lemma 3 tells us that $\mathrm{OPT} \geq \mathrm{UB} - \frac{16(1+\alpha)}{\sqrt{\beta(T+2)\alpha}}$. □

Let us now apply the previous findings to a concrete setting, in which we assume that the objective function f is L-Lipschitz, i.e., $|f(x) - f(y)| \leq L \|x - y\|_2$ for all $x, y \in \mathbb{R}^n$.

Theorem 2. *Let $K \subset \mathbb{R}^n$ be a convex body satisfying $z + r\mathbb{B}_2^n \subset K \subset R\mathbb{B}_2^n$, given by a separation oracle $\mathcal{A}$, and let $f \colon \mathbb{R}^n \to \mathbb{R}$ be an L-Lipschitz convex function given by a subgradient oracle.*

Apply Algorithm 1 to the function $\frac{1}{RL}f$ using norm $\|(x,y)\| := \sqrt{2}\|(x/R, y)\|_2$ and potential $\Phi(a,b) := \frac{1}{4}\|(Ra,b)\|_2^2$. Then, for every $\varepsilon > 0$, after

$$T = O\left(\frac{R^2}{r^2} \cdot \frac{R^2 L^2}{\varepsilon^2}\right)$$

iterations we have $\mathrm{UB} \geq \min_{x \in K} f(x) \geq \mathrm{UB} - \varepsilon$.

Proof. By replacing $f(x)$ by $f(Rx)/(RL)$, K by K/R, ε by $\varepsilon/(RL)$, r by r/R, and z by z/R, we may assume that $R = L = 1$, that $r \in (0,1]$. After this rescaling, note $\|(x,y)\| := \sqrt{2}\|(x,y)\|_2$ and $\Phi(a,b) := \frac{1}{4}\|(a,b)\|_2^2 = \frac{1}{2}\|(a,b)\|_*^2$. Crucially, note that Algorithm 1 is invariant under the above replacement.

We now claim that our choice of input satisfies the conditions of Theorem 1 with $\beta = 1/2$ and $\alpha = r/4$. Given the claim, Theorem 1 directly proves the result. To prove the claim, apart from verifying that the bounds on β and α hold, we must verify smoothness of Φ with respect to the dual norm, a bound of 2 on the norm of $(-x, 1)$ for $x \in K$, as well as a dual norm bound of 1 on $(\nabla f(x), \langle \nabla f(x), x \rangle)$ for $x \in K$.

The setting $\beta = 1/2$ is direct by definition of Φ. Since $\|\cdot\|_*$ is a Euclidean norm, it is immediate that Φ is 1-smooth with respect to $\|\cdot\|_*$. For each $x \in K$, using that $R = L = 1$, we may also verify that

$$\|(x,1)\| = \sqrt{2}\|(x,1)\|_2 = \sqrt{2}\sqrt{\|x\|_2^2 + 1} \leq \sqrt{2}\sqrt{R^2 + 1} = 2,$$

and

$$\|(\nabla f(x), \langle \nabla f(x), x \rangle)\|_* = \frac{1}{\sqrt{2}}\|(\nabla f(x), \langle \nabla f(x), x \rangle)\|_2$$

$$\leq \frac{1}{\sqrt{2}}\sqrt{\|\nabla f(x)\|_2^2 + \|\nabla f(x)\|^2\|x\|^2}$$

$$\leq \frac{1}{\sqrt{2}}\sqrt{L^2 + L^2 R^2} = 1.$$

We now show the lower bound $\alpha \geq r/4$. Firstly, since $\|(-z,1)\|\|(0,1)\|_* = \|(-z,1)\|_2\|(0,1)\|_2 \leq \sqrt{2}$, we see that $\langle (-z,1), (0,1) \rangle = 1 \geq \frac{1}{2}\|(-z,1)\|\|(0,1)\|_*$. Next, any (a,b) returned by the oracle is normalized so that $\|(a,b)\|_* = 1 \Leftrightarrow \|(a,b)\|_2 = \sqrt{2}$. Note then that $\|(-z,1)\|\|(a,b)\|_* \leq 2$. From here, we observe that

$$\langle (a,b), (-z,1) \rangle = b - \langle a, z \rangle = b - \langle a, z + ra/\|a\|_2 \rangle + r\|a\|_2 \geq r\|a\|_2,$$

since $z + ra/\|a\|_2 \in K$ by assumption. Furthermore, $b - \langle a, z \rangle \geq b - \|a\|_2\|z\|_2 \geq b - \|a\|_2$ and $0 \leq b - \langle a, z \rangle \leq b + \|a\|_2$. Thus, $b - \langle a, z \rangle \geq \max\{r\|a\|_2, b - \|a\|_2\}$. We now examine two cases. If $\|a\|_2 \geq 1/2$, then $b - \langle a, z \rangle \geq r/2 \geq r/4 \cdot \|(-z,1)\|\|(a,b)\|_*$. If $\|a\|_2 \leq 1/2$, then $|b| \geq 1$ since $\|(a,b)\|_2^2 = 2$. This gives $b - \langle a, z \rangle \geq b - \|a\|_2 \geq 1/2 \geq r/2$. Thus, $\alpha \geq r/4$, as needed. $\qquad\square$

3 Computational Experiments

In this section, we provide a computational comparison of our method with the standard cut loop, the ellipsoid method, and the analytic center cutting plane method on a testbed of linear optimization instances. For comparison purposes, all four methods are embedded into a common cutting plane framework such that the same termination criteria apply.

Framework. Each method has access to a separation oracle that is equipped with a set of initial linear inequalities valid for K (such as bounds on variables), which are incorporated within each method in a straightforward way. For instance, we initialize our algorithm by adding these constraints to the set A_1. Moreover, for each instance, we will be given a finite upper bound UB and incorporate the linear inequality $f(x) \leq$ UB in a similar way. This upper bound gets updated whenever a feasible solution of better objective value was found. Our framework collects all inequalities queried by the current method and computes the resulting lower bound on the optimum value in every iteration. Each method is stopped whenever the difference of upper and lower bound is below 10^{-3}.

We will also inspect the possibility of a *smart* oracle that, regardless of whether a given point x is feasible, may still provide a valid inequality as well as a feasible solution (for instance, by modifying x in a simple way so that it becomes feasible). Such an oracle is often automatically available and can have a positive impact on the performance of the considered algorithms. For the problems we consider, the actual implementation of a smart oracle will be specified below.

Implementation. The framework has been implemented in `julia 1.6.2` using JuMP and `Gurobi 9.1.1`. To guarantee a fair comparison, all four methods have been implemented in a straightforward fashion. We use the textbook implementation of the ellipsoid method, and Badenbroek's implementation of the analytic center cutting plane method [2]. Our method is implemented[2] in the spirit of Theorem 2, where p_t is computed using `Gurobi`.

Test Sets. We use three problem classes in our experiments: linear programming formulations of the maximum-cardinality matching problem, semidefinite relaxations of the maximum cut problem, and LPBoost instances for classification problems.

For the maximum-cardinality matching problem, we consider the linear program

$$\max\left\{\sum_{e \in E} x_e : x \in [0,1]^E, \sum_{e \in \delta(v)} x_e \leq 1 \text{ for all } v \in V, \right.$$
$$\left. \sum_{e \in E[U]} x_e \leq \frac{|U|-1}{2} \text{ for all } U \subseteq V \text{ with } |U| \text{ odd}\right\},$$

---

[2] https://github.com/christopherhojny/supplement_simple-iterative-methods-linopt-convex-sets.

due to Edmonds [18], where $G = (V, E)$ is a given undirected graph, $\delta(v)$ is the set of all edges incident to v, and $E[U]$ is the set of all edges with both endpoints in U. The latter constraints are handled within an oracle that computes an inequality minimizing $(|U| - 1)/2 - \sum_{e \in E[U]} x_e$, whereas the other inequalities are provided as initial constraints. For the above problem, the smart version of the oracle does not provide a feasible point since there is no obvious way of transforming a given point into a feasible one. However, the smart version always provides the minimizing inequality.

We consider 16 random instances with 500 nodes, generated as follows. For each $r \in \{30, 33, \ldots, 75\}$ we build an instance by sampling r triples of nodes $\{u, v, w\}$ and adding the edges of the induced triangles to the graph, forming the test set `matching`. We believe that these instances are interesting because the r triangles give rise to many constraints to be added by the oracle. Moreover, we selected all 13 instances from the Color02 symposium [12] with less than 300 edges, yielding the test set `matching02`.

Our second set of instances is based on the semidefinite relaxation of Goemans and Williamson [21] for the maximum cut problem

$$\max \left\{ \sum_{\{v,w\} \in E} c(v, w)(1 - X_{v,w})/2 : X_{v,w} = X_{w,v} \text{ for all } v, w \in V, \right.$$
$$\left. X_{v,v} = 1 \text{ for all } v \in V, X \succeq 0 \right\}$$

where c are edge weights on the edges of (V, E). We add the box constraints $X \in [-1, 1]^{V \times V}$ to the initial constraints and handle the semidefiniteness constraint by a separation oracle that, given X, computes an eigenvector h of X of minimum eigenvalue and returns the inequality $\langle hh^\intercal, X \rangle \geq 0$.

Within the smart version of the oracle, this constraint is returned regardless of the feasibility of X. If X is not feasible, the semidefinite matrix $\frac{1}{\lambda - 1} X - \frac{\lambda}{\lambda - 1} I$ is returned, where λ denotes the minimum eigenvalue and I the identity matrix. We generated 10 complete graphs on 10 nodes with edge weights chosen uniformly at random in $[0, 1]$.

Our third set of instances arises from LPBoost [16], a classifier algorithm based on column generation. To solve the pricing problem in column generation, the following linear program is solved:

$$\max \left\{ \gamma : (\gamma, \lambda) \in [-1, 1] \times [0, D]^n, \langle \mathbf{1}, \lambda \rangle = 1, \sum_{i=1}^{m} y_i h(x^i, \omega) \lambda_i \leq -\gamma \text{ for } \omega \in \Omega \right\},$$

where Ω is a set of parameters, for $i \in [m]$, x^i is a data point labeled as $y_i = \pm 1$, $h(\cdot, \omega)$ is a classifier parameterized by $\omega \in \Omega$ that predicts the label of x^i as $h(x^i, \omega) \in \{-1, +1\}$, and $D > 0$ is a parameter. In our experiments, we restrict $h(\cdot, \omega)$ to be a decision tree of height 1, so-called tree stumps, and choose $D = \frac{5}{n}$. To separate a point (γ', λ'), we use `julia`'s `DecisionTree` module to compute a decision stump with score function λ' that weights the data points, whose corresponding inequality classifies (γ', λ') as feasible or not. A smart oracle always returns the computed inequality and decreases γ' until (γ', λ') becomes feasible according to the found decision stump.

Table 1. Comparison of iterations and dual/primal integral without smart oracles.

instance	#iterations				dual integral				primal integral		
	LP	ellipsoid	analytic	our	LP	ellipsoid	analytic	our	ellipsoid	analytic	our
matching	175.44	500.00	500.00	99.81	48.34	473.02	22.13	21.10	52.12	9.29	4.40
matching02	283.77	460.77	491.69	47.15	257.76	339.67	194.26	21.64	23.41	5.91	2.13
maxcut	265.30	500.00	500.00	193.30	7.72	44.32	3.48	6.14	21.15	9.04	6.32
LPboost	91.94	489.06	479.12	278.06	3.15	13.62	20.65	53.15	459.97	100.71	64.08

We extracted all data sets from the UC Irvine Machine Learning Repository [35] that are labeled as multivariate, classification, ten-to-hundred attributes, hundred-to-thousand instances. Data sets with alpha-numeric values or too many missing values have been discarded.

Results. In what follows, we report on the number of iterations, i.e., oracle calls, each method needs to obtain a gap (upper bound minus lower bound) below 10^{-3}. We impose a limit of 500 iterations per instance. Since we are testing naive implementations of each method, we do not report on running time.

To get more insights on the primal and dual performance of the tested methods, we also report on their *primal and dual integrals*. Note that we are solving maximization problems in this section, as opposed to minimization problems in Sect. 2. That is, primal (dual) solutions provide lower (upper) bounds on OPT. If ℓ_i is the lower bound on the optimal objective value OPT in iteration i, the *primal integral* is $\sum_{i=1}^{500} \frac{\text{OPT}-\ell_i}{\text{OPT}-\ell_1}$. The *dual integral* is computed analogously. If an integral is small, this indicates quick progress in finding the correct value of the corresponding bound.

Table 1 summarizes our results without smart oracles, where all numbers are average values. Here, "matching" refers to the random instances and "matching02" to the instances from the Color02 symposium. The standard cut loop is referred to as "LP", the ellipsoid method as "ellipsoid", the analytic center method as "analytic", and Algorithm 1 as "our". Note that Table 1 does not report on the primal integral of "LP" since the standard cut loop is a dual method.

We see that the ellipsoid and analytic center methods are struggling with solving any instance within 500 iterations independent from the problem class. Our algorithm solves the instances of the matching and max-cut problem much faster than the standard cut loop. Only for LPBoost, the standard cut loop clearly dominates our algorithm. To better understand this behavior, the integrals reveal that our algorithm is better in improving the primal bound than the dual bound, with the only exception being LPBoost. The analytic center method, however, performs significantly worse than our algorithm in improving the primal bound. Regarding the dual bound, it performs better than our algorithm (with the exception of matching02). The ellipsoid method is much worse in improving the primal bound in comparison with the analytic center method and our algorithm. Regarding the dual bound, a similar trend can be observed with LPBoost being an exception.

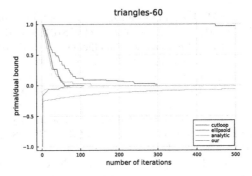

Fig. 1. Typical primal/dual bounds for a random matching instance.

In summary, the analytic center cutting plane method improves the dual bound more quickly than our algorithm. It can find a good primal solution early as the primal integral is small, however it fails to close the remaining gap within the iteration limit. Our algorithm is able to close the primal gap faster, with the trade-off of a slightly slower dual convergence. A typical plot of the of the relative primal and dual gaps is given in Fig. 1.

In a second experiment, we investigate the effect of smart oracles. As Table 2 shows, the algorithms mostly benefit from having access to a smart oracle in the case of LPBoost. A reason might be in the particular structure of these instances: the objective just consists of γ and every truncated convex combination λ is feasible. However, there is no impact of smart oracles on the matching and maxcut instances, respectively.

Table 2. Comparison of iterations and dual/primal integral with smart oracles.

instance	#iterations				dual integral				primal integral		
	LP	ellipsoid	analytic	our	LP	ellipsoid	analytic	our	ellipsoid	analytic	our
matching	175.44	500.00	500.00	99.81	48.34	473.02	22.13	21.10	52.12	9.29	4.40
matching02	283.77	460.77	491.69	47.15	257.76	339.67	194.26	21.64	23.41	5.91	2.13
maxcut	265.30	500.00	500.00	231.00	7.72	42.90	3.48	6.15	20.42	8.91	5.59
LPboost	86.94	346.38	88.00	127.00	3.04	13.50	5.54	5.46	25.41	6.83	6.95

Acknowledgments. We would like to thank Robert Luce and Sebastian Pokutta for their very valuable feedback on our work.

References

1. Atkinson, D.S., Vaidya, P.M.: A cutting plane algorithm for convex programming that uses analytic centers. Math. Program. **69**(1), 1–43 (1995)
2. Badenbroek, R., de Klerk, E.: An analytic center cutting plane method to determine complete positivity of a matrix. INFORMS J. Comput. **34**(2), 1115–1125 (2021)

3. Beck, A.: First-Order Methods in Optimization. Society for Industrial and Applied Mathematics (2017). https://doi.org/10.1137/1.9781611974997

4. Belloni, A., Freund, R.M., Vempala, S.: An efficient rescaled perceptron algorithm for conic systems. Math. Oper. Res. **34**(3), 621–641 (2009)

5. Ben-Tal, A., Nemirovski, A.: Lectures on Modern Convex Optimization. Society for Industrial and Applied Mathematics (2001). https://doi.org/10.1137/1.9780898718829

6. Betke, U.: Relaxation, new combinatorial and polynomial algorithms for the linear feasibility problem. Discrete Comput. Geom. **32**(3), 317–338 (2004). https://doi.org/10.1007/s00454-004-2878-4

7. Chekuri, C., Quanrud, K.: Approximating the held-karp bound for metric TSP in nearly-linear time. In: 2017 IEEE 58th Annual Symposium on Foundations of Computer Science (FOCS). IEEE, October 2017. https://doi.org/10.1109/focs.2017.78

8. Chekuri, C., Quanrud, K.: Near-linear time approximation schemes for some implicit fractional packing problems. In: Proceedings of the Twenty-Eighth Annual ACM-SIAM Symposium on Discrete Algorithms. Society for Industrial and Applied Mathematics, January 2017. https://doi.org/10.1137/1.9781611974782.51

9. Cheney, E.W., Goldstein, A.A.: Newton's method for convex programming and Tchebycheff approximation. Numer. Math. **1**(1), 253–268 (1959)

10. Chubanov, S.: A strongly polynomial algorithm for linear systems having a binary solution. Math. Program. **134**(2), 533–570 (2011). https://doi.org/10.1007/s10107-011-0445-3

11. Chubanov, S.: A polynomial algorithm for linear feasibility problems given by separation oracles. Optim. Online (2017)

12. Color02 - computational symposium: Graph coloring and its generalizations (2002). http://mat.gsia.cmu.edu/COLOR02

13. Dadush, D., Hojny, C., Huiberts, S., Weltge, S.: A simple method for convex optimization in the oracle model. arXiv:2011.08557 (2021). https://arxiv.org/abs/2011.08557

14. Dadush, D., Végh, L.A., Zambelli, G.: Rescaling algorithms for linear conic feasibility. Math. Oper. Res. **45**(2), 732–754 (2020). https://doi.org/10.1287/moor.2019.1011

15. Dantzig, G.B.: Converting a converging algorithm into a polynomially bounded algorithm. Technical report, Stanford University, 1992. 5.6, 6.1, 6.5 (1991)

16. Demiriz, A., Bennett, K.P., Shawe-Taylor, J.: Linear programming boosting via column generation. Mach. Learn. **46**(1), 225–254 (2002)

17. Dunagan, J., Vempala, S.: A simple polynomial-time rescaling algorithm for solving linear programs. Math. Program. **114**(1), 101–114 (2007). https://doi.org/10.1007/s10107-007-0095-7

18. Edmonds, J.: Maximum matching and a polyhedron with 0,1-vertices. J. Res. Natl. Bur. Stand. **69B**(1–2), 125–130 (1964)

19. Frank, M., Wolfe, P.: An algorithm for quadratic programming. Naval Res. Logist. Q. **3**(1–2), 95–110 (1956)

20. Garg, N., Könemann, J.: Faster and simpler algorithms for multicommodity flow and other fractional packing problems. SIAM J. Comput. **37**(2), 630–652 (2007). https://doi.org/10.1137/s0097539704446232

21. Goemans, M.X., Williamson, D.P.: Improved approximation algorithms for maximum cut and satisfiability problems using semidefinite programming. J. ACM **42**(6), 1115–1145 (1995). https://doi.org/10.1145/227683.227684

22. Gomory, R.E.: Outline of an algorithm for integer solutions to linear programs. Bull. Amer. Math. Soc. **64**, 275–278 (1958)
23. Grötschel, M., Lovász, L., Schrijver, A.: Geometric Algorithms and Combinatorial Optimization, vol. 2. Springer, Heidelberg (1988). https://doi.org/10.1007/978-3-642-78240-4
24. Jaggi, M.: Revisiting Frank-Wolfe: projection-free sparse convex optimization. In: Proceedings of Machine Learning Research, vol. 28, pp. 427–435. PMLR, Atlanta, 17–19 June 2013. http://proceedings.mlr.press/v28/jaggi13.html
25. Jiang, H., Lee, Y.T., Song, Z., Wong, S.C.W.: An improved cutting plane method for convex optimization, convex-concave games, and its applications. In: Proceedings of the 52nd Annual ACM SIGACT Symposium on Theory of Computing, STOC 2020, pp. 944–953. Association for Computing Machinery, New York (2020). https://doi.org/10.1145/3357713.3384284
26. Kelley, J.E., Jr.: The cutting-plane method for solving convex programs. J. Soc. Ind. Appl. Math. **8**(4), 703–712 (1960)
27. Khachiyan, L.G.: A polynomial algorithm in linear programming (in russian). Doklady Akademiia Nauk SSSR 224 224, 1093–1096 (1979). English Translation: Soviet Mathematics Doklady **20**, 191–194
28. Lee, Y.T., Sidford, A., Wong, S.C.: A faster cutting plane method and its implications for combinatorial and convex optimization. In: 2015 IEEE 56th Annual Symposium on Foundations of Computer Science, pp. 1049–1065 (2015). https://doi.org/10.1109/FOCS.2015.68
29. Nemirovsky, A., Yudin, D.: Informational complexity and efficient methods for solution of convex extremal problems. Ékon. Math. Metody **12** (1983)
30. Nesterov, Y.: Cutting plane algorithms from analytic centers: efficiency estimates. Math. Program. **69**(1), 149–176 (1995)
31. Plotkin, S.A., Shmoys, D.B., Tardos, É.: Fast approximation algorithms for fractional packing and covering problems. Math. Oper. Res. **20**(2), 257–301 (1995). https://doi.org/10.1287/moor.20.2.257
32. Rosenblatt, F.: The perceptron: a probabilistic model for information storage and organization in the brain. Psychol. Rev. **65**(6), 386–408 (1958). https://doi.org/10.1037/h0042519
33. Shahrokhi, F., Matula, D.W.: The maximum concurrent flow problem. J. ACM **37**(2), 318–334 (1990). https://doi.org/10.1145/77600.77620, http://doi.acm.org/10.1145/77600.77620
34. Sonnevend, G.: New algorithms in convex programming based on a notion of "centre" (for systems of analytic inequalities) and on rational extrapolation. In: Hoffmann, K.H., Zowe, J., Hiriart-Urruty, J.B., Lemarechal, C. (eds.) Trends in Mathematical Optimization: 4th French-German Conference on Optimization, pp. 311–326. Birkhäuser Basel, Basel (1988)
35. UC Irvine Machine Learning Repository. https://archive-beta.ics.uci.edu/ml/datasets. Accessed 3 Sept 2021
36. Vaidya, P.M.: A new algorithm for minimizing convex functions over convex sets. Math. Program. **73**(3), 291–341 (1996). https://doi.org/10.1007/bf02592216

On the Complexity of Separation
from the Knapsack Polytope

Alberto Del Pia, Jeff Linderoth, and Haoran Zhu$^{(\boxtimes)}$

Department of Industrial and Systems Engineering,
University of Wisconsin-Madison, Madison, USA
{delpia,linderoth,hzhu94}@wisc.edu

Abstract. We close three open problems on the separation complexity
of valid inequalities for the knapsack polytope. Specifically, we establish that the separation problems for extended cover inequalities, $(1, k)$-
configuration inequalities, and weight inequalities are all $\mathcal{NP}$-complete.
We also show that, when the number of constraints of the LP relaxation
is constant and its optimal solution is an extreme point, then the separation problems of both extended cover inequalities and weight inequalities
can be solved in polynomial time.

Keywords: Knapsack polytope · Separation problem · Complexity
theory

1 Introduction

The *multi-dimensional knapsack problem* is the integer programming (IP) problem

$$\max\{c^\top x \mid Ax \le d, \ x \in \{0,1\}^n\}, \tag{1}$$

where $A \in \mathbb{Z}_+^{m \times n}$, $c \in \mathbb{Z}_+^n$, and $d \in \mathbb{Z}_+^m$. When the constraint matrix A only has
one row a and the right-hand side vector is a positive integer b, problem (1) is
referred to as *knapsack problem*, and the convex hull of the associated feasible
region, $\text{conv}(\{x \in \{0,1\}^n \mid a^\top x \le b\})$, is referred to as the *knapsack polytope*.

The multi-dimensional knapsack problem is a fundamental problem in discrete optimization, and valid inequalities for the feasible region have been widely
studied, see, e.g., [4,14,19] and paper [11] provides a modern survey. In this
paper, we study the complexity of the separation problem for well-known families of valid inequalities for (1).

A. Del Pia is partially funded by ONR grant N00014-19-1-2322. Any opinions, findings, and conclusions or recommendations expressed in this material are those of the
authors and do not necessarily reflect the views of the Office of Naval Research. The
work of J. Linderoth and H. Zhu is supported by the Department of Energy, Office
of Science, Office of Advanced Scientific Computing Research, Applied Mathematics
program under Contract Number DE-AC02-06CH11347.

K. Aardal and L. Sanitá (Eds.): IPCO 2022, LNCS 13265, pp. 168–180, 2022.
https://doi.org/10.1007/978-3-031-06901-7_13

A standard and computationally useful way for generating cuts for (1) is to generate cuts for the knapsack polytope defined by its individual constraints. Suppose a is a row of the constraint matrix A, and let b be the corresponding coordinate of the right-hand side d. We denote the associated knapsack polytope by $K := \text{conv}(\{x \in \{0,1\}^n \mid a^\mathsf{T} x \leq b\})$.

Many families of valid inequalities for K are based on the notion of a *cover*, which is a subset C of $\{1, 2, \ldots, n\}$ such that $\sum_{i \in C} a_i > b$. Given a cover C, the inequality

$$\sum_{i \in C} x_i \leq |C| - 1$$

is valid for K, and it is called a *cover inequality* (CI). Cover inequalities can often be strengthened through a process called lifting, and the resulting inequalities are called *lifted cover inequalities* (LCIs) [3,8,16,17,22].

Balas [2] gave one family of LCIs known as *extended cover inequality* (ECI), which have the form

$$\sum_{j \notin C : a_j \geq \max_{i \in C} a_i} x_j + \sum_{i \in C} x_i \leq |C| - 1.$$

A *minimal cover* is a cover C such that $\sum_{i \in C \setminus \{j\}} a_i \leq b$ for any $j \in C$. A set $N \cup \{t\}$ with $N \subsetneq \{1, \ldots, n\}$ and $t \notin N$ is called a $(1, k)$-*configuration* for $k \in \{2, \ldots, |N|\}$ if $\sum_{i \in N} a_i \leq b$ and $Q \cup \{t\}$ is a minimal cover for every $Q \subseteq N$ with $|Q| = k$. Padberg [18] showed that for any $(1, k)$-configuration $N \cup \{t\}$, the inequality

$$(|S| - k + 1)x_t + \sum_{i \in S} x_i \leq |S|$$

is valid for K for every $|S| \subseteq N$ with $|S| \geq k$. This inequality is called a $(1, k)$-*configuration inequality*.

Other valid inequalities for the knapsack polytope K arise from the concept of a pack. For the knapsack polytope K, a set $P \subseteq \{1, \ldots, n\}$ is a *pack* if $\sum_{i \in P} a_i \leq b$. Given a pack P, the corresponding *pack inequality* $\sum_{i \in P} a_i x_i \leq \sum_{i \in P} a_i$ is trivially valid for K, as it is implied by the upper bound constraints $x_i \leq 1$. However, pack inequalities can be lifted in several different ways to obtain more interesting *lifted pack inequalities (LPIs)* [1]. Weismantel [21] derived the weight-inequalities, which are LPIs. To define the weight inequalities, let $r(P) := b - \sum_{i \in P} a_i$ be the *residual capacity* of the pack P. The indices $j \notin P$ with $a_j > r(P)$ are lifted to obtain the *weight inequality (WI)*:

$$\sum_{i \in P} a_i x_i + \sum_{j \notin P} \max\{a_j - r(P), 0\} x_j \leq \sum_{i \in P} a_i.$$

Consider the *linear programming (LP) relaxation* of (1):

$$\max\{c^\mathsf{T} x \mid Ax \leq d,\ x \in [0,1]^n\}. \tag{2}$$

For a given family $\mathscr{F}$ of valid inequalities for (1), the associated *separation problem* is defined as follows: "Let x^* be a feasible solution to (2), does there exist an inequality in $\mathscr{F}$ that is violated by x^*? If so, return one such inequality from $\mathscr{F}$." In this paper, we are mainly interested in the weaker decision version of the separation problem where we do not have to return a separating inequality even if it does exist, and we assume that x^* is an optimal solution to (2). In fact, the separation of optimal solution is no harder than the separation of general feasible solution, and from the computational point of view, x^* almost always comes from the optimal solution to some linear relaxation.

The separation problem for several families of valid inequalities for the knapsack polytope has been shown to be $\mathcal{NP}$-complete, including CIs [15], and LCIs [9]. On the other hand the complexity of the separation problem for extended cover inequalities, $(1, k)$-configuration inequalities, and weight inequalities are, to the best of our knowledge, unknown. Kaparis and Letchford stated that the separation problem seems likely to be $\mathcal{NP}$-hard for ECIs in [12]. It was conjectured explicitly in [6] that the separation problem for $(1, k)$-configuration inequalities is $\mathcal{NP}$-hard. Moreover, the complexity of the separation problem for WIs is also open, as mentioned in [11]. In this paper we provide positive answers to all these conjectures. Namely, we show that the separation problems for ECIs, for $(1, k)$-configuration inequalities, and for WIs are all $\mathcal{NP}$-complete. The first two results are proven via a reduction from the separation problem for CIs, and the separation complexity for WIs is given via the reduction from the Subset Sum Problem (SSP).

Along with this $\mathcal{NP}$-hard results, we also present some positive results about the separation problems of those cutting-planes. Specifically, we show that when the number of constraints of the LP relaxation (2) is constant, and the optimal solution x^* is an extreme point, then the seapration problems for ECIs and WIs are both polynomial-time solvable. See Corollary 1 and Corollary 2.

We remark that several heuristics and exact separation algorithms are present in the literature for these families of cuts. Both Gabrel and Minoux [7] and Kaparis and Letchford [12] provide an exact separation algorithm for ECIs that runs in pseudo-polynomial time. Ferreira *et al.* [6] presented simple heuristics for the separation problem of $(1, k)$-configuration inequalities. For the separation problem for WIs, Weismantel [21] proposed an exact algorithm that runs in pseudo-polynomial time. Helmberg and Weismantel [10] further presented a fast separation heuristic for WIs that simply inserts items into the pack P in non-increasing order of x^* value. Kaparis and Letchford [12] gave two exact algorithms and a heuristic for separating WIs and show how to convert these methods into heuristics for separating LPIs.

The separation problems considered in this paper can be defined formally as follows:

Problem CI-SP
Input: $(A, d, c) \in (\mathbb{Z}_+^{m \times n}, \mathbb{Z}_+^m, \mathbb{Z}_+^n)$ and an optimal solution x^* to the LP relaxation (2).
Question: Is there a cover C with respect to some row constraint $a^\mathsf{T} x \leq b$ of (2), such that $\sum_{i \in C} x_i^* > |C| - 1$?

Problem ECI-SP
Input: $(A, d, c) \in (\mathbb{Z}_+^{m \times n}, \mathbb{Z}_+^m, \mathbb{Z}_+^n)$ and an optimal solution x^* to the LP relaxation (2).
Question: Is there a cover C with respect to some row constraint $a^\mathsf{T} x \leq b$ of (2), such that $\sum_{j \notin C : a_j \geq \max_{i \in C} a_i} x_j + \sum_{i \in C} x_i > |C| - 1$?

Problem CONFIG-SP
Input: $(A, d, c) \in (\mathbb{Z}_+^{m \times n}, \mathbb{Z}_+^m, \mathbb{Z}_+^n)$ and an optimal solution x^* to the LP relaxation (2).
Question: Is there a $(1, k)$-configuration $N \cup \{t\}$ and a subset $S \subseteq N$ with $|S| \geq k$ with respect to some row constraint $a^\mathsf{T} x \leq b$ of (2), such that $(|S| - k + 1) x_t^* + \sum_{i \in S} x_i^* > |S|$?

Problem WI-SP
Input: $(A, d, c) \in (\mathbb{Z}_+^{m \times n}, \mathbb{Z}_+^m, \mathbb{Z}_+^n)$ and an optimal solution x^* to the LP relaxation (2).
Question: Is there a pack P with respect to some row constraint $a^\mathsf{T} x \leq b$ of (2), such that $\sum_{i \in P} a_i x_i^* + \sum_{j \notin P} \max\{a_j - r(P), 0\} x_j^* > \sum_{i \in P} a_i$?

For CI-SP, we have the following classic results.

Theorem 1 ([15]).

- *CI-SP is $\mathcal{NP}$-complete, even if $m = 1$.*
- *CI-SP is $\mathcal{NP}$-complete, even if x^* is an extreme point.*

We will show the other three problems, ECI-SP, CONFIG-SP, and WI-SP are all $\mathcal{NP}$-Complete.

Clearly, the $\mathcal{NP}$-hardness of the above problems imply the $\mathcal{NP}$-hardness of the more general separation problem where x^* is a feasible, and not necessarily optimal, solution to (2). We should also remark that, since verifying if a given point violates a given inequality can be obviously done in polynomial time with respect to the input size of such point and inequality, the separation problems for these families of cuts are clearly in class $\mathcal{NP}$. Therefore, when we talk about the separation complexity for those cuts, we do not distinguish between $\mathcal{NP}$-hard and $\mathcal{NP}$-complete throughout this paper.

Notation. For an integer n we set $[n] := \{1, 2, \ldots, n\}$. We define e_n as n-dimensional vector of ones, where we often repress the n if the vector dimension may be implied by the context. For a vector $x \in \mathbb{R}^n$ and $S \subseteq [n]$, we set $x(S) := \sum_{i \in S} x_i$. So for a vector $a \in \mathbb{R}^n, a([n]) = a^\mathsf{T} e = \sum_{i=1}^n a_i$.

2 Extended Cover Inequality Separation

In this section, we establish the complexity of extended cover inequality separation with a simple reduction from the cover inequality separation problem.

In the case where the point to be separated has a small number of fractional components, then extended cover inequality separation can be accomplished in polynomial time.

Theorem 2. *Problem ECI-SP is $\mathcal{NP}$-complete, even if $m = 1$. Furthermore, Problem ECI-SP is $\mathcal{NP}$-complete, even if x^* is an extreme point solution to the LP relaxation* (2).

Proof. We transform CI-SP to ECI-SP. Let $(A, d, c, x^*) \in (\mathbb{Z}_+^{m \times n}, \mathbb{Z}_+^m, \mathbb{Z}_+^n, [0, 1]^n)$ be the input to CI-SP. We construct input to ECI-SP with the property that there is a yes-certificate to CI-SP with input (A, d, c, x^*) if and only if there is a yes-certificate to ECI-SP with input $(A', d', c', y^*) \in (\mathbb{Z}_+^{m \times (n+1)}, \mathbb{Z}_+^m, \mathbb{Z}_+^{n+1}, [0, 1]^{n+1})$.

The data for the ECI-SP instance are constructed as follows:

$$A'_{ij} = A_{ij} \ \forall i \in [m], \forall j \in [n] \qquad A'_{i,n+1} = \sum_{j=1}^{n} A_{ij} \ \forall i \in [m]$$

$$c'_j = c_j \ \forall j \in [n] \qquad c'_{n+1} = M$$

$$d'_i = d_i + \sum_{j=1}^{n} A_{ij} \ \forall i \in [m].$$

The constant M is chosen to be large enough so that if x^* is an optimal solution to the linear program (2), then $y^* = (x^*, 1)$ is an optimal solution to the linear program

$$\max\{(c')^\mathsf{T} y \mid A'y \le d', y \in [0, 1]^{n+1}\}. \tag{3}$$

It is a consequence of linear programming duality that selecting $M \ge (\pi^*)^\mathsf{T} Ae$, where π^* are optimal dual multipliers for the inequality constraints in (2), will ensure the optimality of y^*. Since there is an optimal solution π^* whose encoding length is of polynomial size [20], the encoding size of M is a polynomial function of the input size of CI-SP.

Let $C \subseteq [n]$ be a cover with respect to a row constraint $a^\mathsf{T} x \le b$ of $Ax \le d$ such that the associated CI does not hold at x^*, so $x^*(C) > |C| - 1$. Then $C' := C \cup \{n+1\}$ is a cover with respect to the constraint $(a^\mathsf{T}, a^\mathsf{T} e) \cdot y \le b + a^\mathsf{T} e$ within $A'y \le d'$, and the associated ECI cuts off y^*, since $y^*(C') = 1 + x^*(C) > |C| = |C'| - 1$.

On the other hand, assume that C' is a cover with respect to some row constraint $a'^\mathsf{T} y = (a^\mathsf{T}, a^\mathsf{T} e) \cdot y \le b + a^\mathsf{T} e = b'$ within $A'y \le d'$ such that the associated ECI cuts off y^*. Note that if $n + 1 \notin C'$, then $\sum_{j \in C'} a'_j \le a'([n]) = a^\mathsf{T} e < b + a^\mathsf{T} e$, and C' cannot be a cover with respect to that row constraint. Thus, $n + 1 \in C'$, and the ECI of C' is just its cover inequality $y(C') \le |C'| - 1$. By construction, the set $C := C' \setminus \{n + 1\}$ is a cover with respect to the constraint $a^\mathsf{T} x \le b$ within $Ax \le d$. The ECI of C' cuts off y^*, $y^*(C') = 1 + x^*(C) > |C'| - 1 = |C|$, so $x^*(C) > |C| - 1$, and the CI from C cuts off x^*.

We have shown that there is a yes-certificate to CI-SP with input (A, d, c, x^*) if and only if there is a yes-certificate to ECI-SP with input (A', d', c', y^*).

Together with Theorem 1, this establishes that ECI-SP is $\mathcal{NP}$-complete, even if $m = 1$, and that ECI-SP is $\mathcal{NP}$-complete, even if x^* is an extreme point to the LP relaxation (2). For the second statement of the theorem, it suffices to realize that the input $y^* = (x^*, 1)$ for ECI-SP will be an extreme point of $\{y \in [0,1]^{n+1} \mid A'y \leq d'\}$ if x^* is an extreme point of $\{x \in [0,1]^n \mid Ax \leq d\}$. $\square$

If the fractional support of the input vector x^* is "sparse", then we can separate ECI in polynomial time.

Theorem 3. *Let x^* be the input solution to ECI-SP. If $|\{i \in [n] : x_i^* \in (0,1)\}|$ is a constant, then a separating ECI can be obtained in polynomial time if one exists.*

Proof. For a given point x^* and constraint $a^\mathsf{T} x \leq b$ of $Ax \leq d$, there exists a separating ECI from the constraint if and only if for some $t \in [n]$, there exists a cover C with $\max_{i \in C} a_i = a_t$, such that

$$\sum_{i \in [n]: a_i \geq a_t} x_i^* + \sum_{i \in C: a_i < a_t} x_i^* > |C| - 1. \tag{4}$$

We partition C into four sets, $C = T_1 \cup T_f \cup T_0 \cup T$, with $T_1 = \{i \in C \mid a_i < a_t, x_i^* = 1\}, T_f = \{i \in C \mid a_i < a_t, x_i^* \in (0,1)\}, T_0 = \{i \in C \mid a_i < a_t, x_i^* = 0\}$, and $T = \{i \in C \mid a_i = a_t\}$. With this definition, (4) can be equivalently stated as

$$\sum_{i \in [n]: a_i \geq a_t} x_i^* > \sum_{i \in T_f} (1 - x_i^*) + |T_0| + |T| - 1. \tag{5}$$

The algorithm loops over all $t \in [n]$ and enumerates all $T_f \subseteq \{i \in [n] \mid a_i < a_t, x_i^* \in (0,1)\}$. By our assumption on the cardinality of fractional support of x^*, this is a polynomial number of iterations. For a fixed $t \in [n]$ and $T_f \subseteq C$, the separation problem then amounts to completing the cover C so that

$$|T_0| + |T| < \sum_{i \in [n]: a_i \geq a_t} x_i^* - \sum_{i \in T_f} (1 - x_i^*) + 1. \tag{6}$$

The right-hand side of (6) is a constant, so separation for a fixed index t and subset T_f amounts to solving the knapsack problem

$$\min_{z \in \{0,1\}^{|S_t|}} \left\{ \sum_{i \in S_t} z_i \;\Big|\; \sum_{i \in S_t} a_i z_i \geq b_{t,T_f} \right\}, \tag{7}$$

where $S_t = \{i \in [n] \mid a_i = a_t \text{ or } a_i < a_t, x_i^* = 0\}$, and $b_{t,T_f} = b + 1 - \sum_{i \in T_f} a_i - \sum_{i: a_i < a_t, x_i^* = 1} a_i$. As the non-zero objective coefficients of the knapsack problem (7) are all the same, the problem can be solved in polynomial time by a simple greedy procedure. $\square$

Theorem 3 immediately implies the following corollary.

Corollary 1. *If the number of constraints in the LP relaxation (2) is a constant and x^* is an extreme point solution to (2), then the separating ECI can be obtained in polynomial time if one exists.*

Proof. Let the number of constraints be a constant α. Since x^* is an extreme point, we know that at most α components of x^* are fractional. Hence $|\{i \in [n] : x_i^* \in (0,1)\}| \leq \alpha$. The result then follows from Theorem 3. □

3 $(1, k)$-Configuration Inequality Separation

In this section we establish that the separation problem for $(1, k)$-configuration inequalities is $\mathcal{NP}$-complete using a similar reduction as in the proof of Theorem 2.

Theorem 4. *Problem CONFIG-SP is $\mathcal{NP}$-complete, even if $m = 1$. Furthermore, Problem CONFIG-SP is $\mathcal{NP}$-complete, even if x^* is an extreme point solution to the LP relaxation (2).*

Proof. The proof is very similar to that of Theorem 2. The reduction is from the $\mathcal{NP}$-Complete CI-SP to CONFIG-SP, and details of the reduction are given in the full version of the paper [5]. □

We have settled the complexity of separation for $(1, k)$-configuration inequalities for an input solution x^* that is an extreme point to the LP-relaxation (2), but the complexity of separation for points x^* with a small number of fractional components is still open. In fact, we conjecture it to be $\mathcal{NP}$-Complete.

Conjecture 1. There exists a constant α such that CONFIG-SP is $\mathcal{NP}$-complete, even if the input solution x^* satisfies $|\{i \in [n] \mid x_i^* \in (0,1)\}| \leq \alpha$.

4   Weight Inequality Separation

In this section we show that WI-SP is $\mathcal{NP}$-hard and present special cases where it can be solved in polynomial time. For a pack P of a given knapsack constraint $a^\mathsf{T} x \leq b$, we denote by $C(P) := \{i \in [n] \backslash P \mid a_i > r(P)\}$. With this notation, the WI associated with P takes the form

$$\sum_{i \in P} a_i x_i + \sum_{j \in C(P)} (a_j - r(P)) x_j \leq a(P),$$

where we remind the reader that $r(P) := b - a(P)$. First, we will need the following auxiliary result.

Lemma 1. *Let $(a, b) \in \mathbb{Z}_+^{n+1}$ with $a([n])/b \notin \mathbb{Z}$, and let $x_1^* = \ldots = x_n^* = b/a([n])$. Then there exists a pack P of $a^\mathsf{T} x \leq b$ whose associated WI separates x^* if and only if there exists a pack P' of $a^\mathsf{T} x \leq b$ such that $r(P') > 0$, $P' \cup C(P') = [n]$, and $|C(P')| = \lfloor a([n])/b \rfloor$.*

Proof. The proof of Lemma 1 is self-contained and can be found in the full version of the paper [5].   □

To prove that the separation problem WI-SP is $\mathcal{NP}$-hard, we establish a reduction from the *Subset Sum Problem (SSP)* to WI-SP.

Problem SSP
Input: $\alpha \in \mathbb{Z}_+^n$ and $w \in \mathbb{Z}_+$.
Question: Is there a subset $S \subseteq [n]$ such that $\alpha(S) = w$?

The SSP is among Karp's 21 $\mathcal{NP}$-complete problems [13]. It is simple to check that SSP is $\mathcal{NP}$-complete even if $w > \max(\alpha)$. We are now ready to prove that WI-SP is $\mathcal{NP}$-hard.

Theorem 5. *Problem WI-SP is $\mathcal{NP}$-complete, even if $m = 1$. Furthermore, Problem WI-SP is $\mathcal{NP}$-complete, even if x^* is an extreme point solution to the LP relaxation* (2).

Proof. First, we prove the first part of the statement. We show that WI-SP is $\mathcal{NP}$-hard even in case of a single knapsack constraint. Given an instance $(\alpha, w) \in \mathbb{Z}_+^{n+1}$ of SSP with $w > \max(\alpha)$, we construct a knapsack problem $\max\{c^\mathsf{T}x \mid a^\mathsf{T}x \le b, x \in \{0,1\}^{2n+2}\}$ and give an optimal solution x^* to the associated LP relaxation. The data a, b, c of the constructed knapsack problem is defined as follows:

$$
\begin{aligned}
a_i &:= \alpha_i + 2, \qquad \forall i = 1, \dots, n, \\
a_{n+1} &:= w \cdot (n+1) + 2(n+1)^2 - 3n - \alpha([n]), \\
a_{n+1+j} &:= 2, \qquad \forall j = 1, \dots, n+1, \\
b &:= w + 2n + 3, \\
c &:= a, \\
x_1^* &:= \dots := x_{2n+2}^* := \frac{w + 2n + 3}{w \cdot (n+1) + 2n^2 + 5n + 4}.
\end{aligned}
\tag{8}
$$

It is simple to check that a, b, c are all integral, that (a, b, c, x^*) has polynomial encoding size with respect to that of (α, w), and that $a^\mathsf{T}x^* = b$. Furthermore, x^* is an optimal solution to the knapsack problem described by (8), since $c^\mathsf{T}x^* = a^\mathsf{T}x^* = b$. Hence (a, b, c, x^*) is a feasible input to WI-SP where $m = 1$. Note that $(w \cdot (n+1) + 2n^2 + 5n + 4)/(w + 2n + 3) = n + 1 + 1/(w + 2n + 3) \notin \mathbb{Z}$. Hence, we can apply Lemma 1 and obtain that there exists a separating WI for x^* if and only if there exists a pack P such that:

$$
\begin{aligned}
r(P) &> 0, \qquad P \cup C(P) = [2n + 2], \\
|C(P)| &= \left\lfloor \frac{w \cdot (n+1) + 2n^2 + 5n + 4}{w + 2n + 3} \right\rfloor = n + 1.
\end{aligned}
\tag{9}
$$

Claim 1. There exists a WI from constraint $a^\mathsf{T}x \le b$ that separates x^* if and only if there exists a subset $S \subseteq [n]$ such that $\alpha(S) = w$.

Proof of claim. It suffices to show that there exists pack P such that (9) holds if and only if there exists a subset $S \subseteq [n]$ such that $\alpha(S) = w$.

First, we assume that P is a pack such that (9) holds. The two equations in (9) imply $|P| = 2n+2 - |C(P)| = n+1$. If $\{n+2, n+3, \ldots, 2n+2\} \cap C(P) = \emptyset$, then $P \cup C(P) = [2n+2]$ implies that $\{n+2, n+3, \ldots, 2n+2\} \subseteq P$, which means $P = \{n+2, n+3, \ldots, 2n+2\}$ since $|P| = n+1$. However, since $w > \max(\alpha)$, we know that $2 + \max(\alpha) + 2(n+1) \leq w + 2n + 3 = b$, which implies that $P \cup \{i'\}$ is a pack for any $i' \in [n]$, and this contradicts the assumption $C(P) = [2n+2] \backslash P$ of (9). Therefore, there must exist some $i' \in \{n+2, n+3, \ldots, 2n+2\} \cap C(P)$. Hence $r(P) = b - a(P) < a_{i'} = 2$. Moreover, because $r(P) > 0$, we have $r(P) = 1$, which implies $a(P) = b - 1 = w + 2n + 2$. Since $a_{n+1} = w \cdot (n+1) + 2(n+1)^2 - 3n - \alpha([n]) \geq w + 2(n+1)^2 - 3n + (w \cdot n - \alpha([n])) > w + 2n + 2$, we know $n + 1 \notin P$. Let $S := P \cap [n]$. We then obtain $a(S) = 2|S| + \alpha(S)$ and $a(P \backslash S) = 2(|P| - |S|) = 2(n+1 - |S|)$. Therefore, $w + 2n + 2 = a(P) = a(S) + a(P \backslash S) = \alpha(S) + 2n + 2$, which gives us $\alpha(S) = w$.

Next, we assume that S is a subset of $[n]$ with $\alpha(S) = w$. Clearly, $n + 1 \notin S$. Then we define the set $\tilde{S}$ containing $n + 1 - |S|$ arbitrary indices from $\{n + 2, \ldots, 2n + 2\}$. Then $P := S \cup \tilde{S}$ is a pack such that (9) holds. In fact, we have

$$r(P) = b - a(P)$$
$$= w + 2n + 3 - a(S) - a(\tilde{S})$$
$$= w + 2n + 3 - (2|S| + \alpha(S)) - 2(n + 1 - |S|)$$
$$= 1.$$

This further implies $C(P) = [2n+2] \backslash P$ and $|C(P)| = 2n+2 - |P| = n+1$, since $a_i > 1$ for all $i \in [2n+2]$. Hence (9) is satisfied by pack P. ◇

Claim 1 completes the proof of the first part of the statement, since SSP itself is $\mathcal{NP}$-hard.

Next, we prove the second part of the statement. We show that WI-SP is $\mathcal{NP}$-hard, even if x^* is an extreme point solution to the LP relaxation (2). Given an instance $(\alpha, w) \in \mathbb{Z}_+^{n+1}$ of SSP with $w > \max(\alpha)$, we construct an instance of the multi-dimensional knapsack problem $\max\{c^\mathsf{T} x \mid Ax \leq d, x \in \{0, 1\}^{2N}\}$ and give an optimal solution x^* to the associated LP relaxation, where $N = 2n+2$. Let G be a node-node adjacency matrix of a cycle on N nodes. The constraints of the constructed multi-dimensional knapsack problem are then defined as follows:

$$a^\mathsf{T} y \leq b, \qquad Gz \leq e_N,$$
$$y_i + 2z_1 + 2z_2 + 2z_3 \leq 3 + \epsilon, \qquad \forall i \in [N]. \tag{10}$$

Here $(a, b) \in \mathbb{Z}_+^{N+1}$ is defined as in (8), $\epsilon := (w + 2n + 3)/(w \cdot (n + 1) + 2n^2 + 5n + 4)$ and e_N is the N-dimensional vector with all components equal to one. Now we define the objective vector $c := (a, e_N)$, and we let $x^* := (y^*, z^*) := (\epsilon e_N, e_N/2)$. Note that we can multiply all the rows of (10) by $w \cdot (n + 1) + 2n^2 + 5n + 4$ to get an instance of WI-SP with integral data. The instance defined here clearly has polynomial encoding size with respect to that of (α, w).

First, we verify that this is a valid input for WI-SP. Clearly x^* is feasible. Furthermore, by summing all inequalities in $Gz \leq e_N$, it follows that x^* is an optimal solution to the LP relaxation.

Next we show that x^* is an extreme point of the polyhedron given by (10). Since $N = 2n + 2$ is even, then G is a square matrix with rank $N - 1$. We can further verify that the first $2N$ constraints in (10) give a system of $2N$ linearly independent constraints in $2N$ variables, and the only vector that satisfies all of them at equality is x^*.

Claim 2. There exists a WI from (10) that separates x^* if and only if there exists a WI from the constraint $a^\mathsf{T}y \leq b$ that separates y^*.

Proof of claim. First, we assume that P is a pack with respect to some constraint $a'^\mathsf{T}x \leq b'$ of (10) such that its corresponding WI separates x^*. If such constraint $a'^\mathsf{T}x \leq b'$ comes from the subsystem $Gz \leq e_N$, say $z_1 + z_2 \leq 1$, then the only WI is $z_1 + z_2 \leq 1$, which cannot be violated by x^* since x^* is a feasible point. If $a'^\mathsf{T}x \leq b'$ is $y_i + 2z_1 + 2z_2 + 2z_3 \leq 3 + \epsilon$ for some $i \in [N]$, then all the nonempty packs that do not include variables with zero coefficient are $\{i\}, \{i, N+1\}, \{i, N+2\}, \{i, N+3\}, \{N+1\}, \{N+2\}, \{N+3\}$. The corresponding WIs are $y_i \leq 1$ and:

$$y_i + 2z_1 + (2 - \epsilon)(z_2 + z_3) \leq 3, \qquad 2z_1 + (1 - \epsilon)(z_2 + z_3) \leq 2,$$
$$y_i + 2z_2 + (2 - \epsilon)(z_1 + z_3) \leq 3, \qquad 2z_2 + (1 - \epsilon)(z_1 + z_3) \leq 2,$$
$$y_i + 2z_3 + (2 - \epsilon)(z_1 + z_3) \leq 3, \qquad 2z_3 + (1 - \epsilon)(z_1 + z_2) \leq 2.$$

It is simple to check that none of the above inequalities is violated by $x^* = (\epsilon e_N, e_N/2)$. Hence the constraint $a'^\mathsf{T}x \leq b'$ is just $a^\mathsf{T}y \leq b$. In other words, we have shown that if (10) admits a separating WI that separates x^*, then the constraint $a^\mathsf{T}y \leq b$ admits a separating WI that separates y^*.

On the other hand, any WI from the constraint $a^\mathsf{T}y \leq b$ is also an a WI from the entire linear system (10). We have thereby proven this claim.   ◇

Note that $y^* = \epsilon e_N$ in this proof coincides with the x^* in Claim 1. From Claim 2 and Claim 1, we have completed the proof for the second part of the statement of this theorem, since SSP is $\mathcal{NP}$-hard.   □

Even though the problem WI-SP is $\mathcal{NP}$-hard in general, in the next theorem we provide a special case where it can be solved in polynomial time, and such separating WI can be obtained in polynomial time if one exists.

Theorem 6. *Let x^* be the input solution to WI-SP. If $\max\{|S| : x_i^* \in (0, 1) \forall i \in S, x^*(S) < 1\}$ is a constant, then the separating WI can be obtained in polynomial time if one exists.*

Proof. We assume without loss of generality that $Ax \leq d$ has only a single constraint $a^\mathsf{T}x \leq b$, since we can always solve WI-SP with input (A, d, c, x^*) by

solving the corresponding WI-SP problems for each single constraint individu-
ally. For any $P \subseteq [n]$, let $f(P) := \sum_{i \in P} a_i x_i^* + \sum_{j \in C(P)} (a_j - r(P)) x_j^* - a(P)$.
Then $f(P) > 0$ implies that

$$\sum_{j \in C(P)} x_j^* < \frac{\sum_{i \in P \cup C(P)} a_i x_i^* - a(P)}{r(P)} \leq \frac{b - a(P)}{r(P)} = 1.$$

Among all the packs with the largest $f(P)$ value, let P' be one that is inclusion-
wise maximal. In other words, $f(P') \geq f(P)$ for any pack P, and $f(P) = f(P')$
implies that P' is not contained in P. Let $C := C(P')$. Note that for any $i' \in P'$
we have

$$f(P') - f(P' \setminus \{i'\}) = a_{i'} \left(\sum_{j \in C(P' \setminus \{i'\})} x_j^* + x_{i'}^* - 1 \right) + \sum_{j \in C \setminus C(P' \setminus \{i'\})} (a_j - r(P')) x_j^*$$

$$\in \left[a_{i'} \left(\sum_{j \in C(P' \setminus \{i'\})} x_j^* + x_{i'}^* - 1 \right), a_{i'} \left(\sum_{j \in C} x_j^* + x_{i'}^* - 1 \right) \right].$$

Here the last inequality $f(P') - f(P' \setminus \{i'\}) \leq a_{i'} \left(\sum_{j \in C} x_j^* + x_{i'}^* - 1 \right)$ is simply
because $a_j \leq a_{i'} + r(P')$ for any $j \in C \setminus C(P' \setminus \{i'\})$. Since $P' \setminus \{i'\}$ is also a pack
and $f(P') \geq f(P' \setminus \{i'\})$, we have

$$\sum_{j \in C} x_j^* + x_{i'}^* \geq 1, \quad \forall i' \in P'. \tag{11}$$

On the other hand, for any $i' \in [n] \setminus (C \cup P')$:

$$f(P' \cup \{i'\}) - f(P') = a_{i'} \left(\sum_{j \in C} x_j^* + x_{i'}^* - 1 \right) + \sum_{j \in C(P' \cup \{i'\}) \setminus C} (a_j - r(P' \cup \{i'\})) x_j^*$$

$$\in \left[a_{i'} \left(\sum_{j \in C} x_j^* + x_{i'}^* - 1 \right), a_{i'} \left(\sum_{j \in C(P' \cup \{i'\})} x_j^* + x_{i'}^* - 1 \right) \right].$$

Since $i' \in [n] \setminus (C \cup P')$, the set $P' \cup \{i'\}$ is still a pack, hence $f(P' \cup \{i'\}) -
f(P') \leq 0$. Furthermore, since P' is an inclusion-wise maximal pack with the
largest $f(P')$ value, we have $f(P' \cup \{i'\}) - f(P') < 0$. Therefore,

$$\sum_{j \in C} x_j^* + x_{i'}^* < 1, \quad \forall i' \in [n] \setminus (C \cup P'). \tag{12}$$

From (11) and (12), we obtain

$$P' = \{i \in [n] \setminus C \mid x^*(C) + x_i^* \geq 1\}. \tag{13}$$

We have thereby shown that there exists a WI from knapsack constraint
$a^\mathsf{T} x \leq b$ which separates x^*, if and only if there exists $C \subseteq [n]$, such that the

corresponding P', as defined in (13), is a pack satisfying $f(P') > 0$. Therefore, WI-SP can be solved by checking whether the set $P' = \{i \in [n]\backslash C \mid x^*(C) + x_i^* \geq 1\}$ is a pack with $f(P') > 0$, for any possible $C \subseteq [n]$ with $x^*(C) < 1$.

Let $I_0 := \{i \in [n] \mid x_i^* = 0\}$ and $I_f := \{i \in [n] \mid x_i^* \in (0,1)\}$. From the assumptions of this theorem, we know that $\alpha := \max\{|S| : x^*(S) < 1, S \subseteq I_f\}$ is a constant. For any $T \subseteq I_0$ and $S \subseteq I_f$ with $x^*(S) < 1$, it is easy to see that

$$\{i \in [n]\backslash S \mid x^*(S) + x_i^* \geq 1\} = \{i \in [n]\backslash(S \cup T) \mid x^*(S \cup T) + x_i^* \geq 1\}.$$

Hence, $\{i \in [n]\backslash C \mid x^*(C) + x_i^* \geq 1\}$ is a pack with positive f value for some $C \subseteq [n]$ with $x^*(C) < 1$, if and only if $\{i \in [n]\backslash(C\backslash I_0) \mid x^*(C\backslash I_0) + x_i^* \geq 1\}$ is a pack with positive value, where $C\backslash I_0 \subseteq I_f$ and $x^*(C\backslash I_0) = x^*(C) < 1$. Therefore, WI-SP can be solved by the following procedure:

1. For any $S \subseteq I_f$ with $x^*(S) < 1$, construct the corresponding $P' = \{i \in [n]\backslash S \mid x^*(S) + x_i^* \geq 1\}$.
2. Check if P' is a pack with $f(P') > 0$.
3. If the answer to the previous check is yes for some $S \subseteq I_f$ with $x^*(S) < 1$, then the corresponding P' works as a yes-certificate to WI-SP, and its corresponding WI separates x^*; If the answer is no for all $S \subseteq I_f$ with $x^*(S) < 1$, then x^* cannot be separated by any WI from the knapsack constraint $a^\mathsf{T} x \leq b$.

Since $\alpha = \max\{|S| : x^*(S) < 1, S \subseteq I_f\}$, we have

$$|\{S \mid x^*(S) < 1, S \subseteq I_f\}| \leq \sum_{k=0}^{\alpha} \binom{n}{k} = O(n^\alpha).$$

So this above procedure can be implemented in polynomial time, and we complete the proof. □

In particular, Theorem 6 implies that, if x^* has a constant number of fractional components, then WI-SP can be solved in polynomial time. Following the same logic as in Corollary 1, we directly obtain the following corollary.

Corollary 2. *If the number of constraints in the LP relaxation* (2) *is a constant and x^* is an extreme point solution to* (2), *then the separating WI can be obtained in polynomial time if one exists.*

References

1. Atamtürk, A.: Cover and pack inequalities for (mixed) integer programming. Ann. Oper. Res. **139**(1), 21–38 (2005)
2. Balas, E.: Facets of the knapsack polytope. Math. Program. **8**(1), 146–164 (1975)
3. Balas, E., Zemel, E.: Facets of the knapsack polytope from minimal covers. SIAM J. Appl. Math. **34**(1), 119–148 (1978)

4. Del Pia, A., Linderoth, J., Zhu, H.: Multi-cover inequalities for totally-ordered multiple knapsack sets. In: Singh, M., Williamson, D.P. (eds.) IPCO 2021. LNCS, vol. 12707, pp. 193–207. Springer, Cham (2021). https://doi.org/10.1007/978-3-030-73879-2_14

5. Del Pia, A., Linderoth, J., Zhu, H.: On the complexity of separation from the knapsack polytope. Optimization Online Preprint (2021). http://www.optimization-online.org/DB_FILE/2021/11/8682.pdf

6. Ferreira, C.E., Martin, A., Weismantel, R.: Solving multiple knapsack problems by cutting planes. SIAM J. Optim. **6**(3), 858–877 (1996)

7. Gabrel, V., Minoux, M.: A scheme for exact separation of extended cover inequalities and application to multidimensional knapsack problems. Oper. Res. Lett. **30**(4), 252–264 (2002)

8. Gu, Z., Nemhauser, G.L., Savelsbergh, M.W.P.: Lifted cover inequalities for $0 - 1$ integer programs: computation. INFORMS J. Comput. **10**(4), 427–437 (1998)

9. Gu, Z., Nemhauser, G.L., Savelsbergh, M.W.P.: Lifted cover inequalities for 0-1 integer programs: complexity. INFORMS J. Comput. **11**(1), 117–123 (1999)

10. Helmberg, C., Weismantel, R.: Cutting plane algorithms for semidefinite relaxations. In: Topics in Semidefinite and Interior-Point Methods (Toronto, ON, 1996), Volume 18 of Fields Institute Communications, pp. 197–213. American Mathematical Society, Providence (1998)

11. Hojny, C., et al.: Knapsack polytopes: a survey. Ann. Oper. Res. 1–49 (2019)

12. Kaparis, K., Letchford, A.N.: Separation algorithms for 0-1 knapsack polytopes. Math. Program. **124**(1–2), 69–91 (2010)

13. Karp, R.M.: Reducibility among combinatorial problems. In: Miller, R.E., Thatcher, J.W., Bohlinger, J.D. (eds.) Complexity of computer computations. The IBM Research Symposia Series, pp. 85–103. Springer, Boston (1972). https://doi.org/10.1007/978-1-4684-2001-2_9

14. Kellerer, H., Pferschy, U., Pisinger, D.: Multidimensional knapsack problems. In: Kellerer, H., Pferschy, U., Pisinger, D. (eds.) Knapsack Problems, pp. 235–283. Springer, Heidelberg (2004). https://doi.org/10.1007/978-3-540-24777-7_9

15. Klabjan, D., Nemhauser, G.L., Tovey, C.: The complexity of cover inequality separation. Oper. Res. Lett. **23**(1–2), 35–40 (1998)

16. Letchford, A.N., Souli, G.: On lifted cover inequalities: a new lifting procedure with unusual properties. Oper. Res. Lett. **47**(2), 83–87 (2019)

17. Padberg, M.W.: A note on zero-one programming. Oper. Res. **23**(4), 833–837 (1975)

18. Padberg, M.W.: (1, k)-configurations and facets for packing problems. Math. Program. **18**(1), 94–99 (1980)

19. Puchinger, J., Raidl, G.R., Pferschy, U.: The multidimensional knapsack problem: structure and algorithms. INFORMS J. Comput. **22**(2), 250–265 (2010)

20. Schrijver, A.: Theory of Linear and Integer Programming. Wiley, Hoboken (1986)

21. Weismantel, R.: On the 0/1 knapsack polytope. Math. Program. **77**(3), 49–68 (1997)

22. Wolsey, L.A.: Faces for a linear inequality in 0–1 variables. Math. Program. **8**(1), 165–178 (1975)

Simple Odd β-Cycle Inequalities
for Binary Polynomial Optimization

Alberto Del Pia[1] and Matthias Walter[2($\boxtimes$)]

[1] Department of Industrial and Systems Engineering, Wisconsin Institute
for Discovery, University of Wisconsin-Madison, Madison, USA
delpia@wisc.edu
[2] Department of Applied Mathematics, University of Twente,
Enschede, The Netherlands
m.walter@utwente.nl

Abstract. We consider the multilinear polytope which arises naturally in binary polynomial optimization. Del Pia and Di Gregorio introduced the class of odd β-cycle inequalities valid for this polytope, showed that these generally have Chvátal rank 2 with respect to the standard relaxation and that, together with flower inequalities, they yield a perfect formulation for cycle hypergraph instances. Moreover, they describe a separation algorithm in case the instance is a cycle hypergraph. We introduce a weaker version, called simple odd β-cycle inequalities, for which we establish a strongly polynomial-time separation algorithm for arbitrary instances. These inequalities still have Chvátal rank 2 in general and still suffice to describe the multilinear polytope for cycle hypergraphs.

Keywords: Binary polynomial optimization · Cutting planes · Separation algorithm

1 Introduction

In *binary polynomial optimization* our task is to find a binary vector that maximizes a given multivariate polynomial function. In order to give a mathematical formulation, it is useful to use a hypergraph $G = (V, E)$, where the node set V represents the variables in the polynomial function, and the edge set E represents the monomials with nonzero coefficients. In a binary polynomial optimization problem, we are then given a hypergraph $G = (V, E)$, a profit vector $p \in \mathbb{R}^{V \cup E}$, and our goal is to solve the optimization problem

$$\max \left\{ \sum_{v \in V} p_v z_v + \sum_{e \in E} p_e \prod_{v \in e} z_v : z \in \{0, 1\}^V \right\}. \tag{1}$$

A. Del Pia is partially funded by ONR grant N00014-19-1-2322. Any opinions, findings, and conclusions or recommendations expressed in this material are those of the authors and do not necessarily reflect the views of the Office of Naval Research. M. Walter acknowledges funding support from the Dutch Research Council (NWO) on grant number OCENW.M20.151.

K. Aardal and L. Sanitá (Eds.): IPCO 2022, LNCS 13265, pp. 181–194, 2022.
https://doi.org/10.1007/978-3-031-06901-7_14

Using Fortet's linearization [13,15], we introduce binary auxiliary variables z_e, for $e \in E$, which are linked to the variables z_v, for $v \in V$, via the linear inequalities

$$z_v - z_e \geq 0 \qquad\qquad \forall e \in E, \forall v \in e \qquad (2a)$$

$$(z_e - 1) + \sum_{v \in e}(1 - z_v) \geq 0 \qquad\qquad \forall e \in E. \qquad (2b)$$

It is simple to see that

$$\left\{ z \in \{0,1\}^{V \cup E} : z_e = \prod_{v \in e} z_v \ \forall e \in E \right\} = \left\{ z \in \{0,1\}^{V \cup E} : (2) \right\}.$$

Hence, we can reformulate (1) as the integer linear optimization problem

$$\max \left\{ \sum_{v \in V} p_v z_v + \sum_{e \in E} p_e z_e : (2), \ z \in \{0,1\}^{V \cup E} \right\}. \qquad (3)$$

We define the *multilinear polytope* $ML(G)$ [6], which is the convex hull of the feasible points of (3), and its *standard relaxation* $SR(G)$:

$$ML(G) := \mathrm{conv}\left\{ z \in \{0,1\}^{V \cup E} : (2) \right\},$$

$$SR(G) := \left\{ z \in [0,1]^{V \cup E} : (2) \right\}.$$

Recently, several classes of inequalities valid for $ML(G)$ have been introduced, including 2-link inequalities [4], flower inequalities [7], running intersection inequalities [8], and odd β-cycle inequalities [5]. On a theoretical level, these inequalities fully describe the multilinear polytope for several hypergraph instances: flower inequalities for γ-acyclic hypergraphs, running intersection inequalities for kite-free β-acyclic hypergraphs, and flower inequalities together with odd β-cycle inequalities for cycle hypergraphs. Furthermore, these cutting planes greatly reduce the integrality gap of (3) [5,8] and their addition leads to a significant reduction of the runtime of the state-of-the-art solver BARON [9]. Unfortunately, we are not able to separate efficiently over most of these inequalities. In fact, while the simplest 2-link inequalities can be trivially separated in polynomial time, there is no known polynomial-time algorithm to separate the other classes of cutting planes, and it is known that separating flower inequalities is NP-hard [9].

Contribution. In this paper we introduce a novel class of cutting planes called *simple odd β-cycle inequalities*. As the name suggests, these inequalities form a subclass of the odd β-cycle inequalities introduced in [5]. The main result of this paper is that simple odd β-cycle inequalities can be separated in strongly polynomial time. While our inequalities form a subclass of the inequalities introduced in [5], they still inherit the two most interesting properties of the odd β-cycle inequalities. First, simple odd β-cycle inequalities can have Chvátal rank 2. To the best of our knowledge, our algorithm is the first known polynomial-time separation algorithm over an exponential class of inequalities with Chvátal rank 2. Second, simple odd β-cycle inequalities, together with standard linearization inequalities and flower inequalities with at most two neighbors, provide a perfect formulation of the multilinear polytope for cycle hypergraphs. Finally, we believe that our separation algorithm could lead to significant speedups in solving several applications that can be formulated as (1) with a hypergraph that contains β-cycles. These applications include the image restoration problem in computer vision [4,5], and the low auto-correlation binary sequence problem in theoretical physics [2,5,16,18,19].

Outline. We first introduce certain simple inequalities in Sect. 2 that are then combined to form the simple odd β-cycle inequalities in Sect. 3. Section 4 is dedicated to the polynomial-time separation algorithm. Finally, Sect. 5 relates the simple odd β-cycle inequalities to the general (non-simple) odd β-cycle inequalities in [5].

2   Building Block Inequalities

We consider certain affine linear functions $s : \mathbb{R}^{V \cup E} \to \mathbb{R}$ defined as follows. For each $e \in E$ and each $v \in e$ we define

$$s_{e,v}^{\text{inc}}(z) := z_v - z_e \qquad (s_{e,v}^{\text{inc}})$$

For each $e \in E$ and all $U, W \subseteq e$ with $U, W \neq \varnothing$ and $U \cap W = \varnothing$ we define

$$s_{e,U,W}^{\text{odd}}(z) := 2z_e - 1 + \sum_{u \in U}(1 - z_u) + \sum_{w \in W}(1 - z_w) + \sum_{v \in e \backslash (U \cup W)}(2 - 2z_v) \quad (s_{e,U,W}^{\text{odd}})$$

For all $e, f \in E$ with $e \cap f \neq \varnothing$ and all $U \subseteq e$ with $U \neq \varnothing$ and $U \cap f = \varnothing$ we define

$$s_{e,U,f}^{\text{one}}(z) := 2z_e - 1 + \sum_{u \in U}(1 - z_u) + (1 - z_f) + \sum_{v \in e \backslash (U \cup f)}(2 - 2z_v) \qquad (s_{e,U,f}^{\text{one}})$$

For all $e, f, g \in E$ with $e \cap f \neq \varnothing$, $e \cap g \neq \varnothing$ and $e \cap f \cap g = \varnothing$ we define

$$s_{e,f,g}^{\text{two}}(z) := 2z_e - 1 + (1 - z_f) + (1 - z_g) + \sum_{v \in e \backslash (f \cup g)}(2 - 2z_v) \qquad (s_{e,f,g}^{\text{two}})$$

In this paper we often refer to $s_{e,v}^{\text{inc}}$, $s_{e,U,W}^{\text{odd}}$, $s_{e,U,f}^{\text{one}}$, $s_{e,f,g}^{\text{two}}$ as *building blocks*. Although in these definitions U and W can be arbitrary subsets of an edge e, in the following U and W will always correspond to the intersection of e with another edge. In the next lemma we will show that all building blocks are nonnegative on a relaxation of $\text{ML}(G)$ obtained by adding some flower inequalities [7] to

$SR(G)$, which we will define now. For ease of notation, in this paper, we denote by $[m]$ the set $\{1, \ldots, m\}$, for any nonnegative integer m.

Let $f \in E$ and let e_i, $i \in [m]$, be a collection of distinct edges in E, adjacent to f, such that $f \cap e_i \cap e_j = \emptyset$ for all $i, j \in [m]$ with $i \neq j$. Then the *flower inequality* [5,7] centered at f with neighbors e_i, $i \in [m]$, is defined by

$$(z_f - 1) + \sum_{i \in [m]} (1 - z_{e_i}) + \sum_{v \in f \setminus \cup_{i \in [m]} e_i} (1 - z_v) \geq 0.$$

We denote by $FR(G)$ the polytope obtained from $SR(G)$ by adding all flower inequalities with at most two neighbors. Clearly $FR(G)$ is a relaxation of $ML(G)$. Furthermore, $FR(G)$ is defined by a number of inequalities that is bounded by a polynomial in $|V|$ and $|E|$.

Lemma 1. *Let $G = (V, E)$ be a hypergraph and let s be one of $s_{e,v}^{inc}$, $s_{e,U,W}^{odd}$, $s_{e,U,f}^{one}$, $s_{e,f,g}^{two}$. Then $s(z) \geq 0$ is valid for $FR(G)$. Furthermore, if $z \in ML(G) \cap \mathbb{Z}^{V \cup E}$ and $s(z) = 0$, then the corresponding implication below holds.*

(i) If $s_{e,v}^{inc}(z) = 0$ then $z_v = z_e$.
(ii) If $s_{e,U,W}^{odd}(z) = 0$ then $\prod_{u \in U} z_u + \prod_{w \in W} z_w = 1$.
(iii) If $s_{e,U,f}^{one}(z) = 0$ then $z_f + \prod_{u \in U} z_u = 1$.
(iv) If $s_{e,f,g}^{two}(z) = 0$ then $z_f + z_g = 1$.

Proof. First, $s_{e,v}^{inc}(z) \geq 0$ is part of the standard relaxation and implication (i) is obvious.

Second, $s_{e,U,W}^{odd}(z) \geq 0$ is the sum of the following inequalities from the standard relaxation: $z_e \geq 0$, $1 - z_v \geq 0$ for all $v \in e \setminus (U \cup W)$, and $(z_e - 1) + \sum_{v \in e}(1 - z_v) \geq 0$. If $z \in ML(G) \cap \mathbb{Z}^{V \cup E}$ and $s_{e,U,W}^{odd}(z) = 0$, then each of these inequalities must be tight, thus $z_e = 0$, $z_v = 1$ for each $v \in e \setminus (U \cup W)$. The last (tight) inequality yields $-1 + \sum_{v \in U \cup W}(1 - z_v) = 0$, i.e., precisely one variable z_v, for $v \in U \cup W$, is 0, while all others are 1, which yields implication (ii).

Third, $s_{e,U,f}^{one}(z) \geq 0$ is the sum of the following inequalities: $z_e \geq 0, 1 - z_v \geq 0$ for all $v \in e \setminus (U \cup f)$ and $(z_e - 1) + (1 - z_f) + \sum_{v \in e \setminus f}(1 - z_v) \geq 0$. The latter is the flower inequality centered at e with neighbor f. If $z \in ML(G) \cap \mathbb{Z}^{V \cup E}$ and $s_{e,U,f}^{one}(z) = 0$, then each of these inequalities must be tight, thus $z_e = 0$, $z_v = 1$ for each $v \in e \setminus (U \cup f)$. The last (tight) inequality yields $-1 + (1 - z_f) + \sum_{u \in U}(1 - z_u) = 0$, i.e., either $z_f = 1$ and $z_u = 0$ for exactly one $u \in U$, or $z_f = 0$ and $z_u = 1$ holds for all $u \in U$. Both cases yield implication (iii).

Fourth, we consider $s_{e,f,g}^{two}(z) \geq 0$. Note that due to $e \cap f \neq \emptyset$, $e \cap g \neq \emptyset$ and $e \cap f \cap g = \emptyset$, the three edges e, f, g must all be different. Thus, $s_{e,f,g}^{two}(z) \geq 0$ is the sum of $z_e \geq 0$, $1 - z_v \geq 0$ for all $v \in e \setminus (f \cup g)$ and of $(z_e - 1) + (1 - z_f) + (1 - z_g) + \sum_{v \in e \setminus (f \cup g)}(1 - z_v) \geq 0$. The latter is the flower inequality centered at e with neighbors f and g. If $z \in ML(G) \cap \mathbb{Z}^{V \cup E}$ and $s_{e,f,g}^{two}(z) = 0$ holds, then each of the involved inequalities must be tight, thus $z_e = 0$ and $z_v = 1$ for each $v \in e \setminus (f \cup g)$. The last (tight) inequality implies $-1 + (1 - z_f) + (1 - z_g) = 0$, i.e., $z_f + z_g = 1$. Hence, implication (iv). holds. $\qquad \square$

3   Simple Odd β-Cycle Inequalities

We will consider signed edges by associating either a "+" or a "−" with each edge. We denote by $\{\pm\}$ the set $\{+, -\}$ and by $-p$ a sign change for $p \in \{\pm\}$. In order to introduce simple odd β-cycle inequalities, we first present some more definitions.

Definition 1. *A* closed walk *in G of length $k \geq 3$ is a sequence $C = v_1\text{-}e_1\text{-}v_2\text{-}e_2\text{-}v_3\text{-}\cdots\text{-}v_{k-1}\text{-}e_{k-1}\text{-}v_k\text{-}e_k\text{-}v_1$, where we have $e_i \in E$ as well as $v_i \in e_{i-1} \cap e_i$ and $e_{i-1} \cap e_i \cap e_{i+1} = \varnothing$ for each $i \in [k]$, where we denote $e_0 := e_k$ and $e_{k+1} := e_1$ for convenience. A* signature *of C is a map $\sigma : [k] \to \{\pm\}$. A* signed closed walk *in G is a pair (C, σ) for a closed walk C and a signature σ of C. Similarly, we denote $v_0 := v_k$, $v_{k+1} := v_1$, $\sigma(0) := \sigma(k)$ and $\sigma(k + 1) := \sigma(1)$. We say that (C, σ) is* odd *if there is an odd number of indices $i \in [k]$ with $\sigma(i) = -$; otherwise we say that (C, σ) is* even. *Finally, for any signed closed walk (C, σ) in G, its* length function *is the map $\ell_{(C,\sigma)} : \mathrm{FR}(G) \to \mathbb{R}$ defined by*

$$
\ell_{(C,\sigma)}(z) := \sum_{i \in I_{(+,+,+)}} \left(s^{\mathrm{inc}}_{e_i, v_i}(z) + s^{\mathrm{inc}}_{e_i, v_{i+1}}(z) \right) + \sum_{i \in I_{(-,-,-)}} s^{\mathrm{odd}}_{e_i, e_i \cap e_{i-1}, e_i \cap e_{i+1}}(z)
$$
$$
+ \sum_{i \in I_{(+,+,-)}} s^{\mathrm{inc}}_{e_i, v_i}(z) + \sum_{i \in I_{(-,-,+)}} s^{\mathrm{one}}_{e_i, e_i \cap e_{i-1}, e_{i+1}}(z) + \sum_{i \in I_{(-,+,+)}} s^{\mathrm{inc}}_{e_i, v_{i+1}}(z)
$$
$$
+ \sum_{i \in I_{(+,-,-)}} s^{\mathrm{one}}_{e_i, e_i \cap e_{i+1}, e_{i-1}}(z) + \sum_{i \in I_{(+,-,+)}} s^{\mathrm{two}}_{e_i, e_{i-1}, e_{i+1}}(z),
$$

where $I_{(a,b,c)}$ is the set of edge indices i for which e_{i-1}, e_i and e_{i+1} have sign pattern $(a, b, c) \in \{\pm\}^3$, i.e., $I_{(a,b,c)} := \{i \in [k] : \sigma(i - 1) = a,\ \sigma(i) = b,\ \sigma(i + 1) = c\}$.

We remark that the definition of $\ell_{(C,\sigma)}(z)$ is independent of where the closed walk starts and ends. Namely, if instead of C we consider $C' = v_i\text{-}e_i\text{-}\cdots\text{-}v_k\text{-}e_k\text{-}v_1\text{-}e_1\text{-}\cdots\text{-}v_{i-1}\text{-}e_{i-1}\text{-}v_i$, and we define σ' accordingly, then we have $\ell_{(C,\sigma)}(z) = \ell_{(C',\sigma')}(z)$. Moreover, if $\sigma(i - 1) = -$ or $\sigma(i) = -$, then $\ell_{(C,\sigma)}(z)$ is independent of the choice of $v_i \in e_{i-1} \cap e_i$.

By Lemma 1, the length function of a signed closed walk is nonnegative. We will show that for odd signed closed walks, the length function evaluated in each integer solution is at least 1. Hence, we define the *simple odd β-cycle inequality* corresponding to the odd signed closed walk (C, σ) as

$$
\ell_{(C,\sigma)}(z) \geq 1. \tag{4}
$$

We first establish that this inequality is indeed valid for $\mathrm{ML}(G)$.

Theorem 1. *Simple odd β-cycle inequalities (4) are valid for $\mathrm{ML}(G)$.*

Proof. Let $z \in \mathrm{ML}(G) \cap \{0, 1\}^{V \cup E}$ and assume, for the sake of contradiction, that z violates inequality (4) for some odd signed closed walk (C, σ). Since the coefficients of $\ell_{(C,\sigma)}$ are integer, we obtain $\ell_{(C,\sigma)} \leq 0$. From Lemma 1, we have that $s(z) = 0$ holds for all involved functions $s(z)$. Moreover, edge variables

z_{e_i} for all edges e_i with $\sigma(i) = +$, node variables z_{v_i} for all nodes v_i with $\sigma(i-1) = \sigma(i) = +$, and the expressions $\prod_{v \in e_{i-1} \cap e_i} z_v$ for all nodes i with $\sigma(i-1) = \sigma(i) = -$ are either equal or complementary, where the latter happens if and only if the corresponding edge e_i satisfies $\sigma(i) = -1$. Since the signed closed walk C is odd, this yields a contradiction $z_e = 1 - z_e$ for some edge e of C or $z_v = 1 - z_v$ for some node v of C or $\prod_{v \in e \cap f} z_v = 1 - \prod_{v \in e \cap f} z_v$ for a pair e, f of subsequent edges of C. □

Next, we provide an example of a simple odd β-cycle inequality.

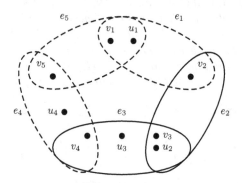

Fig. 1. Figure of the closed walk considered in Example 1. The solid edges have sign $+$ and the dashed edges have sign $-$.

Example 1. We consider the closed walk of length 5 given by the sequence $C = v_1\text{-}e_1\text{-}v_2\text{-}e_2\text{-}v_3\text{-}\cdots\text{-}v_5\text{-}e_5\text{-}v_1$ with signature $(\sigma(1), \sigma(2), \ldots, \sigma(5)) = (-, +, +, -, -)$ depicted in Fig. 1. We have $1 \in I_{(-,-,+)}$, $2 \in I_{(-,-,+)}$, $3 \in I_{(+,+,-)}$, $4 \in I_{(+,-,-)}$, $5 \in I_{(-,-,-)}$. The corresponding simple odd β-cycle inequality is $\ell_{(C,\sigma)}(z) \geq 1$. Using Definition 1, we write $\ell_{(C,\sigma)}(z)$ in terms of the building blocks as

$$\ell_{(C,\sigma)}(z) = s^{\text{one}}_{e_1, e_1 \cap e_5, e_2}(z) + s^{\text{inc}}_{e_2, v_3}(z) + s^{\text{inc}}_{e_3, v_3}(z) + s^{\text{one}}_{e_4, e_4 \cap e_5, e_3}(z) + s^{\text{odd}}_{e_5, e_5 \cap e_4, e_5 \cap e_1}(z).$$

Using the definition of the building blocks, we obtain

$$\ell_{(C,\sigma)}(z) = +2z_{e_1} - 1 + \sum_{u \in e_1 \cap e_5} (1 - z_u) + (1 - z_{e_2}) + \sum_{v \in e_1 \setminus (e_1 \cap e_5 \cup e_2)} (2 - 2z_v)$$

$$+ (z_{v_3} - z_{e_2}) + (z_{v_3} - z_{e_3})$$

$$+ 2z_{e_4} - 1 + \sum_{u \in e_4 \cap e_5} (1 - z_u) + (1 - z_{e_3}) + \sum_{v \in e_4 \setminus (e_4 \cap e_5 \cup e_3)} (2 - 2z_v)$$

$$+ 2z_{e_5} - 1 + \sum_{u \in e_5 \cap e_4} (1 - z_u) + \sum_{w \in e_5 \cap e_1} (1 - z_w) + \sum_{v \in e_5 \setminus (e_5 \cap e_4 \cup (e_5 \cap e_1))} (2 - 2z_v).$$

We write the sums explicitly and obtain

$$
\begin{aligned}
\ell_{(C,\sigma)}(z) &= +2z_{e_1} - 1 + (1 - z_{v_1}) + (1 - z_{u_1}) + (1 - z_{e_2}) \\
&\quad + (z_{v_3} - z_{e_2}) + (z_{v_3} - z_{e_3}) \\
&\quad + 2z_{e_4} - 1 + (1 - z_{v_5}) + (1 - z_{e_3}) + (2 - 2z_{u_4}) \\
&\quad + 2z_{e_5} - 1 + (1 - z_{v_5}) + (1 - z_{v_1}) + (1 - z_{u_1}) \\
&= 2(z_{e_1} - z_{e_2} - z_{e_3} + z_{e_4} + z_{e_5} - z_{v_1} - z_{u_1} + z_{v_3} - z_{u_4} - z_{v_5}) + 7.
\end{aligned}
$$

$\diamond$

Example 1 suggests that, when the function is written explicitly, the coefficients in the function $\ell_{(C,\sigma)}(z)$ exhibit a certain pattern. This different expression of $\ell_{(C,\sigma)}(z)$ is formalized in the next lemma.

Lemma 2. *Given a signed closed walk (C,σ) in G with $k \geq 3$, we have*

$$
\ell_{(C,\sigma)}(z) = \sum_{\substack{i \in [k] \\ \sigma(i)=-}} (2z_{e_i} + 1) - \sum_{\substack{i \in [k] \\ \sigma(i)=+}} 2z_{e_i} + \sum_{\substack{i \in [k] \\ \sigma(i-1)=\sigma(i)=+}} 2z_{v_i} + \sum_{\substack{i \in [k] \\ \sigma(i-1)=\sigma(i)=- \\ v \in e_{i-1} \cap e_i}} 2(1 - z_v)
$$

$$
+ \sum_{\substack{i \in [k]:\sigma(i)=- \\ v \in e_i \setminus (e_{i-1} \cup e_{i+1})}} 2(1 - z_v) - 2|\{i \in [k] : \sigma(i-1) = \sigma(i) = -\}|. \tag{5}
$$

Using Definition 2, we obtain the following result.

Proposition 1. *Simple odd β-cycle inequalities are Chvátal-Gomory inequalities for $\mathrm{FR}(G)$ and can be written in the form*

$$
\sum_{\substack{i \in [k] \\ \sigma(i)=-}} z_{e_i} - \sum_{\substack{i \in [k] \\ \sigma(i)=+}} z_{e_i} + \sum_{\substack{i \in [k] \\ \sigma(i-1)=\sigma(i)=+}} z_{v_i} - \sum_{\substack{i \in [k] \\ \sigma(i-1)=\sigma(i)=- \\ v \in e_{i-1} \cap e_i}} (z_v - 1) - \sum_{\substack{i \in [k]:\sigma(i)=- \\ v \in e_i \setminus (e_{i-1} \cup e_{i+1})}} (z_v - 1)
$$

$$
\geq \frac{1 - |\{i \in [k] : \sigma(i) = -\}|}{2} - |\{i \in [k] : \sigma(i-1) = \sigma(i) = -\}|. \tag{6}
$$

Proof. Let (C,σ) be an odd signed closed walk in a hypergraph G. From Lemma 1 we obtain that $\ell_{(C,\sigma)}(z) \geq 0$ holds for each $z \in \mathrm{FR}(G)$. Lemma 2 reveals that in the inequality $\ell_{(C,\sigma)}(z) \geq 0$, all variables' coefficients are even integers, while the constant term is an odd integer. Hence, the inequality divided by 2 has integral variable coefficients, and we can obtain the corresponding Chvátal-Gomory inequality by rounding the constant term up. The resulting inequality is the simple odd β-cycle inequality (4) scaled by $1/2$ and has the form (6). This shows that simple odd β-cycle inequalities are Chvátal-Gomory inequalities for $\mathrm{FR}(G)$. □

It follows from Proposition 1 that, under some conditions on (C,σ), simple odd β-cycle inequalities are in fact $\{0, 1/2\}$-cuts (see [3]) with respect to $\mathrm{FR}(G)$. Some classes of such cutting planes can be separated in polynomial time, in particular if the involved inequalities only have two odd coefficients. In such a case, these inequalities are patched together such that odd coefficients cancel out and eventually all coefficients are even. We want to emphasize that this generic separation approach does not work in our case since our building block inequalities may have more than 2 odd-degree coefficients. Nevertheless, the separation algorithm presented in the next section is closely related to the idea of cancellation of odd-degree coefficients.

4   Separation Algorithm

The main goal of this section is to show that the separation problem over simple odd β-cycle inequalities can be solved in strongly polynomial time (Theorem 2). This will be achieved by means of an auxiliary undirected graph in which several shortest-path computations must be carried out. The auxiliary graph is inspired by the one for the separation problem of odd-cycle inequalities for the maximum cut problem [1]. However, to deal with our different problem and the more general hypergraphs we will extend it significantly.

Let $G = (V, E)$ be a hypergraph and let $\hat{z} \in \mathrm{FR}(G)$. Define $\mathcal{T} := \{(e, f, g) \in E : e \cap f \neq \varnothing,\ f \cap g \neq \varnothing,\ e \cap f \cap g = \varnothing\}$ to be the set of potential subsequent edge triples. We define the auxiliary graph

$$\bar{G} = (\bar{V}, \bar{E}) = (\bar{V}_+ \cup \bar{V}_- \cup \bar{V}_{\mathrm{E}},\ \bar{E}^{-,-,-} \cup \bar{E}^{+,-,+} \cup \bar{E}^{+,-,-} \cup \bar{E}^{+,+,\pm})$$

and length function $\bar{\ell} : \bar{E} \to \mathbb{R}$ as follows.

$$\bar{V}_+ := V \times \{\pm\}$$
$$\bar{V}_- := \{e \cap f : e, f \in E,\ e \neq f,\ e \cap f \neq \varnothing\} \times \{\pm\}$$
$$\bar{V}_{\mathrm{E}} := E \times \{\pm\}$$
$$\bar{E}^{-,-,-} := \{\{(e \cap f, p), (f \cap g, -p)\} : (e, f, g) \in \mathcal{T},\ p \in \{\pm\}\}$$
$$\bar{\ell}_{\{(U,p),(W,-p)\}} := \min_{e,f,g}\{s_{f,U,W}^{\mathrm{odd}}(\hat{z}) : U = e \cap f,\ W = f \cap g \text{ for some } (e, f, g) \in \mathcal{T}\}$$
$$\bar{E}^{+,-,+} := \{\{(e, p), (g, -p)\} : e, g \in E, e \cap f \neq \varnothing \text{ and } f \cap g \neq \varnothing$$
$$\text{for some } f \in E \text{ with } e \cap f \cap g = \varnothing,\ p \in \{\pm\}\}$$
$$\bar{\ell}_{\{(e,p),(g,-p)\}} := \min_f\{s_{e,f,g}^{\mathrm{two}}(\hat{z}) : f \in E,\ e \cap f \neq \varnothing,\ f \cap g \neq \varnothing,\ e \cap f \cap g = \varnothing\}$$
$$\bar{E}^{+,-,-} := \{\{(e, p), (f \cap g, -p)\} : (e, f, g) \in \mathcal{T},\ p \in \{\pm\}\}$$
$$\bar{\ell}_{\{(e,p),(U,-p)\}} := \min_{f,g}\{s_{f,U,e}^{\mathrm{one}}(\hat{z}) : (e, f, g) \in \mathcal{T},\ U = f \cap g\}$$
$$\bar{E}^{+,+,\pm} := \{\{(v, p), (e, p)\} : v \in e \in E,\ p \in \{\pm\}\}$$
$$\bar{\ell}_{\{(v,p),(e,p)\}} := s_{e,v}^{\mathrm{inc}}(\hat{z})$$

We point out that the graph $\bar{G}$ can have parallel edges, possibly with different lengths. We immediately obtain the following corollary from Lemma 1.

Corollary 1. *The edge lengths $\bar{\ell} : \bar{E} \to \mathbb{R}$ are nonnegative.*

We say that two nodes $\bar{u}, \bar{v} \in \bar{V}$ are *twins* if they only differ in the second component, i.e., the sign. We call a walk $\bar{W}$ in the graph $\bar{G}$ a *twin walk* if its end nodes are twin nodes. For a walk $\bar{W}$ in $\bar{G}$, we denote by $\bar{\ell}(\bar{W})$ the total *length*, i.e., the sum of the edge lengths $\bar{\ell}_e$ along the edges e in $\bar{W}$. In the next two lemmas we study the relationship between odd signed closed walks in G and twin walks in $\bar{G}$.

Lemma 3. *For each odd signed closed walk (C, σ) in G there exists a twin walk $\bar{W}$ in $\bar{G}$ of length $\bar{\ell}(\bar{W}) \leq 1 + s$, where s is the slack of the simple odd β-cycle inequality (4) induced by (C, σ) with respect to $\hat{z}$. In particular, if the inequality is violated by $\hat{z}$, then we have $\bar{\ell}(\bar{W}) < 1$.*

Proof. Let (C, σ) be an odd signed closed walk with $C = v_1\text{-}e_1\text{-}v_2\text{-}e_2\text{-}v_3\text{-}\cdots\cdots v_{k-1}\text{-}v_{k-1}\text{-}v_k\text{-}e_k\text{-}v_1$. For $i \in [k]$, let $p_i := \prod_{j=1}^{i} \sigma(j)$ be the product of signs of all edges up to e_i. Moreover, define $p_0 := \sigma(0) = \sigma(k)$. For each $i \in [k]$, we determine a walk $\bar{W}_i$ in $\bar{G}$ of length at most 2, and construct $\bar{W}$ by going along all these walks in their respective order. The walk $\bar{W}_i$ depends on $\sigma(i-1)$, $\sigma(i)$ and $\sigma(i+1)$:

$$
\bar{W}_i := \begin{cases}
(v_i, p_{i-1}) \to (e_i, p_i) \to (v_{i+1}, p_i) & \text{if } i \in I_{+,+,+} \\
(v_i, p_{i-1}) \to (e_i, p_i) & \text{if } i \in I_{+,+,-} \\
(e_i, p_i) \to (v_{i+1}, p_i) & \text{if } i \in I_{-,+,+} \\
(e_i, p_i) \quad\quad\quad\quad (\text{length } 0) & \text{if } i \in I_{-,+,-} \\
(e_{i-1}, p_{i-1}) \to (e_i \cap e_{i+1}, p_i) & \text{if } i \in I_{+,-,-} \\
(e_{i-1} \cap e_i, p_{i-1}) \to (e_{i+1}, p_i) & \text{if } i \in I_{-,-,+} \\
(e_{i-1} \cap e_i, p_{i-1}) \to (e_i \cap e_{i+1}, p_i) & \text{if } i \in I_{-,-,-} \\
(e_{i-1}, p_{i-1}) \to (e_{i+1}, p_i) & \text{if } i \in I_{+,-,+}.
\end{cases}
$$

The walks $\bar{W}_i$ help to understand the meaning of the different node types: the walk $\bar{W}_i$ starts at a node from $\bar{V}_+$ if $\sigma(i-1) = \sigma(i) = +$, it starts at a node from $\bar{V}_-$ if $\sigma(i-1) = \sigma(i) = -$, and it starts at a node from $\bar{V}_E$ if $\sigma(i-1) \neq \sigma(i)$ holds. Similarly, the walk $\bar{W}_i$ ends at a node from $\bar{V}_+$ if $\sigma(i) = \sigma(i+1) = +$, it ends at a node from $\bar{V}_-$ if $\sigma(i) = \sigma(i+1) = -$, and it ends at a node from $\bar{V}_E$ if $\sigma(i) \neq \sigma(i+1)$ holds.

Note that all edges traversed by each $\bar{W}_i$ are indeed in $\bar{E}$. It is easily verified that, for each $i \in [k-1]$, the walk $\bar{W}_i$ ends at the same node at which the walk $\bar{W}_{i+1}$ starts. Hence $\bar{W}$ is indeed a walk in $\bar{G}$. Since $v_{k+1} = v_1$ holds, C is closed and (C, σ) is odd, it can be checked that $\bar{W}$ is a twin walk. Finally, by construction, $\bar{\ell}(\bar{W}) \leq \ell_{(C,\sigma)}(\hat{z})$ holds, where the inequality comes from the fact that the minima in the definition of $\bar{\ell}$ need not be attained by the edges from C. By definition of s we have $\ell_{(C,\sigma)}(\hat{z}) = 1 + s$, thus $\bar{\ell}(\bar{W}) \leq 1 + s$. $\square$

Lemma 4. *For each twin walk $\bar{W}$ in $\bar{G}$ there exists an odd signed closed walk (C, σ) in G whose induced simple odd β-cycle inequality (4) has slack $\bar{\ell}(\bar{W}) - 1$ with respect to $\hat{z}$. In particular, if $\bar{\ell}(\bar{W}) < 1$ holds, then the inequality is violated by $\hat{z}$.*

Proof. Let $\bar{W}$ be a twin walk in $\bar{G}$. We first construct the signed closed walk (C, σ) by processing the edges of $\bar{W}$ in their order. Throughout the construction we maintain the index i of the next edge to be constructed, which initially is $i :=$ 1. Since the construction depends on the type of the current edge $\bar{e} = \{\bar{u}, \bar{v}\} \in \bar{W}$ (where $\bar{W}$ visits $\bar{u}$ first), we distinguish the relevant cases:

Case 1: $\bar{e} \in \bar{E}^{+,+,\pm}$ and $\bar{u} \in \bar{V}_E$. Hence, $\bar{u} = (e, p)$ and $\bar{v} = (v, p)$ for some $v \in e \in E$ and some $p \in \{\pm\}$. We define $v_i := v$ and continue.

Case 2: $\bar{e} \in \bar{E}^{+,+,\pm}$ and $\bar{u} \in \bar{V}_+$. Hence, $\bar{u} = (v, p)$ and $\bar{v} = (e, p)$ for some $v \in e \in E$ and some $p \in \{\pm\}$ as well as $\ell_{\bar{e}} = s_{e,v}^{inc}$. We define $e_i := e$ and $\sigma(i) := +$. We then increase i by 1 and continue.

Case 3: $\bar{e} \in \bar{E}^{+,-,-}$ and $\bar{u} \in V_E$. Hence, $\bar{u} = (e, p)$ and $\bar{v} = (f \cap g, -p)$ for some $(e, f, g) \in \mathcal{T}$ as well as $\ell_{\bar{e}} = s_{e,U,f}^{one}(\hat{z})$. We define v_i (resp. v_{i+1}) to be any node in $e \cap f$ (resp. $f \cap g$), $e_i := f$ and $\sigma(i) := -$. We then increase i by 1 and continue.

Case 4: $\bar{e} \in \bar{E}^{-,-,-}$. Hence, $\bar{u} = (e \cap f, p)$ and $\bar{v} = (f \cap g, -p)$ for some $(e, f, g) \in \mathcal{T}$ as well as $\ell_{\bar{e}} = s_{e,U,W}^{odd}(\hat{z})$. We define $e_i := f$, $\sigma(i) := -$ and v_{i+1} to be any node in $f \cap g$. We then increase i by 1 and continue.

Case 5: $\bar{e} \in \bar{E}^{+,-,-}$ and $\bar{u} \in V_-$. Hence, $\bar{u} = (e \cap f, p, -)$ and $\bar{v} = (g, -p)$ for some $(e, f, g) \in \mathcal{T}$ with $\ell_{\bar{e}} = s_{e,U,f}^{one}(\hat{z})$. We define $e_i := f$, $\sigma(i) := -$ and v_{i+1} to be any node in $f \cap g$. We then increase i by 1 and continue.

Case 6: $\bar{e} \in \bar{E}^{+,-,+}$. Hence, $\bar{u} = (e, p)$ and $\bar{v} = (g, -p)$ for some $(e, f, g) \in \mathcal{T}$ as well as $\ell_{\bar{e}} = s_{e,f,g}^{two}(\hat{z})$. We define v_i (resp. v_{i+1}) to be any node in $e \cap f$ (resp. $f \cap g$), $e_i := f$, $\sigma(i) := -$, $e_{i+1} := g$ and $\sigma(i+1) := +$. We then increase i by 2 and continue.

After processing all edges of $\bar{W}$, the last defined edge is e_{i-1} and thus we define $k := i - 1$ and $C := v_1\text{-}e_1\text{-}v_2\text{-}e_2\text{-}v_3\text{-}\cdots\text{-}v_{k-1}\text{-}v_k\text{-}e_k\text{-}v_1$. By checking pairs of edges of $\bar{W}$ that arise consecutively, one verifies that for each $i \in [k]$, we also have $v_i \in e_{i-1} \cap e_i$.

To see that (C, σ) is odd, we use the fact that the endnodes of $\bar{W}$ are twin nodes. When traversing an edge $\bar{e}$ from $\bar{u}$ to $\bar{v}$, the second entries of $\bar{u}$ and $\bar{v}$ differ if and only if we set a σ-entry to $-$. Note that in Case 6 we set two such entries, but only one to $-$. We conclude that $\sigma(i) = -$ holds for an odd number of indices $i \in [k]$.

By construction we have $\bar{\ell}(\bar{W}) = \ell_{(C,\sigma)}(\hat{z})$. The slack of the simple odd β-cycle inequality induced by (C, σ) with respect to $\hat{z}$ is then $\ell_{(C,\sigma)}(\hat{z}) - 1 = \bar{\ell}(\bar{W}) - 1$.   □

Theorem 2. *Let $G = (V, E)$ be a hypergraph and let $\hat{z} \in \mathrm{FR}(G)$. The separation problem for simple odd β-cycle inequalities (4) can be solved in time $\mathcal{O}(|E|^5 + |V|^2 \cdot |E|)$.*

Proof. Let $n := |V|$ and $m := |E|$ and assume $m \geq \log(n)$ since otherwise we can merge nodes that are incident to exactly the same edges. First note that, regarding the size of the auxiliary graph $\bar{G}$, we have $|\bar{V}| = \mathcal{O}(m^2 + n)$ and

$|\bar{E}| = \mathcal{O}(mn + m^3)$. For the construction of $\bar{G}$ and the computation of $\bar{\ell}$ we need to inspect all triples $(e, f, g) \in \mathcal{T}$ of edges. This can be done in time $\mathcal{O}(m^3 n)$ since for each of the m^3 edge triples (e, f, g) we have to inspect at most n nodes to check the requirements on the intersections of e, f and g.

According to Lemmas 3 and 4 we only need to check for the existence of a twin walk $\bar{W}$ in $\bar{G}$ with $\ell(\bar{W}) < 1$. This can be accomplished with $|\bar{V}|/2 = \mathcal{O}(m^2 + n)$ runs of Dijkstra's algorithm [12] on $\bar{G}$, each of which takes

$$\mathcal{O}(|\bar{E}| + |\bar{V}| \cdot \log(|\bar{V}|)) = \mathcal{O}((mn + m^3) + (m^2 + n) \cdot \log(m^2 + n))$$

time when implemented with Fibonacci heaps [14]. If $m^2 \geq n$, then the total running time simplifies to $\mathcal{O}(m^5)$, and otherwise we obtain $\mathcal{O}(n^2 m)$. $\qquad \square$

The main reason for this large running time bound is the fact that $|\bar{V}_-|$ can be quadratic in $|E|$.

Clearly, our separation algorithm requires that the edge lengths $\bar{\ell}$ of the auxiliary graph $\bar{G}$ are nonnegative. This in turn requires $\hat{z} \in \mathrm{FR}(G)$, i.e., that the flower inequalities with at most two neighbors are satisfied. As we already mentioned, the number of these flower inequalities is bounded by a polynomial in $|V|$ and $|E|$. We like to point out that one can combine the separation of these flower inequalities with the construction of $\bar{G}$, i.e., one can determine violated inequalities while constructing the auxiliary graph.

5   Relation to Non-simple Odd β-Cycle Inequalities

In this section we relate our simple odd β-cycle inequalities to the odd β-cycle inequalities in [5].

A *cycle hypergraph* is a hypergraph $G = (V, E)$, with $E = \{e_1, \ldots, e_m\}$, where $m \geq 3$, and every edge e_i has nonempty intersection only with e_{i-1} and e_{i+1} for every $i \in \{1, \ldots, m\}$, where, for convenience, we define $e_{m+1} := e_1$ and $e_0 := e_m$. If $m = 3$, it is also required that $e_1 \cap e_2 \cap e_3 = \emptyset$. Given a closed walk $C = v_1\text{-}e_1\text{-}v_2\text{-}e_2\text{-}\cdots\text{-}v_k\text{-}e_k\text{-}v_1$ in a hypergraph $G = (V, E)$, the *support hypergraph* of C is the hypergraph $G(C) = (V(C), E(C))$, where $E(C) := \{e_1, e_2, \ldots, e_k\}$ and $V(C) := e_1 \cup e_2 \cup \cdots \cup e_k$.

Lemma 5. *Let (C, σ) be a signed closed walk in a hypergraph G and assume that the support hypergraph of C is a cycle hypergraph. Let $E^- := \{e_i : i \in [k], \sigma(i) = -\}$, $E^+ := \{e_i : i \in [k], \sigma(i) = +\}$, $S_1 := (\bigcup_{e \in E^-} e) \setminus \bigcup_{e \in E^+} e$, and $S_2 := \{v_1, \ldots, v_k\} \setminus \bigcup_{e \in E^-} e$. Then*

$$\ell_{(C,\sigma)}(z) = -\sum_{v \in S_1} 2z_v + \sum_{e \in E^-} 2z_e + \sum_{v \in S_2} 2z_v - \sum_{e \in E^+} 2z_e + 2|S_1|$$
$$- 2|\{i \in [k] : e_{i-1}, e_i \in E^-\}| + |E^-|.$$

In particular, the simple odd β-cycle inequality corresponding to (C, σ) coincides with the odd β-cycle inequality corresponding to (C, σ). Furthermore, in a cycle hypergraph, every odd β-cycle inequality is a simple odd β-cycle inequality.

Proof. It suffices to observe that

$$\sum_{\substack{i\in[k]\\\sigma(i-1)=\sigma(i)=+}} 2z_{v_i} = \sum_{v\in S_2} 2z_v, \qquad \sum_{\substack{i\in[k]\\\sigma(i-1)=\sigma(i)=-\\v\in e_{i-1}\cap e_i}} 2z_v + \sum_{\substack{i\in[k]:\sigma(i)=-\\v\in e_i\backslash(e_{i-1}\cup e_{i+1})}} 2z_v = \sum_{v\in S_1} 2z_v,$$

$$\sum_{\substack{i\in[k]\\\sigma(i-1)=\sigma(i)=-\\v\in e_{i-1}\cap e_i}} 2 + \sum_{\substack{i\in[k]:\sigma(i)=-\\v\in e_i\backslash(e_{i-1}\cup e_{i+1})}} 2 = 2|S_1| \qquad \text{and} \qquad \sum_{\substack{i\in[k]\\\sigma(i)=-}} 1 = |E^-|.$$

The statement for cycle hypergraphs G follows by inspecting the definition of the odd β-cycle inequalities.   □

As a consequence, we can use the two following known results in order to gain insights about simple odd β-cycle inequalities.

Proposition 2 (Example 2 in [5]). *There exists a cycle hypergraph for which the Chvátal rank of odd β-cycle inequalities can be equal to 2.*

Proposition 3 (Implied by Theorem 1 in [5]). *Flower inequalities are Chvátal-Gomory cuts for* $\mathrm{SR}(G)$.

Theorem 3. *Simple odd β-cycle inequalities can have Chvátal rank 2 with respect to* $\mathrm{SR}(G)$.

Proof. Combining Proposition 3 with Proposition 1 shows that simple odd β-cycle inequalities have Chvátal rank at most 2. Lemma 5 and Proposition 2 show that the Chvátal rank of simple odd β-cycle inequalities for cycle hypergraphs can be equal to 2.   □

For the second insight, we consider a strengthened form of Theorem 5 in [5].

Proposition 4 (Theorem 5 in [5], strengthened). *Let $G = (V,E)$ be a cycle hypergraph. Then $\mathrm{ML}(G)$ is described by all odd β-cycle inequalities and all inequalities from* $\mathrm{FR}(G)$.

The strengthening lies in the fact that in the original statement of Theorem 5 in [5] all flower inequalities are used rather than only those with at most two neighbors. This strengthening of the original statement can be seen by inspecting its proof in [5]. By applying Lemma 5 to Proposition 4 we immediately obtain the following result.

Theorem 4. *Let $G = (V,E)$ be a cycle hypergraph. Then*

$$\mathrm{ML}(G) = \{x \in \mathrm{FR}(G) : x \text{ satisfies all simple odd } \beta\text{-cycle inequalities}\}.$$

Future Research. We would like to conclude this paper with a couple of open questions that could be investigated. An interesting research direction is a computational investigation of simple odd β-cycle inequalities, especially in relation to the applications discussed in Sect. 1, i.e., the image restoration problem in computer vision and the low auto-correlation binary sequence problem in theoretical physics.

The next research direction has a more theoretical flavor. The LP relaxations defined by odd-cycle inequalities [1] for the cut polytope and the affinely isomorphic correlation polytope (see [11]) have the following property: when maximizing a specific objective vector, then one can remove a subset of the odd-cycle inequalities upfront without changing the optimum. More precisely, the removal is based only on the sign pattern of the objective vector (see Theorem 2 in [17]). Since the simple odd β-cycle inequalities can be seen as an extension of the odd cycle inequalities for the cut polytope, the research question is whether a similar property can be proven for simple odd β-cycle inequalities.

The final research direction is that of redundancy of simple odd β-cycle inequalities for which we provide some insight in the full version of the paper [10].

References

1. Barahona, F., Mahjoub, A.R.: On the cut polytope. Math. Program. **36**(2), 157–173 (1986)
2. Bernasconi, J.: Low autocorrelation binary sequences: statistical mechanics and configuration space analysis. J. Phys. **141**(48), 559–567 (1987)
3. Caprara, A., Fischetti, M.: {0, 1/2}-Chvátal-Gomory cuts. Math. Program. **74**(3), 221–235 (1996)
4. Crama, Y., Rodríguez-Heck, E.: A class of valid inequalities for multilinear 0–1 optimization problems. Discret. Optim. **25**, 28–47 (2017)
5. Del Pia, A., Di Gregorio, S.: Chvátal rank in binary polynomial optimization. INFORMS J. Optim. **3**, 315–349 (2021)
6. Del Pia, A., Khajavirad, A.: A polyhedral study of binary polynomial programs. Math. Oper. Res. **42**(2), 389–410 (2017)
7. Del Pia, A., Khajavirad, A.: The multilinear polytope for acyclic hypergraphs. SIAM J. Optim. **28**(2), 1049–1076 (2018)
8. Del Pia, A., Khajavirad, A.: The running intersection relaxation of the multilinear polytope. Math. Oper. Res. **46**, 1008–1037 (2021)
9. Del Pia, A., Khajavirad, A., Sahinidis, N.V.: On the impact of running intersection inequalities for globally solving polynomial optimization problems. Math. Program. Comput. **12**(2), 165–191 (2019). https://doi.org/10.1007/s12532-019-00169-z
10. Del Pia, A., Walter, M.: Simple odd β-cycle inequalities for binary polynomial optimization. arXiv Online Preprint arXiv:2111.04858 (2021)
11. Deza, M.M., Laurent, M.: Geometry of Cuts and Metrics, 1st edn. Springer, Heidelberg (2009)
12. Dijkstra, E.W.: A note on two problems in connexion with graphs. Numer. Math. **1**(1), 269–271 (1959)
13. Fortet, R.: Applications de l'algèbre de boole en recherche opérationelle. Rev. Française D'autom. Inform. Recherche Opér. **4**, 17–26 (1960)
14. Fredman, M.L., Tarjan, R.E.: Fibonacci heaps and their uses in improved network optimization algorithms. J. ACM **34**(3), 596–615 (1987)

15. Glover, F., Woolsey, E.: Converting the 0–1 Polynomial Programming Problem to a 0–1 Linear Program. Oper. Res. **22**(1), 180–182 (1974)
16. Mertens, S., Bessenrodt, C.: On the ground states of the Bernasconi model. J. Phys. A: Math. Gen. **31**(16), 3731–3749 (1998)
17. Michini, C.: Forbidden minors for tight cycle relaxations. Optimization Online Preprint, 5483 (2018)
18. MINLPLib. A library of mixed-integer and continuous nonlinear programming instances (2020)
19. POLIP. Library for polynomially constrained mixed-integer programming (2014)

Combinatorial Algorithms for Rooted Prize-Collecting Walks and Applications to Orienteering and Minimum-Latency Problems

Sina Dezfuli[1], Zachary Friggstad[1], Ian Post[3], and Chaitanya Swamy[2(✉)]

[1] Department of Computer Science, University of Alberta, Edmonton, Canada
{dezfuli,zacharyf}@ualberta.ca
[2] Combinatorics and Optimization, University of Waterloo, Waterloo, Canada
cswamy@uwaterloo.ca
[3] Montain View, USA
ian@ianpost.org

Abstract. We consider the rooted *prize-collecting walks* (PCW) problem, wherein we seek a collection $\mathcal{C}$ of rooted walks having minimum *prize-collecting cost*, which is the (total cost of walks in $\mathcal{C}$) + (total node-reward of the nodes not visited by any walk in $\mathcal{C}$). This problem arises naturally as the Lagrangian relaxation of both *orienteering* (find a length-bounded walk of maximum reward), and the *ℓ-stroll problem* (find a minimum-length walk covering at least ℓ nodes). Our main contribution is to devise a *simple, combinatorial algorithm* for the PCW problem that returns a rooted tree whose prize-collecting cost is at most the *optimum value* of the prize-collecting walks problem. This result applies also to directed graphs, and holds for arbitrary nonnegative edge costs.

We present two applications of our result. We utilize our algorithm to develop combinatorial approximation algorithms for two fundamental vehicle-routing problems (VRPs): (1) orienteering; and (2) *k-minimum-latency problem* (k-MLP), wherein we seek to cover all nodes using k paths starting at a prescribed root node, so as to minimize the sum of the node visiting times. Our combinatorial algorithm allows us to sidestep the part where we solve a preflow-based LP in the LP-rounding algorithms of [13] for orienteering, and in the state-of-the-art 7.183-approximation algorithm for k-MLP in [17]. Consequently, we obtain combinatorial implementations of these algorithms (with the same approximation factors). Compared to algorithms that achieve the current-best approximation factors for orienteering and k-MLP, our algorithms have substantially improved running time, and achieve approximation guarantees that match (k-MLP), or are slightly worse (orienteering) than the current-best approximation factors for these problems. We report various computational results for our resulting orienteering algorithms showing that they perform quite well in practice.

© Springer Nature Switzerland AG 2022
K. Aardal and L. Sanitá (Eds.): IPCO 2022, LNCS 13265, pp. 195–209, 2022.
https://doi.org/10.1007/978-3-031-06901-7_15

1   Introduction

Vehicle-routing problems (VRPs) are a rich class of optimization problems that find various applications, and have been extensively studied in the Operations Research and Computer Science literature (see, e.g., [19].) Broadly speaking, we can distinguish between two types of vehicle-routing problems: one where resource constraints require us to select which set of nodes or clients to visit *and* plan a suitable route(s) for visiting these clients; and the other, where we have a *fixed set* of clients, and seek the most effective route(s) for visiting these clients.

We consider two prominent and well-motivated problems in these two categories: (1) *orienteering* [2,4,7,13], belonging to the first category, wherein nodes have associated rewards for visiting them, and we seek a length-bounded path that collects maximum reward; and (2) *minimum-latency problems* (MLPs) [3,6,17], belonging to the second category, wherein, we seek one or more rooted paths to visit a given set of clients so as to minimize the sum of the client visiting times (i.e., the total latency). Besides its appeal as a natural and clean way of capturing resource constraints in a VRP, the fundamental nature of orienteering stems from the fact that it often naturally arises as a subroutine when solving other VRPs, both in approximation algorithms—e.g., for MLPs (see [5,10,17]), VRPs with time windows [2], distance bounds [15], and regret bounds [12]—as also in computational methods, where orienteering corresponds to the "pricing" problem encountered in solving set covering/partitioning LPs (a.k.a configuration LPs) for VRPs via a column-generation or branch-cut-and-price method (see, e.g., [8]). In particular, we can often formulate the VRP as one of covering clients using suitable paths; solving this covering problem, approximately via a set-cover approach, or its corresponding configuration-LP relaxation, then entails solving an orienteering problem.

Some recent work on orienteering [13] and MLPs [17], has led to promising LP-based approaches for tackling these problems, yielding, for multi-vehicle MLPs, the current-best approximation factors. This approach is based on moving to a bidirected version of the underlying metric and considering a preflow-based LP-relaxation for rooted walk(s) (with in-degree $\geq$ out-degree constraints), and using a powerful arborescence-packing result of Bang-Jensen et al. [1] to decompose an (optimal) LP solution into a convex combination of arborescences that is "at least as good" as the LP solution. Viewing these arborescences as rooted trees in the undirected graph, one can convert the tree into a rooted path/cycle by doubling and shortcutting, and the above works show how to leverage the resulting convex combination of paths/cycles to extract a good solution.

Our Contributions and Related Work. We study the *prize-collecting walks* (PCW) problem, which is the problem of finding a collection $\mathcal{C}$ of r-rooted walks in a digraph $G = (V, E)$ with nonnegative edge costs and node rewards, having minimum *prize-collecting cost*, which is the total cost of the walks in $\mathcal{C}$ + the total node-reward of the nodes not visited by any walk in $\mathcal{C}$. This problem arises as the Lagrangian relaxation of orienteering, and a subroutine encountered in

MLP algorithms, namely that of finding a rooted path of minimum cost covering a certain number of nodes.

Our main contribution is to devise a *simple, combinatorial algorithm* for the PCW problem that returns a *directed tree* (more precisely, an out-arborescence) rooted at r *whose prize-collecting cost is at most the optimal value of the PCW problem*. Let $G = (V, A)$ be a directed graph with arc-set A, arc lengths $c_a \geq 0$ for all $a \in A$, and root r. Let each node $v \in V$ have a reward or penalty $\pi_v \geq 0$. For a multiset of arcs T, define $c(T) = \sum_{a \in A} c_a \cdot$ (number of occurrences of a in T). Define $\pi(S) = \sum_{v \in S} \pi_v$ for any set of nodes S. An *out-arborescence rooted at* r is a subgraph T whose undirected version is a tree containing r, and where every node spanned by T except r has exactly one incoming arc in T; we will often abbreviate this to an out-arborescence. For any subgraph T of G where all nodes in $V(T)$ are reachable from r in T (e.g., out-arborescence rooted at r), define the *prize-collecting cost* of T to be $\mathsf{PCC}(T) := c(T) + \pi(V \backslash V(T))$.

We give a combinatorial polytime algorithm ITERPCA (see Sect. 3) that finds an out-arborescence T whose prize-collecting cost is at most that of any collection of r-rooted walks, i.e.,

$$c(T) + \pi(V \setminus V(T)) \leq O^* := \min_{\substack{\text{collections } \mathcal{C} \text{ of} \\ r\text{-rooted walks}}} \left[\sum_{P \in \mathcal{C}} c(P) + \pi\left(V \setminus \bigcup_{P \in \mathcal{C}} V(P)\right) \right].$$

We actually obtain the stronger guarantee that $\mathsf{PCC}(T)$ is at most the optimal value OPT of a preflow-based LP-relaxation (P) for the PCW problem.

We briefly discuss the ideas underlying our combinatorial algorithm ITER-PCA. Our algorithm and analysis is quite simple, and resembles Edmonds' algorithm for finding a minimum-cost arborescence. It is based on three main ideas for iteratively simplifying the instance. We observe that if we modify the instance by picking any non-root node v, and subtracting a common value θ from the cost of all incoming arcs of v and from π_v while ensuring that the new costs and rewards are nonnegative, then it suffices to prove the desired guarantee for the modified instance. Next, by choosing a suitable θ_v for all non-root nodes, and modifying costs and rewards as above, we ensure that in the modified instance, either: (a) there is a node $v \neq r$ with zero reward; (b) there is a (directed) cycle Z consisting of zero cost arcs; or (c) there is an out-arborescence consisting of zero cost arcs. If (c) applies, then we are done. If (a) or (b) apply, we further simplify the instance as follows: in case (a), we shortcut past v by merging every pair of incoming and outgoing arcs of v and deleting v; in case (b), we contract the cycle Z and set the reward of the contracted node to be the sum of the (modified) rewards of nodes in Z. We then recurse on the simplified instance.

We believe that the above result, and the techniques underlying it, are of independent interest, and will find various applications. We present two applications of our result (Sects. 4, 5), where use our combinatorial algorithm for PCW to give combinatorial implementations of the LP-rounding algorithm for orienteering in [13], and for k-MLP in [17]. We emphasize that our contribution and focus here is to demonstrate how our PCW-algorithm can be utilized to give a more-efficient implementation of existing algorithms; in particular, our resulting algorithms inherit the performance guarantees of the LP-rounding algorithms that they implement. We now discuss these applications, and in doing so, place our result in the context of some extant work. We say that $x \in \mathbb{R}_+^A$ is an r-preflow (or simply preflow), if we have $x(\delta^{\text{in}}(v)) \geq x(\delta^{\text{out}}(v))$ for all $v \neq r$.

- Friggstad and Swamy [13] proposed a novel LP-based approach for orienteering, wherein the LP-relaxation searches for a (r-) preflow of large reward. The first step (and key insight) in their rounding algorithm is to utilize the arborescence-packing result of [1] to cast the LP-solution x as a convex combination of arborescences whose expected reward is at least the LP-optimum and whose expected cost is at most the length bound, say B. They leverage this to show that one can then extract a rooted path having reward at least (LP-optimum)/3 via a simple combinatorial procedure.

  We show (see Sect. 4) that one can utilize our algorithm ITERPCA, in conjunction with binary search, to obtain the desired convex combination combinatorially, that is, without having to solve their LP-relaxation, and thereby obtain a combinatorial 3-approximation. This follows because the PCW problem is obtained by Lagrangifying the "length at most B" constraint. A standard fine tuning of the Lagrangian variable (which affects the node rewards) via binary search then yields the desired distribution (over at most two rooted trees). The same ideas also apply and yield combinatorial approximation algorithms for other variants of orienteering, such as P2P-orienteering (where we seek an r-t path) and cycle orienteering (where we seek a cycle containing r). As noted by [13], while their approximation factor of 3 does not as yet beat the $(2 + \epsilon)$-approximation factor for orienteering [7], their LP-rounding approach is significantly simpler than prior dynamic-programming (DP) based algorithms for orienteering [2,4,7]; with our combinatorial implementation, we also obtain significantly faster algorithms.[1] Moreover, an added subtle benefit of the algorithms in [13] is that they also yield an upper bound on the optimum, which can be used to evaluate the approximation factor of the solution computed on a per-instance basis; our combinatorial algorithms inherit this benefit.

  Our combinatorial algorithm and the associated upper bound may also find use in the context of computational methods for solving other VRPs, since (as mentioned earlier) orienteering corresponds to the pricing problem that needs to be solved in these contexts. Indeed [8] utilizes our combinatorial algorithm to obtain near-optimal solutions to distance-constrained vehicle routing.

  In Sect. 6, we perform a computational study of our combinatorial orienteering algorithms, to better understand the performance of our algorithms in practice. Our computational experiments show that our algorithms perform

---

[1] A straightforward implementation of our orienteering algorithm takes $O(n^4 \cdot K)$ time, where K is the time for binary search. In contrast, the algorithm in [7] has running time at least $O(n^{1/\epsilon^2} \cdot K)$ for obtaining a $\frac{2}{1-\epsilon}$-approximation; thus, $O(n^9 \cdot K)$ time for returning a 3-approximation. The DP-algorithm of Blum et al. [4] has running time at least $O(n^5 \cdot K)$, and its approximation guarantee is no better than 4.

fairly well in practice—in terms of both the solution and the upper bound computed—and much better than that indicated by the theoretical analysis.

- Post and Swamy [17] consider multi-vehicle MLPs. For k-MLP, wherein we seek k rooted paths of minimum total latency that together visit all nodes, they devise two 7.183-approximation algorithms. One of their algorithms (Algorithm 3 in §6.2 [17]) utilizes a subroutine for computing a distribution of rooted trees covering at least k nodes in expectation, whose expected cost is at most that of any collection of rooted walks that together cover at least k nodes. Lagrangifying the coverage constraint again yields a PCW problem. Post and Swamy [17] devised an LP-rounding algorithm for this PCW problem, by considering its LP-relaxation (P), using arborescence packing to obtain a rooted tree with $\mathsf{PCC}(T)$ at most the LP-optimum OPT, and then fine-tuning the node rewards via binary search to obtain the desired distribution. In particular, they obtain the same guarantee that we do, but via solving the LP (P). While not a combinatorial algorithm, they dub their resulting k-MLP algorithm a "more combinatorial" algorithm (as opposed to their other 7.183-approximation algorithm, which needs to explicitly solve a configuration LP).

We can instead utilize *our combinatorial algorithm* to produce the rooted tree T (see Sect. 5); incorporating this within the "more combinatorial" algorithm of [17] yields a fully and truly combinatorial 7.183-approximation algorithm for k-MLP, which is the state-of-the-art for this problem.

We remark that our result bounding the prize-collecting cost of the tree T by the prize-collecting cost of *any* collection of rooted walks is a substantial generalization of an analogous result in [6], who compare against the prize-collecting cost of a *single* walk (and specifically in undirected graphs). As noted in [17], this stronger guarantee is essential for obtaining guarantees for k-MLP.

2   LP-Relaxation for the Prize-Collecting-Walks Problem

Recall that we are given a directed graph $G = (V, A)$, arc costs $c_a \geq 0$ for all $a \in A$, root node $r \in V$, and a reward or penalty $\pi_v \geq 0$ for each node v. (Note that π_r does not affect the prize-collecting cost of any rooted object; so it will sometimes be convenient notationally to assume that $\pi_r = 0$.)

Our LP-relaxation (P) for prize-collecting walks has a variable x_a for each arc a, which represents the multiplicity of arc a in the walk-collection, and a variable p_v for each node $v \neq r$, which indicates whether node v is not covered.

(P)

$$\min \ \sum_{a \in A} c_a x_a + \sum_{v \in V} \pi_v p_v \quad \text{s.t.} \quad x\big(\delta^{\mathrm{in}}(S)\big) + p_v \geq 1 \quad \forall S \subseteq V \backslash \{r\}, v \in S$$

$$x\big(\delta^{\mathrm{in}}(v)\big) \geq x\big(\delta^{\mathrm{out}}(v)\big) \quad \forall v \in V \backslash \{r\}, \qquad x, p \geq 0.$$

The first constraint encodes that for every set $S \subseteq V'$ and every $v \in S$, either S has an incoming arc or we pay the penalty π_v for not visiting v; the second encodes that every node $v \neq r$ has in-degree at least out-degree, so that the solution corresponds to a collection of walks rather than a tree. (Variable p_r does not appear in any constraint, so we may assume that p_r is always 0).

3   A Combinatorial Algorithm

We now present a combinatorial algorithm for prize-collecting walks based on iteratively simplifying the instance. Recall that O^* is the minimum value of $\left[\sum_{P \in \mathcal{C}} c(P) + \pi(V \backslash \bigcup_{P \in \mathcal{C}} V(P)) \right]$ over all collections $\mathcal{C}$ of r-rooted walks. (Recall that a walk may have repeated nodes and arcs, and $c(T) = \sum_{a \in A} c_a \cdot$ (number of occurrences of a in T) for a multiset of arcs T). Throughout this section, the root will remain r, so we drop r from the notation used to refer to an instance. Since we will modify the instance (G, c, π) during the course of our algorithm, we use $O^*(G, c, \pi)$ to denote the above quantity. We use $(P_{(G,c,\pi)})$ to refer to the LP-relaxation (P) for the instance (G, c, π), and $OPT(G, c, \pi)$ to denote its optimal value. We use $\mathsf{PCC}(T; G, c, \pi) := c(T) + \pi(V \backslash V(T))$ to denote the prize-collecting value of T under arc costs c and penalties π, where T is a subgraph of G such that all nodes in $V(T)$ are reachable from r in T. Whenever we say optimal solution below, we mean the optimal walk-collection (i.e., an optimal integral solution to (P)).

Our algorithm ITERPCA resembles Edmond's algorithm for finding a minimum-cost arborescence, and is based on three main ideas for simplifying the instance. However, unlike in the case of min-cost spanning arborescences, our simplifications do *not* leave the problem unchanged; we really exploit the asymmetry that we seek an out-arborescence but are comparing its value against the best collection of r-rooted walks in (G, c, π).

Let $V' = V \setminus \{r\}$. We observe that we may modify the instance by picking a node $v \in V'$, and subtracting a common value θ from the cost of all incoming arcs of v and from π_v, while ensuring that the new values of these quantities are nonnegative (see step (7)). That is, it suffices to prove the desired guarantee for the modified instance $(G, \tilde{c}, \tilde{\pi})$: if T is an out-arborescence with $\mathsf{PCC}(T; G, \tilde{c}, \tilde{\pi}) \leq O^*(G, \tilde{c}, \tilde{\pi})$, then $\mathsf{PCC}(T; G, c, \pi) \leq O^*(G, c, \pi)$ (Lemma 2). By choosing a suitable θ_v for all $v \in V'$ and modifying costs and penalties as above, we may assume that either: (a) there is a node $v \in V'$ with $\tilde{\pi}_v = 0$; (b) there is a (directed) cycle Z consisting of zero $\tilde{c}$-cost arcs; or (c) there is an out-arborescence consisting of zero $\tilde{c}$-cost arcs. If (c) applies, then we are done. If (a) or (b) apply, we further simplify the instance: in case (a), we shortcut past v by merging every pair of incoming and outgoing arcs of v to create a new arc, and delete v (see steps (9)–(15), Lemma 3); in case (b), we contract Z and set the penalty of the contracted node to be $\sum_{v \in V(Z)} \tilde{\pi}_v$ (see steps –, Lemma 4). We then recurse on the simplified instance.

An additional feature of our algorithm is that, by aggregating the θ_v values computed by our algorithm across all recursive calls and translating them suitably to the original graph G, we obtain a certificate $y = (y_S)_{S \subseteq V'}$ such that the quantity $Y = \sum_{S \subseteq V'} y_S$ is sandwiched between the prize-collecting value $\mathsf{PCC}(T; G, c, \pi)$ of our solution, and $O^*(G, c, \pi)$ (which is NP-hard to compute). (We can in fact strengthen the upper bound on Y to $Y \leq OPT(G, c, \pi)$.) This is especially useful when we utilize ITERPCA to implement approximation algorithms for orienteering (see Sect. 4), because there we can utilize Y to obtain a suitable *upper bound* on the optimum value of the orienteering problem (and in fact, the optimal value of the LP-relaxation for orienteering proposed by [13]). This allows us to obtain an *instance-wise approximation guarantee* i.e., an instance-specific bound on the approximation factor of the solution computed for each instance. This instance-wise approximation guarantee is often significantly better than the worst-case approximation guarantee, as is demonstrated by our computational results (Sect. 6). Our computational results also show that our upper bound is a fairly good (over-)estimate of the orienteering optimum. We remark that having both (good) lower and upper bounds on the optimum can be quite useful also for *exact* computational methods for orienteering.

Theorem 1. *On any input* (G, c, π), *algorithm* ITERPCA *runs in polynomial time and returns an out-arborescence* T *and vector* y *such that* $\mathsf{PCC}(T; G, c, \pi) \leq \sum_{S \subseteq V \setminus \{r\}} y_S \leq O^*(G, c, \pi)$. *Furthermore,* $\sum_{S \subseteq V'} y_S \leq OPT(G, c, \pi)$.

The proof of the stronger bound on $\sum_{S \subseteq V'} y_S$ above is a bit technical, so we focus on proving the remainder of Theorem 1 here. Given the recursive nature of ITERPCA, it is natural to use induction (on $|V(G)|$). First, Lemma 2 argues that it suffices to show the inequalities stated in Theorem 1 hold for the instance $(G, \widetilde{c}, \widetilde{\pi})$ specified in step (7)(with "simpler" edge costs and penalties), the out-arborescence T, and the vector $\widetilde{y}$ returned in step (28) or (16). Next, Lemmas 3 and 4 supply essentially the induction step. They show that if the output $(\overline{T}, \overline{y})$ of ITERPCA when it is called recursively on the smaller instance $(\overline{G}, \overline{c}, \overline{\pi})$ in step (12) or (23) satisfies the inequalities stated in Theorem 1, then $(T, \widetilde{y})$ satisfies $\mathsf{PCC}(T; G, \widetilde{c}, \pi) \leq \sum_{S \subseteq V'} \widetilde{y}_S \leq O^*(G, \widetilde{c}, \widetilde{\pi})$. The key observation underlying both proofs is that $O^*(\overline{G}, \overline{c}, \overline{\pi}) \leq O^*(G, \widetilde{c}, \widetilde{\pi})$. Combining this with Lemma 2 finishes the proof. Proofs omitted below appear in the full version [9].

Lemma 2. *Consider the PCW instance* $(G, \widetilde{c}, \widetilde{\pi})$ *obtained after step (7). If the out-arborescence* T *computed in step (15), (26), (27), or (29), and the vector* $\widetilde{y}$ *satisfy* T *and the final vector* y *returned satisfy* $\mathsf{PCC}(T; G, c, \pi) \leq \sum_{S \subseteq V'} y_S \leq O^*(G, c, \pi)$.

Proof. We show that $\mathsf{PCC}(T; G, c, \pi) = \mathsf{PCC}(T; G, \widetilde{c}, \widetilde{\pi}) + \sum_{v \in V'} \theta_v$, and $O^*(G, \widetilde{c}, \widetilde{\pi}) \leq O^*(G, c, \pi) - \sum_{v \in V'} \theta_v$. These inequalities, along with $\sum_{S \subseteq V'} y_S = \sum_{S \subseteq V'} \widetilde{y}_S + \sum_{v \in V'} \theta_v$, yield the lemma.

The first equality follows easily, since every node $v \in V'$ covered by T has exactly one incoming edge whose cost increases by θ_v when going from $\widetilde{c}$ to c, and the penalty of every node $v \in V'$ not covered by T increases by θ_v when going from $\widetilde{\pi}$ to π. (Here we crucially exploit that T is an *out-arborescence*; if $|\delta_T^{in}(v)| > 1$ then $\mathsf{PCC}(T)$ increases by more than θ_v when going from $\widetilde{c}$ to c.)

Algorithm. IterPCA(G, c, π, r): iterative simplification

Input: PCW instance $(G = (V, A), c, \pi, r)$
Output: r-rooted out-arborescence T in G; $y = (y_S)_{S \subseteq V \setminus \{r\}}$

1 Let $V' = V \setminus \{r\}$,   initialize $y \leftarrow \varepsilon c0$, $\widetilde{y} \leftarrow \varepsilon c0$
2 **if** $|V| = 1$ **then return** $(T = \emptyset,$ *null vector*$)$
3 **if** $|V| = 2$, *say* $V = \{r, v\}$ **then**
4 $\quad$ Set $y_{\{v\}} \leftarrow \min\{c_{r,v}, \pi_v\}$
5 $\quad$ **if** $\pi_v > c_{r,v}$ **then return** $(T = \{(r, v)\}, y)$ **else return** $(T = \emptyset, y)$
6 Set $\theta_v \leftarrow \min\{\min_{(u,v) \in A} c_{u,v}, \pi_v\}$ for all $v \in V'$
7 For all $v \in V'$, set $\widetilde{c}_{u,v} \leftarrow c_{u,v} - \theta_v$ for all $(u, v) \in A$, and $\widetilde{\pi}_v \leftarrow \pi_v - \theta_v$; set $\widetilde{\pi}_r \leftarrow 0$

8 **if** *there exists* $v \in V'$ *with* $\widetilde{\pi}_v = 0$ **then**
9 $\quad$ $\overline{G} \leftarrow \big(V \setminus \{v\}, A \setminus (\delta^{\text{ in}}(v) \cup \delta^{\text{ out}}(v)) \cup \{(u, w) : u \in V \setminus \{v\}, w \in V \setminus \{r, v\}\}\big)$
10 $\quad$ For all $u \in V \setminus \{v\}$, $w \in V \setminus \{r, v\}$, set $\overline{c}_{u,w} \leftarrow \min\{\widetilde{c}_{u,w}, \widetilde{c}_{u,v} + \widetilde{c}_{v,w}\}$
11 $\quad$ Set $\overline{\pi} \leftarrow \{\widetilde{\pi}_u\}_{u \in V(\overline{G})}$
12 $\quad$ $(\overline{T}, \overline{y}) \leftarrow$ IterPCA$(\overline{G}, \overline{c}, \overline{\pi}, r)$
13 $\quad$ $\overline{A} \leftarrow \{(u, w) \in \overline{T} : \overline{c}_{u,w} < \widetilde{c}_{u,w}\}$ // note that $\overline{c}_{u,w} = \widetilde{c}_{uv} + \widetilde{c}_{v,w} \, \forall (u, w) \in \overline{A}$
14 $\quad$ $T' \leftarrow \overline{T} \setminus \overline{A} \cup \bigcup_{(u,w) \in \overline{A}}\{(u, v), (v, w)\}$
15 $\quad$ $T \leftarrow$ minimum $\widetilde{c}$-cost spanning arborescence in $(V(T'), A(T'))$
16 $\quad$ Set $\widetilde{y}_S \leftarrow \overline{y}_S$ for all $S \subseteq V \setminus \{r, v\}$

17 **else if** *there exists a cycle* Z *with* $r \notin V(Z)$ *and* $\widetilde{c}_{u,v} = 0$ *for all* $(u, v) \in A(Z)$
$\quad$ **then**
18 $\quad$ Set $\overline{G} \leftarrow$ digraph obtained from G by contracting Z into a single supernode
$\quad\quad$ u_Z, removing self-loops, and replacing parallel (incoming or outgoing) arcs
$\quad\quad$ incident to u_z by a single arc
19 $\quad$ Set $\overline{c}_{u,v} \leftarrow \widetilde{c}_{u,v}$ for all $u \in V \setminus V(Z), v \in V \setminus V(Z)$
20 $\quad$ For all $u \in V \setminus V(Z)$ such that $\delta^{\text{ out}}(u) \cap \delta^{\text{ in}}(Z) \neq \emptyset$, set
$\quad\quad$ $\overline{c}_{u,u_z} \leftarrow \min_{(u,v) \in \delta^{\text{ in}}(Z)} \widetilde{c}_{u,v}$
21 $\quad$ For all $u \in V' \setminus V(Z)$ such that $\delta^{\text{ in}}(u) \cap \delta^{\text{ out}}(Z) \neq \emptyset$, set
$\quad\quad$ $\overline{c}_{u_z,u} \leftarrow \min_{(v,u) \in \delta^{\text{ out}}(Z)} \widetilde{c}_{v,u}$
22 $\quad$ Set $\overline{\pi}_{u_z} \leftarrow \sum_{v \in V(Z)} \widetilde{\pi}_v, \overline{\pi}_u \leftarrow \widetilde{\pi}_u$ for all $u \in V \setminus V(Z)$
23 $\quad$ $(\overline{T}, \overline{y}) \leftarrow$ IterPCA$(\overline{G}, \overline{c}, \overline{\pi}, r)$
24 $\quad$ **if** $u_z \in V(\overline{T})$ **then**
25 $\quad\quad$ Obtain T' from $\overline{T}$ as follows: replace every arc $a \in \overline{T}$ entering or leaving
$\quad\quad\quad$ u_z by the corresponding arc in G whose $\widetilde{c}$-cost defines $\overline{c}_a$; also add (the
$\quad\quad\quad$ nodes and edges of) Z
26 $\quad\quad$ $T \leftarrow$ minimum $\widetilde{c}$-cost spanning arborescence in $(V(T'), A(T'))$
27 $\quad$ **else** $T \leftarrow \overline{T}$
28 $\quad$ For each set $\overline{S} \subseteq V(\overline{G}) \setminus \{r\}$, consider the corresponding set $S \subseteq V'$, which is
$\quad\quad$ $\overline{S}$ if $u_Z \notin \overline{S}$, and $\overline{S} \setminus \{u_z\} \cup V(Z)$ otherwise; set $\widetilde{y}_S \leftarrow \overline{y}_{\overline{S}}$

29 **else** Let $T \leftarrow$ arborescence spanning V with $\widetilde{c}_{u,v} = 0$ for all $(u, v) \in A(T)$

30 Set $y_{\{v\}} \leftarrow \widetilde{y}_{\{v\}} + \theta_v$ for all $v \in V'$, and $y_S \leftarrow \widetilde{y}_S$ for all other subsets $S \subseteq V'$.
31 **return** (T, y)

To see the second inequality, let $\mathcal{C}$ be an optimal solution to the (G, c, π) instance. So for every node $v \in V'$, if v is covered by $\mathcal{C}$, it has at least one incoming edge in this collection of paths, whose cost decreases by θ_v when moving from c to $\tilde{c}$; if v' is not covered, its penalty decreases by θ_v when moving from π to $\tilde{\pi}$. Hence, $O^*(G, \tilde{c}, \tilde{\pi}) \leq O^*(G, c, \pi) - \sum_{v \in V'} \theta_v$.   □

Lemma 3. *Consider a recursive call* ITERPCA (G, c, π, r), *where steps* (9)–(16) *are executed. If* $(\overline{T}, \overline{y})$ *obtained in step* (12) *satisfies* $\mathsf{PCC}(\overline{T}; \overline{G}, \overline{c}, \overline{\pi}) \leq \sum_{S \subseteq V(\overline{G}) \setminus \{r\}} \overline{y}_S \leq O^*(\overline{G}, \overline{c}, \overline{\pi})$, *then the tuple* $(T, \tilde{y})$ *obtained in steps* (15), (16) *satisfy* $\mathsf{PCC}(T; G, \tilde{c}, \tilde{\pi}) \leq \sum_{S \subseteq V'} \tilde{y}_S \leq O^*(G, \tilde{c}, \tilde{\pi})$.

Lemma 4. *Consider a recursive call* ITERPCA (G, c, π, r), *where steps* (18)–(28) *are executed. If* $(\overline{T}, \overline{y})$ *obtained in step* (23) *satisfies* $\mathsf{PCC}(\overline{T}; \overline{G}, \overline{c}, \overline{\pi}) \leq \sum_{S \subseteq V(\overline{G}) \setminus \{r\}} \overline{y}_S \leq O^*(\overline{G}, \overline{c}, \overline{\pi})$, *then the tuple* $(T, \tilde{y})$ *obtained in steps* (26), (28) *or* (27), (28) *satisfies* $\mathsf{PCC}(T; G, \tilde{c}, \tilde{\pi}) \leq \sum_{S \subseteq V'} \tilde{y}_S \leq O^*(G, \tilde{c}, \tilde{\pi})$.

4   Applications for Orienteering Problems

We now show that ITERPCA can be used to obtain fast, combinatorial implementations of the LP-rounding based approximation algorithms devised by [13] for orienteering. The input here consists of a (rational) metric space (V, c), root $r \in V$, a distance bound $B \geq 0$, and nonnegative node rewards $\{\pi_v\}_{v \in V}$. Let $G = (V, E)$ denote the complete graph on G. Three versions of orienteering are often considered in the literature.

- *Rooted orienteering*: find an r-rooted path of cost at most B that collects the maximum reward.
- *Point-to-point (P2P) orienteering*: we are also given an end node t, and we seek an r-t path of cost at most B that collects maximum reward.
- *Cycle orienteering*: find a cycle containing r of cost at most B that collects maximum reward.

By merging nodes at zero distance from each other, and scaling, we may assume that all distances, and B, are positive integers. Friggstad and Swamy [13] propose an LP-relaxation for rooted orienteering, and show that an optimal LP-solution can be rounded to an integer solution losing a factor of 3. This is obtained by decomposing an LP-optimal solution into a convex combination of out-arborescences, and then extracting a rooted path from these arborescences. They adapt their approach to also obtain a 6-approximation for P2P orienteering. Their rounding theorem is stated below, suitably paraphrased. The *regret* (or *excess* [2,4]) of a u-v path P is $c^{\mathsf{reg}}(P) = c(P) - c_{uv}$.

Theorem 5 ([13]). *Fix* $w \in V$. *Let* $T_1, \ldots, T_k$ *be rooted trees in* G, *and* $\gamma_1, \ldots, \gamma_k \geq 0$ *be such that: (i)* $\sum_{i=1}^k \gamma_i = 1$; *(ii)* $\sum_{i=1}^k \gamma_i c(T_i) \leq B$; *and (iii)* $w \in V(T_i)$ *for all* $i = 1, \ldots, k$. *Then, for each* $i = 1, \ldots, k$, *we can extract a rooted path* P_i *from* T_i *(visiting some subset of* $V(T_i)$*) with* $c^{\mathsf{reg}}(P_i) \leq B - c_{rw}$, *such that* $\max_{i=1,\ldots,k} \pi\big(V(P_i)\big) \geq \frac{1}{3} \cdot \sum_{i=1}^k \gamma_i \pi\big(V(T_i)\big)$.

We show that one can utilize IterPCA to obtain combinatorial algorithms for rooted- and P2P- orienteering with the above approximation factors. The high level idea is that Lagrangifying the "cost at most B" constraint for rooted orienteering yields a prize-collecting walks problem, and by fine-tuning the value of the Lagrangian variable, we can leverage IterPCA to obtain a distribution of r-rooted trees having expected cost at most B, and expected reward (essentially) at least the optimum of the rooted orienteering problem.

Theorem 6. *Let $\epsilon > 0$, $N \subseteq V$, $r, w \in N$, and $L \geq c_{rw}$. Let Π^* be the maximum reward of an r-rooted path Q with $\{w\} \subseteq V(Q) \subseteq N$. There is a procedure* BinSearchPCA $(N, L, r, w; \epsilon)$ *that utilizes* IterPCA *to find in polytime rooted trees T_1, T_2, and $\gamma_1, \gamma_2 \geq 0$ with $\gamma_1 + \gamma_2 = 1$ such that: (a) $\{w\} \subseteq V(T_i) \subseteq N$ for $i = 1, 2$; (b) $\sum_{i=1}^{2} \gamma_i c(T_i) \leq L$; and (c) $\sum_{i=1}^{2} \gamma_i \pi(V(T_i)) \geq (1 - \epsilon)\Pi^*$.*

We then apply the rounding algorithm in [13] (i.e., Theorem 5) to the output of Theorem 6 to obtain the stated approximation factors. Our algorithms can thus be seen as a combinatorial implementation of the LP-rounding algorithms in [13]. For cycle orienteering, we adapt the above idea and the analysis in [13], to obtain a combinatorial 4-approximation algorithm. We also leverage the certificate y returned by IterPCA to provide *upper bounds on the optimal value of the {rooted, P2P, cycle}- orienteering problem.* This is quite useful as it allows to assess the approximation guarantee on an instance-by-instance basis. Indeed, our computational experiments in Sect. 6 show that the instance-wise ratio is much better than the worst-case approximation ratio.

Rooted Orienteering. For $w \in V$ with $c_{rw} \leq B$, let Π_w^* be the maximum reward of a rooted path that visits w, and only visits nodes in $\overline{V}_w = \{u \in V : c_{rv} \leq c_{rw}\}$. Let $\epsilon \in (0, 1)$. We consider each $w \in V$ with $c_{rw} \leq B$, apply Theorem 5 on the output of BinSearchPCA $(\overline{V}_w, B, r, w; \epsilon)$, and return the best solution found. This yields a combinatorial $3/(1 - \epsilon)$-approximation algorithm.

For a given guess w (and any L), BinSearchPCA $(\overline{V}_w, L, r, w; \epsilon)$ varies $\lambda \geq 0$, and calls IterPCA on inputs where the reward of each $v \in \overline{V}_w \setminus \{w\}$ is set to $\lambda \pi_v$; define $Y(\lambda) := \sum_{S \subseteq V'} y_S^{(\lambda)}$, where $y^{(\lambda)}$ is the certificate returned by IterPCA for this instance. We have $Y(\lambda) \leq B + \lambda(\pi(\overline{V}_w) - \Pi_w^*)$ by Theorem 1, and rearranging gives $\Pi_w^* \leq \text{UB1}(w, B; \lambda) := \pi(\overline{V}_w) + \frac{B - Y(\lambda)}{\lambda}$. Thus, the optimal value for rooted orienteering is at most $\text{UB1}(B) := \max_{w \in V : c_{rw} \leq B} \min_{\lambda \geq 0} \text{UB1}(w, B; \lambda)$.

P2P Orienteering. Recall that here we seek an r-t path of cost at most B that achieves maximum reward. In [13], a 6-approximation is obtained for this problem by (essentially) guessing the node w on the optimal r-t path with largest $c_{ru} + c_{ut}$ value, and utilizing Theorem 5 on two suitable weighted collections of trees. While they obtain the two collections from an optimal solution to their P2P-orienteering LP, as with rooted orienteering, we can utilize BinSearchPCA instead: we return BinSearchPCA $(\overline{V}_w^{P2P}, B - c_{wt}, r, w; \epsilon)$ and

BINSEARCHPCA $(\overline{V}_w^{\mathsf{P2P}}, B - c_{rw}, t, w; \epsilon)$ as the two collections, where $\overline{V}_w^{\mathsf{P2P}} = \{u \in V : c_{ru} + c_{ut} \le c_{rw} + c_{wt}\}$ (and w is a guess with $c_{rw} + c_{wt} \le B$).

For a given guess w, let $Y_r(\lambda)$ denote $\sum_{S \subseteq N'} y_S^{(\lambda)}$ in BINSEARCHPCA $(\overline{V}_w^{\mathsf{P2P}}, B - c_{wt}, r, w; \epsilon)$, and $Y_t(\lambda)$ denote $\sum_{S \subseteq N'} y_S^{(\lambda)}$ in BINSEARCHPCA $(\overline{V}_w^{\mathsf{P2P}}, B - c_{rw}, t, w; \epsilon)$. For any $\lambda \ge 0$, and any walks Q_1, Q_2 rooted at r and t respectively, such that $\{w\} \subseteq V(Q_1), V(Q_2) \subseteq \overline{V}_w^{\mathsf{P2P}}$, we have $Y_r(\lambda) \le c(Q_1) + \lambda \left[\pi(\overline{V}_w^{\mathsf{P2P}}) - \pi(V(Q_1))\right]$, and $Y_t(\lambda) \le c(Q_2) + \lambda \left[\pi(\overline{V}_w^{\mathsf{P2P}}) - \pi(V(Q_2))\right]$. We obtain an upper bound on the optimal value for P2P orienteering by considering all w with $c_{rw} + c_{wt} \le B$, taking $Q_1 \in \{P_{w,rw}^*, P_w^*\}$, and $Q_2 \in \{P_{w,wt}^*, P_w^*\}$, and collecting all the reward upper bounds resulting from the above inequalities.

Cycle Orienteering. Recall that here we seek a cycle containing r of cost at most B that achieves maximum reward. Taking $t = r$ in our approach for P2P-orienteering yields a combinatorial 6-approximation algorithm. But we can refine this approach and utilize BINSEARCHPCA to obtain a 4-approximation, as also refine our upper-bounding strategy, by leveraging the fact that the tree returned by ITERPCA has prize-collecting cost at most the optimal value of (P).

For any $w \in V$ with $c_{rw} \le B/2$, let C_w^* be the maximum-reward cycle that visits w, and only visits nodes in $\overline{V}_w = \{u \in V : c_{ru} \le c_{rw}\}$, and let $\Pi_w^{*\mathsf{Cyc}} = \pi(V(C_w^*))$. The distribution output by BINSEARCHPCA $(\overline{V}_w, B/2, r, w; \epsilon)$ has expected reward at least $\Pi_w^{*\mathsf{Cyc}}/2 - \epsilon \cdot \Pi_w^{*\mathsf{Cyc}}$, since its expected prize-collecting cost is (essentially) at most that of the fractional solution to (P) where we send a $\frac{1}{2}$-unit of flow from r to w along the two r-w paths in C_w^*. One can extract from this distribution a feasible solution of reward at least $\frac{1-2\epsilon}{4} \cdot \Pi_w^{*\mathsf{Cyc}}$.

For the upper bound, for a given guess w, we compare $Y(\lambda)$ (which denotes $\sum_{S \subseteq N'} y_S^{(\lambda)}$ in BINSEARCHPCA $(\overline{V}_w, L, r, w; \epsilon)$) with the prize-collecting cost of C_w^*, and the above fractional solution. Collecting the reward upper bounds, the optimal value for cycle orienteering is at most

$$\mathsf{UB\text{-}Cyc}(B) := \max_{w \in V : c_{rw} \le B/2} \min_{\lambda \ge 0} \min \left\{\mathsf{UB1}(w, B; \lambda), \mathsf{UB4}(w, B; \lambda)\right\}$$

where $\mathsf{UB4}(w, B; \lambda) := 2 \cdot \pi(\overline{V}_w) - \pi_r - \pi_w + \frac{B - 2 \cdot Y(\lambda)}{\lambda}$.

5   Applications for the k Minimum-Latency Problem

Recall that in the k minimum-latency problem (k-MLP), we have a metric space (V, c) and root $r \in V$. The goal is to find (at most) k-rooted paths that together cover every node, so as to minimize the sum of visiting times of the nodes. The current-best approximation ratio for k-MLP is 7.183, due to Post and Swamy [17], who devise two algorithms for k-MLP having (roughly) this approximation ratio. Their "more combinatorial" algorithm (see Algorithm 3 in §6.2 [17]) relies on a procedure that "solves" the problem of finding a minimum-cost collection of

r-rooted walks that together cover at least k nodes, by returning a distribution of r-rooted trees that in expectation covers at least k nodes and has cost at most the optimal walk-collection cost. In [17], this distribution is obtained by applying the arborescence-packing result of Bang-Jensen et al. [1] to the optimal solution to (P) with node rewards λ, and then varying λ in a binary-search procedure (as we do). We can instead utilize ITERPCA within a binary-search procedure to obtain the desired distribution (over at most two trees). Incorporating this in the more-combinatorial algorithm of [17] yields a fully (and truly) combinatorial 7.183-approximation algorithm for k-MLP.

6   Computational Results for Orienteering

We now present various computational results on the performance of our orienteering algorithms (from Sect. 4) in order to assess the performance of our algorithms in practice. Our experiments demonstrate the effectiveness of our algorithms and upper bounds. They show that the instance-wise approximation ratios, for both the solution returned and the computed upper bound, are much better than the theoretical worst-case bounds, and in fact fairly close to 1.

We implemented our algorithms in C++11, and ran the code on a 2019 Mac-Book Pro with 2.3 GHz Intel Core i9 processor (8 cores) & 16 GB RAM. Our implementation essentially matches the description in Sect. 4, with the following differences. (1) We terminate the binary search (in BINSEARCHPCA) when the interval $[\lambda_1, \lambda_2]$ has width $\lambda_2 - \lambda_1 \leq 10^{-6}$, the precision of the double data type in C++; (2) We extract a solution from the tree T_λ returned by ITERPCA for each λ value encountered in the binary search (as opposed to using only the trees T_{λ_1}, T_{λ_2}), and return the best of these solutions; (3) When extracting a rooted path of a given regret bound R from a path P (obtained from a tree), instead of using a greedy procedure (Lemma 5.1 in [4], or Lemma 2.2 in [12]), we find the maximum-reward subpath Q of P meeting the specifications; (4) When computing the upper bounds on the orienteering optimum, we consider only the λs encountered in the binary search.

We discuss cycle orienteering, as this is computationally the most well-studied version of orienteering, and detail other computational results in [9]. For each guess w of furthest node, we run two binary search procedures to find r-rooted PCA solutions, with target budgets $B/2$ and $B - c_{rw}$. We use the 45 TSP instances with at most 400 nodes from the TSPLIB 2.1 library [18]. These are the instances considered by Fischetti et al. [11] (and by [14]), and three additional datasets from [14]. For each dataset, [11,14] generate node rewards in three ways:

- **Gen 1 - Uniform Rewards:** All nodes apart from the root r have reward 1.
- **Gen 2 - Pseudo-Random Rewards:** The reward of the j-th node is $1 + (7141 \cdot j + 73) \pmod{100}$ apart from the root, which has reward 0.
- **Gen 3 - Far Away Rewards:** The reward of a node $v \neq r$ is $1 + \lfloor 99 \cdot c_{rv}/ \max_w c_{rw} \rfloor$. This is meant to create more challenging instances where the high-reward nodes are further from r.

The distance bound used in each case is $\lceil \mathsf{TSPOpt}/2 \rceil$, where TSPOpt is the cost of the optimal TSP-tour for that dataset (which is provided in TSPLIB). Optimal values are known for all these datasets; most of these were computed in [11], and the rest are from [14]. This allows us to evaluate the instance-wise approximation guarantee of our algorithm, and the quality of our upper bound.

The plots below give an overview of our results: Val is the reward of our solution, Opt is the optimal value, and UB is the upper bound that we compute. The histograms specify the distribution of the $\mathsf{Opt}/\mathsf{Val}$ and UB/Opt ratios across the instances used in the computational experiments. Each histogram bar corresponds to a range of values (for a particular ratio) as indicated on the x-axis, and its height specifies the number of instances where the achieved ratio lies in the range. Detailed results of our experiments appear in the full version [9]. For the (supposedly more challenging) Gen 3 data sets, we also report the instance-wise approximation ratio $\mathsf{Opt}/\mathsf{Val}$, and the ratio UB/Opt. As our results show, our algorithm performs fairly well in practice.

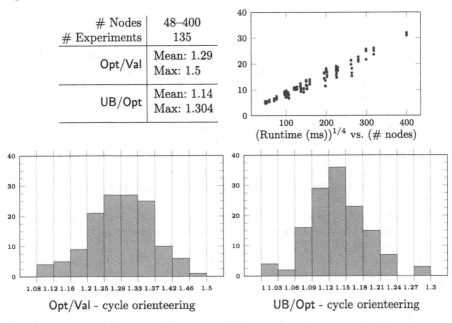

Opt/Val - cycle orienteering   UB/Opt - cycle orienteering

Next, we consider the work of Paul et al. [16], which is the only other work that performs a computational evaluation of an (polytime) approximation algorithm for orienteering. They develop a 2-approximation algorithm for cycle orienteering, and report computational results for the *unrooted* problem (where no root node is specified). Instead of a direct comparison of their algorithm vs. ours (by trying all possible roots), we use our two algorithms in *combination* to see if this yields improved solutions for the underlying instances. We run our algorithm as a *postprocessing* step, taking the root r and the guess w of the furthest node to be nodes from the solution returned by [16]. The histograms below show that

this postprocessing almost always yields improvements, sometimes by a significant factor, on both the TSPLIB and the Citi Bike data sets considered in [16].
The improvement factors are on the x-axis; the height of a bar is the number of
instances for which we achieve this factor.

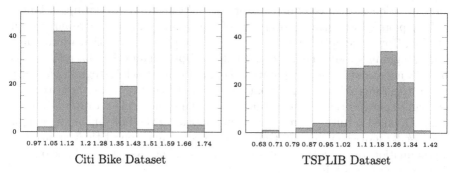

Citi Bike Dataset   TSPLIB Dataset

References

1. Bang-Jensen, J., Frank, A., Jackson, B.: Preserving and increasing local edge-
   connectivity in mixed graphs. SIAM J. Discret. Math. **8**(2), 155–178 (1995)
2. Bansal, N., Blum, A., Chawla, S., Meyerson, A.: Approximation algorithms for
   deadline-TSP and vehicle routing with time windows. In: Proceedings of 36th
   STOC, pp. 166–174 (2004)
3. Blum, A., Chalasani, P., Coppersmith, D., Pulleyblank, B., Raghavan, P., Sudan,
   M.: The minimum latency problem. In: Proceedings of 26th STOC, pp. 163–171
   (1994)
4. Blum, A., Chawla, S., Karger, D.R., Lane, T., Meyerson, A., Minkoff, M.: Approximation algorithms for orienteering and discount-reward TSP. SIAM J. Comput.
   37, 653–670 (2007)
5. Chakrabarty, D., Swamy, C.: Facility location with client latencies: linear-
   programming based techniques for minimum-latency problems. Math. Oper. Res.
   41(3), 865–883 (2016)
6. Chaudhuri, K., Godfrey, P.B., Rao, S., Talwar, K.: Paths, trees and minimum
   latency tours. In: Proceedings of 44th FOCS, pp. 36–45 (2003)
7. Chekuri, C., Korula, N., Pál, M.: Improved algorithms for orienteering and related
   problems. ACM Trans. Algorithms **8**(3), 1–27 (2012)
8. Dezfuli, S.S.: Solving the LP relaxation of distance-constrained vehicle routing
   problem using column generation. Master's thesis, University of Alberta (2019)
9. Dezfuli, S.S., Friggstad, Z., Post, I., Swamy, C.: Combinatorial algorithms for
   rooted prize-collecting walks and applications to orienteering and minimum-latency
   problems. Manuscript posted on the CS arXiv, November 2021
10. Fakcharoenphol, J., Harrelson, C., Rao, S.: The k-traveling repairman problem.
    ACM Trans. Algorithms **3**(4), Article 40 (2007)
11. Fischetti, M., Salazar-Gonzáles, J.J., Toth, P.: Solving the orienteering problem
    through branch and cut. INFORMS J. Comput. **10**, 133–148 (1998)
12. Friggstad, Z., Swamy, C.: Approximation algorithms for regret-bounded vehicle
    routing and applications to distance-constrained vehicle routing. In: Proceedings
    of 46th STOC, pp. 744–753 (2014)

13. Friggstad, Z., Swamy, C.: Compact, provably-good LPs for orienteering and regret-bounded vehicle routing. In: Eisenbrand, F., Koenemann, J. (eds.) IPCO 2017. LNCS, vol. 10328, pp. 199–211. Springer, Cham (2017). https://doi.org/10.1007/978-3-319-59250-3_17
14. Kobeaga, G., Merino, M., Lozano, J.A.: A revisited branch-and-cut algorithm for large-scale orienteering problems. CoRR: 2011.02743 (2021)
15. Nagarajan, V., Ravi, R.: Approximation algorithms for distance constrained vehicle routing problems. Networks **59**(2), 209–214 (2012)
16. Paul, A., Freund, D., Ferber, A., Shmoys, D.B., Williamson, D.B.: Budgeted prize-collecting traveling salesman and minimum spanning tree problems. Math. Oper. Res. **45**(2), 576–590 (2019)
17. Post, I., Swamy, C.: Linear-programming based techniques for multi-vehicle minimum latency problems. In: Proceedings of 26th SODA, pp. 512–531 (2015)
18. Reinelt, G.: TSPLIB - a traveling salesman library. ORSA J. Comput. **3**, 376–384 (1991)
19. Toth, P., Vigo, D. (eds.): The Vehicle Routing Problem. SIAM Monographs on Discrete Mathematics and Applications, Philadelphia (2002)

Intersecting and Dense Restrictions
of Clutters in Polynomial Time

Martin Drees[(✉)]

Research Institute for Discrete Mathematics, University of Bonn, Bonn, Germany
math@mdrees.de

Abstract. A clutter is a family of sets, called members, such that no
member contains another. It is called intersecting if every two members
intersect, but not all members have a common element. Dense clutters
additionally do not have a fractional packing of value 2. We are looking
at certain substructures of clutters, namely minors and restrictions.

For a set of clutters we introduce a general sufficient condition such
that for every clutter we can decide whether the clutter has a restriction
in that set in polynomial time. It is known that the sets of intersecting
and dense clutters satisfy this condition. For intersecting clutters we
generalize the statement to k-wise intersecting clutters using a much
simpler proof.

We also give a simplified proof that a dense clutter with no proper
dense minor is either a delta or the blocker of an extended odd hole.
This simplification reduces the running time of the algorithm for finding
a delta or the blocker of an extended odd hole minor from previously
$\mathcal{O}(n^4)$ to $\mathcal{O}(n^3)$ filter oracle calls.

Keywords: Clutters · Clutter minors · Deltas · Odd holes

1 Introduction

A clutter is a family of sets, called members, over a finite ground set V such that
no member contains another [7]. Clutters are *isomorphic* if they can be obtained
from each other by relabeling the ground set. A *cover* of $\mathcal{C}$ is a set $B \subseteq V$ such
that $B \cap C \neq \emptyset$ for all members $C \in \mathcal{C}$. It is called *minimal* if it does not contain
another cover. The *covering number* $\tau(\mathcal{C})$ is the minimum cardinality of a cover.
A *packing* of a clutter is a set of pairwise disjoint members. The *packing number*
$\nu(\mathcal{C})$ is the maximum cardinality of a packing. Clearly, $\tau(\mathcal{C}) \geq \nu(\mathcal{C})$. The *blocker*
$b(\mathcal{C})$ is the clutter given by the minimal covers of $\mathcal{C}$. Edmonds and Fulkerson [7]
observed that $b(b(\mathcal{C})) = \mathcal{C}$.

Consider the following example of a clutter. Take the edge set of a graph
as the ground set and the s-t-paths as members. The minimal covers are the
inclusionwise minimal s-t-cuts and $\tau(\mathcal{C}) = \nu(\mathcal{C})$ by Menger's theorem [10].

A clutter $\mathcal{C}$ is *ideal* if the polyhedron $\{\mathbf{x} \geq 0 \colon x(C) \geq 1 \ \forall C \in \mathcal{C}\}$ is integral.
The notion of idealness was introduced by Cornuéjols and Novick [5]. The above

© Springer Nature Switzerland AG 2022
K. Aardal and L. Sanitá (Eds.): IPCO 2022, LNCS 13265, pp. 210–222, 2022.
https://doi.org/10.1007/978-3-031-06901-7_16

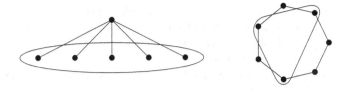

Fig. 1. Δ_6 and an extended odd hole of dimension 7

clutter is ideal, which can be shown by using the famous Max-Flow-Min-Cut theorem by Ford and Fulkerson [8]. Lehman [9] proved that a clutter is ideal if and only if its blocker is ideal.

As one might expect, not every clutter is ideal. In fact, Ding, Feng and Zang [6] showed that the problem to decide whether a clutter is ideal is co-NP-complete. A simple reason why a clutter is not ideal is that it has covering number at least 2, but no fractional packing of value two. Clutters with this property are called *dense* [2].

An important notion needed to study clutters is that of minors.

Definition 1 ([11]). *Let C be a clutter over a ground set V and $I, J \subseteq V$ disjoint subsets. The* minor *of C obtained after* deleting *I and* contracting *J is the clutter $C\backslash I/J$ over the ground set $V - (I \cup J)$ whose members are the inclusion-wise minimal sets of $\{C - J : C \in C, C \cap I = \emptyset\}$. If $I \cup J \neq \emptyset$ the minor is called* proper.

Seymour [12] showed that $b(C\backslash I/J) = b(C)\backslash J/I$. He also proved that every minor of an ideal clutter is ideal. It therefore suffices to find a non-ideal minor to certify that the clutter is not ideal. This motivates the problem to decide whether a clutter has a dense minor. Abdi, Lee and Cornuéjols [2] showed that this problem can be solved in polynomial time. This is quite surprising as from the hardness result of Ding, Feng and Zang [6], one can conclude that this problem is NP-complete if the input is the blocker of the clutter [2].

For the study of dense minors, the following two examples (see Fig. 1) are fundamental. Let $n \geq 3$ be an integer. Take the clutter over the ground set $\{1, 2, \ldots, n\}$ with members $\{1, 2\}, \{1, 3\}, \ldots, \{1, n\}$ and $\{2, 3, \ldots, n\}$. This clutter is called Δ_n and a clutter isomorphic to this clutter is called a *delta of dimension n*. Observe that $b(\Delta_n) = \Delta_n$.

Let $n \geq 5$ be an odd integer. Consider the clutter C over the ground set $\{1, 2 \ldots, n\}$ with minimum cardinality members $\{1, 2\}, \{2, 3\}, \ldots, \{n-1, n\}$ and $\{n, 1\}$. A clutter that is isomorphic to C is an *extended odd hole of dimension n*.

Abdi and Lee [4] proved that every dense clutter has a delta or the blocker of an extended odd hole minor. They also showed that this minor can be found in polynomial time.

A nontrivial clutter is called k-*wise intersecting* if it has no cover of size 1 and every k (not necessarily different) members have a common element. Abdi, Cornuéjols, Huynh and Lee [1] conjectured that for $k \geq 4$ these clutters are non-ideal.

The 2-wise intersecting clutters are also called intersecting clutters. They satisfy $\tau(\mathcal{C}) \geq 2$ and $\nu(\mathcal{C}) = 1$. Note that dense clutters are intersecting. Abdi, Cornuéjols and Lee [2] showed that one can decide in polynomial time whether a clutter has an intersecting minor.

As we do not allow covers of size 1 in many cases, we occasionally consider only special kinds of minors.

Definition 2 ([2]). *Let $\mathcal{C}$ be a clutter over a ground set V and $I \subseteq V$. Let*

$$J := \{u \in V - I : \{u\} \text{ is a cover of } \mathcal{C} \backslash I\} . \tag{1}$$

The minor $\mathcal{C} \backslash I / J$ is the restriction *of $\mathcal{C}$ after restricting I.*

The following are our main contributions:

- We introduce the concept for a set of clutters $\mathcal{S}$ to be k-unifying (Sect. 2). This gives a sufficient condition to decide whether a clutter has a restriction in $\mathcal{S}$ in polynomial time. The abstract unifying theorem could also be applied to other sets of clutters and might therefore be of independent interest.
- Abdi, Cornuéjols and Lee [2] showed that the set of intersecting clutters is 3-unifying. We generalize this result and show that k-wise intersecting clutters are $(k + 1)$-unifying (Sect. 3). This is achieved by a much simpler proof.
- We give a simplified proof that dense clutters with no proper dense minors are deltas or the blocker of extended odd holes (Sect. 4). This is also the main step to show that the set of dense clutters is 3-unifying.
- This simplification allows us to formulate a faster algorithm to find a delta or the blocker of an extended odd hole minor (Sect. 5). The running time is improved from previously $\mathcal{O}(n^4)$ [4] to $\mathcal{O}(n^3)$ filter oracle calls.

2   The Unifying Theorem

We will present a sufficient condition to decide whether a clutter has a restriction with a certain property in polynomial time.

Definition 3. *Let $\mathcal{S}$ be a set of clutters. A clutter $\mathcal{C} \in \mathcal{S}$ is called* restriction-minimal *in $\mathcal{S}$ if no proper restriction of $\mathcal{C}$ is in $\mathcal{S}$.*

Definition 4. *A set of clutters is called k-unifying if every restriction-minimal clutter in this set has k members whose union is the ground set of that clutter.*

We are ready to formulate the main theorem of this section.

Theorem 1 (Unifying theorem). *Let $\mathcal{S}$ be a k-unifying set of clutters and $\mathcal{C}$ be a clutter over a ground set V. Then the following two statements are equivalent:*

(i) $\mathcal{C}$ has a restriction in $\mathcal{S}$;
(ii) there are k members $C_1, C_2, \ldots, C_k$ of $\mathcal{C}$ such that the clutter obtained from $\mathcal{C}$ after restricting $V - \bigcup_{i=1}^{k} C_i$ is in $\mathcal{S}$.

Proof. ($\Longleftarrow$) is immediate. ($\Longrightarrow$) Since clutters have a finite ground set, $\mathcal{C}$ has a restriction-minimal restriction $\mathcal{C}\backslash I/J$ in $\mathcal{S}$. As $\mathcal{S}$ is k-unifying, we find k members $C'_1, C'_2, \ldots, C'_k$ in this restriction such that $\bigcup_{i=1}^{k} C'_i = V - (I \cup J)$. Let $C_i = C'_i \cup J$, which are members of $\mathcal{C}$ by the definition of a restriction. That yields $I = V - \bigcup_{i=1}^{k} C_i$.   □

Corollary 1. *Let $\mathcal{C}$ be a clutter over a ground set V with n elements and m members. Let $\mathcal{S}$ be a k-unifying set of clutters for a fixed $k \geq 2$. Furthermore, there is an oracle given, which decides whether a given clutter is in $\mathcal{S}$ in polynomial time. Then one can decide in polynomial time depending on n and m whether $\mathcal{C}$ has a restriction in $\mathcal{S}$.*

In the next sections we will discuss examples of k-unifying sets of clutters.

3 k-Wise Intersecting Clutters

Theorem 2. *For $k \geq 2$ the set of k-wise intersecting clutters is $(k+1)$-unifying.*

Proof. Let $\mathcal{C}$ be k-wise intersecting such that no proper restriction is k-wise intersecting. We have to show that $\mathcal{C}$ has $(k+1)$ members whose union is the ground set. Choose $(k+1)$ members $C_1, C_2, \ldots, C_{k+1}$ of $\mathcal{C}$ such that $|\bigcap_{i=1}^{k+1} C_i|$ is minimal. Assume for a contradiction there is a $v \in V$ such that $v \notin C_i$ for $i = 1, 2, \ldots, k+1$. Consider the restriction $\mathcal{C}' = \mathcal{C}\backslash v/J$. We get $J \subseteq \bigcap_{i=1}^{k+1} C_i$.

Since $\mathcal{C}'$ is not k-wise intersecting and $\mathcal{C}' = \{\emptyset\}$ or $\tau(\mathcal{C}') \geq 2$, we find k not necessarily different members $C'_1, \ldots, C'_k$ of $\mathcal{C}'$ with empty intersection. They imply k members $C_1^*, C_2^*, \ldots, C_k^*$ in $\mathcal{C}$ with $\bigcap_{i=1}^{k} C_i^* \subseteq J$. Note that the intersection of these k members is not empty since $\mathcal{C}$ is k-wise intersecting. We find an element $u \in \bigcap_{i=1}^{k} C_i^*$. Since $\{u\}$ is not a cover, there is a member C_{k+1}^* with $u \notin C_{k+1}^*$. We conclude

$$\bigcap_{i=1}^{k+1} C_i^* \subsetneq J \subseteq \bigcap_{i=1}^{k+1} C_i, \tag{2}$$

contradicting the minimality assumption.   □

Corollary 2 (Reformulation of [2], Proposition 3.3). *The set of intersecting clutters is 3-unifying.*

We can therefore decide in polynomial time whether a clutter has an intersecting restriction.

Remark 1 ([2], Remark 1.2). A clutter has an intersecting restriction if and only if it has an intersecting minor.

As a consequence thereof, given a clutter explicitly with its members, we can decide in polynomial time whether the clutter has an intersecting minor. It is conjectured that ideal clutters with no intersecting minor also have the max-flow min-cut property [2].

4 Dense Clutters

We will prove that the set of dense clutters is 3-unifying. In the process, we will also prove that minimally dense clutters are deltas or blockers of extended odd holes.

Abdi and Lee [4] considered clutters with $\min\{|C| : C \in \mathcal{C}\} = 2$ and the graph with vertex set V with the members of cardinality 2 as edges. We will consider this graph of the blocker.

Definition 5. *Let $\mathcal{C}$ be a clutter over a ground set V with $\tau(\mathcal{C}) = 2$. The covering graph of $\mathcal{C}$ is the graph with vertex set V and the covers of size two of $\mathcal{C}$ as edges.*

The covering graph of a delta is a single vertex connected to all other vertices. The covering graph of an extended odd hole is an odd cycle. Conversely, a clutter with a covering graph of an odd cycle is an extended odd hole.

The following lemma analyzes the structure of covers of size 2 in clutters with a fractional packing of value 2.

Lemma 1. *Let $\mathcal{C}$ be a clutter over a ground set V with $\tau(\mathcal{C}) = 2$, connected covering graph, and a fractional packing of value 2. Then the covering graph is bipartite and $\mathcal{C}$ has two members representing the colour classes. In particular there are members L and K of $\mathcal{C}$ with $K \cap L = \emptyset$ and $K \cup L = V$.*

Proof. For $C \in \mathcal{C}$ let x_C be the value assigned to C in the fractional packing of value 2. Let $B = \{b_1, b_2\}$ be an arbitrary cover of size 2 of $\mathcal{C}$ and C be an arbitrary member of $\mathcal{C}$ with $x_C > 0$. Since B is a cover, we have $|B \cap C| \geq 1$. Assume $|B \cap C| > 1$. We conclude

$$2 \geq \sum_{b_1 \in C'} x_{C'} + \sum_{b_2 \in C'} x_{C'} = \sum_{b_1 \in C' \text{ or } b_2 \in C'} x_{C'} + \sum_{b_1 \in C' \text{ and } b_2 \in C'} x_{C'} \geq 2 + x_C > 2 ,$$

(3)

a contradiction. Therefore $|B \cap C| = 1$.

Since the cover of size 2 was arbitrary, each member C with $x_C > 0$ has exactly one element with each of these covers in common. In the covering graph G, such a member is a stable set and vertex cover. Let s and t be two arbitrary vertices of G. As G is connected, there is an s-t-path in G. Such a path has to be alternating between vertices in C and vertices not in C. In particular, the covering graph cannot contain an odd cycle, since that would result in two paths between the same vertices of different parity in length. Thus, the covering graph is bipartite and we get two colour classes.

In a connected bipartite graph, the only sets of vertices that are a stable set and a vertex cover, are the two colour classes.

It is impossible that all members with $x_C > 0$ are only one of the two colour classes, because that would be the only member in the fractional packing and a value of 2 would not be possible. Therefore, each colour class is represented by a member, so there are members K and L with $K \cap L = \emptyset$ and $K \cup L = V$. □

Definition 6. *A clutter is called* minimally dense *if it is dense and no proper minor is dense. A clutter is called* strictly dense *if it is dense and no proper restriction is dense.*

Given a minimally dense clutter, we will consider proper minors with covering number at least 2. Such a minor then has a fractional packing of value 2. The idea is to construct minors with a connected covering graph and then apply Lemma 1 to deduce specific members of that minor. They will imply members of the original clutter such that in total we get a delta or the blocker of an extended odd hole.

The first step is to show that the covering graph of the original clutter is actually connected.

We will use the following lemma to get a certificate for a dense clutter.

Lemma 2 ([2], Lemma 1.6). *Let $\mathcal{C}$ be a clutter with $\tau(\mathcal{C}) \geq 2$ over a ground set V. Then the following are equivalent:*

(i) $\mathcal{C}$ is dense,
(ii) there is a $w \in \mathbb{R}_{\geq 0}^{V}$ with $\mathbf{1}^{T} w = 1$ such that $\sum_{u \in C} w_u > \frac{1}{2}$ for all $C \in \mathcal{C}$.

Lemma 3. *A minimally dense clutter has $\tau(\mathcal{C}) = 2$ and connected covering graph.*

Proof. If an element $v \in V$ does not appear in a cover of size two of $\mathcal{C}$, the proper minor $\mathcal{C} \backslash v$ has covering number at least 2. Thus, this minor has a fractional packing of value 2 which is also a fractional packing of value 2 for $\mathcal{C}$, a contradiction. Therefore, $\tau(\mathcal{C}) = 2$ and each element of the ground set appears in a cover of size 2.

Let G be the covering graph of $\mathcal{C}$. Assume G is not connected. Let A be the vertex set of one component of G and $B = V - A$. Let H be the subgraph of G induced by A. Note that A is a cover of $\mathcal{C}$ since it contains at least one cover.

Consider the minor $\mathcal{C}' = \mathcal{C}/B$. If $\mathcal{C}'$ has a cover of size 1, this would also be a cover of $\mathcal{C}$, since no member is entirely contracted as A is a cover. Hence, $\mathcal{C}'$ has a fractional packing of value 2. The covering graph H' of $\mathcal{C}'$ contains the edges of H and is thus connected. By applying Lemma 1 on $\mathcal{C}'$, we get that H' is bipartite and the two colour classes K and L are members in $\mathcal{C}'$, implying that K and L are not covers in $\mathcal{C}'$ and $\mathcal{C}$.

Since $\mathcal{C}$ is dense, by Lemma 2 there is a $w \in \mathbb{R}_{\geq 0}^{V}$ such that $\sum_{u \in C} w_u > \frac{w(V)}{2}$ for all $C \in \mathcal{C}$. Let without loss of generality $w(\overline{K}) \geq w(L)$. Each member of $\mathcal{C} \backslash K / L$ then has weight greater than

$$\frac{w(V)}{2} - w(L) \geq \frac{w(V) - w(K) - w(L)}{2} = \frac{w(V - A)}{2} . \tag{4}$$

Hence, the certificate for $\mathcal{C}$ also implies a certificate for $\mathcal{C} \backslash K / L$. Since that minor cannot be dense, we have $\tau(\mathcal{C} \backslash K / L) < 2$. Thus, this minor has a cover of size 1. Let this cover be $\{b\}$. The proper minor $\mathcal{C}'' = \mathcal{C} \backslash b / (B - \{b\})$ has no cover of size 1 since there is no edge between the vertex sets A and B. Using the same argument as for $\mathcal{C}'$, the covering graph of $\mathcal{C}''$ contains the edges of H, is bipartite, and has the colour classes as members. Therefore, K and L are members of $\mathcal{C}''$, but not covers. That yields that $K \cup \{b\}$ is not a cover of $\mathcal{C}$, a contradiction. □

We will use the following tool to find delta minors.

Lemma 4 ([3], **see also** [4], **Theorem 5**). *Let $\mathcal{C}$ be a clutter over a ground set V. Let $u, v, w \in V$ be distinct elements and such that $\{u, v\}$ and $\{u, w\}$ are members of $\mathcal{C}$. Let $C \in \mathcal{C}$ such that $\{u, v, w\} \cap C = \{v, w\}$. Then $\mathcal{C}$ has a delta minor.*

Proof. Let $I = V - (C \cup \{u\})$. Let $C_1 = \{x \in C : \{u, x\} \in \mathcal{C}\}$ and let $C_2 = C - C_1$. Note that $|C_1| \geq 2$ as $v, w \in C_1$. Starting with the minor $\mathcal{C} \backslash I$, contract elements x of C_2 one by one as long as $\{u, x\}$ is not a member of the current minor. We get a clutter $\mathcal{C}' = \mathcal{C} \backslash I / C_2'$. Note that due to the definition of C_2, $\{u\}$ is not a member of this clutter as it is not a member of $\mathcal{C} \backslash I$. As C is a member of $\mathcal{C}$, we get that $C - C_2$ contains a member of $\mathcal{C}'$. Actually, $C - C_2'$ is a member in $\mathcal{C}'$ because there is no member $C' \subsetneq C - C_2'$ as it would imply a member $C^* \subsetneq C$ of $\mathcal{C}$. Therefore, $\mathcal{C}'$ has the members $C - C_2'$ and $\{u, x\}$ for all $x \in C - C_2'$. There cannot be further members due to the definition of a clutter. Therefore, $\mathcal{C}'$ is a delta and $\mathcal{C}$ has a delta minor.

We are now ready to prove the following fundamental result.

Theorem 3 (Reformulation of [4], Theorem 3). *A minimally dense clutter is a delta or the blocker of an extended odd hole.*

Proof. By the previous lemma, the covering graph G of the minimally dense clutter $\mathcal{C}$ is connected. If G contains an odd cycle, contracting all other elements leads to a minor with covering number at least 2, but the covering graph is not bipartite. If the minor has a fractional packing of value 2, Lemma 1 implies a contradiction. Therefore, the minor is dense and thus not proper. In conclusion, G does not properly contain a cycle and thus is a cycle with no additional chords. Therefore, $\mathcal{C}$ is the blocker of an extended odd hole or a delta if $|V| = 3$.

We can now assume that G is bipartite. Let X and Y be the colour classes of G.

One of them is a cover, otherwise we would have disjoint members. So let without loss of generality Y be a cover and $B \subseteq Y$ be a minimal cover. We get $|Y| \geq 3$ as there is no edge in $G[Y]$.

Remove an $x \in X$ from G such that the number of vertices from X in the same component of the resulting graph is maximal. Let this maximal component be M_x. If this number is not $|X| - 1$, take an $x' \in X$ that is not in M_x. Removing x' instead of x does not disconnect M_x. Furthermore, x is connected to M_x, a contradiction to the maximality. Therefore, there is an $x \in X$ such that removing x from G does not disconnect the other vertices in X.

Let G' be the graph resulting from G after removing x. The components of G' consist of one component containing all other vertices in X and some vertices from Y. All other components have to be a single vertex from Y, because there are no edges between vertices in Y. Let this set of isolated vertices be Z.

Since G is connected, $\{x, z\}$ is an edge of G for all $z \in Z$. Assume that there is a $y \in Z - B$. Consider the minor $\mathcal{C}/y$. The covering graph of this minor is connected. Therefore, this minor has a member $C' \subseteq X$. It implies a member $C \subseteq X \cup \{y\}$ in $\mathcal{C}$, a contradiction to B being a cover. So $Z \subseteq B$.

If $|Z| \geq 2$, the blocker of $\mathcal{C}$ has members $\{x, z_1\}, \{x, z_2\}$. As B is a minimal cover of $\mathcal{C}$, it is a member of $b(\mathcal{C})$. Furthermore, we have $B \cap \{x, z_1, z_2\} = \{z_1, z_2\}$. Applying Lemma 4 yields that $b(\mathcal{C})$ has a delta minor. Let $b(\mathcal{C})\backslash I/J = \Delta_n$. We get $\mathcal{C}\backslash J/I = b(\Delta_n) = \Delta_n$. Thus, $\mathcal{C}$ has a delta minor. As deltas are dense, $\mathcal{C}$ is a delta itself.

If $|Z| < 2$ consider the minor $\mathcal{C}\backslash Z/x$. This minor has no cover of size 1 because even if $|Z| = 1$ there is no edge incident to $z \in Z$ other than $\{x, z\}$. Furthermore, this minor is not trivial as we would have $|V| \leq 2$. As that minor also has a connected covering graph, we find a member $C' \subseteq X$. This implies a member $C \subseteq X$ of $\mathcal{C}$, a contradiction to Y being a cover. $\qquad\square$

In the rest of this section, we will prove that the set of dense clutters is 3-unifying.

Lemma 5. *A minimally dense clutter over a ground set V has three members C_1, C_2 and C_3 with empty intersection whose union is the ground set.*

Proof. We will prove that such a clutter $\mathcal{C}$ has two members C_1 and C_2 whose union is the ground set and $|C_1 \cap C_2| = 1$. We can then choose an arbitrary C_3 not containing the common element to complete the proof.

If $\mathcal{C}$ is a delta, choose $C_1 = \{1, 2\}$ and $C_2 = \{2, 3, \ldots, n\}$.

If $\mathcal{C}$ is the blocker of an extended odd hole, take an arbitrary $v \in V$ and consider the minor $\mathcal{C}/v$. The covering graph of this minor is connected and bipartite since it is a path. By Lemma 1, we find two disjoint members K and L of this minor with $K \cup L = V - \{v\}$. Since $\mathcal{C}$ has no disjoint members, they both contain v in $\mathcal{C}$. $\qquad\square$

The following lemma bridges the gap from minimally dense to strictly dense clutters.

Lemma 6. *Let $\mathcal{C}$ be a strictly dense clutter over a ground set V and $J \subseteq V$ such that $\mathcal{C}/J$ is dense. Then for each $v \in J$, there is a $w \in V - J$ such that $\{v, w\}$ is a cover of size 2 in $\mathcal{C}$.*

Proof. Assume there is a $v \in J$ such that no such cover of size 2 exists. Then the minor $\mathcal{C}\backslash v/(J - \{v\})$ has no cover of size 1 and thereby covering number at least 2. Furthermore, this minor has no fractional packing of value 2 since it contains only a subset of the members in $\mathcal{C}/J$. Therefore this minor is dense. The restriction obtained from $\mathcal{C}$ after restricting v contracts a subset of $J - \{v\}$ since $\tau(\mathcal{C}\backslash v/(J - \{v\})) \geq 2$. This implies that restriction is also dense, a contradiction. $\qquad\square$

Theorem 4 ([2], Proposition 4.5). *The set of dense clutters is 3-unifying.*

Proof. Let C be a strictly dense clutter over a ground set V. Choose U such that C/U is dense but no proper contraction minor is. Let $C \backslash I/(U \cup U')$ be a proper minor of C/U with covering number at least 2. If $I \neq \emptyset$, the restriction $C \backslash I/J$ is not dense and therefore has a fractional packing of value 2. Since $J \subseteq (U \cup U')$, the minor $C \backslash I/(U \cup U')$ also has a fractional packing of value 2. If $I = \emptyset$, we get the same result by the definition of U.

Hence, C/U is minimally dense. By Corollary 5, we find three members C_1', C_2' and C_3' in C/U with empty intersection and union $V - U$.

Let $C_1, C_2, C_3 \in C$ such that $C_i' \subseteq C_i \subseteq C_i' \cup U$ for $i = 1, 2, 3$. Suppose there is a $v \in V - (C_1 \cup C_2 \cup C_3)$. Clearly, $v \in U$. By Lemma 6, there is a cover $\{v, w\}$ with $w \in V - U$. Since none of the three members contains U, they all contain w, a contradiction. $\qquad\square$

By the unifying theorem, given a clutter explicitly with its members, we can find a dense restriction in polynomial time or state that there is none. Similarly to Remark 1, having a dense restriction is equivalent to having a dense minor. Furthermore, this is also equivalent to having a delta or the blocker of an extended odd hole minor.

Remark 2 ([2], Remark 4.4). Let C be a clutter over a ground set V. Then the following statements are equivalent:

(i) C has a dense restriction.
(ii) C has a delta or the blocker of an extended odd hole minor.
(iii) C has a dense minor.

5   Finding Delta or Blocker of Extended Odd Hole Minors

We will discuss an algorithm which finds a delta or the blocker of an extended odd hole minor of a dense clutter.

The clutter C over the ground set V is given by a filter oracle [13]. This oracle returns in time θ whether a set $A \subseteq V$ contains a member of C. The algorithm implied by the proofs given in Sect. 4 has a running time of $\mathcal{O}(n^3(\theta + n))$. This improves the running time of the algorithm by Abdi and Lee in [4] with a running time of $\mathcal{O}(n^4)$ oracle calls.

We will start with some basic properties about filter oracles.

Lemma 7 ([4], Remark 9). *Let C be a clutter over a ground set V given by a filter oracle. Let $n = |V|$ and θ be the time required to run the oracle. Then the following statements hold:*

(i) The members of cardinality 1 can be computed in $\mathcal{O}(n\theta)$ time.
(ii) The members of cardinality 2 can be computed in $\mathcal{O}(n^2\theta)$ time.
(iii) Given a set $C \subseteq V$, it can be checked in $\mathcal{O}(n\theta)$ time, whether $C \in C$.
(iv) Given a set $A \subseteq V$ that contains a member, such a member can be found in $\mathcal{O}(n^2\theta)$ time.

Lemma 8 ([13], **Theorem 5.1**). *Let $\mathcal{C}$ be a clutter over a ground set V given by a filter oracle, which runs in time θ. Then the filter oracle also implies a filter oracle for the blocker $b(\mathcal{C})$ running in $\theta + \mathcal{O}(|V|)$ time.*

Lemma 9 ([13], **Theorem 5.2**). *Let $\mathcal{C}$ be a clutter over a ground set V given by a filter oracle, which runs in time θ. Let $I, J \subseteq V$ be disjoint. Then the filter oracle implies a filter oracle for the minor $\mathcal{C} \backslash I / J$, which runs in time $\theta + \mathcal{O}(|V|)$.*

The input of the final algorithm is a dense clutter including a certificate $w \in \mathbb{R}_{\geq 0}^V$ as in Lemma 2. We allow a scaled certificate, that means it only has to satisfy $w(C) > \frac{w(V)}{2}$ for all members $C \in \mathcal{C}$. Note that the certificate can be calculated with a linear program, if the members are explicitly given.

We will formulate an algorithm that, given a dense clutter including a certificate, computes a delta, the blocker of an extended odd hole minor, or a proper dense minor. If the output is a proper dense minor, we will call the algorithm again. We need to make sure that a certificate for the proper dense minor can be computed easily.

Lemma 10. *Let $\mathcal{C}$ be a dense clutter over a ground set V with $\tau(\mathcal{C}) = 2$ and connected bipartite covering graph. The clutter is inputted via a filter oracle running in time θ. Let $n = |V|$. Then a certificate $w \in \mathbb{R}_{\geq 0}^V$ such that $w(C) > \frac{w(V)}{2}$ for all members $C \in \mathcal{C}$ can be computed in $\mathcal{O}(n^2(\theta + n))$ time.*

Proof. Since all members of cardinality 2 of the blocker can be computed in $\mathcal{O}(n^2(\theta + n))$ time, the covering graph G of $\mathcal{C}$ can be computed in that time. Let X and Y be the colour classes of the bipartition, which can also be calculated in quadratic time. For $v \in V$ let $w(v)$ be the degree of v in G. This defines an initial certificate with $w(V) = 2m$ where m is the number of edges. Each member $C \in \mathcal{C}$ contains at least one element of each cover of size 2. Thus, C as a vertex set is incident to all edges, implying $w(C) \geq m$. If a member contains more than one element of a cover, at least one edge has to be counted twice and therefore $w(C) \geq m+1$. By the same argument as in Lemma 1, the only possible members incident to at most one element of each cover of size 2 are X and Y. Check whether X or Y are members of $\mathcal{C}$. This can be done in linear time. Only one of them can be a member since $\mathcal{C}$ is dense. If none of them is a member, return w as the certificate. If without loss of generality X is a member, let $x \in X$ and $y \in Y$. Increase $w(x)$ by $\frac{1}{2}$ and decrease $w(y)$ by $\frac{1}{2}$ to get a certificate w'. This does not change the total sum and each member then fulfils $w'(C) \geq m+\frac{1}{2}$, so w' is indeed a certificate and can be returned.

We are now ready to formulate the main algorithm, see Algorithm 1.

Theorem 5. *The algorithm works correctly in $\mathcal{O}(n^2(\theta + n))$ time. If it returns a proper dense minor, a certificate for this minor being dense can be calculated in $\mathcal{O}(n^2(\theta + n))$ time.*

Algorithm 1: Finding a delta or the blocker of an extended odd hole minor

---

Input   : Dense clutter C over a ground set V with certificate w inputted via a filter oracle

Output: Delta or extended odd hole minor or a proper dense minor of C

1 **if** *there is $v \in V$ such that v does not appear in a cover of size 2* **then**
2 $\quad$ **return** $C \backslash v$ *as proper dense minor*

3 Compute covering graph G of C;
4 **if** *G is not bipartite* **then**
5 $\quad$ Find an odd cycle O with no chords in G;
6 $\quad$ **return** $C/(V - O)$ *as delta or the blocker of an extended odd hole minor*;

7 **if** *G is not connected* **then**
8 $\quad$ Let A be the vertex set of one component, $B = V - A$;
9 $\quad$ Let K and L be colour classes of this component with $w(K) \geq w(L)$;
10 $\quad$ **if** *K and L are not members of C/B* **then**
11 $\quad\quad$ **return** C/B *as proper dense minor*;
12 $\quad$ **if** *$\tau(C \backslash K/L) \geq 2$* **then**
13 $\quad\quad$ **return** $C \backslash K/L$ *as proper dense minor*;
14 $\quad$ Find cover $\{b\}$ of $C \backslash K/L$;
15 $\quad$ **return** $C \backslash b/(B - \{b\})$ *as proper dense minor*;

16 Let X and Y be the colour classes of G such that Y is a cover;
17 Compute minimal cover $B \subseteq Y$;
18 Compute $x \in X$ such that removing x from G does not disconnect $X - \{x\}$;
19 Compute set of isolated vertices $Z \subseteq Y$ after removing x;
20 **if** $Z \not\subseteq B$ **then**
21 $\quad$ Let $y \in Z - B$;
22 $\quad$ **return** C/y *as proper dense minor*;

23 **if** $|Z| \geq 2$ **then**
24 $\quad$ Find a delta minor by Lemma 4;
25 $\quad$ **return** *that delta minor*;

26 **return** $C \backslash Z/x$ *as proper dense minor*;

Proof. The correctness of the algorithm is given by the proof of Lemma 3 and Theorem 3. Whenever we get a contradiction or an excluded case in the proof due to the clutter being minimally dense, the algorithm outputs the proper dense minor. Note that in almost all cases where a proper dense minor is returned, we have the situation of Lemma 10 with the minor $C \backslash v$ in the first step and $C \backslash K/L$ as the only exceptions.

In both cases, we can take the projection of the certificate of the original clutter on the new ground set. For $C \backslash v$ this is a certificate because $w(C)$ does not change for members C, but $w(V) \geq w(V - \{v\})$. For $C \backslash K/L$ the argument is given in the proof of Lemma 3. The running time for the certificate is satisfied in both cases. It remains to show that the running time of the algorithm is indeed $\mathcal{O}(n^2(\theta + n))$.

The covers of size 2 can be computed in $\mathcal{O}(n^2(\theta + n))$ time since they are minimal covers and members of cardinality 2 in the blocker can be computed in that time. Therefore, we find a $v \in V$ that does not appear in a cover of size 2 or obtain the covering graph G in $\mathcal{O}(n^2(\theta + n))$ time.

In each connected component of G, we can find an odd cycle or a bipartition in quadratic time.

Checking whether K and L are members of $\mathcal{C}/B$ can be done in a linear number of oracle calls. By finding all minimal covers of size 1 in the blocker, we can decide whether $\tau(\mathcal{C} \backslash K/L) \geq 2$ is satisfied in a linear number of oracle calls. If there is a cover of size 1, we can use it as the cover $\{b\}$ of $\mathcal{C} \backslash K/L$.

We can decide in one oracle call whether X or Y is a cover. The minimal cover $B \subseteq Y$ can be obtained in $\mathcal{O}(n^2(\theta + n))$ time by Lemma 7. The $x \in X$ such that removing x from G does not disconnect $X - \{x\}$ can be computed in quadratic time as we can start with any element and find a better one given by the proof in Theorem 3 in linear time.

The set Z is calculated in linear time since these are just components of a graph. Checking $Z \subseteq B$ and possibly finding $y \in Z - B$ can also be done in linear time as an order of the elements can be assumed.

The computation of the delta minor in Lemma 4 can also be implemented to run in $\mathcal{O}(n(\theta + n))$ time. The sets I, C_1 and C_2 of the proof can be determined in that time. Let X denote the set of elements of C_2 that are already contracted when considering $x \in C_2$. We query the oracle whether $\{u, x\} \cup X$ contains a member in $\mathcal{C}$. As C is a member of $\mathcal{C}$ and $\{u\}$ is not a member of $\mathcal{C} \backslash I/X$, this is by construction of X equivalent to $\{u, x\}$ being a member of $\mathcal{C} \backslash I/X$. Hence, each element $x \in C_2$ can be processed in $\mathcal{O}(\theta + n)$ time. In conclusion, every single step of the algorithm can be implemented in $\mathcal{O}(n^2(\theta + n))$ time, concluding the proof. $\square$

By applying the algorithm recursively and calculating the new certificate, the cardinality of the ground set decreases in each iteration. As an immediate consequence, we get the following corollary as the main result of this section.

Corollary 3. *There is an algorithm that, given a dense clutter over the ground set V by a filter oracle and a certificate w for that clutter being dense, finds a delta or the blocker of an extended odd hole minor in $\mathcal{O}(|V|^3(\theta + |V|))$ time, where θ is the time required for an oracle call.*

In conclusion, given a clutter $\mathcal{C}$ over a ground set V explicitly with its members, we can decide in polynomial time (in $|\mathcal{C}|$ and $|V|$) whether $\mathcal{C}$ has a dense restriction. Recall that this is equivalent to $\mathcal{C}$ having a delta or the blocker of an extended odd hole minor.

If $\mathcal{C}$ has a dense restriction, we can find it and compute a certificate in polynomial time.

Acknowledgement. I would like to thank Ahmad Abdi, Gérard Cornuéjols and Stephan Held for helpful discussions.

References

1. Abdi, A., Cornuéjols, G., Huynh, T., Lee, D.: Idealness of k-wise intersecting families. Math. Program. 1–22 (2020). https://doi.org/10.1007/s10107-020-01587-x
2. Abdi, A., Cornuéjols, G., Lee, D.: Intersecting restrictions in clutters. Combinatorica **40**(5), 605–623 (2020). https://doi.org/10.1007/s00493-020-4076-2
3. Abdi, A., Cornuéjols, G., Pashkovich, K.: Ideal clutters that do not pack. Math. Oper. Res. **43**(2), 533–553 (2018). https://doi.org/10.1287/moor.2017.0871
4. Abdi, A., Lee, D.: Deltas, extended odd holes and their blockers. J. Combin. Theory **136**, 193–206 (2019). https://doi.org/10.1016/j.jctb.2018.10.006
5. Cornuejols, G., Novick, B.: Ideal 0, 1 matrices. J. Combin. Theory Ser. B **60**(1), 145–157 (1994). https://doi.org/10.1006/jctb.1994.1009
6. Ding, G., Feng, L., Zang, W.: The complexity of recognizing linear systems with certain integrality properties. Math. Program. **114**, 321–334 (2008). https://doi.org/10.1007/s10107-007-0103-y
7. Edmonds, J., Fulkerson, D.: Bottleneck extrema. J. Combin. Theory **8**(3), 299–306 (1970). https://doi.org/10.1016/S0021-9800(70)80083-7
8. Ford, L., Fulkerson, D.: Flows in networks. Amer. Math. Monthly **71**, 1059–1060 (1962). https://doi.org/10.2307/2311955
9. Lehman, A.: On the width-length inequality. Math. Program. **17**, 403–417 (1979). https://doi.org/10.1007/BF01588263
10. Menger, K.: Zur allgemeinen Kurventheorie. Fundam. Math. **10**, 96–115 (1927). https://doi.org/10.4064/FM-10-1-96-115
11. Seymour, P.: The forbidden minors of binary clutters. J. London Math. Soc. **s2–12**(3), 356–360 (1976). https://doi.org/10.1112/jlms/s2-12.3.356
12. Seymour, P.: The matroids with the max-flow min-cut property. J. Combin. Theory Ser. B **23**(2), 189–222 (1977). https://doi.org/10.1016/0095-8956(77)90031-4
13. Seymour, P.: On Lehman's width-length characterization. In: Polyhedral Combinatorics (1990). https://doi.org/10.1090/dimacs/001/11

LP-Based Approximations for Disjoint Bilinear and Two-Stage Adjustable Robust Optimization

Omar El Housni[1], Ayoub Foussoul[2(✉)], and Vineet Goyal[2]

[1] ORIE, Cornell Tech, New York, USA
oe46@cornell.edu
[2] IEOR, Columbia University, New York, USA
{af3209,vg2277}@columbia.edu

Abstract. We consider the class of disjoint bilinear programs $\max \{\mathbf{x}^T\mathbf{y} \mid \mathbf{x} \in \mathcal{X}, \mathbf{y} \in \mathcal{Y}\}$ where $\mathcal{X}$ and $\mathcal{Y}$ are packing polytopes. We present an $O(\frac{\log\log m_1}{\log m_1} \frac{\log\log m_2}{\log m_2})$-approximation algorithm for this problem where m_1 and m_2 are the number of packing constraints in $\mathcal{X}$ and $\mathcal{Y}$ respectively. In particular, we show that there exists a near-optimal solution $(\mathbf{x},\mathbf{y})$ such that $\mathbf{x}$ and $\mathbf{y}$ are "near-integral". We give an LP relaxation of this problem from which we obtain the near-optimal near-integral solution via randomized rounding. As an application of our techniques, we present a tight approximation for the two-stage adjustable robust optimization problem with covering constraints and right-hand side uncertainty where the separation problem is a bilinear optimization problem. In particular, based on the ideas above, we give an LP restriction of the two-stage problem that is an $O(\frac{\log n}{\log\log n} \frac{\log L}{\log\log L})$-approximation where L is the number of constraints in the uncertainty set. This significantly improves over state-of-the-art approximation bounds known for this problem.

Keywords: Disjoint bilinear programming · Two-stage robust optimization · Approximation algorithms

1 Introduction

We consider the following class of disjoint bilinear programs,

$$z_{\mathsf{PDB}} = \max_{\mathbf{x},\mathbf{y}} \{\mathbf{x}^T\mathbf{y} \mid \mathbf{x} \in \mathcal{X}, \mathbf{y} \in \mathcal{Y}\}, \tag{PDB}$$

where $\mathcal{X}$ and $\mathcal{Y}$ are packing polytopes given by an intersection of knapsack constraints. Specifically,

$$\mathcal{X} := \{\mathbf{x} \geq \mathbf{0} \mid \mathbf{P}\mathbf{x} \leq \mathbf{p}\} \quad \text{and} \quad \mathcal{Y} := \{\mathbf{y} \geq \mathbf{0} \mid \mathbf{Q}\mathbf{y} \leq \mathbf{q}\},$$

where $\mathbf{P} \in \mathbb{R}_+^{m_1 \times n}$, $\mathbf{Q} \in \mathbb{R}_+^{m_2 \times n}$, $\mathbf{p} \in \mathbb{R}_+^{m_1}$ and $\mathbf{q} \in \mathbb{R}_+^{m_2}$. We refer to this problem as a packing disjoint bilinear program PDB. This is a subclass of the well-studied

© Springer Nature Switzerland AG 2022
K. Aardal and L. Sanità (Eds.): IPCO 2022, LNCS 13265, pp. 223–236, 2022.
https://doi.org/10.1007/978-3-031-06901-7_17

disjoint bilinear problem: $\max\limits_{\mathbf{x},\mathbf{y}} \{\mathbf{x}^T\mathbf{M}\mathbf{y} \mid \mathbf{x} \in \mathcal{X},\ \mathbf{y} \in \mathcal{Y}\}$, where $\mathbf{M}$ is a general $n \times n$ matrix.

Disjoint bilinear programming is NP-hard in general (Chen et al. [9]). We show that it is NP-hard to even approximate within any finite factor. Several heuristics have been studied for this problem including cutting-planes algorithms (Konno et al. [24]), polytope generation methods (Vaish et al. [30]), Benders decomposition (Geoffrion [18]), reduction to concave minimization (Thieu [29]) and two-stage robust optimization (Zhen et al. [33]). Algorithms for non-convex quadratic optimization can also be used to solve disjoint bilinear programs.

Many important applications can be formulated as a disjoint bilinear program including fixed charge network flows (Rebennack et al. [27]), concave cost facility location (Soland [28]), bilinear assignment problems (Ćustić et al. [34]), non-convex cutting-stock problems (Harjunkoski et al. [23]), multicommodity flow network interdiction problems (Lim and Smith [25]), bimatrix games (Mangasarian and Stone [26], Firouzbakht et al. [17]) pooling problems (Gupte et al. [22]).

One important application closely related to disjoint bilinear optimization that we focus on in this paper, is the two-stage adjustable robust optimization. In particular, the separation problem of a two-stage adjustable robust problem can be formulated as a disjoint bilinear optimization problem. More specifically, we consider the following two-stage adjustable robust problem,

$$z_{\mathsf{AR}} = \min_{\mathbf{x},t} \{\mathbf{c}^T\mathbf{x} + t \mid t \geq \mathcal{Q}(\mathbf{x}),\ \mathbf{x} \in \mathcal{X}\}, \tag{AR}$$

where for all $\mathbf{x} \in \mathcal{X}$,

$$\mathcal{Q}(\mathbf{x}) = \max_{\mathbf{h}\in\mathcal{U}} \min_{\mathbf{y}\geq 0} \{\mathbf{d}^T\mathbf{y} \mid \mathbf{By} \geq \mathbf{h} - \mathbf{Ax}\}.$$

Here $\mathbf{A} \in \mathbb{R}^{m\times n}$, $\mathbf{B} \in \mathbb{R}_+^{m\times n}$, $\mathbf{c} \in \mathbb{R}_+^n$, $\mathbf{d} \in \mathbb{R}_+^n$, $\mathcal{X} \subset \mathbb{R}_+^n$ is a polyhedral cone, and $\mathcal{U}$ is a polyhedral uncertainty set. The separation problem of AR is the following: given a candidate solution $(\mathbf{x},t)$, decide if it is feasible, i.e., $\mathbf{x} \in \mathcal{X}$ and $t \geq \mathcal{Q}(\mathbf{x})$ or give a separating hyperplane. This is equivalent to solving $\mathcal{Q}(\mathbf{x})$. We will henceforth refer to $\mathcal{Q}(\mathbf{x})$ as the separation problem. For ease of notation, we use $\mathcal{Q}(\mathbf{x})$ to refer to both the problem and its optimal value. In this two-stage problem, the adversary observes the first-stage decision $\mathbf{x}$ and reveals the worst-case scenario of $\mathbf{h} \in \mathcal{U}$. Then, the decision maker selects a second-stage recourse decision $\mathbf{y}$ such that $\mathbf{By}$ covers $\mathbf{h} - \mathbf{Ax}$. The goal is to select a first-stage decision such that the total cost in the worst-case is minimized. This model has been widely considered in the literature (Dhamdhere et al. [11], Feige et al. [16], Gupta et al. [21], Bertsimas and Goyal [4], Bertsimas and Bidkhori [5], Bertsimas and de Ruiter [7], Xu et al. [31], Zhen et al. [32], El Housni and Goyal [12], El Housni et al. [14,15]), and has many applications including set cover, capacity planning and network design problems under uncertain demand.

Several uncertainty sets have been considered in the literature including polyhedral uncertainty sets, ellipsoids and norm balls (see Bertsimas et al. [6]). Some of the most important uncertainty sets are budget of uncertainty sets (Bertsimas and Sim [8], Gupta et al. [20], El Housni and Goyal [13]) and intersections of budget of uncertainty sets such as CLT sets (see Bandi and Bertsimas [1]) and inclusion-constrained budgeted sets (see Gounaris et al. [19]). These have been widely used in practice. Following this motivation, we consider in this paper the following uncertainty set,

$$\mathcal{U} := \{\mathbf{h} \geq \mathbf{0} \mid \mathbf{Rh} \leq \mathbf{r}\},$$

where $\mathbf{R} \in \mathbb{R}_+^{L \times m}$ and $\mathbf{r} \in \mathbb{R}_+^L$. This is a generalization of the previously mentioned sets. We refer to this as a packing uncertainty set.

Feige et al. [16] show that AR is hard to approximate within any factor better than $\Omega(\frac{\log n}{\log \log n})$ even in the special case of a single budget of uncertainty set. Bertsimas and Goyal [4] give an $O(\sqrt{m})$-approximation in the case where the first-stage matrix $\mathbf{A}$ is non-negative. Recently, El Housni and Goyal [13] give an $O(\frac{\log n}{\log \log n})$-approximation in the case of a single budget of uncertainty set and an $O(\frac{\log^2 n}{\log \log n})$-approximation in the case of an intersection of disjoint budgeted sets. In general, they show an $O(\frac{L \log n}{\log \log n})$-approximation in the case of a packing uncertainty set with L constraints. However, this bound scales linearly with L. The two-stage robust covering problem was also considered in the discrete case where the first and second stage solutions $\mathbf{x}$ and $\mathbf{y}$ are restricted to be in $\{0, 1\}^m$. For this problem, Feige et al. [16] and Gupta et al. [21] give an $O(\log n \log m)$-approximation and an $O(\log n + \log m)$-approximation respectively in the case where $\mathbf{A} = \mathbf{B} \in \{0, 1\}^{m \times n}$, $\mathbf{d} = \lambda \mathbf{c}$ for some $\lambda > 0$ and the uncertainty set $\mathcal{U}$ is a budget of uncertainty set with equal weights, i.e., $\mathcal{U} = \{\mathbf{h} \in [0, 1]^m \mid \sum_{i=1}^m h_i \leq k\}$. Gupta et al. [20] consider a more general uncertainty set, namely, intersection of *p-system* and *q-knapsack* and give an $O(pq \log n)$-approximation of the two-stage problem.

The goal of this paper is to provide LP-based approximation algorithms with provable guarantees for the packing disjoint bilinear program as well as the two-stage adjustable robust problem that improve over the approximation bounds known for these problems.

1.1 Our Contributions

A Polylogarithmic Approximation Algorithm for PDB. We present an LP-rounding based randomized approximation algorithm for PDB. Our algorithm relies on a new idea that might be of independent interest. In particular, we show the existence of a near-optimal near-integral solution of this problem. That is, a near-optimal solution $(\hat{\mathbf{x}}, \hat{\mathbf{y}})$ such that $\hat{x}_i \in \{0, \max_{\mathbf{x} \in \mathcal{X}} x_i/\zeta_1\}$ and $\hat{y}_i \in \{0, \max_{\mathbf{y} \in \mathcal{Y}} y_i/\zeta_2\}$ for some logarithmic factors ζ_1 and ζ_2. We give an LP relaxation of PDB, i.e., a linear program whose optimal cost is greater than the optimum of PDB, from which we obtain such $(\hat{\mathbf{x}}, \hat{\mathbf{y}})$ via randomized rounding. More specifically, we have the following theorem,

Theorem 1. *There exists an LP-rounding based randomized algorithm that gives an $O(\frac{\log \log m_1}{\log m_1} \frac{\log \log m_2}{\log m_2})$-approximation to PDB.*

Approximating the Two-Stage Problem AR. We present an LP-based approximation for AR. The separation problem for AR is a variant of PDB. However, the objective is a difference of a bilinear and a linear term making it challenging to approximate. Our approach approximates AR directly. In particular, using ideas from our approximation of PDB, we give a compact linear restriction of AR, that is, a linear program whose optimal objective is greater than the optimum of AR, and show that it is a polylogarithmic approximation of AR. In particular, we have the following theorem.

Theorem 2. *There exists an LP restriction of AR that gives an* $O(\frac{\log n}{\log \log n} \frac{\log L}{\log \log L})$-*approximation to AR.*

Our bound improves significantly over the prior approximation bound of $O(\frac{L \log n}{\log \log n})$ [13] known for this problem. It also shows a striking contrast between the fractional and the discrete case of the two-stage robust covering problem. In fact, the discrete two-stage robust covering problem under a packing uncertainty set with L constraints (*L-knapsack*) considered in [20] is hard to approximate within any factor better than L^δ, for some $\delta > 0$. This follows from the hardness of the maximum independent set problem.

We compare the performance of our approximation to affine policies. Affine policies are widely used approximate policies in dynamic robust optimization where the second-stage decision **y** is restricted to be an affine function of the uncertain right-hand side **h**. It is known that the optimal affine policy can be computed efficiently (Ben-Tal et al. [2]). We show that our algorithm is significantly faster than finding the optimal affine policy while providing good approximate solutions. Specifically, in randomly generated instances with $n = m = L = 100$, the cost of our solution is within 30% of the cost of the optimal affine policy in all of the instances we consider. However, our algorithm is significantly faster terminating in less than 0.1s for all instances. In contrast, it takes $1000s$ or larger on average to compute the optimal affine policy for $n > 80$.

2   A Polylogarithmic Approximation for PDB

In this section, we present an $O(\frac{\log \log m_1}{\log m_1} \frac{\log \log m_2}{\log m_2})$-approximation for PDB (Theorem 1). To prove this theorem, we show an interesting structural property of PDB. In particular, we show that there exists a near-optimal solution of PDB that is "near-integral". Let us define for all $i \in [n]$,

$$\theta_i = \max_{\mathbf{x} \in \mathcal{X}} x_i, \quad \gamma_i = \max_{\mathbf{y} \in \mathcal{Y}} y_i, \quad \zeta_1 = \frac{3 \log m_1}{\log \log m_1} + 2 \quad \text{and} \quad \zeta_2 = \frac{3 \log m_2}{\log \log m_2} + 2.$$

We formally state our structural property in the following lemma.

Lemma 1. (Structural Property). *There exists a feasible solution* $(\hat{\mathbf{x}}, \hat{\mathbf{y}})$ *of PDB whose objective value is within* $O(\frac{\log \log m_1}{\log m_1} \frac{\log \log m_2}{\log m_2})$ *of the optimum and such that* $\hat{x}_i \in \{0, \frac{\theta_i}{\zeta_1}\}$ *and* $\hat{y}_i \in \{0, \frac{\gamma_i}{\zeta_2}\}$ *for all* $i \in [n]$.

We obtain such a solution satisfying the above property using an LP relaxation of PDB via a randomized rounding approach.

LP Relaxation and Rounding. We consider the following linear program,

$$
z_{\mathsf{LP-PDB}} = \max_{\omega \geq 0} \left\{ \sum_{i=1}^{n} \theta_i \gamma_i \omega_i \; \middle| \; \begin{array}{c} \sum_{i=1}^{n} \theta_i \mathbf{P}_i \omega_i \leq \mathbf{p} \\ \sum_{i=1}^{n} \gamma_i \mathbf{Q}_i \omega_i \leq \mathbf{q} \end{array} \right\}, \tag{LP-PDB}
$$

where $\mathbf{P}_i$ is the i-th column of $\mathbf{P}$ and $\mathbf{Q}_i$ is the i-th column of $\mathbf{Q}$. We first show that LP-PDB is a relaxation of PDB.

Lemma 2. $z_{\mathsf{PDB}} \leq z_{\mathsf{LP-PDB}}$.

Proof. Let $(\mathbf{x}^*, \mathbf{y}^*)$ be an optimal solution of PDB. Let $\boldsymbol{\omega}^*$ be such that $\omega_i^* = \frac{x_i^*}{\theta_i} \cdot \frac{y_i^*}{\gamma_i}$ for all $i \in [n]$. By definition, we have $x_i^* \leq \theta_i$ and $y_i^* \leq \gamma_i$ for all $i \in [n]$. Hence,

$$
\sum_{i=1}^{n} \theta_i \mathbf{P}_i \omega_i^* = \sum_{i=1}^{n} \theta_i \mathbf{P}_i \frac{x_i^*}{\theta_i} \frac{y_i^*}{\gamma_i} \leq \sum_{i=1}^{n} \mathbf{P}_i x_i^* \leq \mathbf{p},
$$

and

$$
\sum_{i=1}^{n} \gamma_i \mathbf{Q}_i \omega_i^* = \sum_{i=1}^{n} \gamma_i \mathbf{Q}_i \frac{x_i^*}{\theta_i} \frac{y_i^*}{\gamma_i} \leq \sum_{i=1}^{n} \mathbf{Q}_i y_i^* \leq \mathbf{q}.
$$

Note that we use the fact that $\mathbf{P}$ and $\mathbf{Q}$ are non-negative in the above inequalities. Therefore, $\boldsymbol{\omega}^*$ is feasible for LP-PDB with objective value

$$
\sum_{i=1}^{n} \theta_i \gamma_i \omega_i^* = \sum_{i=1}^{n} x_i^* y_i^* = z_{\mathsf{PDB}},
$$

which concludes the proof. □

Now, to construct our near-optimal near-integral solution, we consider the randomized rounding approach described in Algorithm 1. Note that by definition of θ_i, $\max_{\omega}\{\omega_i \mid \sum_{j=1}^{n} \theta_j \mathbf{P}_j \omega_j \leq \mathbf{p}, \boldsymbol{\omega} \geq 0\} = 1$, for all $i \in [n]$. Hence, for all $i \in [n]$, ω_i^* defined in Algorithm 1 is such that $\omega_i^* \leq 1$. In our proof of Lemma 1, we use the following variant of Chernoff bounds.

Lemma 3. (Chernoff Bounds [10]).
(a) *Let* $\chi_1, \ldots, \chi_r$ *be independent Bernoulli trials. Denote* $\Xi := \sum_{i=1}^{r} \epsilon_i \chi_i$ *where* $\epsilon_1, \ldots, \epsilon_r$ *are reals in [0,1]. Let* $s > 0$ *such that* $\mathbb{E}(\Xi) \leq s$. *Then for any* $\delta > 0$ *we have,*

$$
\mathbb{P}(\Xi \geq (1+\delta)s) \leq \left(\frac{e^\delta}{(1+\delta)^{1+\delta}} \right)^s.
$$

(b) *Let* $\chi_1, \ldots, \chi_r$ *be independent Bernoulli trials. Denote* $\Xi := \sum_{i=1}^{r} \epsilon_i \chi_i$ *where* $\epsilon_1, \ldots, \epsilon_r$ *are reals in (0,1]. Then for any* $0 < \delta < 1$,

$$
\mathbb{P}(\Xi \leq (1-\delta)\mathbb{E}(\Xi)) \leq e^{-\frac{1}{2}\delta^2 \mathbb{E}(\Xi)}.
$$

Algorithm 1.

Input: $\epsilon > 0$.

Output: feasible solution verifying lemma 1 with probability at least $1 - \epsilon - o(1)$.

1: Let $\boldsymbol{\omega}^*$ be an optimal solution of LP-PDB and let $T = 8\lceil \log \frac{1}{\epsilon} \rceil$.

2: Initialize $\mathbf{x}^* = \mathbf{0}$, $\mathbf{y}^* = \mathbf{0}$ and $\mathtt{max} = 0$.

3: **for** $t = 1, \ldots, T$ **do**

4:     let $\tilde{\omega}_1, \ldots, \tilde{\omega}_n$ be i.i.d. Bernoulli variables with $\mathbb{P}(\tilde{\omega}_i = 1) = \omega_i^*$ for $i \in [n]$.

5:     let $\hat{x}_i = \theta_i \tilde{\omega}_i / \zeta_1$ and $\hat{y}_i = \gamma_i \tilde{\omega}_i / \zeta_2$ for $i \in [n]$.

6:     **if** $(\hat{\mathbf{x}}, \hat{\mathbf{y}})$ is feasible for PDB and $\hat{\mathbf{x}}^T \hat{\mathbf{y}} \geq \mathtt{max}$ **then**

7:         set $x_i^* = \hat{x}_i$, $y_i^* = \hat{y}_i$ for all $i \in [n]$ and $\mathtt{max} = \hat{\mathbf{x}}^T \hat{\mathbf{y}}$.

8:     **end if**

9: **end for**

10: **return** $(\mathbf{x}^*, \mathbf{y}^*)$

Proof of Lemma 1. It is sufficient to prove that, with constant probability, $(\hat{\mathbf{x}}, \hat{\mathbf{y}})$ constructed at each iteration of Algorithm 1 is a feasible solution of PDB verifying the structural property. In particular, let $\boldsymbol{\omega}^*$ be an optimal solution of LP-PDB as defined in Algorithm 1. Consider some iteration $t \in [T]$. Let $\tilde{\omega}_1, \ldots, \tilde{\omega}_n$ and $(\hat{\mathbf{x}}, \hat{\mathbf{y}})$ be as defined in iteration t of the main loop. We show that the following properties hold with a constant probability,

$$\sum_{i=1}^n \mathbf{P}_i \hat{x}_i \leq \mathbf{p}, \qquad \sum_{i=1}^n \mathbf{Q}_i \hat{y}_i \leq \mathbf{q}, \qquad \sum_{i=1}^n \hat{x}_i \hat{y}_i \geq \frac{z_{\mathrm{LP-PDB}}}{2\zeta_1\zeta_2}. \qquad (1)$$

First, we have,

$$\mathbb{P}(\sum_{i=1}^n \mathbf{P}_i \hat{x}_i > \mathbf{p}) = \mathbb{P}\left(\sum_{i=1}^n \theta_i \mathbf{P}_i \frac{\tilde{\omega}_i}{\zeta_1} > \mathbf{p} \right) \leq \sum_{j=1}^{m_1} \mathbb{P}\left(\sum_{i=1}^n \theta_i P_{ji} \frac{\tilde{\omega}_i}{\zeta_1} > p_j \right)$$

$$= \sum_{j \in [m_1] : p_j > 0} \mathbb{P}\left(\sum_{i=1}^n \frac{\theta_i P_{ji}}{p_j} \tilde{\omega}_i > \zeta_1 \right)$$

$$\leq \sum_{j \in [m_1] : p_j > 0} \frac{e^{\zeta_1 - 1}}{(\zeta_1)^{\zeta_1}}$$

$$\leq m_1 \frac{e^{\zeta_1 - 1}}{(\zeta_1)^{\zeta_1}},$$

where the first inequality follows from a union bound on m_1 constraints. The second equality holds because for all $j \in [m_1]$ such that $p_j = 0$, we have $\mathbb{P}\left(\sum_{i=1}^n \theta_i P_{ji} \frac{\tilde{\omega}_i}{\zeta_1} > p_j \right) = 0$. In fact, $p_j = 0$ implies $\sum_{i=1}^n \theta_i P_{ji} \frac{\omega_i^*}{\zeta_1} = 0$, by feasibility of $\boldsymbol{\omega}^*$ in LP-PDB. Therefore, we have $\sum_{i=1}^n \theta_i P_{ji} \frac{\tilde{\omega}_i}{\zeta_1} = 0$ almost surely. The second inequality follows from the Chernoff bounds (a) with $\delta = \zeta_1 - 1$ and $s = 1$. In particular, $\frac{\theta_i P_{ji}}{p_j} \in [0, 1]$ by definition of θ_i for all $i \in [n]$ and $j \in [m_1]$ such that $p_j > 0$ and $\mathbb{E}\left[\sum_{i=1}^n \frac{\theta_i P_{ji}}{p_j} \tilde{\omega}_i \right] = \sum_{i=1}^n \frac{\theta_i P_{ji}}{p_j} \omega_i^* \leq 1$, for all $j \in [m_1]$ such that $p_j > 0$, which holds by feasibility of $\boldsymbol{\omega}^*$. Next, note that $\frac{e^{\zeta_1 - 1}}{(\zeta_1)^{\zeta_1}} = O(\frac{1}{m_1^2})$, therefore, there exists a constant $c > 0$ such that,

$$\mathbb{P}(\sum_{i=1}^{n} \mathbf{P}_i \hat{x}_i > \mathbf{p}) \leq \frac{c}{m_1}. \tag{2}$$

By a similar argument there exists a constant $c' > 0$, such that

$$\mathbb{P}(\sum_{i=1}^{n} \mathbf{Q}_i \hat{y}_i > \mathbf{q}) \leq \frac{c'}{m_2}, \tag{3}$$

Finally we have,

$$\mathbb{P}\left(\sum_{i=1}^{n} \hat{x}_i \hat{y}_i < \frac{1}{2\zeta_1\zeta_2} z_{\mathsf{LP-PDB}}\right) = \mathbb{P}\left(\sum_{i=1}^{n} \frac{\theta_i \gamma_i \tilde{\omega}_i^2}{\zeta_1\zeta_2} < \frac{1}{2\zeta_1\zeta_2} \sum_{i=1}^{n} \theta_i \gamma_i \omega_i^*\right)$$
$$= \mathbb{P}\left(\sum_{i=1}^{n} \frac{\theta_i \gamma_i}{\sum_{j=1}^{n} \theta_j \gamma_j \omega_j^*} \tilde{\omega}_i < \frac{1}{2}\right) \leq e^{-\frac{1}{8}}, \tag{4}$$

where the last inequality follows from Chernoff bounds (b) with $\delta = 1/2$. In particular, $\frac{\theta_i \gamma_i}{\sum_{j=1}^{n} \theta_j \gamma_j \omega_j^*} \leq 1$ for all $i \in [n]$, this is because the unit vector $\mathbf{e}_i$ is feasible for LP-PDB for all $i \in [n]$ such as $\theta_i \gamma_i \leq z_{\mathsf{LP-PDB}} = \sum_{j=1}^{n} \theta_j \gamma_j \omega_j^*$, and we also have, $\mathbb{E}\left[\sum_{i=1}^{n} \frac{\theta_i \gamma_i}{\sum_{j=1}^{n} \theta_j \gamma_j \omega_j^*} \tilde{\omega}_i\right] = 1$. Combining inequalities (2), (3) and (4) we get that $(\hat{\mathbf{x}}, \hat{\mathbf{y}})$ verifies the properties (1) with probability at least $1 - \frac{c}{m_1} - \frac{c'}{m_2} - e^{-\frac{1}{8}} = 1 - e^{-\frac{1}{8}} - o(1)$, which is greater than a constant for m_1 and m_2 large enough. This implies in particular that with positive probability, $(\hat{\mathbf{x}}, \hat{\mathbf{y}})$ is feasible and has an objective value that is greater than $\frac{1}{2\zeta_1\zeta_2} z_{\mathsf{LP-PDB}}$. From Lemma 2, this is greater than $\frac{1}{2\zeta_1\zeta_2} z_{\mathsf{PDB}}$. Therefore, with constant probability, $(\hat{\mathbf{x}}, \hat{\mathbf{y}})$ is a feasible solution of PDB that verifies the structural property. □

Proof of theorem 1. Let $(\mathbf{x}^*, \mathbf{y}^*)$ be the output solution of Algorithm 1. Then $(\mathbf{x}^*, \mathbf{y}^*)$ is such that

$$\mathbf{x}^* \in \mathcal{X}, \qquad \mathbf{y}^* \in \mathcal{Y}, \qquad \mathbf{x}^{*T}\mathbf{y}^* \geq \frac{1}{2\zeta_1\zeta_2} \cdot z_{\mathsf{PDB}},$$

if and only if

$$\hat{\mathbf{x}} \in \mathcal{X}, \qquad \hat{\mathbf{y}} \in \mathcal{Y}, \qquad \hat{\mathbf{x}}^T\hat{\mathbf{y}} \geq \frac{1}{2\zeta_1\zeta_2} \cdot z_{\mathsf{PDB}},$$

for some iteration t of the main loop. From our proof of lemma 1, this happens with probability at least $1 - (e^{-\frac{1}{8}} - o(1))^T \geq 1 - \epsilon - o(1)$. Therefore, with probability at least $1 - \epsilon - o(1)$, Algorithm 1 outputs a feasible solution of PDB whose objective value is within $O(\frac{\log\log m_1}{\log m_1} \frac{\log\log m_2}{\log m_2})$ of z_{PDB}. □

Hardness of the General Disjoint Bilinear Program. Like packing linear programs, the covering linear programs are known to have logarithmic integrality gaps. Hence, a natural question to ask would be whether similar results can be proven for an equivalent covering version of PDB, i.e., a disjoint bilinear program of the form,

$$z_{\text{cdb}} = \min_{\mathbf{x},\mathbf{y}}\{\mathbf{x}^T\mathbf{y} \mid \mathbf{Px} \geq \mathbf{p}, \ \mathbf{Qy} \geq \mathbf{q}, \ \mathbf{x}, \mathbf{y} \geq \mathbf{0}\} \qquad \text{(CDB)}$$

where $\mathbf{P} \in \mathbb{R}_+^{m_1 \times n}$, $\mathbf{Q} \in \mathbb{R}_+^{m_2 \times n}$, $\mathbf{p} \in \mathbb{R}_+^{m_1}$ and $\mathbf{q} \in \mathbb{R}_+^{m_2}$. However, the previous analysis does not extend to the covering case. In particular, we have the following inapproximability result.

Theorem 3. *The covering disjoint bilinear program CDB is NP-hard to approximate within any finite factor.*

The proof of Theorem 3 uses a polynomial time transformation from the *Monotone Not-All-Equal 3-Satisfiability* (MNAE3SAT) NP-complete problem and is deferred to the full version of the paper.

3   From Disjoint Bilinear Optimization to Two-Stage Adjustable Robust Optimization

In this section, we present a polylogarithmic approximation algorithm for AR. In particular, we give a compact linear restriction of AR that provides near-optimal first-stage solutions with cost that is within a factor of $O(\frac{\log n}{\log \log n} \frac{\log L}{\log \log L})$ of z_{AR}. Our proof uses ideas from our approximation of PDB applied to the separation problem $\mathcal{Q}(\mathbf{x})$.

Recall the two-stage adjustable problem AR,

$$\min_{\mathbf{x} \in \mathcal{X}} \quad \mathbf{c}^T\mathbf{x} + \mathcal{Q}(\mathbf{x}),$$

where for all $\mathbf{x} \in \mathcal{X}$,

$$\mathcal{Q}(\mathbf{x}) = \max_{\mathbf{h} \in \mathcal{U}} \min_{\mathbf{y} \geq \mathbf{0}} \{\mathbf{d}^T\mathbf{y} \mid \mathbf{By} \geq \mathbf{h} - \mathbf{Ax}\}.$$

Let us write $\mathcal{Q}(\mathbf{x})$ in its bilinear form. In particular, we take the dual of the inner minimization problem on $\mathbf{y}$ to get,

$$\mathcal{Q}(\mathbf{x}) = \max_{\mathbf{h},\mathbf{z} \geq \mathbf{0}} \left\{ \mathbf{h}^T\mathbf{z} - (\mathbf{Ax})^T\mathbf{z} \ \middle| \ \begin{array}{l} \mathbf{B}^T\mathbf{z} \leq \mathbf{d} \\ \mathbf{Rh} \leq \mathbf{r} \end{array} \right\}.$$

For the special case where $\mathbf{A} = \mathbf{0}$, the optimal first-stage solution is $\mathbf{x} = \mathbf{0}$ and AR reduces to an instance of PDB. Therefore, our algorithm for PDB gives an $O(\frac{\log n}{\log \log n} \frac{\log L}{\log \log L})$-approximation algorithm of AR in this special case.

In the general case, the separation problem $\mathcal{Q}(\mathbf{x})$ is the difference of a bilinear and a linear term. This makes it challenging to approximate $\mathcal{Q}(\mathbf{x})$. Instead, we attempt to approximate AR directly. In particular, for any $\mathbf{x} \in \mathcal{X}$ and $\mathbf{y}_0 \geq \mathbf{0}$ such that $\mathbf{Ax} + \mathbf{By}_0 \geq \mathbf{0}$, we consider the following linear program:

$$
\mathcal{Q}^{\mathsf{LP}}(\mathbf{x}, \mathbf{y}_0) = \max_{\omega \geq 0} \left\{ \sum_{i=1}^{m} (\theta_i \gamma_i - \theta_i \mathbf{a}_i^T \mathbf{x} - \theta_i \mathbf{b}_i^T \mathbf{y}_0) \omega_i \;\middle|\; \begin{array}{l} \sum_{i=1}^{m} \theta_i \mathbf{b}_i \omega_i \leq \mathbf{d} \\[2mm] \sum_{i=1}^{m} \gamma_i \mathbf{R}_i \omega_i \leq \mathbf{r} \end{array} \right\},
$$

where $\theta_i := \max_{\mathbf{z}}\{z_i \mid \mathbf{Bz} \leq \mathbf{d}, \mathbf{z} \geq \mathbf{0}\}$, $\gamma_i := \max_{\mathbf{h}}\{h_i \mid \mathbf{Rh} \leq \mathbf{r}, \mathbf{h} \geq \mathbf{0}\}$, $\mathbf{a}_i$ and $\mathbf{b}_i$ are the i-th row of $\mathbf{A}$ and $\mathbf{B}$ respectively and $\mathbf{R}_i$ is the i-th column of $\mathbf{R}$. The role of $\mathbf{y}_0$ here is to handle the case when some of the entries of $\mathbf{A}$ are negative. In fact, our approximation relies on the non-negativity of $\mathbf{Ax}$. Since this is not the case in general, we add a second-stage covering term $\mathbf{By}_0$ to $\mathbf{Ax}$ for some static second-stage solution $\mathbf{y}_0 \geq \mathbf{0}$ such that $\mathbf{Ax}+\mathbf{By}_0 \geq \mathbf{0}$. For ease of notation, we use $\mathcal{Q}^{\mathsf{LP}}(\mathbf{x}, \mathbf{y}_0)$ to refer to both the problem and its optimal value. Let $\eta := \frac{3 \log n}{\log \log n} + 2$ and $\beta := \frac{3 \log L}{\log \log L} + 2$. Similar to PDB, we show the following structural property of the separation problem.

Structural Property. For $\mathbf{x} \in \mathcal{X}$ and $\mathbf{y}_0 \geq \mathbf{0}$ such that $\mathbf{Ax} + \mathbf{By}_0 \geq \mathbf{0}$, there exists a near-integral solution $(\mathbf{h}, \mathbf{z}) \in \{0, \frac{\gamma_i}{\beta}\}^m \times \{0, \frac{\theta_i}{\eta}\}^m$ of $\mathcal{Q}(\mathbf{x})$ such that,

$$
\sum_{i=1}^{m} \mathbf{b}_i z_i \leq \mathbf{d}, \qquad \sum_{i=1}^{m} \mathbf{R}_i h_i \leq \mathbf{r},
$$

$$
\sum_{i=1}^{m} h_i z_i - (\mathbf{a}_i^T \mathbf{x} + \mathbf{b}_i^T \mathbf{y}_0) z_i \geq \frac{\mathcal{Q}^{\mathsf{LP}}(\beta \mathbf{x}, \beta \mathbf{y}_0)}{2 \eta \beta}. \tag{5}
$$

We construct such solution following a similar procedure as in Algorithm 1. In particular, let ω^* be an optimal solution of $\mathcal{Q}^{\mathsf{LP}}(\beta \mathbf{x}, \beta \mathbf{y}_0)$, consider $\tilde{\omega}_1, \ldots, \tilde{\omega}_m$ i.i.d. Bernoulli random variables such that $\mathbb{P}(\tilde{\omega}_i = 1) = \omega_i^*$ for all $i \in [m]$ and let $(\mathbf{h}, \mathbf{z})$ and let $h_i = \frac{\gamma_i \tilde{\omega}_i}{\beta}$ and $z_i = \frac{\theta_i \tilde{\omega}_i}{\eta}$ for all $i \in [m]$. Such $(\mathbf{h}, \mathbf{z})$ satisfies the properties (5) with a constant probability. The proof of this fact is similar to the proof of Lemma 1 and is omitted due of lack of space. The proof is given in the full version of the paper for completeness.

Because of the linear term, the solution given by this structural property is not necessarily a near-optimal solution of $\mathcal{Q}(\mathbf{x})$ anymore. However, the existence of such solution allows us to bound $\mathcal{Q}(\mathbf{x})$ as follows.

Lemma 4. *For $\mathbf{x} \in \mathcal{X}$ and $\mathbf{y}_0 \geq \mathbf{0}$ such that $\mathbf{Ax} + \mathbf{By}_0 \geq \mathbf{0}$ we have,*

$$
\frac{1}{2 \eta \beta} \mathcal{Q}^{\mathsf{LP}}(\beta \mathbf{x}, \beta \mathbf{y}_0) \leq \mathcal{Q}(\mathbf{x}) \leq \mathcal{Q}^{\mathsf{LP}}(\mathbf{x}, \mathbf{y}_0) + \mathbf{d}^T \mathbf{y}_0.
$$

Our Linear Restriction. Before proving Lemma 4, let us discuss how we derive our linear restriction of AR. In particular, consider the following problem where $\mathcal{Q}(\mathbf{x})$ is replaced by $\mathcal{Q}^{\mathsf{LP}}(\mathbf{x}, \mathbf{y}_0)$ in the expression of AR:

$$z_{\mathsf{LP-AR}} = \min_{\mathbf{x} \in \mathcal{X}, \mathbf{y}_0 \geq 0} \left\{ \mathbf{c}^T \mathbf{x} + \mathbf{d}^T \mathbf{y}_0 + \mathcal{Q}^{\mathsf{LP}}(\mathbf{x}, \mathbf{y}_0) \mid \mathbf{Ax} + \mathbf{By}_0 \geq 0 \right\}. \quad (6)$$

Note for given $\mathbf{x}, \mathbf{y}_0$, $\mathcal{Q}^{\mathsf{LP}}(\mathbf{x}, \mathbf{y}_0)$ is a maximization LP. Taking its dual and substituting in (6), we get the following LP:

$$z_{\mathsf{LP-AR}} = \min_{\mathbf{x}, \mathbf{y}_0, \mathbf{y}, \boldsymbol{\alpha}} \quad \mathbf{c}^T \mathbf{x} + \mathbf{d}^T \mathbf{y}_0 + \mathbf{d}^T \mathbf{y} + \mathbf{r}^T \boldsymbol{\alpha}$$

$$\text{s.t.} \quad \theta_i \mathbf{a}_i^T \mathbf{x} + \theta_i \mathbf{b}_i^T \mathbf{y}_0 + \theta_i \mathbf{b}_i^T \mathbf{y} + \gamma_i \mathbf{R}_i^T \boldsymbol{\alpha} \geq \theta_i \gamma_i \quad \forall i, \quad \text{(LP-AR)}$$

$$\mathbf{Ax} + \mathbf{By}_0 \geq 0,$$

$$\mathbf{x} \in \mathcal{X}, \mathbf{y}_0, \mathbf{y} \geq 0, \boldsymbol{\alpha} \geq 0.$$

We claim that LP-AR is a restriction of AR and gives an $O(\frac{\log L}{\log \log L} \frac{\log n}{\log \log n})$-approximation for . We first give the proof of Lemma 4.

Proof of Lemma 4. First, let $(\mathbf{h}^*, \mathbf{z}^*)$ be an optimal solution of $\mathcal{Q}(\mathbf{x})$. Define $\boldsymbol{\omega}^*$ such that $\omega_i^* = \frac{z_i^*}{\theta_i} \cdot \frac{h_i^*}{\gamma_i}$ for all $i \in [m]$. Then $\boldsymbol{\omega}^*$ is feasible for $\mathcal{Q}^{\mathsf{LP}}(\mathbf{x}, \mathbf{y}_0)$ with objective value,

$$\sum_{i=1}^{m} (\theta_i \gamma_i - \theta_i \mathbf{a}_i^T \mathbf{x} - \theta_i \mathbf{b}_i^T \mathbf{y}_0) \omega_i^* = \sum_{i=1}^{m} h_i^* z_i^* - (\mathbf{a}_i^T \mathbf{x} + \mathbf{b}_i^T \mathbf{y}_0) \frac{h_i^*}{\gamma_i} z_i^*$$

$$\geq \sum_{i=1}^{m} h_i^* z_i^* - (\mathbf{a}_i^T \mathbf{x} + \mathbf{b}_i^T \mathbf{y}_0) z_i^*$$

$$= \sum_{i=1}^{m} h_i^* z_i^* - \mathbf{a}_i^T \mathbf{x} z_i^* - (\sum_{i=1}^{m} \mathbf{b}_i z_i^*)^T \mathbf{y}_0$$

$$\geq \mathcal{Q}(\mathbf{x}) - \mathbf{d}^T \mathbf{y}_0,$$

where the first inequality follows from the fact that $\frac{h_i}{\gamma_i} \leq 1$ and $\mathbf{a}_i^T \mathbf{x} + \mathbf{b}_i^T \mathbf{y}_0 \geq 0$ for all $i \in [m]$, and the last inequality follows from the fact that $\sum_{i=1}^{m} \mathbf{b}_i z_i^* \leq \mathbf{d}$. Hence $\mathcal{Q}(\mathbf{x}) - \mathbf{d}^T \mathbf{y}_0 \leq \mathcal{Q}^{\mathsf{LP}}(\mathbf{x}, \mathbf{y}_0)$.

Now, consider $(\mathbf{h}, \mathbf{z}) \in \{0, \frac{\gamma_i}{\beta}\}^m \times \{0, \frac{\theta_i}{\eta}\}^m$ satisfying properties (5). The first two properties imply that $(\mathbf{h}, \mathbf{z})$ is a feasible solution for $\mathcal{Q}(\mathbf{x})$. The objective value of this solution is given by,

$$\sum_{i=1}^{m} h_i z_i - \mathbf{a}_i^T \mathbf{x} z_i \geq \sum_{i=1}^{m} h_i z_i - (\mathbf{a}_i^T \mathbf{x} + \mathbf{b}_i^T \mathbf{y}_0) z_i \geq \frac{1}{2\eta\beta} \mathcal{Q}^{\mathsf{LP}}(\beta \mathbf{x}, \beta \mathbf{y}_0).$$

The first inequality holds because $\mathbf{b}_i^T \mathbf{y}_0 \geq 0$ for all $i \in [m]$, and the second inequality follows from the properties (5). Hence, $\mathcal{Q}(\mathbf{x}) \geq \frac{1}{2\eta\beta} \mathcal{Q}^{\mathsf{LP}}(\beta \mathbf{x}, \beta \mathbf{y}_0)$. □

Now, we are ready to prove Theorem 2.

Proof of Theorem 2. We prove the following:

$$z_{AR} \leq z_{LP-AR} \leq 3\eta\beta z_{AR}.$$

Let $\mathbf{x}_{LP}^*$, $\mathbf{y}_{0,LP}^*$ denote an optimal solution of (6). We have,

$$z_{LP-AR} = \mathbf{c}^T\mathbf{x}_{LP}^* + \mathbf{d}^T\mathbf{y}_{0,LP}^* + \mathcal{Q}^{LP}(\mathbf{x}^*, \mathbf{y}_{0,LP}^*) \geq \mathbf{c}^T\mathbf{x}_{LP}^* + \mathcal{Q}(\mathbf{x}_{LP}^*) \geq z_{AR},$$

where the first inequality follows from Lemma 4 and the last inequality follows from the feasibility of $\mathbf{x}_{LP}^*$ in AR.

To prove the upper bound for z_{LP-AR}, let $\mathbf{x}^*$ denote an optimal first-stage solution of AR and let $\mathbf{y}_0^* \in \arg\min_{\mathbf{y}\geq 0}\{\mathbf{d}^T\mathbf{y} \mid \mathbf{A}\mathbf{x}^* + \mathbf{B}\mathbf{y} \geq 0\}$. Since $0 \in \mathcal{U}$ is a feasible second-stage scenario,

$$\mathbf{c}^T\mathbf{x}^* + \mathbf{d}^T\mathbf{y}_0^* \leq z_{AR}. \tag{7}$$

Now, we have

$$z_{AR} \geq \mathcal{Q}(\mathbf{x}^*)$$
$$\geq \frac{1}{2\eta\beta}\mathcal{Q}^{LP}(\beta\mathbf{x}^*, \beta\mathbf{y}_0^*)$$
$$= \frac{1}{2\eta\beta}(\beta\mathbf{c}^T\mathbf{x}^* + \beta\mathbf{d}^T\mathbf{y}_0^* + \mathcal{Q}^{LP}(\beta\mathbf{x}^*, \beta\mathbf{y}_0^*)) - \frac{1}{2\eta}(\mathbf{c}^T\mathbf{x}^* + \mathbf{d}^T\mathbf{y}_0^*)$$
$$\geq \frac{1}{2\eta\beta}(\beta\mathbf{c}^T\mathbf{x}^* + \beta\mathbf{d}^T\mathbf{y}_0^* + \mathcal{Q}^{LP}(\beta\mathbf{x}^*, \beta\mathbf{y}_0^*)) - \frac{1}{2}z_{AR}$$
$$\geq \frac{1}{2\eta\beta}z_{LP-AR} - \frac{1}{2}z_{AR},$$

where the second inequality follows from Lemma 4, the third inequality follows from (7) and the fact that $\eta \geq 1$. For the last inequality, note that $\beta\mathbf{x}^* \in \mathcal{X}$, $\beta\mathbf{y}_0^* \geq 0$ and $\beta\mathbf{A}\mathbf{x}^* + \beta\mathbf{B}\mathbf{y}_0^* \geq 0$. Therefore, $\beta\mathbf{x}^*$, $\beta\mathbf{y}_0^*$ is a feasible solution for (6). This implies that $z_{LP-AR} \leq 3\eta\beta z_{AR}$. □

4   Numerical Experiments

In this section, we compare our approximation to finding the optimal affine policy. Affine policies are widely used approximation policy for AR. Ben-Tal et al. [2] show that the optimal affine policy can be found in polynomial time by solving a linear program with polynomially many constraints and variables. We show that our algorithm is significantly faster and provides good approximate solutions. The results of the experiment are given in Table 1.

Experimental Setup. We consider the same instances as in Ben-Tal et al. [3], namely, we consider instances of AR where $n = m$, $\mathbf{c} = \mathbf{d} = \mathbf{e}$ and $\mathbf{A} = \mathbf{B} = \mathbf{I}_m + \mathbf{G}$, where $\mathbf{I}_m$ is the identity matrix and $\mathbf{G}$ is a random normalized Gaussian matrix. We consider the case where $\mathcal{X} = \mathbb{R}_+^m$ and $\mathcal{U}$ is an intersection of L budget of uncertainty sets of the form $\mathcal{U} = \left\{ \mathbf{h} \in [0,1]^m \mid \boldsymbol{\omega}_l^T \mathbf{h} \le 1 \; \forall l \in [L] \right\}$, where the weight vectors $\boldsymbol{\omega}_l$ are normalized Gaussian vectors, i.e., $\omega_{l,i} = \dfrac{|G_{l,i}|}{\sqrt{\sum_i (G_{l,i})^2}}$ for $\{G_{l,i}\}$ i.i.d. standard Gaussian variables.

We compare the running time of our algorithm in seconds denoted by $T_{\mathsf{LP-AR}}$ with the running time needed to compute the optimal affine policy denoted by T_{aff}, for different values of $n = m$ and L. We also compare the ratio between the optimal cost of LP-AR denoted by $z_{\mathsf{LP-AR}}$ and the cost of the optimal affine policy denoted by z_{aff}. The results are given in Table 1 and were obtained using Gurobi v9.1.2 on a dual-core laptop with 8 Go of RAM and 1.8 GHz processor.

Results. Table 1 shows that solving LP-AR significantly faster than finding the optimal affine policy. For example, when $n = m = 100$ computing the optimal affine policy is more than 10000 times slower than LP-AR for all considered values of L. Furthermore, the cost of LP-AR stays within approximately 30% of the cost of the optimal affine policy. We also observe numerically that this gap gets smaller when we increase the dimension of our problem and therefore our algorithm gets close to the optimal affine policy for large instances, which are usually the computationally challenging instances for the optimal affine policy.

Table 1. Comparison of the optimal value and the running time in seconds between our algorithm and the optimal affine policy, for different values of $n = m$ and L.

n	T_{aff}	$T_{\mathsf{LP-AR}}$	$\frac{z_{\mathsf{LP-AR}}}{z_{\mathsf{aff}}}$	n	T_{aff}	$T_{\mathsf{LP-AR}}$	$\frac{z_{\mathsf{LP-AR}}}{z_{\mathsf{aff}}}$	n	T_{aff}	$T_{\mathsf{LP-AR}}$	$\frac{z_{\mathsf{LP-AR}}}{z_{\mathsf{aff}}}$
20	0.57	0.07	1.28	20	1.15	0.05	1.32	20	1.58	0.07	1.38
30	3.70	0.08	1.25	30	3.03	0.03	1.30	30	3.62	0.07	1.35
40	12.5	0.06	1.25	40	8.58	0.03	1.31	40	11.5	0.05	1.32
50	33.2	0.10	1.23	50	28.2	0.02	1.28	50	29.6	0.08	1.33
60	76.0	0.03	1.23	60	78.7	0.04	1.27	60	78.7	0.05	1.30
70	222	0.03	1.22	70	193	0.10	1.27	70	175	0.09	1.27
80	430	0.04	1.21	80	508	0.10	1.26	80	386	0.10	1.28
90	768	0.06	1.21	90	1116	0.06	1.25	90	657	0.13	1.27
100	1790	0.17	1.22	100	1714	0.04	1.22	100	1354	0.12	1.27

(a) L=20   (b) L=50   (c) L=100

5   Conclusion

In this paper, we consider the class of packing disjoint bilinear programs PDB and present an LP-rounding based randomized approximation algorithm for this problem. In particular, we show the existence of a near-optimal near-integral solution for PDB. We give an LP relaxation from which we obtain such solution using a randomized rounding of an optimal solution. We apply our ideas to the two-stage adjustable problem AR whose separation problem is a variant of

PDB. While a direct application of the approximation algorithm for PDB does not work for AR, we derive an LP restriction of AR, based on similar ideas, that gives a polylogarithmic approximation of AR. We compare our algorithm with the widely used affine policies and show that it is significantly faster and provides near-optimal solutions.

References

1. Bandi, C., Bertsimas, D.: Tractable stochastic analysis in high dimensions via robust optimization. Math. Program. **134**(1), 23–70 (2012). https://doi.org/10.1007/s10107-012-0567-2
2. Ben-Tal, A., Goryashko, A., Guslitzer, E., Nemirovski, A.: Adjustable robust solutions of uncertain linear programs. Math. Program. **99**, 351–376 (2004). https://doi.org/10.1007/s10107-003-0454-y
3. Ben-Tal, A., El Housni, O., Goyal, V.: A tractable approach for designing piecewise affine policies in two-stage adjustable robust optimization. Math. Program. **182**, 57–102 (2019). https://doi.org/10.1007/s10107-019-01385-0
4. Bertsimas, D., Goyal, V.: On the power and limitations of affine policies in two-stage adaptive optimization. Math. Program. **134**, 491–531 (2012). https://doi.org/10.1007/s10107-011-0444-4
5. Bertsimas, D., Bidkhori, H.: On the performance of affine policies for two-stage adaptive optimization: a geometric perspective. Math. Program. **153**(2), 577–594 (2014). https://doi.org/10.1007/s10107-014-0818-5
6. Bertsimas, D., Brown, D.B., Caramanis, C.: Theory and applications of robust optimization. SIAM Rev. **53**(3), 464–501 (2011)
7. Bertsimas, D., Ruiter, F.: Duality in two-stage adaptive linear optimization: faster computation and stronger bounds. INFORMS J. Comput. **28**, 500–511 (2016)
8. Bertsimas, D., Sim, M.: The price of robustness. Oper. Res. **52**, 35–53 (2004)
9. Chen, X., Deng, X., Teng, S.H.: Settling the complexity of computing two-player nash equilibria. J. ACM **56**(3), 1–57 (2009)
10. Chernoff, H.: A measure of asymptotic efficiency for tests of a hypothesis based on the sum of observations. Ann. Math. Stat. **23**(4), 493–507 (1952)
11. Dhamdhere, K., Goyal, V., Ravi, R., Singh, M.: How to pay, come what may: approximation algorithms for demand-robust covering problems. In: 46th Annual IEEE Symposium on Foundations of Computer Science (FOCS'05), pp. 367–376 (2005)
12. El Housni, O., Goyal, V.: Beyond worst-case: a probabilistic analysis of affine policies in dynamic optimization. In: Proceedings of the 31st International Conference on Neural Information Processing Systems, pp. 4759–4767 (2017)
13. El Housni, O., Goyal, V.: On the optimality of affine policies for budgeted uncertainty sets. Math. Oper. Res. **46**(2), 674–711 (2021)
14. EL Housni, O., Goyal, V., Hanguir, O., Stein, C.: Matching drivers to riders: a two-stage robust approach. In: Approximation, Randomization, and Combinatorial Optimization. Algorithms and Techniques (APPROX/RANDOM 2021), vol. 207, pp. 12:1–12:22 (2021)
15. El Housni, O., Goyal, V., Shmoys, D.: On the power of static assignment policies for robust facility location problems. In: Singh, M., Williamson, D.P. (eds.) IPCO 2021. LNCS, vol. 12707, pp. 252–267. Springer, Cham (2021). https://doi.org/10.1007/978-3-030-73879-2_18

16. Feige, U., Jain, K., Mahdian, M., Mirrokni, V.: Robust combinatorial optimization with exponential scenarios. In: Fischetti, M., Williamson, D.P. (eds.) IPCO 2007. LNCS, vol. 4513, pp. 439–453. Springer, Heidelberg (2007). https://doi.org/10.1007/978-3-540-72792-7_33

17. Firouzbakht, K., Noubir, G., Salehi, M.: On the power of static assignment policies for robust facility location problems. IEEE Trans. Commun. **64**(1), 429–440 (2016). https://doi.org/10.1007/978-3-030-73879-2_18

18. Geoffrion, A.: Generalized benders decomposition. J. Optim. Theory Appl. **10**, 237–260 (1972)

19. Gounaris, C., Repoussis, P., Tarantilis, C., Wiesemann, W., Floudas, C.: An adaptive memory programming framework for the robust capacitated vehicle routing problem. Transp. Sci. **50**, 1139–1393 (2014)

20. Gupta, A., Nagarajan, V., Ravi, R.: Robust and maxmin optimization under matroid and knapsack uncertainty sets. ACM Trans. Algorithms **12**(1), 1–21 (2015)

21. Gupta, A., Nagarajan, V., Ravi, R.: Thresholded covering algorithms for robust and max-min optimization. Math. Program. **146**, 583–615 (2014). https://doi.org/10.1007/s10107-013-0705-5

22. Gupte, A., Ahmed, S., Dey, S.S., Cheon, M.S.: Relaxations and discretizations for the pooling problem. J. Glob. Optim. **67**(3), 631–669 (2016). https://doi.org/10.1007/s10898-016-0434-4

23. Harjunkoski, I., Westerlund, T., Pörn, R., Skrifvars, H.: Different transformations for solving non-convex trim-loss problems by minlp. Eur. J. Oper. Res. **105**, 594–603 (1998)

24. Konno, H.: A cutting plane algorithm for solving bilinear programs. Math. Program. **11**, 14–27 (1976). https://doi.org/10.1007/BF01580367

25. Lim, C., Smith, J.C.: Algorithms for discrete and continuous multicommodity flow network interdiction problems. IIE Trans. **39**(1), 15–26 (2007)

26. Mangasarian, O., Stone, H.: Two-person nonzero-sum games and quadratic programming. J. Math. Anal. Appl. **9**(3), 348–355 (1964)

27. Rebennack, S., Nahapetyan, A., Pardalos, P.: Bilinear modeling solution approach for fixed charge network flow problems. Optim. Lett. **3**, 347–355 (2009). https://doi.org/10.1007/s11590-009-0114-0

28. Soland, R.: Optimal facility location with concave costs. Oper. Res. **22**, 373–382 (1974)

29. Thieu, T.V.: A note on the solution of bilinear programming problems by reduction to concave minimization. Math. Program. **41**(1–3), 249–260 (1988). https://doi.org/10.1007/BF01580766

30. Vaish, H., Shetty, C.M.: The bilinear programming problem. Nav. Res. Logistics Q. **23**(2), 303–309 (1976)

31. Xu, G., Burer, S.: A copositive approach for two-stage adjustable robust optimization with uncertain right-hand sides. Comput. Optim. Appl. **70**(1), 33–59 (2017). https://doi.org/10.1007/s10589-017-9974-x

32. Zhen, J., den Hertog, D., Sim, M.: Adjustable robust optimization via fourier-motzkin elimination. Oper. Res. **66**(4), 1086–1100 (2018)

33. Zhen, J., Marandi, A., den Hertog, D., Vandenberghe, L.: Disjoint bilinear programming: a two-stage robust optimization perspective. Optimization Online (2018). www.optimization-online.org/DB_HTML/2018/06/6685.html

34. Ćustić, A., Sokol, V., Punnen, A.P., Bhattacharya, B.: The bilinear assignment problem: complexity and polynomially solvable special cases. Math. Program. **166**(1), 185–205 (2017). https://doi.org/10.1007/s10107-017-1111-1

A Constant-Factor Approximation for Generalized Malleable Scheduling Under $M^\natural$-Concave Processing Speeds

Dimitris Fotakis[1] , Jannik Matuschke[2]([✉]) , and Orestis Papadigenopoulos[3]

[1] National Technical University of Athens, Athens, Greece
fotakis@cs.ntua.gr
[2] KU Leuven, Leuven, Belgium
jannik.matuschke@kuleuven.be
[3] The University of Texas at Austin, Austin, USA
papadig@cs.utexas.edu

Abstract. In generalized malleable scheduling, jobs can be allocated and processed simultaneously on multiple machines so as to reduce the overall makespan of the schedule. The required processing time for each job is determined by the joint processing speed of the allocated machines. We study the case that processing speeds are job-dependent $M^\natural$-concave functions and provide a constant-factor approximation for this setting, significantly expanding the realm of functions for which such an approximation is possible. Further, we explore the connection between malleable scheduling and the problem of fairly allocating items to a set of agents with distinct utility functions, devising a black-box reduction that allows to obtain resource-augmented approximation algorithms for the latter.

1 Introduction

Parallel execution of a job on multiple machines is often used to optimize the overall makespan in time-critical task scheduling systems. Practical applications are numerous and diverse, varying from task scheduling in production and logistics, such as quay crane allocation in naval logistics [4,15] and cleaning activities on trains [3], to optimizing the performance of computationally demanding tasks, such as web search index update [28] and training neural networks [12] (see also [9,10] for further references and examples).

The model of *malleable* (a.k.a. *moldable*) jobs, introduced by Du and Leung [8], captures the algorithmic aspects of scheduling jobs that can be executed simultaneously on multiple machines. A malleable job can be assigned to an arbitrary subset of machines to be processed *non-preemptively* and in *unison*, i.e., with the same starting and completion time on each of the allocated machines. Importantly, the scheduler decides on the degree of parallelization for each job by choosing the set of machines allocated to each job (in contrast to non-malleable parallel machine models, where a single machine is allocated to each task).

Despite the significant interest in the model, most of the work on scheduling malleable jobs considers the case of identical machines, where the processing

© Springer Nature Switzerland AG 2022
K. Aardal and L. Sanitá (Eds.): IPCO 2022, LNCS 13265, pp. 237–250, 2022.
https://doi.org/10.1007/978-3-031-06901-7_18

time of a job only depends on the number of allocated machines. A common assumption is that a job's processing time is non-increasing in the number k of allocated machines, while a job's work (i.e., k times the job's processing time on k machines) is non-decreasing in k. This is usually referred to as the *monotone work assumption* and accounts for communication and coordination overhead due to parallelization. The approximability of makespan minimization in the setting of malleable job scheduling on identical machines is very well understood. Constant-factor approximation algorithms are known since the work of Turek et al. [27]. Following a line of successive improvements in the approximation factor [17, 21,22], two recent results by Jansen and Thöle [18] and Jansen and Land [16] implied a polynomial-time approximation scheme for malleable scheduling on identical machines.

On the other hand, scheduling malleable jobs on non-identical machines has received much less attention. As a natural first step, building on previous work by Correa et al. [7] on the closely related *splittable job* model, Fotakis et al. [9] introduced the setting of *speed-implementable processing-time* functions, where each machine i has an unrelated "speed" s_{ij} for each job j and a job's processing time is a non-decreasing function of the total allocated speed fulfilling a natural generalization of the non-decreasing work assumption. They devised an LP-based 3.16-approximation for this setting.

However, as recently observed in [10], the aforementioned models, in which the processing power of a heterogeneous set of machines is expressed by a single scalar, cannot capture the (possibly complicated) combinatorial interaction effects arising among different machines processing the same job. Practical settings where such complicated interdependencies among machines may arise include modern heterogeneous parallel computing systems, typically consisting of CPUs, GPUs, and I/O nodes [5], and highly distributed processing systems, where massive parallelization is subject to constraints imposed by the underlying communication network [1]; see [10] for further references and examples. Having such practical settings in mind, Fotakis et al. [10] introduced a *generalized malleable scheduling* model, where the processing time $f_j(S) = 1/g_j(S)$ of a job j depends on a job-specific processing speed function $g_j(S)$ of the set of machines S allocated to j. In addition to motivating and introducing the model, they derived an LP-based 5-approximation for scaled matroid rank processing speeds, and a $O(\log \min\{n, m\})$-approximation algorithm for submodular processing speeds, where n is the number of jobs and m is the number of machines.

Fotakis et al. [10] left open whether there are processing speed functions more general than scaled matroid rank functions for which generalized malleable scheduling can be approximated within a constant factor. In this work, employing notions and techniques from the field of Discrete Convexity [23], we present a constant-factor approximation algorithm for job-dependent $M^\natural$-*concave* (a.k.a. *gross substitute*) processing speed functions, thus significantly expanding the realm of functions for which such an approximation is possible. We further point out a connection between malleable scheduling and the so-called max-min fair allocation problem, devising a black-box reduction that allows to obtain resource-augmented approximation algorithms for the latter.

1.1 Generalized Malleable Scheduling and Main Results

To discuss our contribution in more detail, we need to formally introduce the *Generalized Malleable Scheduling* problem. We are given a set of *jobs* J to be assigned to a set of *machines* M. Each job $j \in J$ is equipped with a *processing time function* $f_j : 2^M \to \mathbb{R}_{\geq 0}$ that specifies the time $f_j(S)$ needed for the completion of j, when assigned to a subset of machines $S \subseteq M$. We assume that functions f_j are accessed through a *value oracle* that, given $S \subseteq M$, returns the value of $f_j(S)$. A *schedule* consists of two parts: (i) an *assignment* $\mathbf{S} = (S_j)_{j \in J}$ of each job $j \in J$ to a non-empty set of machines $S_j \subseteq M$; and (ii) a *starting time* vector $\mathbf{t} = (t_j)_{j \in J}$, specifying the time t_j at which jobs in S_j start to jointly process job j. A schedule is *feasible*, if $S_j \cap S_{j'} = \emptyset$ for all $j, j' \in J$ with $t_j < t_{j'} < t_j + f_j(S_j)$, i.e., while a machine is involved in processing a job j, it cannot start processing any other job j'. The objective is to find a feasible schedule of minimum *makespan* $C(\mathbf{S}, \mathbf{t}) := \max_{j \in J} \{t_j + f_j(S_j)\}$.

An interesting relaxation of the above SCHEDULING problem is the ASSIGNMENT problem, asking for an assignment $\mathbf{S}$ that minimizes the *load* $L(\mathbf{S}) := \max_{i \in M} \sum_{j \in J : i \in S_j} f_j(S_j)$. Clearly, the load of an assignment is a lower bound on the makespan of any feasible schedule using that same assignment.

The *processing speed* of a set of machines S for a job j is $g_j(S) := 1/f_j(S)$. Under the assumption that for each job $j \in J$, the processing speed function g_j is submodular, [10, Theorem 1] shows that any assignment of maximum machine load C can be transformed in polynomial time into a so-called *well-structured* schedule, where each machine shares at most one job with another machine, of makespan at most $\frac{2e}{e-1}C$. Thus, suffering a small constant-factor loss in the approximation ratio, we can approximate the optimal makespan in Generalized Malleable Scheduling by approximating the ASSIGNMENT problem.

Our main contribution is an $O(1)$-approximation for the ASSIGNMENT problem with $M^\natural$-concave processing speeds (see Sect. 1.2 for definitions of submodularity and $M^\natural$-concavity). Because all $M^\natural$-concave functions are submodular, our results, together with the aforementioned transformation [10], imply the following theorem.

Theorem 1. *There is a polynomial time constant-factor approximation algorithm for* ASSIGNMENT *and* SCHEDULING *with* $M^\natural$*-concave processing speeds.*

1.2 Submodularity, $M^\natural$-Concavity, and Matroids

Let $g : 2^M \to \mathbb{R}$ be a function. The *demand set* for g and a vector $p \in \mathbb{R}^M$ is defined by $\mathcal{D}(g, p) := \operatorname{argmax}_{S \subseteq M} g(S) - \sum_{i \in S} p_i$. We say that g is

- *submodular* if $g(S \cup T) + g(S \cap T) \leq g(S) + g(T)$ for all $S, T \subseteq M$.
- $M^\natural$-*concave* if for any $p', p'' \in \mathbb{R}^M$ with $p' \leq p''$ and any $S' \in \mathcal{D}(g, p')$ there is an $S'' \in \mathcal{D}(g, p'')$ with $S' \cap \{i \in M : p'_i = p''_i\} \subseteq S''$.

Submodularity of g is equivalent to the following diminishing returns property: $g(S \cup \{i\}) - g(S) \geq g(T \cup \{i\}) - g(T)$ for all $S \subseteq T \subseteq M$ and $i \in M \backslash T$.

A *matroid* on the ground set M is a non-empty set family $\mathcal{F} \subseteq 2^M$ such that (i) $T \in \mathcal{F}$ implies $S \in \mathcal{F}$ for all $S \subseteq T$ and (ii) for every $S, T \in \mathcal{F}$ with $|S| < |T|$ there is an $i \in T \backslash S$ such that $S \cup \{i\} \in \mathcal{F}$. It is well known that the the rank function $r(S) := \max_{S \subseteq T : S \in \mathcal{F}} |S|$ for $S \subseteq M$ of any matroid $\mathcal{F}$ is submodular.

$M^\natural$-concavity, also known as the *gross substitutability*, defines an important subclass of submodular functions. Gross-substitute functions have been widely studied, receiving particular attention in economics and operations research, due to their applications in diverse fields such as labor and housing markets [14, 19], inventory management [6], or structural analysis for engineering systems [24]; see the survey by Paes Leme [25] and the textbook by Murota [23] for an overview of the rich collection of algorithmic results for these class of functions.

In the following, we provide an example for a subclass of $M^\natural$-concave functions, known as *matroid-based valuations* [25] or *Rado valuations* [13], which naturally arise in the context of malleable scheduling.

Example: Rado Valuations. Assume that each $j \in J$ is equipped with a set of *processing slots* V_j together with a matroid $\mathcal{F}_j$ on V_j. Each slot represents a role that a machine can take to speed up the completion of job j. For each machine $i \in M$ and each slot $v \in V_j$, a weight $w_{iv} \geq 0$ specifies how much i would contribute to the processing of job j when assigned to v. When j is processed by a set of machines S, each machine in S can fill at most one of j's processing slots and each slot can be taken by at most one machine, i.e., the machines in S are matched to the slots in V_j. The matroid $\mathcal{F}_j$ on V_j specifies which of its slots can be used simultaneously (e.g., the slots could be partitioned into groups and from each group only a limited number of slots may be used). Thus, a feasible matching of the machines in S to the slots in V_j is a function $\pi : S \to V_j \cup \{\emptyset\}$ (where $\emptyset$ denotes the machine not being used) such that $\pi(i) = \pi(i') \neq \emptyset$ implies $i = i'$ and such that $\{\pi(i) : i \in S, \pi(i) \neq \emptyset\} \in \mathcal{F}_j$. Denoting the set of feasible matchings from S to V_j by $\Pi_{S,j}$, the processing speed of a set of machines S for job j is given by $g_j(S) = \max_{\pi \in \Pi_{S,j}} \sum_{i \in S, \pi(i) \neq \emptyset} w_{i\pi(i)}$.

1.3   Organization of the Paper

In the following Sects. 2 to 4 we discuss our main result in detail. Our constant-factor approximation is based on a linear programming relaxation of the ASSIGN-MENT problem that we discuss in Sect. 2, where we show that a pair of optimal primal and dual solution to the LP generate a weighted matroid that can be used to approximate the processing speed functions. This is exploited in the remaining three steps of the algorithm, that, based on the LP solution, partition the set of jobs into three groups, each of which is assigned separately. We first give an overview of these steps in Sect. 3 and then discuss them in detail in Sect. 4. In Sect. 5, we finally turn our attention to the MAX-MIN FAIR ALLOCATION (MMFA) problem and present a blackbox reduction that turns approximation algorithms for generalized malleable scheduling into resource-augmented approximation algorithms for MMFA. All missing proofs can be found in the full version of the paper [11].

2   The Configuration LP

In the following we introduce a *configuration LP*, which features a fractional variable $x(S,j) \geq 0$ for each non-empty $S \subseteq M$ and each $j \in J$ (and an auxiliary slack variable $s_i \geq 0$ for each machine $i \in M$). We will show that this LP, for a given target bound C, is feasible if there is an assignment of load at most C.

$$\textbf{maximize:} \qquad \sum_{i \in M} s_i \qquad\qquad \text{(GS)}$$

$$\textbf{s.t.:} \quad \sum_{S \subseteq M : S \neq \emptyset} \left(2 - \tfrac{1}{C g_j(S)}\right) x(S,j) \geq 1 \quad \forall j \in J$$

$$\sum_{j \in J} \sum_{S \subseteq M : i \in S} \tfrac{1}{g_j(S)} x(S,j) + s_i \leq C \quad \forall i \in M$$

$$x, s \geq 0$$

The dual of (GS) is the following LP.

$$\textbf{minimize:} \qquad -\sum_{j \in J} \lambda_j + C \sum_{i \in M} \mu_i \qquad\qquad \text{(GS-D)}$$

$$\textbf{s.t.:} \quad \left(2g_j(S) - \tfrac{1}{C}\right)\lambda_j - \sum_{i \in S} \mu_i \leq 0 \quad \forall j \in J, S \subseteq M, S \neq \emptyset$$

$$\mu_i \geq 1 \qquad\qquad \forall i \in M$$

$$\lambda_j \geq 0 \qquad\qquad \forall j \in J$$

The following lemma reveals that (GS) is indeed a relaxation of ASSIGNMENT.

Lemma 1. *If there exists an assignment* $\mathbf{S}$ *with* $L(\mathbf{S}) \leq C$, *then (GS) is feasible. In addition, if* λ, μ *is an optimal solution to (GS-D), then* $\lambda_j > 0$ *for all* $j \in J$.

Henceforth, let (x,s) and (λ, μ) be primal-dual optimal solutions to (GS) and (GS-D), respectively. For $j \in J$, define

$$g_j^*(S) := 2g_j(S) - \sum_{i \in S} \frac{\mu_i}{\lambda_j} \quad \text{and} \quad \mathcal{D}_j := \operatorname{argmax}_{S \subseteq M} g_j^*(S).$$

Note that the first constraint of (GS-D) is equivalent to $g_j^*(S) \leq 1/C$ for all $S \subseteq M$ and $j \in J$. Thus complementary slackness implies that sets in the support of x are maximizers of g_j^*, as formalized in the next lemma.

Lemma 2. *If* $x(S,j) > 0$ *for* $S \subseteq M$ *and* $j \in J$, *then* $g_j(S) \geq \frac{1}{2C}$ *and* $S \in \mathcal{D}_j$.

We now observe that the sets $\mathcal{D}_j$ defined in the preceding lemma induce a matroid. This is a consequence of $M^\natural$-concavity, which g_j^* inherits from g_j.

Lemma 3. *The system $\mathcal{F}_j := \{S \subseteq M : S \subseteq T \text{ for some } T \in \mathcal{D}_j\}$ is a matroid for any $j \in J$*

Our final lemma in the analysis of (GS) shows that the values μ_i/λ_j provide an approximation for g_j on the matroid $\mathcal{F}_j$.

Lemma 4. *Let $S \in \mathcal{F}_j$ and $j \in J$. Then $g_j(S) \geq \frac{1}{2}\sum_{i \in S} \frac{\mu_i}{\lambda_j}$.*

Proof. Because $S \in \mathcal{F}_j$ there is $T \in \mathcal{D}_j$ with $S \subseteq T$. Let $i_1, \ldots, i_\ell$ be an arbitrary ordering of the elements in S and let $S_k = \{i_1, \ldots, i_k\}$ for $k \in [\ell]$. We obtain

$$2g_j(S) = \sum_{k=1}^{\ell} 2g_j(S_k) - 2g_j(S_k\backslash\{i_k\}) \geq \sum_{k=1}^{\ell} 2g_j(T) - 2g_j(T\backslash\{i_k\}) \geq \sum_{k=1}^{\ell} \frac{\mu_{i_k}}{\lambda_j},$$

where the first inequality follows from submodularity and the second follows from the fact that $T \in \mathcal{D}_j$ implies $2g_j(T) - \sum_{i \in T} \frac{\mu_i}{\lambda_j} \geq 2g_j(T \setminus \{i'\}) - \sum_{i \in T\setminus\{i'\}} \frac{\mu_i}{\lambda_j}$ for all $i' \in T$, as T is a maximizer of g_j^*. □

3 Overview of the Algorithm

Given a target makespan C, our algorithm starts from computing an optimal primal-dual solution (x, s) and (λ, μ) to (GS) and (GS-D). Note that such solutions can be computed via dual separation, as the separation problem for (GS-D) can be solved via a greedy algorithm for $M^\natural$-concave functions.

If (GS) turns out to be infeasible, we conclude that there is no assignment of maximum load C by Lemma 1. Otherwise, we continue by partitioning the job set into three types. For each of these types, we will show independently how to turn the corresponding part of the solution to (GS) into an assignment whose load can be bounded by a constant factor times C. Binary search for the smallest C for which (GS) is feasible then yields a constant-factor approximation.

The first type are the jobs that are assigned by our algorithm to a single-machine. For $j \in J$ define $M_j^+ := \{i \in M : g_j(\{i\}) \geq \frac{1}{16C}\}$. Let

$$J_1 := \Big\{j \in J : \sum_{i \in M_j^+} \sum_{S \subseteq M : i \in S} \frac{g_j(\{i\})}{g_j(S)} x(S, j) \geq \frac{1}{16}\Big\}.$$

In Sect. 4.1, we show how to obtain an assignment for the jobs in J_1 using the LP rounding algorithm of Lenstra et al. [20] for non-malleable unrelated machine scheduling. This establishes the following result.

Lemma 5. *Step 1 of the algorithm computes in polynomial time an assignment for the jobs in J_1 with a maximum machine load of at most $32C$.*

The second type of jobs are those that are assigned predominantly to groups of machines with a relatively low total weight. Formally, define

$$\mathcal{S}_j^2 := \Big\{S \subseteq M : \sum_{i \in S} \mu_i/\lambda_j \leq \frac{4}{C}\Big\} \text{ and } J_2 := \Big\{j \in J \setminus J_1 : \sum_{S \in \mathcal{S}_j^2} x(S, j) \geq \frac{1}{8}\Big\}.$$

In Sect. 4.2, we show how to assign the jobs in J_2 via the so-called WELFARE problem, which can be solved optimally in polynomial time for gross-substitute functions. This establishes the following result.

Lemma 6. *Step 2 of the algorithm computes in polynomial time an assignment for the jobs in J_2 with a maximum machine load of at most $40C$.*

Finally, we consider the jobs in $J_3 := J \setminus (J_1 \cup J_2)$. Assigning these jobs is more involved than in the preceding cases.

In Sect. 4.3, we modify the fractional LP solution for the jobs in J_3 in such a way that the sum of fractional assignments for each machine is bounded by a constant. This transformation uses the fact that we already filtered out jobs using predominantly fast machines or slow assignments in Steps 1 and 2.

In Sect. 4.4, we partition the set of machines for each job according to their weights μ_i/λ_j into classes whose weight differs by a factor of 2. We then use the approximation of the functions g_j via matroids described in Lemma 4 to reformulate the problem of assigning a sufficient number of machines from each weight class to each job in J_3, while assigning to each machine only a constant number of jobs, as an intersection of two polymatroids. The transformed LP solution derived before guarantees the existence of a feasible solution to this polymatroid intersection, from which we can derive an assignment of the jobs in J_3, establishing the following result.

Lemma 7. *Step 3 of the algorithm computes in polynomial time an assignment for the jobs in J_3 with a maximum machine load of at most $121C$.*

By concatenating the assignments for the individual job types we obtain the following result, which implies Theorem 1.

Theorem 2. *There exists a polynomial-time 193-approximation algorithm for* ASSIGNMENT *when processing speeds are $M^\natural$-concave.*

4   Description and Analysis of the Intermediate Steps

4.1   Step 1: Single-machine Assignments for J_1

Recalling the definitions of $M_j^+ = \left\{ i \in M : g_j(\{i\}) \geq \frac{1}{16C} \right\}$ for $j \in J$ and of $J_1 = \left\{ j \in J : \sum_{i \in M_j^+} \sum_{S \subseteq M : i \in S} \frac{g_j(\{i\})}{g_j(S)} x(S, j) \geq \frac{1}{16} \right\}$, consider the following assignment LP:

$$\sum_{i \in M_j^+} y_{ij} \geq 1 \qquad\qquad \forall j \in J_1 \qquad\qquad \text{(GS-A)}$$

$$\sum_{j \in J_1} \frac{1}{g_j(\{i\})} y_{ij} \leq 16C \qquad\qquad \forall i \in M$$

$$y_{ij} = 0 \qquad \forall j \in J, i \in M \setminus M_j^+$$

$$y \geq 0$$

Note that (GS-A) corresponds to an instance of the classic makespan minimization problem on unrelated machines. It can be shown that the solution x to (GS) induces a feasible solution to (GS-A). By applying the rounding algorithm of Lenstra et al. [20] to an extreme point solution of (GS-A), we get an assignment of the jobs in J_1 in which each machine $i \in M$ receives a load of at most $16C + 1/g_j(\{i\})$ for some $j \in J_1$ with $i \in M_j^+$. Because $g_j(\{i\}) \geq \frac{1}{16C}$ for $j \in J_1$ and $i \in M_j^+$, the load of this assignment is at most $32C$ and Lemma 5 follows.

4.2 Step 2: Assignments with Low Total Speed for J_2

Recalling the definitions of $\mathcal{S}_j^2 = \{S \subseteq M : \sum_{i \in S} \mu_i/\lambda_j \leq 4/C\}$ and of $J_2 = \{j \in J \setminus J_1 : \sum_{S \in \mathcal{S}_j^2} x(S,j) \geq \frac{1}{8}\}$ consider the following LP:

$$\text{maximize:} \qquad \sum_{S \subseteq M} \sum_{j \in J_2} g_j^*(S) z(S,j) \qquad\qquad \text{(GS-W)}$$

$$\text{s.t.:} \qquad \sum_{S \subseteq M} z(S,j) \;\leq\; 1 \qquad \forall j \in J_2$$

$$\sum_{j \in J_2} \sum_{S \subseteq M : i \in S} z(S,j) \;\leq\; 20 \qquad \forall i \in M$$

$$z \;\geq\; 0$$

Again, it can be shown that the solution x to (GS) induces a feasible solution to (GS-W). Moreover, (GS-W) corresponds to an LP relaxation of the so-called WELFARE MAXIMIZATION problem. Because the functions g_j^* are $M^\natural$-concave, (GS-W) is integral, and integer optimal solutions can be found in polynomial time [26]. Thus, let z be such an integer optimal solution. Lemma 8 below guarantees that for each $j \in J_2$ there is a set $S_j \subseteq M$ with $z(S_j, j) = 1$ and $g_j(S_j) \geq 1/2C$. Since each machine participates in the execution of at most 20 jobs, each of processing time $2C$, we obtain an assignment with load at most $40C$ for the jobs in J_2, and Lemma 6 follows.

Lemma 8. *Let z be an integral optimal to (GS-W). Then for each $j \in J_2$ there is a set S_j with $z(S_j, j) = 1$ and $g_j(S_j) \geq \frac{1}{2C}$.*

Proof (sketch). Using the construction of J_2, it can be shown that setting $z'(S,j) := x(S,j)/\sum_{S' \in \mathcal{S}_j^2} x(S',j)$ for $j \in J_2$ and $S \in \mathcal{S}_j^2$ yields a solution of value $\sum_{j \in J_2} \sum_{S \subseteq M} g_j^*(S) z'(S,j) = |J_2|/C$ to (GS-W). Now let z be an integral optimal solution. Note that $\sum_{j \in J_2} \sum_{S \subseteq M} g^*(S) z(S,j) \geq |J_2|/C$ by total dual integrality. Because $g_j^*(S) \leq \frac{1}{C}$ for $S \subseteq M$ and $j \in J$ by the constraints of (GS-D), we conclude that for each $j \in J_2$ there is a set S_j with $z(S_j, j) = 1$ and $g_j^*(S_j) = \frac{1}{C}$. The latter implies $g_j(S_j) = \left(g_j^*(S_j) + \sum_{i \in S_j} \mu_i/\lambda_j\right)/2 \geq \frac{1}{2C}$. □

4.3 Step 3A: Splitting Assignments with High Total Speed

Recall that $J_3 = J \setminus (J_1 \cup J_2)$. For $j \in J_3$ define

$$\mathcal{S}_j^3 := \left\{ S \subseteq M : x(S,j) > 0, \; S \notin \mathcal{S}_j^2, \; \sum_{i \in S \setminus M_j^+} \frac{\mu_i}{\lambda_j} > \sum_{i \in S \cap M_j^+} \frac{\mu_i}{\lambda_j} \right\}.$$

We construct a new fractional assignment $x' : 2^M \times J_3 \to \mathbb{R}_{\geq 0}$ as follows. For each $j \in J_3$ and each $T \in \mathcal{S}_j^3$, find a partition $\mathcal{A}_j(T)$ of $T \setminus M_j^+$ such that

- $\sum_{i \in A} \mu_i/\lambda_j \leq 4/C$ for all $A \in \mathcal{A}_j(T)$.
- $\sum_{i \in A} \mu_i/\lambda_j + \sum_{i \in A'} \mu_i/\lambda_j > 4/C$ for all distinct sets $A, A' \in \mathcal{A}_j(T)$.

Notice that if $i \in T \setminus M_j^+$ for some $T \in \mathcal{S}_j^3$ then $\{i\} \in \mathcal{F}_j$ because $x(T,j) > 0$ and thus $\mu_i/\lambda_j \leq 2g_j(\{i\}) < 1/(8C) \leq 4/C$ by Lemma 4. Therefore, a partition as the one described above exists and can be constructed greedily. For $j \in J_3$ and $S \subseteq M$ let $\mathcal{T}_j(S) := \{T \in \mathcal{S}_j^3 : S \in \mathcal{A}_j(T)\}$ and define

$$\bar{x}(S,j) := \sum_{T \in \mathcal{T}_j(S)} \frac{g_j(S)}{\sum_{A \in \mathcal{A}_j(T)} g_j(A)} x(T,j) \quad \text{and} \quad x'(S,j) := \frac{\bar{x}(S,j)}{\gamma_j},$$

where $\gamma_j := \sum_{S \subseteq M} \bar{x}(S,j)$. We further define $x'_{ij} := \sum_{S \subseteq M : i \in S} x'(S,j)$ for ease of notation.

The modified fractional assignment x' exhibits several properties that will be useful in the construction of an integral assignment for J_3.

Lemma 9. *The assignment x' described above fulfills the following properties:*
1. *$\sum_{S \subseteq M} x'(S,j) = 1$ for all $j \in J_3$,*
2. *$S \cap M_j^+ = \emptyset$ and $S \in \mathcal{F}_j$ for all $S \subseteq M$ and $j \in J_3$ with $x'(S,j) > 0$.*
3. *$\sum_{j \in J_3} x'_{ij} \leq 26$ for all $i \in M$,*
4. *$\sum_{i \in M} \frac{\mu_i}{\lambda_j} x'_{ij} \geq \frac{79}{40C}$ for all $j \in J_3$,*

Proof (sketch). The first two properties of the lemma follow directly from construction of x' and from the fact that $S \in \mathcal{F}_j$ when $x(S,j) > 0$. Intuitively, the construction of x' splits assignments of high total weight into assignments of moderate weight, then scaling the solution by $1/\gamma_j$ so as to compensate for the fact that machines in M_j^+ and assignments in $\mathcal{S}_j^2$ are ignored. The main part of the proof is to show that this scaling factor and thus the blow-up in makespan can be bounded by a small constant. Once this is established, Property 3 follows from the fact that $x'(S,j) > 0$ implies $g_j(S) \leq \frac{1}{2}(\frac{1}{C} + \sum_{i \in S} \frac{\mu_i}{\lambda_j}) \leq \frac{5}{2C}$ by construction and hence the total sum of fractional assignments for every machine is bounded by a constant. Finally, Property 4 can be derived from the fact that $\sum_{i \in S \cap M_j^+} \frac{\mu_i}{\lambda_j} > \frac{1}{2} \sum_{i \in S} \frac{\mu_i}{\lambda_j} \geq \frac{2}{C}$ for all $S \in \mathcal{S}_j^3$. $\square$

4.4   Step 3B: Constructing an Integer Assignment for J_3

For $j \in J_3$ and $k \in \mathbb{Z}$ define

$$M_{jk} := \left\{ i \in M : \frac{1}{2^{k+1}C} < \frac{\mu_i}{\lambda_j} \leq \frac{1}{2^k C} \right\} \quad \text{and} \quad d_{jk} := \left\lfloor \sum_{i \in M_{jk}} x'_{ij} \right\rfloor.$$

Note that there are only polynomially many $k \in \mathbb{Z}$ with $d_{jk} > 0$.

Furthermore, define $r'_j(U) := \sum_{k \in \mathbb{Z}} \min\{r_j(U \cap M_{jk}), d_{jk}\}$ for $j \in J_3$ and $U \subseteq M$, where r_j is the rank function of the matroid $\mathcal{F}_j$. Consider the LP:

$$\textbf{maximize:} \quad \sum_{i \in M} \sum_{j \in J_3} y_{ij} \qquad\qquad\qquad \text{(GS-P)}$$

$$\textbf{s.t.:} \quad \sum_{j \in J_3} y_{ij} \leq 26 \qquad\qquad \forall i \in M$$

$$\sum_{i \in U} y_{ij} \leq r'_j(U) \quad \forall U \subseteq M, j \in J_3$$

$$y \geq 0$$

Let y be an integer optimal solution to the following LP (such a point exists and can be computed in polynomial time due to Lemmas 10 and 11 below). The assignment for J_3 is constructed by setting $S_j := \{i \in M : y_{ij} > 0\}$ for $j \in J_3$.

Because y is integral, $|\{j \in J_3 : i \in S_j\}| \leq 26$ for each $i \in M$. Moreover, Lemma 13 guarantees that $f_j(S_j) \leq 320C/69$ for each $j \in J_3$. We have thus found an assignment of the jobs in J_3 with maximum load $26 \cdot 320C/69 < 121C$. This completes the proof of Lemma 7 and the description of the algorithm.

To complete the analysis, it remains to prove Lemmas 10 and 13 invoked above. We first observe that the function r'_j is submodular for each $j \in J_3$ and hence (GS-P) is indeed the intersection of two polymatroids. As a consequence, we obtain the following lemma.

Lemma 10. *All extreme points of (GS-P) are integral. If (GS-P) is feasible, an optimal extreme point can be computed in polynomial time.*

We next show that (GS-P) has a feasible solution that attains the bound $\sum_{i \in M_{jk}} y_{ij} \leq d_{jk}$ implicit in the definition of r'_j for each k and j with equality.

Lemma 11. *(GS-P) has a feasible solution of value $\sum_{j \in J_3} \sum_{k \in \mathbb{Z}} d_{jk}$.*

Proof (sketch). For $j \in J_3$ and $i \in M_{jk}$, let $y'_{ij} := \frac{d_{jk}}{\sum_{i' \in M_{jk}} x'_{i'j}} x'_{ij}$ if $d_{jk} > 0$ and $y'_{ij} := 0$ otherwise. By construction, $\sum_{i \in M} \sum_{j \in J_3} y'_{ij} = \sum_{j \in J_3} \sum_{k \in \mathbb{Z}} d_{jk}$.

To see that y' is also feasible, observe that $\sum_{j \in J_3} y'_{ij} \leq \sum_{j \in J_3} x'_{ij} \leq 26$ by Property 3 of Lemma 9 and the fact that $y'_{ij} \leq x'_{ij}$ by construction. Moreover, Properties 1 and 2 imply $\sum_{i \in U \cap M_{jk}} y'_{ij} \leq \min\{d_{jk}, r(U \cap M_{jk})\}$ for all $U \subseteq M$ and all $k \in \mathbb{Z}$. As $M_{jk} \cap M_{jk'} = \emptyset$ for $k \neq k'$, we conclude that $\sum_{i \in U} y'_{ij} \leq r'(U)$. Hence y' is a feasible solution to (GS-P). ☐

To prove Lemma 13, we use the following consequence of the properties of x' described in Lemma 9.

Lemma 12. $\sum_{k \in \mathbb{Z}} \frac{1}{2^k C} d_{jk} \geq \frac{69}{40C}$ *for every $j \in J_3$.*

Proof. Note that $x'(S, j) > 0$ implies $S \cap M_j^+ = \emptyset$ by Property 2 of Lemma 9 and hence $\frac{\mu_i}{\lambda_j} \leq 2g_j(\{i\}) < \frac{1}{8C}$ for all $i \in S$. Hence $d_{jk} = 0$ for $k < 3$ and thus

$$\sum_{k \in \mathbb{Z}} \frac{1}{2^k C} d_{jk} = \sum_{k=3}^{\infty} \frac{1}{2^k C} \left\lfloor \sum_{i \in M_{jk}} x'_{ij} \right\rfloor \geq \sum_{i \in M} \frac{\mu_i}{\lambda_j} x'_{ij} - \sum_{k=3}^{\infty} \frac{1}{2^k C}$$

$$\geq \sum_{S \subseteq M} \sum_{i \in S} \frac{\mu_i}{\lambda_j} x'(S, j) - \frac{1}{4C} \geq \frac{79}{40C} - \frac{1}{4C} = \frac{69}{40C},$$

where the last inequality follows from Property 4 of Lemma 9. ☐

We are now ready to combine Lemmas 11 and 12 and the fact that $S_j \in \mathcal{F}_j$ to show that each job in J_3 is indeed assigned sufficient processing speed.

Lemma 13. *For $j \in J_3$ let $S_j := \{i \in M : y_{ij} > 0\}$. Then $g_j(S_j) \geq \frac{69}{320C}$.*

Proof. Note that

$$\sum_{j \in J_3} \sum_{k \in \mathbb{Z}} d_{jk} \leq \sum_{j \in J_3} \sum_{i \in M} y_{ij} = \sum_{j \in J_3} \sum_{k \in \mathbb{Z}} \sum_{i \in M_{jk}} y_{ij} \leq \sum_{j \in J_3} \sum_{k \in \mathbb{Z}} r'_j(M_{jk}) \leq \sum_{j \in J_3} \sum_{k \in \mathbb{Z}} d_{jk}$$

where the first identity follows from by Lemma 11 and the final two inequalities follow from feasibility of y and definition of r'_j, respectively. We conclude that all inequalities are fulfilled with equality, which is only possible if $\sum_{i \in M_{jk}} y_{ij} = d_{jk}$ for all $j \in J_3$ and $k \in \mathbb{Z}$.

Let $T_j \subseteq S_j$ with $T_j \in \mathcal{F}_j$ maximizing $\sum_{i \in T_j} \frac{\mu_i}{\lambda_j}$ computed by the matroid greedy algorithm. The greedy algorithm ensures $\left| T_j \cap \bigcup_{k=0}^{\ell} M_{jk} \right| \geq r_j(S_j \cap M_{j\ell})$ for all $\ell \in \mathbb{Z}$. We conclude that

$$\sum_{i \in T_j} \frac{\mu_i}{\lambda_j} \geq \frac{1}{2} \sum_{k=0}^{\infty} r_j(S_j \cap M_{jk}) \frac{1}{2^{k+1}C} \geq \frac{1}{2} \sum_{k=0}^{\infty} \frac{1}{2^{k+1}C} \sum_{i \in M_{jk}} y_{ij}$$

$$\geq \frac{1}{4} \sum_{k=0}^{\infty} \frac{1}{2^k C} d_{jk} \geq \frac{69}{160C},$$

where the last inequality follows from Lemma 12. Because $T_j \in \mathcal{F}_j$, we conclude that $g_j(S_j) \geq g_j(T_j) \geq \frac{69}{320C}$ by Lemma 4. □

5 Generalized Malleable Jobs and Fair Allocations

In this section, we explore an interesting relation between generalized malleable scheduling and MAX-MIN FAIR ALLOCATION (MMFA). In this problem, we are given a set of *items* I and a set of *agents* A. Each agent $j \in A$ has a *utility function* $u_j : 2^I \to \mathbb{R}_{\geq 0}$ on the items. Our goal is to assign the items to the agents in a way to maximize the minimum utility, that is, to find an assignment **T** with $|\{j \in A : i \in T_j\}| \leq 1$ for all $i \in I$ so as to maximize $\min_{j \in A} u_j(T_j)$.

We show that any approximation algorithm for malleable scheduling implies a *resource-augmented* approximation for MMFA in which some items may be assigned to a small number of agents (this can be interpreted as a moderate way of splitting these items, e.g., multiple agents sharing a resource by taking turns).

To formalize this result, we establish two definitions. A β-augmented α-approximation algorithm for MMFA is an algorithm that given an MMFA instance computes in polynomial time an assignment **T** with $\min_{j \in A} u_j(T_j) \geq \frac{1}{\alpha} V^*$ and $|\{j \in A : i \in T_j\}| \leq \beta$ for all $i \in I$, where V^* is the optimal solution value of the MMFA instance. Moreover, we say that a class of set functions $\mathcal{C}$ is *closed under truncation* if for any $h \in \mathcal{C}$ and any $t \in \mathbb{R}_{\geq 0}$, the function h^t defined by $h^t(S) := \min\{h(S), t\}$ is contained in $\mathcal{C}$.

Theorem 3. *Let $\mathcal{C}$ be a class of set functions closed under truncation. If there is an α-approximation algorithm for ASSIGNMENT with processing speeds from $\mathcal{C}$, then there is an $\lfloor \alpha \rfloor$-augmented α-approximation algorithm for MMFA with utilities from $\mathcal{C}$.*

Proof. Given an instance of MMFA and a target value V for the minimum utility (to be determined by binary search), let $M := I$ and $J := A$, i.e., we introduce a machine for each item and a job for each agent. Define processing speeds g_j for $j \in J$ by $g_j(S) := \min\{u_j(S), V\}$ for $S \subseteq M$. Now apply the α-approximation algorithm to this ASSIGNMENT instance and obtain an assignment **S**.

If $\max_{i \in M} \sum_{j \in J: i \in S_j} 1/g_j(S_j) \leq \alpha/V$, then return the assignment **S** as a solution to the MMFA instance (note that in this case, $g_j(S_j) \geq V/\alpha$ for each $j \in J$ and $|\{j \in J : i \in S_j\}| \leq \lfloor \alpha \rfloor$ for all $i \in M$, because $g_j(S_j) \leq V$ for all $j \in J$). If, on the other hand, $\max_{i \in M} \sum_{j \in J: i \in S_j} 1/g_j(S_j) > \alpha/V$, then we conclude that the MMFA instance does not allow for a solution of value V (because such a solution would have load $1/V$ in the ASSIGNMENT instance). $\square$

This black-box reduction, together with the 3.16-approximation for speed-implementable functions [9] implies a 3-augmented 3.16-approximation for the well-known SANTA CLAUS problem [2] (the special case of MMFA with linear utilities). A more careful analysis delivers the following stronger result:

Corollary 1. *There is a 2-augmented 2-approximation for* SANTA CLAUS.

Although $M^\natural$-concave functions are not closed under truncation as defined above, a slightly different form of truncating processing speeds allows us to apply the reduction on the constant-factor approximation for $M^\natural$-concave processing speeds presented in this paper. We thus obtain the following result.

Corollary 2. *There exists a $\mathcal{O}(1)$-augmentation $\mathcal{O}(1)$-approximation algorithm for MMFA with $M^\natural$-concave utilities.*

6   Conclusion

In this paper we have presented a constant-factor approximation for generalized malleable scheduling under $M^\natural$-concave processing speeds. To achieve a constant approximation guarantee, our algorithm makes extensive use of structural results from discrete convex analysis. We think that some of the techniques, such as the rounding technique for weighted polymatroids in Sect. 4.4, might be of independent interest and applicable in other contexts as well. We have not made any attempt to optimize the constant in the approximation ratio, but we expect that significant additional insights are required to achieve a reasonably small (single-digit) approximation guarantee.

An intriguing open question is whether there exists a constant-factor approximation for generalized malleable scheduling under *submodular* processing speeds, for which only a logarithmic approximation is known to date, together with a strong inapproximability result for the more general XOS functions [10]. Our present work is a significant progress in this direction, but several steps of our algorithm, such as the approximation of processing speeds via weighted matroids and the elimination of low-speed assignments crucially use structural properties of $M^\natural$-concavity not present in submodular functions. Overcoming these issues is an interesting direction for future research.

Acknowledgments. Dimitris Fotakis was supported by the Hellenic Foundation for Research and Innovation (H.F.R.I.) under the "First Call for H.F.R.I. Research Projects to support Faculty members and Researchers and the procurement of high-cost research equipment grant", project BALSAM, HFRI-FM17-1424. Orestis Papadigenopoulos was partially supported by the NSF Institute for Machine Learning, Award number 2019844.

References

1. Bampis, E., Dogeas, K., Kononov, A., Lucarelli, G., Pascual, F.: Scheduling malleable jobs under topological constraints. In: 2020 IEEE International Parallel and Distributed Processing Symposium (IPDPS), pp. 316–325. IEEE (2020)
2. Bansal, N., Sviridenko, M.: The santa claus problem. In: Proceedings of the Thirty-Eighth Annual ACM Symposium on Theory of Computing, pp. 31–40 (2006)
3. Bartolini, E., Dell'Amico, M., Iori, M.: Scheduling cleaning activities on trains by minimizing idle times. J. Sched. **20**, 1–14 (2017). https://doi.org/10.1007/s10951-017-0517-1
4. Blazewicz, J., Cheng, T.E., Machowiak, M., Oguz, C.: Berth and quay crane allocation: a moldable task scheduling model. J. Oper. Res. Soc. **62**(7), 1189–1197 (2011). https://doi.org/10.1057/jors.2010.54
5. Bleuse, R., Hunold, S., Kedad-Sidhoum, S., Monna, F., Mounié, G., Trystram, D.: Scheduling independent moldable tasks on multi-cores with GPUs. IEEE Trans. Parallel Distrib. Sys. **28**(9), 2689–2702 (2017)
6. Chen, X., Li, M.: $M^\natural$-Convexity and its applications in operations. Oper. Res. (2021)
7. Correa, J., et al.: Strong LP formulations for scheduling splittable jobs on unrelated machines. Math. Program. **154**(1), 305–328 (2014). https://doi.org/10.1007/s10107-014-0831-8
8. Du, J., Leung, J.Y.T.: Complexity of scheduling parallel task systems. SIAM J. Discrete Math. **2**(4), 473–487 (1989)
9. Fotakis, D., Matuschke, J., Papadigenopoulos, O.: Malleable scheduling beyond identical machines. In: Approximation, Randomization, and Combinatorial Optimization. Algorithms and Techniques (APPROX/RANDOM 2019). Leibniz International Proceedings in Informatics (LIPIcs), vol. 145, pp. 17:1–17:14. Dagstuhl (2019)
10. Fotakis, D., Matuschke, J., Papadigenopoulos, O.: Assigning and scheduling generalized malleable jobs under subadditive or submodular processing speeds. Technical report (2021). arXiv:2111.06225
11. Fotakis, D., Matuschke, J., Papadigenopoulos, O.: A constant-factor approximation for generalized malleable scheduling under $M^\natural$-concave processing speeds. Technical report (2021). arXiv:2111.06733
12. Fujiwara, I., Tanaka, M., Taura, K., Torisawa, K.: Effectiveness of moldable and malleable scheduling in deep learning tasks. In: 2018 IEEE 24th International Conference on Parallel and Distributed Systems (ICPADS), pp. 389–398. IEEE (2018)
13. Garg, J., Husić, E., Végh, L.A.: Approximating nash social welfare under rado valuations. In: Proceedings of the 53rd Annual ACM SIGACT Symposium on Theory of Computing, pp. 1412–1425 (2021)
14. Gul, F., Stacchetti, E.: Walrasian equilibrium with gross substitutes. J. Econ. Theory **87**(1), 95–124 (1999)

15. Imai, A., Chen, H.C., Nishimura, E., Papadimitriou, S.: The simultaneous berth and quay crane allocation problem. Transp. Res. Part E Logistics Transp. Rev. **44**(5), 900–920 (2008)
16. Jansen, K., Land, F.: Scheduling monotone moldable jobs in linear time. In: 2018 IEEE International Parallel and Distributed Processing Symposium (IPDPS), pp. 172–181. IEEE (2018)
17. Jansen, K., Porkolab, L.: Linear-time approximation schemes for scheduling malleable parallel tasks. Algorithmica **32**(3), 507–520 (2002). https://doi.org/10.1007/s00453-001-0085-8
18. Jansen, K., Thöle, R.: Approximation algorithms for scheduling parallel jobs. SIAM J. Comput. **39**(8), 3571–3615 (2010)
19. Kelso, A.S., Crawford, V.P.: Job matching, coalition formation, and gross substitutes. Econometrica J. Econometric Soc. **50**(6), 1483–1504 (1982)
20. Lenstra, J.K., Shmoys, D.B., Tardos, E.: Approximation algorithms for scheduling unrelated parallel machines. Math. Program. **46**(3), 259–271 (1990). https://doi.org/10.1007/BF01585745
21. Mounié, G., Rapine, C., Trystram, D.: A $\frac{3}{2}$-approximation algorithm for scheduling independent monotonic malleable tasks. SIAM J. Comput. **37**(2), 401–412 (2007)
22. Mounie, G., Rapine, C., Trystram, D.: Efficient approximation algorithms for scheduling malleable tasks. In: Proceedings of the Eleventh Annual ACM symposium on Parallel Algorithms and architectures, pp. 23–32 (1999)
23. Murota, K.: Discrete Convex Analysis, vol. 10. SIAM (2003)
24. Murota, K.: Matrices and matroids for systems analysis, vol. 20. Springer, Heidelberg (2009). https://doi.org/10.1007/978-3-642-03994-2
25. Paes Leme, R.: Gross substitutability: an algorithmic survey. Games Econ. Behav. **106**, 294–316 (2017)
26. Paes Leme, R., Wong, S.C.: Computing Walrasian equilibria: fast algorithms and structural properties. Math. Program. **179**(1), 343–384 (2020). https://doi.org/10.1007/s10107-018-1334-9
27. Turek, J., Wolf, J.L., Yu, P.S.: Approximate algorithms scheduling parallelizable tasks. In: Proceedings of the Fourth Annual ACM Symposium on Parallel Algorithms and Architectures, pp. 323–332 (1992)
28. Wu, X., Loiseau, P.: Algorithms for scheduling deadline-sensitive malleable tasks. In: 2015 53rd Annual Allerton Conference on Communication, Control, and Computing (Allerton), pp. 530–537. IEEE (2015)

Improved Approximations for Capacitated Vehicle Routing with Unsplittable Client Demands

Zachary Friggstad, Ramin Mousavi[(✉)], Mirmahdi Rahgoshay,
and Mohammad R. Salavatipour

Department of Computing Science, University of Alberta, Edmonton, Canada
{zacharyf,mousavih,rahgosha,mrs}@ualberta.ca

Abstract. In this paper, we present improved approximation algorithms for the (unsplittable) Capacitated Vehicle Routing Problem (CVRP) in general metrics. In CVRP, introduced by Dantzig and Ramser (1959), we are given a set of points (clients) V together with a depot r in a metric space, with each $v \in V$ having a demand $d_v > 0$, and a vehicle of bounded capacity Q. The goal is to find a minimum cost collection of tours for the vehicle, each starting and ending at the depot, such that each client is visited at least once and the total demands of the clients in each tour is at most Q. In the unsplittable variant we study, the demand of a node must be served entirely by one tour. We present two approximation algorithms for unsplittable CVRP: a combinatorial $(\alpha + 1.75)$-approximation, where α is the approximation factor for the Traveling Salesman Problem, and an approximation algorithm based on LP rounding with approximation guarantee $\alpha + \ln(2) + \delta \approx 3.194 + \delta$ in $n^{O(1/\delta)}$ time. Both approximations can further be improved by a small amount when combined with recent work by Blauth, Traub, and Vygen (2021), who obtained an $(\alpha + 2 \cdot (1 - \varepsilon))$-approximation for unsplittable CVRP for some constant ε depending on α ($\varepsilon > 1/3000$ for $\alpha = 1.5$).

Keywords: Capacitated vehicle routing · Combinatorial optimization · Approximation algorithm

1 Introduction

Vehicle routing problems are among the most well known and well studied problems in Combinatorial Optimization. The goal is generally to find cost-efficient delivery routes for delivering items from depots to clients in a network using vehicles. The Capacitated Vehicle Routing Problem (CVRP), introduced by Dantzig and Ramser in 1959 [12], generalizes the classic Traveling Salesman Problem and has numerous applications. In CVRP, we are given as input a complete graph $G = (V, E)$ with metric edge weights (also referred to as costs)

Z. Friggstad—Supported by an NSERC Discovery Grant and Accelerator Supplement
M.R. Salavatipour—Supported by NSERC

ⓒ Springer Nature Switzerland AG 2022
K. Aardal and L. Sanitá (Eds.): IPCO 2022, LNCS 13265, pp. 251–261, 2022.
https://doi.org/10.1007/978-3-031-06901-7_19

$c(e) \in \mathbb{R}_{\geq 0}$, a depot $r \in V$, and a vehicle with capacity $Q > 0$, and wish to compute a minimum weight/cost collection of tours, each starting and ending at the depot and visiting at most Q customers, whose union covers all the customers. In the more general setting, each node v is given along with a demand $d(v) \in \mathbb{Z}_{\geq 1}$ and the goal is to find a set of tours of the minimum total cost, each of which includes r, such that the union of the tours covers the demand at every client and every tour serves at most Q demand.

There are three common versions of CVRP: *unit, splittable*, and *unsplittable*. In the splittable variant, the demand of a node can be delivered using multiple tours so each tour must also specify how much demand it serves at each client[1]. However, in the unsplittable variant the entire demand of a client must be delivered by a single tour (*e.g.* each demand is an indivisible good of a certain size). This obviously requires that $d_v \leq Q$ for all clients v. The unit demand case is a special case of the unsplittable case where every node has a unit demand, and the demand of a client must be delivered by a single tour. It is easy to see that the splittable demand case can be reduced to the unit demand case in pseudo-polynomial time using multiple collocated clients of unit demands. However, the unsplittable version is more challenging. For example, it contains the bin-packing problem as a special case; when all clients are have distance 1 from r and distance 0 from each other.

CVRP has also been referred to as the k-tours problem [3,4]. Both the splittable and unsplittable versions admit constant factor approximation algorithms in polynomial-time. Haimovich and Kan [17] showed that a heuristic, called iterative partitioning, yields an $(\alpha + 1(1 - 1/Q))$-approximation for the unit demand case if one uses an α-approximation for the Traveling Salesman Problem (TSP). A similar approach produces a $2 + (1 - 2/Q)\alpha)$-approximation for the unsplittable variant [2]. Despite their simplicity, these remained the best approximations for these two variants for over 35 years. Recently, Blauth et al. [10] improved these approximations giving an $(\alpha + 2 \cdot (1 - \varepsilon))$-approximation algorithm for unsplittable CVRP and a $(\alpha + 1 - \varepsilon)$-approximation algorithm for unit demand CVRP and splittable CVRP where ε is a constant depending only on α. For $\alpha = 3/2$, they showed $\varepsilon > 1/3000$. All the variants are APX-hard in general metric spaces [25].

In this paper we make significant progress on improving the approximation guarantee for unsplittable CVRP. More specifically we present a simple combinatorial algorithm with ratio 3.25, and then a 3.194-approximation algorithm based on linear programming (LP). Our algorithms are completely independent of the improvements by Blauth et al. [10]. By incorporating their approach, we can further improve both ratios by a small constant $\varepsilon' > 0$. However, for the sake of simplicity we prefer to present our main results without factoring in this last improvement.

Theorem 1. *There is an approximation algorithm for the unsplittable CVRP with ratio $\alpha + 1.75$, where α is the best approximation ratio for TSP.*

---

[1] One can show using that restricting the demand served to each client by each tour to integer quantities does not change the optimum solution cost.

The running time of this algorithm is dominated by computing two α-approximate TSP tours and a minimum cost matching. For example, using the simple (combinatorial) Christofides-Serdyukov 1.5-approximation we get a combinatorial 3.25-approximation for unsplittable CVRP whose running time is dominated by computing $O(1)$ perfect matchings in graphs with $O(|V|)$ nodes. Computing a perfect matching in a graph with n nodes can be done in $O(n^3)$ time [15]; hence our algorithm runs in $O(|V|^3)$ time.

If we allow greater running time, we can improve the approximation guarantee further by using linear programming.

Theorem 2. *For any $\delta > 0$, there is an approximation algorithm for unsplittable CVRP with ratio $\ln(2) + \alpha + \frac{1}{1-\delta}$ and running time $n^{O(\frac{1}{\delta})}$, where α is the best approximation ratio for TSP.*

Finally, we show how combining these two results with the approach in [10] actually yields further improvements: a combinatorial $(\alpha + 1.75 - \varepsilon')$-approximation and an LP-based $(\alpha + \ln(2) + \frac{1}{1-\delta} - \varepsilon')$-approximation in time $n^{O(\frac{1}{\delta})}$, where $\varepsilon' > 0$ is an absolute constant.

It is worth noting as the classical results on CVRP [2,17], our results also can be extended to the asymmetric metric where $c(u,v)$ is not necessarily equal to $c(v,u)$. For example, the analogous of Theorem 1 in the asymmetric metric is a $(\beta + 1.75)$-approximation where β is the best approximation factor for Asymmetric Traveling Salesman Problem.

We discuss these further improvements and the extension to the asymmetric metric in more details in the full version of the paper [14].

1.1 Related Work

CVRP captures classic TSP when Q, the vehicle capacity, is at least the total demand of all clients. For general metrics, Haimovich and Kan [17] considered a simple heuristic, called tour partitioning, which starts from a TSP tour and then splits it into tours of size at most Q by making back-and-forth trips to r at certain points along the TSP tour. They showed this gives a $(1 + (1 - 1/Q)\alpha)$-approximation for splittable CVRP, where α is the approximation ratio for TSP. Essentially the same algorithm yields a $(2 + (1 - 2/Q)\alpha)$-approximation for unsplittable CVRP [2]. These stood as the best-known bounds until recently, when Blauth et al. [10] showed that given a TSP approximation α, there is an $\varepsilon > 0$ such that there is an $(\alpha + 2 \cdot (1 - \varepsilon))$-approximation algorithm for CVRP. For $\alpha = 3/2$, they showed $\varepsilon > 1/3000$. They also describe a $(\alpha + 1 - \varepsilon)$-approximation algorithm for unit demand CVRP and splittable CVRP.

For the case of trees, Labbé et al. [23] showed splittable CVRP is NP-hard, and Golden et al. [16] showed unsplittable version is hard to approximate better than 1.5. This is via a simple reduction from bin packing. For splittable CVRP (again on trees), Hamaguchi et al. [18] defined a lower bound for the cost of the optimal solution and gave a 1.5 approximation with respect to the lower bound. Asano et al. [4] improved the approximation to $(\sqrt{41} - 1)/4$ with respect to the same lower bound and also showed the existence of instances whose optimal cost is exactly 4/3 times the lower bound. Later, Becker [5] gave a 4/3-approximation with respect to the lower bound. Becker and Paul [9] showed a $(1, 1 + \varepsilon)$-bicriteria polynomial-time approximation scheme for splittable CVRP in trees, i.e. a PTAS but every tour serves at most $(1 + \varepsilon)Q$ demand. Recently, Jayaprakash and Salavatirpour [19] presented a QPTAS for unit-demand CVRP for trees and more generally graphs of bounded treewidth, bounded doubling metrics, or bounded highway dimension. Even more recently, building upon ideas of [9] and [19], Mathieu and Zhou [24] have presented a PTAS for splittable CVRP on trees.

Das and Mathieu [13] gave a quasi-polynomial-time approximation scheme (QPTAS) for CVRP in the Euclidean plane ($\mathbb{R}^2$). A PTAS for when Q is $O(\log n / \log \log n)$ or Q is $\Omega(n)$ was shown by Asano et al. [4]. A PTAS for Euclidean plane $\mathbb{R}^2$ for moderately large values of Q, i.e. $Q \leq 2^{\log^\delta n}$ where $\delta = \delta(\varepsilon)$, was shown by Adamaszek et al. [1], building on the work of Das and Mathieu [13]. For high dimensional Euclidean spaces $\mathbb{R}^d$, Khachay et al. [20] showed a PTAS when Q is $O(\log \log^{1/d} n)$. For graphs of bounded doubling dimension, Khachay et al. [21] gave a QPTAS when the optimal number of tours is polylog(n) and Khachay et al. [22] gave a QPTAS when Q is polylog(n).

The next results we summarize are all for the case $Q = O(1)$. CVRP remains APX-hard in general metrics in this case but is polynomial-time solvable on trees. There exists a PTAS for CVRP in the Euclidean plane ($\mathbb{R}^2$) (again for when Q is fixed) as shown by Khachay et al. [20]. A PTAS for planar graphs was given by Becker et al. [8] and a QPTAS for planar and bounded-genus graphs was then given by Becker et al. [6]. A PTAS for graphs of bounded highway dimension and an exact algorithm for graphs with treewidth tw with running time $O(n^{tw \cdot Q})$ was shown by Becker et al. [7]. Cohen-Addad et al. [11] showed an efficient PTAS for graphs of bounded-treewidth, an efficient PTAS for bounded highway dimension, an efficient PTAS for bounded genus metrics and a QPTAS for minor-free metrics.

Organization of the Paper: We start with definitions and preliminaries in Sect. 2. The proof of Theorem 1 is presented in Sect. 3 and the proof of Theorem 2 is presented in Sect. 4.

2   Preliminaries

For ease of exposition, we assume we have scaled all the demands and the capacity of the vehicle so that the capacity is 1 and each $d(v) \in (0, 1]$ (so demands can be rational numbers). Also, we treat r as a separate node from the rest of the nodes. Formally:

Definition 1 (CAPACITATED VEHICLE ROUTING). *An instance (V, r, c, d) of* CAPACITATED VEHICLE ROUTING *(CVRP) consists of:*

– a set of clients V, where $|V| = n$,

- *a depot r, not in V,*
- *metric travel costs/distances $c : (V \cup \{r\}) \times (V \cup \{r\}) \to \mathbb{R}_{\geq 0}$,*
- *a demand $d_v \in (0, 1]$ for each customer $v \in V$.*

A feasible solution is a collection of tours $\mathcal{T}$ such that

- *every tour $T \in \mathcal{T}$ is a cycle containing r,*
- *every client belongs to exactly one tour,*
- *$\sum_{v \in T} d_v \leq 1$ for all $T \in \mathcal{T}$.*

The goal is to find a feasible solution with minimum cost where the cost is the sum of costs of the edges in the solution and denoted by $c(\mathcal{T}) := \sum_{T \in \mathcal{T}} c(T) :=$
$$\sum_{T \in \mathcal{T}} \sum_{(u,v) \in T} c(u, v).$$

Observe we are viewing a tour T as both a set of edges comprising a cycle plus the set of endpoints of these edges, so we may use notation like $v \in T$ for a location v and also $(u, v) \in T$ for a pair of locations (u, v) appearing consecutively along the tour T. It is convenient to view the depot r as having $d_r = 0$, for example when we sum the demand of all locations on a tour.

Fix an unsplittable CVRP instance $\mathcal{I} = (V, r, c, d)$ for the rest of this paper. We use OPT to denote an optimal solution for $\mathcal{I}$ and opt the value of this optimal solution.

Definition 2 (Feasible tours). *A tour T that spans r and some clients is called feasible for $\mathcal{I}$ if the total demand of the clients in T is at most 1, i.e., $\sum_{v \in T} d_v \leq 1$.*

Clients are partitioned into *small* and *big* clients based on a parameter $\delta \in [0, \frac{1}{2}]$, which will be chosen differently for our two algorithms.

Definition 3 (Small and big clients). *For a fixed $\delta \in [0, \frac{1}{2}]$, we say a client v is small if $d_v \in [0, \delta]$, and big otherwise.*

Let $\mathcal{D} := \sum_{v \in V} 2 \cdot d_v \cdot c(r, v)$. This is historically referred to as the radial lower bound and the following simple well-known lemma has been used often in previous work.

Lemma 1 (Haimovich and Kan [17]). $\mathcal{D} \leq$ opt.

We also define a similar sum for small and big clients separately, i.e., $\mathcal{D}_{small} := \sum_{\substack{v \in V: \\ v \text{ is small}}} 2 \cdot d_v \cdot c(r, v)$, and $\mathcal{D}_{big} := \sum_{\substack{v \in V: \\ v \text{ is big}}} 2 \cdot d_v \cdot c(r, v)$. Also define $\mathcal{D}'_{big} := \sum_{\substack{v \in V: \\ v \text{ is big}}} 2 \cdot c(r, v)$, which is the cost of serving all big clients using a separate tour for each client.

Given a TSP tour, the algorithm by Haimovich and Kan has the vehicle begin by randomly filling the "tank" of demand it carries with some value

$\theta \sim (0, 1]$. It then travels about the TSP tour: if the tank has insufficient demand to serve a client it travels to the depot to get enough demand to serve the client, returns to serve the client, and then returns to the depot to refill the tank appropriately before resuming the tour. The probability that such a resupply trip is performed when trying to serve a client v is d_v, so the total cost of performing these round trips is at most $2 \cdot \mathcal{D} \le 2 \cdot \text{opt}$ in expectation.

One of the main driving forces behind our improvements is the following idea. For a small client, if we think of the vehicle's tank as only holding $1 - \delta$ demand and keep a reserved tank holding demand δ, then if we cannot serve a client with the demand in the main tank, we can serve it using the reserve tank and only make one round trip to the depot to refill both tanks before proceeding. Both of our main algorithms balance this idea with approaches to handling big clients.

We formalize this notion of using a reserve tank in Lemma 2 below. When $\delta = 0$ this gives the same result as in [2,17].

Lemma 2 (δ-tank Lemma). *Let $\mathcal{A}$ be a TSP tour on $V \cup \{r\}$ and define small and big clients based on a fixed $\delta \le 1/2$. There is an algorithm that turns $\mathcal{A}$ into a feasible solution for the CVRP instance with cost*

$$c(\mathcal{A}) + \frac{1}{1 - \delta} \cdot \mathcal{D}_{small} + \frac{2}{1 - \delta} \cdot \mathcal{D}_{big} - \frac{\delta}{1 - \delta} \cdot \mathcal{D}'_{big}, \tag{1}$$

and running time $O(n^2)$.

Proof We sketch the high level idea behind the proof. See the full version of the paper [14] for the complete proof.

The idea is to reserve δ portion of the vehicle's tank and fill out the rest with a random amount. Then, the vehicle visits the vertices in the same order as they appear in $\mathcal{A}$. As the vehicle visits the clients (vertices), it serves their demand. However, the vehicle might need to make some round trips to the depot to refill. Using the reserved tank and the initial random filling, we bound the cost of these round trips to the depot against different parts of $\mathcal{D}$ as shown in (1). $\square$

3   A Combinatorial 3.25-Approximation

In this section, we set $\delta := \frac{1}{3}$. So v is a small client if $d_v \le \frac{1}{3}$ and big if $d_v > \frac{1}{3}$. Note that in any feasible solution, there are at most two big clients in any single tour. Our algorithm tries two things: the first serves only big clients by pairing them up optimally to form these tours and then runs the classic 3.5-approximation on the small clients but using our δ-tank procedure (see Lemma 2) for performing the tour splitting. The other simply runs the 3.5-approximation using δ-tank tour splitting on all clients.

Let us first explain how we use matching. Consider an auxiliary graph $G_{aux} = (V_{big}, E_{aux})$ where $V_{big} \subseteq V$ is the set of all big clients and E_{aux} constructed as follows: for any pair of big clients u, v where $d_u + d_v \le 1$ we add and edge between u and v with cost $c(r,u) + c(u,v) + c(v,r)$. Furthermore, for every big client v there is a loop in G_{aux} with cost equal to $2 \cdot c(r,v)$. We compute a min-cost perfect matching[2] which corresponds to the cheapest way to select tours to serve only the big clients. The precise details are presented in Algorithm 1.

Algorithm 1. $(\alpha + 1.75)$-approximation

1: The first solution is constructed as follows:
2:    Compute a min-cost perfect matching M on G_{aux}. Let $\mathcal{T}'$ be the tours corresponding to the edges in M.
3:    Compute a TSP tour $\mathcal{A}$ on small clients and r.
4:    Apply Lemma 2 to $\mathcal{A}$ with $\delta = \frac{1}{3}$ and let $\mathcal{T}''$ be the resulting solution.
5:    $\mathcal{T} \leftarrow \mathcal{T}' \cup \mathcal{T}''$
6: The second solution is constructed as follows:
7:    Compute a TSP tour $\mathcal{A}$ on $V \cup \{r\}$.
8:    Apply Lemma 2 to $\mathcal{A}$ with $\delta = \frac{1}{3}$ and let $\mathcal{F}$ be the resulting solution.
9: Return the cheaper of the two solutions $\mathcal{T}$ and $\mathcal{F}$.

3.1 Analysis

We begin with two simple observations.

Lemma 3 $cost(M) \leq$ opt.

Proof Each tour in any feasible solution contains at most two big clients. So, after shortcutting all tours in OPT past small clients, we get tours corresponding to a perfect matching G_{aux} with cost at most opt.   □

Lemma 4 $cost(M) \leq \mathcal{D}'_{big}$.

Proof Consider all the loops in G_{aux}. The cost of all the loops is exactly $\mathcal{D}'_{big}$ and this is a matching so it is an upper bound on the minimum cost of a perfect matching.   □

Next, we compute the cost of the first solution in the algorithm. Note that $c(\mathcal{T}') = cost(M)$. Using an α-approximation for TSP, the cost of $\mathcal{A}$ is at most $\alpha \cdot$ opt: again we are using the metric property which shows opt upper bounds the optimum TSP tour since the union of all tours in OPT is connected and Eulerian. Finally, applying the δ-tank lemma to $\mathcal{A}$ results in a solution of cost at most $c(\mathcal{A}) + \frac{1}{1-\delta} \cdot \mathcal{D}_{small}$ since there is no big client on $\mathcal{A}$. Overall, we have

$$c(\mathcal{T}) = c(\mathcal{T}') + c(\mathcal{T}'') \leq cost(M) + \alpha \cdot \text{opt} + \frac{3}{2} \cdot \mathcal{D}_{small}. \qquad (2)$$

---

[2] A set of edges M that may contain loops is a perfect matching if each node lies in precisely one edge: so a node is either matched with another node via a normal edge or with itself via a loop.

Next, we compute the cost of the second solution. From the δ-tank lemma,

$$c(\mathcal{F}) = \alpha \cdot \text{opt} + \frac{3}{2} \cdot \mathcal{D}_{small} + 3 \cdot \mathcal{D}_{big} - \frac{1}{2} \cdot \mathcal{D}'_{big}. \tag{3}$$

Combining these, we bound the cost of the solution output by the algorithm as follows:

$$
\begin{aligned}
\min\{c(\mathcal{T}), c(\mathcal{F})\} &\leq \frac{c(\mathcal{T}) + c(\mathcal{F})}{2} \\
&= \frac{2 \cdot \alpha \cdot \text{opt} + 3 \cdot (\mathcal{D}_{small} + \mathcal{D}_{big}) + cost(M) - \frac{1}{2} \cdot \mathcal{D}'_{big}}{2} \\
&\leq \frac{2 \cdot \alpha \cdot \text{opt} + 3 \cdot \mathcal{D} + \frac{1}{2} \cdot cost(M)}{2} \\
&\leq \alpha \cdot \text{opt} + 1.5 \cdot \text{opt} + 0.25 \cdot \text{opt} \\
&= (\alpha + 1.75) \cdot \text{opt},
\end{aligned}
\tag{4}
$$

where the second inequality follows from Lemma 4 and the last inequality follows from Lemmas 1 & 3. This finishes the proof of Theorem 1.

4 An Improved LP-Based Approximation

In this section let δ be a fixed constant in the range $(0, 1/2]$. Smaller δ lead to better approximations with increased, but still polynomial, running times.

Define the small and big clients for this value δ as in Definition 3. Let V_{big} be the set of big clients. We consider the following configuration LP for big clients: Let $\mathcal{J}$ be the set of all feasible tours where each tour consists of some big clients and the depot. Note $|\mathcal{J}|$ is bounded by $n^{O(\frac{1}{\delta})}$ as there can be at most $\frac{1}{\delta}$ big clients in each tour. For each $T \in \mathcal{J}$ let $c(T)$ be the cost of tour T. For each tour $T \in \mathcal{J}$, we have a variable x_T indicating this tour is chosen by the algorithm.

$$\text{minimize:} \quad \sum_{T \in \mathcal{J}} c(T) \cdot x_T \qquad \textbf{(Configuration-LP)}$$

$$\text{subject to:} \quad \sum_{\substack{T \in \mathcal{J}: \\ v \in T}} x_T \geq 1 \quad \forall v \in V_{big} \tag{5}$$

$$x \geq 0$$

By shortcutting all tours in the optimum solution past small clients and discarding tours with no big clients, we see there is an integer solution to **(Configuration-LP)** with cost at most opt. Thus, the optimum LP value provides a lower bound on opt.

Our algorithm independently samples tours spanning large clients using an optimal LP solution. After this, some large clients and all small clients remain uncovered, we cover them using the classic 3.5-approximation but use the δ-tank tour splitting approach. Algorithm 2 contains the full description of our approach. With foresight, we set $\gamma := \ln(2)$.

Algorithm 2. $(3.194 + \frac{1}{1-\delta})$-approximation

1: $\mathcal{T} \leftarrow \emptyset$. {This will be a collection of tours.}
2: Compute an optimal solution x^* of (**Configuration-LP**).
3: **for** $T \in \mathcal{J}$ **do**
4:   with probability $\min\{1, \gamma \cdot x_T\}$ add T to $\mathcal{T}$.
5: Approximate a TSP tour $\mathcal{A}$ spanning $\{r\} \cup (V \setminus V(\mathcal{T}))$ where $V(\mathcal{T})$ is the vertices covered in $\mathcal{T}$.
6: Apply the δ-tank lemma to $\mathcal{A}$ and let $\mathcal{T}'$ be the resulting collection of tours.
7: Return $\mathcal{T} \cup \mathcal{T}'$.

It could be that some clients lie on multiple tours due to the randomized rounding step. One can shortcut the tours past repeated occurrences of clients so each client lies on exactly one tour.

4.1 Analysis

We first bound the probability of a big client not being covered in the randomized rounding step of Algorithm 2 (steps 3–4).

Lemma 5 *For a* $v \in V_{big}$, $\Pr[v$ *is not covered by* $\mathcal{T}] \leq e^{-\gamma}$.

Proof The event that a big client v is not covered is if we do not sample any tour T that contains v in the randomized rounding step. So

$$\Pr[v \text{ is not covered by } \mathcal{T}] = \prod_{T \in \mathcal{T}} (1 - \gamma \cdot x_T) \leq e^{-\gamma \cdot \sum_{T \in \mathcal{T}: v \in T} x_T} \leq e^{-\gamma},$$

where the last bound follows from the constraint in (**Configuration-LP**) for v. □

Next, we bound the expected costs of $\mathcal{T}$ and $\mathcal{T}'$, separately. The cost of $\mathcal{T}$ is bounded as follows:

$$\mathbb{E}[\mathcal{T}] = \gamma \cdot cost(x^*) \leq \gamma \cdot \text{opt.} \tag{6}$$

Using the δ-tank lemma, we bound the expected cost of $\mathcal{T}'$ but with the following changes: in (1), we drop the negative term and we incorporate the fact that a big client is on $\mathcal{A}$ with probability at most $e^{-\gamma}$, see Lemma 5.

$$\mathbb{E}[c(\mathcal{T}')] \leq c(\mathcal{A}) + \frac{1}{1-\delta} \cdot \mathcal{D}_{small}$$

$$+ 2 \cdot \sum_{v \in V_{big}} \Pr[v \text{ is not covered by } \mathcal{T}] \cdot \frac{d_v}{1-\delta} \cdot 2 \cdot c(r, v)$$

$$= c(\mathcal{A}) + \frac{1}{1-\delta} \cdot \mathcal{D}_{small} + e^{-\gamma} \cdot \frac{2}{1-\delta} \cdot \mathcal{D}_{big} \tag{7}$$

$$= c(\mathcal{A}) + \frac{1}{1-\delta} \cdot \mathcal{D}_{small} + \frac{1}{1-\delta} \cdot \mathcal{D}_{big}$$

$$= c(\mathcal{A}) + \frac{1}{1-\delta} \cdot \mathcal{D} \leq \alpha \cdot \text{opt} + \frac{1}{1-\delta} \cdot \mathcal{D}.$$

The second equality follows from our choice of $\gamma = \ln 2$. From (6) and (7), the expected cost of the solution returned by Algorithm 2 is at most

$$
\begin{aligned}
\mathbb{E}[c(\mathcal{T} \cup \mathcal{T}')] &\leq \ln 2 \cdot \text{opt} + \alpha \cdot \text{opt} + \frac{1}{1-\delta} \cdot \mathcal{D} \\
&\leq (\ln 2 + \alpha + \frac{1}{1-\delta}) \cdot \text{opt,}
\end{aligned}
\tag{8}
$$

where the last inequality follows from Lemma 1. This finishes the proof of Theorem 2. We briefly comment that this algorithm can be derandomized efficiently using the method of conditional expectation since the probability a big client is covered and its expected contribution to the δ-tank upper bound can be computed efficiently even if some tours have been sampled or rejected so far. Note there is a numerical issue in that $\gamma \cdot x_T$ may not be a rational number, but this error can be absorbed in the $\frac{1}{1-\delta}$ part of the guarantee by choosing δ to be slightly smaller.

Acknowledgements. We thank an anonymous reviewer for pointing out that our approaches could be extended to the asymmetric metric setting.

References

1. Adamaszek, A., Czumaj, A., Lingas, A.: PTAS for k-tour cover problem on the plane for moderately large values of k. In: Dong, Y., Du, D.-Z., Ibarra, O. (eds.) ISAAC 2009. LNCS, vol. 5878, pp. 994–1003. Springer, Heidelberg (2009). https://doi.org/10.1007/978-3-642-10631-6_100
2. Altinkemer, K., Gavish, B.: Heuristics for unequal weight delivery problems with a fixed error guarantee. Oper. Res. Lett. **6**(4), 149–158 (1987). https://www.sciencedirect.com/science/article/pii/0167637787900125
3. Arora, S.: Polynomial time approximation schemes for euclidean traveling salesman and other geometric problems. J. ACM **45**(5), 753–782 (1998)
4. Asano, T., Katoh, N., Tamaki, H., Tokuyama, T.: Covering points in the plane by k-tours: Towards a polynomial time approximation scheme for general k. In: 29th ACM Symposium on the Theory of Computing (STOC), pp. 275–283 (1997)
5. Becker, A.: A tight 4/3 approximation for capacitated vehicle routing in trees. In: 21st International Conference on Approximation Algorithms for Combinatorial Optimization Problems (APPROX), pp. 3:1–3:15 (2018)
6. Becker, A., Klein, P.N., Saulpic, D.: A quasi-polynomial-time approximation scheme for vehicle routing on planar and bounded-genus graphs. In: 25th Annual European Symposium on Algorithms (ESA), pp. 12:1–12:15 (2017)
7. Becker, A., Klein, P.N., Saulpic, D.: Polynomial-time approximation schemes for k-center, k-median, and capacitated vehicle routing in bounded highway dimension. In: 26th Annual European Symposium on Algorithms (ESA), pp. 8:1–8:15 (2018)
8. Becker, A., Klein, P.N., Schild, A.: A PTAS for bounded-capacity vehicle routing in planar graphs. In: 16th International Algorithms and Data Structures Symposium (WADS), pp. 99–111 (2019)
9. Becker, A., Paul, A.: A framework for vehicle routing approximation schemes in trees. In: Friggstad, Z., Sack, J.-R., Salavatipour, M.R. (eds.) WADS 2019. LNCS, vol. 11646, pp. 112–125. Springer, Cham (2019). https://doi.org/10.1007/978-3-030-24766-9_9

10. Blauth, J., Traub, V., Vygen, J.: Improving the approximation ratio for capacitated vehicle routing. In: Singh, M., Williamson, D.P. (eds.) IPCO 2021. LNCS, vol. 12707, pp. 1–14. Springer, Cham (2021). https://doi.org/10.1007/978-3-030-73879-2_1

11. Cohen-Addad, V., Filtser, A., Klein, P.N., Le, H.: On light spanners, low-treewidth embeddings and efficient traversing in minor-free graphs. In: 61st IEEE Annual Symposium on Foundations of Computer Science (FOCS), pp. 589–600 (2020)

12. Dantzig, G.B., Ramser, J.H.: The truck dispatching problem. Manag. Sci. **6**(1), 80–91 (1959)

13. Das, A., Mathieu, C.: A quasipolynomial time approximation scheme for euclidean capacitated vehicle routing. Algorithmica **73**(1), 115–142 (2015). https://doi.org/10.1007/s00453-014-9906-4

14. Friggstad, Z., Mousavi, R., Rahgoshay, M., Salavatipour, M.R.: Improved approximations for cvrp with unsplittable demands (2021). arXiv preprint arXiv:2111.08138

15. Gabow, H.N.: An efficient implementation of edmonds' algorithm for maximum matching on graphs. J. ACM (JACM) **23**(2), 221–234 (1976)

16. Golden, B.L., Wong, R.T.: Capacitated arc routing problems. Networks **11**(3), 305–315 (1981)

17. Haimovich, M., Kan, A.H.G.R.: Bounds and heuristics for capacitated routing problems. Math. Oper. Res. **10**(4), 527–542 (1985)

18. Hamaguchi, S., Katoh, N.: A capacitated vehicle routing problem on a tree. In: Chwa, K.-Y., Ibarra, O.H. (eds.) ISAAC 1998. LNCS, vol. 1533, pp. 399–407. Springer, Heidelberg (1998). https://doi.org/10.1007/3-540-49381-6_42

19. Jayaprakash, A., Salavatipour, M.R.: Approximation schemes for capacitated vehicle routing on graphs of bounded treewidth, bounded doubling, or highway dimension. In: Proceedings of the 2022 ACM-SIAM Symposium on Discrete Algorithms SODA (2022)

20. Khachay, M., Dubinin, R.: PTAS for the euclidean capacitated vehicle routing problem in R^d. In: Kochetov, Y., Khachay, M., Beresnev, V., Nurminski, E., Pardalos, P. (eds.) DOOR 2016. LNCS, vol. 9869, pp. 193–205. Springer, Cham (2016). https://doi.org/10.1007/978-3-319-44914-2_16

21. Khachay, M., Ogorodnikov, Y.: QPTAS for the CVRP with a moderate number of routes in a metric space of any fixed doubling dimension. In: Kotsireas, I.S., Pardalos, P.M. (eds.) LION 2020. LNCS, vol. 12096, pp. 27–32. Springer, Cham (2020). https://doi.org/10.1007/978-3-030-53552-0_4

22. Khachay, M., Ogorodnikov, Y., Khachay, D.: An extension of the das and mathieu QPTAS to the case of polylog capacity constrained CVRP in metric spaces of a fixed doubling dimension. In: Kononov, A., Khachay, M., Kalyagin, V.A., Pardalos, P. (eds.) MOTOR 2020. LNCS, vol. 12095, pp. 49–68. Springer, Cham (2020). https://doi.org/10.1007/978-3-030-49988-4_4

23. Labbé, M., Laporte, G., Mercure, H.: Capacitated vehicle routing on trees. Oper. Res. **39**(4), 616–622 (1991)

24. Mathiue, C., Zhou, H.: A PTAS for Capacitated Vehicle Routing on Trees (2020). CoRR arXiv:2111.03735, https://arxiv.org/abs/2111.03735

25. Papadimitriou, C.H., Yannakakis, M.: The traveling salesman problem with distances one and two. Math. Oper. Res. **18**(1), 1–11 (1993)

SOCP-Based Disjunctive Cuts for a Class of Integer Nonlinear Bilevel Programs

Elisabeth Gaar[1]([envelope]) [iD], Jon Lee[2] [iD], Ivana Ljubić[3] [iD], Markus Sinnl[1,4] [iD], and Kübra Tanınmış[1] [iD]

[1] Institute of Production and Logistics Management,
Johannes Kepler University Linz, Linz, Austria
{elisabeth.gaar,markus.sinnl,kuebra.taninmis_ersues}@jku.at
[2] University of Michigan, Ann Arbor, MI, USA
jonxlee@umich.edu
[3] ESSEC Business School of Paris, Cergy, France
ljubic@essec.edu
[4] JKU Business School, Johannes Kepler University Linz, Linz, Austria

Abstract. We study a class of bilevel integer programs with second-order cone constraints at the upper level and a convex quadratic objective and linear constraints at the lower level. We develop disjunctive cuts to separate bilevel infeasible points using a second-order-cone-based cut-generating procedure. To the best of our knowledge, this is the first time disjunctive cuts are studied in the context of discrete bilevel optimization. Using these disjunctive cuts, we establish a branch-and-cut algorithm for the problem class we study, and a cutting plane method for the problem variant with only binary variables. We present a preliminary computational study on instances with no second-order cone constraints at the upper level and a single linear constraint at the lower level. Our study demonstrates that both our approaches outperform a state-of-the-art generic solver for mixed-integer bilevel linear programs that is able to solve a linearized version of our test instances, where the non-linearities are linearized in a McCormick fashion.

Keywords: Bilevel optimization · Disjunctive cuts · Conic optimization · Nonlinear optimization · Branch-and-cut

1 Introduction

Bilevel programs (BPs) are challenging hierarchical optimization problems, in which the feasible solutions of the upper level problem depend on the optimal

This research was funded in whole, or in part, by the Austrian Science Fund (FWF) [P 35160-N]. For the purpose of open access, the author has applied a CC BY public copyright licence to any Author Accepted Manuscript version arising from this submission. It is also supported by the Johannes Kepler University Linz, Linz Institute of Technology (Project LIT-2019-7-YOU-211) and the JKU Business School. J. Lee was supported on this project by ESSEC and by ONR grant N00014-21-1-2135.

© Springer Nature Switzerland AG 2022
K. Aardal and L. Sanitá (Eds.): IPCO 2022, LNCS 13265, pp. 262–276, 2022.
https://doi.org/10.1007/978-3-031-06901-7_20

solution of the lower level problem. BPs allow to model two-stage two-player Stackelberg games in which two rational players (often called *leader* and *follower*) compete in a sequential fashion. BPs have applications in many different domains such as machine learning [1], logistics [15], revenue management [25], the energy sector [17,31] and portfolio optimization [16]. For more details about BPs see, e.g., the book by Dempe and Zemkoho [10] and two recent surveys [21,34].

In this work, we consider the following integer nonlinear bilevel programs with convex leader and follower objective functions (IBNPs)

$$\min \ c'x + d'y \tag{1a}$$
$$\text{s.t. } Mx + Ny \geq h \tag{1b}$$
$$\tilde{M}x + \tilde{N}y - \tilde{h} \in \mathcal{K} \tag{1c}$$
$$y \in \arg\min \{q(y) : Ax + By \geq f, \ y \in \mathcal{Y}, \ y \in \mathbb{Z}^{n_2}\} \tag{1d}$$
$$x \in \mathbb{Z}^{n_1}, \tag{1e}$$

where the decision variables x and y are of dimension n_1 and n_2, respectively, and $n = n_1 + n_2$. Moreover, we have $c \in \mathbb{R}^{n_1}$, $d \in \mathbb{R}^{n_2}$, $M \in \mathbb{R}^{m_1 \times n_1}$, $N \in \mathbb{R}^{m_1 \times n_2}$, $h \in \mathbb{R}^{m_1}$, $\tilde{M} \in \mathbb{R}^{\tilde{m}_1 \times n_1}$, $\tilde{N} \in \mathbb{R}^{\tilde{m}_1 \times n_2}$, $\tilde{h} \in \mathbb{R}^{\tilde{m}_1}$, $A \in \mathbb{Z}^{m_2 \times n_1}$, $B \in \mathbb{Z}^{m_2 \times n_2}$, and $f \in \mathbb{Z}^{m_2}$. We assume that each row of A and B has at least one non-zero entry and the constraints $Ax + By \geq f$ are referred to as *linking constraints*. Furthermore, $q(y)$ is a convex quadratic function of the form $q(y) = y'Ry + g'y$ with $R = V'V$ and $V \in \mathbb{R}^{n_3 \times n_2}$ with $n_3 \leq n_2$, $\mathcal{K}$ is a given cross-product of second-order cones, and $\mathcal{Y}$ is a polyhedron.

Note that even though we formulate the objective function (1a) as linear, we can actually consider any convex objective function which can be represented as a second-order cone constraint and whose optimal value is integer when $(x, y) \in \mathbb{Z}^n$ (e.g., a convex quadratic polynomial with integer coefficients). To do so, we can use an epigraph reformulation to transform it into a problem of the form (1).

Our work considers the *optimistic* case of bilevel optimization. This means that whenever there are multiple optimal solutions for the follower problem, the one which is best for the leader is chosen, see, e.g., [27]. We note that already mixed-integer bilevel linear programming (MIBLP) is Σ_2^p-hard [26].

The *value function reformulation* (VFR) of the bilevel model (1) is given as

$$\min \ c'x + d'y \tag{2a}$$
$$\text{s.t. } Mx + Ny \geq h \tag{2b}$$
$$\tilde{M}x + \tilde{N}y - \tilde{h} \in \mathcal{K} \tag{2c}$$
$$Ax + By \geq f \tag{2d}$$
$$q(y) \leq \Phi(x) \tag{2e}$$
$$y \in \mathcal{Y} \tag{2f}$$
$$(x, y) \in \mathbb{Z}^n, \tag{2g}$$

where the so-called *value function* $\Phi(x)$ of the *follower problem*

$$\Phi(x) = \min \{q(y) : Ax + By \geq f, \ y \in \mathcal{Y}, \ y \in \mathbb{Z}^{n_2}\} \tag{3}$$

is typically non-convex and non-continuous. Note that the VFR is equivalent to the original bilevel model (1). The *high point relaxation* (HPR) is obtained when dropping (2e), i.e., the optimality condition of y for the follower problem, from the VFR (2). We denote the continuous relaxation (i.e., replacing the integer constraint (2g) with the corresponding variable bound constraints) of the HPR as $\overline{\text{HPR}}$. A solution (x^*, y^*) is called *bilevel infeasible*, if it is feasible for $\overline{\text{HPR}}$, but not feasible for the original bilevel model (1).

Our Contribution. Since the seminal work of Balas [3], and more intensively in the past three decades, disjunctive cuts (DCs) have been successfully exploited for solving mixed-integer (nonlinear) programs (MI(N)LPs) [4]. While there is a plethora of work on using DCs for MINLPs [5], we are not aware of any previous applications of DCs for solving IBNPs. In this work we demonstrate how DCs can be used within in a branch-and-cut (B&C) algorithm to solve (1). This is the first time that DCs are used to separate bilevel infeasible points, using a cut-generating procedure based on second-order cone programming (SOCP). Moreover, we also show that our DCs can be used in a finitely-convergent cutting plane procedure for 0–1 IBNPs, where the HPR is solved to optimality before separating bilevel infeasible points. Our computational study is conducted on instances in which the follower minimizes a convex quadratic function, subject to a covering constraint linked with the leader. We compare the proposed B&C and cutting plane approaches with a state-of-the-art solver for MIBLPs (which can solve our instances after applying linearization in a McCormick fashion), and show that the latter one is outperformed by our new DC-based methodologies.

Literature Overview. For MIBLPs with integrality restrictions on (some of) the follower variables, state-of-the-art computational methods are usually based on B&C (see, e.g., [11–13, 35]). Other interesting concepts are based on multi-branching, see [36, 38].

   Considerably less results are available for non-linear BPs, and in particular with integrality restrictions at the lower level. In [29], Mitsos et al. propose a general approach for non-convex follower problems which solves nonlinear optimization problems to compute upper and lower bounds in an iterative fashion. In a series of papers on the so-called *branch-and-sandwich* approach, tightened bounds on the optimal value function and on the leader's objective function value are calculated [22–24]. A solution algorithm for mixed-IBNPs proposed in [28] by Lozano and Smith approximates the value function by dynamically inserting additional variables and big-M type of constraints. Recently, Kleinert et al. [20] considered bilevel problems with a mixed-integer convex-quadratic upper level and a continuous convex-quadratic lower level. The method is based on outer approximation after the problem is reformulated into a single-level one using the strong duality and convexification. In [8], Byeon and Van Hentenryck develop a solution algorithm for bilevel problems, where the leader problem can be modeled as a mixed-integer second-order conic problem and the follower problem can be modeled as a second-order conic problem. The algorithm is based on a dedicated Benders decomposition method. In [37], Weninger et al. propose a methodology that can tackle any kind of a MINLP at the upper level which can be handled by an off-the-shelf solver. The mixed-integer lower level problem has to be convex, bounded, and satisfy Slater's condition for the continuous variables. This exact method is derived from a previous approach proposed in [39] by Yue et al. for finding bilevel feasible solutions. For a more detailed overview of the recent literature on computational bilevel optimization we refer an interested reader to [9,21,34].

The only existing application of DCs in the context of bilevel *linear* optimization is by Audet et al., [2] who derive DCs from LP-complementarity conditions. In [18], Júdice et al. exploit a similar idea for solving mathematical programs with equilibrium constraints. DCs are frequently used for solving MINLPs (see, e.g., [4], and the many references therein, and [32,33]). In [19], Kılınç-Karzan and Yıldız derive closed-form expressions for inequalities describing the convex hull of a two-term disjunction applied to the second-order cone.

2   Disjunctive Cut Methodology

The aim of this section is to derive DCs for the bilevel model (1) with the help of SOCP, so we want to derive DCs that are able to separate bilevel infeasible points from the convex hull of bilevel feasible ones. Toward this end, we assume throughout this section that we have a second-order conic convex set $\mathcal{P}$, such that the set of feasible solutions of the VFR is a subset of $\mathcal{P}$, and such that $\mathcal{P}$ is a subset of the set of feasible solutions of the $\overline{\text{HPR}}$. This implies that $\mathcal{P}$ fulfills (2b), (2c), (2d) and (2f) and potentially already some DCs. Moreover, we assume that (x^*, y^*) is a bilevel infeasible point in $\mathcal{P}$. The point (x^*, y^*) is an *extreme point* of $\mathcal{P}$, if it is not a convex combination of any other two points of $\mathcal{P}$.

2.1   Preliminaries

For clarity of exposition in what follows, we consider only one linking constraint of problem (1), i.e., $m_2 = 1$ and thus $A = a'$ and $B = b'$ for some $a \in \mathbb{Z}^{n_1}$, $b \in \mathbb{Z}^{n_2}$ and $f \in \mathbb{Z}$. Note however that our methodology can be generalized for multiple linking constraints leading to one additional disjunction for every additional linking constraint. Moreover, our DCs need the following assumptions.

Assumption 1. *All variables are bounded in the HPR and $\mathcal{Y}$ is bounded.*

Assumption 1 ensures that the HPR is bounded. We note that in a bilevel-context already for the linear case of MIBLPs, unboundedness of the $\overline{\text{HPR}}$ does not imply anything for the original problem, all three options (infeasible, unbounded, and existence of an optimum) are possible. For more details see, e.g., [13].

Assumption 2. *For every x, such that there exists a y with (x, y) being feasible for the $\overline{\text{HPR}}$, the follower problem (3) is feasible.*

Assumption 3. *$\overline{\text{HPR}}$ has a feasible solution satisfying its nonlinear constraint (2c) strictly, and its dual has a feasible solution.*

Assumption 3 ensures that we have strong duality between $\overline{\text{HPR}}$ and its dual, and so we can solve the $\overline{\text{HPR}}$ (potentially with added cuts) to arbitrary accuracy.

2.2   Deriving Disjunctive Cuts

To derive DCs we first examine bilevel feasible points. It is easy to see and also follows from the results by Fischetti et al. [12], that for any $\hat{y} \in \mathcal{Y} \cap \mathbb{Z}^{n_2}$ the set

$$S(\hat{y}) = \{(x,y) : a'x \geq f - b'\hat{y}, \; q(y) > q(\hat{y})\}$$

does not contain any bilevel feasible solutions, as for any $(x,y) \in S(\hat{y})$ clearly $\hat{y}$ is a better follower solution than y for x. Furthermore, due to the integrality of our variables and of a and b, the extended set

$$S^+(\hat{y}) = \{(x,y) : a'x \geq f - b'\hat{y} - 1, \; q(y) \geq q(\hat{y})\}$$

does not contain any bilevel feasible solutions in its interior, because any bilevel feasible solution in the interior of $S^+(\hat{y})$ is in $S(\hat{y})$. Based on this observation intersection cuts have been derived in [12], however $S^+(\hat{y})$ is not convex in our case, so we turn our attention to DCs. As a result, for any $\hat{y} \in \mathcal{Y} \cap \mathbb{Z}^{n_2}$ any bilevel feasible solution is in the disjunction $\mathcal{D}_1(\hat{y}) \vee \mathcal{D}_2(\hat{y})$, where

$$\mathcal{D}_1(\hat{y}) : a'x \leq f - b'\hat{y} - 1 \qquad \text{and} \qquad \mathcal{D}_2(\hat{y}) : q(y) \leq q(\hat{y}).$$

To find a DC, we want to generate valid linear inequalities for

$$\{(x,y) \in \mathcal{P} : \mathcal{D}_1(\hat{y})\} \vee \{(x,y) \in \mathcal{P} : \mathcal{D}_2(\hat{y})\}, \tag{4}$$

so in other words we want to find a valid linear inequality that separates the bilevel infeasible solution (x^*, y^*) from

$$\mathcal{D}(\hat{y}) = \text{conv}\left(\{(x,y) \in \mathcal{P} : \mathcal{D}_1(\hat{y})\} \cup \{(x,y) \in \mathcal{P} : \mathcal{D}_2(\hat{y})\}\right).$$

Toward this end, we first derive a formulation of $\mathcal{P}$. If we have already generated some DCs of the form $\alpha'x + \beta'y \geq \tau$, then they create a bunch of constraints $\mathcal{A}x + \mathcal{B}y \geq \mathcal{T}$. We take these cuts, together with $Mx + Ny \geq h$ and $a'x + b'y \geq f$ and also $y \in \mathcal{Y}$, which can be represented as $\mathcal{C}y \geq \mathcal{U}$, and we bundle them all together as

$$\bar{M}x + \bar{N}y \geq \bar{h}, \tag{5}$$

such that $\mathcal{P}$ is represented by (5) and (2c), and where

$$\bar{M} = \begin{pmatrix} M \\ a' \\ \mathcal{A} \\ 0 \end{pmatrix}, \qquad \bar{N} = \begin{pmatrix} N \\ b' \\ \mathcal{B} \\ \mathcal{C} \end{pmatrix}, \qquad \bar{h} = \begin{pmatrix} h \\ f \\ \mathcal{T} \\ \mathcal{U} \end{pmatrix}.$$

The representation of $\mathcal{D}_1(\hat{y})$ is straightforward. It is convenient to write $\mathcal{D}_2(\hat{y})$ in SOCP-form using a standard technique. Indeed, $\mathcal{D}_2(\hat{y})$ is equivalent to the standard second-order (Lorentz) cone constraint $z^0 \geq \|(z^1, z^2)\|$ with

$$z^0 = \frac{1 - (g'y - q(\hat{y}))}{2}, \qquad z^1 = Vy, \qquad z^2 = \frac{1 + (g'y - q(\hat{y}))}{2}.$$

Because z^0, z^1 and z^2 are linear in y, we can as well write it in the form

$$\tilde{D}y - \tilde{c} \in \mathcal{Q}, \tag{6}$$

where $\mathcal{Q}$ denotes a standard second-order cone, which is self dual, and

$$\tilde{D} = \begin{pmatrix} -\frac{1}{2}g' \\ V \\ \frac{1}{2}g' \end{pmatrix} \quad \text{and} \quad \tilde{c} = \begin{pmatrix} \frac{-1-q(\hat{y})}{2} \\ 0 \\ \frac{-1+q(\hat{y})}{2} \end{pmatrix}.$$

We employ a scalar dual multiplier σ for the constraint $\mathcal{D}_1(\hat{y})$ and we employ a vector $\rho \in \mathcal{Q}^*$ of dual multipliers for the constraint (6), representing $\mathcal{D}_2(\hat{y})$. Furthermore, we employ two vectors $\bar{\pi}_k$, $k = 1, 2$, of dual multipliers for the constraints (5) and we employ two vectors $\tilde{\pi}_k$, $k = 1, 2$, of dual multipliers for the constraints (2c), both together representing $\mathcal{P}$. Then every (α, β, τ) corresponding to a valid linear inequality $\alpha'x + \beta'y \geq \tau$ for $\mathcal{D}(\hat{y})$ corresponds to a solution of

$$\alpha' = \bar{\pi}_1'\bar{M} + \tilde{\pi}_1'\tilde{M} + \sigma a' \tag{7a}$$

$$\alpha' = \bar{\pi}_2'\bar{M} + \tilde{\pi}_2'\tilde{M} \tag{7b}$$

$$\beta' = \bar{\pi}_1'\bar{N} + \tilde{\pi}_1'\tilde{N} \tag{7c}$$

$$\beta' = \bar{\pi}_2'\bar{N} + \tilde{\pi}_2'\tilde{N} + \rho'\tilde{D} \tag{7d}$$

$$\tau \leq \bar{\pi}_1'\bar{h} + \tilde{\pi}_1'\tilde{h} + \sigma(f - 1 - b'\hat{y}) \tag{7e}$$

$$\tau \leq \bar{\pi}_2'\bar{h} + \tilde{\pi}_2'\tilde{h} + \rho'\tilde{c} \tag{7f}$$

$$\bar{\pi}_1 \geq 0, \ \bar{\pi}_2 \geq 0, \ \tilde{\pi}_1 \in \mathcal{K}^*, \ \tilde{\pi}_2 \in \mathcal{K}^*, \ \sigma \leq 0, \ \rho \in \mathcal{Q}^*, \tag{7g}$$

where $\mathcal{K}^*$ and $\mathcal{Q}^*$ are the dual cones of $\mathcal{K}$ and $\mathcal{Q}$, respectively (see, e.g., Balas [4, Theorem 1.2]).

To attempt to generate a valid inequality for $\mathcal{D}(\hat{y})$ that is violated by the bilevel infeasible solution (x^*, y^*), we solve

$$\max \ \tau - \alpha'x^* - \beta'y^* \qquad \text{(CG-SOCP)}$$
$$\text{s.t. } (7a)\text{—}(7g).$$

A positive objective value for a feasible (α, β, τ) corresponds to a valid linear inequality $\alpha'x + \beta'y \geq \tau$ for $\mathcal{D}(\hat{y})$ violated by (x^*, y^*), i.e. the inequality gives a DC separating (x^*, y^*) from $\mathcal{D}(\hat{y})$.

Finally, we need to deal with the fact that the feasible region of (CG-SOCP) is a cone. So (CG-SOCP) either has its optimum at the origin (implying that (x^*, y^*) cannot be separated), or (CG-SOCP) is unbounded, implying that there is a violated inequality, which of course we could scale by any positive number so as to make the violation as large as we like. The standard remedy for this is to introduce a normalization constraint to (CG-SOCP). A typical good choice (see [14]) is to impose $\|(\bar{\pi}_1, \bar{\pi}_2, \tilde{\pi}_1, \tilde{\pi}_2, \sigma, \rho)\|_1 \leq 1$, but in our context, because we are using a conic solver, we can more easily and efficiently impose

$\|(\bar{\pi}_1, \bar{\pi}_2, \tilde{\pi}_1, \tilde{\pi}_2, \sigma, \rho)\|_2 \leq 1$, which is just one constraint for a conic solver. Thus, we will from now on consider normalization as part of (CG-SOCP).

To be able to derive DCs we make the following additional assumption.

Assumption 4. *The dual of* (CG-SOCP) *has a feasible solution in its interior and we have an exact solver for* (CG-SOCP).

We have the following theorem, which allows us to use DCs in solution methods.

Theorem 1. *Let $\mathcal{P}$ be a second-order conic convex set, such that the set of feasible solutions of the VFR is a subset of $\mathcal{P}$, and such that $\mathcal{P}$ is a subset of the set of feasible solutions of the $\overline{HPR}$. Let (x^*, y^*) be bilevel infeasible and be an extreme point of $\mathcal{P}$. Let $\hat{y}$ be a feasible solution to the follower problem for $x = x^*$ (i.e., $\hat{y} \in \mathcal{Y} \cap \mathbb{Z}^{n_2}$ and $a'x^* + b'\hat{y} \geq f$) such that $q(\hat{y}) < q(y^*)$.*

Then there is a DC that separates (x^*, y^*) from $\mathcal{D}(\hat{y})$ and it can be obtained by solving (CG-SOCP).

Proof. Assume that there is no cut that separates (x^*, y^*) from $\mathcal{D}(\hat{y})$, then (x^*, y^*) is in $\mathcal{D}(\hat{y})$. However, due to the definition of $\hat{y}$, the point (x^*, y^*) does not fulfill $\mathcal{D}_1(\hat{y})$ and does not fulfill $\mathcal{D}_2(\hat{y})$. Therefore, in order to be in $\mathcal{D}(\hat{y})$, the point (x^*, y^*) must be a convex combination of one point in $\mathcal{P}$ that fulfills $\mathcal{D}_1(\hat{y})$, and another point in $\mathcal{P}$ that fulfills $\mathcal{D}_2(\hat{y})$. This is not possible due to the fact that (x^*, y^*) is an extreme point of $\mathcal{P}$.

Thus, there is a cut that separates (x^*, y^*) from $\mathcal{D}(\hat{y})$. By construction of (CG-SOCP) and due to Assumption 4, we can use (CG-SOCP) to find it.   □

Note that there are two reasons why a feasible $\overline{HPR}$ solution (x^*, y^*) is bilevel infeasible: it is not integer or y^* is not the optimal follower response for x^*. Thus, in the case that (x^*, y^*) is integer, there is a better follower response $\tilde{y}$ for x^*. Then Theorem 1 with $\hat{y} = \tilde{y}$ implies that (x^*, y^*) can be separated from $\mathcal{D}(\hat{y})$. We present solution methods based on this observation in Sect. 3.2.

2.3   Separation Procedure for Disjunctive Cuts

We turn our attention to describing how to computationally separate our DCs for a solution $(x^*, y^*) \in \mathcal{P}$ now. Note that we do not necessarily need the optimal solution of the follower problem (3) for $x = x^*$ to be able to cut off a bilevel infeasible solution (x^*, y^*), as any $\hat{y}$ that is feasible for the follower problem with $q(\hat{y}) < q(y^*)$ gives a violated DC as described in Theorem 1. Thus, we implement two different strategies for separation which are described in Algorithm 1.

In the first one, denoted as O, we solve the follower problem to optimality, and use the optimal $\hat{y}$ in (CG-SOCP). In the second strategy, denoted as G, for each feasible integer follower solution $\hat{y}$ with a better objective value than $q(y^*)$ obtained during solving the follower problem, we try to solve (CG-SOCP). The procedure returns the first found significantly violated cut, i.e., it finds a DC greedily. A cut $\alpha'x + \beta'y \geq \tau$ is considered to be *significantly violated* by (x^*, y^*) if $\tau - \alpha'x^* - \beta'y^* > \varepsilon$ for some $\varepsilon > 0$.

Algorithm 1: separation

Input : A feasible $\overline{\text{HPR}}$ solution (x^*, y^*), a separation *strategy* O or G, a set $\mathcal{P}$

Output: A significantly violated disjunctive cut or nothing

---

1 **while** *the follower problem is being solved for $x = x^*$ by an enumeration based method* **do**

2 $\quad$ **for** *each feasible integer $\hat{y}$ with $q(\hat{y}) < q(y^*)$* **do**

3 $\quad\quad$ **if** *strategy = G or (strategy = O and $\hat{y}$ is optimal)* **then**

4 $\quad\quad\quad$ solve (CG-SOCP) for (x^*, y^*), $\hat{y}$ and $\mathcal{P}$;

5 $\quad\quad\quad$ **if** $\tau - \alpha' x^* - \beta' y^* > \varepsilon$ **then**

6 $\quad\quad\quad\quad$ **return** $\alpha' x + \beta' y \geq \tau$;

If (x^*, y^*) is a bilevel infeasible solution satisfying integrality constraints, Algorithm 1 returns a violated cut with both strategies. Otherwise, i.e., if (x^*, y^*) is not integer, a cut may not be obtained, because it is possible that there is no feasible $\hat{y}$ for the follower problem with $q(\hat{y}) < q(y^*)$.

3   Solution Methods Using Disjunctive Cuts

We now present two solution methods based on DCs, one applicable for the general bilevel model (1), one dedicated to a binary version of (1).

3.1   A Branch-and-Cut Algorithm

We propose to use the DCs in a B&C algorithm to solve the bilevel model (1). The B&C can be obtained by modifying any given continuous-relaxation-based B&B algorithm to solve the HPR (assuming that there is an off-the-shelf solver for $\overline{\text{HPR}}$ that always returns an extreme optimal solution (x^*, y^*) like e.g., a simplex-based B&B for a linear $\overline{\text{HPR}}$[1]).

The algorithm works as follows: Use $\overline{\text{HPR}}$ as initial relaxation $\mathcal{P}$ at the root-node of the B&C. Whenever a solution (x^*, y^*) which is integer is encountered in a B&C node, call the DC separation. If a violated cut is found, add the cut to the set $\mathcal{P}$ (which also contains, e.g., variable fixing by previous branching decisions, previously added globally or locally valid DCs, ...) of the current B&C node, otherwise the solution is bilevel feasible and the incumbent can be updated. Note that DCs are only locally valid except the ones from the root node, since $\mathcal{P}$ includes branching decisions. If $\mathcal{P}$ is empty or optimizing over $\mathcal{P}$ leads to an objective function value that is larger than the objective function value of the current incumbent, we fathom the current node. In our implementation, we also use DC separation for fractional (x^*, y^*) as described in Sect. 2.3 for strengthening the relaxation.

---

[1] This assumption is without loss of generality, as we can outer approximate second-order conic constraints of $\mathcal{P}$ and get an extreme optimal point by a simplex method.

Theorem 2. *The B&C solves the bilevel model (1) in a finite number of B&C-iterations under our assumptions.*

Proof. First, suppose the B&C terminates, but the solution (x^*, y^*) is not bilevel feasible. This is not possible, as by Theorem 1 and the observations thereafter the DC generation procedure finds a violated cut to cut off the integer point (x^*, y^*) in this case.

Next, suppose the B&C terminates and the solution (x^*, y^*) is bilevel feasible, but not optimal. This is not possible, since by construction the DCs never cut off any bilevel feasible solution.

Finally, suppose the B&C never terminates. This is not possible, as all variables are integer and bounded, thus there is only a finite number of nodes in the B&C tree. Moreover, this means there is also a finite number of integer points (x^*, y^*), thus we solve the follower problem and (CG-SOCP) a finite number of times. The follower problem is discrete and can therefore be solved in a finite number of iterations. □

3.2 An Integer Cutting Plane Algorithm

The DCs can be directly used in a cutting plane algorithm under the following assumption.

Assumption 5. *All variables in the bilevel model (1) are binary variables.*

The algorithm is detailed in Algorithm 2. It starts with the HPR as initial relaxation of VFR, which is solved to optimality. Then the chosen DC separation routine (either O or G) is called to check if the obtained integer optimal solution is bilevel feasible. If not, the obtained DC is added to the relaxation to cut off the optimal solution, and the procedure is repeated with the updated relaxation.

Due to Assumption 5 each obtained integer optimal solution is an extreme point of the convex hull of $\overline{\text{HPR}}$, and thus due to Theorem 1 a violated cut will be produced by the DC separation if the solution is not bilevel feasible.

Algorithm 2: cutting plane

   Input : An instance of problem (1) where all variables are binary
   Output: An optimal solution (x^*, y^*)
1 $\mathcal{R} \leftarrow$ HPR; $\mathcal{P} \leftarrow$ set of feasible solutions of $\overline{\text{HPR}}$; violated $\leftarrow$ *True*;
2 **do**
3   violated $\leftarrow$ *False*;
4   solve $\mathcal{R}$ to optimality, let (x^*, y^*) be the obtained optimal solution;
5   call **separation** for (x^*, y^*) and $\mathcal{P}$ with strategy O or G;
6   **if** *a violated cut is found for* (x^*, y^*) **then**
7 | violated $\leftarrow$ *True*; add the violated cut to $\mathcal{R}$ and to $\mathcal{P}$;
8 **while** *violated*;
9 **return** (x^*, y^*)

---

4   Computational Analysis

In this section we present preliminary computational results.

4.1   Instances

In our computations, we consider the quadratic bilevel covering problem

$$\min\ \hat{c}'x + \hat{d}'y \tag{8a}$$

$$\text{s.t.}\ \hat{M}'x + \hat{N}'y \geq \hat{h} \tag{8b}$$

$$y \in \arg\min\{y'\hat{R}y : \hat{a}'x + \hat{b}'y \geq \hat{f},\ y \in \{0,1\}^{n_2}\} \tag{8c}$$

$$x \in \{0,1\}^{n_1}, \tag{8d}$$

where $\hat{c} \in \mathbb{R}^{n_1}$, $\hat{d} \in \mathbb{R}^{n_2}$, $\hat{M} \in \mathbb{R}^{m_1 \times n_1}$, $\hat{N} \in \mathbb{R}^{m_1 \times n_2}$, $\hat{h} \in \mathbb{R}^{m_1}$, $\hat{R} = \hat{V}'\hat{V} \in \mathbb{Z}^{n_2 \times n_2}$, $\hat{a} \in \mathbb{Z}^{n_1}$, $\hat{b} \in \mathbb{Z}^{n_2}$, and $\hat{f} \in \mathbb{Z}$. This problem can be seen as the covering-version of the quadratic bilevel knapsack problem studied by Zenarosa et al. in [40] (there it is studied with a quadratic non-convex leader objective function, only one leader variable and no leader constraint (8b)). The linear variant of such a bilevel knapsack-problem is studied in, e.g., [6,7]. We note that [6,7,40] use problem-specific solution approaches to solve their respective problem. The structure of (8) allows an easy linearization of the nonlinear follower objective function using a standard McCormick-linearization to transform the problem into an MIBLP. Thus we can compare the performance of our algorithm against a state-of-the-art MIBLP-solver MIX++ from Fischetti et al. [12] to get a first impression of whether our development of a dedicated solution approach for IBNPs exploiting nonlinear techniques is a promising endeavour.

We generated random instances in the following way. We consider $n_1 = n_2$ for $n_1 + n_2 = n \in \{20, 30, 40, 50\}$ and we study instances with no (as in [40]) and with one leader constraint (8b), so $m_1 \in \{0, 1\}$. For each n we create five random instances for $m_1 = 0$ and five random instances for $m_1 = 1$. Furthermore, we chose all entries of $\hat{c}$, $\hat{d}$, $\hat{M}$, $\hat{N}$, $\hat{a}$, and $\hat{b}$ randomly from the interval $[0, 99]$. The values of $\hat{h}$ and $\hat{f}$ are set to the sum of the entries of the corresponding rows in the constraint matrices divided by four. The matrix $\hat{V} \in \mathbb{R}^{n_2 \times n_2}$ has random entries from the interval $[0, 9]$.

4.2   Computational Environment

All experiments are executed on a single thread of an Intel Xeon E5-2670v2 machine with 2.5 GHz processor with a memory limit of 8 GB and a time limit of 600 s. Our B&C algorithm and our cutting plane algorithm both are implemented in C++. They make use of IBM ILOG CPLEX 12.10 (in its default settings) as branch-and-cut framework in our B&C algorithm and as solver for $\mathcal{R}$ in our cutting plane algorithm. During the B&C, CPLEX's internal heuristics are allowed and a bilevel infeasible heuristic solution is just discarded if a violated cut cannot be obtained. For calculating the follower response $\hat{y}$ for a given x^*, we also use CPLEX. For solving (CG-SOCP), we use MOSEK [30].

Table 1. Results for the quadratic bilevel covering problem.

n	Setting	t	Gap	RGap	Nodes	nICut	nFCut	t_F	t_S	nSol
20	I-0	1.6	0.0	42.9	158.9	44.0	0.0	0.7	0.4	10
	IF-0	7.0	0.0	46.1	82.8	13.5	151.3	5.0	1.5	10
	I-G	1.1	0.0	42.6	192.4	56.7	0.0	0.2	0.4	10
	IF-G	3.3	0.0	42.1	102.4	17.3	183.9	0.7	2.0	10
30	I-0	26.1	0.0	40.4	2480.0	325.5	0.0	22.3	2.2	10
	IF-0	246.5	9.6	45.9	522.6	24.9	2104.1	216.0	20.3	8
	I-G	2.7	0.0	48.6	1630.6	226.1	0.0	0.4	1.7	10
	IF-G	55.2	0.0	39.6	669.6	29.9	1631.7	5.9	40.5	10
40	I-0	262.0	3.6	70.4	9209.4	1308.8	0.0	233.5	18.6	8
	IF-0	439.9	35.7	66.8	391.3	30.1	1751.9	390.7	43.8	4
	I-G	82.3	0.0	67.9	14225.5	1379.1	0.0	4.2	47.1	10
	IF-G	387.1	6.1	64.0	1039.8	53.4	3783.1	22.0	331.4	8
50	I-0	537.6	46.3	72.5	10921.1	1553.6	0.0	458.4	67.5	2
	IF-0	600.0	71.6	72.7	156.3	24.8	1272.5	545.2	51.5	0
	I-G	417.9	20.2	71.8	93621.8	6928.2	0.0	17.6	102.5	4
	IF-G	519.8	40.5	72.8	2537.6	56.0	12548.1	45.4	244.9	3

4.3   Numerical Results

In this preliminary computational study we use a simplified version of both the the B&C and the cutting plane algorithm, namely we always use the initial $\mathcal{P}$, i.e., the $\overline{\text{HPR}}$, as input for the separation of DCs and do not update it.

While executing our B&C algorithm, we consider four different settings for the separation of cuts. I and IF denote the settings where only integer solutions are separated and where both integer and fractional solutions are separated, respectively. For each of them, we separate the cuts using the routine separation with strategies 0 and G, which is indicated with an "-0" or "-G" next to the relevant setting name. The resulting four settings are I-0, IF-0, I-G and IF-G. Similarly, the cutting plane algorithm is implemented with both separation strategies, leading to the settings CP-0 and CP-G. We determine the minimum acceptable violation $\varepsilon = 10^{-6}$ for our experiments. During the integer separation of (x^*, y^*), while solving the follower problem, we make use of the follower objective function value $q(y^*)$, by setting it as an upper cutoff value. This is a valid approach because a violated DC exists only if $\Phi(x^*) < q(y^*)$.

The results of the B&C algorithm are presented in Table 1, as averages of the problems with the same size n. We provide the solution time t (in seconds), the optimality gap Gap at the end of time limit (calculated as $100(z^* - LB)/z^*$, where z^* and LB are the best objective function value and the lower bound, respectively), the root gap RGap (calculated as $100(z_R^* - LB_r)/z^*$, where z_R^* and LB_r are the best objective function value and the lower bound at the end of the

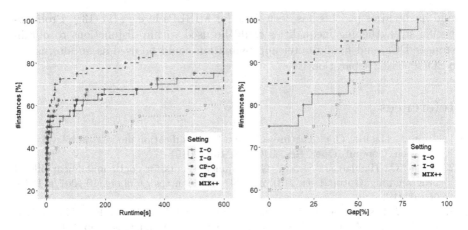

Fig. 1. Runtimes and final optimality gaps for the quadratic bilevel covering problem under our different settings and the benchmark solver MIX++.

root node, respectively), the number of B&C nodes Nodes, the numbers of integer nICut and fractional cuts nFCut, the time t_F to solve the follower problems, the time t_S to solve (CG-SOCP), and the number of optimally solved instance nSol out of 10. I-G is the best performing setting in terms of solution time and final optimality gaps. Although IF-O and IF-G yield smaller trees, they are inefficient because of invoking the separation routine too often, which is computationally costly. Therefore, they are not included in further comparisons.

In Fig. 1, we compare the B&C results with the results obtained by the cutting plane algorithm as well as a state-of-the-art MIBLP solver MIX++ of Fischetti et al. [12], which is able to solve the linearized version of our instances. Figure 1 shows the cumulative distributions of the runtime and the optimality gaps at the end of the time limit. It can be seen that settings with G perform better than their O counterparts. While CP-O and CP-G perform close to I-O, they are significantly outperformed by I-G. The solver MIX++ is also outperformed by both the cutting plane algorithm and the B&C.

5   Conclusions

In this article we showed that SOCP-based DCs are an effective and promising methodology for solving a challenging family of discrete BPs with a convex quadratic objective and linear constraints at the lower level.

There are still many open questions for future research. From the computational perspective, dealing with multiple linking constraints at the lower level requires an implementation of a SOCP-based separation procedure based on multi-disjunctions. The implementation can also be extended to deal with second-order cone constraints at the upper level. Moreover, the proposed B&C could be enhanced by bilevel-specific preprocessing procedures, or bilevel-specific valid inequalities (as this has been done for MIBLPs in e.g., [11–13]). Problem-specific strengthening inequalities could be used within disjunctions to obtain stronger DCs, and finally outer-approximation could be used as an alternative to SOCP-based separation.

References

1. Agor, J., Özaltın, O.Y.: Feature selection for classification models via bilevel optimization. Comput. Oper. Res. **106**, 156–168 (2019)
2. Audet, C., Haddad, J., Savard, G.: Disjunctive cuts for continuous linear bilevel programming. Optim. Lett. **1**(3), 259–267 (2007). https://doi.org/10.1007/s11590-006-0024-3
3. Balas, E.: Disjunctive programming. In: Hammer, P.L., Johnson, E.L., Korte, B.H. (eds.) Annals of Discrete Mathematics 5: Discrete Optimization, pp. 3–51. North Holland (1979)
4. Balas, E.: Disjunctive Programming. Springer, Cham (2018). https://doi.org/10.1007/978-3-030-00148-3
5. Belotti, P., Liberti, L., Lodi, A., Nannicini, G., Tramontani, A., et al.: Disjunctive inequalities: applications and extensions. In: Cochran, J.J., Cox, L.A., Keskinocak, P., Kharoufeh, J.P., Smith, J.C. (eds.) Wiley Encyclopedia of Operations Research and Management Science, vol. 2, pp. 1441–1450. Wiley, Hoboken (2011)
6. Brotcorne, L., Hanafi, S., Mansi, R.: A dynamic programming algorithm for the bilevel knapsack problem. Oper. Res. Lett. **37**(3), 215–218 (2009)
7. Brotcorne, L., Hanafi, S., Mansi, R.: One-level reformulation of the bilevel knapsack problem using dynamic programming. Discrete Optim. **10**(1), 1–10 (2013)
8. Byeon, G., Van Hentenryck, P.: Benders subproblem decomposition for bilevel problems with convex follower. INFORMS J. Comput. (2022)
9. Cerulli, M.: Bilevel optimization and applications. Ph.D. Thesis. École Polytechnique, Paris (2021)
10. Dempe, S., Zemkoho, A. (eds.): Bilevel Optimization. SOIA, vol. 161. Springer, Cham (2020). https://doi.org/10.1007/978-3-030-52119-6
11. Fischetti, M., Ljubić, I., Monaci, M., Sinnl, M.: Intersection cuts for bilevel optimization. In: Louveaux, Q., Skutella, M. (eds.) IPCO 2016. LNCS, vol. 9682, pp. 77–88. Springer, Cham (2016). https://doi.org/10.1007/978-3-319-33461-5_7
12. Fischetti, M., Ljubić, I., Monaci, M., Sinnl, M.: A new general-purpose algorithm for mixed-integer bilevel linear programs. Oper. Res. **65**(6), 1615–1637 (2017)
13. Fischetti, M., Ljubić, I., Monaci, M., Sinnl, M.: On the use of intersection cuts for bilevel optimization. Math. Program. **172**(1), 77–103 (2018). https://doi.org/10.1007/s10107-017-1189-5
14. Fischetti, M., Lodi, A., Tramontani, A.: On the separation of disjunctive cuts. Math. Program. **128**(1), 205–230 (2011). https://doi.org/10.1007/s10107-009-0300-y
15. Fontaine, P., Crainic, T.G., Gendreau, M., Minner, S.: Population-based risk equilibration for the multimode hazmat transport network design problem. Eur. J. Oper. Res. **284**(1), 188–200 (2020)
16. González-Díaz, J., González-Rodríguez, B., Leal, M., Puerto, J.: Global optimization for bilevel portfolio design: economic insights from the Dow Jones index. Omega **102**, 102353 (2021)

17. Grimm, V., Orlinskaya, G., Schewe, L., Schmidt, M., Zöttl, G.: Optimal design of retailer-prosumer electricity tariffs using bilevel optimization. Omega **102**, 102327 (2021)
18. Júdice, J.J., Sherali, H.D., Ribeiro, I.M., Faustino, A.M.: A complementarity-based partitioning and disjunctive cut algorithm for mathematical programming problems with equilibrium constraints. J. Global Optim. **36**(1), 89–114 (2006). https://doi.org/10.1007/s10898-006-9001-8
19. Kılınç-Karzan, F., Yıldız, S.: Two-term disjunctions on the second-order cone. In: Lee, J., Vygen, J. (eds.) IPCO 2014. LNCS, vol. 8494, pp. 345–356. Springer, Cham (2014). https://doi.org/10.1007/978-3-319-07557-0_29
20. Kleinert, T., Grimm, V., Schmidt, M.: Outer approximation for global optimization of mixed-integer quadratic bilevel problems. Math. Program. **188**(2), 461–521 (2021). https://doi.org/10.1007/s10107-020-01601-2
21. Kleinert, T., Labbé, M., Ljubić, I., Schmidt, M.: A survey on mixed-integer programming techniques in bilevel optimization. EURO J. Comput. Optim. **9**, 100007 (2021)
22. Kleniati, P.M., Adjiman, C.S.: Branch and sandwich: a deterministic global optimization algorithm for optimistic bilevel programming problems. Part II: convergence analysis and numerical results. J. Global Optim. **60**(3), 459–481 (2014). https://doi.org/10.1007/s10898-013-0121-7
23. Kleniati, P.M., Adjiman, C.S.: A generalization of the branch-and-sandwich algorithm: from continuous to mixed-integer nonlinear bilevel problems. Comput. Chem. Eng. **72**, 373–386 (2015)
24. Kleniati, P.M., Adjiman, C.S.: Branch-and-sandwich: a deterministic global optimization algorithm for optimistic bilevel programming problems. Part I: theoretical development. J. Global Optim. **60**(3), 425–458 (2014). https://doi.org/10.1007/s10898-013-0121-7
25. Labbé, M., Violin, A.: Bilevel programming and price setting problems. Ann. Oper. Res. **240**(1), 141–169 (2015). https://doi.org/10.1007/s10479-015-2016-0
26. Lodi, A., Ralphs, T.K., Woeginger, G.J.: Bilevel programming and the separation problem. Math. Program. **146**(1), 437–458 (2014). https://doi.org/10.1007/s10107-013-0700-x
27. Loridan, P., Morgan, J.: Weak via strong stackelberg problem: new results. J. Global Optim. **8**(3), 263–287 (1996). https://doi.org/10.1007/BF00121269
28. Lozano, L., Smith, J.C.: A value-function-based exact approach for the bilevel mixed-integer programming problem. Oper. Res. **65**(3), 768–786 (2017)
29. Mitsos, A., Lemonidis, P., Barton, P.I.: Global solution of bilevel programs with a nonconvex inner program. J. Global Optim. **42**(4), 475–513 (2008). https://doi.org/10.1007/s10898-007-9260-z
30. MOSEK ApS: MOSEK Fusion API for C++ manual. Version 9.2. (2021). http://docs.mosek.com/9.2/toolbox/index.html
31. Plein, F., Thürauf, J., Labbé, M., Schmidt, M.: Bilevel optimization approaches to decide the feasibility of bookings in the European gas market. Math. Method. Oper. Res. (2021). https://doi.org/10.1007/s00186-021-00752-y
32. Saxena, A., Bonami, P., Lee, J.: Convex relaxations of non-convex mixed integer quadratically constrained programs: extended formulations. Math. Program. **124**, 383–411 (2010). https://doi.org/10.1007/s10107-010-0371-9
33. Saxena, A., Bonami, P., Lee, J.: Convex relaxations of non-convex mixed integer quadratically constrained programs: projected formulations. Math. Program. **130**, 359–413 (2010). https://doi.org/10.1007/s10107-010-0340-3

34. Smith, J.C., Song, Y.: A survey of network interdiction models and algorithms. Eur. J. Oper. Res. **283**(3), 797–811 (2020)
35. Tahernejad, S., Ralphs, T.K., DeNegre, S.T.: A branch-and-cut algorithm for mixed integer bilevel linear optimization problems and its implementation. Math. Program. Comput. **12**(4), 529–568 (2020). https://doi.org/10.1007/s12532-020-00183-6
36. Wang, L., Xu, P.: The watermelon algorithm for the bilevel integer linear programming problem. SIAM J. Optimiz. **27**(3), 1403–1430 (2017)
37. Weninger, D., Orlinskaya, G., Merkert, M.: An exact projection-based algorithm for bilevel mixed-integer problems with nonlinearities (2020). http://www.optimization-online.org/DB_HTML/2020/12/8153.html
38. Xu, P., Wang, L.: An exact algorithm for the bilevel mixed integer linear programming problem under three simplifying assumptions. Comput. Oper. Res. **41**, 309–318 (2014)
39. Yue, D., Gao, J., Zeng, B., You, F.: A projection-based reformulation and decomposition algorithm for global optimization of a class of mixed integer bilevel linear programs. J. Global Optim. **73**(1), 27–57 (2018). https://doi.org/10.1007/s10898-018-0679-1
40. Zenarosa, G.L., Prokopyev, O.A., Pasiliao, E.L.: On exact solution approaches for bilevel quadratic 0–1 knapsack problem. Ann. Oper. Res. **298**(1), 555–572 (2021). https://doi.org/10.1007/s10479-018-2970-4

Non-adaptive Stochastic Score Classification and Explainable Halfspace Evaluation

Rohan Ghuge[1]([✉]), Anupam Gupta[2], and Viswanath Nagarajan[1]

[1] University of Michigan, Ann Arbor, MI, USA
rghuge@umich.edu
[2] Carnegie Mellon University, Pittsburgh, PA, USA

Abstract. We consider the stochastic score classification problem. There are several binary tests, where each test i is associated with a probability p_i of being positive, a cost c_i, and a weight a_i. The score of an outcome is a weighted sum of all positive tests, and the range of possible scores is partitioned into intervals corresponding to different classes. The goal is to perform tests sequentially (and possibly adaptively) so as to identify the class at the minimum expected cost. We provide the first constant-factor approximation algorithm for this problem, which improves over the previously-known logarithmic approximation ratio. Moreover, our algorithm is *non adaptive*: it just involves performing tests in a *fixed* order until the class is identified. Our approach also extends to the d-dimensional score classification problem and the "explainable" stochastic halfspace evaluation problem (where we want to evaluate some function on d halfspaces). We obtain an $O(d^2 \log d)$-approximation algorithm for both these extensions. Finally, we perform computational experiments that demonstrate the practical performance of our algorithm for score classification. We observe that, for most instances, the cost of our algorithm is within 50% of an information-theoretic lower bound on the optimal value.

Keywords: Stochastic optimization · Approximation algorithms · Stochastic probing · Adaptivity

1 Introduction

The problem of diagnosing complex systems often involves running a large number of tests for each component of such a system. One option to diagnose such systems is to perform tests on *all* components, which can be prohibitively expensive and slow. Therefore, we are interested in a policy that tests components one by

---

The full version of the paper is available at [13]. R. Ghuge and V. Nagarajan were supported in part by NSF grants CMMI-1940766 and CCF-2006778. A. Gupta was supported in part by NSF awards CCF-1907820, CCF-1955785, and CCF-2006953.

K. Aardal and L. Sanitá (Eds.): IPCO 2022, LNCS 13265, pp. 277–290, 2022.
https://doi.org/10.1007/978-3-031-06901-7_21

one, and minimizes the average cost of testing. (See [23] for a survey.) Concretely, we consider a setting where the goal is to test various components, in order to assign a risk class to the system (e.g., whether the system has low/medium/high risk).

The *stochastic score classification* (SSClass) problem introduced by [14] models such situations. There are n components in a system, where each component i is "working" with independent probability p_i. While the probabilities p_i are known *a priori*, the random outcomes $X_i \in \{0, 1\}$ are initially unknown. The outcome X_i of each component i can be determined by performing a test of cost c_i: $X_i = 1$ if i is working and $X_i = 0$ otherwise. The overall status of the system is determined by a linear score $r(X) := \sum_{i=1}^{n} a_i X_i$, where the coefficients $a_i \in \mathbb{Z}$ are input parameters. We are also given a collection of intervals $I_1, I_2, \ldots, I_k$ that partition the real line (i.e., all possible scores). The goal is to determine the interval I_j (also called the *class*) that contains $r(X)$, while incurring minimum expected cost. A well-studied special case is when there are just two classes, which corresponds to evaluating a halfspace or linear-threshold-function [11].

Example: Consider a system which must be assigned a risk class of *low, medium,* or *high*. Suppose there are five components in the system, each of which is working with probability $\frac{1}{2}$. The score for the entire system is the number of working components. A score of 5 corresponds to the "Low" risk class, scores between 2 and 4 correspond to "Medium" risk, and a score of at most 1 signifies "High" risk. Suppose that after testing components $\{1, 2, 3\}$, the system has score 2 (which occurs with probability $\frac{3}{8}$) then it will be classified as medium risk irrespective of the remaining two components: so testing can be stopped. Instead, if the system has score 3 after testing components $\{1, 2, 3\}$ (which occurs with probability $\frac{1}{8}$) then the class of the system cannot be determined with certainty (it may be either medium or low), and so further testing is needed.

A related problem is the *d-dimensional stochastic score classification problem* (*d*-SSClass), which models the situation when a system has d different functions, each with an associated linear score (as above). We must now perform tests on the underlying components to *simultaneously* assign a class to each of the d functions.

In another related problem, a system again has d different functions. Here, the status (working or failed) of each function is determined by some halfspace, and the overall system is considered operational if all d functions are working. The goal of a diagnosing policy is to decide whether the system is operational, and if not, to return at least one function that has failed (and therefore needs maintenance). This is a special case of a problem we call *explainable stochastic halfspace evaluation* (EX-SFE).

Solutions for all these problems (SSClass, *d*-SSClass, and EX-SFE) are *sequential decision processes*. At each step, a component is tested and its outcome (working or failed) is observed. The information from all previously tested components can then be used to decide on the next component to test; this makes the process *adaptive*. This process continues until the risk class can be determined with certainty from the tested components. One simple class of solutions are *non-adaptive* solutions, which are simply described by a priority list:

we then test components in this fixed order until the class can be uniquely determined. Such solutions are simpler and faster to implement, compared to their adaptive counterparts: the non-adaptive testing sequence needs to be constructed just once, after which it can be used for all input realizations. However, non-adaptive solutions are weaker than adaptive ones, and our goal is to bound the *adaptivity gap*, the multiplicative ratio between the performance of the non-adaptive solution to that of the optimal adaptive one. Our main result shows that SSClass has a constant adaptivity gap, thereby answering an open question posed by [14]. Additionally, we show an adaptivity gap of $O(d^2 \log d)$ for both the d-SSClass and EX-SFE problems.

Before we present the results and techniques, we formally define the problem. For any integer m, we use $[m] := \{1, 2, \ldots, m\}$. An instance of SSClass consists of n independent $\{0, 1\}$ random variables $X = X_1, \ldots, X_n$, where variable X_i has $\Pr[X_i = 1] = \mathbb{E}[X_i] = p_i$. The cost to probe/query X_i is $c_i \in \mathbb{R}_+$; both p_i, c_i are known to us. We are also given non-negative weights $a_i \in \mathbb{Z}_+$, and the *score* of the outcome $X = (X_1, \ldots, X_n)$ is $r(X) = \sum_{i=1}^{n} a_i X_i$. In addition, we are given $B + 1$ integers $\alpha_1, \ldots, \alpha_{B+1}$ such that *class j* corresponds to the interval $I_j := \{\alpha_j, \ldots, \alpha_{j+1} - 1\}$. The *score classification function* $h : \{0, 1\}^n \to \{1, \ldots, B\}$ assigns $h(X) = j$ precisely when $r(X) \in I_j$. The goal is to determine $h(X)$ at minimum expected cost. We assume non-negative weights only for simplicity: any instance with positive and negative weights can be reduced to an equivalent instance with all positive weights (see full version for details). Let $W := \sum_{i=1}^{n} a_i$ denote the total weight. In our algorithm, we associate two numbers $(\beta_j^0, \beta_j^1) \in \mathbb{Z}_+^2$ with each class j, where $\beta_j^0 = W - \alpha_{j+1} + 1$ and $\beta_j^1 = \alpha_j$.

1.1 Results and Techniques

Our main result is the following algorithm (and adaptivity gap).

Theorem 1. *There is a polynomial-time non-adaptive algorithm (called NaCl) for stochastic score classification with expected cost at most a constant factor times that of the optimal adaptive policy.*

This result improves on the prior work from [11,14] in several ways. Firstly, we get a constant-factor approximation, improving upon the previous $O(\log W)$ and $O(B)$ ratios, where W is the sum of weights, and B the number of classes. Secondly, our algorithm is non-adaptive in contrast to the previous adaptive ones. Finally, our algorithm has nearly-linear runtime, which is faster than the previous algorithms.

An added benefit of our approach is that we obtain a "universal" solution that is simultaneously $O(1)$-approximate for all class-partitions. Indeed, the non-adaptive list produced by NaCl only depends on the probabilities, costs, and weights, and not on the class boundaries $\{\alpha_j\}$; these α_j values are only needed in the stopping condition for probing.

Algorithm Overview. To motivate our algorithm, suppose that we have probed a subset $S \subseteq [n]$ of variables, and there is a class j such that $\sum_{i \in S} a_i X_i \geq \alpha_j$ and $\sum_{i \in S} a_i (1 - X_i) \geq W - \alpha_{j+1} + 1$. The latter condition can be rewritten as $\sum_{i \in S} a_i X_i + \sum_{i \notin S} a_i \leq \alpha_{j+1} - 1$. So, we can conclude that the final score lies in $\{\alpha_j, \ldots, \alpha_{j+1} - 1\}$ irrespective of the outcomes of variables in $[n] \setminus S$. This means that $h(X) = j$. On the other hand, if the above condition is not satisfied for *any* class j, we must continue probing. Towards this end, we define two types of rewards for each variable $i \in [n]$: $R_0(i) = a_i \cdot (1 - X_i)$ and $R_1(i) = a_i \cdot X_i$. (See Fig. 1.) The total R_0-reward and R_1-reward from the probed variables correspond to upper and lower bounds on the score, respectively. Our non-adaptive algorithm NACL probes the variables in a predetermined order until $\sum_{i \in S} R_0(i) \geq \beta_j^0 = W - \alpha_{j+1} + 1$ and $\sum_{i \in S} R_1(i) \geq \beta_j^1 = \alpha_j$ for *some* class j (at which point it determines $h(X) = j$).

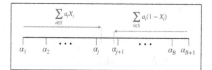 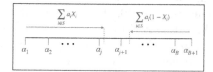

(a) The R_0 reward (upper bound) and R_1 reward (lower bound) lie in class j and hence probing can be stopped.

(b) Here $f(X)$ could be j or $j + 1$, so probing must continue.

Fig. 1. Illustration of non-adaptive approach

To get this ordering we first build two separate lists: list L_b for the R_b-rewards (for $b = 0, 1$) minimizes the cost required to cover some target amount of R_b-reward. Finally, interleaving lists L_0 and L_1 gives the final list. The idea behind list L_0 is as follows: if we only care about a *single* class j, we can set a target of β_j^0 and use the non-adaptive algorithm for stochastic knapsack cover [21]. Since the class j is unknown, so is the target β_j^0 on the R_0-reward. Interestingly, we show how to construct a "universal" non-adaptive list L_0 that works for *all* targets simultaneously. The construction proceeds in phases: in each phase $\ell \geq 0$, the algorithm adds a subset of variables with cost $O(2^\ell)$ that (roughly) maximizes the *expected* R_0 reward. Naïvely using the expected rewards can lead to poor performance, so a natural idea is to use rewards *truncated* at logarithmically-many scales (corresponding to the residual target); see for example, [12]. Moreover, to get a constant-factor approximation, we use the *critical scaling* idea from [21]. Roughly speaking, this identifies a *single* scale κ such that with constant probability (1) the algorithm obtains large reward (truncated at scale κ), and (2) any subset of cost 2^ℓ has small reward.

Analysis Overview. The analysis of Theorem 1 relates the "non-completion" probabilities of our algorithm after cost $\gamma \cdot 2^\ell$ to that of the optimal adaptive algorithm after cost 2^ℓ, for each phase $\ell \geq 0$. The factor γ corresponds to the approximation ratio of the algorithm. In order to relate these non-completion probabilities, we consider the R_0 and R_1-rewards obtained by an optimal adaptive algorithm, and argue that the non-adaptive algorithm obtains a higher R_0 as *well as* R_1 reward (with constant probability). Thus, if the optimal adaptive algorithm decides $h(X)$, so does the non-adaptive algorithm. Our algorithm/analysis also use various properties of the fractional knapsack problem.

Extensions. Next, we consider the d-dimensional score classification problem (d-SSClass) and obtain the following result:

Theorem 2. *There is a non-adaptive $O(d^2 \log d)$-approximation algorithm for d-dimensional stochastic score classification.*

We achieve this by extending the above approach (for $d = 1$). We now define two rewards (corresponding to R_0 and R_1) for *each* dimension d. Then, we apply the list building algorithm for each of these rewards, resulting in $2d$ separate lists. Finally, we interleave these lists to obtain the non-adaptive probing sequence. The analysis is also an extension of the $d = 1$ case. The main differences are as follows. Just accounting for the $2d$ lists results in an extra $O(d)$ factor in the approximation. Furthermore, we need to ensure (for each phase ℓ) that with constant probability, our non-adaptive algorithm achieves more reward than the optimum for *all* the $2d$ rewards. We incur another $O(d \log d)$ factor in the approximation in order to achieve this stronger property.

In a similar vein, our main result for EX-SFE is the following:

Theorem 3. *There is a non-adaptive $O(d^2 \log d)$-approximation algorithm for explainable stochastic halfspace evaluation.*

The non-adaptive list for EX-SFE is constructed in the same manner as for d-SSClass, but the stopping rule relies on the oracle for verifying witnesses of $f \circ h$. As a special case, we obtain:

Corollary 1. *There is a non-adaptive $O(d^2 \log d)$-approximation algorithm for the explainable stochastic intersection of half-spaces problem.*

The stochastic intersection of halfspaces problem (in a slightly different model) was studied previously by [6], where an $O(\sqrt{n \log d})$-approximation algorithm was obtained assuming all probabilities $p_i = \frac{1}{2}$. The main difference from our model is that [6] do not require a witness at the end. So their policy can stop if it concludes that there exists a violated halfspace (even without knowing which one), whereas our policy can only stop after it identifies a violated halfspace (or determines that all halfspaces are satisfied). We note that our approximation ratio is independent of the number of variables n and holds for arbitrary probabilities. The proofs of Theorems 2 and 3 are omitted in this extended abstract and appear in the full version of the paper.

Computational Results. Finally, we evaluate the empirical performance of our algorithm for score classification. In these experiments, our non-adaptive algorithm performs nearly as well as the previous-best *adaptive* algorithms, while being an order of magnitude faster. In fact, on many instances, our algorithm provides an *improvement* in both the cost as well as the running time. On most instances, the cost of our algorithm is within 50% of an information-theoretic lower bound on the optimal value.

1.2   Related Work

The special case of SSClass with $B = 2$ classes is the well-studied stochastic Boolean function evaluation for *linear threshold functions* (SBFT). Here, the goal is to identify whether a single halfspace is satisfied (i.e., the score is above or below a threshold). [11] gave an elegant 3-approximation algorithm for SBFT using an adaptive dual greedy approach. Prior to their work, only an $O(\log W)$-approximation was known, based on the more general stochastic submodular cover problem [15, 20].

The general SSClass problem was introduced by [14], who showed that it can be formulated as an instance of stochastic submodular cover. Then, using general results such as [15, 20], they obtained an adaptive $O(\log W)$-approximation algorithm. Furthermore, [14] obtained an adaptive $3(B-1)$-approximation algorithm for SSClass by extending the approach of [11] for SBFT; recall that B is the number of classes. A main open question from this work was the possibility of a constant approximation for the general SSClass problem. We answer this in the affirmative. Moreover, our algorithm is non-adaptive: so we also bound the adaptivity gap.

The stochastic knapsack cover problem (SKC) is closely related to SBFT. Given a set of items with random rewards and a target k, the goal is to (adaptively) select a subset of items having total reward at least k. The objective is to minimize the expected cost of selected items. [11] gave an adaptive 3-approximation algorithm for SKC. Later, [21] gave a *non-adaptive* $O(1)$-approximation algorithm for SKC. In fact, the result in [21] applied to the more general stochastic k-TSP problem [12]. Our algorithm and analysis use some ideas from [12, 21]. We use the notion of a "critical scale" from [21] to identify the correct reward truncation threshold. The approach of using non-completion probabilities in the analysis is similar to [12]. There are also a number of differences: we exploit additional structure in the (fractional) knapsack problem and obtain a simpler and nearly-linear time algorithm.

More generally, non-adaptive solutions (and adaptivity gaps) have been used in solving various other stochastic optimization problems such as max-knapsack [5, 10], matching [2, 4], probing [18, 19] and orienteering [3, 16, 17]. Our result shows that this approach is also useful for SSClass.

SSClass and EX-SFE also fall under the umbrella of designing query strategies for "priced information", where one wants to evaluate a function by sequentially querying variables (that have costs). There are two lines of work here: comparing to an optimal strategy (as in our model) [1, 6, 14, 22], and comparing to the min-cost solution in hindsight (i.e., competitive analysis) [7–9]. We note that the "explainable" requirement in the EX-SFE problem (that we solve) is similar to the requirement in [22].

2   Preliminaries

We first state some basic results for the deterministic knapsack problem. In an instance of the knapsack problem, we are given a set T of items with non-negative costs $\{c_i : i \in T\}$ and rewards $\{r_i : i \in T\}$, and a budget D on the total cost. The goal is to select a subset of items of total cost at most D that maximizes the total reward. The LP relaxation is the following:

$$g(D) = \max\left\{\sum_{i \in T} r_i \cdot x_i \mid \sum_{i \in T} c_i \cdot x_i \leq D,\, x \in [0,1]^T\right\}, \qquad \forall D \geq 0.$$

The following algorithm $\mathcal{A}_{\mathsf{KS}}$ solves the *fractional* knapsack problem and also obtains an approximate integral solution. Assume that the items are ordered so that $\frac{r_1}{c_1} \geq \frac{r_2}{c_2} \geq \cdots$. Let t index the first item (if any) so that $\sum_{i=1}^{t} c_i \geq D$. Let $\psi := \frac{1}{c_t}(D - \sum_{i=1}^{t-1} c_i)$ which lies in $(0,1]$. Define

$$x_i = \begin{cases} 1 & \text{if } i \leq t-1 \\ \psi & \text{if } i = t \\ 0 & \text{if } i \geq t+1 \end{cases}.$$

Return x as the optimal fractional solution and $Q = \{1, \cdots, t\}$ as an integer solution. We use the following well-known result (proved in the full version, for completeness).

Theorem 4. *Consider algorithm $\mathcal{A}_{\mathsf{KS}}$ on any instance of the knapsack problem with budget D.*

1. *$\langle r, x \rangle = \sum_{i=1}^{t-1} r_i + \psi \cdot r_t = g(D)$ and so x is an optimal LP solution.*
2. *The derivative $g'(D) = \frac{r_t}{c_t}$.*
3. *Solution Q has cost $c(Q) \leq D + c_{max}$ and reward $r(Q) \geq g(D)$.*
4. *$g(D)$ is a concave function of D.*

3 The Stochastic Score Classification Algorithm

Our non-adaptive algorithm creates two lists L_0 and L_1 separately. These lists are based on the R_0 and R_1 rewards of the variables, where $R_0(i) = a_i(1 - X_i)$ and $R_1(i) = a_i X_i$. It interleaves lists L_0 and L_1 together (by power-of-2 costs) and then probes the variables in this non-adaptive order until the class is identified.

3.1 The Algorithm

We first explain how to build the lists L_0 and L_1. We only consider list L_0 below (the algorithm/analysis for L_1 are identical). The list building algorithm operates in phases. For each phase $\ell \geq 0$ it gets a budget of $O(2^\ell)$, and it solves several instances of the deterministic knapsack problem, where rewards are truncated expectations of R_0. We will use the following truncation values, also called *scales*.

$$\mathcal{G} := \left\{\theta^\ell \;:\; 0 \leq \ell \leq 1 + \log_\theta W\right\}, \text{ where } \theta > 1 \text{ is a constant.}$$

For each scale $\tau \in \mathcal{G}$, we find a deterministic knapsack solution with reward $\mathbb{E}[\min\{R_0/\tau, 1\}]$ (see Equation 1 for the formal definition) and budget $\approx C2^\ell$

(where $C > 1$ is a constant). Including solutions for each scale would lead to an $O(\log W)$ loss in the approximation factor. Instead, as in [21], we identify a "critical scale" and only include solutions based on the critical scale. To this end, each scale τ is classified as either *rich* or *poor*. Roughly, in a rich scale, the knapsack solution after budget $C2^\ell$ still has large "incremental" reward (formalized by the derivative of g being at least some constant δ). The *critical scale* is the smallest scale κ that is poor, and so represents a transition from rich to poor. For our analysis, we will choose constant parameters C, δ and θ so that $\frac{C\delta}{\theta} > 1$. We note however, that our algorithm achieves a constant approximation ratio for any constant values $C > 1$, $\delta \in (0, 1)$ and $\theta > 0$.

Algorithm 1. PICKREPS$(\ell, \tau, \mathbf{r})$

1: let $T \subseteq [n]$ denote the variables with non-zero reward and cost at most 2^ℓ
2: run algorithm $\mathcal{A}_{\text{KS}}$ (Theorem 4) on the knapsack instance with items T and budget $D = C2^\ell$
3: let $f = g'(D)$ be the derivative of the LP value and $Q \subseteq T$ the integral solution from $\mathcal{A}_{\text{KS}}$
4: **if** $f > \delta 2^{-\ell}$ **then**
5:     scale τ is **rich**
6: **else**
7:     scale τ is **poor**
8: **return** Q

Subroutine PICKREPS (Algorithm 1) computes the knapsack solution for each scale τ, and classifies the scale as rich/poor. The subroutine BUILDLIST(R) (Algorithm 2) builds the list for any set of random rewards $\{R(i) : i \in [n]\}$. List L_b (for $b = 0, 1$) is obtained by running BUILDLIST(R_b). Finally, the non-adaptive algorithm NACL involves interleaving the variables in lists L_0 and L_1; this is described in Algorithm 3. The resulting policy probes variables in the order given by NACL until the observed upper and lower bounds on the score lie within the same class. Note that there are $O(\log(nc_{max}))$ phases and $O(\log W)$ scales: so the total number of deterministic knapsack instances solved is polylogarithmic. Moreover, the knapsack algorithm $\mathcal{A}_{\text{KS}}$ runs in $O(n \log n)$ time. So the overall runtime of our algorithm is nearly linear.

3.2 The Analysis

Lemma 1. *The critical scale κ in Step 6 of Algorithm 2 is always well defined.*

Proof. To prove that there is a smallest poor scale, it suffices to show that not all scales can be rich. We claim that the last scale $\tau \geq W$ cannot be rich. Suppose (for a contradiction) that scale τ is rich. Then, by concavity of g (see property 4 in Theorem 4), we have $g(D) \geq D \cdot g'(D) > D \cdot \delta 2^{-\ell} = C\delta \geq 1$. On the other hand, the total deterministic reward at this scale, $\sum_{i=1}^n r_i^\tau \leq \frac{W}{\tau} \leq 1$. Thus, $g(D) \leq 1$, a contradiction.

Algorithm 2. BUILDLIST($\{R(i) : i \in [n]\}$)

1: list $\Pi \leftarrow \emptyset$
2: **for** phase $\ell = 0, 1, \ldots$ **do**
3:    **for** each scale $\tau \in \mathcal{G}$ **do**
4:       define truncated rewards

$$r_i^\tau = \begin{cases} \mathbb{E}\left[\min\left\{\frac{R(i)}{\tau}, 1\right\}\right], & \text{if } i \notin \Pi \\ 0, & \text{otherwise} \end{cases} \tag{1}$$

5:       $S_{\ell,\tau} \leftarrow \text{PICKREPS}(\ell, \tau, \mathbf{r}^\tau)$
6:    let κ be the smallest **poor** scale in $\mathcal{G}$ (this is called the critical scale)
7:    $\Pi_\ell \leftarrow S_{\ell,\kappa}$ and $\Pi \leftarrow \Pi \circ \Pi_\ell$
8: **return** list Π

Algorithm 3. NACL (NON-ADAPTIVE CLASSIFIER)

1: list $L_b \leftarrow \text{BUILDLIST}(\{R_b(i) : i \in [n]\})$ for $b = 0, 1$
2: let L_b^ℓ denote the variables in phase ℓ for list L_b
3: for each phase ℓ, set $S_\ell \leftarrow L_0^\ell \cup L_1^\ell$
4: **return** list $S_0, S_1, \cdots, S_\ell \cdots$

Lemma 2. *The cost* $c(S_{\ell,\tau}) \leq (C+1)2^\ell$ *for any phase* ℓ. *Hence, the cost incurred in phase* ℓ *of* NACL *is at most* $(C+1)2^{\ell+1}$.

Proof. Consider any call to PICKREPS in phase ℓ. We have $S_{\ell,\tau} = Q$ where Q is the integer solution from Theorem 4. It follows that $c(S_{\ell,\tau}) = c(Q) \leq C2^\ell + \max_{i \in T} c_i \leq (C+1)2^\ell$; note that we only consider variables of cost at most 2^ℓ (see Step 1 of Algorithm 1). Finally, the variables S_ℓ in phase ℓ of NACL consist of the phase-ℓ variables of both L_0 and L_1. So the total cost of these variables is at most $(C+1)2^{\ell+1}$. □

We now analyze the cost incurred by our non-adaptive strategy NACL. We denote by OPT an optimal adaptive solution for SSClass. To analyze the algorithm, we use the following notation.

- u_ℓ: probability that NACL is not complete by end of phase ℓ.
- u_ℓ^*: probability that OPT costs at least 2^ℓ.

We can assume by scaling that the minimum cost is 1. So $u_0^* = 1$. For ease of notation, we use OPT and NA to denote the *random* cost incurred by OPT and NACL respectively. We also divide OPT into phases: phase ℓ corresponds to variables in OPT after which the cumulative cost is between $2^{\ell-1}$ and 2^ℓ. The following lemma forms the crux of the analysis.

Lemma 3. *For any phase* $\ell \geq 1$, *we have* $u_\ell \leq q \cdot u_{\ell-1} + u_\ell^*$ *where* $q \leq 0.3$.

Given Lemma 3, the proof of Theorem 1 is standard (see, for example, [12]).

3.3   Proof of Lemma 3

Recall that NACL denotes the non-adaptive algorithm, and NA its random cost. Fix any phase $\ell \geq 1$, and let σ denote the realization of the variables probed in the first $\ell - 1$ phases of NACL. We further define the following *conditioned on* σ:

- $u_\ell(\sigma)$: probability that NACL is not complete by end of phase ℓ.
- $u_\ell^*(\sigma)$: probability that OPT costs at least 2^ℓ, i.e., OPT is not complete by end of phase ℓ.

If NACL does not complete before phase ℓ then $u_{\ell-1}(\sigma) = 1$, and we will prove

$$u_\ell(\sigma) \leq u_\ell^*(\sigma) + 0.3. \tag{2}$$

We can complete the proof using this. Note that $u_{\ell-1}(\sigma)$ is either 0 or 1. If $u_{\ell-1}(\sigma) = 0$ then $u_\ell(\sigma) = 0$ as well. So, Eq. (2) implies that $u_\ell(\sigma) \leq u_\ell^*(\sigma) + 0.3u_{\ell-1}(\sigma)$ for all σ. Taking expectation over σ gives Lemma 3. It remains to prove Eq. (2).

We denote by $\mathcal{R}_0$ and $\mathcal{R}_0^*$ the total R_0 reward obtained in the first ℓ phases by NACL and OPT respectively. We similarly define $\mathcal{R}_1$ and $\mathcal{R}_1^*$. To prove Equation (2), we will show that the probabilities (conditioned on σ) $\mathbf{P}(\mathcal{R}_0^* > \mathcal{R}_0)$ and $\mathbf{P}(\mathcal{R}_1^* > \mathcal{R}_1)$ are small. Intuitively, this implies that with high probability, if OPT finishes in phase ℓ, then so does NACL. Formally, we prove the following key lemma.

Lemma 4 (Key Lemma). *For $b \in \{0, 1\}$, we have $\mathbf{P}(\mathcal{R}_b < \mathcal{R}_b^* \mid \sigma) \leq 0.15$.*

Using these lemmas, we prove Equation (2).

Proof (Proof of Equation (2)). Recall that we associate a pair (β_j^0, β_j^1) with every class j. If OPT finishes in phase ℓ, then there exists some j such that $\mathcal{R}_0^* \geq \beta_j^0$ and $\mathcal{R}_1^* \geq \beta_j^1$. Thus,

$$\mathbf{P}(\text{OPT finishes in phase } \ell \mid \sigma) = 1 - u_\ell^*(\sigma) = \mathbf{P}(\exists j : \mathcal{R}_0^* \geq \beta_j^0 \text{ and } \mathcal{R}_1^* \geq \beta_j^1 \mid \sigma).$$

From Lemma 4 and union bound, we have $\mathbf{P}(\mathcal{R}_0 < \mathcal{R}_0^* \text{ or } \mathcal{R}_1 < \mathcal{R}_1^* \mid \sigma) \leq 0.3$. Then, we have

$$
\begin{aligned}
1 - u_\ell(\sigma) &= \mathbf{P}(\text{NA finishes in phase } \ell \mid \sigma) \\
&\geq \mathbf{P}\left((\text{OPT finishes in phase } \ell) \bigwedge \mathcal{R}_0 \geq \mathcal{R}_0^* \bigwedge \mathcal{R}_1 \geq \mathcal{R}_1^* \mid \sigma\right) \\
&\geq \mathbf{P}(\text{OPT finishes in phase } \ell \mid \sigma) - \mathbf{P}(\mathcal{R}_0 < \mathcal{R}_0^* \text{ or } \mathcal{R}_1 < \mathcal{R}_1^* \mid \sigma) \\
&\geq (1 - u_\ell^*(\sigma)) - 0.3
\end{aligned}
$$

Upon rearranging, this gives $u_\ell(\sigma) \leq u_\ell^*(\sigma) + 0.3$ as desired.   □

Proof Sketch of the Key Lemma. We now provide an intuition for the proof of Lemma 4 with $b = 0$ (the case $b = 1$ is identical). Henceforth, reward will only refer to R_0. Observe that in phase ℓ of Algorithm 2, the previously probed variables $\Pi \subseteq \sigma$. For ease of notation, let σ also represent the set of variables probed in the first $\ell - 1$ phases. Recall that S_ℓ is the set of variables probed by NACL in phase ℓ. Let O_ℓ be the variables probed by OPT in phase ℓ; so the total cost of O_ℓ is at most 2^ℓ. Note that O_ℓ can be a random subset as OPT is adaptive. Also, S_ℓ is a deterministic subset as NACL is a non-adaptive list. Roughly, we show that (conditioned on σ) the probability that O_ℓ has more reward than S_ℓ is small. The key idea is to use the *critical scale* κ (in phase ℓ) to argue that the following hold with constant probability (1) reward of $O_\ell \setminus (S_\ell \cup \sigma)$ is at most κ, and (2) reward of $S_\ell \setminus O_\ell$ is at least κ. This would imply that with constant probability, NACL gets at least as much reward as OPT by the end of phase ℓ. Formally, we show:

Lemma 5. *If $\mathcal{A}$ is any adaptive policy of selecting variables from $[n] \setminus (S_\ell \cup \sigma)$ with total cost $\leq 2^\ell$ then* $\mathbf{P}\left[R_0(\mathcal{A}) < \kappa\right] \geq 1 - \delta$. *Hence,*

$$\mathbf{P}\left[R_0(O_\ell \setminus (S_\ell \cup \sigma)) \geq \kappa\right] \leq \delta.$$

Lemma 6. *We have* $\mathbf{P}\left(R_0(S_\ell \setminus O_\ell) < \kappa\right) \leq e^{-(\mu - \ln \mu - 1)}$. *Here,* $\mu := (C-1)\delta/\theta$.

Combining these two lemmas with an appropriate choice of constants C, δ and θ, we obtain Lemma 4. We defer the proofs to the full version.

4   Computational Results

We provide a summary of computational results of our non-adaptive algorithm for the stochastic score classification problem. We conducted all of our computational experiments using Python 3.8 with a 2.3 GHz Intel Core i5 processor and 16 GB 2133MHz LPDDR3 memory. We use synthetic data to generate instances of SSClass for our experiments.

Instance Generation. We test our algorithm on synthetic data generated as follows. We first set $n \in \{100, 200, \ldots, 1000\}$. Given n, we generate n Bernoulli variables, each with probability chosen uniformly from $(0, 1)$. We set the costs of each variable to be an integer in $[10, 100]$. To select cutoffs (when $B \neq 2$), we first select $B \in \{5, 10, 15\}$ and then select the cutoffs (based on the value of B) uniformly at random in the score interval. We provide more details and plots in the full version of the paper. For each n we generate 10 instances. For each instance, we sample 50 realizations in order to calculate the average cost and average runtime.

Algorithms. We compare our non-adaptive SSClass algorithm (Theorem 1) against a number of prior algorithms. For SBFT instances, we compare to the *adaptive* 3-approximation algorithm from [11]. For *unweighted* SSClass instances, we compare to the non-adaptive $2(B - 1)$-approximation algorithm from [14]. For general SSClass instances, we compare to the *adaptive* $O(\log W)$-approximation algorithm from [14]. As a benchmark, we also compare to a naive non-adaptive algorithm that probes variables in a random order. We also compare to an information-theoretic lower bound (no adaptive policy can do better than this lower bound). We obtain this lower bound by using an integer linear program to compute the (offline) optimal probing cost for a given realization (see full version for details), and then taking an average over 50 realizations.

Parameters C, δ, and θ. As noted in Sect. 3, our algorithm achieves a constant factor approximation guarantee for any constant $C > 1$, $\delta \in (0, 1)$ and $\theta > 1$. For our final computations, we (arbitrarily) choose values $C = 2$, $\delta = 0.01$, and $\theta = 2$.

Reported Quantities. For every instance, we compute the cost and runtime of each algorithm by taking an average over 50 independent realizations. For the non-adaptive algorithms, note that we only need *one* probing sequence for each instance. On the other hand, adaptive algorithms need to find the probing sequence afresh for each realization. Thus, the non-adaptive algorithms are significantly faster (see full version for corresponding runtime plots).

In Table 1, we report the average performance ratio (cost of the algorithm divided by the information-theoretic lower bound) of the various algorithms. For each instance type (SBFT, Unweighted SSClass and SSClass), we report the performance ratio averaged over *all* values of n (10 choices) and all instances (10 each). Note that values closer to 1 demonstrate better performance.

Table 1. Average performance ratios relative to the lower bound.

Instance type	Our Alg.	GGHK Alg.	Random list
Unweighted SSClass, $B = 5$	1.50	1.48	1.80
Unweighted SSClass, $B = 10$	1.25	1.24	1.33
Unweighted SSClass, $B = 15$	1.13	1.13	1.19
SSClass, $B = 5$	1.59	1.94	2.43
SSClass, $B = 10$	1.34	1.45	1.73
SSClass, $B = 15$	1.22	1.39	1.47

Instance type	Our Alg.	DHK Alg.	Random list
SBFT	2.18	1.74	5.63

We observe that for unweighted SSClass instances, our algorithm performs nearly as well as the $2(B - 1)$-approximation algorithm. For general (weighted) SSClass instances, our algorithm performs considerably better than the adaptive $O(\log W)$-approximation algorithm. For SBFT instances, the performance of our algorithm is about 25% worse than the adaptive 3-approximation algorithm while its runtime is an order of magnitude faster.

References

1. Allen, S., Hellerstein, L., Kletenik, D., Ünlüyurt, T.: Evaluation of monotone DNF formulas. Algorithmica **77** (2015)
2. Bansal, N., Gupta, A., Li, J., Mestre, J., Nagarajan, V., Rudra, A.: When LP is the cure for your matching woes: improved bounds for stochastic matchings. Algorithmica **63**(4), 733–762 (2012). https://doi.org/10.1007/s00453-011-9511-8
3. Bansal, N., Nagarajan, V.: On the adaptivity gap of stochastic orienteering. Math. Program. **154**(1–2), 145–172 (2015). https://doi.org/10.1007/s10107-015-0927-9
4. Behnezhad, S., Derakhshan, M., Hajiaghayi, M.: Stochastic matching with few queries: (1-ε) approximation. In: Proccedings of the 52nd Annual ACM SIGACT Symposium on Theory of Computing, pp. 1111–1124 (2020)
5. Bhalgat, A., Goel, A., Khanna, S.: Improved approximation results for stochastic knapsack problems. In: SODA, pp. 1647–1665 (2011)
6. Blanc, G., Lange, J., Tan, L.Y.: Query strategies for priced information, revisited. In: Proceedings of the Thirty-Second Annual ACM-SIAM Symposium on Discrete Algorithms (SODA 2021), pp. 1638–1650 (2021)
7. Charikar, M., Fagin, R., Guruswami, V., Kleinberg, J., Raghavan, P., Sahai, A.: Query strategies for priced information (extended abstract), pp. 582–591 (2000)
8. Cicalese, F., Laber, E.S.: A new strategy for querying priced information. In: Proceedings of the Thirty-Seventh Annual ACM Symposium on Theory of Computing, STOC 2005, pp. 674–683 (2005)
9. Cicalese, F., Laber, E.S.: On the competitive ratio of evaluating priced functions. J. ACM, **58**(3) (2011)
10. Dean, B.C., Goemans, M.X., Vondrák, J.: Approximating the stochastic knapsack problem: the benefit of adaptivity. Math. Oper. Res. **33**(4), 945–964 (2008)
11. Deshpande, A., Hellerstein, L., Kletenik, D.: Approximation algorithms for stochastic submodular set cover with applications to Boolean function evaluation and min-knapsack. ACM Trans. Algorithms **12**(3), 1–28 (2016)
12. Ene, A., Nagarajan, V., Saket, R.: Approximation algorithms for stochastic k-TSP. In: 37th IARCS Annual Conference on Foundations of Software Technology and Theoretical Computer Science, pp. 27:14–27:27 (2017)
13. Ghuge, R., Gupta, A., Nagarajan, V.: Non-adaptive stochastic score classification and explainable halfspace evaluation. CoRR abs/2111.05687 (2021)
14. Gkenosis, D., Grammel, N., Hellerstein, L., Kletenik, D.: The stochastic score classification problem. In: 26th Annual European Symposium on Algorithms (ESA), vol. 112, pp. 36:1–36:14 (2018)
15. Golovin, D., Krause, A.: Adaptive submodularity: a new approach to active learning and stochastic optimization. CoRR abs/1003.3967 (2017)
16. Guha, S., Munagala, K.: Multi-armed bandits with metric switching costs. In: Albers, S., Marchetti-Spaccamela, A., Matias, Y., Nikoletseas, S., Thomas, W. (eds.) ICALP 2009. LNCS, vol. 5556, pp. 496–507. Springer, Heidelberg (2009). https://doi.org/10.1007/978-3-642-02930-1_41
17. Gupta, A., Krishnaswamy, R., Nagarajan, V., Ravi, R.: Running errands in time: approximation algorithms for stochastic orienteering. Math. Oper. Res. **40**(1), 56–79 (2015)
18. Gupta, A., Nagarajan, V.: A stochastic probing problem with applications. In: Goemans, M., Correa, J. (eds.) IPCO 2013. LNCS, vol. 7801, pp. 205–216. Springer, Heidelberg (2013). https://doi.org/10.1007/978-3-642-36694-9_18

19. Gupta, A., Nagarajan, V., Singla, S.: Adaptivity gaps for stochastic probing: submodular and XOS functions. In: Proceedings of the Twenty-Eighth Annual ACM-SIAM Symposium on Discrete Algorithms (SODA), pp. 1688–1702 (2017)
20. Im, S., Nagarajan, V., van der Zwaan, R.: Minimum latency submodular cover. ACM Trans. Algorithms **13**(1), 13:1–13:28 (2016)
21. Jiang, H., Li, J., Liu, D., Singla, S.: Algorithms and adaptivity gaps for stochastic k-TSP. In: 11th Innovations in Theoretical Computer Science Conference (ITCS), vol. 151, pp. 45:1–45:25 (2020)
22. Kaplan, H., Kushilevitz, E., Mansour, Y.: Learning with attribute costs. In: Proceedings of the Thirty-Seventh Annual ACM Symposium on Theory of Computing, STOC 2005, pp. 356–365 (2005)
23. Ünlüyurt, T.: Sequential testing of complex systems: a review. Discrete Appl. Math. **142**(1), 189–205 (2004)

On the Complexity of Finding Shortest Variable Disjunction Branch-and-Bound Proofs

Max Gläser$^{(\boxtimes)}$ and Marc E. Pfetsch

Department of Mathematics, TU Darmstadt, Darmstadt, Germany
{glaeser,pfetsch}@mathematik.tu-darmstadt.de

Abstract. We investigate the complexity of finding small branch-and-bound trees using variable disjunctions. We first show that it is not possible to approximate the size of a smallest branch-and-bound tree within a factor of $2^{\frac{1}{5}n}$ in time $2^{\delta n}$ with $\delta < \frac{1}{5}$, unless the strong exponential time hypothesis fails. Similarly, for any $\varepsilon > 0$, no polynomial time $2^{(\frac{1}{2}-\varepsilon)n}$-approximation is possible, unless P = NP. We then discuss that finding small branch-and-bound trees generalizes finding short treelike resolution refutations. Therefore, hardness results, in particular non-automatizability results, transfer from this setting. Finally, we show that computing the size of a smallest branch-and-bound tree is #P-hard. Similar results hold for estimating the size of the tree produced by branching rules like most-infeasible branching.

Keywords: Binary program · Branch-and-bound proof · Counting problem

1 Introduction

The currently dominant strategy to solve mixed-integer linear programs is the branch-and-cut method. Besides the addition of cutting planes, the core is formed by branch-and-bound, based on linear programming (LP) relaxations. The nodes are usually created by variable branching, i.e., a disjunction on the variable bounds. Many branching rules, i.e., methods to choose the branching variable at each node, are known, see, e.g., [1,2,8] to mention a few.

This raises the desire to evaluate the performance of branching rules in comparison to the best possible, i.e., to the smallest size of a tree. In this paper, we investigate the theoretical complexity of computing the size of a smallest branch-and-bound tree based on variable branching. We concentrate on proving optimality by assuming that an objective cut using the optimal value is integrated into the system. Then the goal is to estimate the size of a smallest tree to prove infeasibility of a linear inequality system with integrality requirements on the variables. Moreover, we restrict attention to pure branch-and-bound algorithms, i.e., no cutting plane separation, domain reduction, presolving is used.

© Springer Nature Switzerland AG 2022
K. Aardal and L. Sanitá (Eds.): IPCO 2022, LNCS 13265, pp. 291–304, 2022.
https://doi.org/10.1007/978-3-031-06901-7_22

We first prove in Sect. 3 that it is not possible to approximate the size of a smallest branch-and-bound tree using variable branching within a factor of $2^{\frac{1}{5}n}$ in time $2^{\delta n}$ with $\delta < \frac{1}{5}$, where n is the number of variables, unless the strong exponential time hypothesis fails (Theorem 2). The same argument can be used to show that unless P = NP, no polynomial time algorithm can approximate the size of -a- smallest tree within -a factor of $-2^{\frac{1}{2}-\varepsilon}$ for any $\varepsilon > 0$ (Theorem 1). This result significantly strengthens the hardness of approximation within a factor of 2 in Hendel et al. [15].

However, can the smallest size be approximated, if polynomial time in the size of the instance is spent for every node of the smallest tree? This is captured by the notion of automatizability of proof systems, i.e., a proof can be produced in polynomial time in the size of the instance and of a shortest proof. In Sect. 4, we transfer hardness results from the literature to show that branch-and-bound with variable branching is not automatizable under reasonable hardness assumptions. However, it is quasi-automatizable, i.e., a proof can be produced in quasi-polynomial time in the size of the instance and smallest proof.

In Sect. 5, we prove that computing the size of a smallest branch-and-bound tree is #P-hard (Theorem 6). To the best of our knowledge, such a hardness result is novel, even across all commonly considered proof systems. Furthermore, for a binary program it is #P-hard to compute the size of the tree produced by many branching rules from practice (e.g., the most-infeasible branching rule), given suitable tie-breaking. Since one can construct a SAT-formula for which satisfying assignments correspond to leaves of the resulting tree [18, 22], this task is actually in #P (when solving LPs with a polynomial time algorithm). Due to the famous theorem of Toda [23], neither of the two aforementioned estimation problems can be solved in polynomial time, even when given access to an oracle from the polynomial hierarchy, unless the polynomial hierarchy collapses.

Note that Valiant and Vazirani [26] show that every problem in #P has a polynomial-time randomized approximation scheme, assuming access to an NP-oracle, making a result by Sipser [21] and Stockmeyer [22] explicit. However, we do not have a reason to believe that computing the size of a smallest branch-and-bound tree lies in #P. We note that Le Bodic and Nemhauser [20] consider an abstract model in which branching on a particular variable improves the value of the LP-bound for the two children by a fixed amount and show that computing the size of a shortest tree in their model is (weakly) NP-hard.

Branch-and-bound algorithms in practice almost always cut off the optimal LP solution of the currently considered node by branching on a fractional variable. Thus, our results also hold for LP-based branch-and-bound trees.

Note that pure branch-and-bound described above produces mixed results: On the one hand, it solves random binary programs (with a fixed number of constraints) in polynomial time [12], on the other hand, it takes exponential time to prove infeasibility of matching polytopes with an objective cut [6, Theorem 2.2], even though matching is solvable in polynomial time.

2   Preliminaries

2.1   Binary Programs and Branch-and-Bound Proofs

We consider binary programs (BP) of the following form

$$\max\{c^\top x \; : \; Ax \leq b, \; x \in \{0,1\}^n\},$$

where $n \in \mathbb{N}$, $A \in \mathbb{Q}^{m \times n}$, $b \in \mathbb{Q}^m$, $c \in \mathbb{Q}^n$. We define $[n] := \{1, \ldots, n\}$.

To concentrate the investigation on the ability of branch-and-bound to prove optimality (or infeasibility), we assume that the problems are bounded and we know the optimal values of feasible instances, which are integrated by an objective cut into the problem. As a consequence, we are aiming for proving infeasibility of the system

$$Ax \leq b, \quad x \in \{0,1\}^n. \tag{P}$$

To fix notation, we formalize bound-and-bound proofs. A *(variable disjunction) branch-and-bound proof (of infeasibility)* of (P) or *branch-and-bound refutation* of (P) is a rooted binary directed tree T with the following properties:

- Every non-leaf node N is labeled with some variable x_i (the *branching variable* for N), one of the outgoing edges from N is labeled by the *variable fixing* $x_i = 0$ and is called the *left child* (or $x_i = 0$-*child*) of N and the other outgoing edge from N is labeled by $x_i = 1$ and is called the *right child* (or $x_i = 1$-*child*) of N.
- To each node N there corresponds a linear program (P^N), which is the LP-relaxation of (P) strengthened by all the variable fixings which occur as labels on the path from the root to N and (P^N) is infeasible if and only if N is a leaf. We say N is *(in)feasible*, if (P^N) is.

The $x_i = a$-branch at a node N, $a \in \{0,1\}$, is the directed subtree rooted at the $x_i = a$-child of N. Without loss of generality, we assume every variable occurs at most once as a label of a node on any root-leaf path in T.

We are then interested in $T(I)$, which for an infeasible instance I of problem (P) is the size of a shortest branch-and-bound proof; the *size* of a proof is the number of its nodes.

2.2   Boolean Formulas and the Exponential Time Hypothesis

A *literal* is a Boolean variable or its negation. A *clause* C is a disjunction of literals. A Boolean formula φ in *conjunctive normal form (CNF)* is a conjunction of clauses. Let $\mathcal{C}$ denote the set of clauses in φ. For a clause $C \in \mathcal{C}$ denote by $\mathcal{P}(C)$ the set of unnegated variables in C and by $\mathcal{N}(C)$ the set of negated variables in C. A CNF φ is a *k-CNF*, if every clause of φ contains at most k literals. Without loss of generality, every k-CNF contains at most $\binom{n}{k}2^k$ clauses.

For some results we will rely on the (strong) exponential time hypothesis formulated by Impagliazzo and Paturi [16] as a hardness assumption:
Let $(k$-$)$SAT denote the problem of deciding whether a $(k$-$)$CNF is satisfiable and

$$s_k := \inf\{s \in \mathbb{R}_+ : k - \text{SAT can be decided in time } O(2^{sn})\}.$$

The *exponential time hypothesis (ETH)* states $s_3 > 0$ and the *strong exponential time hypothesis (SETH)* states $s_\infty := \lim_{k \to \infty} s_k = 1$.

3   Hardness of Approximation

We first prove that the size $\mathcal{T}$ of a smallest branch-and-bound proof cannot be approximated within any exponential factor within subexponential time, unless ETH fails.

Consider $J_n = \{x \in [0,1]^{2n} : \sum_{i=1}^{2n} 2x_i = 2n + 1\}$, a classical example by Jeroslow [17]. Note that $J_n \cap \mathbb{Z}^{2n}$ is empty, but J_n is not. It is easy to see that one needs to fix at least n variables before attaining LP-infeasibility. Thus, any proof of $J_n \cap \mathbb{Z}^{2n} = \varnothing$ must contain at least $2^{n+1} - 1$ nodes.

Furthermore, for a given CNF φ with variables $x_1, \ldots, x_n$ and clauses $\mathcal{C}$, consider the following binary problem:

$$\sum_{x_i \in \mathcal{P}(C)} x_i + \sum_{x_i \in \mathcal{N}(C)} (1 - x_i) \geq \tfrac{1}{2} \quad \forall C \in \mathcal{C}, \qquad x_1, \ldots, x_n \in \{0, 1\}. \qquad (Q_\varphi)$$

Satisfying assignments of φ correspond to integer points in (Q_φ).

Given a set of instances $\mathcal{I}$ and $\alpha \geq 1$, an *α-approximation algorithm* for a function $f : \mathcal{I} \mapsto \mathbb{N}$ is an algorithm A which satisfies $A(I)/\alpha \leq f(I) \leq A(I) \cdot \alpha$ for all $I \in \mathcal{I}$. Our first result is

Theorem 1. *There is no polynomial time α-approximation algorithm for $\mathcal{T}$ with $\alpha < 2^{(\frac{1}{2} - \varepsilon)n}$ for any $\varepsilon > 0$, unless $P = NP$.*

The proof is analogous to that of the next result:

Theorem 2. *For any parameter $\lambda \in \mathbb{N}$ with $\lambda > 1$, there is no α-approximation algorithm for $\mathcal{T}$ with $\alpha < 2^{(\frac{1}{2} - \frac{1.5}{2\lambda+1})n}$ running in time $O(2^{\delta n})$ with $\delta < \frac{1}{1+2\lambda}$, unless SETH fails. Furthermore, for every $\varepsilon > 0$, there is some $\delta > 0$ such that there is no α-approximation algorithm for $\mathcal{T}$ with $\alpha < 2^{(\frac{1}{2} - \varepsilon)n}$ running in time $O(2^{\delta n})$, unless ETH fails.*

In particular, for $\lambda = 2$, we obtain that it is not possible to approximate $\mathcal{T}$ within a factor of $2^{\frac{1}{5}n}$ in time $2^{\delta n}$ with $\delta < \frac{1}{5}$, unless SETH fails. Moreover, Theorem 1 corresponds to $\lambda \to \infty$.

Proof. We prove the theorem by contraposition: Assume such an algorithm exists for some λ. We will show that we can then give a family of algorithms for SAT falsifying (S)ETH. Let a CNF φ with variables $y_1, \ldots, y_m$ and clauses $\mathcal{C}$ be given, and let $\ell := \lambda m$. Consider the following binary program (J) with $n = 2\ell + m$ variables:

$$\sum_{i=1}^{2\ell} 2x_i = 2\ell + 1, \tag{Ja}$$

$$\sum_{y_i \in \mathcal{P}(C)} y_i + \sum_{y_i \in \mathcal{N}(C)} 1 - y_i \geq \tfrac{1}{2} \qquad \forall C \in \mathcal{C}, \tag{Jb}$$

$$x_i, \, y_j, \, \in \{0,1\} \qquad \forall i \in [2\ell], \, j \in [m].$$

Assume that φ has no satisfying assignment. Then exhaustively branching on all y_i yields infeasibility due to (Jb) and thus a proof of size at most $2^{m+1} - 1$.

Conversely, assume there is an assignment satisfying φ and we are given a branch-and-bound proof T of (J). It is easy to see that we can obtain a proof T' for $J_\ell \cap \mathbb{Z}^{2\ell} = \varnothing$ with $|T'| \leq |T|$ from T and thus we must have $|T| \geq 2^{\ell+1} - 1$. Since $\ell \geq m$, we have

$$\frac{2^{\ell+1} - 1}{2^{m+1} - 1} \geq \frac{2^{\ell+1}}{2^{m+1}} = 2^{\ell-m} = 2^{(\lambda-1)m} = 2^{\frac{\lambda-1}{2\lambda+1}(2\ell+m)} = 2^{\frac{\lambda-1}{2\lambda+1}n}.$$

Therefore, approximating the size of a shortest branch-and-bound proof for (J) within factor $2^{\frac{\lambda-1}{2\lambda+1}n}$ decides whether φ is satisfiable. Further, an $O(2^{\delta n})$-time algorithm with $\delta < \frac{1}{2\lambda+1}$ for this task is an $O(2^{(2\lambda+1)\delta m})$-time algorithm for k-SAT for arbitrary k with $(2\lambda+1)\delta < 1$ and thus contradicts SETH.

Furthermore, choose λ with $\frac{1.5}{2\lambda+1} < \varepsilon$ for any fixed $\varepsilon > 0$. Then an $O(2^{\delta n})$-time algorithm approximating T within factor $2^{(\frac{1}{2}-\varepsilon)n}$, yields an $O(2^{\delta' m})$-time algorithm for 3-SAT for $\delta' := (2\lambda+1)\delta$. If such an algorithm exists for every $\delta > 0$ (and therefore every δ'), this violates ETH, which yields the second part of the statement. $\qquad \square$

Remark 3. In the above proof, we can always branch on a fractional variable, i.e., Theorem 1 and 2 hold for LP-based branch-and-bound trees.

Remark 4. Note that for BPs with an exponential number of constraints, we can replace J_n by $P_n := \{x \in [0,1]^n : \sum_{i \in R} x_i + \sum_{i \notin R}(1 - x_i) \geq \tfrac{1}{2} \, \forall R \subseteq [n]\}$, for which any branch-and-bound proof with variable disjunctions must be a complete binary tree of depth n. We can then strengthen the factor $\frac{1}{2} - \frac{2}{2\lambda+1}$ in the exponent of the approximation ratio in the first part of Theorem 2 to $1 - \frac{2}{\lambda+1}$ and the factor of $\frac{1}{2} - \varepsilon$ in the second part to $1 - \varepsilon$. Note P_n is separable in polynomial time: Given a point $\hat{x} \in [0,1]^n$, the set $\hat{R} := \{i \in [n] : \hat{x}_i \leq \tfrac{1}{2}\}$ minimizes the left hand side. Thus, $\hat{x} \in P_n$ if and only if $\hat{x}$ satisfies $\sum_{i \in \hat{R}} \hat{x}_i + \sum_{i \notin \hat{R}}(1 - \hat{x}_i) \geq \tfrac{1}{2}$.

Cook et al. [10] showed that there is no cutting plane proof of $P_n \cap \mathbb{Z}^n = \varnothing$ of size at most $2^n/n$, even if we replace $\frac{1}{2}$ on the right hand side by 1. Dadush and Tiwari [11] showed that there also does not exist a branch-and-bound proof of size $2^n/n$ even if we allow branching on general disjunctions.

4 (Non-)Automatizability of Branch-and-Bound

In this section we sketch how results about automatizing finding shortest so-called treelike resolution refutations transfer to the case of branch-and-bound proofs.

Treelike resolution is a proof system which certifies the infeasibility of Boolean formulas in CNF. Treelike resolution can be seen as a branch-and-bound procedure, where we branch by fixing a variable to either 1 (true) or 0 (false) and prune only nodes in which the already fixed variables *falsify a clause*, i.e., we prune a node N, if there is a clause C with fixings $x = 0$ for all $x \in \mathcal{P}(C)$ and $x = 1$ for all $x \in \mathcal{N}(C)$ in the subproblem associated with N [19, Section 5.2]. A *treelike resolution proof* or *refutation* is a tree produced by this procedure.

Automatizability. It is easy to see that not all infeasible instances of an NP-complete problem can have proofs of size bounded by some polynomial, unless NP = coNP. Therefore the reason for the inability of polynomial-time algorithms to find shortest proofs might be the size of these proofs instead of the computational difficulty of finding them. To investigate this possibility, Bonet et al. [9] introduced the concept of *automatization*. A proof system is called *automatizable (quasi-automatizable)*, if given an infeasible instance I, an infeasibility proof in this system can be produced in time polynomial (quasipolynomial) in $|I|$ and L, where $|I|$ is the encoding length of I and L is the size of a shortest proof of I.

Beame and Pitassi [7] proved the positive result that branch-and-bound using variable disjunctions is quasi-automatizable for binary programs, for which we reproduce the proof for completeness:

Consider the following recursive procedure $R(S, n)$, which generates a branch-and-bound proof for a given binary program (P) with n variables, if there is a proof of size at most S. Among the $2n$ possible fixings of a variable in (P), find a fixing $x_i = a \in \{0, 1\}$, such that $R(S/2, n-1)$ succeeds on $(P \cap \{x_i = a\})$ by trial-and-error. This works if there exists a proof T of size at most S, because one of the branches at the root node of T has size at most $S/2$. Then apply $R(S, n-1)$ to $(P \cap \{x_i = 1-a\})$, which succeeds if there exists a proof of size at most S and because $\mathcal{T}$ is non-increasing with respect to fixing variables. The resulting running time recursion $t(S, n) = 2n \cdot t(S/2, n-1) + t(S, n-1)$ solves to roughly $n^{\log S}$. We can then try $R(S, n)$ for increasing values of S until it succeeds.   □

We now observe that the treelike resolution refutations of φ are the branch-and-bound proofs using variable disjunctions of (Q_φ). Indeed, in both cases a node is infeasible if and only if the already fixed variables falsify a clause.

Thus, we can replace "treelike resolution" by "branch-and-bound using variable disjunctions" in a result by Alekhnovich and Razborov [3], which has been strengthened by Eickmeyer et al. [14] to obtain: Branch-and-bound using variable disjunctions is not automatizable unless FPT = W[P]. Here, FPT ⊆ W[P] may be seen as an analogue of the question whether NP ⊆ P in the world of parameterized problems.

Notably, the recent breakthrough of Atserias and Müller [5], showing non-automatizability of general resolution under the weaker and optimal assumption P $\neq$ NP, likely does not transfer to branch-and-bound using variable disjunctions (or treelike resolution). In fact, their approach – to show that it is NP-hard to distinguish between instances with a refutation of size bounded by some polynomial and instances with refutations of at least exponential size – likely does not transfer; otherwise, quasi-automatizability of branch-and-bound using variable disjunction would cause the exponential time hypothesis to fail [3,5].

5   #P-Hardness of Exact Computation

We first review #P-hardness as introduced by Valiant [24,25], which we will use to bound the hardness of finding a shortest branch-and-bound proof from below:

Definition 5. *A function* $g \colon \{0,1\}^* \to \mathbb{N}$ *is in #P if there exists a polynomial* $p \colon \mathbb{N} \to \mathbb{N}$ *and a polynomial time Turing machine* M *such that for every* $x \in \{0,1\}^*$ *we have*

$$g(x) = |\{y \in \{0,1\}^{p(|x|)} : M \text{ accepts on input } (x, y)\}|.$$

So #P is the class of functions which count certificates of some NP-problem. For example, #SAT is the function that on input of a Boolean formula returns the number of satisfying assignments.

We will identify natural numbers with their binary representation and let FP^f denote the class of functions $g \colon \{0,1\}^* \to \mathbb{N}$ which are computable in polynomial time by a Turing machine with access to an oracle computing f. A function f is #P-*hard*, if #P $\subseteq$ FP^f. Because the Cook-Levin reduction preserves the number of certificates, #SAT is #P-complete, see, e.g., [4, Theorem 17.10].

Our main result is the following:

Theorem 6. *Computing* $\mathcal{T}$ *for binary programs is #P-hard.*

Our proof is based on a highly symmetrical problem with additional constraints on a subset of variables, ensuring that a Boolean formula φ is satisfied. Due to the symmetry without additional constraints, it is optimal to first branch on variables occuring in additional constraints. Then the number of surviving branches after fixing all the variables subject to additional constraints – i.e. the number of assignments satisfying φ – is roughly proportional to the length of a smallest branch-and-bound proof.

Let a formula φ in CNF with variables $x_1 \ldots, x_n$ be given. We then introduce additional variables $x_{n+1}, \ldots, x_{3n}, y_1, \ldots, y_{3n}, z_1, \ldots, z_{3n}$ and consider the following BP (P_φ):

$$\sum_{x_i \in \mathcal{P}(C)} x_i + \sum_{x_i \in \mathcal{N}(C)} (1 - x_i) \geq \tfrac{1}{2} \qquad \forall C \in \mathcal{C}, \qquad (P_\varphi a)$$

$$\sum_{i=1}^{3n} (y_i + z_i) \geq 3n + \tfrac{1}{2}, \qquad\qquad\qquad (P_\varphi b)$$

$$y_i \leq \tfrac{3}{2} - x_i, \ z_i \leq \tfrac{1}{2} + x_i \qquad \forall i \in [3n], \qquad (P_\varphi c)$$

$$(x, y, z) \in \{0,1\}^{3 \cdot 3n}.$$

In particular, none of the variables $x_{n+1}, \dots, x_{3n}$ are contained in a constraint of type $(P_\varphi a)$. Note that because of $(P_\varphi b)$ and $(P_\varphi c)$, this problem is always infeasible, even when ignoring the constraints $(P_\varphi a)$. The following Lemmas 7 and 8 will give sufficiently tight bounds on $\#\mathrm{SAT}(\varphi)$ depending on $T(P_\varphi)$ to see that knowledge of $T(P_\varphi)$ suffices to compute $\#\mathrm{SAT}(\varphi)$ in polynomial time. This then proves Theorem 6.

Lemma 7. $T(P_\varphi) \leq 6 \cdot 2^{2n} \cdot \#\mathrm{SAT}(\varphi) - 5 + 6(2^n - \#\mathrm{SAT}(\varphi))$

Proof. Construct a branch-and-bound proof T by choosing the branching variable in every node N as follows:

1. If N is a $x_i = 0$ child, choose z_i and if N is a $x_i = 1$-branch choose y_i,
2. otherwise choose an unbranched x_i with smallest index.

We say a node is at depth ℓ, if its distance from the root is ℓ. By construction, each node at even depth is labeled with some x_i, whereas nodes at odd depth are labeled with either some y_i or z_i.

Since we are proving an upper bound, we can assume that constraint $(P_\varphi a)$ is only added at depth $2n$. Therefore, the problem never becomes infeasible after fixing a variable x_i. Thus, all nodes at odd depths are feasible. At even depths except for $2n$ and $6n$, exactly half of the nodes are feasible, namely the $y_i = 0$ and $z_i = 0$ nodes.

Together, this yields 2^{i+1} nodes at odd depth $2i + 1$ for $i = 0, \dots, n - 1$ and $2 \cdot \#\mathrm{SAT}(\varphi)$ nodes at depth $2n+1$. For odd depth $2(n+i)+1$ for $i = 0, \dots, 2n-1$, we have 2^{i+1} nodes for each solution of φ. Moreover, assigning to every node at odd depth the set consisting of its two children and itself, yields a partition of the node set without the root. This yields:

$$T(P_\varphi) \leq 3 \cdot |\{\text{nodes labeled with } y \text{ or } z \text{ before level } 2n\}|$$
$$+ 3 \cdot |\{\text{nodes labeled with } y \text{ or } z \text{ after level } 2n\}| + 1$$
$$= 3\left(2\sum_{i=0}^{n-1} 2^i + 2 \cdot \#\mathrm{SAT}(\varphi) \cdot \sum_{i=0}^{2n-1} 2^i\right) + 1$$
$$= 6(2^n - 1) + 6 \cdot \#\mathrm{SAT}(\varphi) \cdot (2^{2n} - 1) + 1,$$

which shows the claim. $\qquad\qquad\qquad\qquad\qquad\qquad\qquad\qquad\qquad \square$

Lemma 8. $T(P_\varphi) \geq 6 \cdot 2^{2n} \cdot \#\mathrm{SAT}(\varphi) - 5.$

Proof. Let $\mathcal{S}$ denote the set of points $(x, y, z) \in \{0, 1\}^{3 \cdot 3n}$ such that

1. $x_1, \ldots, x_n$ satisfy φ,
2. $x_{n+1}, \ldots, x_{3n}$ can have any value and
3. $z_i = x_i$ and $y_i = 1 - x_i$ for all $i \in [3n]$ (i.e. y_i and z_i have the maximal binary value which is consistent with x_i with respect to $(P_\varphi c)$).

Evidently, $|\mathcal{S}| = \#\mathrm{SAT}(\varphi) \cdot 2^{2n}$.

Let T denote a smallest branch-and-bound proof of (P_φ) using variable disjunctions. We will show that every point $s \in \mathcal{S}$ must correspond to a unique leaf of T (Claim 2) and that the problems associated to these leaves must contain many variable fixings of y and z variables (Claim 1). This allows us to lower-bound the number of nodes of T labeled with these variables in terms of $|\mathcal{S}|$ (Claim 3). Together with the fact that T can be chosen exhibiting certain structure (Claim 4), this will allow us to isolate sufficiently many nodes of T for our desired bound.

We say that a 0/1-point $p = (x, y, z) \in \{0, 1\}^{3 \cdot 3n}$ is *ruled out* by a leaf N of T, if p satisfies all variable fixings labeling the edges on the path from the root to N. Such a leaf N certifies that p is not feasible for (P_φ). Every $p \in \{0, 1\}^{3 \cdot 3n}$ is ruled out by some leaf.

Claim 1. The subproblem $P_\varphi(N)$ associated to a leaf N ruling out $s = (x, y, z) \in \mathcal{S}$ must contain one of the variable fixings $y_i = 0$ or $z_i = 0$ for every $i = 1, \ldots, 3n$.

Proof of claim. Assume there is $i = 1, \ldots 3n$, such that $P_\varphi(N)$ neither contains $y_i = 0$ nor $z_i = 0$. Thus, if we define for $j = 1, \ldots, 3n$,

$$\hat{x}_j = x_j, \quad \hat{y}_j = \begin{cases} \frac{1}{2}, & \text{if } j = i \text{ and } x_j = 1, \\ 1, & \text{if } j = i \text{ and } x_j = 0, \\ y_j, & \text{otherwise,} \end{cases} \quad \hat{z}_j = \begin{cases} 1, & \text{if } j = i \text{ and } x_j = 1, \\ \frac{1}{2}, & \text{if } j = i \text{ and } x_j = 0, \\ z_j, & \text{otherwise,} \end{cases}$$

then $(\hat{x}, \hat{y}, \hat{z})$ is a feasible solution to $P_\varphi(N)$, contradicting infeasibility of N. ∎

Claim 2. A leaf can rule out at most one point from $\mathcal{S}$.

Proof of claim. Assume $(\tilde{x}, \tilde{y}, \tilde{z})$, $(x', y', z') \in \mathcal{S}$ with $\tilde{x} \neq x'$ are both ruled out by a leaf N. (Note that differences in y and z imply differences in x.) Thus, there is $i \in [3n]$ such that $\tilde{x}_i \neq x'_i$ and the subproblem $P_\varphi(N)$ associated to N does not contain variable fixings on x_i, y_i or z_i. This is impossible by Claim 1. ∎

Claim 3. There are at least $2(|\mathcal{S}| - 1)$ nodes in T labeled either with some y_i or z_i, which have a descendant ruling out a point from $\mathcal{S}$.

Proof of claim. Let X denote the set of inner nodes N in T for which both branches contain a node ruling out a point from $\mathcal{S}$. By Claim 2, we have $|X| = |\mathcal{S}| - 1$. For every $N \in X$, we will choose a set $\mathcal{H}_N$ of two descendants labeled with either y_i or z_i, which have a descendant ruling out a point from $\mathcal{S}$. To this end, we emphasize that both branches in T starting at N contain a leaf ruling out a point from S and consider the following cases:

1. If N is labeled by some x_i, then no node on the path from the root to N can be labeled with y_i or z_i, since otherwise not both the $x_i = 1$- and the $x_i = 0$-branch can contain a leaf ruling (since only one is consistent with the earlier branching decision). Since both branches at N contain a leaf ruling out a point from $\mathcal{S}$, both of these branches must contain a node labeled with either y_i or z_i by Claim 1. Choose $\mathcal{H}_N$ to be any two such nodes.
2. Otherwise, N is labeled with some y_i or z_i, say y_i. Once again, no node on the path from the root to N can be labeled with x_i or z_i. Furthermore, the $y_i = 1$-branch must contain a node ruling out a point from $\mathcal{S}$. Again by Claim 1, a node U in this branch must be labeled with z_i. Choose $\mathcal{H}_N = \{N, U\}$.

We claim that $\mathcal{H}_N \cap \mathcal{H}_M = \emptyset$ for distinct nodes N, $M \in X$. For the sake of contradiction assume $\mathcal{H}_N \cap \mathcal{H}_M \neq \emptyset$. By construction, the labels of N and M must be distinct elements of $\{x_i, y_i, z_i\}$ for some index i. Furthermore, since $\mathcal{H}_N$ and $\mathcal{H}_M$ contain only descendants of N and M, there must be a ancestor relationship between N and M in T, say N is a descendant of M. But then not both branches at N can contain a leaf ruling out a point from $\mathcal{S}$, as only for one of these branches the branching decision at N is consistent with the branching decision at M, contradicting our choice of X. Thus, the set $\mathcal{H} = \bigcup_{N \in X} \mathcal{H}_N$ is a collection of $2(|\mathcal{S}| - 1)$ nodes labeled with either y_i or z_i, which have a descendant ruling out a point from $\mathcal{S}$, as desired.   ∎

Claim 4. T can be chosen to satisfy the following properties:

1. Let be N be a node labeled with some y_i, then N either has an ancestor A labeled with x_i and is in the $x_i = 1$-branch at A or a descendant in the $y_i = 1$-branch at N is labeled with x_i. Analogously, if N is labeled with some z_i, then N either has an ancestor A labeled with x_i and is in the $x_i = 0$-branch at A or a descendant in the $z_i = 1$-branch at N is labeled with x_i.
2. If a variable y_i or z_i is branched after x_i, then it is branched immediately after, i.e., if there is a node N labeled with some y_i or z_i which is a descendant of some A labeled with x_i then N is A's child.

Proof of claim. 1. Assume there is a node N labeled with y_i that is neither in the $x_i = 1$ branch of an ancestor labeled with x_i nor has a descendant in the $y_i = 1$-branch labeled with x_i. We claim that the subtree S rooted at N can be replaced by the $y_i = 1$-branch S' at N as shown in Figure 1, removing the branching on y_i. Since S' is a subtree of S, every node of S' corresponds to a node of S in the obvious way. Let T' denote the tree obtained by cutting off S from T and attaching S' to T at N instead. It remains to show that all of the problems $P_\varphi(L')$ for a leaf L' of S' are infeasible, where $P_\varphi(L')$ is taken with respect to T'. Let L be the node corresponding to L' in S and $P_\varphi(L)$ be the associated subproblem (with respect to T). Then $P_\varphi(L)$ is $P_\varphi(L')$ strengthened by the variable fixing $y_i = 1$, but does not contain the variable fixing $x_i = 1$. It is obvious that if $P_\varphi(L')$ has a feasible solution, then it has a feasible solution with $y_i = 1$ (and $x_i = \frac{1}{2}$) and therefore the infeasibility of $P_\varphi(L)$ implies the infeasibility of $P_\varphi(L')$. The case where N is labeled with some z_i is analogous.

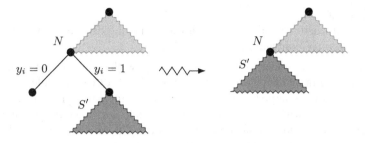

Fig. 1. The modifications to T to prove part 1 of Claim 4. We see the case where N is labeled with y_i, the other case is analogous. The fixing of y_i does not contribute to the infeasibility of the leaves in the $y_i = 1$-branch and therefore can be removed.

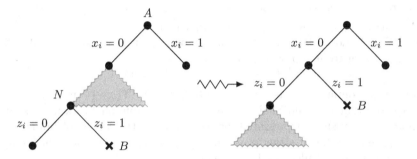

Fig. 2. The modifications to T to prove part 2 of Claim 4. We see the case where N is labeled with z_i, the other case is analogous. The gray triangle is the $x_i = 0$-branch at A without the subtree rooted at N. The node B is infeasible because of the incompatible choices of z_i and x_i. The only leaf where fixed variables become free is B, thus all leaves remain infeasible.

2. Assume a node N which is labeled with some y_i or z_i is a descendant, but not a child, of a node A labeled with x_i. Then by Part 1, either N is labeled with y_i and in the $x_i = 1$-branch at A, or N is labeled with z_i and in the $x_i = 0$-branch at A. In either case, we can find another branch-and-bound proof T' of the same size, in which the number of times a variable y_i or z_i is branched after, but not immediately after x_i is reduced by one, which suffices by induction. This can be achieved by reattaching certain subtrees of T as shown in Fig. 2 and verifying that the leaves remain infeasible. ∎

Finally, we assign to every node N labeled with y_i which has a descendant ruling out a point from $\mathcal{S}$ the following nodes:

1. If N's parent M is labeled with x_i:
   - Half of M and
   - N's infeasible child O, in which the branching decisions for x_i in M and for y_i in N contradict.
2. Otherwise, by Claim 4, there is a node M in the $y_i = 1$-branch at N which is labeled with x_i and we assign to N:

– Half of M and
– M's infeasible child O, in which the branching decisions for x_i in M and for y_i in N contradict.

We assign nodes to nodes labeled with some z_i with a descendant ruling out a point from S analogously. Note that in both cases, O cannot rule out a point from S. We have assigned $\frac{3}{2}$ nodes to every N labeled with y_i or z_i, none of which are labeled with y_i or z_i and none of which can rule out a point from S. Moreover, every node is assigned at most once: The only case in which a node could be assigned more than once appears if a node N labeled with x_i is assigned with $\frac{1}{2}$ to its two children and an ancestor each. Assume N is in the $y_i = 1$ branch of that ancestor. Then no node in its $x_i = 1$ branch rules out a node from S and N is not assigned to both of its children. All other cases are analogous.

In total, T contains at least

(a) $|S|$ nodes ruling out a point from S by Claim 2,
(b) $2(|S| - 1)$ nodes labeled with either y_i or z_i and have a descendant ruling out a point from S by Claim 3,
(c) and $\frac{3}{2} \cdot 2(|S| - 1)$ nodes assigned to the nodes from (b).

Therefore T contains at least $6|S| - 5 = 6 \cdot 2^{2n} \cdot \#\mathrm{SAT}(\varphi) - 5$ nodes.   □

We note that Dey and Shah [13] independently from our work use a similar technique: They show exponential lower bounds on the length of branch-and-bound proofs for the lot-sizing problem by giving a large set S for which only a unique point can be ruled out by any leaf.

Finally, we can prove Theorem 6:

Proof (of Theorem 6). The intervals for $T(P_\varphi)$ given by the bounds from Lemmas 7 and 8

$$[6 \cdot 2^{2n} \cdot \#\mathrm{SAT}(\varphi) - 5, \ 6 \cdot 2^{2n} \cdot \#\mathrm{SAT}(\varphi) - 5 + 6(2^n - \#\mathrm{SAT}(\varphi))]$$

are disjoint for different values of $\#\mathrm{SAT}(\varphi)$. Therefore, we can use $T(P_\varphi)$ to infer $\#\mathrm{SAT}(\varphi)$. Since (P_φ) can be constructed in polynomial time and $\#\mathrm{SAT}$ is $\#\mathrm{P}$-hard, T is $\#\mathrm{P}$-hard as well.   □

If we use the left hand side of constraint $(P_\varphi b)$ as an objective function, the branch-and-bound proof constructed in Lemma 7 always branches on a variable which is fractional in the current unique LP-optimum. Thus, finding the length of a shortest proof which eliminates the current LP-optimum with each branching decision is also $\#\mathrm{P}$-hard.

Furthermore, the branching rule described in Lemma 7 is most-infeasible branching with the tie-breaking rule $x_1 > y_1 > z_1 > \cdots > x_{3n} > y_{3n} > z_{3n}$, because all fractional values of the unique optimum solution are $\frac{1}{2}$. Thus, if $\mathcal{M}(I)$ denotes the size of the branch-and-bound proof produced by variable branching using most-infeasible branching with the above tie-breaking rule, we have:

Corollary 9. *Computing $\mathcal{M}$ for binary programs is $\#\mathrm{P}$-hard.*

Using Theorem 6, it is not hard to derive hardness results similar to Theorems 1 and 2 for $\mathcal{M}$. To this end, note that the number of variables of (P_φ) is nine times the number of variables of φ.

6 Outlook

Firstly, we currently know $P^{\#SAT} \subseteq P^{\mathcal{T}} \subseteq \text{PSPACE}$. It remains to determine the exact complexity of $\mathcal{T}$. Secondly, when trying to adapt Theorem 6 for the case of treelike resolution in the straightforward way, the analogous formula for (P_φ) suffers from exponential blow-up. It would be interesting to know if this can be remedied.

Acknowledgements. We thank three reviewers for their helpful comments that improved the presentation of the paper. This work was supported by the DFG, Project A01 in CRC/Transregio 154 (Project # 239904186).

References

1. Achterberg, T., Berthold, T.: Hybrid branching. In: van Hoeve, W.-J., Hooker, J.N. (eds.) CPAIOR 2009. LNCS, vol. 5547, pp. 309–311. Springer, Heidelberg (2009). https://doi.org/10.1007/978-3-642-01929-6_23
2. Achterberg, T., Koch, T., Martin, A.: Branching rules revisited. Oper. Res. Lett. **33**(1), 42–54 (2005). https://doi.org/10.1016/j.orl.2004.04.002
3. Alekhnovich, M., Razborov, A.A.: Resolution is not automatizable unless W[P] is tractable. SIAM J. Comput. **38**(4), 1347–1363 (2008). https://doi.org/10.1137/06066850X
4. Arora, S., Barak, B.: Computational Complexity: A Modern Approach. Cambridge University Press, Cambridge (2009). https://doi.org/10.1017/CBO9780511804090
5. Atserias, A., Müller, M.: Automating resolution is NP-hard. J. ACM **67**(5), 1–17 (2020). https://doi.org/10.1145/3409472
6. Basu, Amitabh, Conforti, Michele, Di Summa, Marco, Jiang, Hongyi: Complexity of branch-and-bound and cutting planes in mixed-integer optimization - II. In: Singh, Mohit, Williamson, David P.. (eds.) IPCO 2021. LNCS, vol. 12707, pp. 383–398. Springer, Cham (2021). https://doi.org/10.1007/978-3-030-73879-2_27
7. Beame, P., Pitassi, T.: Simplified and improved resolution lower bounds. In: Proceedings of 37th Conference on Foundations of Computer Science (FOCS), pp. 274–282. IEEE (1996). https://doi.org/10.1109/SFCS.1996.548486
8. Berthold, T., Salvagnin, D.: Integration of AI and OR techniques in constraint programming for combinatorial optimization problems. In: Gomes, C., Sellmann, M. (eds.) Cloud Branching. LNCS, vol. 7874, pp. 28–43. Springer, Cham (2013). https://doi.org/10.1007/978-3-642-38171-3_3
9. Bonet, M.L., Pitassi, T., Raz, R.: On interpolation and automatization for Frege systems. SIAM J. Comput. **29**(6), 1939–1967 (2000). https://doi.org/10.1137/S0097539798353230
10. Cook, W., Coullard, C.R., Turán, G.: On the complexity of cutting-plane proofs. Discrete Appl. Math. **18**(1), 25–38 (1987). https://doi.org/10.1016/0166-218X(87)90039-4
11. Dadush, D., Tiwari, S.: On the complexity of branching proofs. In: Proceedings of the 35th Computational Complexity Conference. Schloss Dagstuhl, Germany (2020). https://doi.org/10.4230/LIPIcs.CCC.2020.34
12. Dey, S.S., Dubey, Y., Molinaro, M.: Branch-and-bound solves random binary IPs in polytime. In: Proceedings of the 2021 ACM-SIAM Symposium on Discrete Algorithms (SODA), pp. 579–591. SIAM (2021). https://doi.org/10.1137/1.9781611976465.35

13. Dey, S.S., Shah, P.: Lower bound on size of branch-and-bound trees for solving lot-sizing problem. arXiv preprint arXiv:2112.03965 (2021)
14. Eickmeyer, K., Grohe, M., Grüber, M.: Approximation of natural W[P]-complete minimisation problems is hard. In: 23rd Annual IEEE Conference on Computational Complexity, pp. 8–18. IEEE (2008). https://doi.org/10.1109/CCC.2008.24
15. Hendel, G., Anderson, D., Le Bodic, P., Pfetsch, M.E.: Estimating the size of branch-and-bound trees. INFORMS J. Comput. **34**, 934–952 (2021). https://doi.org/10.1287/ijoc.2021.1103
16. Impagliazzo, R., Paturi, R.: On the complexity of k-SAT. J. Comput. Syst. Sci. **62**(2), 367–375 (2001). https://doi.org/10.1006/jcss.2000.1727
17. Jeroslow, R.G.: Trivial integer programs unsolvable by branch-and-bound. Math. Program. **6**(1), 105–109 (1974). https://doi.org/10.1007/BF01580225
18. Knuth, D.E.: Estimating the efficiency of backtrack programs. Math. comput. **29**(129), 122–136 (1975). https://doi.org/10.1090/S0025-5718-1975-0373371-6
19. Krajíček, J.: Proof Complexity. Cambridge University Press., Cambridge (2019). https://doi.org/10.1017/9781108242066
20. Le Bodic, P., Nemhauser, G.: An abstract model for branching and its application to mixed integer programming. Math. Program. **166**(1), 369–405 (2017). https://doi.org/10.1007/s10107-016-1101-8
21. Sipser, M.: A complexity theoretic approach to randomness. In: Proceedings of the Fifteenth Annual ACM Symposium on Theory of computing, pp. 330–335 (1983). https://doi.org/10.1145/800061.808762
22. Stockmeyer, L.: On approximation algorithms for #P. SIAM J. Comput. **14**(4), 849–861 (1985). https://doi.org/10.1137/0214060
23. Toda, S.: PP is as hard as the polynomial-time hierarchy. SIAM J. Comput. **20**(5), 865–877 (1991). https://doi.org/10.1137/0220053
24. Valiant, L.G.: The complexity of computing the permanent. Theor. Comput. Sci. **8**(2), 189–201 (1979). https://doi.org/10.1016/0304-3975(79)90044-6
25. Valiant, L.G.: The complexity of enumeration and reliability problems. SIAM J. Comput. **8**(3), 410–421 (1979). https://doi.org/10.1137/0208032
26. Valiant, L.G., Vazirani, V.V.: NP is as easy as detecting unique solutions. In: Proceedings of the Seventeenth Annual ACM Symposium on Theory of Computing, pp. 458–463 (1985). https://doi.org/10.1016/0304-3975(86)90135-0

Matroid-Based TSP Rounding
for Half-Integral Solutions

Anupam Gupta[1], Euiwoong Lee[3], Jason Li[2], Marcin Mucha[4],
Heather Newman[1(✉)], and Sherry Sarkar[1]

[1] Carnegie Mellon University, Pittsburgh, USA
hanewman@andrew.cmu.edu
[2] Simons Institute for the Theory of Computing and UC Berkeley,
Berkeley, USA
[3] University of Michigan, Ann Arbor, USA
[4] University of Warsaw, Warsaw, Poland

Abstract. We show how to round any half-integral solution to the subtour-elimination relaxation for the TSP, while losing a less-than-1.5 factor. Such a rounding algorithm was recently given by Karlin, Klein, and Oveis Gharan based on sampling from max-entropy distributions. We build on an approach of Haddadan and Newman to show how sampling from the matroid intersection polytope, and a new use of max-entropy sampling, can give better guarantees.

1 Introduction

The (symmetric) traveling salesman problem asks: given an graph $G = (V, E)$ with edge-lengths $c_e \geq 0$, find the shortest tour that visits all vertices at least once. The Christofides-Serdyukov algorithm [1,10] gives a $3/2$-approximation to this APX-hard problem; this was recently improved to a $(3/2 - \varepsilon)$-approximation by the breakthrough work of Karlin, Klein, and Oveis Gharan, where $\varepsilon > 0$ [7]. A related question is: *what is the integrality gap of the subtour-elimination polytope relaxation for the TSP?* Wolsey had adapted the Christofides-Serdyukov analysis to show an upper bound of $3/2$ [12] (also [11]), and there exists a lower bound of $4/3$. Building on their above-mentioned work, Karlin, Klein, and Oveis Gharan gave an integrality gap of $1.5 - \varepsilon'$ for another small constant $\varepsilon' > 0$ [5], thereby making the first progress towards the conjectured optimal value of $4/3$ in nearly half a century.

Both these recent results are based on a randomized version of the Christofides-Serdyukov algorithm proposed by Oveis Gharan, Saberi, and Singh [8]. This algorithm first samples a spanning tree (plus perhaps one edge) from the *max-entropy distribution* with marginals matching the LP solution, and adds an O-join on the odd-degree vertices O in it, thereby getting an Eulerian spanning subgraph. Since the first step has expected cost equal to that of the LP solution, these works then bound the cost of this O-join by strictly less than half the optimal value, or the LP value. The proof uses a cactus-like decomposition of the min-cuts of the graph with respect to the values x_e, like in [8].

© Springer Nature Switzerland AG 2022
K. Aardal and L. Sanitá (Eds.): IPCO 2022, LNCS 13265, pp. 305–318, 2022.
https://doi.org/10.1007/978-3-031-06901-7_23

Given the 3/2 barrier has been broken, we can ask: what other techniques can be effective here? How can we make further progress? These questions are interesting even for cases where the LP has additional structure. The half-integral cases (i.e., points for which $x_e \in \{0, 1/2, 1\}$ for all e) are particularly interesting due to the Schalekamp, Williamson, and van Zuylen conjecture, which says that the integrality gap is achieved on instances where the LP has optimal half-integral solutions [9]. The team of Karlin, Klein, and Oveis Gharan first used their max-entropy approach to get an integrality gap of 1.49993 for half-integral LP solutions [6], before they moved on to the general case in [7] and obtained an integrality gap of $1.5 - \varepsilon$; the latter improvement is considerably smaller than in the half-integral case. It is natural to ask: can we do better for half-integral instances?

In this paper, we answer this question affirmatively. We show how to get tours of expected cost of ≈ 1.499 times the linear program value using an algorithm based just on matroid intersection techniques. Moreover, some of these ideas can also strengthen the max-entropy sampling approach in the half-integral case. While the matroid intersection approach and the strengthened max-entropy approach each separately yield improvements over the bound in [6], the improvement obtained by combining these two approaches is slightly better:

Theorem 1.1. *Let x be a half-integral solution to the subtour elimination polytope with cost $c(x)$. There is a randomized algorithm that rounds x to an integral solution whose cost is at most $(1.5 - \varepsilon) \cdot c(x)$, where $\varepsilon = 0.001695$.*

We view our work as showing a proof-of-concept of the efficacy of combinatorial techniques (matroid intersection, and flow-based charging arguments) in getting an improvement for the half-integral case. We hope that these techniques, ideally combined with max-entropy sampling techniques, can give further progress on this central problem.

Our Techniques. The algorithm is again in the Christofides-Serdyukov framework. It is easiest to explain for the case where the graph (a) has an even number of vertices, and (b) has no (non-trivial) proper min-cuts with respect to the LP solution values x_e—specifically, the only sets for which $x(\partial S) = 2$ correspond to the singleton cuts. Here, our goal is that each edge is "even" with some probability: i.e., both of its endpoints have even degree with probability $p > 0$. In this case we use an idea due to Haddadan and Newman [3]: we *shift* and get a $\{1/3, 1\}$-valued solution y to the subtour elimination polytope K_{TSP}. Specifically, we find a random perfect matching M in the support of x, and set $y_e = 1$ for $e \in M$, and $1/3$ otherwise, thereby ensuring $\mathbb{E}[y] = x$. To pick a random tree from this shifted distribution y, we do one of the following:

1. We pick a random "independent" set M' of matching edges (so that no edge in E is incident to two edges of M'). For each $e' \in M'$, we place partition matroid constraints enforcing that exactly one edge is picked at each endpoint—which, along with e' itself, gives degree 2 and thereby makes the edge even as desired. Finding spanning trees subject to another matroid constraint can be implemented using matroid intersection.

2. Or, instead we sample a random spanning tree from the max-entropy distribution, with marginals being the shifted value y. (In contrast, [6] sample trees

from x itself; our shifting allows us to get stronger notions of evenness than they do: e.g., we can show that every edge is "even-at-last" with constant probability, as opposed to having at least one even-at-last edge in each tight cut with some probabiltiy.)

(Our algorithm randomizes between the two samplers to achieve the best guarantees.) For the O-join step, it suffices to give fractional values z_e to edges so that for every odd cut in T, the z-mass leaving the cut is at least 1. In the special case we consider, each edge only participates in two min-cuts—those corresponding to its two endpoints. So set $z_e = {}^{x_e}/3$ if e is even, and ${}^{x_e}/2$ if not; the only cuts with $z(\partial S) < 1$ are minimum cuts, and these cuts will not show up as O-join constraints, due to evenness. For this setting, if an edge is even with probability p, we get a $(3/2 - {}^p/6)$-approximation!

It remains to get rid of the two simplifying assumptions. To sample trees when $|V|$ is odd (an open question from [3]), we add a new vertex to fix the parity, and perform local surgery on the solution to get a new TSP solution and reduce to the even case. The challenge here is to show that the losses incurred are small, and hence each edge is still even with constant probability.

Finally, what if there are proper tight sets S, i.e., where $x(\partial S) = 2$? We use the cactus decomposition of a graph (also used in [6,8]) to sample spanning trees from pieces of G with no proper min-cuts, and stitch these trees together. These pieces are formed by contracting sets of vertices in G, and have a hierachical structure. Moreover, each such piece is either of the form above (a graph with no proper min-cuts) for which we have already seen samplers, or else it is a double-edged cycle (which is easily sampled from). Since each edge may now lie in many min-cuts, we no longer just want an edge to have both endpoints be even. Instead, we use an idea from [6] that uses the hierarchical structure on the pieces considered above. Every edge of the graph is "settled" at exactly one of these pieces, and we ask for both of its endpoints to have even degree *in the piece at which it is settled*. The z_e value of such an edge may be lowered from an initial value of ${}^{x_e}/2$ in the O-join without affecting constraints corresponding to cuts *in the piece at which it is settled*.

Since cuts at other levels of the hierarchy may now be deficient because of the lower values of z_e, we may need to increase the z_f values for other "lower" edges f to satisfy these deficient cuts. This last part requires a charging argument, showing that each edge e has z_e that is strictly smaller than ${}^{x_e}/2$ in expectation. For our samplers, the naïve approach of distributing charge uniformly as in [6] does not work, so we instead formulate this charging as a flow problem.

Due to lack of space we present the simpler samplers and the main algorithm here, and defer many of the proofs and the details to the full version.

2   Notation and Preliminaries

Given a multigraph $G = (V, E)$, and a set $S \subseteq V$, let ∂S denote the cut consisting of the edges connecting S to $V \setminus S$; S and $\bar{S} := V \setminus S$ are called *shores* of the cut. A subset $S \subseteq V$ is *proper* if $1 < |S| < |V| - 1$; a cut ∂S is called *proper* if the set S is a proper subset. A set S is *tight* if $|\partial S|$ equals the size of the minimum edge-cut in G. Two sets S and S' are *crossing* if $S \cap S'$, $S \setminus S'$, $S' \setminus S$, and $V \setminus (S \cup S')$ are all non-empty.

Define the *subtour elimination polytope* $K_{TSP}(G) \subseteq \mathbb{R}^{|E|}$:

$$\{x \geq 0 \mid x(\partial v) = 2 \; \forall v \in V, x(\partial S) \geq 2 \; \forall \text{ proper } S\}. \tag{1}$$

Let x be half-integral and feasible for (1). W.l.o.g. we can focus on solutions with $x_e = 1/2$ for each $e \in E$, doubling edges if necessary. The support graph G is then a 4-regular 4-edge-connected (henceforth 4EC) multigraph.

Fact 2.1. *If $x \in K_{TSP}(G)$, then $x|_{E(V(G)\setminus\{r\})}$ is in the spanning tree polytope $K_{spT}(G[V(G) \setminus \{r\}])$ for any $r \in V$, and $x/2$ is in the perfect matching polytope $K_{PM}(G)$ (when $|V(G)|$ is even) and in the O-join dominator polytope $K_{join}(G, O)$, $O \subseteq V(G)$, $|O|$ even, given by:*

$$\{z \geq 0 \mid z(\partial S) \geq 1 \; \forall S \subseteq V, |S \cap O| \text{ odd.}\}.$$

Lemma 2.1. *Consider a sub-partition $\mathcal{P} = \{P_1, P_2, \ldots, P_t\}$ of the edge set of G. Let x be a fractional solution to the spanning tree polytope that satisfies $x(P_i) \leq 1$ for all $i \in [t]$. The integrality of the matroid intersection polytope implies that we can efficiently sample from a probability distribution $\mathcal{D}$ over spanning trees which contain at most one edge from each of the parts P_i, such that $\mathbb{P}_{T \leftarrow \mathcal{D}}[e \in T] = x_e$.*

Let z be in the relative interior of the spanning tree polytope. The *max-entropy distribution* is a distribution μ of spanning trees that maximizes the entropy of μ subject to $\mathbb{P}(e \in T) = z_e$ [8]. It is a λ-uniform spanning tree distribution and thus is *strongly Rayleigh (SR)*.

Theorem 2.2 (Negative Correlation [8]). *Let μ be an SR distribution on spanning trees.*

1. *Let S be a set of edges and $X_S = |S \cap T|$, where $T \sim \mu$. Then, $X_S \sim \sum_{i=1}^{|S|} Y_i$, where the Y_i are independent Bernoulli random variables with success probabilities p_i and $\sum_i p_i = \mathbb{E}[X_S]$.*
2. *For any set of edges S and $e \notin S$,*
   (i) $\mathbb{E}_\mu[X_S] \leq \mathbb{E}_\mu[X_S \mid X_e = 0] \leq \mathbb{E}_\mu[X_S] + \mathbb{P}_\mu(e \in T)$, and
   (ii) $\mathbb{E}_\mu[X_S] - 1 + \mathbb{P}_\mu(e \in T) \leq \mathbb{E}_\mu[X_S \mid X_e = 1] \leq \mathbb{E}_\mu[X_S]$.

Theorem 2.3 ([4], Corollary 2.1). *Let $g : \{1, \ldots, m\} \to \mathbb{R}$ and $0 \leq p \leq m$. Let $B_1, \ldots, B_m$ be Bernoulli r.v.s with probabilities $p_1^*, \ldots, p_m^*$ that maximize (or minimize) $\mathbb{E}[g(B_1 + \ldots + B_m)]$ over all possible success probabilities p_i for B_i for which $p_1 + \cdots + p_m = p$. Then $\{p_1^*, \ldots, p_m^*\} \in \{0, x, 1\}$ for some $x \in (0, 1)$.*

3   Samplers

We now describe the MaxEnt and MatInt samplers for graphs that contain no proper min-cuts, and give bounds on certain correlations between edges that will be used in Sect. 5 to prove that every edge is "even" with constant probability.

For lack of space, we focus on the case where $|V|$ is even; the case of $|V|$ being odd is slightly more technical; please see the full version of the paper.

Suppose the graph $H = (V, E)$ is 4-regular and 4EC, contains at least four vertices, and has no proper min-cuts. H is a simple graph, because parallel edges between u, v would mean that $\partial(\{u, v\})$ is a proper min-cut. Also, all proper cuts have six or more edges. We are given a dedicated *external* vertex $r \in V(H)$; the vertices $I := V \setminus \{r\}$ are called *internal*. (In future sections, r will be given by a cut hierarchy.) Call the edges in ∂r *external edges*; all other edges are *internal*. An internal vertex is called a *boundary vertex* if it is adjacent to r. An edge is said to be *special* if both of its endpoints are non-boundary vertices.

We show two ways to sample a spanning tree on $H[I]$, the graph induced on the internal vertices, being faithful to the marginals x_e, i.e., $\mathbb{P}_T(e \in T) = x_e$ for all $e \in E(H) \setminus \partial r$. Moreover, we want that for each internal edge, both its endpoints have even degree in T with constant probability. This property will allow us to lower the cost of the O-join in Sect. 6. While both samplers will satisfy this property, each will do better in certain cases. The MATINT sampler targets special edges; it allows us to randomly "hand-pick" edges of this form and enforce that both of its endpoints have degree 2 in the tree. The MAXENT sampler, on the other hand, relies on maximizing the randomness of the spanning tree sampled (subject to being faithful to the marginals); negative correlation properties allow us to obtain the evenness property, specifically, better probabilities than MATINT for non-special edges, and a worse one for the special edges.

Our samplers will depend on the parity of $|V|$: when $|V|$ is even, the MATINT sampler in Sect. 3.1 was given by [3, Theorem 13]. They left the case of odd $|V|$ as an open problem, which we solve.

3.1 Samplers for Even $|V(H)|$

Since H is 4-regular and 4EC and $|V(H)|$ is even, setting a value of $1/4 = x_e/2$ on each edge gives a solution to $K_{PM}(H)$ by Fact 2.1.

1. Sample a perfect matching M s.t. $\mathbb{P}(e \in M) = 1/4 = x_e/2$ for all $e \in E(H)$.
2. Define a new fractional solution y (that depends on M): set $y_e = 1$ for $e \in M$, and $y_e = 1/3$ otherwise. We have $y \in K_{TSP}(H)$ (and hence $y|_I \in K_{spT}(H[I])$ by Fact 2.1): indeed, each vertex has $y(\partial v) = 1 + 3 \cdot 1/3 = 2$ because M is a perfect matching. Moreover, every proper cut U in H has at least six edges, so $y(\partial U) \geq |\partial U| \cdot 1/3 \geq 2$. Furthermore,

$$\mathbb{E}_M[y_e] = 1/4 \cdot 1 + 3/4 \cdot 1/3 = 1/2 = x_e. \tag{2}$$

3. Sample a spanning tree faithful to the marginals y, using one of two samplers:
   (a) MAXENT Sampler: Sample from the max-entropy distribution on spanning trees with marginals y. (Since y may not be in the relative interior of the spanning tree polytope, contract the 1-valued edges to obtain

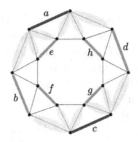

Fig. 1. The matching M consists of the green and brown edges; one possible choice of M' has edges $\{a, c\}$ in brown, and the constraints are placed on the edges adjacent to those in M' (marked in gray).

a 6-regular, $1/3$-uniform solution. This may have nontrivial min-cuts, so once again use a cactus hierarchy to decompose the graph into pieces (see Sect. 4.1); the induced solution on each piece is in the relative interior of the spanning tree polytope. For each piece, sample a λ-uniform spanning tree that preserves marginals, and then stitch these trees together.)

(b) MATINT Sampler:

i. Color the edges of M using 7 colors such that no edge of H is adjacent to two edges of M having the same color; e.g., greedily 7-color the 6-regular graph H/M. Let M' be one of these color classes picked uniformly at random. So $\mathbb{P}(e \in M') = 1/28$, and $\mathbb{P}(\partial v \cap M' \neq \varnothing) = 1/7$ see Fig. 1.

ii. For each edge $e = uv \in M'$, let L_{uv} and R_{uv} be the sets of edges incident at u and v other than e. Note that $|L_{uv}| = |R_{uv}| = 3$. Place partition matroid constraints $y(L_{uv}) \leq 1$ and $y(R_{uv}) \leq 1$ on each of these sets. Finally, restrict the partition constraints to the internal edges of H; this means some of these constraints are no longer tight for the solution y.

(c) Given the sub-matching $M' \subseteq M$, and the partition matroid $\mathcal{M}$ on the internal edges defined using M', use Lemma 2.1 to sample a tree on $H[V \setminus \{r\}]$ (i.e., on the internal vertices and edges of H) with marginals y_e, subject to this partition matroid $\mathcal{M}$.

Conditioned on the matching M, we have $\mathbb{P}(e \in T \mid M) = y_e$; now using (2), we have $\mathbb{P}(e \in T) = x_e$ for all $e \in (E \setminus \partial r)$.

The main idea for the odd case is to duplicate the external vertex with a pair of parallel edges between these copies. Since this gives a graph with proper min-cuts, we cannot apply shifting naively. Instead we perform "local surgery" on the LP solution to get a feasible fractional spanning tree. Showing that these changes still give us a tree with good evenness properties requires some care, and the ideas are deferred to the full version for lack of space.

3.2 Correlation Properties of Samplers

Let T be a tree sampled using either the MATINT or the MAXENT sampler. The following claims will be used to prove the evenness property in Sect. 5. Each table gives lower bounds on the corresponding probabilities for each sampler.

Lemma 3.1. *If f, g are internal edges incident to a vertex v, then*

Probability Statement	MATINT	MAXENT
$\mathbb{P}(|T \cap \{f, g\}| = 2)$	$1/9$	$1/9$
$\mathbb{P}(T \cap \{f, g\} = \{f\})$	$1/9$	$12/72$

Lemma 3.2. *If edges e, f, g, h incident to a vertex v are all internal, then*

Probability Statement	MATINT	MAXENT
$\mathbb{P}(|T \cap \{e, f, g, h\}| = 2)$	$2/21$	$8/27$

Lemma 3.3. *For an internal edge $e = uv$:*

(a) if both endpoints are non-boundary vertices, then

Probability Statement	MATINT	MAXENT
$\mathbb{P}(|\partial_T(u)| = |\partial_T(v)| = 2)$	$1/36$	$128/6561$

(b) if both u, v are boundary vertices, then

Probability Statement	MATINT	MAXENT
$\mathbb{P}(\text{exactly one of } u, v \text{ has odd degree in } T)$	$1/9$	$5/18$

To give a sense of the techniques, we give the proof for the last statement above when $|V(H)|$ is even. The other proofs are similar in flavor, please see the full version.

Proof (Lemma 3.3a, Even Case). The MATINT claims: The event happens when $e \in M'$, which happens w.p. $1/28$, which is at least $1/36$.

The MAXENT claims: Condition on $e \in M$. Let $S_1 = \partial(u) \setminus e$ and $S_2 = \partial(v) \setminus e$. Denote $S_1 = \{a, b, c\}$. Lower bound $\mathbb{P}(|S_1 \cap T| = 1)$ using Theorem 2.3: $\mathbb{E}[|S_1 \cap T|] = 3 \cdot 1/3 = 1$, so $\mathbb{P}(|S_1 \cap T| = 1) \geq 3 \cdot 1/3 \cdot (2/3)^{2/3} = 4/9$. Consider the distribution over the edges in S_2 conditioned on $a \in T$; this distribution is also SR. By Theorem 2.2, $1/3 \leq \mathbb{E}[X_{S_2} \mid X_a = 1] \leq 1$. Applying Theorem 2.2 twice more,

$$1/3 \leq \mathbb{E}[X_{S_2} \mid X_a = 1, X_{b,c} = 0] \leq 1 + 1/3 + 1/3 = 5/3.$$

By Theorem 2.3, $\mathbb{P}(X_{S_2} = 1 \mid X_a = 1, X_{b,c} = 0) \geq 3 \cdot 1/9 \cdot (8/9)^2 = 64/243$. Using symmetry, we obtain $\mathbb{P}(X_{S_2} = 1 \wedge X_{S_1} = 1 \wedge e \in M) \geq 64/243 \cdot 4/9 \cdot 1/4 \geq 128/6561$.

4 Sampling Algorithm for General Solutions

Now that we can sample a spanning tree from a graph with no proper min-cuts, we introduce the algorithm to sample a spanning tree plus one edge (an r_0 tree) from a 4-regular, 4EC graph, perhaps with proper min-cuts.

W.l.o.g., assume that the graph $G = (V, E)$ has a set of three special vertices $\{r_0, u_0, v_0\}$, with each pair r_0, u_0 and r_0, v_0 having a pair of edges between them (used in line 18).(We can introduce dummy nodes to ensure this property, which is for simplicity—it guarantees that the top set in the cut hierarchy is a cycle set.) Define a *double cycle* to be a cycle graph in which each edge is replaced by a pair of parallel edges, and call each such pair *partner edges*.

Algorithm 1. Sampling Algorithm for a Half-Integral Solution

1: **let** G be the support graph of a half-integral solution x.
2: **let** $T = \varnothing$.
3: **while** $\exists$ a proper tight set of G not crossed by another proper tight set **do**
4:     **let** S be a minimal such set (and choose S such that $r_0 \notin S$).
5:     Define $G' = G/(V \setminus S)$.
6:     **if** G' is a double cycle **then**
7:         Label S a *cycle set*.
8:         **sample** a random edge from each set of partner edges in $G[S]$; add these edges to T.
9:     **else** // *G' has no proper min-cuts (Lemma 4.1).*
10:        Label S a *degree set*.
11:        **if** $G' = K_5$ **then**
12:            **sample** a random path on $G[S]$
13:        **else**
14:            W.p. λ, let μ be the MaxEnt distribution over $E(S)$
15:            W.p. $1 - \lambda$, let μ be the MatInt distribution.
16:            **sample** a spanning tree on $G[S]$ from μ and add its edges to T.
17:    **let** $G = G/S$
18: Due to r_0, u_0, v_0, at this point G is a double cycle (Lemma 4.1). Sample one edge between each pair of adjacent vertices in G.

As in [6], we refer to the sets in line 4 as *critical sets*. The algorithm samples from the same pieces as in [6], with the key differences being randomizing between the MatInt and MaxEnt samplers as well as a critical optimization for K_5's (the latter will become clear later and in the full version of the paper).

Lemma 4.1. *Algorithm 1 is well-defined: In every iteration of Algorithm 1, G' is either a double cycle or a graph with no proper min-cuts, and graph remaining at the end of the algorithm (line 18) is a double cycle.*

We will prove the following theorem in Sect. 6. This in turn gives Theorem 1.1.

Theorem 4.1. *Let* T *be the* r_0*-tree chosen from Algorithm 1, and* O *be the set of odd degree vertices in* T. *The expected cost of the minimum cost* O*-join for* T *is at most* $(1/2 - \varepsilon) \cdot c(x)$.

4.1   The Cut Hierarchy

To lower the cost of the O-join, we need a complete description of the min-cuts of G, which will be achieved by the implicit hierarchy of sets Algorithm 1 induces. This hierarchical decomposition is the same as the one used in [6]; however, here we give an explicit way to construct the hierarchy of tight cuts. The hierarchy is given by a rooted tree $\mathcal{T} = (V_{\mathcal{T}}, E_{\mathcal{T}})$.[1] The node set $V_{\mathcal{T}}$ corresponds to all critical sets found by the algorithm, along with a root node and leaf nodes labelled the vertices in $V_G \setminus \{r_0\}$. If S is a critical set, we label the node in $V_{\mathcal{T}}$ with S, where we view $S \subseteq V_G$ and not $V_{G'}$. The root node is labelled $V_G \setminus \{r_0\}$. A node S is a child of S' if $S \subset S'$ and S' is the first superset of S contracted after S in the algorithm. Also, the root node is a parent of all nodes corresponding to critical sets that are not strictly contained in any other critical set. Each leaf node is a child of the smallest critical set that contains it. Observe that vertex sets labelling the children of a node are a partition of the vertex set labelling that node. A node in $V_{\mathcal{T}}$ is a *cycle* or *degree* node if the corresponding critical set labelling it is a cycle or degree set. (We take the root node as a cycle node. The leaf nodes are not labelled as degree or cycle nodes see Fig. 2.)

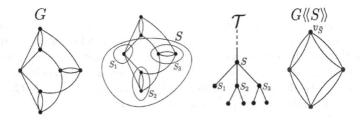

Fig. 2. A portion of the cut hierarchy $\mathcal{T}$ and the local multigraph $G\langle\langle S\rangle\rangle$.

Let $S \subseteq V_G$ be a set labelling a node in $\mathcal{T}$. Define the *local multigraph* $G\langle\langle S\rangle\rangle$ to be the following graph: take G and contract the subsets of V_G labelling the children of S in $\mathcal{T}$ down to single vertices and contract $\bar{S}$ to a single vertex $v_{\bar{S}}$. Remove any self-loops. The vertex $v_{\bar{S}}$ is called the *external vertex*; all other vertices are called *internal vertices*. An internal vertex is called a *boundary vertex* if it is adjacent to the external vertex. The edges in $G\langle\langle S\rangle\rangle \setminus v_{\bar{S}}$ are called *internal edges*. Observe $G\langle\langle S\rangle\rangle$ is precisely the graph G' in line 5 of Algorithm 1 when S is a critical set, and is a double cycle when $S = V_G \setminus \{r_0\}$.

---

[1] Since there are several graphs under consideration, the vertex set of G is called V_G. Moreover, for clarity, we refer to elements of V_G as vertices, and elements of $V_{\mathcal{T}}$ as nodes.

Properties of T

1. Let $S \subseteq V_G$ be a set labelling a node in T. If S is a degree node in T, then $G\langle\langle S \rangle\rangle$ has at least five vertices and no proper min-cuts. If S is a cycle node in T, then $G\langle\langle S \rangle\rangle$ is exactly a double cycle. These are by Lemma 4.1.
2. Algorithm 1 can be restated: For each non-leaf node S in T, sample a random path on $G\langle\langle S \rangle\rangle \setminus v_{\bar{S}}$ if it is a double cycle or K_5; otherwise use the MaxEnt or MatInt samplers w.p. λ and $1 - \lambda$, respectively, on $G\langle\langle S \rangle\rangle \setminus v_{\bar{S}}$.
3. For a degree set S, the graph $G\langle\langle S \rangle\rangle$ having no proper min-cuts implies that it has no parallel edges. In particular, no vertex has parallel edges to the external vertex in $G\langle\langle S \rangle\rangle$. Hence we get the following:

Corollary 4.1. *For a set S labeling a non-leaf node in T and any internal vertex $v \in G\langle\langle S \rangle\rangle$: if S is a cycle set then $|\partial v \cap \partial S| \in \{0, 2\}$, and if S is a degree set then $|\partial v \cap \partial S| \in \{0, 1\}$.*

The cactus representation of min-cuts [2] translates to the following complete characterization of the min-cuts of G in terms of local multigraphs.

Lemma 4.2. *Any min-cut in G is either (a) ∂S for some node S in T, or (b) ∂X where X is obtained as follows: for some cycle set S in T, X is the union of vertices corresponding to some contiguous segment of the cycle $G\langle\langle S \rangle\rangle$.*

5   Analysis Part I: The Even-at-Last Property

We now define a notion of evenness for every edge in G that will allow us to reduce the cost of the O-join in Sect. 6. In the case where G has no proper min-cuts, we called an edge *even* if both of its endpoints were even in T. The general definition of evenness extends this idea, but now depends on where an edge belongs in the hierarchy T. Specifically, we say an edge $e \in E(G)$ is *settled* at S if S is the (unique) set such that e is an internal edge of $G\langle\langle S \rangle\rangle$; call S the *last set* of e. If S is a degree or cycle set, we call e a *degree edge* or *cycle edge*.

Let S be the last set of e, and $T\langle\langle S \rangle\rangle$ be the restriction of T to $G\langle\langle S \rangle\rangle$.

1. A degree edge e is called *even-at-last (EAL)* if both its endpoints have even degree in $T\langle\langle S \rangle\rangle$.
2. For a cycle edge $e = uv$, the graph $G\langle\langle S \rangle\rangle \setminus \{v_{\bar{S}}\}$ is a chain of vertices $v_\ell, \dots, u, v, \dots, v_r$, with consecutive vertices connected by two parallel edges. Let $C := \{v_\ell, \dots, u\}$, and $C' := \{v, \dots, v_r\}$ be a partition of this chain. The cuts ∂C and $\partial C'$ are called the *canonical cuts* for e. Cycle edge e is called *even-at-last (EAL)* if both canonical cuts are crossed an even number of times by $T\langle\langle S \rangle\rangle$; in other words, if there is exactly one edge in $T\langle\langle S \rangle\rangle$ from each of the two pairs of external partner edges leaving v_ℓ and v_r.

Informally, a degree edge is EAL in the general case if it is even in the tree at the level at which it is settled. Let e be settled at a *degree* set S. We say that e is *special* if both of its endpoints are non-boundary vertices in $G\langle\langle S\rangle\rangle$ and *half-special* if exactly one of its endpoints is a boundary vertex in $G\langle\langle S\rangle\rangle$. The key property used in Sect. 6 to reduce each z_e in the fractional O-join is:

Theorem 5.1 (The Even-at-Last Property). *The table below gives lower bounds on the probability that special, half-special, and all other types of degree edges are EAL in each of the two samplers.*

	special	*half-special*	*other degree edges*
MATINT	$1/36$	$1/21$	$1/18$
MAXENT	$128/6561$	$4/27$	$12/144$

Moreover, a cycle edge is EAL w.p. at least $\lambda \cdot 12/144 + (1 - \lambda) \cdot 1/18$.

Proof. Let e be settled at S. Let T_S be the spanning tree sampled on the internal vertices of $G\langle\langle S\rangle\rangle$ (in Algorithm 1, the spanning tree sampled on $G[S]$). We show the proof when S is a degree set.

1. If e is special, then Lemma 3.3(a) gives the bounds in the table.
2. Suppose one of the endpoints of $e = uv$ (say u) is a boundary vertex in S, with edge f incident to u leaving S. By Lemma 3.2, the other endpoint v is even in T_S w.p. $2/21$ for the MATINT sampler and $8/27$ for the MAXENT sampler. Moreover, the edge f is chosen at a higher level than S and is therefore independent of T_S, and hence can make the degree of u even w.p. $1/2$. Thus e is EAL w.p. $1/21$ for the MATINT sampler and $4/27$ for the MAXENT sampler.
3. Suppose both endpoints of e are boundary vertices of S, with edges f, g leaving S. Let $q_=$ be the probability that the degrees of vertices u, v in the tree T_S chosen within S have the same parity, and $q_\neq = 1 - q_=$. Now, when S is contracted and we choose a r_0-tree T' on the graph G/S consistent with the marginals, let $p_=$ be the probability that either both or neither of f, g are chosen in T', and $p_\neq = 1 - p_=$. Hence

$$\mathbb{P}(e\ EAL) = q_{oo}p_{11} + q_{oe}p_{10} + q_{eo}p_{01} + q_{ee}p_{00} = 1/2(p_= q_= + p_\neq q_\neq), \quad (3)$$

where q_{oo}, q_{oe}, q_{eo} and q_{ee} correspond to different parity combinations of u and v in T_S and $p_{00}, p_{01}, p_{10}, p_{11}$ correspond to whether f and g are chosen in T'. The second equality follows from symmetry.
   (a) If f, g are settled at different levels, then they are independent. This gives $p_= = p_\neq = 1/2$, and hence $\mathbb{P}(e\ EAL) = 1/4$ regardless of the sampler.
   (b) If f, g have the same last set which is a degree set, then by Lemma 3.1 $p_{11}, p_{01}, p_{10} \geq 1/9$. By symmetry, $p_{00} \geq 1/9$. So (3) gives $\mathbb{P}(e\ EAL) \geq 1/9$.
   (c) If f, g have the same last set which is a cycle set, consider the case where f, g are partners, in which case $p_\neq = 1$. Now (3) implies that $\mathbb{P}(e\ EAL) = q_\neq/2$, which by Lemma 3.3(b) is $\geq 1/2 \cdot 1/9 = 1/18$ in the MATINT sampler, and $1/2 \cdot 5/18 = 5/36$ in the MAXENT sampler. If f, g are not partners, then they are chosen independently, in which case again $p_= = p_\neq = 1/2$, and hence $\mathbb{P}(e\ EAL) = 1/4$.

The proof for cycle sets follows similar lines, and is in the full version.

6 Analysis Part II: The O-Join and Charging

To prove Theorem 4.1, we construct an O-join for the random tree T, and bound its expected cost via a charging argument. The structure of here is similar to [6]; however, we use a flow-based approach to perform the charging instead of the naive one, and also use our stronger property that *every* edge is EAL with constant probability (versus the weaker property obtained in [6] that every tight cut contains an EAL edge with constant probability).

Let O denote the (random set of) odd-degree vertices in T. The dominant of the O-join polytope $K_{join}(G,O)$ is given by

$$\{x(\partial S) \geq 1 \mid \forall S \subseteq V, |\partial S \cap T| \text{ odd.}\}.$$

This polytope is integral, so it suffices to show the *existence* of a fractional O-join solution $z \in K_{join}(G,O)$ with low expected cost. (The expectation is taken over O.)

To construct the fractional O-join z, we begin with $z = {}^x/2$. Notice that $z(\partial S) \geq 1$ is a tight constraint when S is a min-cut. For any e that is EAL in T, we first flip a biased coin to know whether to reduce z_e: the purpose of the coin flips is to "flatten" the probability of reducing z_e to the bound given by Theorem 5.1 on the probability that e is EAL. Now the amount of the reduction in z_e depends on whether e is a degree or cycle edge; the amount is later optimized by the solution to a linear program. In the case of the cycle edge, we are able to reduce an edge by the full $1/12$; degree edges cannot be reduced as drastically.

However, these reductions may make z infeasible, and we need to fix that. Indeed, suppose f is EAL and that we reduce z_f. Say f is settled at S. If S is a degree set, then the only min-cuts of $G\langle\!\langle S \rangle\!\rangle$ are the degree cuts. So the only min-cuts that the edge f is part of in $G\langle\!\langle S \rangle\!\rangle$ are the degree cuts of its endpoints; call them U, V, in $G\langle\!\langle S \rangle\!\rangle$ (U and V are vertices in $G\langle\!\langle S \rangle\!\rangle$ representing sets U and V in G). But since $|\partial U \cap T|$ and $|\partial V \cap T|$ are both even by definition of EAL, we need not worry that reducing z_f causes $z(\partial U) \geq 1$ and $z(\partial V) \geq 1$ to be violated. Likewise, if S is a cycle set, then by definition of EAL all min-cuts S' in $G\langle\!\langle S \rangle\!\rangle$ containing e have $|\partial S' \cap T|$ even, so again we need not worry.

Since f is only an internal edge for its last set S, the only cuts S' for which the constraint $z(\partial S') \geq 1$, $|\partial S' \cap T|$ odd, may be violated as a result of reducing z_f are cuts represented in lower levels of the hierarchy. Specifically, lef f be an external edge for some $G\langle\!\langle X \rangle\!\rangle$ (meaning X is lower in the hierarchy T than S) and S' be a min-cut of $G\langle\!\langle X \rangle\!\rangle$ (either a degree cut or a canonical cut). By Lemma 4.2, cuts of the form S' are the only cuts that may be deficient as a result of reducing z_f. Call the internal edges of $\partial S'$ *lower edges*. When z_f is reduced and $|\partial S' \cap T|$ is odd, we must distribute an increase (charge) over the lower edges totalling the amount by which z_f is reduced, so that $z(\partial S') = 1$.

Fix edge e, say with last set S. We need to show that in expectation the charge it receives from external edges in $G\langle\!\langle S \rangle\!\rangle$ is strictly less than the initial expected reduction to z_e. External edges bring in charge to internal edges and Corollary 4.1 says that every vertex in a critical set can have at 0, 1, or 2 external edges (and there are 4 external edges). When e is a cycle edge, we distribute charge from an external edge evenly between e and its partner. When e is a degree edge, charge will be optimally distributed according to a maximum-flow solution. Specifically, in order to minimize the maximum charge given on any edge, we bipartize $G\langle\!\langle S\rangle\!\rangle$ into $H = (L, R)$: the vertices in L and R represent external and internal edges, respectively, and the edges of H are those pairs of edges in $G\langle\!\langle S\rangle\!\rangle$ which are adjacent. Each vertex in L releases a unit of charge and each vertex in R absorbs at most c units of charge.

The problem of optimally distributing charge reduces to finding the smallest constant c such that there exists a flow with capacity at most c. An argument based on Hall's condition characterizes precisely when a flow distributing c units of charge to internal edges exists. In order to optimize the constant c (found to be $1/2$), the case where $G\langle\!\langle S\rangle\!\rangle = K_5$ happens to be a bottleneck; hence, we treat the K_5 case separately in order to gain from the max-flow formulation, by choosing a very natural sampling method for it, and then reducing its edges differently from those of other degree sets.

Acknowledgments. AG supported in part by NSF awards CCF-1907820, CCF-1955785, and CCF-2006953. MM partially supported by the ERC CoG grant TUg-bOAT no 772346.

References

1. Christofides, N.: Worst-case analysis of a new heuristic for the travelling salesman problem. Technical report 338, Graduate School of Industrial Administration, Carnegie-Mellon University, Pittsburgh (1976)
2. Fleiner, T., Frank, A.: A quick proof for the cactus representation of mincuts. Technical report Quick-proof No. 2009–03, EGRES (2009)
3. Haddadan, A., Newman, A.: Towards improving Christofides algorithm for half-integer TSP. In: ESA, pp. 56:1–56:12 (2019). https://doi.org/10.4230/LIPIcs.ESA.2019.56
4. Hoeffding, W.: On the distribution of the number of successes in independent trials. Ann. Math. Stat. **27**(3), 713–721 (1956)
5. Karlin, A., Klein, N., Oveis Gharan, S.: A (slightly) improved bound on the integrality gap of the subtour LP for TSP. CoRR abs/2105.10043 (2021). https://arxiv.org/abs/2105.10043
6. Karlin, A.R., Klein, N., Oveis Gharan, S.: An improved approximation algorithm for TSP in the half integral case. In: STOC, pp. 28–39. ACM (2020). https://doi.org/10.1145/3357713.3384273
7. Karlin, A.R., Klein, N., Oveis Gharan, S.: A (slightly) improved approximation algorithm for metric TSP. CoRR abs/2007.01409 (2020). https://arxiv.org/abs/2007.01409
8. Oveis Gharan, S., Saberi, A., Singh, M.: A randomized rounding approach to the traveling salesman problem. In: FOCS, pp. 550–559 (2011). https://doi.org/10.1109/FOCS.2011.80
9. Schalekamp, F., Williamson, D.P., van Zuylen, A.: 2-matchings, the traveling salesman problem, and the subtour LP: a proof of the boyd-carr conjecture. Math. Oper. Res. **39**(2), 403–417 (2014). https://doi.org/10.1287/moor.2013.0608

10. Serdyukov, A.: O nekotorykh ekstremal'nykh obkhodakh v grafakh. Upravlyaemye sistemy, **17**, 76–79 (1978). http://nas1.math.nsc.ru/aim/journals/us/us17/us17_007.pdf
11. Shmoys, D.B., Williamson, D.P.: Analyzing the Held-Karp TSP bound: a monotonicity property with application. Info. Proc. Let. **35**(6), 281–285 (1990)
12. Wolsey, L.A.: Heuristic analysis, linear programming and branch and bound. In: Combinatorial Optimization II, Mathematical Programming Studies, vol. 13, pp. 121–134. Springer (1980). https://doi.org/10.1007/BFb0120913

The Two-Stripe Symmetric Circulant TSP is in P

Samuel C. Gutekunst[1], Billy Jin[2(✉)], and David P. Williamson[2]

[1] Departments of Computer Science and Mathematics, Bucknell University,
Lewisburg, PA, USA
s.gutekunst@bucknell.edu
[2] School of Operations Research and Information Engineering, Cornell University,
Ithaca, NY, USA
{bzj3,davidpwilliamson}@cornell.edu

Abstract. The symmetric circulant TSP is a special case of the traveling salesman problem in which edge costs are symmetric and obey circulant symmetry. Despite the substantial symmetry of the input, remarkably little is known about the symmetric circulant TSP, and the complexity of the problem has been an often-cited open question. Considerable effort has been made to understand the case in which only edges of two lengths are allowed to have finite cost: the two-stripe symmetric circulant TSP (see Greco and Gerace [8] and Gerace and Greco [6]). In this paper, we resolve the complexity of the two-stripe symmetric circulant TSP. To do so, we reduce two-stripe symmetric circulant TSP to the problem of finding certain minimum-cost Hamiltonian paths on *cylindrical graphs*. We then solve this Hamiltonian path problem. Our results show that the two-stripe symmetric circulant TSP is in P. Note that the input of a two-stripe symmetric circulant TSP instance consists of a constant number of inputs (including n, the number of cities), so that a polynomial-time algorithm for the decision problem must run in time polylogarithmic in n, and a polynomial-time algorithm for the optimization problem cannot output the tour. We address this latter difficulty by showing that the optimal tour must fall into one of two parameterized classes of tours, and that we can output the class and the parameters in polynomial time. Thus we make a substantial contribution to the set of polynomial-time solvable special cases of the TSP [7,10], and take an important step towards resolving the complexity of the general symmetric circulant TSP.

1 Introduction

The traveling salesman problem (TSP) is one of the most famous problems in combinatorial optimization. An input to the TSP consists of a set of n cities

S. C. Gutekunst—Supported in part by NSF grants DGE-1650441 and CCF-1908517.
B. Jin—Supported in part by NSERC fellowship PGSD3-532673-2019 and by NSF grant CCF-2007009.
D. P. Williamson—Supported in part by NSF grants CCF-1908517 and CCF-2007009.

© Springer Nature Switzerland AG 2022
K. Aardal and L. Sanitá (Eds.): IPCO 2022, LNCS 13265, pp. 319–332, 2022.
https://doi.org/10.1007/978-3-031-06901-7_24

$[n] := \{1, 2, ..., n\}$ and edge costs c_{ij} for each pair of distinct $i, j \in [n]$, representing the cost of traveling from city i to city j. Given this information, the TSP is to find a minimum-cost tour visiting every city exactly once. Throughout this paper, we assume that the edge costs are *symmetric* (so that $c_{ij} = c_{ji}$ for all distinct $i, j \in [n]$) and interpret the n cities as vertices of the complete undirected graph K_n with edge costs $c_e = c_{ij}$ for edge $e = \{i, j\}$. In this setting, the TSP is to find a minimum-cost Hamiltonian cycle on K_n.

With just this set-up, the TSP is well known to be NP-hard. An algorithm that could approximate TSP solutions in polynomial time to within any factor $\alpha > 1$ would imply P=NP (see, e.g., Theorem 2.9 in Williamson and Shmoys [14]). Thus it is common to consider special cases, such as requiring costs to obey the triangle inequality (i.e. requiring costs to be *metric*, so that $c_{ij} + c_{jk} \geq c_{ik}$ for all $i, j, k \in [n]$), or costs that are distances between points in Euclidean space.

In this paper, we consider a different class of instances: circulant TSP. In **circulant TSP**, the matrix of edge costs $C = (c_{ij})_{i,j=1}^n$ is circulant; the cost of edge $\{i, j\}$ only depends on $i - j$ mod n. Our assumption that the edge costs are symmetric and that K_n is a simple graph implies that, for symmetric circulant TSP instances, we can write our cost matrix in terms of $\lfloor \frac{n}{2} \rfloor$ parameters:

$$C = (c_{(j-i) \bmod n})_{i,j=1}^n = \begin{pmatrix} 0 & c_1 & c_2 & c_3 & \cdots & c_1 \\ c_1 & 0 & c_1 & c_2 & \cdots & c_2 \\ c_2 & c_1 & 0 & c_1 & \ddots & c_3 \\ \vdots & \vdots & \vdots & \vdots & \ddots & \vdots \\ c_1 & c_2 & c_3 & c_4 & \cdots & 0 \end{pmatrix}, \tag{1}$$

with $c_0 = 0$ and $c_i = c_{n-i}$ for $i = 1, ..., \lfloor \frac{n}{2} \rfloor$. Importantly, in circulant TSP we do not assume that the edge costs satisfy the triangle inequality.

Circulant TSP is a compelling open problem because of the intriguing structure which circulant symmetry broadly provides to combinatorial optimization problems. It is unclear whether a given combinatorial optimization problem should remain hard or become easy when restricted to circulant instances. Some classic combinatorial optimization problems are easy when restricted to circulant instances. For example, in the late 80's, Burkard and Sandholzer [3] showed that the decision problem for whether or not a symmetric circulant graph (i.e. a graph whose adjacency matrix is circulant) is Hamiltonian can be solved in polynomial time and showed that bottleneck TSP is polynomial-time solvable on symmetric circulant graphs. Bach, Luby, and Goldwasser (cited in Gilmore, Lawler, and Shmoys [7]) showed that one could find minimum-cost Hamiltonian paths in (not-necessarily-symmetric) circulant graphs in polynomial time. In contrast, Codenotti, Gerace, and Vigna [4] show that Max Clique and Graph Coloring remain NP-hard when restricted to circulant graphs and do not admit constant-factor approximation algorithms unless P=NP.

Because of this ambiguity, the complexity of circulant TSP has often been cited as an open problem (see, e.g., Burkhard [1], Burkhard, Deĭneko, Van Dal, Van der Veen, and Woeginger [2], and Lawler, Lenstra, Rinnooy Kan, and Shmoys [11]). It is not known if the circulant TSP is solvable in polynomial-time or is NP-hard. Prior to this work, the complexity of circulant TSP was not understood even when restricted to instances where only two of the edge costs $c_1, \ldots, c_{\lfloor \frac{n}{2} \rfloor}$ are finite: the *symmetric two-stripe circulant TSP* (which we will henceforth abbreviate as *two-stripe TSP*) (see Greco and Gerace [8] and Gerace and Greco [6]). Yang, Burkard, Çela, and Woeginger [15] provide a polynomial-time algorithm for *asymmetric* TSP in circulant graphs with only two stripes having finite edge costs. The symmetric two-stripe circulant TSP is not, however, a special case of the asymmetric two-stripe version.[1]

Despite substantial structure and symmetry, the complexity of two-stripe circulant TSP had previously remained elusive. General upper- and lower-bounds for circulant TSP stem from Van der Veen, Van Dal, and Sierksma [13]; Greco and Gerace [8] and Gerace and Greco [6] focus specifically on the two-stripe TSP, and prove sufficient (but not necessary) conditions for these upper- and lower-bounds to apply. Van der Veen, Van Dal, and Sierksma [13] and Gerace and Greco [5] give a general heuristic for circulant TSP that provides a tour within a factor of two of the optimal solution; no improvements to this general heuristic have been made when constrained to the two-stripe version.

In this paper, we take the first step toward resolving the polynomial-time solvability of circulant TSP by showing that symmetric two-stripe circulant TSP is solvable in polynomial time. We need to be clear on what we mean by "solvable in polynomial time" for this problem. The input is the number n (represented in binary by $O(\log n)$ bits), the indices of the two finite cost edges, and the corresponding costs. So to run in polynomial time in this case, we should run in time *polylogarithmic* in n. We will show how to compute the cost of the optimal tour in $O(\log^2 n)$ time, and thus the decision problem of whether the cost of the optimal tour is at most a bound given as input is solvable in polynomial time; this places the decision version of the problem in the class P. Notice, however, time polynomial in the input size is not sufficient to output the complete sequence of n vertices to visit in the tour. Nevertheless, given two parameterized classes of tours that we will later describe, we are able to compute in polynomial time to which class the optimal tour belongs, as well as the values of the parameters.

Thus our main contribution is to make a substantial addition to the set of polynomial-time solvable special cases of the TSP [7,10], and to take an important step towards resolving the complexity of the symmetric circulant TSP.

In Sect. 2, we begin by providing background on the structure of circulant graphs, and previous work on the two-stripe TSP. We then state our main result, Theorem 1, which characterizes the cost of an optimal tour. In Sect. 3, we reduce the two-stripe TSP to the problem of finding certain minimum-cost Hamiltonian paths on cylinder graphs. In Sect. 4 we prove Theorem 1 assuming a characterization theorem for Hamiltonian paths between the first and last column in cylinder graphs. We formally state our algorithm in Sect. 5, and show that it runs in polynomial time. In Sect. 6, we present a proof sketch of the aforementioned characterization theorem. We conclude in Sect. 7.

---

[1] In the symmetric case, edges $\{v, v + i\}$ and $\{v, v - i\}$ of cost c_i connect v to both $v + i$ and $v - i$; in the asymmetric case, there are edge costs $c_1, \ldots, c_{n-1}$ and an edge $(v, v + i)$ of cost c_i only connects from v to $v + i$. To encode two general symmetric circulant edges would require four asymmetric circulant edges.

Because of space limitations, many results in the body of the paper are stated without proofs, and many figures are suppressed. A full version of the paper can be accessed at http://people.orie.cornell.edu/dpw/2stripe.pdf.

2 Background and Notation

In this section, we formalize our notation for the two-stripe TSP, and describe pertinent background results we will use. Throughout this paper, we use $\equiv_n$ to denote equivalence mod n. All calculations are implicitly mod n, unless indicated otherwise. For example, we use $v + a_1$ to denote the vertex reachable from v by following an edge of length a_1; the label $v + a_1$ is implicitly taken mod n.

2.1 Circulant Graphs

Recall that a **circulant graph** is a simple graph whose adjacency matrix is circulant. In symmetric circulant TSP, all edges $\{i, j\}$ such that $i - j \equiv_n k$ or $i - j \equiv_n (n - k)$ have the same cost c_k. Such edges are typically referred to as in the k-**th stripe**, or as of **length** k. In the two-stripe TSP, only two of the edge costs $c_1, c_2, ..., c_{\lfloor \frac{n}{2} \rfloor}$ are finite. We refer to those two edge lengths as a_1 and a_2, so that $0 \le c_{a_1} \le c_{a_2} < \infty$, and for $i \notin \{a_1, a_2\}$, $c_i = \infty$.

We use the following definition to describe circulant graphs including exactly the edges associated with some set of stripes S. In the two-stripe TSP, we are generally interested in subsets S of size 1 or 2.

Definition 1. *Let $S \subset \{1, ..., \lfloor \frac{n}{2} \rfloor\}$. The **circulant graph** $C\langle S \rangle$ is the (simple, undirected, unweighted) graph including exactly the edges associated with the stripes S. I.e., the graph with adjacency matrix $A = (a_{ij})_{i,j=1}^n$, with $a_{ij} = 1$ if $(i - j) \bmod n \in S$ or $(j - i) \bmod n \in S$ and 0 otherwise.*

Provided that $C\langle \{a_1, a_2\} \rangle$ is Hamiltonian, the **two-stripe TSP** is to find a minimum-cost Hamiltonian cycle in the graph $C\langle \{a_1, a_2\} \rangle$. Since $c_{a_1} \le c_{a_2}$, the problem is to find a Hamiltonian cycle in $C\langle \{a_1, a_2\} \rangle$ using as few edges of length a_2 as possible.

Early work from Burkard and Sandholzer [3] provides necessary and sufficient conditions for $C\langle \{a_1, a_2\} \rangle$ to be Hamiltonian: $\gcd(n, a_1, a_2)$ must equal 1. That this condition is necessary follows directly: If $\gcd(n, a_1, a_2) := g_2 > 1$, then for any $n, m \in \mathbb{Z}$, $na_1 + ma_2 \equiv_{g_2} 0$. Thus any combination of edges of length a_1 and a_2 will remain within the vertices $\{v \in [n] : v \equiv_{g_2} 0\}$, which is a strict subset of $[n]$. We will constructively show tours demonstrating sufficiency for the two-stripe case. Throughout this paper, we let $g_1 := \gcd(n, a_1)$ and $g_2 := \gcd(n, a_1, a_2)$. As argued above, an instance to the two-stripe TSP has a solution with finite cost if and only if $g_2 = 1$.

Circulant graphs have a rich structure that allows us to understand $C\langle \{a_1, a_2\} \rangle$ in terms of $C\langle \{a_1\} \rangle$. From a more general result of Burkard and Sandholzer [3], the graph $C\langle \{a_1\} \rangle$ consists of g_1 components. Each component is a cycle of size n/g_1: Start at some vertex i for $0 \le i < g_1$, and continue

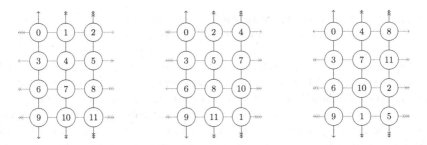

Fig. 1. Three drawings of two-stripe instances on $n = 12$ vertices with $a_1 = 3$. In the left instance $a_2 = 1$, in the middle instance $a_2 = 2$, and on the right instance $a_2 = 4$. Black edges are of length a_1 and red edges are of length a_2. Arrows denote horizontal/vertical "wrapping around," and same-colored arrows with the same number of arrowheads wrap to each other. In all instances, e.g., an edge of length a_1 wraps around vertically, connecting 0 (top-left) to 9 (bottom-left).

following edges of length a_1 until reaching $i + \left(\frac{n}{g_1} - 1\right) a_1$. Taking one more length-a_1 edge wraps back to vertex i, since $i + \frac{n}{g_1} a_1 = i + n\frac{a_1}{g_1} \equiv_n i$ (where the final equivalence follows because g_1 divides a_1 by definition)[2].

Provided $g_2 = 1$, the result of Burkard and Sandholzer indicates $C\langle\{a_1, a_2\}\rangle$ is Hamiltonian (and therefore connected). Thus edges of length a_2 connect the components of $C\langle\{a_1\}\rangle$. They do so, moreover, in an extremely structured manner. Consider, e.g., Fig. 1, which shows $C\langle\{a_1, a_2\}\rangle$ for three different two-stripe instances on $n = 12$ vertices. In all cases, $a_1 = 3$. Since $g_1 = GCD(12, 3) = 3$, there are three components of $C\langle\{a_1\}\rangle$ in all instances. The instances are drawn so that each component is in its own column. While the columns are drawn in slightly different orders for each instance, and while the specific red length-a_2 edges connect different pairs of vertices, all three instances are extremely structured. For example, let v be any vertex in the first column. Regardless of the instance or which vertex v is within the first column, $v + a_2$ is always in the second column. Similarly, if v is in the second column, $v + a_2$ is always in the third; if v is in the third column, $v + a_2$ is always in the first.

The following claim from Gutekunst and Williamson [9] makes this structure precise: Consider the graph $C\langle\{a_1, a_2\}\rangle$, contract each component of $C\langle\{a_1\}\rangle$ into a single vertex, and remove extra copies of parallel edges. Then, provided $g_2 = 1$, the resulting graph is a cycle.

This structure suggests a convention for drawing two-stripe circulant graphs (used in Fig. 1), where each column corresponds to a component of $C\langle\{a_1\}\rangle$. We further arrange the columns (and the ordering of vertices within each column) so that edges of a_2 can be generally drawn horizontally. Specifically, we take the convention that 0 is the top-left vertex. The first column will proceed vertically-down as $0, a_1, 2a_1, 3a_1, ..., \left(\frac{n}{g_1} - 1\right) a_1$. Our second column "translates" the first

---

[2] In the special case that n is even and $a_1 = n/2$, we think of each of the $g_1 = GCD(n, n/2) = n/2$ components of $C\langle\{a_1\}\rangle$ as a cycle on a pair of vertices.

column by a_2, so that the top vertex is a_2, the second vertex is $a_1 + a_2$, the third vertex is $2a_1 + a_2$, and so on. Our third column translates by another a_2, and so on.

Returning to Fig. 1, recall that the three instances have $n = 12$ and $a_1 = 3$. In the left instance $a_2 = 1$, in the middle instance $a_2 = 2$, and in the right instance $a_2 = 4$. Up to a re-labelling of vertices, the only structural change occurs in how the right-most column is connected to the left-most column: the edges that wrap around horizontally, connecting the last column back to the first. On the left, taking an edge of length a_2 from a vertex in the last column wraps around to the first column, but one row lower (2 in the top row connects to 3 in the second row, 5 in the second row connects to 6 in the third row, etc.). In the middle, wrapping around shifts down two rows (4 in the top row connects to 6 in the third row, etc.). On the right, edges wrap back to the same row.

2.2   Cylinder Graphs

Part of the challenge of two-stripe TSP is the differing ways that horizontal edges wrap around between the first and last column. Our first result, in Sect. 3, removes this difficulty, and allows us to work on "cylinder graphs": graphs that are similar to those in Fig. 1, but without any horizontal edges wrapping around between the last and first column.

Definition 2. *Let $n = r \times c$. A $r \times c$ **cylinder graph** is a graph with n vertices arranged into an $r \times c$ grid. For $0 \le i \le r - 1$ and $0 \le j \le c - 1$, a vertex in row i and column j is adjacent to:*

- *The vertex in row i and column $j + 1$, provided $j < c - 1$,*
- *The vertex in row i and column $j - 1$, provided $j > 0$,*
- *The vertex in row $i - 1 \mod r$ and column j, and the vertex in row $i + 1$ $\mod r$ and column j.*

It will often be helpful to refer to a vertex in row x and column y as (x, y), indexed by its column and row, starting from 0. We take the convention that the top-left vertex is $(0, 0)$. Hence, the bottom-right vertex is $(r - 1, c - 1)$. In general, a vertex $v = xa_1 + ya_2$ will have cylindrical coordinates (x, y).

2.3   Trivial Cases

We note several cases where the two-stripe TSP is trivial:

- If $g_2 > 1$, then $G\langle\{a_1, a_2\}\rangle$ is not Hamiltonian.
- If $g_1 = 1$, then the graph $G\langle\{a_1\}\rangle$ is Hamiltonian. Hence, since $c_{a_1} \le c_{a_2}$, a cheapest Hamiltonian cycle costs nc_{a_1} and consists of a Hamiltonian cycle on $G\langle\{a_1\}\rangle$. Note that any time n is prime, $g_1 = 1$, so that two-stripe (and indeed, general circulant) TSP is trivial any time n is prime.
- If $c_{a_1} = c_{a_2}$ and $g_2 = 1$, then $G\langle\{a_1, a_2\}\rangle$ is Hamiltonian and any Hamiltonian cycle costs $nc_{a_1} = nc_{a_2}$.

These observations are also made in [8] and [6].

Following the example of Fig. 1, we think of the "expensive" length-a_2 edges as "horizontal edges." The "cheap" length-a_1 edges are analogously considered "vertical edges." Then the goal is to minimize the total number of expensive (horizontal) edges. From here on we assume without loss of generality that $c_{a_1} = 0$ and $c_{a_2} = 1$, so every vertical edge has cost 0 and every horizontal edge has cost 1.

2.4   Main Result

The main result of our paper, stated below, is a characterization of the optimal tour in any non-trivial two-stripe circulant TSP instance. Section 4 contains its proof, and Sect. 5 shows how this result naturally leads to a polynomial-time algorithm for two-stripe TSP.

Theorem 1. *Let $r = \frac{n}{g_1}$ and $c = g_1$. Suppose the cylindrical coordinates of $-a_2$ are $(x, c-1)$. Let m^* be the smallest integer value of m such that $m \geq -\frac{c}{2}$, and $x \in \{2m + c, -(2m + c)\} \mod r$. Then:*

- *If $m^* \leq 0$, the cost of the optimal tour is c.*
- *If $0 < 2m^* < c - 2$, the cost of the optimal tour is $c + 2m^*$.*
- *If $2m^* \geq c - 2$ or m^* does not exist, the cost of the optimal tour is $2c - 2$.*

2.5   Previous Results on Two-Stripe TSP

Despite the restrictive structure of two-stripe TSP, many of the main previous results for it are inherited from the more-general circulant TSP. From an approximation algorithms perspective, the state-of-the-art remains a heuristic from Van der Veen, Van Dal, and Sierksma [13]; on any two-stripe instance, this heuristic provides a tour of cost within a factor of two of the optimal cost (see also Gerace and Greco [5], for a more general 2-approximation algorithm). The performance guarantee uses a combinatorial lower bound of Van der Veen, Van Dal, and Sierksma [13], which when specialized to the two-stripe case states that any Hamiltonian cycle must cost at least g_1; this is clear, since there must be at least $c = g_1$ edges to join the c columns in a cycle. If a tour of cost g_1 exists, we will refer to it as a "lower bound tour."

We can exhibit an upper bound of $2(g_1 - 1)$ on the optimal solution to any two-stripe instance by providing a feasible tour (as usual, provided that $g_2 = 1$); such tours are illustrated in Fig. 2. Henceforth, we will refer to these tours as "upper bound tours". Such tours immediately give a 2-approximation algorithm: they provides a feasible solution costing at most $2(g_1 - 1) < 2g_1$, while g_1 is a lower bound on the optimal cost.

More specific results on the two-stripe TSP come from Greco and Gerace [8] and Gerace and Greco [6]. They provide an algebraic test that can identify certain instances where the upper- or lower-bounds occur, but leave as open fully characterizing either extreme, as well as determining the optimal solution to any instance where the optimal value is between those bounds. They further propose

Fig. 2. Feasible Hamiltonian cycles of cost $2(g_1 - 1)$ when g_1 is even (left) and odd (right).

Fig. 3. Two examples of GG paths, both using 3 horizontal edges between the first and second column. Because GG paths use exactly one edge between each pair of consecutive columns from the second onward, they are extremely structured.

a particular construction of a tour (described in the next section), and conjecture that it is the optimal tour in any case where the lower- or upper-bound is not optimal. Our resolution of the two-stripe TSP confirms this conjecture.

3 GG Paths and a Reduction to Hamiltonian Paths on Cylinder Graphs

In this section, we reduce the two-stripe TSP to the problem of finding a minimum-cost Hamiltonian path from 0 to $-a_2$ on an associated cylinder graph. We then introduce specific Hamiltonian paths on cylinder graphs, which we will call *GG paths*, and characterize the set of vertices that they can reach. We name these paths after Gerace and Greco ([6], [8]), who conjectured that they are sufficient to describe an optimal tour in settings where the upper bound is not optimal.

3.1 Reduction of 2-Stripe Problem to Cylinder Graph Problem

We first reduce two-stripe TSP instances to Hamiltonian paths on cylinder graphs that start in the first column and end in the last column. As previously stated, we can always attain a tour of cost $2(g_1 - 1)$. Theorem 2 says that we can find a cheaper tour costing $k < 2(g_1 - 1)$ if and only if there there is a corresponding Hamiltonian path on a cylinder graph using $k - 1$ horizontal edges.

Theorem 2. *Consider a 2-stripe instance with n vertices and edges of length a_1 and a_2. There exists a Hamiltonian cycle costing $k < 2(g_1 - 1)$ if and only if there is a Hamiltonian path on an $r \times c$ cylinder graph with $c = g_1$ and $r = \frac{n}{g_1}$ using $k - 1$ horizontal edges, starting at $(0, 0)$, and ending at $(x, c - 1)$, where x is the unique solution to $xa_1 \equiv_n -g_1 a_2$ in $\{0, 1, ..., \frac{n}{g_1} - 1\}$.*

The idea of the proof is that a tour of cost less than $2(g_1 - 1) = 2(c - 1)$ must have some pair of consecutive columns such that only one horizontal edge of the tour crosses between the two columns. We can "shift" the tour so that this edge wraps around from column $c - 1$ to 0, and so that this edge passes through $(0, 0)$. Removing the wraparound edge leaves a Hamiltonian path from $(0,0)$ to $(x, c - 1)$, where the endpoint is equivalent to the vertex $-a_2$.

3.2 GG Paths

Theorem 2 reduces two-stripe TSP to the problem of finding a minimum-cost Hamiltonian path between 0 and $-a_2$ in a cylinder graph. Next, we introduce a specific class of Hamiltonian paths: *GG paths*. We will later show (in Theorem 3), that there is always a minimum-cost Hamiltonian path that is a GG path.

Definition 3. *A **GG path** with $2k + 1 + (c - 2)$ horizontal edges in a cylinder graph is a Hamiltonian path that*

1. *Starts at $(0, 0)$ and ends in the last column,*
2. *Uses $2k + 1$ horizontal edges between the first pair of columns, and*
3. *Uses 1 horizontal edge between each of the $c - 2$ remaining pairs of columns.*

See Fig. 3 for two examples of GG paths. Because GG paths only use one horizontal edge between consecutive columns starting from the second column onward, they are extremely structured. Moreover, when moving to a new column (starting at the third column), a GG path has exactly two options: the first edge in that column vertically is either vertically "up" (going from (i, j) to $(i - 1, j)$) or vertically "down" (going from (i, j) to $(i + 1, j)$). The net effect is that, if the last vertex visited in the second column is $(i, j - 1)$, the last vertex visited in the third column will be $(i + 1, j)$ or $(i - 1, j)$.

This structure of GG paths applies inductively. Starting in the third column, each time a GG path enters a new column, it has the exact same two options. Tracing out this process allows us to quickly determine where a GG path can end, based on its path through the first two columns.

In Theorem 3, we will show that when solving the minimum-cost Hamiltonian path problem on a cylinder graph between $(0, 0)$ and a vertex in the last column, it suffices to consider GG paths: If a minimum-cost Hamiltonian path between $(0,0)$ and $(i, c - 1)$ requires k horizontal edges, there is a GG path from $(0, 0)$ to $(i, c - 1)$ using k horizontal edges. Hence, we characterize exactly where a GG path, with some fixed number of horizontal edges, can end up. We define this set below.

Definition 4. *Let $A_{r,c,m}$ denote the set of row indices of vertices in the last column reachable on a GG path in an $r \times c$ cylinder graph using **at most** $2m + (c-1)$ total horizontal edges.*

In the definition of $A_{r,c,m}$, it is useful to think of the m as the number of extra "expensive pairs" of horizontal edges used between the first and second column. Note also that a GG path can have at most $\lfloor \frac{r-1}{2} \rfloor$ extra horizontal pairs, because the extra horizontal edges are all between the first two columns.

Proposition 1. *Let $m \le \lfloor \frac{r-1}{2} \rfloor$. Then $A_{r,c,m} = \{c + 2m - 2i \mod r : 0 \le i \le c + 2m\}$.*

Our proof follows the structure of GG paths discussed above: we need only identify the vertices reachable when $c = 2$. Then we follow the "triangular growth" of the reachable vertices, where a GG path ending in some row k of column c corresponds to a GG path ending in column $c+1$ in row $k-1$ or $k+1$ (taken mod r).

4   Proof of Main Result

In this section, we prove our main result, Theorem 1, which characterizes the cost of an optimal tour for the two-stripe TSP. In what follows, we will sometimes use the notation $\{a_1, a_2, \ldots, a_k\} \mod r$ as a shorthand for $\{a_1 \mod r, a_2 \mod r, \ldots, a_k \mod r\}$.

Our proof of Theorem 1 relies on Theorem 3. This is one of the main technical results of the paper. We give a sketch of its proof in Sect. 6.

Theorem 3. *Consider a cylinder graph on $n = r \times c$ vertices, with r rows and c columns. Suppose we have a Hamiltonian path, starting at 0 and ending in the last column, and suppose it uses at most $(c-1) + 2m$ horizontal edges. Then it must end at a row in $A_{r,c,m}$.*

Proof (of Theorem 1). Suppose $m^* \le 0$. This implies $x \in \{-c, -c+2, \ldots, c-2, c\} \mod r$. Note that this set is equal to $A_{r,c,0}$ by Proposition 1. By definition of $A_{r,c,0}$, there exists a GG path, using exactly $c-1$ horizontal edges, from 0 to $-a_2$. Appending the edge from $-a_2$ to 0 gives a Hamiltonian cycle with cost c. This cycle is optimal, since c is a lower bound on the cost of any tour.

Next, suppose $0 < 2m^* < c - 2$. By Proposition 1, we have that $x \in A_{r,c,m^*}$. Moreover, the minimality of m^* implies that m^* is the *smallest* value of m for which $x \in A_{r,c,m}$. By definition of A_{r,c,m^*}, there is a GG path from 0 to $-a_2$ with cost $(c-1)+2m^*$, and appending the wraparound edge from $-a_2$ to 0 gives a tour of cost $c + 2m^*$. To show that this is optimal, note first that since $2m^* < c - 2$, we have $c+2m^* < 2(c-1)$. Therefore by Theorem 2, the optimal tour consists of a cheapest Hamiltonian path P from 0 to $-a_2$, plus the wraparound edge from $-a_2$ to 0. Now, Theorem 3 says that we can always take P to be a GG path. Since m^* is the smallest value of m for which $x \in A_{r,c,m}$, it follows that the cost of P is $(c-1) + 2m^*$, so we are done.

Finally, suppose that $2m^* \geq c - 2$ or m^* does not exist. By Proposition 1, this implies that $x \notin A_{r,c,m}$ for any m with $0 \leq 2m < c - 2$. This, together with Theorems 2 and 3, implies that the upper bound tour is optimal, the cost of which is $2c - 2$.

5   The Algorithm

In this section, we show how Theorem 1 naturally leads to an algorithm for two-stripe circulant TSP. We will describe the algorithm formally and prove that it runs in polynomial time. As mentioned in the introduction and Sect. 2, it is possible to represent the input to the two-stripe TSP with 3 numbers: n, the number of nodes, and a_1 and a_2, the lengths of the edges with finite cost. (Recall that we assume without loss of generality that $c_{a_1} = 0$ and $c_{a_1} = 1$.)

It is important to recall what we mean by "polynomial time". We take "polynomial time" to mean polynomial time in the bit size of the input, which in our case is *polylogarithmic* in n. Note then that we cannot, strictly speaking, output the entire tour as a sequence of vertices, because this would require listing n numbers, taking $\Omega(n)$ time. Instead, we will show a guarantee on our algorithm that is similar in spirit to the statement of Theorem 1: There are two parametrized classes of tours to which the optimal tour can belong, and we can determine the class as well as output a set of parameters that describe the tour in polynomial time. In particular, we are able to compute the *cost* of the optimal tour in polynomial time.

Below is the algorithm. It mimics the statement of Theorem 1. Throughout the remainder of this section, we will let $r := \frac{n}{g_1}$ and $c := g_1$.

1. Calculate the row index of $-a_2$. That is, calculate the value of $x \in \{0, 1, \ldots, r - 1\}$ such that the cylindrical coordinates of $-a_2$ are $(x, c - 1)$. (We will explain how to do this step in $O(\log^2 n)$ time.)
2. Compute m^*, the smallest integer value of m such that $m \geq -\frac{c}{2}$ and $x \in \{c+2m, -(c+2m)\} \mod r$. (We will explain how to do this step in $O(\log^2 n)$ time.)
3. If $2m^* < c - 2$, the cost of the optimal tour is $c + \max\{0, 2m^*\}$. It is achieved by taking a cheapest GG path from 0 to $-a_2$ (which will use exactly $2\max\{m^*, 0\} + (c-1)$ horizontal edges), and appending the wraparound edge from $-a_2$ to 0.
4. Otherwise, if $2m^* \geq c - 2$ or m^* does not exist, the cost of the optimal tour is $2c - 2$. In this case, the upper bound tour is optimal.

Theorem 4 (Correctness of the Algorithm).   *The algorithm correctly determines a min-cost tour.*

Theorem 5. *The algorithm calculates the cost of the optimal tour in $O(\log^2 n)$ time. Moreover, it can output a set of parameters that describe an optimal tour in an additional $O(\log^2 n)$ time, and it can return the sequence of n vertices in the tour in $O(n)$ time.*

Proof. To implement the algorithm, we need to first describe how to find the row index x of $-a_2$. We will show how to do this in $O(\log^2 n)$ time. By Theorem 2, we know that $0 \leq x < \frac{n}{g_1}$ is the unique solution to $a_1 x \equiv -g_1 a_2 \mod n$. Since $g_1 = \gcd(a_1, n)$, we can divide both sides of the above relation by g_1 to obtain the equivalent relation $\left(\frac{a_1}{g_1}\right) x \equiv -a_2 \mod \frac{n}{g_1}$. Since $\gcd(\frac{a_1}{g_1}, \frac{n}{g_1}) = 1$ this congruence has a unique solution $x \in \{0, 1, \ldots, \frac{n}{g_1}\}$, and it can be found using the extended Euclidean algorithm in $O(\log^2 n)$ time. (See, for example, Theorem 44 in [12].)

In the second step of the algorithm, we need to compute m^*, the smallest integer value of m such that $m \geq -\frac{c}{2}$ and $x \in \{c + 2m, -(c + 2m)\} \mod r$. Doing this essentially involves solving two congruences, each of which can be solved with the extended Euclidean algorithm in $O(\log^2 n)$ time. We omit the argument for space reasons.

In the interest of space, we have omitted the remaining part of the proof (which shows how to output a small number of parameters which uniquely describe the optimal tour in $O(\log^2 n)$ time, and how to output the entire tour in $O(n)$ time). Essentially, the argument observes that the optimal tour is either an upper bound tour, or a GG path from 0 to $-a_2$ plus the wraparound edge from $-a_2$ to 0. The former tour is easy to describe. For the latter tour, one needs to reason about the GG path P which forms a part of the optimal tour. Since GG paths are extremely structured, we can show that there is a natural, compact way to describe P, and that this description can be computed in $O(\log^2 n)$ time. Again, this running time comes from solving linear congruences using the extended Euclidean algorithm.

6   Proof Sketch of Theorem 3

Proof (Sketch). Our proof involves a minimal counterexample argument. Specifically, we will consider a counterexample to Theorem 3 that is minimal in two senses. Suppose that there exists an $r \times c$ cylinder graph with a Hamiltonian path P from $(0,0)$ to a vertex v in last column. Suppose further that P uses at most $(c-1) + 2m$ horizontal edges, but the row index of v is not in $A_{r,c,m}$. Among all such cylinder graphs and corresponding Hamiltonian paths, we specifically consider an instantiation where:

1. r and c are minimal with respect to first $r + c$, then r, then c.
2. Among all counterexamples with r rows and c columns, consider a counterexample path P that is minimal with respect to the **reverse-lexicographic ordering** of horizontal edges: That is, the lexicographic ordering of $(h_{c-2}, h_{c-3}, \ldots, h_1, h_0)$, where h_i is the number of horizontal edges used in P between the ith and $(i+1)$st columns.

The intuition for considering a counterexample that is minimal with respect to the reverse-lexicographic ordering is as follows: Among all Hamiltonian paths that use $(c-1)+2m$ horizontal edges, GG paths are minimal with respect to this ordering. Our strategy for arriving at a contradiction builds on this intuition. At a high level, the idea of the proof is to devise a sequence of transformations that are applied to P, with the intent of creating a Hamiltonian path P' with $\text{cost}(P') \leq \text{cost}(P)$, and such that the reverse lexicographic order of P' is smaller than that of P.

Our transformations will generate a sequence of subgraphs $P = H_0, H_1, \ldots, H_k$, such that H_{i+1} is obtained by applying a transformation on H_i. By properties of the transformations, we will be able to show that H_i satisfies the following three invariants: (1) H_i uses at most $(c - 1) + 2m$ horizontal edges, (2) H_i is either a 0-v Hamiltonian path or the disjoint union of a 0-v path and a cycle, and (3) H_i has a smaller reverse-lexicographic order than that of P.

Observe that if any H_i is a 0-v Hamiltonian path, then this immediately contradicts the minimality of P. Hence, we may assume that after applying these transformations, each H_i is the disjoint union of a 0-v path P_i and a cycle C_i.

Let $(P_1, C_1), \ldots, (P_k, C_k)$ be the sequence of path/cycle pairs generated by the sequence of transformations. The next part of the proof shows that each C_i must be a doubled vertical edge (i.e. a 2-cycle). We do this via a "backward induction" argument: We first show that C_k is a 2-cycle. Then, we show inductively that assuming $C_k, C_{k-1}, \ldots, C_{i+1}$ are 2-cycles, C_i must also be a 2-cycle. This inductive step is itself a mini proof by contradiction: Assuming C_i is *not* a 2-cycle, we show that the structure of the minimal counterexample implies that that many edges in the graph of (P_i, C_i) are forced. We then use the presence of these forced edges to derive a contradiction.

Finally, the last step of the proof completes the argument by showing that in a minimal counterexample, it is not possible for $C_1, \ldots, C_k$ to all be 2-cycles. This is the place where we use the assumption that $r + c$ is minimal: By deleting two specific rows in the cylinder graph, we reduce to a graph with two fewer rows. Then, the fact that Theorem 3 holds on this smaller instance will enable us to show that, in fact the row index of v is in $A_{r,c,m}$ after all.

7   Conclusion

The natural next question is to consider general symmetric circulant TSP, where all the edge lengths can potentially have finite cost. For that problem, a 2-approximation algorithm is known, but it is open whether it is polynomial-time solvable. As an intermediate step, one might consider a variant with some constant number of edge lengths having finite costs. One might wonder, for example, if there is a dynamic programming approach that extends work from this paper to the constant-stripe case.

References

1. Burkard, R.E.: Efficiently solvable special cases of hard combinatorial optimization problems. Math. Prog. **79**(1), 55–69 (1997). https://doi.org/10.1007/BF02614311
2. Burkard, R. E., Deĭneko, V.G., van Dal, R., van der Veen, J.A., Woeginger, G.J.: Well-solvable special cases of the traveling salesman problem: a survey. SIAM Rev. **40**(3), 496–546 (1998). http://dx.doi.org/10.1137/S0036144596297514. https://doi.org/10.1137/S0036144596297514
3. Burkard, R.E., Sandholzer, W.: Efficiently solvable special cases of bottleneck travelling salesman problems. Discrete Appl. Math. **32**(1), 61–76 (1991). http://www.sciencedirect.com/science/article/pii/0166218X9190024Q. https://doi.org/10.1016/0166-218X(91)90024-Q
4. Codenotti, B., Gerace, I., Vigna, S.: Hardness results and spectral techniques for combinatorial problems on circulant graphs. Linear Algebra Appl. **285**(1), 123–142 (1998). http://www.sciencedirect.com/science/article/pii/S002437959810126X. https://doi.org/10.1016/S0024-3795(98)10126-X
5. Gerace, I., Greco, F.: Bounds for the symmetric circulant traveling salesman problem. Rapporto Tecnico 4/2006, Dipartimento de Matematica e Informatica, Università di Perugia (2006)
6. Gerace, I., Greco, F.: The travelling salesman problem in symmetric circulant matrices with two stripes. Math. Struct. Comput. Sci. **18**(1), 165–175 (2008)
7. Gilmore, P.C., Lawler, E.L., Shmoys, D.B.: Well-solved special cases. In: Lawler, E.L., Lenstra, J. K., Rinnooy Kan, A.H.G., Shmoys, D.B., (eds.) The Traveling Salesman Problem: A Guided Tour of Combinatorial Optimization, New York, pp. 87–143. Wiley (1985)
8. Greco, F., Gerace, I.: The traveling salesman problem in circulant weighted graphs with two stripes. Electron. Notes Theor. Comput. Sci. **169**, 99–109 (2007). http://www.sciencedirect.com/science/article/pii/S1571066107000497. https://doi.org/10.1016/j.entcs.2006.07.032
9. Gutekunst, S.C., Williamson, D.P.: Characterizing the integrality gap of the subtour LP for the circulant traveling salesman problem. SIAM J. Discrete Math. **33**(4), 2452–2478 (2019)
10. Kabadi, S.N.: Polynomially solvable cases of the TSP. In: Gutin, G., Punnen, A.P. (eds.) The Traveling Salesman Problem and Its Variations. Combinatorial Optimization, vol. 12. Springer, Boston (2007). https://doi.org/10.1007/0-306-48213-4_11
11. Lawler, E.L., Lenstra, J.K., Rinnooy Kan, A.H.G., Shmoys, D.B.: The Traveling Salesman Problem: A Guided Tour of Combinatorial Optimization. Wiley (1985)
12. Shoup, V.: A Computational Introduction to Number Theory and Algebra. Cambridge University Press, Cambridge (2009)
13. van der Veen, J.A.A., van Dal, R., Sieksma, G.: The symmetric circulant traveling salesman problem. Research Memorandum 429, Institute of Economic Research, Faculty of Economics, University of Groningen, (1991)
14. Williamson, D.P., Shmoys, D.B.: The Design of Approximation Algorithms. Cambridge University Press, New York (2011)
15. Yang, Q.F., Burkard, R.E., Çela, E., Woeginger, G.J.: Hamiltonian cycles in circulant digraphs with two stripes. Discrete Math. **176**(1), 233–254 (1997)

An Abstract Model for Branch-and-Cut

Aleksandr M. Kazachkov[1]([✉])(ID), Pierre Le Bodic[2](ID),
and Sriram Sankaranarayanan[3](ID)

[1] University of Florida, Gainesville, USA
akazachkov@ufl.edu
[2] Monash University, Melbourne, Australia
pierre.lebodic@monash.edu
[3] Indian Institute of Management, Ahmedabad, India
srirams@iima.ac.in

Abstract. Branch-and-cut is the dominant paradigm for solving a wide range of mathematical programming problems—linear or nonlinear—combining intelligent search (via branch-and-bound) and relaxation-tightening procedures (via cutting planes, or cuts). While there is a wealth of computational experience behind existing cutting strategies, there is simultaneously a relative lack of theoretical explanations for these choices, and for the tradeoffs involved therein. Recent papers have explored abstract models for branching and for comparing cuts with branch-and-bound. However, to model practice, it is crucial to understand the impact of jointly considering branching and cutting decisions. In this paper, we provide a framework for analyzing how cuts affect the size of branch-and-cut trees, as well as their impact on solution time. Our abstract model captures some of the key characteristics of real-world phenomena in branch-and-cut experiments, regarding whether to generate cuts only at the root or throughout the tree, how many rounds of cuts to add before starting to branch, and why cuts seem to exhibit nonmonotonic effects on the solution process.

Keywords: Integer programming · Branch-and-bound · Cutting planes

1 Introduction

The branch-and-cut (B&C) paradigm is a hybrid of the branch-and-bound (B&B) [20] and cutting plane methods [12–14]. It is central to a wide range of modern global optimization approaches [2,6], particularly mixed-integer linear and nonlinear programming solvers [1]. Cutting planes, or *cuts*, tighten the relaxation of a given optimization problem and are experimentally known to significantly improve a B&B process [1], but determining which cuts to add is currently based on highly-engineered criteria and computational insights. An outstanding open problem is a rigorous underpinning for the choices involved in branch-and-cut. While there has been active exploration of the theory of branching [3,8–10,21] and comparing cutting and branching [4,22], the interaction of

© Springer Nature Switzerland AG 2022
K. Aardal and L. Sanitá (Eds.): IPCO 2022, LNCS 13265, pp. 333–346, 2022.
https://doi.org/10.1007/978-3-031-06901-7_25

the two together remains poorly understood. Most recently, Basu et al. [5] showed that using B&C can outperform either branching or cutting alone.

This paper introduces a theoretical framework for analyzing the practical challenges involved in making B&C decisions. We build on recent work by Le Bodic and Nemhauser [21], which provides an abstract model of B&B, based on how much bound improvement is gained by branching on a variable at a node of the B&B tree. This model not only is theoretically useful, but also can improve branching decisions in solvers [3].

Specifically, we add a cuts component to the abstract B&B model from Le Bodic and Nemhauser [21]. We apply this enhanced model to account for both the utility of the cuts in proving bounds, as well as the additional time taken to solve the nodes of a B&C tree after adding cuts.

In this abstract model, given the relative strengths of cuts, branching, and the rate at which node-processing time grows with additional cuts, we quantify (i) the number of cuts, and (ii) cut positioning (at the root or deeper in the tree) to minimize both the tree size and the solution time of an instance. This thereby captures some of the main tradeoffs between cutting and branching, in that cuts can improve the bound or even the size of a B&C tree, but meanwhile slow down the solution time overall. We use a *single-variable* abstract B&C model, where every branching variable has identical effect on the bound, and we only address the *dual* side of the problem, i.e., we are only interested in proving a good *bound* on the optimal value, as opposed to generating better integer-feasible solutions.

We emphasize that our motivation is to advance a theoretical understanding of empirically-observed phenomena in solving optimization problems, and our results show that some of the same challenges that solvers encounter in applying cuts do arise in theory. While we also state some prescriptive recommendations in our abstract model, these are not intended to be immediately computationally viable. Instead, the intent of the prescriptive results is to see whether our abstraction affords enough simplicity to make precise theoretical statements.

Summary of contributions and paper structure. We provide a generic view of B&C in Sect. 2. Section 3 introduces our abstract B&C model, in which the quality of cuts and branching remains fixed throughout the tree. In Sect. 4, we analyze the effect of cuts on tree size; we prove that in this case it is never necessary to add cuts after the root node, and we provide a lower bound on the optimal number of cutting plane rounds that will minimize the B&C tree size.

We then study how cuts affect solving time for a tree. In Sect. 5.1, we show that cuts are guaranteed to be helpful for sufficiently hard instances. In contrast to the case of tree size, in this more general setting, adding cuts after the root node may be better. However, one of our main results is Theorem 13, which shows that if the solution time for nodes in a B&C tree grows linearly with the number of added cuts and branching yields the same bound improvement in both children, then it is optimal to add all cuts at the root. The proof of this theorem makes use of independently-interesting conditions, presented in Sect. 5.2, on optimally applying a fixed number of cuts. We conclude in Sect. 6 with a discussion of open problems and potential directions for future work.

All missing proofs can be found in the publicly-accessible full paper [17].

2   Preliminaries

We are given a generic optimization problem (*OP*)—linear or nonlinear, with or without integers—which is to be solved using a B&C algorithm. For convenience, we assume that the OP is a minimization problem. We also assume that we already have a feasible solution to the OP, so that our only goal is to efficiently certify the optimality or quality of that solution.

The B&C approach involves creating a computationally tractable relaxation of the original problem, which we call the *root* of the B&C tree and assume is provided to us. For example, when the OP is a mixed-integer linear program, we usually start with its linear programming relaxation. The value of the solution to this relaxation provides a lower bound on the optimal value to the OP. B&C proceeds by either (1) tightening the relaxation through adding *valid* cuts, which will remove parts of the current relaxation but no OP-feasible points, or (2) splitting the feasible region, creating two subproblems, which we call the *children* of the original (*parent*) relaxation. Both of these operations improve the lower bound with respect to the original relaxation. The B&C procedure repeats on the new relaxation with cuts added in the case of (1), and recursively on the children in the case of (2); we assume that tractability is maintained in either case. Moreover, we assume that all children remain OP-feasible. We now formally define a *B&C tree* as used in this paper.

Definition 1 (B&C tree). *A B&C tree T is a rooted binary tree with node set $\mathcal{V}_T$ that is node-labeled by a function $g_T : \mathcal{V}_T \to \mathbb{R}_{\geq 0}$, indicating the bound improvement at each node with respect to the bound at the root node, such that*

1. *The root v_0 has label $g_T(v_0) = 0$.*
2. *A node v with exactly one child v' is a cut node, and we say that a cut or round of cuts is added at node v. The bound at v' is $g_T(v') = g_T(v) + c_v$, where c_v is the nonnegative value associated with the round of cuts at v.*
3. *A node v with exactly two children v_1 and v_2 is a branch node, and we say that we branch at node v. The bounds at the children of v are $g_T(v_1) = g_T(v) + \ell_v$ and $g_T(v_2) = g_T(v) + r_v$, where (ℓ_v, r_v) is the pair of bound improvement values associated with branching at v.*

We say that T proves a bound of G if $g_T(v) \geq G$ for all leaves $v \in \mathcal{V}_T$.

While Definition 1 is generic, the abstraction we study is restricted to the *single-variable* version in which ℓ_v and r_v is the same for every branch node $v \in \mathcal{V}_T$, and c_v is constant for each cut node $v \in \mathcal{V}_T$, so we drop the subscript v.

3   The Abstract Branch-and-Cut Model

This section introduces the *Single Variable Branch-and-Cut* (SVBC) model, an abstraction of a B&C tree as presented in Definition 1. First, we define a formal notion of the time taken to process a B&C tree as the sum of the node processing times, which in turn depends on the following definition of a time-function.

Definition 2 (Time-function). *A function* w $: \mathbb{Z}_{\geq 0} \rightarrow [1, \infty)$ *is a time-function if it is nondecreasing and* w$(0) = 1$.

Definition 3 (Node time and tree time). *Given a B&C tree* T, *a node* $v \in \mathcal{V}_T$, *and a time-function* w,

(i) *the* (node) *time of* v, *representing the time taken to process node* v, *is* w(z), *where* z *is the number of* cut *nodes in the path from the root of* T *to* v.
(ii) *the* (tree) *time of* T, *denoted by* $\tau_w(T)$, *is the sum of the node times of all the nodes in the tree.*

We simply say $\tau(T)$ *when the time-function* w *is clear from context.*

Definition 3 models the observation that cuts generally make the relaxation harder to solve, and hence applying more cuts increases node processing time. Note that (i) if w $= \mathbf{1}$, i.e., w$(z) = 1$ for all $z \in \mathbb{Z}_{\geq 0}$, we obtain the regular notion of size of a tree, which counts the number of nodes in the tree, and (ii) the time of a pure cutting tree with t cuts (i.e., $t + 1$ nodes) is $\sum_{i=0}^{t}$ w(i).

Finally, we state the SVBC model in Definition 4. In this model, the relative bound improvement at every cut node is always the same constant c, and every branch node is associated to the same (ℓ, r) pair of bound improvement values. We also assume that the time to solve a node depends on the number of cuts added to the relaxation up to that node.

Definition 4 (Single Variable Branch-and-Cut (SVBC)). *A B&C tree is a* Single Variable Branch-and-Cut (SVBC) *tree with parameters* $(\ell, r; c, w)$ *if the bound improvement value associated with each branch node is* (ℓ, r), *the bound improvement by each cut node is* c, *and the time-function is* w. *We say such a tree is an* SVBC$(\ell, r; c, w)$ *tree.*

Without loss of generality, we assume $0 \leq \ell \leq r$.

Definition 5 (τ-minimality). *Given a function* w $: \mathbb{Z}_{\geq 0} \rightarrow [1, \infty)$, *we say that a B&C tree* T *that proves bound* G *is* τ-minimal *if, for any other B&C tree* T' *that also proves bound* G *with the same* $(\ell, r; c, w)$, *it holds that* $\tau(T') \geq \tau(T)$.

It is often the case that applying a round of cuts at a node may not improve the bound as much as branching at that node, but the advantage is that cutting adds only one node to the tree, while branching creates two subproblems. A first question is whether there always exists a minimal-size tree with *only* branch nodes or *only* cut nodes. We address this in Example 6, which illustrates our notation, shows that cut nodes can help reduce the size of a B&C tree despite improving the bound less than branch nodes, and highlights the fact that finding a minimum-sized B&C tree proving a particular bound G involves strategically using both branching and cutting.

Example 6 (Branch-and-cut can outperform pure branching or pure cutting). Figure 1 shows three B&C trees that prove the bound $G = 6$. The tree in panel (a) only has branch nodes, (b) only has cut nodes, and (c) has both branch and cut nodes. As seen in the figure, branching and cutting together can strictly outperform pure branching or pure cutting methods, in terms of tree size. ■

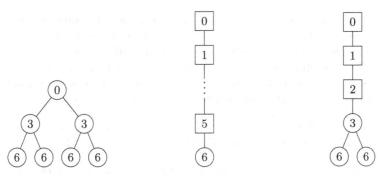

(a) Pure branching: 7 nodes (b) Pure cutting: 7 nodes (c) Branch-and-cut: 6 nodes

Fig. 1. Three B&C trees proving $G = 6$, with $\ell = r = 3$, $c = 1$, and $\text{w} = 1$.

4   Optimizing Tree Size

In this section, we examine the number of cuts that minimize the *size* $|\mathcal{V}_T|$ of a B&C tree T, i.e., optimizing $\tau(T)$ when $\text{w} = 1$. In Lemma 7, we first address the location of these cuts—should they be at the root or deeper in the tree?

Lemma 7. *For any target bound $G > 0$ to prove and a fixed set of parameters $(\ell, r; c, 1)$, for any minimal $SVBC(\ell, r; c, 1)$ tree T with $k \geq 1$ cut nodes, there exists a minimal $SVBC(\ell, r; c, 1)$ tree T' such that there are exactly k cut nodes in T' that form a path starting at the root of T'.*

The proof follows from observing that if a cut node is moved from deeper in the tree to the root, then the bound at every leaf node weakly improves.

Next, we present Theorem 8, which provides the optimal number of rounds of cuts for an $SVBC(\ell, r; c, 1)$ tree when $c \leq \ell = r$, as a function of the tree parameters and the target bound. The theorem also implies that the bound proved by branching is at most $r \log_2 \lceil r/c \rceil$, and the rest is proved by cutting.

Theorem 8. *Let $\delta^* := \lfloor \log_2 \lceil r/c \rceil \rfloor$. When $0 < c \leq \ell = r$, the number of rounds of cuts at the root to minimize the size of an $SVBC(\ell, r; c, 1)$ tree proving bound G is precisely enough to make the remaining bound before branching at most*

$$
G^* := \begin{cases}
r(\delta^* - 1) & \text{if } G \geq r\delta^* \text{ and } \lceil (G - c \lceil (G - r\delta^*)/c \rceil - r(\delta^* - 1))/c \rceil < 2^{\delta^*} \\
r\delta^* & \text{if } G \geq r\delta^* \text{ and } \lceil (G - c \lceil (G - r\delta^*)/c \rceil - r(\delta^* - 1))/c \rceil \geq 2^{\delta^*} \\
r \lfloor G/r \rfloor & \text{if } G < r\delta^* \text{ and } \lceil (G - r \lfloor G/r \rfloor)/c \rceil < 2^{r\lceil G/r \rceil} \\
G & \text{otherwise,}
\end{cases}
$$

achieved with $k^* := \lceil (G - G^*)/c \rceil$ cut rounds. Moreover, the size of any minimal $SVBC(\ell, r; c, 1)$ tree that proves bound G is at least $2^{G^*/r + 1} - 1 + k^*$.

Although Theorem 8 prescribes an optimal number k^* of cut nodes when $\ell = r$, it is not true that tree size monotonically decreases as the number of cuts goes from 0 to k^*. For example, if $G = G^* + 2c$, using one cut node would *increase* the overall tree size, while two cut rounds would reduce tree size by $2^{1+G^*/r} - 2$. This phenomenon, present even when $w = 1$, might indicate one of the practical challenges in determining whether a cut family benefits an instance.

In Theorem 9, we show that even in general, for $\ell \neq r$, it is always optimal to add at least one cut round for sufficiently large target bounds.

Theorem 9. *If* $0 < c \leq \ell \leq r$ *and* $G > r \lfloor \log_2 \lceil r/c \rceil \rfloor$, *then the minimal* $SVBC(\ell, r; c, 1)$ *tree proving a bound* G *has at least one cut node.*

Proof. Consider a pure branching tree that proves G. The number of leaf nodes of this tree is at least $2^{\lceil G/r \rceil}$, since $\ell \leq r$, and all the parents of each of these leaf nodes have a remaining bound in $(0, r]$ that needs to be proved. Now suppose we add $\lceil r/c \rceil$ rounds of cuts. All of the leaf nodes of the pure branch-and-bound tree would then be pruned, since the parent nodes would already prove the desired target bound of G. As a result, there is benefit to cutting when $2^{\lceil G/r \rceil} > \lceil r/c \rceil$, which holds when $G > r \lfloor \log_2 \lceil r/c \rceil \rfloor$. $\qquad\square$

Corollary 10. *If* $0 < c \leq \ell \leq r$, *then every minimal* $SVBC(\ell, r; c, 1)$ *tree proving a bound* $G > \overline{G} := r \lfloor \log_2 \lceil r/c \rceil \rfloor$ *has at least* $\lceil (G - \overline{G})/c \rceil$ *cut nodes, which can always be repositioned next to the root node due to Lemma 7.*

Example 11. The following example, from Basu et al. [5], shows that SVBC trees with constant $c \leq \ell = r$ have been studied in the literature, and that cuts not only decrease the size of a branch-and-bound tree, but in fact can lead to an exponential improvement.

Consider the independent set problem, defined on a graph G with vertices V and edge set E, in which G consists of m disjoint triangles (cliques of size three): $\max_x \{\sum_{v \in V} x_v : x \in \{0, 1\}^{|V|}; \ x_u + x_v \leq 1, \ \forall \{u, v\} \in E\}$. The optimal value is m, using $x_v = 1$ for exactly one vertex of every clique, while the linear relaxation has optimal value $3m/2$, obtained by setting $x_v = 1/2$ for all $v \in V$.

Suppose we branch on x_v, $v \in V$, where v belongs to a clique with vertices u and w. One can verify that the objective value of the relaxation decreases by exactly $\ell = 1/2$ with respect to the parent for the left ($x_v \leq 0$) branch, and similarly, by $r = 1/2$ for the right ($x_v \geq 1/2$) branch. Notice that once we branch on x_v, the remaining problem can be seen as fixing the values of the three variables corresponding to vertices in the triangle that v belongs to, while keeping the remaining variables unchanged. In other words, it is a subproblem with exactly the same structure as the original one, except removing the decision variables for the vertices of a single clique.

Finally, we look at families of cutting planes that we can derive. By adding up the three constraints corresponding to the edges of any triangle $\{u, v, w\}$, we obtain the implication $2(x_u + x_v + x_w) \leq 3$. Since all variables are integer-restricted, we can infer that $x_u + x_v + x_w \leq \lfloor 3/2 \rfloor = 1$ for every clique. Each such cut corresponds to a change of $c = 1/2$ in the objective, and there exists one such cut for every clique of three vertices.

Hence, by Theorem 8, we have that, not counting cut nodes, the optimal depth of the $SVBC(\ell, r; c, \mathbf{1})$ tree that proves the bound $G = m/2$ is $\delta^* = \lfloor \log_2 \lceil r/c \rceil \rfloor = 0$. This implies that the optimal number of cut rounds is $k^* = \lceil G/c \rceil = \lceil (m/2)/(1/2) \rceil = m$, with a corresponding tree with m total nodes, compared to a pure branching tree, which would have depth $\lceil G/r \rceil = m$ and thus $2^{m+1} - 1$ nodes—exponentially many more than if cut nodes are used. ■

5   Optimizing Tree Time

Although the previous section's focus on decreasing the *size* of a branch-and-cut tree is useful, in practice the quantity of interest is the *time* it takes to solve an instance. The two notions do not intersect: it can be that one tree is smaller than another, but because the relaxations at each node solve more slowly in the smaller tree, the smaller tree ultimately solves in more time than the larger one. This plays prominently into cut selection criteria, as strong cuts can be dense, and adding such cuts to the relaxation slows down the solver.

5.1   Time-Functions Bounded by a Polynomial

We first show that if the time-function is bounded above by a polynomial, then for sufficiently large G, it is optimal to use at least one cut node.

Theorem 12. *Suppose we have an $SVBC(\ell, r; c, \mathrm{w})$ tree T and the values of w are bounded above by a polynomial. Then, there exists $\overline{G} > 0$ such that every τ-minimal SVBC tree proves a bound of $G > \overline{G}$ has at least one cut node.*

Proof. Let $\mathrm{w}(z) \leq \alpha z^d + 1$ for some $\alpha, d > 0$ be the polynomial upper bound for each $z \in \mathbb{Z}_{\geq 0}$. A pure branching tree T_B proving a bound G has at least $2^{\lceil G/r \rceil + 1} - 1$ nodes. The same lower bound holds for $\tau(T_B)$.

Now consider a pure cutting tree T_C proving bound G. Such a tree has exactly $k = \lceil G/c \rceil + 1$ nodes. The tree time for T_C is $\tau(T_C) = \sum_{i=0}^{t-1} \mathrm{w}(i) \leq t + \alpha \sum_{i=1}^{k-1} i^d \leq k + \alpha(k-1)^{d+1} < p(k)$, where p is some polynomial. For sufficiently large values of G, $2^{\lceil G/r \rceil + 1} - 1 > p\left(\lceil \frac{G}{c} \rceil\right)$ for any polynomial p, implying that a τ-minimal tree has at least one cut node. □

Next, we observe that an analogue of Lemma 7 does *not* hold for τ-minimality. Figure 2 provides an example where the unique τ-minimal B&C tree has no cuts at the root. Despite that, for the special case where $\ell = r$ and w is a *linear* time-function, we prove in Theorem 13 that there is a τ-minimal tree having only root cuts. We emphasize that the result may not hold if either of these assumptions is relaxed. For instance, with a general time-function, one can construct an example of a unique τ-minimal symmetric tree in which not all cuts are at the root.

Theorem 13. *If $c \leq \ell = r$ and the time-function is linear, then there is a τ-minimal tree with all cuts at the root.*

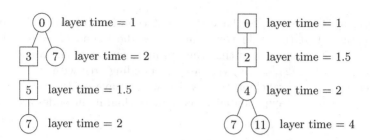

Fig. 2. Consider the SVBC tree that must prove a bound $G = 7$, with parameters $\ell = 3$, $r = 7$, $c = 2$, and $w(k) = k/2 + 1$. The time of the first tree is 6.5 and of the second tree is 8.5. Thus, cutting at the root node is strictly inferior to cutting at the leaf. One can also check that the pure branching tree has a time of 7 and the pure cutting tree has a time of 10, showing that the unique τ-minimal B&C tree is the tree in the left panel.

We prove Theorem 13 in Sect. 5.3. On the way, we present several intermediate results of independent interest in Sect. 5.2, which relate properties of general time-functions to the optimal number and location of cuts in the tree.

5.2   Adding k Cuts Along Every Root-to-Leaf Path

We first analyze adding k cuts to a generic $SVBC(\ell, r; c, w)$ tree and prescribe how many should be placed before the first branch node.

Lemma 14. *Consider a B&C tree in which each root-to-leaf path has exactly k cut nodes, and each cut node can only be located either before or immediately after the first branching node. Then the time of the tree is minimized by adding*

$$t^\star \in \underset{0 \le t \le k}{\arg\min} \left\{ w(t) - \sum_{i=0}^{t-1} w(i) \right\}$$

cut nodes before the first branch node, and $k - t^\star$ cut nodes in a path starting at each child of the first branch node.

Next we provide a technical condition that is sufficient to allow all cut nodes to be added to the root of the tree.

Lemma 15. *For any $c \le \ell = r$, among all $SVBC(\ell, r; c, w)$ trees in which there are exactly $k^\star$ cut nodes along the path from the root to every leaf node, if the time-function w satisfies $w(z + 1) \le 2w(z)$ for all $z \in \mathbb{Z}_{\ge 0}$, then there is a τ-minimal SVBC tree in which all cuts are at the root node.*

Proof. Let T denote the tree in which all $k^\star$ cuts are attached to the root node. Let T' be the tree in which $k^\star - 1$ cuts are at the root, and one cut node is added after the first branch node. Both of these trees are shown in Fig. 3.

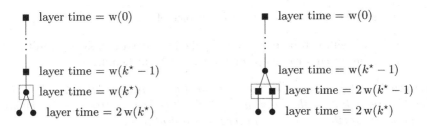

Fig. 3. The effect of moving the last cut at the root down one level is splitting one node with node time $w(k^\star)$ into two nodes that each have node time $w(k^\star - 1)$.

The time of T is no greater than that of T' if and only if $w(k^\star) \leq 2\,w(k^\star - 1)$. Recursively applying the same principle shows that tree T is weakly better than a tree with an arbitrary number of the $k^\star$ cut nodes moved to the second level. Similarly, it does not help to add cuts deeper in the tree either, as the same logic applies to each subtree rooted at a node at higher depth. □

Lemma 15 assumes a fixed number of cuts are added along every root-to-leaf path. In contrast, the next set of results gives conditions under which a given number of cuts can (or cannot) be τ-minimal.

Lemma 16. *If $c \leq \ell = r$ and the time-function satisfies $w(1) > 2$, then a τ-minimal $SVBC(\ell, r; c, w)$ tree proving bound G cannot have exactly one cut along each root-to-leaf path.*

Proof. We show that the pure branching tree has strictly lower tree time than a tree that also proves bound G and has exactly one cut along every path from the root node to every leaf node. We refer to the latter as the "one-cut tree", which we assume without loss of generality is a τ-minimal one-cut tree and has all cuts at the same depth (by symmetry, due to $\ell = r$).

Denote by T_0 the size, and by δ the depth, of the pure branching tree. Suppose we place a single cut at δ_1 along every root-to-leaf path for the one-cut tree. Let T'_0 denote the nodes of the one-cut tree up to depth δ_1; these nodes all have time $w(0) = 1$. Let T'_1 denote the subtree rooted after the cut node and let δ_2 be the depth of tree T'_1; there are 2^{δ_1} of these trees—one for each leaf of tree T'_0—and each node of these trees has time $w(1)$. See Fig. 4 for reference. Note that if the addition of a cut does not reduce the total number of nodes with respect to T_0, then the lemma statement is trivially true; hence, we can assume that $\delta_1 + \delta_2 = \delta - 1$.

The tree time of the one-cut tree is

$$w(0) \cdot (2^{\delta_1 + 1} - 1) + w(1) \cdot 2^{\delta_1} \cdot (2^{\delta_2 + 1} - 1) = (2^{\delta_1 + 1} - 1) + w(1)(2^{\delta_1 + \delta_2 + 1} - 2^{\delta_1}).$$

The tree time of the pure branching tree is $2^{\delta + 1} - 1$, which is less than that of the one-cut tree if and only if

$$w(1) > \frac{2^{\delta + 1} - 2^{\delta_1 + 1}}{2^{\delta_1 + \delta_2 + 1} - 2^{\delta_1}} = \frac{2^{\delta + 1} - 2^{\delta_1 + 1}}{2^{\delta} - 2^{\delta_1}} = 2.$$

Observe that this holds regardless of the choice of δ_1. $\square$

The above result can be recursively applied to give a necessary condition for a larger number of cuts to be optimal, as we prove in Lemma 17.

Lemma 17. *If $c \leq \ell = r$ and the time-function satisfies $\mathrm{w}(k) > 2\,\mathrm{w}(k-1)$ for a given integer $k \geq 1$, then a τ-minimal $SVBC(\ell, r; c, \mathrm{w})$ tree proving bound G cannot have exactly k cuts along every root-to-leaf path.*

Proof. Define the "k-cut tree" as a τ-minimal tree in which every root-to-leaf path has exactly k cuts. Since $\ell = r$ and we are minimizing tree time, we can assume without loss of generality that the k-cut tree is symmetric. For a given k-cut tree, let δ_1 denote the depth at which the penultimate—$(k-1)^{\mathrm{st}}$—cut occurs along every root-to-leaf path. Let T_0' be the subtree of the k-cut tree consisting of nodes up to and including those at depth δ_1. For any cut node at depth δ_1, let T_1' be the subtree rooted at that cut node's child, and let G' be the remaining bound that needs to be proved after T_0'. The tree T_1' has exactly one cut along every root-to-leaf path, and its root has time $\mathrm{w}(k-1)$. Define $\overline{\mathrm{w}}(z) := \mathrm{w}(z+k-1)/\mathrm{w}(k-1)$ for all $z \in \mathbb{Z}_{\geq 0}$, which satisfies the requirements of a time-function from Definiton 2. Observe that the tree time of T_1' with respect to $\overline{\mathrm{w}}$ is precisely $1/\mathrm{w}(k-1)$ multiplied by the tree time of T_1' in the k-cut tree.

We have $\overline{\mathrm{w}}(1) = \mathrm{w}(k)/\mathrm{w}(k-1) > 2$ by assumption. By Lemma 16, there does not exist a τ-minimal tree with respect to $\overline{\mathrm{w}}$ that has exactly one cut along every root-to-leaf path. It follows that there exists a one-cut tree proving bound G with less tree time than the k-cut tree, by replacing T_1' with a tree that proves the same bound G' without any cut nodes. $\square$

Lemma 17 states that a tree with k cuts along every root-to-leaf path is τ-minimal only if the tree with $k-1$ cuts does not have lower tree time. For the special case of $k = 2$, in Lemma 19, we provide another necessary condition for two cuts to be optimal, by comparing the two-cut tree with a pure branching

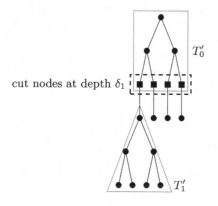

Fig. 4. A tree with a single cut located at depth δ_1 along every root-leaf path.

tree. We will apply the following lemma, which asserts that if adding two cuts is optimal, then those cuts can be adjacent.

Lemma 18. *If $c \leq \ell = r$ and there exists a τ-minimal tree proving bound G with exactly two cuts along every root-to-leaf path, then there is a τ-minimal $SVBC(\ell, r; c, \mathrm{w})$ tree in which the two cuts are adjacent along every path.*

Proof. Consider a symmetric (without loss of generality, because $\ell = r$) tree T with two cuts along every root-to-leaf path, in which the second cut node is not adjacent to the first. Now let T' be nearly identical to T, except that the second cut nodes are "shifted up" one level in the tree (this operation can be seen in Fig. 3, where the first panel corresponds to shifting up cuts compared to the second panel). This is equivalent to merging every pair of these second cut nodes, each with time $\mathrm{w}(1)$, into a single node with time $\mathrm{w}(2)$. By Lemma 17 with $k = 2$, it must hold that $\mathrm{w}(2) \leq 2\,\mathrm{w}(1)$. Hence, T' weighs no more than T. The shifting up operation can be repeated, without increasing tree time, until the second cut node is adjacent to the first. $\qquad\square$

Lemma 19. *If $c \leq \ell = r$ and the time-function satisfies $\mathrm{w}(2) > 2 + \mathrm{w}(1)$, then a τ-minimal $SVBC(\ell, r; c, \mathrm{w})$ tree proving bound G cannot have exactly two cuts along each root-to-leaf path.*

Proof. Assume for the sake of contradiction that $\mathrm{w}(2) > 2 + \mathrm{w}(1)$ but there exists a symmetric (without loss of generality, as $\ell = r$) τ-minimal tree with two cuts along every root-to-leaf path. By Lemma 18, we can assume the two cut nodes are adjacent along every root-to-leaf path.

For a given root-to-leaf path, let T be the subtree rooted at the first cut node along that path, and observe that T must also be τ-minimal. Define T' by "shifting" the two cut nodes down one "level" with respect to T, as illustrated in Fig. 5, with the following modification in the case that the cut nodes in T are already at the lowest level of the tree, i.e., when T is a path. When T is a path, to prove bound G, the second cut node is not necessary, so the last layer of T' can be removed; that is, T' consists of one branch node whose two children are cut nodes, and the children of the cut nodes are the leaf nodes of T'.

Thus, T' is the same as T except that the second cut node and first branch node in T, with time $\mathrm{w}(1) + \mathrm{w}(2)$, are replaced by four nodes, with total time $2(\mathrm{w}(0) + \mathrm{w}(1)) = 2 + 2\,\mathrm{w}(1) < \mathrm{w}(1) + \mathrm{w}(2)$. This shows that $\tau(T') < \tau(T)$, which contradicts the assumption that T is τ-minimal. $\qquad\square$

5.3   Proof of Theorem 13

Proof (Proof of Theorem 13). Since $\ell = r$, there exists a τ-minimal tree that is symmetric, in which every root-to-leaf path has the same number of cut nodes, and if there is a cut at a certain depth for one path, every other path has a cut at the same depth. Let $k^\star$ denote the number of cuts added along every root-to-leaf path in a τ-minimal B&C tree.

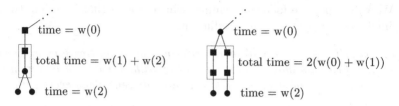

Fig. 5. Illustration of shifting two cuts down in the tree. The boxed nodes cause the only difference in times between the two subtrees.

We assume $w(z) = \alpha z + 1$ for some $\alpha > 0$. It follows that $\sum_{i=0}^{t-1} w(i) = \alpha t(t-1)/2 + t$. As a result, $w(t) - \sum_{i=0}^{t-1} w(i)$ turns out to be a concave quadratic. Thus, its minimum over an interval is achieved at the extreme points.

Thus, by Theorem 14, to decide if cuts should be placed at the root or immediately after the first branch node, we only need to check which one of $t = 0$ (all cuts after the root) or $t = k^\star$ (only root cuts) minimizes

$$w(t) - \sum_{i=0}^{t-1} w(i) = (\alpha t + 1) - \frac{\alpha t(t-1)}{2} - t = \alpha t \left(1 - \frac{t-1}{2}\right) - t + 1.$$

For $t = 0$, this expression is equal to 1. Picking $t = k^\star > 0$ is no worse if and only if

$$\alpha k^\star \left(1 - \frac{k^\star - 1}{2}\right) - k^\star + 1 \leq 1 \iff 1 - \frac{k^\star - 1}{2} \leq \frac{1}{\alpha} \iff k^\star \geq 3 - \frac{2}{\alpha}. \quad (1)$$

If $k^\star = 1$, then inequality (1) holds if and only if $\alpha \leq 1$. By Lemma 16, if $k^\star = 1$ is optimal, then $w(1) \leq 2$, which holds if and only if $\alpha \leq 1$. For $k^\star = 2$, (1) holds if and only if $\alpha \leq 2$. By Lemma 19, if $k^\star = 2$ is optimal, then $w(2) \leq 2 + w(1)$, which holds if and only if $\alpha \leq 2$. If $k^\star \geq 3$, then (1) is true for all $\alpha > 0$. $\qquad\square$

6 Conclusion and Potential Extensions

We analyze a framework capturing several crucial tradeoffs in jointly making branching and cutting decisions for optimization problems. For example, we show that adding cuts can yield nonmonotonic changes in tree size, which can make it difficult to evaluate the effect of cuts computationally. Our results highlight challenges for improving cut selection schemes, in terms of their effect on branch-and-cut tree size and solution time, albeit for a simplified setting in which the strength of cuts and branching is assumed constant and known. There do exist contexts in which the relative strength of cuts compared to branching decisions can be approximated, such as by inferring properties for a family of instances, an idea that has seen recent success with machine learning methods applied to integer programming problems [11,15,18,19,23]. This lends hope to apply our results to improve cut selection criteria for such families of instances and this warrants future computational study, though it is far from straightforward.

This paper focuses on the single-variable version of the abstract branch-and-cut model. Some results extend directly to bounds for a generalization of the model permitting different possible branching variables, by assuming the "single branching variable" corresponds to the *best* possible branching variable at every node, but an in-depth treatment of the general case remains open. Further, an appealing extension of the general time-functions considered in Sect. 5 is to investigate branching on general disjunctions [7,16,22], which has been the subject of recent computational study [24].

One important aspect left as future work is accounting for diminishing cut strength over rounds. This could be analyzed within our model by assuming an exponentially-increasing time-function, as many more cuts would be needed each round to achieve a constant bound improvement; these extra cuts would take more time to generate, and the impact on the node solving time would be progressively worse. An alternative is to expand our model to allow cuts to worsen (in bound improvement) with each round.

Finally, most of the results we present in Sect. 5 for general time-functions assumes that branching on a variable leads to the same bound improvement for both children. The general situation of unequal and/or nonconstant bound improvements remains open, both regarding the best location of cut nodes and the optimal number of cuts to be added, and merits future theoretical and experimental investigation.

Acknowledgements. The authors thank Andrea Lodi, Canada Excellence Research Chair in Data Science for Real-Time Decision Making, for financial support and creating a collaborative environment that facilitated the interactions that led to this paper, as well as Monash University for supporting Pierre's trip to Montréal.

References

1. Achterberg, T., Wunderling, R.: Mixed integer programming: analyzing 12 years of progress. In: Jünger, M., Reinelt, G. (eds.) Facets of Combinatorial Optimization. Springer, Heidelberg (2013). https://doi.org/10.1007/978-3-642-38189-8_18
2. Al-Khayyal, F.A.: An implicit enumeration procedure for the general linear complementarity problem. In: Hoffman, K.L., Jackson, R.H.F., Telgen, J. (eds.) Computation Mathematical Programming. Mathematical Programming Studies, vol. 31. Springer, Heidelberg (1987). https://doi.org/10.1007/BFb0121176
3. Anderson, D., Le Bodic, P., Morgan, K.: Further results on an abstract model for branching and its application to mixed integer programming. Math. Program. **190**(1), 811–841 (2020). https://doi.org/10.1007/s10107-020-01556-4
4. Basu, A., Conforti, M., Di Summa, M., Jiang, H.: Complexity of cutting planes and branch-and-bound in mixed-integer optimization (2020). https://arxiv.org/abs/2003.05023
5. Basu, A., Conforti, M., Di Summa, M., Jiang, H.: Complexity of branch-and-bound and cutting planes in mixed-integer optimization - II. In: Singh, M., Williamson, D.P. (eds.) IPCO 2021. LNCS, vol. 12707, pp. 383–398. Springer, Cham (2021). https://doi.org/10.1007/978-3-030-73879-2_27

6. Burer, S., Vandenbussche, D.: A finite branch-and-bound algorithm for nonconvex quadratic programming via semidefinite relaxations. Math. Program. **113**(2), 259–282 (2008). https://doi.org/10.1007/s10107-006-0080-6

7. Cornuéjols, G., Liberti, L., Nannicini, G.: Improved strategies for branching on general disjunctions. Math. Program. **130**(2, Ser. A), 225–247 (2011). https://doi.org/10.1007/s10107-009-0333-2

8. Dey, S.S., Dubey, Y., Molinaro, M.: Branch-and-bound solves random binary packing IPs in polytime. In: Marx, D. (ed.) Proceedings of the 2021 ACM-SIAM Symposium on Discrete Algorithms, SODA 2021, Virtual Conference January 10–13, 2021, pp. 579–591. SIAM (2021). https://doi.org/10.1137/1.9781611976465.35

9. Dey, S.S., Dubey, Y., Molinaro, M.: Lower bounds on the size of general branch-and-bound trees. Math. Program. (2022). https://doi.org/10.1007/s10107-022-01781-z

10. Dey, S.S., Dubey, Y., Molinaro, M., Shah, P.: A theoretical and computational analysis of full strong-branching (2021)

11. Gasse, M., Chételat, D., Ferroni, N., Charlin, L., Lodi, A.: Exact combinatorial optimization with graph convolutional neural networks. In: Advances in Neural Information Processing Systems, pp. 15580–15592 (2019)

12. Gomory, R.E.: Outline of an algorithm for integer solutions to linear programs. Bull. Amer. Math. Soc. **64**, 275–278 (1958)

13. Gomory, R.E.: Solving linear programming problems in integers. Comb. Anal. **10**, 211–215 (1960)

14. Gomory, R.E.: An algorithm for integer solutions to linear programs. Recent Adv. Math. Program. **64**, 260–302 (1963)

15. Huang, Z., et al.: Learning to select cuts for efficient mixed-integer programming. Pattern Recogn. **123**, 108353 (2022). https://doi.org/10.1016/j.patcog.2021.108353

16. Karamanov, M., Cornuéjols, G.: Branching on general disjunctions. Math. Program. **128**(1–2), 403–436 (2011). https://doi.org/10.1007/s10107-009-0332-3

17. Kazachkov, A.M., Le Bodic, P., Sankaranarayanan, S.: An abstract model of branch-and-cut (2021). https://arxiv.org/abs/2111.09907

18. Khalil, E., Dai, H., Zhang, Y., Dilkina, B., Song, L.: Learning combinatorial optimization algorithms over graphs. In: Advances in Neural Information Processing Systems, pp. 6348–6358 (2017)

19. Khalil, E., Le Bodic, P., Song, L., Nemhauser, G., Dilkina, B.: Learning to branch in mixed integer programming. In: Thirtieth AAAI Conference on Artificial Intelligence (2016)

20. Land, A.H., Doig, A.G.: An automatic method of solving discrete programming problems. Econometrica **28**, 497–520 (1960)

21. Le Bodic, P., Nemhauser, G.: An abstract model for branching and its application to mixed integer programming. Math. Prog. **166**(1–2), 369–405 (2017). https://doi.org/10.1007/s10107-016-1101-8

22. Mahajan, A.: On selecting disjunctions in mixed integer linear programming. Ph.D. thesis, Lehigh University (2009)

23. Tang, Y., Agrawal, S., Faenza, Y.: Reinforcement learning for integer programming: learning to cut. In: Proceedings of the 37th International Conference on Machine Learning (ICML 2020), pp. 9367–9376 (2020)

24. Yang, Y., Boland, N., Savelsbergh, M.: Multivariable branching: a 0-1 knapsack problem case study. INFORMS J. Comput. **33**(4), 1354–1367 (2021). https://doi.org/10.1287/ijoc.2020.1052

Neural Networks with Linear Threshold Activations: Structure and Algorithms

Sammy Khalife$^{(\boxtimes)}$ and Amitabh Basu

Department of Applied Mathematics and Statistics, Johns Hopkins University,
Baltimore, USA
{khalife.sammy,basu.amitabh}@jhu.edu

Abstract. In this article we present new results on neural networks with linear threshold activation functions $x \mapsto 1_{\{x>0\}}$. We precisely characterize the class of functions that are representable by such neural networks and show that 2 hidden layers are necessary and sufficient to represent any function representable in the class. This is a surprising result in the light of recent exact representability investigations for neural networks using other popular activation functions like rectified linear units (ReLU). We also give precise bounds on the sizes of the neural networks required to represent any function in the class. Finally, we design an algorithm to solve the *empirical risk minimization (ERM)* problem to global optimality for these neural networks with a fixed architecture. The algorithm's running time is polynomial in the size of the data sample, if the input dimension and the size of the network architecture are considered fixed constants. The algorithm is unique in the sense that it works for any architecture with any number of layers, whereas previous polynomial time globally optimal algorithms work only for restricted classes of architectures.

1 Introduction

A basic question in a rigorous study of neural networks is a precise characterization of the class of functions representable by neural networks with a certain activation function. The question is of fundamental importance because neural network functions are a popular hypothesis class in machine learning and artificial intelligence. Every aspect of learning using neural networks benefits from a better understanding of the function class: from the statistical aspect of understanding the *bias* introduced in the learning procedure by using a particular neural hypothesis class, to the algorithmic question of training, i.e., finding the "best" function in the class that extrapolates the given sample of data points.

It may seem that the universal approximation theorems for neural networks render this question less relevant, especially since these results apply to a broad class of activation functions [1,5,15]. We wish to argue otherwise. Knowledge of the finer structure of the function class obtained by using a particular activation function can be exploited advantageously. For example, the choice of a certain activation function may lead to much smaller networks that achieve the

© Springer Nature Switzerland AG 2022
K. Aardal and L. Sanitá (Eds.): IPCO 2022, LNCS 13265, pp. 347–360, 2022.
https://doi.org/10.1007/978-3-031-06901-7_26

same bias compared to the hypothesis class given by another activation function, even though the universal approximation theorems guarantee that asymptotically both activation functions achieve arbitrarily small bias. As another example, one can design targeted training algorithms for neural networks with a particular activation function if the structure of the function class is better understood, as opposed to using a generic algorithm like some variant of (stochastic) gradient descent. This has recently led to globally optimal empirical risk minimization algorithms for *rectified linear units (ReLU)* neural networks with specific architecture [2,4,6] that are very different in nature from conventional approaches like (stochastic) gradient descent; see also [7–12,16].

Recent results of this nature have been obtained with ReLU neural networks. Any neural network with ReLU activations clearly gives a piecewise linear function. Conversely, any piecewise linear function $\mathbb{R}^n \to \mathbb{R}$ can be *exactly* represented with at most $\lceil \log_2(n+1) \rceil$ hidden layers [2], thus characterizing the function class representable using ReLU activations. However, it remains an open question if $\lceil \log_2(n+1) \rceil$ are indeed needed. It is conceivable that all piecewise linear functions can be represented by 2 or 3 hidden layers. It is believed this is not the case and there is a strict hierarchy starting from 1 hidden layer, all the way to $\lceil \log_2(n+1) \rceil$ hidden layers. It is known that there are functions representable using 2 hidden layers that cannot be represented with a single hidden layer, but even the 2 versus 3 hidden layer question remains open. Some partial progress on this question can be found in [13].

In this paper, we study the class of functions representable using *threshold activations* (also known as the Heaviside activation, unit step activation, and McCulloch-Pitts neurons). It is easy to see that any function represented by such a neural network is a piecewise constant function. We show that *any* piecewise constant function can be represented by such a neural network, and surprisingly – contrary to what is believed to be true for ReLU activations – there is always a neural network with at most 2 hidden layers that does the job. We also establish that there are functions that cannot be represented by a single hidden layer and thus one cannot do better than 2 hidden layers in general. Our constructions also show that the size of the neural network is at most linear in the number of "pieces" of the function, giving a relatively efficient encoding compared to recent results for ReLU activations which give a polynomial size network only in the case of fixed input dimension [13]. Finally, we use these insights to design an algorithm to solve the empirical risk minimization (training) problem for these neural networks to global optimality whose running time is polynomial in the size of the data sample, assuming the input dimension and the network architecture are fixed. To the best of our knowledge, this is the first globally optimal training algorithm for any family of neural networks that works for arbitrary architectures and has computational complexity that is polynomial in the number of data points, that does not involve a discretization of parameter space or the input space. A very interesting result by Bienstock, Munoz and Pokutta [3] formulates the training problem as a linear programming problem which solves the problem to ϵ-accuracy in time that is *linear* in the number of data points and polynomial in $\frac{1}{\epsilon}$, assuming the input dimension and the network architecture are fixed. This is done via a discretization of the neural network parameter space and input space. For a general convex loss, our algorithm will also have to be content with ϵ-approximate solutions, since this is the best one can do for minimizing general convex functions. However, our running time is polynomial in $\log(1/\epsilon)$, in contrast to $\frac{1}{\epsilon}$. Moreover, for certain loss functions like the ℓ_1 or ℓ_∞, our algorithm will indeed be exact, because the convex optimization problem becomes a linear programming problem, but the algorithm in [3] will still need to rely on discretizations, leading to an approximation. On the other hand, the linear dependence of the algorithm in [3] on the number of data points is much better than our algorithm, although it should be noted that under its current form, their analysis does not formally extend for the linear threshold activation functions, since $x \mapsto 1_{\{x>0\}}$ is not Lipschitz continuous.

We next introduce necessary definitions and notation, followed by a formal statement of our results.

Definitions and Notation. A *polyhedral complex* $\mathcal{P}$ is a collection of convex polyhedra having the following properties:

(A) For every $P, P' \in \mathcal{P}, P \cap P'$ is a face of P and P'.
(B) every face of a polyhedron in $\mathcal{P}$ belongs to $\mathcal{P}$.

We denote by $\dim(P)$ the dimension of a polyhedron and by $\mathring{P}$ the relative interior of P. $|\mathcal{P}|$ will denote the number of polyhedra in a polyhedral complex $\mathcal{P}$ and is called the *size of $\mathcal{P}$*.

Definition 1 (Piecewise constant function). *We say that a function $f : \mathbb{R}^n \to \mathbb{R}$ is* piecewise constant *if there exists a finite polyhedral complex that covers $\mathbb{R}^n$ and f is constant in the relative interior of each polyhedron in the complex. We use PWC_n as a shorthand for the piecewise constant function from $\mathbb{R}^n$ to $\mathbb{R}$.*

Note that there may be multiple polyhedral complexes that correspond to a given piecewise constant function, with possibly different sizes. For example, the indicator function of the nonnegative orthant $\mathbb{R}^n_+$ is a piecewise constant function but there are many different ways to break up the complement of the nonnegative orthant into polyhedral regions. We say that a polyhedral complex $\mathcal{P}$ is *compatible* with a piecewise constant function f if f is constant in the relative interior of every polyhedron in $\mathcal{P}$.

The *threshold activation function* is a map from $\mathbb{R}$ to $\{0,1\}$ given by the indicator of the positive reals, i.e., $x > 0$. By extending this to apply coordinatewise, we get a function $\sigma : \mathbb{R}^d \to \{0,1\}^d$ for any $d \geq 1$, i.e., $\sigma(x)_i$ is 1 if and only if $x_i > 0$ for $i = 1, \ldots, d$. For any subset $X \subseteq \mathbb{R}^n$, 1_X wil denote its *indicator function*, i.e., $1_X(y) = 1$ if $y \in X$ and 0 otherwise.

Definition 2 (Linear threshold deep neural network (DNN)). *For any number of hidden layers $k \in \mathbb{N}$, input and output dimensions $w_0, w_{k+1} \in \mathbb{N}$, a $\mathbb{R}^{w_0} \to \mathbb{R}^{w_{k+1}}$ linear threshold DNN is given by specifying a sequence of k*

natural numbers $w_1, w_2, \cdots, w_k$ representing widths of the hidden layers, a set of $k+1$ affine transformations $T_i : \mathbb{R}^{w_{i-1}} \to \mathbb{R}^{w_i}$, $i = 1, \ldots, k + 1$. Such a linear threshold DNN is called a $(k+1)$-layer DNN, and is said to have k hidden layers. The function $f : \mathbb{R}^{w_0} \to \mathbb{R}^{w_{k+1}}$ computed or represented by this DNN is:

$$f = T_{k+1} \circ \sigma \circ T_k \circ \cdots T_2 \circ \sigma \circ T_1.$$

The size of the DNN, or the number of neurons in the DNN, is $w_1 + \ldots + w_k$. For natural numbers n, k and a tuple $w = (w_1, \ldots, w_k)$, we use $H_n^w(k)$ to denote the family of all possible linear threshold DNNs with input dimension $w_0 = n$, k hidden layers with widths $w_1, \ldots, w_k$ and output dimension $w_{k+1} = 1$. $H_n(k) := \bigcup_{w=(w_1,\ldots,w_k)} H_n^w(k)$ will denote the family of all linear threshold activation neural networks with k hidden layers. We say that a neuron or a neural network computes a subset $X \subseteq \mathbb{R}^n$ if and only if its output is the indicator function 1_X of that subset.

Our Contributions

Any function expressed by a linear threshold neural network is a constant piecewise function (i.e. $H_n(k) \subseteq \mathrm{PWC}_n$ for all natural numbers k), because a composition of piecewise constant functions is piecewise constant. In this work we show that linear threshold neural networks with 2 hidden layers can compute any constant piecewise function, i.e. $H_n(2) = \mathrm{PWC}_n$. We also prove that this is optimal. More formally,

Theorem 1. *For all natural numbers $k \geq 2$,*

$$H_n(1) \subsetneq H_n(2) = H_n(k) = \mathrm{PWC}_n.$$

Equivalently, any piecewise constant function $f : \mathbb{R}^n \to \mathbb{R}$ can be computed with linear threshold DNN with at most 2 hidden layers. Moreover, the DNN needs at most $3|\mathcal{P}|$ neurons, where $\mathcal{P}$ is any polyhedral complex compatible with f.

Next, we show that the bound on the size of the neural network in Theorem 1 is in a sense best possible, up to constant factors.

Proposition 1. *There is a family of piecewise constant functions such that for any function f in the family, any linear threshold DNN representing f has size at least the size of the smallest polyhedral complex compatible with f.*

Finally, we present a new algorithm to perform exact *empirical risk minimization (ERM)* for linear threshold neural networks with fixed architecture, i.e., fixed k and $w = (w_1, \ldots, w_k)$. Given D data points $(x_i, y_i) \in \mathbb{R}^n \times \mathbb{R}$, $\{i = 1 \cdots, D\}$, the ERM problem with hypothesis class $H_n^w(k)$ is

$$\min_{f \in H_n^w(k)} \frac{1}{D} \sum_{i=1}^{D} \ell(f(x_i), y_i) \tag{1}$$

where ℓ is a convex loss function.

Theorem 2. *For natural numbers n, k and tuple $w = (w_1, \ldots, w_k)$, there exists an algorithm that computes the global optimum (1) with running time $O(D^{w_1 n} \cdot 2^{\sum_{i=1}^{k-1} w_i^2 w_{i+1}} \cdot \mathrm{poly}(D, w_1, \ldots, w_k))$. Thus, the algorithm is polynomial in the size of the data sample, if $n, k, w_1, \ldots, w_k$ are considered fixed constants.*

The rest of the article is organized as follows. In Sect. 2 we present our representability results for the class of linear threshold neural networks, including a proof of Theorem 1. In Sect. 3 we prove the lower bound stated in Proposition 1 using the structure of breakpoints of piecewise constant functions. Our ERM algorithm and the proof of Theorem 2 are presented in Sect. 4 with intermediate results. We conclude with a short discussion and open problems in Sect. 5.

2 Representability Results

The following lemma is clear from the definitions.

Lemma 1. *Let $f \in \mathrm{PWC}_n$ and let $\mathcal{P}$ be a compatible polyhedral complex. Then, there exists a unique sequence $(\alpha_1, \cdots, \alpha_{|\mathcal{P}|}) \in \mathbb{R}^{|\mathcal{P}|}$ such that*

$$f = \sum_{P \in \mathcal{P}} \alpha_P 1_{\mathring{P}}$$

Proposition 2. $\mathrm{H}_1(1) = \mathrm{PWC}_1$, *i.e., linear threshold neural networks with a single hidden layer can compute any piecewise constant function $\mathbb{R} \to \mathbb{R}$. Moreover, if $f \in \mathrm{PWC}_1$ and $\mathcal{P}$ is a polyhedral complex of f, then f can be computed with $3|\mathcal{P}| + 1$ neurons.*

Proof. Let $f : \mathbb{R} \to \mathbb{R}$ a piecewise constant function. Then using Lemma 1 there exists a polyhedral complex whose union is $\mathbb{R}$ and such that f is constant on the relative interior of each of the polyhedra. In $\mathbb{R}$, non empty polyhedra are either reduced to a point, or they are the intervals of the form $[a, b],] -\infty, a]$, $[a, +\infty[$ with $a \leq b \in \mathbb{R}$, or $\mathbb{R}$ itself. We first show that we can compute the indicator function on each of the interior of those intervals with at most two neurons. The interior of $[a, +\infty[,] -\infty, b]$ or $\mathbb{R}$ can obviously be computed by one neuron (e.g. $x \mapsto 1_{\{ax < 0\}}$ with $a = 0$ for $\mathbb{R}$). The last cases (singletons and polyhedron of the form $[a, b]$) requires a more elaborate construction. To compute the function $1_{\{x \in]a,b[\}}$, it is sufficient to implement a Dirac function, since $1_{\{x \in]a,b[\}} = 1_{\{b < x\}} - 1_{\{a < x\}} - \delta_a(x)$ where δ_a is the Dirac in $a \in \mathbb{R}$, i.e., $\delta_a : \mathbb{R} \to \mathbb{R}, \ x \mapsto 1_{\{x = a\}}$. δ_a can be computed by a linear combination of three neurons, since $g_a : \mathbb{R} \to \mathbb{R}, \ x \mapsto 1_{\mathbb{R}} - (1_{\{x < a\}} + 1_{\{x > a\}})$ is equal to δ_a. Using a linear combination of the basis functions (polyhedra and faces), we can compute exactly f. To show that $3|\mathcal{P}| + 1$ neurons suffice, $1_{\mathbb{R}}$ is computed with one shared neuron, and then 3 other neurons are needed at most for one polyhedron using our construction. $\qquad\square$

We next show that starting with two dimensions, linear threshold DNNs with a single hidden layer cannot compute every possible piecewise constant function.

Proposition 3. *Let $C_2 := \{(x_1, x_2) \in \mathbb{R}^2 \mid 0 \leq x_1, x_2 \leq 1\}$. Then 1_{C_2} cannot be represented by any linear threshold neural network with one hidden layer.*

Proof. Consider any piecewise constant function on $\mathbb{R}^2$ represented by a single hidden layer neural network, say $g := x \mapsto \sum_{i=1}^{m} \alpha_i 1_{\{x \in \mathbb{R}^2 : \langle a_i, x \rangle + b_i < 0\}}$ with $\alpha_1, \cdots, \alpha_m \in \mathbb{R}$, $a_1, \ldots, a_m \in \mathbb{R}^2$ and $b_1, \ldots, b_m \in \mathbb{R}$. For the sake of clarity, we may suppose that for all $i \neq j$, either $a_i \neq a_j$ or $b_i \neq b_j$ to avoid possible compensations. This implies that the set of nondifferentiable points of g is a union of lines in $\mathbb{R}^2$. However, the set of nondifferentiable points of 1_{C_2} are the sides of the cube, which is a union of finite length line segments. This shows that 1_{C_2} cannot be represented by a single hidden layer linear threshold DNN. □

We will now build towards a proof of Theorem 1 which states that 2 hidden layers actually suffice to represent any piecewise constant function in PWC_n.

Lemma 2. *Let P be a polyhedron in $\mathbb{R}^n$ given by the intersection of m halfspaces. Then, 1_P can be computed with a two hidden layer neural network and $m + 1$ neurons in total.*

Proof. Let P a polyhedron, i.e. $P = \{x \in \mathbb{R}^n \mid Ax \leq b\}$ with $A = (a_1, \cdots, a_m)^T \in \mathbb{R}^{m \times n}$ and $b = (b_1, \cdots, b_m) \in \mathbb{R}^m$. Let us consider the m neurons $(\phi_i : x \mapsto 1_{\{x \in \mathbb{R}^n : \langle a_i, x \rangle > b_i\}})_{1 \leq i \leq m}$, and $\phi : x \mapsto \sum_i \phi_i(x)$. Then for all $x \in \mathbb{R}^n$, $\phi(x) < 1$ if and only if $x \in P$. Now, defining $\psi : y \mapsto 1_{\{y \in \mathbb{R} : y < 1\}}$ yields $\psi \circ \phi = 1_P$. ψ can obviously be computed with a neuron. Therefore, one can compute 1_P with m neurons in the first hidden layer and one neuron in the second, which proves the result. □

Lemma 3. *Let P be a polyhedron in $\mathbb{R}^n$. Then the indicator function of its relative interior can be computed with a two hidden layer neural network, using the indicator of P and the indicators of its faces.*

Proof. Let P be a polyhedron. First, we always have $1_{\mathring{P}} = 1_P - 1_{\text{Union of facets of P}}$. Therefore it is sufficient to prove that we can implement $1_{\text{Union of facets of P}}$ for any P. Using the inclusion exclusion principle on indicator functions, suppose that the facets of P are $f_1, \cdots, f_l$, then:

$$1_{\bigcup_{j=1}^{l} f_k} = \sum_{j=1}^{l} (-1)^{j+1} \sum_{1 \leq i_1 < \cdots < i_j \leq l} 1_{f_{i_1} \cap \cdots \cap f_{i_j}}$$

It should be noted that for any $j \in \{1, \cdots, l\}$, $F = f_{i_1} \cap \cdots \cap f_{i_j}$ is either empty, or a face of P, hence a polyhedron of dimension lower or equal to $\dim(P) - 1$. Therefore, using Lemma 2, we can implement F with a two hidden neural network with at most $m + 1$ neurons, where m is the number of halfspaces in an inequality description of P. If s is the number of faces of P, then there are at most s polyhedra to compute. □

Combining these results, we can now provide a:

Proof of Theorem 1. Thanks to Lemma 1, in order to represent $f \in \mathrm{PWC}_n$, it is sufficient to compute the indicator function of the relative interior of each polyhedron in one of its polyhedral complex $\mathcal{P}$. This can be achieved with just two hidden layers using Lemma 3. This establishes the equalities in the statement of the theorem. The strict containment $H_n(1) \subsetneq H_n(2)$ is given by Proposition 3.

Let m be the total number of halfspaces used in an inequality description of all the polyhedra in the polyhedral complex $\mathcal{P}$. Since all faces are included in the polyhedral complex, there exists an inequality description with $m \leq 2|\mathcal{P}|$. The factor 2 appears because for each facet of a full dimensional polyhedron in $\mathcal{P}$, one may need both directions of the inequality that describes this facet. Then the construction in the proofs of Lemmas 2 and 3 show that one needs at most $m \leq 2|\mathcal{P}|$ neurons in the first hidden layer and at most $|\mathcal{P}|$ neurons in the second hidden layer. $\square$

3   Proof of Proposition 1

Even though it is possible to represent any piecewise constant function using only two hidden layers, one may wonder if there is some advantage of using more hidden layers, for instance to decrease the number of neurons to compute a target function. We are unable to settle this question in general. However, we show that the linear bound in Theorem 1 cannot be improved in general. More precisely, we prove Proposition 1 in this section.

We first introduce the notion of breakpoint for piecewise constant functions.

Definition 3. *Let $f \in \mathrm{PWC}_n$. We say that $x \in \mathbb{R}^n$ is a breakpoint of f if and only if for all $\epsilon > 0$ the ball centered in x and radius ϵ contains a point y such that $f(y) \neq f(x)$.*

Lemma 4. *For any piecewise constant function $f : \mathbb{R}^n \to \mathbb{R}$, the breakpoints of $1_{\{x \in \mathbb{R}^n \,:\, f(x) > 0\}}$ are breakpoints of f.*

Proof. Let $f : \mathbb{R}^n \to \mathbb{R}$ be a piecewise constant function and x a breakpoint of $1_{\{x \in \mathbb{R}^n \,:\, f(x) > 0\}}$. Then for any $\epsilon > 0$, there exists $y \in \mathcal{B}(x, \epsilon)$ such that either $f(x) \leq 0$ and $f(y) > 0$, or $f(x) > 0$ and $f(y) \leq 0$. In both cases, $f(x) \neq f(y)$ so by definition x is a breakpoint of f. $\square$

For any single neuron with a linear threshold activation with k inputs, the output is the indicator of an open halfspace, i.e., $1_{\{x \in \mathbb{R}^k : \langle a, x \rangle + b > 0\}}$ for some $a \in \mathbb{R}^k$ and $b \in \mathbb{R}$. We say that $\{x \in \mathbb{R}^k : \langle a, x \rangle + b = 0\}$ is the *hyperplane associated with this neuron*. This concept is needed in the next proposition.

Proposition 4. *The set of breakpoints of a function represented by a linear threshold DNN with any number of hidden layers is included in the union of hyperplanes associated with the neurons in the first hidden layer.*

Proof. We give a proof by induction on the number k of hidden layers. We remind that if $\alpha_i \in \mathbb{R}^n$ and $\beta_i \in \mathbb{R}$ are the weights and bias of a neuron in the first layer, the corresponding hyperplane is $\{x \in \mathbb{R}^n : \alpha_i x + \beta_i = 0\}$.

<u>Base case:</u> For $k = 1$, let ϕ be a one hidden layer DNN with w_1 neurons, i.e. there exists $\gamma_1, \cdots, \gamma_{w_1} \in \mathbb{R}$ and closed half-spaces $H_1, \cdots, H_{w_1}$ such that:

$$\phi = \sum_{i=1}^{w_1} \gamma_i 1_{H_i^c}$$

Let $P_1, \cdots, P_{w_1}$ be the hyperplanes associated to each $H_1, \cdots, H_{w_1}$. Then we claim that ϕ does not have any breakpoint in $\mathbb{R}^n - \bigcup_{i=1}^{w_1} P_i$ where $P_1, \cdots, P_{w_1}$ are the hyperplanes associated to $H_1, \cdots, H_{w_1}$. To formally prove it, let x be a point of $\mathbb{R}^n - \bigcup_{i=1}^{w_1} P_i$. Then for each i, x is either in $\mathring{H}_i$ or H_i^c. This means that x belongs to a intersection of open sets, say $E = \cap_{i=1}^{w_1} O_i$ where $\forall i$, $O_i = \mathring{H}_i$ or $O_i = H_i^c$. First, E is an open set so there exists $\epsilon > 0$ such that the ball $\mathcal{B}(x, \epsilon)$ centered in x and with radius ϵ is included in E. Furthermore, by definition, ϕ is constant on E, hence ϕ is constant on $\mathcal{B}(x, \epsilon)$ and x cannot be a breakpoint. We proved that the breakpoints of ϕ are in $\bigcup_{i=1}^{w_1} P_i$.

<u>Induction step:</u> Let us suppose the statement is true for all neural networks with k hidden layers, and consider a neural network ϕ with $k + 1$ hidden layers. It should be noted that the output of a neuron in the last hidden layer is of the form

$$1_{\{x \in \mathbb{R}^n \, : \, a_1 \psi_1(x) + \cdots a_{w_k} \psi_{w_k}(x) + \beta > 0\}} = 1_{\{x \in \mathbb{R}^n \, : \, \Psi(x) > 0\}}$$

where $\Psi : x \mapsto a_1 \psi_1(x) + \cdots a_{w_k} \psi_{w_k}(x) + \beta$ is the piecewise function represented by a neural network of depth k. Lemma 4 states that the breakpoints of $1_{\{x \in \mathbb{R}^n \, : \, \Psi(x) > 0\}}$ are breakpoints of Ψ, a DNN of depth k. Using the induction assumption, the breakpoints of Ψ belong to the hyperplanes introduced in the first layer. Hence the breakpoints of ϕ, which is a linear combination of such neurons, are included in the hyperplanes of the first layer.   □

Proof of Proposition 1. Let us construct a family of functions in $\mathbb{R}^n$. Let us consider the sets $P_1 := \{x \in \mathbb{R}^n : x_1 \le 0\}$, $P_i := \{x \in \mathbb{R}^n : (i - 2) < x_1 \le i - 1\}$ for $i \in \{2, \cdots, N - 1\}$, and $P_N := \{x \in \mathbb{R}^n : x_1 > N - 2\}$. Note that $\bigcup_{i=1}^N P_i = \mathbb{R}^n$. Let $f \in \text{PWC}_n$ such that such that $f(x \in P_i) = i$. It is easy to see that f is a piecewise constant function and that the breakpoints of f is a set of $N - 1$ hyperplanes, with empty pairwise intersections. By Proposition 4, any linear threshold neural network must have these hyperplanes associated with neurons in the first hidden layer, and therefore we must have at least N neurons in the first hidden layer.   □

4   Globally Optimal Empirical Risk Minimization

In this section we present a algorithm to train a linear threshold DNN with fixed architecture to optimality. Let us recall the corresponding optimization problem. Given D data points $(x_i, y_i) \in \mathbb{R}^n \times \mathbb{R}$, $\{i = 1 \cdots, D\}$, find the function $f \in \mathrm{H}_n^w(k)$ represented by a k-hidden layer $\mathbb{R}^n \to \mathbb{R}$ DNN with widths $w = (w_1, \cdots, w_k)$ that solves:

$$\min_{f \in \mathrm{H}_n^w(k)} \frac{1}{D} \sum_{i=1}^{D} \ell(f(x_i), y_i) \tag{2}$$

where ℓ is a convex loss function. We first present the idea with 2 hidden layers and then adapt the method for an arbitrary number of layers.

4.1   Preliminaries

Definition 4. *Let $m \geq 1$ be any natural number. We say a collection $\mathcal{A}$ of subsets of $\{1, \ldots, m\}$ is* linearly separable *if there exist $\alpha_1, \ldots, \alpha_m, \beta \in \mathbb{R}$ such that any subset $A \subseteq \{1, \ldots, m\}$ is in $\mathcal{A}$ if and only if $\sum_{s \in A} \alpha_s + \beta > 0$. Define*

$$\mathcal{L}_m := \{\mathcal{A} : \mathcal{A} \text{ is a linearly separable collection of subsets of } \{1, \ldots, m\}\},$$

i.e., $\mathcal{L}_m$ denotes the set of all linearly separable collections of subsets of $\{1, \ldots, m\}$.

Remark 1. We note that given a collection $\mathcal{A}$ of subsets of $\{1, \cdots, m\}$ one can test if $\mathcal{A}$ is linearly separable by checking if the optimum value of the following linear program is strictly positive:

$$\max_{t \in \mathbb{R}, \alpha \in \mathbb{R}^n, \beta \in \mathbb{R}} \quad t$$

$$\text{s.t.} \quad \sum_{s \in A} \alpha_s + \beta \geq t \quad \forall A \in \mathcal{A} \quad \text{and} \quad \sum_{s \in A} \alpha_s + \beta \leq 0 \quad \forall A \notin \mathcal{A}$$

In Algorithm 1 below, we will enumerate through all possible collections in $\mathcal{L}_m$ (for different values of m). We assume this has been done a priori using the linear programs above and this enumeration can be done in time $|\mathcal{L}_m|$ during the execution of Algorithm 1.

Remark 2. In $\mathbb{R}^2$, $\mathcal{A} = \{\emptyset, \{1\}, \{2\}\}$ is linearly separable, but $\{\emptyset, \{1, 2\}\}$ is not linearly separable because the set of inequalities $\beta > 0$, $\alpha_1 + \alpha_2 > 0$, $\alpha_1 + \beta \leq 0$ and $\alpha_2 + \beta \leq 0$ have no solution. Two examples of linearly separable collections in $\mathbb{R}^3$ are given in Fig. 1.

Proposition 5. *Let $k \geq 2$ and $w = (w_1, \ldots, w_k)$, and consider a DNN of $\mathrm{H}_n^w(k)$. Any neuron of this neural network computes some subset of $\mathbb{R}^n$. Suppose we fix the weights of the neural network up to the $(k-1)$-th hidden layer. This fixes the sets $Y_1, \ldots, Y_{w_{k-1}} \subseteq \mathbb{R}^n$ computed by the w_{k-1} neurons in this layer.*

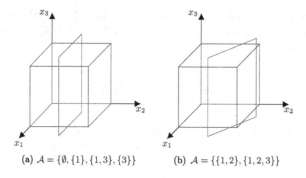

(a) $\mathcal{A} = \{\emptyset, \{1\}, \{1,3\}, \{3\}\}$ **(b)** $\mathcal{A} = \{\{1,2\}, \{1,2,3\}\}$

Fig. 1. Two linearly separable collections of $2^{\{1,2,3\}}$ in $\mathbb{R}^3$. The subsets of $\{1,2,3\}$ are represented by the vertices of $\{0,1\}^3$. The blue hyperplanes represent a possible separation of the corresponding vertices, giving two different linearly separable collections.

Then a neuron in the k-th layer computes $X \subseteq \mathbb{R}^n$ (by adjusting the weights and bias of this neuron) if and only if there exists a linearly separable collection $\mathcal{A}$ of subsets of $\{1, \ldots, w_{k-1}\}$ such that:

$$X = \bigcup_{A \in \mathcal{A}} \bigcap_{s \in A} Y_s$$

Proof. Let $\alpha \in \mathbb{R}^{w_{k-1}}$, $\beta \in \mathbb{R}$ be the weights and bias of the neuron in the k-th layer. By definition, the set represented by this neuron is

$$S_{\alpha,\beta} = \{x \in \mathbb{R}^n : \alpha_1 1_{Y_1}(x) + \cdots + \alpha_{w_{k-1}} 1_{Y_{w_{k-1}}}(x) + \beta > 0\}$$

We define the collection $\mathcal{A} = \{A \subseteq \{1, \cdots, w_{k-1}\} : \sum_{i \in A} \alpha_i + \beta > 0\}$. By definition, $\mathcal{A}$ is a linearly separable collection. Now let us consider the set:

$$O = \bigcup_{A \in \mathcal{A}} \bigcap_{s \in A} Y_s$$

Let $x \in O$. Then there exists $A \subseteq \{1, \cdots, w_{k-1}\}$ such that $x \in \bigcap_{s \in A} Y_s$ and therefore, $\sum_{s \in A} \alpha_s 1_{Y_s}(x) + \beta > 0$. This means that $x \in S_{\alpha,\beta}$, hence $O \subseteq S_{\alpha,\beta}$. Now, let $x \in S_{\alpha,\beta}$. Then $\alpha_1 1_{Y_1}(x) + \cdots + \alpha_{w_{k-1}} 1_{Y_{w_{k-1}}}(x) + \beta > 0$. Let $A = \{s : x \in Y_s\}$ then $\sum_{s \in A} \alpha_s + \beta > 0$ and $x \in \bigcap_{s \in A} Y_s$, hence $S_{\alpha,\beta} \subseteq O$, and $S_{\alpha,\beta} = O$.

Conversely, let $\mathcal{A}$ be a linearly separable collection of subsets of $\{1, \cdots, w_{k-1}\}$. By definition there exists $\alpha \in \mathbb{R}^n$ and $\beta \in \mathbb{R}$ such that $A \in \mathcal{A} \iff \sum_{s \in A} \alpha_s + \beta > 0$. These are then taken as the weights of the neuron in the k-th hidden layer and its output is the function $1_{\{x \in \mathbb{R}^n \,:\, \sum_{i=1}^{w_{k-1}} \alpha_i 1_{Y_i}(x) + \beta > 0\}}$. $\qquad\square$

4.2   Proof of Theorem 2

Consider a neural network with k hidden layers and widths $w = (w_1, \ldots, w_k)$ that implements a function in $H_n^w(k)$. The output of any neuron on these data points is in $\{0,1\}$ and thus each neuron can be thought of as picking out a subset of the set $X := \{x_1, \ldots, x_D\}$. Proposition 5 provides a way to enumerate these subsets of X in a systematic manner.

Definition 5. *For any finite subset $F \subseteq \mathbb{R}^n$, a subset F' of F is said to be linearly separable if there exists $a \in \mathbb{R}^n$, $b \in \mathbb{R}$ such that $F' = \{x \in F : \langle a, x \rangle + b > 0\}$.*

The following is a well-known result in combinatorial geometry [17].

Theorem 3. *For any finite subset $F \subseteq \mathbb{R}^n$, there are at most $2\binom{|F|}{n}$ linearly separable subsets.*

By considering the natural mapping between subsets of $\{1, \ldots, m\}$ and $\{0,1\}^m$, we also obtain the following corollary.

Corollary 1. *For any $m \geq 1$, there are at most $2\binom{2^m}{m}$ linearly separable collections of subsets of $\{1, \ldots, m\}$. In other words, $|\mathcal{L}_m| \leq 2\binom{2^m}{m}$.*

Algorithm 1. Algorithm to solve (2) for linear threshold DNNs with n inputs, k hidden layers and widths $w = (w_1, \ldots, w_k)$.

1: **Input** Dimension n, Dataset $(x_i, y_i)_{i=1}^D$, Integers $w_1, \ldots, w_k$
2: **Output** Solution of Problem 2
3: Define $X = (x_1, \ldots, x_D) \subseteq \mathbb{R}^n$. Let $\mathcal{H}$ be the collection of linearly separable subsets of X.
4: Initialize $OPT = +\infty$, $SOL = \emptyset$.
5: **for** each choice of $H_1, \ldots, H_{w_1} \in \mathcal{H}$, $\mathcal{A}_1^i, \ldots, \mathcal{A}_{w_i}^i \in \mathcal{L}_{w_{i-1}}$ for $i = 2, \ldots, k$ **do**
6:      Define $Y_j^1 = H_j$ for $j = 1, \ldots, w_1$.
7:      Set the weights of the neurons in the first layer to compute Y_j^1 for $j = 1, \ldots, w_1$.
8:      **for** i = 2 to k **do**
9:          **for** j=1 to w_i **do**
10:             Define $Y_j^i = \bigcup_{A \in \mathcal{A}_j^i} \bigcap_{s \in A} Y_s^{i-1}$.
11:             Set the weights of neuron j of layer i in accordance with $\mathcal{A}_j^i$ to compute Y_j^i.
12:         **end for**
13:     **end for**
14:     For each $i = 1, \ldots, D$ and $j = 1, \ldots, w_k$, compute $\delta_{ij} \leftarrow 1_{Y_j^k}(x_i)$, using the neural network constructed so far.
15:     Solve the convex minimization problem in the decision variables $\gamma_1, \ldots, \gamma_{w_k} \in \mathbb{R}$:

$$\text{temp} = \min_{\gamma \in \mathbb{R}^{w_k}} \sum_{i=1}^{D} \ell\left(\sum_{j=1}^{w_k} \gamma_j \delta_{ij}, y_i\right)$$

16:     If temp $< OPT$, then update $OPT = $ temp and SOL to be the current neural network with weights computed in the previous steps.
17: **end for**

Proof of Theorem 2. Algorithm 1 solves (2). The correctness comes from the observation that a recursive application of Proposition 5 shows that the sets $Y_1^k, \ldots, Y_{w_k}^k$ computed by the algorithm are all possible subsets of X computed by the neurons in the last hidden layer. The $\gamma_1, \ldots, \gamma_{w_k}$ are simply the weights of the last layer that combine the indicator functions of these subsets to yield the function value of the neural network on each data point. The convex minimization problem in line 13 finds the optimal γ_j values, for this particular choice of subsets $Y_1^k, \ldots, Y_{w_k}^k$. Selecting the minimum over all these choices solves the problem.

The outermost for loop iterates at most $O(D^{w_1 n} \cdot 2^{\sum_{i=1}^{k-1} w_i^2 w_{i+1}})$ times using Theorem 3 and Corollary 1. The computation of the δ_{ij} values in Step 14 can be done in time $\mathrm{poly}(D, w_1, \ldots, w_k)$. The convex minimization problem in w_k variables and D terms in the sum can be solved in $\mathrm{poly}(D, w_k)$ time. Putting these together gives the overall running time.

We now show that the exponential dependence on the dimension n in Theorem 2 is actually necessary unless $P = NP$. We consider the version of (2) with single neuron and show that it is NP-hard with a direct reduction.

Theorem 4 *(NP-hardness). The One-Node-Linear-Threshold problem is NP-hard when the dimension n is considered part of the input. This implies in particular that Problem 2 is NP-hard when n is part of the input.*

Proof. We here use a result of [14, Theorem 3.1], which showed that the following decision problem is NP-complete.

MinDis(Halfspaces): Given disjoint sets of positive and negative examples of $\mathbb{Z}^n$ and a bound $k \geq 1$, does there exist a separating hyperplane which leads to at most k misclassifications?

MinDis(Halfspaces) is a special case of (2) with a single neuron: given data points $x_1, \cdots, x_D \in \mathbb{R}^n$ and $y_1, \cdots, y_D \in \{0, 1\}$, compute $\alpha \in \mathbb{R}^n, \beta \in \mathbb{R}$ that minimizes $\frac{1}{D} \sum_{i=1}^{D} (1_{\{\langle \alpha, x_i \rangle + \beta > 0\}} - y_i)^2$. □

5   Open Questions

We showed that neural networks with linear threshold activations can represent any piecewise constant function f using at most two hidden layers and a linear number of neurons with respect to the size of any polyhedral complex compatible with f. Furthermore, we provided a family of functions for which this linear dependence cannot be improved. However, it is possible that there are other families of functions where the behaviour is different: by increasing depth, one could possibly represent these functions using an exponentially smaller number of neurons, compared to what is needed with two layers. For instance, in the case of ReLU activations, there exist functions for which depth brings an exponential gain in the size of the neural network [2,18]. We think it is a very interesting open question to determine if such families of functions exist for linear threshold networks.

On the algorithmic side, we solve the empirical risk minimization problem to global optimality with running time that is polynomial in the size of the data sample, assuming that the input dimension and the architecture size are fixed constants. The running time is exponential in terms of these parameters (see Theorem 2). While the exponential dependence on the input dimension cannot be avoided unless $P = NP$ (see Theorem 4), another very interesting open question is to determine if the exponential dependence on the architectural parameters is really needed, or if an algorithm can be designed that has complexity which is polynomial in both the data sample and the architecture parameters. A similar question is also open in the case of ReLU neural networks [2].

References

1. Anthony, M., Bartlett, P.L.: Neural Network Learning: Theoretical Foundations. Cambridge University Press, Cambridge (1999)
2. Arora, R., Basu, A., Mianjy, P., Mukherjee, A.: Understanding deep neural networks with rectified linear units. In: International Conference on Learning Representations (2018)
3. Bienstock, D., Muñoz, G., Pokutta, S.: Principled deep neural network training through linear programming. arXiv preprint arXiv:1810.03218 (2018)
4. Boob, D., Dey, S.S., Lan, G.: Complexity of training ReLU neural network. Discrete Optim. **44**, 100620 (2020)
5. Cybenko, G.: Approximation by superpositions of a sigmoidal function. Math. Control Signals Syst. **2**(4), 303–314 (1989)
6. Dey, S.S., Wang, G., Xie, Y.: Approximation algorithms for training one-node ReLU neural networks. IEEE Trans. Signal Process. **68**, 6696–6706 (2020)
7. Froese, V., Hertrich, C., Niedermeier, R.: The computational complexity of ReLU network training parameterized by data dimensionality. arXiv preprint arXiv:2105.08675 (2021)
8. Goel, S., Kanade, V., Klivans, A., Thaler, J.: Reliably learning the ReLU in polynomial time. In: Conference on Learning Theory, pp. 1004–1042. PMLR (2017)
9. Goel, S., Klivans, A., Manurangsi, P., Reichman, D.: Tight hardness results for training depth-2 ReLU networks. arXiv preprint arXiv:2011.13550 (2020)
10. Goel, S., Klivans, A., Meka, R.: Learning one convolutional layer with overlapping patches. In: International Conference on Machine Learning, pp. 1783–1791. PMLR (2018)
11. Goel, S., Klivans, A.R.: Learning neural networks with two nonlinear layers in polynomial time. In: Conference on Learning Theory, pp. 1470–1499. PMLR (2019)
12. Goel, S., Klivans, A.R., Manurangsi, P., Reichman, D.: Tight hardness results for training depth-2 ReLU networks. In: 12th Innovations in Theoretical Computer Science Conference (ITCS 2021). LIPIcs, vol. 185, pp. 22:1–22:14. Schloss Dagstuhl - Leibniz-Zentrum für Informatik (2021)
13. Hertrich, C., Basu, A., Di Summa, M., Skutella, M.: Towards lower bounds on the depth of ReLU neural networks. arXiv preprint arXiv:2105.14835 (2021). To appear in NeurIPS 2021
14. Hoffgen, K.U., Simon, H.U., Vanhorn, K.S.: Robust trainability of single neurons. J. Comput. Syst. Sci. **50**(1), 114–125 (1995)
15. Hornik, K.: Approximation capabilities of multilayer feedforward networks. Neural Netw. **4**(2), 251–257 (1991)

16. Manurangsi, P., Reichman, D.: The computational complexity of training ReLU(s). arXiv preprint arXiv:1810.04207 (2018)
17. Matousek, J.: Lectures on Discrete Geometry, vol. 212. Springer Science & Business Media (2013)
18. Telgarsky, M.: Benefits of depth in neural networks. In: Conference on Learning Theory, pp. 1517–1539. PMLR (2016)

A PTAS for the Horizontal Rectangle Stabbing Problem

Arindam Khan[1], Aditya Subramanian[1(✉)], and Andreas Wiese[2]

[1] Indian Institute of Science, Bengaluru, India
{arindamkhan,adityasubram}@iisc.ac.in
[2] Vrije Universiteit Amsterdam, Amsterdam, The Netherlands
a.wiese@vu.nl

Abstract. We study rectangle stabbing problems in which we are given n axis-aligned rectangles in the plane that we want to *stab*, i.e., we want to select line segments such that for each given rectangle there is a line segment that intersects two opposite edges of it. In the *horizontal rectangle stabbing problem* (STABBING), the goal is to find a set of horizontal line segments of minimum total length such that all rectangles are stabbed. In *general rectangle stabbing problem*, also known as *horizontal-vertical stabbing problem* (HV-STABBING), the goal is to find a set of rectilinear (i.e., either vertical or horizontal) line segments of minimum total length such that all rectangles are stabbed. Both variants are NP-hard. Chan, van Dijk, Fleszar, Spoerhase, and Wolff [5] initiated the study of these problems by providing $O(1)$-approximation algorithms. Recently, Eisenbrand, Gallato, Svensson, and Venzin [11] have presented a QPTAS and a polynomial-time 8-approximation algorithm for STABBING but it is open whether the problem admits a PTAS.

In this paper, we obtain a PTAS for STABBING, settling this question. For HV-STABBING, we obtain a $(2 + \varepsilon)$-approximation. We also obtain PTASes for special cases of HV-STABBING: (i) when all rectangles are squares, (ii) when each rectangle's width is at most its height, and (iii) when all rectangles are δ-large, i.e., have at least one edge whose length is at least δ, while all edge lengths are at most 1. Our result also implies improved approximations for other problems such as *generalized minimum Manhattan network*.

Keywords: Geometric optimization · Approximation algorithms · Line stabbing · Rectangles

1 Introduction

Rectangle stabbing problems are natural geometric optimization problems. Here, we are given a set of n axis-parallel rectangles $\mathcal{R}$ in the two-dimensional plane. For each rectangle $R_i \in \mathcal{R}$, we are given points $(x_1^{(i)}, y_1^{(i)}), (x_2^{(i)}, y_2^{(i)}) \in \mathbb{R}^2$ that denote its bottom-left and top-right corners, respectively. Also, we denote its *width* and *height* by $w_i := x_2^{(i)} - x_1^{(i)}$ and $h_i := y_2^{(i)} - y_1^{(i)}$, respectively. Our goal

© Springer Nature Switzerland AG 2022
K. Aardal and L. Sanitá (Eds.): IPCO 2022, LNCS 13265, pp. 361–374, 2022.
https://doi.org/10.1007/978-3-031-06901-7_27

is to compute a set of line segments $\mathcal{L}$ that stab all input rectangles. We call a rectangle *stabbed* if a segment $\ell \in \mathcal{L}$ intersects both of its horizontal or both of its vertical edges. We study several variants. In the *horizontal rectangle stabbing problem* (STABBING) we want to find a set of horizontal segments of minimum total length such that each rectangle is stabbed. The *general rectangle stabbing* (HV-STABBING) problem generalizes STABBING and involves finding a set of axis-parallel segments of minimum total length such that each rectangle in $\mathcal{R}$ is stabbed. The *general square stabbing* (SQUARE-STABBING) problem is a special case of HV-STABBING where all rectangles in the input instance are squares. These problems have applications in bandwidth allocation, message scheduling with time-windows on a direct path, and geometric network design [3,5,10].

Note that STABBING and HV-STABBING are special cases of weighted geometric set cover problem, where the rectangles correspond to elements and potential line segments correspond to sets, and the weight of a set equals the length of the corresponding segment. A set contains an element if the corresponding line segment stabs the corresponding rectangle. This already implies an $O(\log n)$-approximation algorithm [9] for HV-STABBING and STABBING.

Chan, van Dijk, Fleszar, Spoerhase, and Wolff [5] initiated the study of STABBING. They proved STABBING to be NP-hard via a reduction from planar vertex cover. Also, they presented a constant[1] factor approximation algorithm using *decomposition techniques* and the *quasi-uniform sampling* method [19] for weighted geometric set cover. In particular, they showed that STABBING instances can be decomposed into two disjoint *laminar* set cover instances of small shallow cell complexity for which the *quasi-uniform sampling* yields an $O(1)$-approximation using techniques from [8].

Recently, Eisenbrand, Gallato, Svensson, and Venzin [11] presented a quasi-polynomial time approximation scheme (QPTAS) for STABBING. This shows that STABBING is not APX-hard unless NP $\subseteq$ DTIME($2^{\text{poly}\log n}$). The QPTAS relies on *the shifting technique* by Hochbaum and Maass [14], applied to a grid, consisting of randomly shifted vertical grid lines that are equally spaced. With this approach, the plane is partitioned into narrow disjoint vertical strips which they then process further. Then, this routine is applied recursively. They also gave a polynomial time dynamic programming based exact algorithm for STABBING for laminar instances (in which the projections of the rectangles to the x-axis yield a laminar family of intervals). Then they provided a simple polynomial-time 8-approximation algorithm by reducing any given instance to a laminar instance. It remains open whether there is a PTAS for the problem.

1.1   Our Results

In this paper, we give a PTAS for STABBING and thus resolve this open question. Also, we extend our techniques to HV-STABBING for which we present a polynomial time $(2 + \varepsilon)$-approximation and PTASes for several special cases: when all

---

[1] The constant is not explicitly stated, and it depends on a not explicitly stated constant in [7].

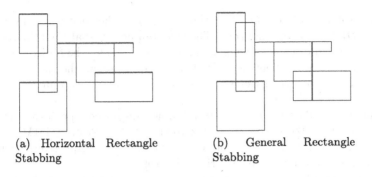

(a) Horizontal Rectangle
Stabbing

(b)   General   Rectangle
Stabbing

Fig. 1. A solution for an instance of STABBING and HV-STABBING.

input rectangles are squares, more generally when for each rectangle its width
is at most its height, and finally for δ-large rectangles, i.e., when each rectangle
has one edge whose length is within $[\delta, 1]$ and 1 is the maximum length of each
edge of any input rectangle (in each dimension).

Our algorithm for STABBING is in fact quite easy to state: it is a dynamic
program (DP) that recursively subdivides the plane into smaller and smaller rectangular regions. In the process, it guesses line segments from OPT. However, its
analysis is intricate. We show that there is a sequence of recursive decompositions that yields a solution whose overall cost is $(1 + \varepsilon)$OPT. Instead of using a
set of equally spaced grid lines as in [11], we use a *hierarchical* grid with several
levels for the decomposition. In each level of our decomposition, we subdivide
the given rectangular region into strips of narrow width and guess $O_\varepsilon(1)$ line
segments from OPT inside them which correspond to the current level. One crucial ingredient is that we slightly extend the segments, such that the guessed
horizontal line segments are aligned with our grid. The key consequence is that
it will no longer be necessary to remember these line segments once we have
advances three levels further in the decomposition. Also, for the guessed vertical
line segments of the current level we introduce additional (very short) horizontal line segments, such that we do not need to remember them either, once we
advanced three levels more in the decomposition. Therefore, the DP needs to
remember previously guessed line segments from only the last three previous
levels and afterwards these line segments *vanish*. This allows us to bound the
number of arising subproblems (and hence of the DP-cells) by a polynomial.

Our techniques easily generalize to a PTAS for SQUARE-STABBING and to
a PTAS for HV-STABBING if for each input rectangle its width is at most its
height, whereas the QPTAS in [11] worked only for STABBING. We use the latter
PTAS as a subroutine in order to obtain a polynomial time $(2+\varepsilon)$-approximation
for HV-STABBING – which improves on the result in [5].

Then, we extend our techniques above to the setting of δ-large rectangles of
HV-STABBING. This is an important subclass of rectangles and are well-studied
for other geometric problems [1,15]. To this end, we first reduce the problem to
the setting in which all input rectangles are contained in a rectangular box that

admits a solution of cost $O(1/\varepsilon^3)$. Then we guess the relatively long line segments in OPT in polynomial time. The key argument is that then the remaining problem splits into two independent subproblems, one for the horizontal and one for the vertical line segments in OPT. For each of those, we then apply our PTAS for STABBING which then yields a PTAS for HV-STABBING if all rectangles are δ-large.

Finally, our PTAS for STABBING implies improved approximation ratios for the GENERALIZED MINIMUM MANHATTAN NETWORK (GMMN) and x-separated 2D-GMMN problems, of $(4+\varepsilon)\log n$ and $4+\varepsilon$, respectively, by improving certain subroutines of the algorithm in [10].

Due to space limitations, many proofs had to be omitted, and we refer the reader to the full version of this paper [16].

1.2   Further Related Work

Finke et al. [12] gave a polynomial time exact algorithm for a special case of STABBING where all input rectangles have their left edges lying on the y-axis. Das et al. [10] studied a related variant in the context of the GENERALIZED MINIMUM MANHATTAN NETWORK (GMMN) problem. In GMMN, we are given a set of n terminal-pairs and the goal is to find a minimum-length rectilinear network such that each pair is connected by a *Manhattan path*. They obtained a 4-approximation for a variant of STABBING where all rectangles intersect a vertical line. Then they used it to obtain a $(6+\varepsilon)$-approximation algorithm for the x-separated 2D-GMMN problem, a special case of 2D-GMMN, and $(6+\varepsilon)(\log n)$-approximation for 2-D GMMN.

Gaur et al. [13] studied the problem of stabbing rectangles by a *minimum number of axis-aligned lines* and gave an LP-based 2-approximation algorithm. Kovaleva and Spieksma [17] considered a weighted generalization of this problem and gave an $O(1)$-approximation algorithm.

STABBING and HV-STABBING are related to geometric set cover which is a fundamental geometric optimization problem. Brönnimann and Goodrich [4] in a seminal paper gave an $O(d\log(d\text{OPT}))$-approximation for unweighted geometric set cover where d is the dual VC-dimension of the set system and OPT is the value of the optimal solution. Using ε-*nets*, Aronov et al. [2] gave an $O(\log\log OPT)$-approximation for hitting set for axis-parallel rectangles. Later, Varadarajan [19] developed quasi-uniform sampling and provided sublogarithmic approximation for weighted set cover where sets are weighted fat triangles or weighted disks. Chan et al. [8] generalized this to any set system with low *shallow cell complexity*. Afterward, Chan and Grant [6] and Mustafa et al. [18] have settled the APX-hardness statuses of all natural weighted geometric set cover problems.

2   Dynamic Program

We present a dynamic program that computes a $(1+\varepsilon)$-approximation to HV-STABBING for the case where $h_i \geq w_i$ for each rectangle $R_i \in \mathcal{R}$. This implies directly a PTAS for the setting of squares for the same problem, and we will argue that it also yields a PTAS for STABBING. Later, we will use it as a subroutine to obtain a $(2 + \varepsilon)$-approximation for HV-STABBING and a PTAS for the setting of δ-large rectangles of HV-STABBING.

For a line segment ℓ, we use the notation $|\ell|$ to represent its length, and for a set of segments $\mathcal{L}$, we use notation $c(\mathcal{L})$ to represent the cost of the set, which is also the total length of the segments contained in it. We use the term OPT interchangeably to refer to the optimal solution to the problem and also to $c(\text{OPT})$, i.e., the cost of the optimal solution.

2.1 Preprocessing Step

First, we show that by some simple scaling and discretization steps we can ensure some simple properties that we will use later. Without loss of generality we assume that $(1/\varepsilon) \in \mathbb{N}$ and we say that a value $x \in \mathbb{R}$ is *discretized* if x is an integral multiple of ε/n.

Lemma 1. *For any positive constant $\varepsilon < 1/3$, by losing a factor $(1 + O(\varepsilon))$ in the approximation ratio, we can assume for each $R_i \in \mathcal{R}$ the following properties hold:*

- *$\varepsilon/n \le w_i \le 1$,*
- *$x_1^{(i)}, x_2^{(i)}$ are discretized and within $[0, n]$,*
- *$y_1^{(i)}, y_2^{(i)}$ are discretized and within $[0, 4n^2]$, and*
- *each horizontal line segment in the optimal solution has width at most $1/\varepsilon$.*

Henceforth in this paper when we refer to the set of input rectangles $\mathcal{R}$, we are referring to a set $\mathcal{R}'$ that has been obtained after applying the preprocessing from Lemma 1 to the input set $\mathcal{R}$, and when we refer to OPT, we are referring to the optimal solution to the set of rectangles $\mathcal{R}'$, which is a $(1 + O(\varepsilon))$ approximation of the optimal solution of the input instance.

2.2 Description of the Dynamic Program

Our algorithm is based on a dynamic program (DP). It has a cell $\text{DP}(S, \mathcal{L})$ for each combination of

- a rectangular region $S \subseteq [0, n] \times [0, 4n^2]$ with discretized coordinates (that is not necessarily equal to an input rectangle in $\mathcal{R}$).
- a set $\mathcal{L}$ of at most $3(1/\varepsilon)^3$ line segments, each of them horizontal or vertical, such that for each $\ell \in \mathcal{L}$ we have that $\ell \subseteq S$, and all coordinates of ℓ are discretized.

This DP-cell corresponds to the subproblem of stabbing all rectangles in $\mathcal{R}$ that are contained in S and that are not already stabbed by the line segments in $\mathcal{L}$. Therefore, the DP stores solution $\text{SOL}(S, \mathcal{L})$ in the cell $\text{DP}(S, \mathcal{L})$ such that $\text{SOL}(S, \mathcal{L}) \cup \mathcal{L}$ stabs all rectangles in $\mathcal{R}$ that are contained in S.

Given a DP-cell $DP(S, \mathcal{L})$, our DP computes a solution for it as follows. If $\mathcal{L}$ already stabs each rectangle from $\mathcal{R}$ that is contained in S, then we simply define a solution $SOL(S, \mathcal{L}) := \emptyset$ for the cell $DP(S, \mathcal{L})$ and do not compute anything further. Another simple case is when there is a line segment $\ell \in \mathcal{L}$ such that $S \setminus \ell$ is divided into two rectangular regions S_1, S_2 (these regions are henceforth referred to as connected components). In this case we define $SOL(S, \mathcal{L}) := SOL(S_1, \mathcal{L} \cap S_1) \cup SOL(S_2, \mathcal{L} \cap S_2) \cup \{\ell\}$, where for any set of line segments $\mathcal{L}'$ and any rectangle S' we define $\mathcal{L}' \cap S' := \{\ell' \cap S' | \ell' \in \mathcal{L}' \wedge \ell' \cap S' \neq \emptyset\}$. In case that there is more than one such line segment $\ell \in \mathcal{L}$ then we pick one according to some arbitrary but fixed global tie-breaking rule. We will later refer to this as *trivial operation*.

Otherwise, we do each of the following operations which produces a set of candidate solutions:

1. *Add operation:* Consider each set $\mathcal{L}'$ of line segments with discretized coordinates such that $|\mathcal{L}| \cup |\mathcal{L}'| \leq 3\varepsilon^{-3}$ and each $\ell \in \mathcal{L}'$ is contained in S and horizontal or vertical. For each such set $\mathcal{L}'$ we define the solution $\mathcal{L}' \cup SOL(S, \mathcal{L} \cup \mathcal{L}')$ as a candidate solution.
2. *Line operation:* Consider each vertical/horizontal line ℓ with a discretized vertical/horizontal coordinate such that $S \setminus \ell$ has two connected components S_1 and S_2. Let $\mathcal{R}_\ell$ denote the rectangles from $\mathcal{R}$ that are contained in S and that are stabbed by ℓ. For the line ℓ we do the following:
   (a) compute an $O(1)$-approximate solution $\mathcal{L}(\mathcal{R}_\ell)$ for the rectangles in $\mathcal{R}_\ell$ using the polynomial time algorithm in [5].
   (b) produce the candidate solution $\mathcal{L}(\mathcal{R}_\ell) \cup SOL(S_1, \mathcal{L} \cap S_1) \cup SOL(S_2, \mathcal{L} \cap S_2)$.

Note that in the line operation we consider entire lines, not just line segments. We define $SOL(S, \mathcal{L})$ to be the solution of minimum cost among all the candidate solutions produced above and store it in $DP(S, \mathcal{L})$.

We do the operation above for each DP-cell $DP(S, \mathcal{L})$. Finally, we output the solution $SOL([0, n] \times [0, 4n^2], \emptyset)$, i.e., the solution corresponding to the cell $DP([0, n] \times [0, 4n^2], \emptyset)$.

We remark that instead of using the $O(1)$-approximation algorithm in [5] for stabbing the rectangles in $\mathcal{R}_\ell$, one could design an algorithm with a better approximation guarantee, using the fact that all rectangles in $\mathcal{R}_\ell$ are stabbed by the line ℓ. However, for our purposes an $O(1)$-approximate solution is good enough.

2.3 Definition of DP-Decision Tree

We want to show that the DP above computes a $(1 + \varepsilon)$-approximate solution. For this, we define a tree T in which each node corresponds to a cell $DP(S, \mathcal{L})$ of the DP and a corresponding solution $\overline{SOL}(S, \mathcal{L})$ to this cell. The root node of T corresponds to the cell $DP([0, n] \times [0, 4n^2], \emptyset)$. Intuitively, this tree represents doing one of the possible operations above, of the DP in the root problem $DP([0, n] \times [0, 4n^2], \emptyset)$ and recursively one of the possible operations in each resulting DP-cell. The corresponding solutions in the nodes are the solutions obtained by choosing exactly these operations in each DP-cell. Since the DP always picks the solution of minimum total cost this implies that the computed solution has a cost that is at most the cost of the root, $c(\overline{\text{SOL}}([0, n] \times [0, 4n^2], \emptyset))$.

Formally, we require T to satisfy the following properties. We require that a node v is a leaf if and only if for the corresponding DP-cell $\text{DP}(S, \mathcal{L})$ the DP directly defined that $\text{DP}(S, \mathcal{L}) = \emptyset$ because all rectangles in $\mathcal{R}$ that are contained in S are already stabbed by the segments in $\mathcal{L}$. If a node v for a DP-cell $\text{DP}(S, \mathcal{L})$ has one child then we require that we reduce the problem for $\text{DP}(S, \mathcal{L})$ to the child by applying the add operation, i.e., there is a set $\mathcal{L}'$ of horizontal/vertical line segments with discretized coordinates such that $|\mathcal{L}| \cup |\mathcal{L}'| \le 3(1/\varepsilon)^3$, the child node of v corresponds to the cell $\text{DP}(S, \mathcal{L} \cup \mathcal{L}')$, and $\overline{\text{SOL}}(S, \mathcal{L}) = \overline{\text{SOL}}(S, \mathcal{L} \cup \mathcal{L}') \cup \mathcal{L}'$.

Similarly, if a node v has two children then we require that we can reduce the problem of $\text{DP}(S, \mathcal{L})$ to these two children by applying the trivial operation or the line operation. Formally, assume that the child nodes correspond to the subproblems $\text{DP}(S_1, \mathcal{L}_1)$ and $\text{DP}(S_2, \mathcal{L}_2)$. If there is a segment $\ell \in \mathcal{L}$ such that $S_1 \cup S_2 \cup \ell = S$, then the applied operation was a trivial operation, and it must also be true that $\mathcal{L}_1 \cup \mathcal{L}_2 \cup \{\ell\} = \mathcal{L}$ and $\overline{\text{SOL}}(S, \mathcal{L}) = \overline{\text{SOL}}(S_1, S_1 \cap \mathcal{L}) \cup \overline{\text{SOL}}(S_2, S_2 \cap \mathcal{L})$. If no such segment exists, then the applied operation was a line operation on a line along the segment ℓ, such that $S_1 \cup S_2 \cup \ell = S$, $\mathcal{L}_1 \cup \mathcal{L}_2 = \mathcal{L}$, and $\overline{\text{SOL}}(S, \mathcal{L}) = \overline{\text{SOL}}(S_1, S_1 \cap \mathcal{L}) \cup \overline{\text{SOL}}(S_2, S_2 \cap \mathcal{L}) \cup \mathcal{L}(\mathcal{R}_\ell)$; where $\mathcal{L}(\mathcal{R}_\ell)$ is a $O(1)$-approximate solution for the set of segments stabbing the set of rectangles intersected by ℓ.

We call a tree T with these properties a *DP-decision-tree*. If there exists a DP-decision-tree with cost $(1+\varepsilon)\text{OPT}$, then our DP computes a solution with at most this cost since the choices in each node of the DP-decision-tree are possible choices of the DP in each node, and in each node the DP makes the choice that minimizes the overall costs.

Lemma 2. *If there is a DP-decision-tree T' for which $c(\overline{\text{SOL}}([0, n] \times [0, 4n^2], \emptyset)) \le (1 + \varepsilon)\text{OPT}$ then the DP is a $(1 + \varepsilon)$-approximation algorithm with a running time of $(n/\varepsilon)^{O(1/\varepsilon^3)}$.*

We define now a DP-decision-tree for which $c(\overline{\text{SOL}}(S, \mathcal{L})) \le (1+\varepsilon)\text{OPT}$. Assume w.l.o.g. that $1/\varepsilon \in \mathbb{N}$. We start by defining a hierarchical grid of vertical lines. Let $a \in \mathbb{N}_0$ be a random offset to be defined later. The grid lines have levels. For each level $j \in \mathbb{N}_0$, there is a grid line $\{a + k \cdot \varepsilon^{j-2}\} \times \mathbb{R}$ for each $k \in \mathbb{N}$. Note that for each $j \in \mathbb{N}_0$ each grid line of level j is also a grid line of level $j + 1$. Also note that any two consecutive lines of some level j are exactly ε^{j-2} units apart.

We say that a line segment $\ell \in \text{OPT}$ is *of level j* if the length of ℓ is in $(\varepsilon^j, \varepsilon^{j-1}]$ (Note that we can have vertical segments which are longer than $1/\varepsilon$, we consider these also to be of level 0). We say that a *horizontal* line segment of some level j is *well-aligned* if both its left and its right x-coordinates lie on a grid line of level $j + 3$, i.e., if both of its x-coordinates are of the form $a + k \cdot \varepsilon^{j+1}$. We say that a *vertical* line segment of some level j is *well-aligned* if both its top and bottom y-coordinates are integral multiples of ε^{j+1}. This would be similar to the segment's end points lying on an (imaginary) horizontal grid line of level $j + 3$. In order to make a line segment from OPT well-aligned, it suffices to extend it by a factor $1 + O(\varepsilon)$, which hence increases the cost by at most this factor.

Lemma 3. *By losing a factor* $1 + O(\varepsilon)$, *we can assume that each line segment* $\ell \in$ OPT *is well-aligned.*

We define the tree T by defining recursively one of the possible operations (trivial operation, add operation, line operation) for each node v of the tree. After applying an operation, we always add children to the processed node v that corresponds to the subproblems that we reduce to, i.e., for a node v corresponding to the subproblem $DP(S, \mathcal{L})$, if we are applying the trivial (resp. line) operation along a segment (resp. line) ℓ, then we add children corresponding to the DP subproblems $DP(S_1, S_1 \cap \mathcal{L})$ and $DP(S_2, S_2 \cap \mathcal{L})$, where S_1 and S_2 are the connected components of $S \backslash \ell$. Similarly if we apply the add operation on v with the set of segments $\mathcal{L}'$ then we add the child node corresponding to the subproblem $DP(S, \mathcal{L} \cup \mathcal{L}')$.

First level. We start with the root $DP([0, n] \times [0, 4n^2], \emptyset)$. We apply the line operation for each vertical line that corresponds to a (vertical) grid line of level 0. Consider one of the resulting subproblems $DP(S, \emptyset)$. Suppose that there are more than ε^{-3} line segments (horizontal or vertical) from OPT of level 0 inside S. We want to partition S into smaller rectangles, such that within each of these rectangles S' at most $O(\varepsilon^{-3})$ of these level 0 line segments start or end. This will make it possible for us to guess them. To this end, we consider the line segments from OPT of level 0 inside S, take their endpoints and order these endpoints non-decreasingly by their y-coordinates. Let $p_1, p_2, ..., p_k$ be these points in this order. For each $k' \in \mathbb{N}$ with $k'/\varepsilon^3 \leq k$, we consider the point p_{k'/ε^3}. Let ℓ' be the horizontal line that contains p_{k'/ε^3}. We apply the line operation to ℓ'.

Lemma 4. *Let* $DP(S', \emptyset)$ *be one of the subproblems after applying the operations above. There are at most* ε^{-3} *line segments* $\mathcal{L}'$ *(horizontal or vertical) from* OPT *of level 0 that have an endpoint inside* S'.

In each resulting subproblem $DP(S', \emptyset)$, for each vertical line segment $\ell \in$ OPT that crosses S', i.e., such that $S' \setminus \ell$ has two connected components, we apply the line operation for the line that contains ℓ. In each subproblem $DP(S'', \emptyset)$ obtained after this step, we apply the add operation to the line segments from OPT of level 0 that intersects S'' (or to be more precise, their intersection with S''), i.e., to the set $\mathcal{L}' := \{\ell \cap S'' \mid \ell \in \text{OPT} \wedge \ell \cap S'' \neq \emptyset \wedge \ell \text{ is of level } 0\}$. Claim 4 implies that $|\mathcal{L}'| \leq \varepsilon^{-3}$. In each obtained subproblem, we apply the trivial operation until it is no longer applicable. We say that all these operations correspond to level 0.

Subsequent levels. Next, we do a sequence of operations that correspond to levels $j = 1, 2, 3, ...$ Assume by induction that for some j each leaf in the current tree

T corresponds to a subproblem $\mathrm{DP}(S, \mathcal{L})$ such that $\ell \cap S \in \mathcal{L}$ for each line segment $\ell \in \mathrm{OPT}$ of each level $j' < j$ for which $\ell \cap S \neq \emptyset$. Take one of these leaves and assume that it corresponds to a subproblem $\mathrm{DP}(S, \mathcal{L})$. We apply the line operation for each vertical line that corresponds to a (vertical) grid line of level j.

Consider a corresponding subproblem $\mathrm{DP}(S', \mathcal{L})$. Suppose that there more than ε^{-3} line segments (horizontal or vertical) from OPT of level j that have an endpoint inside S'. Like above, we consider these endpoints and we order them non-decreasingly by their y-coordinates. Let $p_1, p_2, ..., p_k$ be these points in this order. For each $k' \in \mathbb{N}$ with $k'/\varepsilon^3 \leq k$, we consider the point p_{k'/ε^3} and apply the line operation for the horizontal line ℓ' that contains p_{k'/ε^3}. If for a resulting subproblem $\mathrm{DP}(S'', \mathcal{L})$ there is a vertical line segment $\ell \in \mathcal{L}$ of some level $j' < j - 2$ with an endpoint p inside S'', then we apply the line operation for the horizontal line that contains p.

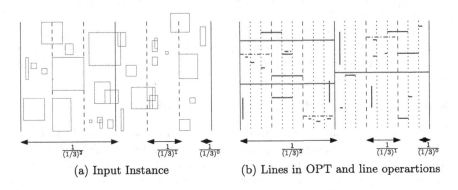

(a) Input Instance   (b) Lines in OPT and line operartions

Fig. 2. Horizontal line operations

Lemma 5. *Let* $\mathrm{DP}(S', \mathcal{L})$ *be one of the subproblems after applying the operations above. There are at most* ε^{-3} *line segments* $\mathcal{L}'$ *(horizontal or vertical) from* OPT *of level* j *that have an endpoint inside* S'.

Consider a resulting subproblem $\mathrm{DP}(S'', \mathcal{L})$. For each line segment $\ell \in \mathrm{OPT}$ such that ℓ crosses S'', i.e., $S'' \setminus \ell$ has two connected components, we apply the line operation to the line that contains ℓ. We apply the trivial operation until it is no longer applicable. In each subproblem $\mathrm{DP}(S''', \mathcal{L})$ obtained after this step, we apply the add operation to the line segments of level j that have an endpoint in S''', i.e., to the set $\mathcal{L}' := \{\ell \cap S''' \mid \ell \in \mathrm{OPT} \land \ell \cap S''' \neq \emptyset \land \ell \text{ is of level } j\}$.

As an example, look at Fig. 2, with $\varepsilon = 1/3$. The solid lines in it are of level 0, the dashed lines of level 1, and dotted lines of level 2. Also the black lines are vertical grid lines, the blue lines are (well-aligned) lines in OPT and the red lines are lines along which horizontal line operations are applied. It can be seen from this example that a segment of level j, by virtue of it being well-aligned, will get removed by a trivial operation of level less than $j + 3$.

2.4   Analysis of DP-Decision Tree

We want to prove that the resulting tree T is indeed a DP-decision-tree corresponding to a solution of cost at most $(1 + \varepsilon)$OPT. To this end, first we need to show that whenever we apply the add operation to a subproblem DP$(S, \mathcal{L})$ for a set $\mathcal{L}'$ then $|\mathcal{L}| + |\mathcal{L}'| \leq 3\varepsilon^{-3}$. The *key insight* for this is that if we added a line segment $\ell \in$ OPT of some level j, then it will not be included in the respective set $\mathcal{L}$ of later subproblems of level $j + 3$ or higher since ℓ is well-aligned. More precisely, if ℓ is horizontal then its x-coordinates are aligned with the grid lines of level $j + 3$. Hence, if ℓ or a part of ℓ is contained in a set $\mathcal{L}$ of some subproblem DP$(S, \mathcal{L})$ for some level $j + 3$, then we applied the trivial operation to ℓ and thus ℓ "disappeared" from $\mathcal{L}$ (note that here by disappear we mean that the segment does not need to be considered in $\mathcal{L}$ anymore, and gets added to the solution of the DP subproblem). If ℓ is vertical and it appears in a DP$(S, \mathcal{L})$ for some level $j + 3$ then we applied the line operation to the horizontal lines that contain the two endpoints of ℓ. Afterwards, we applied the trivial operation to ℓ until ℓ "disappeared" from $\mathcal{L}$.

In particular, for each subproblem DP$(S, \mathcal{L})$ constructed by operations of level j, the set $\mathcal{L}$ can contain line segments of levels $j - 2, j - 1$, and j; but no line segments of a level j' with $j' < j - 2$. Using this, we prove the following lemma.

Lemma 6. *The constructed tree T is a DP-decision-tree.*

We want to show that the cost of the solution corresponding to T is at most $(1 + O(\varepsilon))$OPT. In fact, depending on the offset a this might or might not be true. However, we show that there is a choice for a such that this is true (in fact, we will show that for a random choice for a the cost will be at most $(1 + O(\varepsilon))$OPT in expectation). Intuitively, when we apply the line operation to a vertical grid line ℓ of some level j then the incurred cost is at most $O(1)$ times the cost of the line segments from OPT of level j or larger that stab at least one rectangle intersected by ℓ. A line segment $\ell' \in$ OPT of level j stabs such a rectangle only if ℓ' is intersected by ℓ (if ℓ' is horizontal) or the x-coordinate of ℓ' is close to ℓ (if ℓ' is vertical). Here we use that $h_i \geq w_i$ for each rectangle $R_i \in$ OPT.

Thus, we want to bound the total cost over all levels j of the line segments from OPT that are in level j and that are intersected or close to grid lines of level j or smaller. We will show that if we choose a randomly then the total cost of such grid lines is at most $\varepsilon \cdot$ OPT in expectation. Hence, by using the constant approximation algorithm from [5] in expectation the total cost due to all line operations for vertical line segments is at most $O(\varepsilon) \cdot$ OPT.

When we apply the line operation for a horizontal line, then the cost of stabbing the corresponding rectangles is at most the width of the rectangle S of the current subproblem DP$(S, \mathcal{L})$. We will charge this cost to the line segments of OPT inside S of the current level or higher levels. We will argue that we can charge each such line operation to line segments from OPT whose total width is at least $1/\varepsilon$ times the width of S. This costs another $O(\varepsilon) \cdot$ OPT in total due to all applications of line operations for horizontal segments.

The add operation yields a cost of exactly OPT and the trivial operation does not cost anything. This yields a total cost of $(1 + O(\varepsilon))$OPT.

Lemma 7. *There is a choice for the offset a such that the solution $\overline{\mathrm{SOL}}([0,n] \times [0, 4n^2], \emptyset)$ in T has a cost of at most $(1 + O(\varepsilon))$OPT.*

Proof. In the tree as defined above, the add operations are only applied on segments from OPT, and hence the cost across all such add operations is at most c(OPT). Similarly, the trivial operations are applied on segments which were 'added' before, and hence their cost is also already accounted for. So we are left with analyzing the cost of stabbing the rectangles which are intersected by the lines along which we apply the line operations. We claim that for a random offset a, this cost is $O(\varepsilon \cdot \mathrm{OPT})$, which gives us the required result.

Let us first consider any line operation of level j that is applied to a horizontal line ℓ. This operation would create 2 cells of width at most ε^{j-2}, one of which either contains ε^{-3} endpoints of segments (horizontal or vertical) from OPT of level j; or contains at least one vertical segment from OPT of level $j' < j - 2$, i.e., the cost of the segments from OPT with at least one endpoint in this cell is at least $\varepsilon^{-3} \cdot \varepsilon^j = \varepsilon^{j-3}$. Since a segment of width ε^{j-2} (width of cell) is sufficient to stab all rectangles stabbed by ℓ, we see that this horizontal line takes only ε times the cost of the segments in OPT with at least one endpoint in the cell. We charge the cost of this horizontal segment to these corresponding endpoints. Since each such segment in OPT of level j can be charged at most twice, by summing over all horizontal line operations over all levels we get that the cost of such line operations is at most $2\varepsilon \cdot \mathrm{OPT}$.

Now, let us consider the line operations applied to vertical grid lines. We wish to bound the cost of stabbing all the rectangles intersected or *close to* grid lines (will be formally defined shortly), over all levels j. This can also be stated as bounding the cost, over all levels j, of line segments in level j of OPT (call this set OPT_j) intersected or close to grid lines of level j or smaller. For a horizontal segment $\ell \in \mathrm{OPT}_j$, let I_ℓ be the indicator variable representing the event that a grid line of level j or smaller intersects ℓ ($I_\ell = 0$ for vertical segments). Since $|\ell| \leq \varepsilon^{j-1}$, if we take a random offset a, we obtain that $\mathbb{E}[I_\ell] \leq \varepsilon^{j-1}/\varepsilon^{j-2} = \varepsilon$. For a vertical segment $\ell \in \mathrm{OPT}_j$, let J_ℓ be the indicator variable representing the event that a grid line of level j or smaller intersect the rectangle stabbed by ℓ ($J_\ell = 0$ for horizontal segments). Since for $j > 0$, $|\ell| \leq \varepsilon^{j-1}$, we know that for the rectangle stabbed by ℓ, the dimensions satisfy $w_i \leq h_i \leq \varepsilon^{j-1}$. This means that to stab such a rectangle, ℓ has to lie *close to*, i.e., within $\pm\varepsilon^{j-1}$ of the vertical grid line. So for a random offset a and level $j > 0$ we obtain that $\mathbb{E}[J_\ell] \leq 2\varepsilon^{j-1}/\varepsilon^{j-2} = 2\varepsilon$. For level 0, we note that even though the vertical segments can be very long, the maximum width of a rectangle is at most 1. So ℓ has to lie within ± 1 of the grid line, giving us: $\mathbb{E}[J_\ell] \leq 2/\varepsilon^{-2} = 2\varepsilon^2 \leq 2\varepsilon$. With the expectations computed above, we can upper bound the expected cost of segments in OPT intersected by vertical line operations as:

$$\mathbb{E}\left[\sum_j \sum_{\ell \in \mathrm{OPT}_j} (I_\ell + J_\ell) \cdot |\ell|\right] = \sum_j \sum_{\ell \in \mathrm{OPT}_j} |\ell| \cdot (\mathbb{E}[I_\ell] + \mathbb{E}[J_\ell])$$

$$\leq \sum_j \sum_{\ell \in \mathrm{OPT}_j} |\ell| \cdot (\varepsilon + 2\varepsilon) = 3\varepsilon \cdot \mathrm{OPT}$$

Now, by using the α-approximation algorithm for stabbing from [5], where α is a constant, the solution returned by our algorithm takes an additional cost of $3\alpha \cdot \varepsilon \cdot \mathrm{OPT}$.   □

This gives our main theorem.

Theorem 1. *There is a $(1+\varepsilon)$-approximation algorithm for the general rectangle stabbing problem with a running time of $(n/\varepsilon)^{O(1/\varepsilon^3)}$, assuming that $h_i \geq w_i$ for each rectangle $R_i \in \mathcal{R}$.*

Theorem 1 has some direct implications. First, it yields a PTAS for the general square stabbing problem.

Corollary 1. *There is a PTAS for the general square stabbing problem.*

Also, it yields a $(2 + \varepsilon)$-approximation algorithm for the general rectangle stabbing problem for arbitrary rectangles: we can simply split the input into rectangles R_i for which $h_i \geq w_i$ holds, and those for which $h_i < w_i$ holds, and output the union of these two solutions.

Corollary 2. *There is a $(2 + \varepsilon)$-approximation algorithm for the general rectangle stabbing problem with a running time of $(n/\varepsilon)^{O(1/\varepsilon^3)}$.*

Finally, it yields a PTAS for the horizontal rectangle stabbing problem: we can take the input of that problem and stretch all input rectangles vertically such that it is always very costly to stab any rectangle vertically (so in particular our $(1+\varepsilon)$-approximate solution would never do this). Then we apply the algorithm due to Theorem 1.

Corollary 3. *There is a $(1+\varepsilon)$-approximation algorithm for the horizontal rectangle stabbing problem with a running time of $(n/\varepsilon)^{O(1/\varepsilon^3)}$.*

3 δ-Large Rectangles

We now consider the case of δ-large rectangles for some given constant δ, i.e., where for each input rectangle R_i we assume that $w_i \leq 1$ and $h_i \leq 1$ and additionally $w_i \geq \delta$ or $h_i \geq \delta$. For this case we again give a PTAS in which we use our algorithm due to Theorem 1 as a subroutine.

First, by losing only a factor of $1+\varepsilon$, we divide the instance into independent subproblems which are disjoint rectangular cells. For each cell C_i, we denote by $\mathrm{OPT}(C_i)$ the cells from OPT that are contained in C_i and our routine ensures that $c(\mathrm{OPT}(C_i)) \leq O(1/\varepsilon^3)$. Then for each cell C_i, the number of segments in

$\mathrm{OPT}(C_i)$ with length longer than δ is bounded by $O(1/\delta\varepsilon^3)$. We guess them in polynomial time. Now, the remaining segments in OPT are all of length smaller than δ, and hence they can stab a rectangle only along its shorter dimension. Hence, we can divide the remaining rectangles into two disjoint sets, one with $h_i \geq w_i$ and the other with $w_i > h_i$, and use Theorem 1 to get a $1 + \varepsilon$ approximation of the remaining problem.

Theorem 2. *For* HV-STABBING *with δ-large rectangles, there is a $(1 + \varepsilon)$-approximation algorithm with a running time of $(n/\varepsilon)^{O(1/\delta\varepsilon^3)}$.*

4 Conclusion

In this paper, we have settled the STABBING problem by giving a PTAS for it, and also give a $(2 + \varepsilon)$-approximate solution for the HV-STABBING problem and PTASs for some related special cases of these problems. It is not immediately clear whether these techniques could be extended to obtain a PTAS for the HV-STABBING problem, or even if such a PTAS exists or not, since the question of the APX-hardness of HV-STABBING is still open.

Acknowledgement. Arindam Khan was partly supported by Pratiksha Trust Young Investigator Award, Google India Research Award, and Google ExploreCS Award. Andreas Wiese was partially supported by the FONDECYT Regular grant 1200173.

References

1. Adamaszek, A., Wiese, A.: Approximation schemes for maximum weight independent set of rectangles. In: FOCS, pp. 400–409 (2013)
2. Aronov, B., Ezra, E., Sharir, M.: Small-size \eps-nets for axis-parallel rectangles and boxes. SIAM J. Comput. **39**(7), 3248–3282 (2010)
3. Becchetti, L., Marchetti-Spaccamela, A., Vitaletti, A., Korteweg, P., Skutella, M., Stougie, L.: Latency-constrained aggregation in sensor networks. ACM Trans. Algorithms **6**(1), 1–20 (2009)
4. Brönnimann, H., Goodrich, M.T.: Almost optimal set covers in finite VC-dimension. Discret. Comput. Geom. **14**(4), 463–479 (1995). https://doi.org/10.1007/BF02570718
5. Chan, T.M., van Dijk, T.C., Fleszar, K., Spoerhase, J., Wolff, A.: Stabbing rectangles by line segments - how decomposition reduces the shallow-cell complexity. In: Hsu, W., Lee, D., Liao, C. (eds.) ISAAC, vol. 123, pp. 61:1–61:13 (2018)
6. Chan, T.M., Grant, E.: Exact algorithms and APX-hardness results for geometric packing and covering problems. Comput. Geom. **47**(2), 112–124 (2014). https://doi.org/10.1016/j.comgeo.2012.04.001
7. Chan, T.M., Grant, E., Könemann, J., Sharpe, M.: Weighted capacitated, priority, and geometric set cover via improved quasi-uniform sampling. In: Proceedings of the Twenty-Third Annual ACM-SIAM Symposium on Discrete Algorithms, pp. 1576–1585. SIAM (2012)
8. Chan, T.M., Grant, E., Könemann, J., Sharpe, M.: Weighted capacitated, priority, and geometric set cover via improved quasi-uniform sampling. In: Rabani, Y. (ed.) SODA, pp. 1576–1585 (2012)

9. Chvatal, V.: A greedy heuristic for the set-covering problem. Math. Oper. Res. **4**(3), 233–235 (1979)
10. Das, A., Fleszar, K., Kobourov, S., Spoerhase, J., Veeramoni, S., Wolff, A.: Approximating the generalized minimum manhattan network problem. Algorithmica **80**(4), 1170–1190 (2018)
11. Eisenbrand, F., Gallato, M., Svensson, O., Venzin, M.: A QPTAS for stabbing rectangles. CoRR abs/2107.06571 (2021). https://arxiv.org/abs/2107.06571
12. Finke, G., Jost, V., Queyranne, M., Sebő, A.: Batch processing with interval graph compatibilities between tasks. Discret. Appl. Math. **156**(5), 556–568 (2008)
13. Gaur, D.R., Ibaraki, T., Krishnamurti, R.: Constant ratio approximation algorithms for the rectangle stabbing problem and the rectilinear partitioning problem. J. Algorithms **43**(1), 138–152 (2002)
14. Hochbaum, D.S., Maass, W.: Approximation schemes for covering and packing problems in image processing and VLSI. J. ACM **32**(1), 130–136 (1985)
15. Khan, A., Pittu, M.R.: On guillotine separability of squares and rectangles. In: APPROX/RANDOM. vol. 176, pp. 47:1–47:22 (2020)
16. Khan, A., Subramanian, A., Wiese, A.: A PTAS for the horizontal rectangle stabbing problem. CoRR abs/2111.05197 (2021). https://arxiv.org/abs/2111.05197
17. Kovaleva, S., Spieksma, F.: Approximation algorithms for rectangle stabbing and interval stabbing problems. SIAM J. Discret. Math. **20**(3), 748–768 (2006)
18. Mustafa, N.H., Raman, R., Ray, S.: Quasi-polynomial time approximation scheme for weighted geometric set cover on pseudodisks and halfspaces. SIAM J. Comput. **44**(6), 1650–1669 (2015). https://doi.org/10.1137/14099317X
19. Varadarajan, K.R.: Weighted geometric set cover via quasi-uniform sampling. In: STOC, pp. 641–648 (2010)

Lattice-Free Simplices with Lattice Width $2d - o(d)$

Lukas Mayrhofer, Jamico Schade, and Stefan Weltge[✉]

Technical University of Munich, Munich, Germany
{lukas.mayrhofer,jamico.schade,weltge}@tum.de

Abstract. The Flatness theorem states that the maximum lattice width $\mathrm{Flt}(d)$ of a d-dimensional lattice-free convex set is finite. It is the key ingredient for Lenstra's algorithm for integer programming in fixed dimension, and much work has been done to obtain bounds on $\mathrm{Flt}(d)$. While most results have been concerned with upper bounds, only few techniques are known to obtain lower bounds. In fact, the previously best known lower bound $\mathrm{Flt}(d) \geq 1.138d$ arises from direct sums of a 3-dimensional lattice-free simplex.

In this work, we establish the lower bound $\mathrm{Flt}(d) \geq 2d - O(\sqrt{d})$, attained by a family of lattice-free simplices. Our construction is based on a differential equation that naturally appears in this context.

Additionally, we provide the first local maximizers of the lattice width of 4- and 5-dimensional lattice-free convex bodies.

Keywords: Flatness theorem · Lattice-free · Simplices

1 Introduction

A convex body in $\mathbb{R}^d$ is called *lattice-free* if it does not contain any integer points in its interior. Lattice-free convex bodies appear in many important works concerning the theory of integer programming. They are central objects in cutting plane theory [1,5,8,11,15] and can been used as certificates of optimality in convex integer optimization [4,7,9,10,22,25,26]. Moreover they play a crucial role in Lenstra's algorithm [21] for integer programming in fixed dimension.

A fundamental property of lattice-free convex bodies is that they are "flat" with respect to the integer lattice. The *lattice width* of a convex body $K \subseteq \mathbb{R}^d$ is

$$\mathrm{lw}(K) := \min \left\{ \max_{x \in K} \langle x, y \rangle - \min_{x \in K} \langle x, y \rangle : y \in \mathbb{Z}^d \setminus \{\mathbf{0}\} \right\},$$

where $\langle \cdot, \cdot \rangle$ denotes the standard scalar product. The famous *Flatness theorem*, first proved by Khinchine [20], states that the maximum lattice width $\mathrm{Flt}(d)$ of a d-dimensional lattice-free convex body is finite. It has been applied in several results in mixed-integer programming (e.g., [2,12,14,17,21]) and much work has been done to obtain bounds on $\mathrm{Flt}(d)$. Combining the work of Banaszczyk,

© Springer Nature Switzerland AG 2022
K. Aardal and L. Sanitá (Eds.): IPCO 2022, LNCS 13265, pp. 375–386, 2022.
https://doi.org/10.1007/978-3-031-06901-7_28

Litvak, Pajor, Szarek [6] and Rudelson [23], the currently best upper bound is
$\mathrm{Flt}(d) = O(d^{4/3} \operatorname{polylog}(d))$, and it is conjectured that $\mathrm{Flt}(d) = \Theta(d)$ holds [6].

While most results have been concerned with upper bounds, only few techniques are known to obtain lower bounds. It is easy to see that $\mathrm{Flt}(d) \geq d$ holds by observing that $\Delta := \operatorname{conv}\{\mathbf{0}, d \cdot \mathbf{e}_1, \ldots, d \cdot \mathbf{e}_d\}$ is lattice-free and satisfies $\mathrm{lw}(\Delta) = d$. This bound has been improved for small dimensions: For $d = 2$, Hurkens [19] proved that

$$\mathrm{Flt}(2) = 1 + 2/\sqrt{3} \approx 1.08d$$

holds, and this is already the last dimension for which $\mathrm{Flt}(d)$ is known. For $d = 3$, Codenotti & Santos [13] recently constructed a 3-dimensional lattice-free simplex of lattice width $2 + \sqrt{2}$, showing that

$$\mathrm{Flt}(3) \geq 2 + \sqrt{2} \approx 1.14d$$

holds. Averkov, Codenotti, Macchia & Santos [3] showed that this simplex is a local maximizer of the lattice width among 3-dimensional lattice-free convex bodies. In fact, Codenotti & Santos conjectured that it is actually a global maximizer, i.e., that the above bound is tight.

It is possible to lift these examples to higher dimensions: Using the notion of a direct sum, one can show that $\mathrm{Flt}(d_1 + d_2) \geq \mathrm{Flt}(d_1) + \mathrm{Flt}(d_2)$ holds for all positive integers d_1, d_2, see [13, Prop. 1.4]. However, prior to this work, d-dimensional lattice-free convex bodies with lattice width strictly larger than $1.14d$ were not known.

In this work, we establish new lower bounds on $\mathrm{Flt}(d)$. In terms of small dimensions, we obtain the first local maximizers of the lattice width among d-dimensional lattice-free convex bodies for $d = 4$ and $d = 5$, implying

$$\mathrm{Flt}(4) \geq 2 + 2\sqrt{1 + 2/\sqrt{5}} \approx 1.19d \quad \text{and} \quad \mathrm{Flt}(5) \geq 5 + \frac{2}{\sqrt{3}} \approx 1.23d.$$

Unfortunately, it becomes inherently more difficult to obtain local maximizers of the lattice width in larger dimensions. However, we show that for each d, it is still possible to construct lattice-free simplices with strictly increasing ratios of lattice width to dimension. Our main result is the following:

Theorem 1. *There exist d-dimensional lattice-free simplices $(\Delta_d)_{d \geq 2}$ with lattice width $2d - O(\sqrt{d})$.*

In particular, we see that

$$\mathrm{Flt}(d) \geq 2d - O(\sqrt{d})$$

holds.

In contrast to the previously mentioned lower bounds, Theorem 1 is not based on lifting small-dimensional examples to higher dimensions, but yields explicit constructions of lattice-free simplices for all dimensions. An interesting aspect of our construction is that it arises by considering the problem from an infinite-dimensional point of view. In fact, the facet-defining normal vectors of our simplices are discretizations of the solution to a differential equation that naturally appears within our approach (see Sect. 3).

Our paper is organized as follows. In Sect. 2, we describe the simplices mentioned in Theorem 1, prove that they are lattice-free, and determine their lattice widths. The background on how our construction was obtained is provided in Sect. 3. In Sect. 4, we present the local maximizers of the lattice width in dimensions 4 and 5. We conclude with some open questions in Sect. 5 and comment on difficulties for obtaining local maximizers in dimensions $d \geq 6$.

2   Lattice-Free Simplices of Large Lattice Width

In this part, we will present d-dimensional lattice-free simplices with lattice width $2d - O(\sqrt{d})$. Our construction yields highly symmetric simplices, at least when viewed as subsets of

$$H := \{x \in \mathbb{R}^{d+1} : \langle \mathbf{1}, x \rangle = 1\}.$$

In fact, we will consider a simplex $\Delta \subseteq H$ and see that its projection $\pi(\Delta) \subseteq \mathbb{R}^d$ onto the first d coordinates has the desired properties. The simplex is given by

$$\Delta := \left\{x \in H : \langle \overrightarrow{a}^i, x \rangle \leq a_{d+1} \text{ for } i = 0, 1, \ldots, d \right\},$$

where $a \in \mathbb{R}^{d+1}$ is defined via

$$a_i := \delta^{i-1} - 1$$

for every $i \in [d + 1] := \{1, \ldots, d + 1\}$ with

$$\delta := \left(1 - \sqrt{\frac{2}{d+1}}\right)^{-1},$$

and $\overrightarrow{a}^i$ arises from a by cyclically shifting all entries i positions to the right. During this section, we will prove the following statements.

(A) Δ is a d-dimensional simplex.
(B) Δ does not contain an integer point in its relative interior.
(C) For every $c \in \mathbb{Z}^{d+1} \setminus \{\lambda \cdot \mathbf{1} : \lambda \in \mathbb{R}\}$ we have

$$\max_{x \in \Delta}\langle c, x \rangle - \min_{x \in \Delta}\langle c, x \rangle \geq \frac{d \cdot \delta^{d+2} - \delta^{d+1} - (d+1) \cdot \delta^d + \delta + 1}{(\delta - 1) \cdot (\delta^{d+1} - 1)}. \tag{1}$$

Let us first show that these claims indeed imply our main result.

Proof of Theorem 1. Since Δ is a d-dimensional simplex by (A), the same holds for $\pi(\Delta)$. Suppose that $\pi(\Delta)$ is not lattice-free, i.e., there exists a point $x' \in \mathbb{Z}^d$ in the interior of $\pi(\Delta)$. The integer point $x = (x_1', \ldots, x_d', 1 - x_1' - \cdots - x_d')$ is then contained in the relative interior of Δ, a contradiction to (B).

To obtain a lower bound on the lattice width of $\pi(\Delta)$, let α denote the right-hand side of (1). Consider any $c' \in \mathbb{Z}^d \setminus \{0\}$ and note that $c := (c_1', \ldots, c_d', 0)$ satisfies $c \in \mathbb{Z}^{d+1} \setminus \{\lambda \cdot \mathbf{1} : \lambda \in \mathbb{R}\}$. Thus, by (C) we obtain

$$\max_{x \in \pi(\Delta)} \langle c', x \rangle - \min_{x \in \pi(\Delta)} \langle c', x \rangle = \max_{x \in \Delta} \langle c, x \rangle - \min_{x \in \Delta} \langle c, x \rangle \geq \alpha.$$

In particular, we see that $\mathrm{lw}(\pi(\Delta)) \geq \alpha$ holds. Note that

$$\alpha > \frac{d \cdot \delta^{d+2} - \delta^{d+1} - (d+1) \cdot \delta^d}{(\delta - 1) \cdot \delta^{d+1}} = \frac{d \cdot \delta^2 - \delta - (d+1)}{(\delta - 1) \cdot \delta}$$

holds, where the inequality follows from $\delta > 1$. Substituting $k = \sqrt{(d+1)/2}$, we get $d = 2k^2 - 1$ and $\delta = \frac{k}{k-1}$, and hence

$$\alpha > \frac{(2k^2 - 1) \cdot \left(\frac{k}{k-1}\right)^2 - \frac{k}{k-1} - 2k^2}{\frac{1}{k-1} \cdot \frac{k}{k-1}}$$

$$= 2(2k^2 - 1) - 4k + 3 = 2d - 4k + 3 = 2d - \sqrt{8d + 8} + 3.$$

$\square$

In the proofs of (A) and (B) we will make use of the following auxiliary facts.

Lemma 1. *For every $x \in \mathbb{R}^{d+1}$ with $\langle \mathbf{1}, x \rangle \geq 0$ there exists some ℓ such that $\overrightarrow{x}_1^\ell + \cdots + \overrightarrow{x}_j^\ell \leq \langle \mathbf{1}, x \rangle$ holds for all $j \in [d]$. Moreover, if $\langle \mathbf{1}, x \rangle > 0$, then each of these inequalities is strict.*

Proof. For each $j \in [d+1]$ we define $S_j = x_1 + \cdots + x_j$. Note that $S_{d+1} = \langle \mathbf{1}, x \rangle$. Let $k \in [d+1]$ denote the smallest index such that $S_k = \max\{S_1, \ldots, S_{d+1}\}$ holds. We will show that $\ell := d + 1 - k$ satisfies the claim.

Recall that $\overrightarrow{x}^\ell = (x_{k+1}, x_{k+2}, \ldots, x_{d+1}, x_1, \ldots, x_k)$ and let $j \in [d]$. If $j \leq d + 1 - k$, then we have

$$\overrightarrow{x}_1^\ell + \cdots + \overrightarrow{x}_j^\ell = x_{k+1} + \cdots + x_{k+j} = S_{k+j} - S_k \leq 0 \leq \langle \mathbf{1}, x \rangle,$$

where the inequality $S_{k+j} - S_k \leq 0$ holds due to the choice of k. If $j \geq d + 2 - k$, then

$$\overrightarrow{x}_1^\ell + \cdots + \overrightarrow{x}_j^\ell = x_{k+1} + \cdots + x_{d+1} + x_1 + \cdots + x_{j+k-(d+1)}$$

$$= S_{d+1} - S_k + S_{j+k-(d+1)}.$$

Since $j \leq d$, we have $j + k - (d+1) < k$ and hence by the choice of k we see that $S_{j+k-(d+1)} < S_k$ holds, which yields $\overrightarrow{x}_1^\ell + \cdots + \overrightarrow{x}_j^\ell < S_{d+1} = \langle \mathbf{1}, x \rangle$. $\square$

Lemma 2. *For every $x \in \mathbb{R}^{d+1} \setminus \{0\}$ with $\langle 1, x \rangle = 0$ there exists some i such that $\langle \overrightarrow{a}^i, x \rangle > 0$.*

Proof. Let ℓ be as in Lemma 1. Defining $y := \overrightarrow{x}^\ell$ we have that $S_j := y_1 + \cdots + y_j \leq 0$ holds for all $j \in [d]$. Recall that $S_{d+1} := y_1 + \cdots + y_{d+1} = 0$ and that $y \neq 0$. Thus, we must have $S_{j^*} < 0$ for some $j^* \in [d]$. We obtain

$$\begin{aligned}
\langle a, y \rangle &= a_1 y_1 + a_2 y_2 + a_3 y_3 + \cdots + a_{d+1} y_{d+1} \\
&= a_1 S_1 + a_2 (S_2 - S_1) + a_3 (S_3 - S_2) + \cdots + a_{d+1}(S_{d+1} - S_d) \\
&= S_1(a_1 - a_2) + S_2(a_2 - a_3) + \cdots + S_d(a_d - a_{d+1}) + a_{d+1} S_{d+1} \\
&\geq S_{j^*}(a_{j^*} - a_{j^*+1}) + a_{d+1} S_{d+1} \\
&= S_{j^*}(a_{j^*} - a_{j^*+1}) \\
&> 0,
\end{aligned}$$

where we used the fact that $a_j < a_{j+1}$ holds for every $j \in [d]$. The claim follows since $\langle \overrightarrow{a}^{d+1-\ell}, x \rangle = \langle a, \overrightarrow{x}^\ell \rangle = \langle a, y \rangle$. □

Lemma 3. *For every $x \in H \cap \mathbb{Z}^{d+1}$ there exists some i such that $\langle \overrightarrow{a}^i, x \rangle \geq a_{d+1}$. In particular, this implies that Δ does not contain an integer point in its relative interior.*

Proof. We proceed similarly to the proof of Lemma 2. We pick ℓ as in Lemma 1 and define $y := \overrightarrow{x}^\ell$. Note that $S_j := y_1 + \cdots + y_j < 1$ holds for all $j \in [d]$. Since y is integer, we even have $S_j \leq 0$ for all $j \in [d]$. Setting $S_{d+1} := y_1 + \cdots + y_{d+1} = 1$ we hence obtain

$$\begin{aligned}
\langle a, y \rangle &= a_1 y_1 + a_2 y_2 + a_3 y_3 + \cdots + a_{d+1} y_{d+1} \\
&= a_1 S_1 + a_2 (S_2 - S_1) + a_3 (S_3 - S_2) + \cdots + a_{d+1}(S_{d+1} - S_d) \\
&= S_1(a_1 - a_2) + S_2(a_2 - a_3) + \cdots + S_d(a_d - a_{d+1}) + a_{d+1} S_{d+1} \\
&\geq a_{d+1} S_{d+1} = a_{d+1},
\end{aligned}$$

where the inequality holds since we have $a_j \leq a_{j+1}$ for every $j \in [d]$ and $S_{d+1} = 1$. Again, the claim follows since $\langle \overrightarrow{a}^{d+1-\ell}, x \rangle = \langle a, \overrightarrow{x}^\ell \rangle = \langle a, y \rangle$. □

Lemma 4. *Δ is a d-dimensional simplex.*

Proof. The point $p := \frac{1}{d+1} 1$ is contained in H and satisfies

$$\langle \overrightarrow{a}^i, p \rangle = \frac{a_1 + \cdots + a_{d+1}}{d+1} < a_{d+1}$$

for all i. In particular, we see that Δ contains a ball within H around p, and hence Δ is d-dimensional.

Since Δ is defined by $d + 1$ linear inequalities, it suffices to show that Δ is bounded in order to prove that it is a simplex. The recession cone of Δ is given by

$$\text{recc}(\Delta) = \left\{ x \in \mathbb{R}^{d+1} : \langle 1, x \rangle = 0, \langle \overrightarrow{a}^i, x \rangle \leq 0 \text{ for } i = 0, 1, \ldots, d \right\}.$$

From Lemma 2, it follows directly that $\operatorname{recc}(\varDelta) = \{0\}$, proving that $\varDelta$ is indeed bounded. $\square$

For the proof of (C), let us first determine the vertices of $\varDelta$. To this end, define the vector $v \in \mathbb{R}^{d+1}$ via

$$v_1 = \frac{1 - d \cdot \delta^d + (d-2) \cdot \delta^{d+1} + \delta^{d+2}}{(\delta - 1) \cdot (\delta^{d+1} - 1)},$$

$$v_2 = \cdots = v_d = \frac{(\delta - 1) \cdot \delta^d}{\delta^{d+1} - 1},$$

$$v_{d+1} = \frac{-\delta + \delta^d + (d-1) \cdot \delta^{d+1} - (d-1) \cdot \delta^{d+2}}{(\delta - 1) \cdot (\delta^{d+1} - 1)}.$$

Lemma 5. $\varDelta = \operatorname{conv}\left(\left\{\overrightarrow{v}^0, \overrightarrow{v}^1, \ldots, \overrightarrow{v}^{d+1}\right\}\right).$

Proof. For every i, we will show that $\overrightarrow{v}^i$ is contained in H, satisfies all linear inequalities defining $\varDelta$, and all but one even with equality. This shows that each $\overrightarrow{v}^i$ is a vertex of $\varDelta$. Since $\varDelta$ is a simplex and all $\overrightarrow{v}^i$ are distinct, we obtain the claim. To this end, due to the circulant structure of $\varDelta$, we may assume that $\overrightarrow{v}^i = v$ holds. First, observe that

$$(\delta - 1) \cdot (\delta^{d+1} - 1) \cdot \langle \mathbf{1}, v \rangle = (\delta - 1) \cdot (\delta^{d+1} - 1) \cdot (v_1 + (d-1) \cdot v_2 + v_{d+1})$$
$$= \delta^{d+2} - \delta^{d+1} - \delta + 1 = (\delta - 1) \cdot (\delta^{d+1} - 1)$$

holds, which shows that v is contained in H. Next, note that we have

$$\langle \overrightarrow{a}^0, v \rangle = v_2 \cdot \sum_{k=1}^{d+1} a_i + (v_1 - v_2) \cdot a_1 + (v_{d+1} - v_2) \cdot a_{d+1}$$

and

$$\langle \overrightarrow{a}^i, v \rangle = v_2 \cdot \sum_{k=1}^{d+1} a_i + (v_1 - v_2) \cdot a_{d+2-i} + (v_{d+1} - v_2) \cdot a_{d+1-i}$$

for $i \in [d]$. To evaluate these expressions, observe that we have

$$(\delta - 1) \cdot (\delta^{d+1} - 1) \cdot v_2 \cdot \sum_{k=1}^{d+1} a_i = (\delta - 1)^2 \cdot \delta^d \cdot \sum_{k=0}^{d} (\delta^k - 1)$$
$$= (\delta - 1)^2 \cdot \delta^d \cdot \left(\frac{\delta^{d+1} - 1}{\delta - 1} - (d+1) \right)$$
$$= \delta^d \cdot (\delta - 1) \cdot \left((\delta^{d+1} - 1) - (d+1) \cdot (\delta - 1) \right),$$

as well as

$$(\delta - 1) \cdot (\delta^{d+1} - 1) \cdot (v_1 - v_2) = 1 - (d+1) \cdot \delta^d + d \cdot \delta^{d+1}$$

and

$$(\delta - 1) \cdot (\delta^{d+1} - 1) \cdot (v_{d+1} - v_2) = -\delta + (d+1) \cdot \delta^{d+1} - d \cdot \delta^{d+2}.$$

Thus, for $i = 0$ we obtain

$$\underbrace{(\delta - 1) \cdot (\delta^{d+1} - 1)}_{>0} \cdot \langle \overrightarrow{a}^0, v \rangle = (1 - \delta^{d+1}) \cdot \delta \cdot ((d-1) \cdot \delta^d - d \cdot \delta^{d-1} + 1)$$

$$= (1 - \delta^{d+1}) \cdot \delta \cdot (\delta - 1) \cdot \left((d-1) \cdot \delta^{d-1} - \sum_{k=0}^{d-2} \delta^k \right)$$

$$= \underbrace{(1 - \delta^{d+1})}_{<0} \cdot \delta \cdot (\delta - 1) \cdot \underbrace{\sum_{k=0}^{d-2} (\delta^{d-1} - \delta^k)}_{>0},$$

which implies $\langle \overrightarrow{a}^0, v \rangle < 0 < a_{d+1}$. For $i \in [d]$ we see that

$$(\delta - 1) \cdot (\delta^{d+1} - 1) \cdot \langle \overrightarrow{a}^i, v \rangle = (\delta^d - 1) \cdot (\delta - 1) \cdot (\delta^{d+1} - 1)$$

$$= (\delta - 1) \cdot (\delta^{d+1} - 1) \cdot a_{d+1}$$

holds, and hence $\langle \overrightarrow{a}^i, v \rangle = a_{d+1}$. □

Note that, in order to prove (C), it remains to show the following:

Lemma 6. *For every $c \in \mathbb{Z}^{d+1} \setminus \{\lambda \cdot \mathbf{1} : \lambda \in \mathbb{R}\}$ we have*

$$\max_{x \in \Delta} \langle c, x \rangle - \min_{x \in \Delta} \langle c, x \rangle \geq v_1 - v_{d+1}. \tag{2}$$

Proof. Since not all entries of c are equal, there exists some i such that $\overrightarrow{c}_1^i \geq \overrightarrow{c}_j^i$ for all $j \in [d]$ and $\overrightarrow{c}_1^i > \overrightarrow{c}_{d+1}^i$. Due to the circulant symmetry of Δ, we may replace c by $\overrightarrow{c}^i$ without changing the left-hand side in (2). Thus, we may assume that $c = \overrightarrow{c}^i$ holds. Let $j \in [d]$ denote the smallest index such that $c_j > c_{j+1}$ holds. Since c_1 is a maximal entry of c, it is clear that $c_j = c_1$ holds. Moreover, since c is an integer vector, we see that

$$c_1 = c_j \geq c_{j+1} + 1 \quad \text{and} \quad c_j = c_1 \geq c_{d+1} + 1 \tag{3}$$

hold. We obtain

$$\max_{x \in \Delta} \langle c, x \rangle - \min_{x \in \Delta} \langle c, x \rangle$$

$$\geq \langle c, v \rangle - \langle c, \overrightarrow{v}^j \rangle = \langle c, v - \overrightarrow{v}^j \rangle$$

$$= c_1(v_1 - v_2) + c_j(v_2 - v_{d+1}) + c_{j+1}(v_2 - v_1) + c_{d+1}(v_{d+1} - v_2)$$

$$= (c_1 - c_{j+1})(v_1 - v_2) + (c_j - c_{d+1})(v_2 - v_{d+1}).$$

We claim that $v_1 \geq v_2$ and $v_2 \geq v_{d+1}$, which, together with (3), implies

$$\max_{x \in \Delta} \langle c, x \rangle - \min_{x \in \Delta} \langle c, x \rangle \geq (c_1 - c_{j+1})(v_1 - v_2) + (c_j - c_{d+1})(v_2 - v_{d+1})$$

$$\geq (v_1 - v_2) + (v_2 - v_{d+1}) = v_1 - v_{d+1},$$

in which case we are done. To see that $v_1 \geq v_2$ and $v_2 \geq v_{d+1}$ hold, we use

$$\underbrace{(\delta - 1) \cdot (\delta^{d+1} - 1)}_{>0} \cdot (v_1 - v_2) = 1 - d \cdot \delta^d + (d - 2) \cdot \delta^{d+1} + \delta^{d+2} - (\delta - 1)^2 \cdot \delta^d$$

$$= \underbrace{(\delta - 1)}_{>0} \cdot \underbrace{\sum_{k=0}^{d-1} (\delta^d - \delta^k)}_{>0}$$

and $v_2 - v_{d+1} = \delta(v_1 - v_2)$.   □

3   An Infinite-Dimensional View

In this section, we would like to provide some background on the construction of the simplices in the previous section. It is inspired by the lattice-free simplices in [19] and [13], which attain the largest (known) lattice widths in dimensions 2 and 3, respectively. In fact, they can be also described in the form

$$\left\{ x \in H : \langle \vec{a}^i, x \rangle \leq a_{d+1} \text{ for } i = 0, 1, \ldots, d \right\},   (4)$$

for some vector $a \in \mathbb{R}^{d+1}$.

Such sets were also used by Doolittle, Katthän, Nill, Santos [16] in the context of simplices whose integer points coincide with their vertices. Within this setting, constructions of Sebő [24] were already based on highly-symmetric simplices. Herr, Rehn, Schürmann [18] provide some more background on how symmetry interacts with lattice-freeness.

A convenient property of a set as in (4) is that it is lattice-free whenever the entries of a are non-decreasing, see the proof of Lemma 3. Since we are limited to the hyperplane H, we may assume that $a_1 = 0$ and $a_{d+1} = 1$. Every vertex of the above set is the cyclic shift of some vector $v \in \mathbb{R}^{d+1}$ satisfying

$$\begin{bmatrix} 1 & 1 & \cdots & 1 & 1 \\ a_2 & a_3 & \cdots & a_{d+1} & a_1 \\ a_3 & a_4 & \cdots & a_1 & a_2 \\ \vdots & \vdots & & \vdots & \vdots \\ a_{d+1} & a_1 & \cdots & a_{d-1} & a_d \end{bmatrix} v = \mathbf{1}.   (5)$$

In the examples from [13,19], the lattice width is equal to $v_1 - v_{d+1}$, which is why we particularly focus on $\lambda = \frac{v_1 - v_{d+1}}{2d}$ in what follows. (Note that we obtain simplices with λ close to 1.) Let C arise from the above matrix by deleting the first row as well as the first and last columns, and set $w = (v_2, \ldots, v_d)$. The above system is equivalent to

$$
v_1 \begin{bmatrix} a_2 \\ a_3 \\ \vdots \\ a_{d+1} \end{bmatrix} + Cw + v_{d+1} \begin{bmatrix} a_1 \\ a_2 \\ \vdots \\ a_d \end{bmatrix} = 1, \quad v_1 + \langle 1, w \rangle + v_{d+1} = 1.
$$

Substituting $v_1 = \frac{1}{2}(1 - \langle 1, w \rangle) + d\lambda$ and $v_{d+1} = \frac{1}{2}(1 - \langle 1, w \rangle) - d\lambda$, this leads to

$$
\lambda \cdot d \left(\begin{bmatrix} a_2 \\ a_3 \\ \vdots \\ a_{d+1} \end{bmatrix} - \begin{bmatrix} a_1 \\ a_2 \\ \vdots \\ a_d \end{bmatrix} \right) = 1 - Cw - (1 - \langle 1, w \rangle) \frac{1}{2} \left(\begin{bmatrix} a_1 \\ a_2 \\ \vdots \\ a_d \end{bmatrix} + \begin{bmatrix} a_2 \\ a_3 \\ \vdots \\ a_{d+1} \end{bmatrix} \right).
$$

In order to understand which vectors a lead to a large value λ, it is convenient to think of a as (the discretization of) a function $y \colon [0, 1] \to \mathbb{R}$, w as a function $\omega \colon [0, 1] \to \mathbb{R}$, and regard d to be large. We consider the continuous analogue

$$
\lambda \cdot y' = 1 - \mathrm{C}\omega - \left(1 - \int_0^1 \omega(s) \, ds \right) \cdot y. \tag{6}
$$

The matrix C is replaced by the convolution operator C given by

$$
(\mathrm{C}\,\omega)(t) := \int_0^1 y(t + s) \, \omega(s) \, ds,
$$

where we set $y(t) = y(t - 1)$ whenever $1 < t \leq 2$. The boundary conditions $a_1 = 0$ and $a_{d+1} = 1$ are represented by $y(0) = 0$ and $y(1) = 1$. Applying $\int_0^1 \cdot \, dt$ to both sides of (6) yields

$$
\lambda = 1 - \int_0^1 y(t) \, dt \leq 1,
$$

which suggests that the general approach will not yield simplices of lattice width larger than $2d$. Luckily, choosing ω to be a constant function already yields solutions close to that bound: If ω is constant, taking the derivative of both sides in (6), we see that $y'' = \gamma \cdot y'$ holds for some $\gamma \in \mathbb{R}$, and hence y has to be an exponential function. In the previous section we have seen that, choosing the vector a to be a discretization of an exponential function, we indeed obtain simplices of the desired lattice widths.

4   Local Maximizers in Dimensions 4 and 5

Let us mention that it is possible to construct lattice-free simplices with slightly larger lattice widths than the simplices presented in the Sect. 2. In fact, by optimizing the choice of $a_1, \ldots, a_{d+1}$ to maximize $v_1 - v_{d+1}$ in the construction from

Sect. 3 in dimensions 4 and 5, we obtain the following simplices. For dimension $d = 4$, we define the simplex $\Delta_4 := \mathrm{conv}\{\overrightarrow{v}^1, \ldots, \overrightarrow{v}^5\}$, where $v \in \mathbb{R}^5$ with

$$v_1 = \frac{1}{5}\left(7 - 2\sqrt{5} + 2\sqrt{10 + 2\sqrt{5}}\right) \qquad v_4 = v_2$$

$$v_2 = \frac{1}{5}\left(-3 + 4\sqrt{5} - 4\sqrt{5 - 2\sqrt{5}}\right) \qquad v_5 = \frac{1}{5}\left(-3 - 2\sqrt{5} - 2\sqrt{5 + 2\sqrt{5}}\right).$$

$$v_3 = \frac{1}{5}\left(7 - 4\sqrt{5} + 6\sqrt{5 - 2\sqrt{5}}\right)$$

For dimension $d = 5$, we define $\Delta_5 := \mathrm{conv}\{\overrightarrow{v}^1, \ldots, \overrightarrow{v}^6\}$, where $v \in \mathbb{R}^6$ with

$$v_1 = \frac{1}{18}\left(57 - 7\sqrt{3}\right) \qquad\qquad v_4 = v_3$$

$$v_2 = \frac{1}{3}\left(4\sqrt{3} - 5\right) \qquad\qquad v_5 = v_2$$

$$v_3 = \frac{1}{18}\left(27 - 11\sqrt{3}\right) \qquad\qquad v_6 = \frac{1}{18}\left(-33 - 19\sqrt{3}\right).$$

We can prove the following.

Theorem 2. *Let* $\pi_d : \mathbb{R}^{d+1} \to \mathbb{R}^d$ *denote the projection onto the first d coordinates. Then,* $\pi_4(\Delta_4)$ *and* $\pi_5(\Delta_5)$ *are lattice-free simplices with*

$$\mathrm{lw}(\pi_4(\Delta_4)) = 2 + 2\sqrt{1 + \frac{2}{\sqrt{5}}} \qquad \text{and} \qquad \mathrm{lw}(\pi_5(\Delta_5)) = 5 + \frac{2}{\sqrt{3}}.$$

Both of them are local maximizers of lattice width among lattice-free convex bodies in the respective dimensions.

A precise version of the latter statement is the following: For $d = 4$ and $d = 5$ there is some $\varepsilon > 0$ such that every lattice-free convex body $K \subseteq \mathbb{R}^d$ whose Hausdorff distance to $\pi_d(\Delta_d)$ is at most ε satisfies $\mathrm{lw}(K) \leq \mathrm{lw}(\pi_d(\Delta_d))$.

The fact that $\pi_4(\Delta_4)$ and $\pi_5(\Delta_5)$ are lattice-free can be easily confirmed by calculating their inequality descriptions and using Lemma 3. The lattice widths can be determined using a strategy similar to the proof of Lemma 6. Showing that both simplices are local maximizers can be conducted by directly following all steps used by Averkov, Codenotti, Macchia & Santos in [3] for the three-dimensional case. As in [3], the necessary computations are rather complex but can be verified using a computer algebra system in a straightforward way.

5   Open Questions

We conclude our paper by posing some questions that naturally arise from our result, starting with the following.

Problem 1. Are there any d-dimensional lattice-free convex bodies with lattice width greater than $2d$?

Actually, we do not even know a d-dimensional lattice-free convex body with lattice width equal to $2d$. As indicated in Sect. 3, we cannot exceed this bound with our approach.

Following the previous section, all known (local) maximizers of the lattice width among lattice-free convex bodies for $d \leq 5$ are obtained by maximizing $v_1 - v_{d+1}$ over all vectors $a \in \mathbb{R}^{d+1}$, where v is the unique solution of (5). The question directly arises whether this approach can be used to obtain local maximizers in dimensions $d \geq 6$. Unfortunately, this does not work: While the maximizing vectors a are non-decreasing for $d \leq 5$, and hence result in lattice-free simplices, this is not true anymore for $d \geq 6$.

Problem 2. Determine local maximizers of the lattice width among all lattice-free convex bodies in dimensions $d \geq 6$.

Recall that the known constructions in dimensions $3, 4, 5$ are only known to be *local* maximizers. We still do not know whether any of these simplices is actually a global maximizer.

Problem 3. Do there exist any d-dimensional lattice-free convex bodies whose lattice widths exceed the lattice widths of the known local maximizers for $d = 3, 4, 5$?

Acknowledgements. The authors would like to thank Gennadiy Averkov and Paco Santos for valuable feedback and discussions on earlier stages of this work, and Amitabh Basu for discussions about applications of the Flatness theorem within optimization. This work was supported by the Deutsche Forschungsgemeinschaft (DFG, German Research Foundation), project number 451026932.

References

1. Andersen, K., Louveaux, Q., Weismantel, R., Wolsey, L.A.: Inequalities from two rows of a simplex tableau. In: Fischetti, M., Williamson, D.P. (eds.) IPCO 2007. LNCS, vol. 4513, pp. 1–15. Springer, Heidelberg (2007). https://doi.org/10.1007/978-3-540-72792-7_1
2. Averkov, G., Basu, A., Paat, J.: Approximation of corner polyhedra with families of intersection cuts. SIAM J. Optim. **28**(1), 904–929 (2018)
3. Averkov, G., Codenotti, G., Macchia, A., Santos, F.: A local maximizer for lattice width of 3-dimensional hollow bodies. Discrete Appl. Math. **298**, 129–142 (2021)
4. Baes, M., Oertel, T., Weismantel, R.: Duality for mixed-integer convex minimization. Math. Prog. **158**(1), 547–564 (2015). https://doi.org/10.1007/s10107-015-0917-y
5. Balas, E.: Intersection cuts-a new type of cutting planes for integer programming. Oper. Res. **19**(1), 19–39 (1971)
6. Banaszczyk, W., Litvak, A.E., Pajor, A., Szarek, S.J.: The flatness theorem for nonsymmetric convex bodies via the local theory of banach spaces. Math. oper. res. **24**(3), 728–750 (1999)

7. Basu, A., Conforti, M., Cornuéjols, G., Weismantel, R., Weltge, S.: Optimality certificates for convex minimization and helly numbers. Oper. Res. Lett. **45**(6), 671–674 (2017)

8. Basu, A., Conforti, M., Di Summa, M.: A geometric approach to cut-generating functions. Math. Program. **151**(1), 153–189 (2015). https://doi.org/10.1007/s10107-015-0890-5

9. Blair, C.E., Jeroslow, R.G.: Constructive characterizations of the value-function of a mixed-integer program i. Discrete Appl. Math. **9**(3), 217–233 (1984)

10. Blair, C.E., Jeroslow, R.G.: Constructive characterizations of the value function of a mixed-integer program ii. Discrete Appl. Math. **10**(3), 227–240 (1985)

11. Borozan, V., Cornuéjols, G.: Minimal valid inequalities for integer constraints. Math. Oper. Res. **34**(3), 538–546 (2009)

12. Cevallos, A., Weltge, S., Zenklusen, R.: Lifting linear extension complexity bounds to the mixed-integer setting. In: Proceedings of the Twenty-Ninth Annual ACM-SIAM Symposium on Discrete Algorithms. pp. 788–807. SIAM (2018)

13. Codenotti, G., Santos, F.: Hollow polytopes of large width. Proc. Am. Math. Soc. **148**(2), 835–850 (2020)

14. Dash, S., Dobbs, N.B., Günlük, O., Nowicki, T.J., Świrszcz, G.M.: Lattice-free sets, multi-branch split disjunctions, and mixed-integer programming. Math. Program. **145**(1), 483–508 (2013). https://doi.org/10.1007/s10107-013-0654-z

15. Del Pia, A., Weismantel, R.: Relaxations of mixed integer sets from lattice-free polyhedra. Ann. Oper. Res. **240**(1), 95–117 (2015). https://doi.org/10.1007/s10479-015-2024-0

16. Doolittle, J., Katthän, L., Nill, B., Santos, F.: Empty simplices of large width. arXiv preprint. arXiv:2103.14925 (2021)

17. Fukasawa, R., Poirrier, L., Xavier, Á.S.: The (not so) trivial lifting in two dimensions. Math. Prog. Comp. **11**(2), 211–235 (2018). https://doi.org/10.1007/s12532-018-0146-5

18. Herr, K., Rehn, T., Schürmann, A.: On lattice-free orbit polytopes. Discrete Comput. Geom. **53**(1), 144–172 (2014). https://doi.org/10.1007/s00454-014-9638-x

19. Hurkens, C.: Blowing up convex sets in the plane. Linear Algebra Appl. **134**, 121–128 (1990)

20. Khinchine, A.: A quantitative formulation of kronecker's theory of approximation. Izv. Akad. Nauk SSR Ser. Matem. **12**(2), 113–122 (1948)

21. Lenstra, H.W., Jr.: Integer programming with a fixed number of variables. Math. Oper. Res. **8**(4), 538–548 (1983)

22. Morán, R., Diego, A., Dey, S.S., Vielma, J.P.: A strong dual for conic mixed-integer programs. SIAM J. Optim. **22**(3), 1136–1150 (2012)

23. Rudelson, M.: Distances between non-symmetric convex bodies and the mm^*-estimate. Positivity **24**, 161–178 (2000)

24. Sebő, A.: An introduction to empty lattice simplices. In: Cornuéjols, G., Burkard, R.E., Woeginger, G.J. (eds.) IPCO 1999. LNCS, vol. 1610, pp. 400–414. Springer, Heidelberg (1999). https://doi.org/10.1007/3-540-48777-8_30

25. Wolsey, L.A.: The b-hull of an integer program. Discrete Appl. Math. **3**(3), 193–201 (1981)

26. Wolsey, L.A.: Integer programming duality: price functions and sensitivity analysis. Math. Program. **20**(1), 173–195 (1981). https://doi.org/10.1007/BF01589344

Graph Coloring and Semidefinite Rank

Renee Mirka[(✉)], Devin Smedira, and David P. Williamson

Cornell University, Ithaca, NY 14850, USA
{rem379,dts88,davidpwilliamson}@cornell.edu

Abstract. This paper considers the interplay between semidefinite programming, matrix rank, and graph coloring. Karger, Motwani, and Sudan [10] give a vector program for which a coloring of the graph can be encoded as a semidefinite matrix of low rank. By complementary slackness conditions of semidefinite programming, if an optimal dual solution has sufficiently high rank, any optimal primal solution must have low rank. We attempt to characterize graphs for which we can show that the corresponding dual optimal solution must have sufficiently high rank. In the case of the original Karger, Motwani, and Sudan vector program, we show that any graph which is a k-tree has sufficiently high dual rank, and we can extract the coloring from the corresponding low-rank primal solution. We can also show that if the graph is not uniquely colorable, then no sufficiently high rank dual optimal solution can exist. This allows us to completely characterize the planar graphs for which dual optimal solutions have sufficiently high dual rank, since it is known that the uniquely colorable planar graphs are precisely the planar 3-trees.

We then modify the semidefinite program to have an objective function with costs, and explore when we can create a cost function whose optimal dual solution has sufficiently high rank. We show that it is always possible to construct such a cost function given the graph coloring. The construction of the cost function gives rise to a heuristic for graph coloring which we show works well in the case of planar graphs; we enumerated all maximal planar graphs with a K_4 of up to 14 vertices, and the heuristics successfully colored 99.75% of them.

Our research was motivated by the Colin de Verdière graph invariant [5] (and a corresponding conjecture of Colin de Verdière), in which matrices that have some similarities to the dual feasible matrices must have high rank in the case that graphs are of a certain type; for instance, planar graphs have rank that would imply the 4-colorability of the primal solution. We explore the connection between the conjecture and the rank of the dual solutions.

1 Introduction

Given an undirected graph $G = (V, E)$, a *coloring* of G is an assignment of colors to the vertices V such that for each edge $(i, j) \in E$, i and j receive different colors. The *chromatic number* of G, denoted $\chi(G)$, is the minimum

---

Supported by NSF grant CCF-2007009.

K. Aardal and L. Sanitá (Eds.): IPCO 2022, LNCS 13265, pp. 387–401, 2022.
https://doi.org/10.1007/978-3-031-06901-7_29

number of colors used such that a coloring of G exists. The *clique number* of a graph G, denoted $\omega(G)$, is the size of the largest *clique* in the graph; a set $S \subseteq V$ of vertices is a clique if for every distinct pair $i, j \in S$, $(i, j) \in E$. It is easy to see that $\omega(G) \leq \chi(G)$. Graph colorings have been intensively studied for over a century. One of the most well-known theorems of graph theory, the *four-color theorem*, states that four colors suffice to color any planar graph G; the problem of four-coloring a planar graph can be traced back to the 1850 s, and the computer-assisted proof of the four-color theorem by Appel and Haken [2,3] is considered a landmark in graph theory. See Jensen and Toft [9] and Molloy and Reed [13] for book-length treatments of graph coloring in general. Fritsch and Fritsch [7], Ore [14], and Wilson [17] provide book-length treatments of the four-color theorem in particular, and Robertson, Sanders, Seymour, and Thomas [15] give a simplified computer-assisted proof of the four-color theorem.

This paper considers the use of semidefinite programming in graph coloring. The connection between semidefinite programming and graph coloring was initiated by Lovász [12], who introduced the Lovász theta function, $\theta(\bar{G})$, which is computable via semidefinite programming; $\bar{G}$ is the complement of graph G, in which all edges of G are replaced by nonedges and vice versa. Lovász showed that $\omega(G) \leq \theta(\bar{G}) \leq \chi(G)$; Knuth [11] gives a helpful overview of this result.

Another use of semidefinite programming for graph coloring was introduced by Karger, Motwani, and Sudan [10] (KMS), who showed how to color k-colorable graphs with $O(n^{1-3/(k+1)} \log^{1/2} n)$ colors in polynomial time using semidefinite programming, where n is the number of vertices in the graph. A starting point of the algorithm of KMS is the following vector program, which KMS called the *strict vector chromatic number*; the vector program can be solved via semidefinite programming:

$$\begin{aligned}
& \text{minimize} && \alpha \\
& \text{subject to} && v_i \cdot v_j = \alpha, \ \forall (i, j) \in E, \\
& (SVCN\text{-}P) && v_i \cdot v_i = 1, \ \forall i \in V, \\
& && v_i \in \mathbb{R}^n, \quad \forall i \in V.
\end{aligned}$$

KMS observe that any k-colorable graph has a feasible solution to the vector program with $\alpha = -1/(k-1)$: let $v_1 = (1, 0, \ldots, 0) \in \Re^{k-1}$ and inductively find $v_i \in \Re^{k-1}$ for $1 < i \leq k-1$ by setting $v_i(j) = 0$ for $j > i$ and otherwise solving the system of equations given by $v_l \cdot v_i = -1/(k-1)$ for $1 \leq l \leq i-1$ and $v_i \cdot v_i = 1$. Finally, let $v_k = -\sum_{j=1}^{k-1} v_j$, then assign one color to each of the vectors v_i. This guarantees that each vector v_i has unit length (so $v_i \cdot v_i = 1$) and that for any edge $(i, j) \in E$, $v_i \cdot v_j = -1/(k-1)$. It is important for the following discussion to observe that this solution lies in a $(k-1)$-dimensional space. KMS also observe that there is a natural connection between the strict vector chromatic number and the Lovász theta function. In particular, for the solution α to the vector program above, it is possible to show that $\alpha = 1/(1 - \theta(\bar{G}))$ (see [10, Theorem 8.2]). If the graph G has a k-clique K_k and is k-colorable, then by Lovász's theorem, $\theta(\bar{G}) = k$, and so the feasible solution with $\alpha = -1/(k-1)$ is an optimal solution. It is also possible to argue directly that a graph with a K_k must have $\alpha \geq -1/(k-1)$, again proving that the feasible solution given above is an optimal one. We will call the feasible solution above (in which the vectors are recursively constructed) the *reference solution*.

The goal of this paper is to explore situations in which the reference solution is the unique optimal solution of a semidefinite program (SDP), either the SDP corresponding to the strict vector chromatic number given above, or another that we will give shortly. To do this, we will use complementary slackness conditions for semidefinite programs. Consider the primal and dual SDPs shown in standard form below, where the constraint that X is a positive semidefinite matrix is represented by $X \succeq 0$, and we take the outer product of matrices, so that $C \bullet X$, for instance, denotes $\sum_{i=1}^{\ell} \sum_{j=1}^{\ell} c_{ij} x_{ij}$.

$$
\begin{array}{ll}
\text{minimize } C \bullet X & \text{maximize } b^T y \\
\text{subject to } A_i \bullet X = b_i \text{ for } i = 1, \ldots, m, & \text{subject to } S = C - \sum_{i=1}^{m} y_i A_i, \\
(P) \qquad X \succeq 0, & (D) \qquad S \succeq 0, \\
\qquad X \in \Re^{\ell \times \ell}, & \qquad S \in \Re^{\ell \times \ell}.
\end{array}
$$

Duality theory for semidefinite programs (e.g. Alizadeh [1]) shows that for any feasible primal solution X and any feasible dual solution y, $C \bullet X \geq b^T y$. Furthermore if $C \bullet X = b^T y$, so that the solutions are optimal, then it must be the case that $\text{rank}(X) + \text{rank}(S) \leq \ell$, and $XS = 0$, where we refer to $\text{rank}(X)$ and $\text{rank}(S)$ as the *primal rank* and *dual rank*, respectively. Thus if we want to show that any optimal primal solution has rank at most r, it suffices to show the existence of an optimal dual solution of rank at least $\ell - r$. Turning back to the strict vector chromatic number vector program, the corresponding dual vector program is

$$
\begin{array}{ll}
\text{maximize} & -\sum_i u_i \cdot u_i \\
\text{subject to} & \sum_{i \neq j} u_i \cdot u_j \geq 1, \\
(SVCN\text{-}D) \; u_i \cdot u_j = 0, & \forall (i,j) \notin E, i \neq j \\
u_i \in \mathbb{R}^n, & \forall i \in V.
\end{array}
$$

Thus, given a k-colorable graph G with a K_k, if we can show a dual feasible solution of value $-1/(k-1)$ and rank $n - k + 1$, then we know that the primal solution must have rank at most $k - 1$; in the cases we can show this, we can also show that the reference solution is the unique primal optimal solution. We will for shorthand say that there is an optimal dual solution of *sufficiently high rank*.

Our first result is to partially characterize the set of graphs for which the optimal solution to the strict vector chromatic number vector program is the reference solution. In particular, we can show that if the graph is a $(k-1)$-*tree*, then the reference solution is the unique optimal solution to the SDP. In the opposite direction, if the graph is not *uniquely colorable*, then the dual does not have sufficiently high rank, and there exist optimal primal solutions that are not the reference solution and are at least k-dimensional. A $(k-1)$-tree is a graph constructed by starting with a complete graph on k vertices. We then iteratively add vertices v; for each new vertex v, we add $k - 1$ edges from v to previously added vertices such that v together with these $k - 1$ neighbors form a clique. A k-colorable graph is uniquely colorable if it has only one possible coloring up to a permutation of the colors. The k-tree graphs are easily shown to be uniquely colorable. In the case of planar graphs with a K_4, these results imply a complete characterization of the graphs for which the optimal solution is the reference solution, since it is known that the uniquely 4-colorable planar graphs are exactly the planar 3-trees, also known as the Apollonian networks [6]. We argue that it is not surprising that graphs are not uniquely k-colorable do not have the reference solution as the sole optimal solution; we show that one can find a convex combination of the two different reference solutions corresponding to the two different colorings that gives an optimal SDP solution of rank higher than $k - 1$, and clearly the convex combination is also feasible for the SDP.

To get around the issue of unique colorability, we instead look for minimum-cost feasible solutions to the SDP above. That is, given a cost matrix C, we look to find optimal solutions to the primal SDP

$$
\begin{aligned}
& \text{minimize} \quad C \bullet X \\
& \text{subject to } X_{ij} = -1/(k-1), \ \forall (i,j) \in E, \\
& (CP) \qquad X_{ii} = 1, \qquad \qquad \forall i \in V, \\
& \qquad \quad X \succeq 0.
\end{aligned}
$$

The corresponding dual SDP is

$$
\begin{aligned}
& \text{maximize} \ \sum_{i=1}^{n} y_i - \frac{2}{k-1} \sum_{e \in E} z_e \\
& \text{subject to } S = C - \sum_{i=1}^{n} y_i E_{ii} - \sum_{e \in E} z_e E_e, \\
& (CD) \qquad S \succeq 0,
\end{aligned}
$$

where E_{ii} is the matrix with a 1 at position ii and 0 elsewhere and for $e = (i,j)$, E_e is the matrix with 1 at positions ij and ji and 0 elsewhere. Once again, the reference solution is a feasible solution to the primal SDP. The goal now is to find a cost function C such that there is an optimal dual solution of sufficiently high rank (here rank $n - k + 1$), so that the reference solution is the unique optimal solution to the primal SDP. We show that it is always possible to find a cost function C such that the dual has sufficiently high rank. Our construction of C depends on the coloring of the graph; however, we do show that such a C exists.

Furthermore, the construction of C suggests a heuristic for finding a coloring of the graph, and we show that the heuristic works well. We enumerated all maximal planar graphs of up to 14 vertices containing a K_4. The heuristics successfully colored all graphs of up to 11 vertices, and at least 99.75% of all graphs on 12, 13, and 14 vertices. The heuristics involve repeatedly solving semidefinite programs, and thus are not practical for large graphs (although they still run in polynomial time). However, we view them as a proof of concept that it might be possible to use our framework to reliably 4-color planar graphs.

Our interest in this direction of research was prompted by the Colin de Verdière invariant [5] (see also [16] for a useful survey of the invariant). A *generalized Laplacian* $L = (\ell_{ij})$ of graph G is a matrix such that the entries $\ell_{ij} < 0$

when $(i, j) \in E$, and $\ell_{ij} = 0$ when $(i, j) \notin E$. The Colin de Verdiére invariant, $\mu(G)$, is defined as follows.

Definition 1. *The Colin de Verdière invariant $\mu(G)$ is the largest corank of a generalized Laplacian L of G such that:*

1. *L has exactly one negative eigenvalue of multiplicity one;*
2. *there is no nonzero matrix $X = (x_{ij})$ such that $LX = 0$ and such that $x_{ij} = 0$ whenever $i = j$ or $\ell_{ij} \neq 0$.*

Colin de Verdiére shows that $\mu(G) \leq 3$ if and only if G is planar; in other words, if G is planar *any* generalized Laplacian of G with exactly one negative eigenvalue of multiplicity 1 will have rank at least $n - 3$ (modulo the second condition on the invariant, which we will ignore for the moment). Other results show that G is outerplanar if and only if $\mu(G) \leq 2$, and G is a collection of paths if and only if $\mu(G) \leq 1$. Colin de Verdière [5] conjectures that $\chi(G) \leq \mu(G) + 1$; this result is known to hold for $\mu(G) \leq 4$. We note that the part of the dual matrix $- \sum_{i=1}^{n} y_i E_{ii} - \sum_{e \in E} z_e E_e$ is indeed a generalized Laplacian L of a planar graph when the $z_e \geq 0$ for all $e \in E$, and that if G is connected, then the y_i can be adjusted so that this matrix has a single negative eigenvalue of multiplicity one. Thus this part of the matrix, under these conditions, must have sufficiently high rank, as desired to verify that the optimal primal solution is the reference solution. This would show that if the graph G has a clique on $\mu(G) + 1$ vertices, then indeed $\chi(G) = \mu(G) + 1$. So, for example, this would prove that any planar graph with a K_4 can be four-colored, leading to a non-computer assisted proof of the four-color theorem. However, we do not know how to find the corresponding cost matrix C or show that the dual S we find is optimal. Still, we view our heuristics as a step towards finding a way to construct the cost matrix C without knowledge of the coloring, and without using the proofs of the four-color theorem that have been developed thus far.

The rest of this paper is structured as follows. In Sect. 2, we give some preliminary results on semidefinite programming. In Sect. 3, we show our results for the strict vector chromatic number SDP, and show that $(k-1)$-trees imply dual solutions of sufficiently high rank, while graphs that are not uniquely colorable imply that such dual solutions cannot exist. In Sect. 4, we turn to the SDP with cost matrix C, and show that for any k-colorable graph with a k-clique, a cost matrix C exists that gives rise to a dual of sufficiently high rank. In Sect. 5, we give two heuristics for coloring planar graphs based on our construction of the cost matrix C, and show a case where the heuristic fails to find a 4-coloring of a planar graph. We give some open questions in Sect. 6. For space reasons, many proofs are omitted.

2   Preliminaries

In this section, we recall some basic facts about semidefinite matrices and semidefinite programs that we will use in subsequent sections.

Recall the primal and dual semidefinite programs (P) and (D) from the introduction. We always have weak duality for semidefinite programs, so that the following holds.

Fact 1. *Given any feasible X for (P) and y for (D), $C \bullet X \geq b^T y$.*

Thus if we can produce a feasible X for (P) and a feasible y for (D) such that $C \bullet X = b^T y$, then X must be optimal for (P) and y optimal for (D).

The following is also known, and is the semidefinite programming version of complementary slackness conditions for linear programming.

Fact 2. *[1, Theorem 2.10, Corollary 2.11] For optimal X for (P) and y for (D), $XS = 0$ and $rank(X) + rank(S) \leq \ell$.*

Semidefinite programs and vector programs (such as the strict vector chromatic vector program) are equivalent because a symmetric $X \in \Re^{n \times n}$ is positive semidefinite if and only if $X = QDQ^T$ for a real matrix $Q \in \Re^{n \times n}$ and diagonal matrix D in which the entries of D are the eigenvalues of X, and the eigenvalues are all nonnegative. We can then consider $D^{1/2}$, the diagonal matrix in which each diagonal entry is the square root of the corresponding entry of D. Then $X = (QD^{1/2})(QD^{1/2})^T$. If we let $v_i \in \Re^n$ be the ith row of $QD^{1/2}$, then $x_{ij} = v_i \cdot v_j$, and similarly, given the vectors v_i, we can construct a semidefinite matrix X with $x_{ij} = v_i \cdot v_j$. We also make the following observation based on this decomposition.

Observation 1. *Given a semidefinite matrix $X = QDQ^T \in \Re^{n \times n}$, $rank(X) = d$ if and only if the vectors $v_i \in \Re^n$ with v_i the ith row of $QD^{1/2}$ are supported on just d coordinates.*

3   The Strict Vector Chromatic Number SDP

Recall the strict vector chromatic SDP given in the introduction, labeled (SVCN-P). In what follows we will let the matrix $X = (X_{ij})$ be the SDP matrix such that $X_{ij} = v_i \cdot v_j$ and $S = (S_{ij})$ be the SDP matrix related to the dual solution (SVCN-D) such that $S_{ij} = u_i \cdot u_j$.

Lemma 1. *Given an optimal primal solution X to (SVCN-P) and optimal dual solution S to (SVCN-D), we have that $rank(X) + rank(S) \leq n$.*

Our main result for this section is about graphs that are $(k-1)$-trees.

Definition 2. *A $(k-1)$-tree with n vertices is an undirected graph constructed by beginning with the complete graph on k vertices and repeatedly adding vertices in such a way that each new vertex, v, has $k-1$ neighbors that, together with v, form a k-clique.*

An easy inductive argument shows that these graphs are k-colorable. Also, $(k-1)$-trees are known to be uniquely k-colorable, where uniquely colorable means every coloring produces that same vertex partitioning. Once k colors are assigned to the initial complete graph with k vertices, the color of each new vertex is uniquely determined by its $k-1$ neighbors. This partitioning into color classes is unique up to permuting the colors. Note that by construction, a $(k-1)$-tree contains a K_k. By discussion in the introduction, the optimal value of (SVCN-P) for a $(k-1)$-tree will be exactly $-1/(k-1)$.

Our goal is to show there is a feasible solution to the dual (SVCN-D) with high rank. In particular, given a $(k-1)$-tree with n vertices, we show the existence of a dual solution with rank at least $n-k+1$. This ensures that any primal solution has rank at most $k-1$; we show that the reference solution is the unique optimal primal solution. This is formalized in the following theorem.

Theorem 1. *Given a $(k-1)$-tree G with n vertices, there is an optimal dual solution S to (SVCN-D) with rank at least $n-k+1$, and thus any optimal primal solution X to (SVCN-P) has rank at most $k-1$.*

We subsequently prove that the reference solution is indeed the unique optimal solution in this case.

Theorem 2. *The reference solution is the unique optimal primal solution (up to rotation) for a $(k-1)$-tree $G = (V, E)$.*

To prove Theorem 1, we need a number of supporting lemmas. We begin with the following.

Lemma 2. *Let $tri(G)$ denote the number of triangles in a $(k-1)$-tree, G. Then, for a $(k-1)$-tree G with n vertices,*

$$|E(G)| = (2n - k)\frac{k-1}{2} \tag{1}$$

$$tri(G) = \frac{(3n - 2k)(k-1)(k-2)}{6}. \tag{2}$$

Consider a $(k-1)$-tree G with n vertices. For $v \in V$ we denote the neighborhood of v by $N(v) = \{u : (u, v) \in E\}$. We define the following matrix $S(G)$ which may be referred to as S if G is clear from context.

$$S(G)_{ij} = \begin{cases} \dfrac{|N(i)| - (k-2)}{k(k-1)(n-k+1)} & i = j \\[3mm] \dfrac{|N(i) \cap N(j)| - (k-3)}{k(k-1)(n-k+1)} & (i,j) \in E \\[3mm] 0 & (i,j) \notin E, i \neq j. \end{cases}$$

We will show that $S(G)$ is an optimal dual solution with rank $n - k + 1$. First, we show $S(G)$ is a feasible solution with help from the following lemma.

Lemma 3. *For a $(k-1)$-tree G with n vertices, $S(G)$ is positive semidefinite.*

Proof. Observe that it suffices to show $S'(G) = k(k-1)(n-k+1)S(G)$ is positive semidefinite (PSD) since $k(k-1)(n-k+1) > 0$ for $n \geq k$. We proceed by induction. First consider $(k-1)$-trees with k vertices. There is only one, $G = K_k$. Furthermore, $S'(K_k)$ is equal to the all-ones matrix which has eigenvalues k and 0 with multiplicity $k-1$ and thus is PSD.

Now assume there is some integer n such that for every $(k-1)$-tree, G, with at most n vertices, $S'(G)$ is PSD. Consider a $(k-1)$-tree G with $n+1$ vertices. Since it is a $(k-1)$-tree, it can be constructed from some smaller $(k-1)$-tree G' with n vertices by adding a vertex v and $(k-1)$ edges that form a k clique with the $k-1$ neighbors. By assumption, $S'(G')$ is PSD. Let I be the set of indices of the $k-1$ neighbors of v. Then we observe that $S'(G) = T + v_{n+1}v_{n+1}^T$ where

$$
T = \left[\begin{array}{c|c} S'(G') & \begin{matrix} 0 \\ \vdots \\ 0 \end{matrix} \\ \hline \begin{matrix} 0 & \cdots & 0 \end{matrix} & 0 \end{array} \right]
$$

and

$$
v_{n+1}(i) = \begin{cases} 1 & i \in I \cup \{n+1\} \\ 0 & \text{otherwise} \end{cases}.
$$

Then $x^T S'(G)x = x^T T x + x^T v_{n+1}v_{n+1}^T x \geq x^T v_{n+1}v_{n+1}^T x = (v_{n+1}^T x)^2 \geq 0$ where the first inequality is due to T being PSD since $S'(G')$ is PSD.   □

Lemma 4. *For a $(k-1)$-tree G with n vertices, $S(G)$ is a feasible dual slack matrix.*

Proof. Lemma 3 shows that $S(G)$ is PSD. To complete this claim, we must show that the dual constraints are satisfied. That $S(G)_{ij} = 0$ for $(i,j) \notin E$ is clear by construction. The other constraint requires $\sum_{i \neq j} s_{ij} \geq 1$. Using (1) and (2) from Lemma 2 we can prove that the inequality holds; the algebra is omitted for space reasons.   □

We can now show that $S(G)$ is an optimal dual solution.

Theorem 3. *For a $(k-1)$-tree G with n vertices, $S(G)$ is an optimal dual solution.*

Proof. We remarked earlier that the optimal primal value for a $(k-1)$-tree is $-1/(k-1)$. Thus for $S(G)$ to be an optimal dual solution, it suffices to show that $-\sum_i S_{ii} = -1/(k-1)$. Again using (1) from Lemma 2, we have

$$-\sum_{i=1}^{n} S_{ii} = -\sum_{i=1}^{n} \frac{|N(i)| - (k-2)}{k(k-1)(n-k+1)}$$

$$= -\frac{1}{k(k-1)(n-k+1)} \left[-(k-2)n + \sum_{i=1}^{n} |N(i)| \right]$$

$$= -\frac{-(k-2)n + 2|E|}{k(k-1)(n-k+1)}$$

$$= -\frac{-(k-2)n + ((2n-k)(k-1))}{k(k-1)(n-k+1)}$$

$$= -\frac{-nk + 2n + 2nk - 2n - k^2 + k}{k(k-1)(n-k+1)} = -1/(k-1).$$

$\square$

Finally, we want to show that for a $(k-1)$-tree G with n vertices, $S(G)$ has rank at least $n - k + 1$. This guarantees that any primal solution has rank at most $k - 1$.

Theorem 4. *For a $(k-1)$-tree G with n vertices, $S(G)$ has rank at least $n-k+1$.*

Proof. It again suffices to show the claim is true for $S'(G) = k(k-1)(n-k+1)S(G)$. Proceeding by induction, for $n = k$ we have $rank(S'(G)) = rank(S'(K_k)) = 1 = k - (k-1)$ with $S'(K_k)$ equal to the all-ones matrix. Assuming the claim is true for all $(k-1)$-trees with at most n vertices, we consider a $(k-1)$-tree G with $n+1$ vertices. We again use the decomposition $S'(G) = T + v_{n+1}v_{n+1}^T$ where

$$T = \begin{bmatrix} S'(G') & \begin{matrix} 0 \\ \vdots \\ 0 \end{matrix} \\ \begin{matrix} 0 \cdots 0 \end{matrix} & 0 \end{bmatrix} , v_{n+1}(i) = \begin{cases} 1 & i \in I \cup \{n+1\} \\ 0 & \text{otherwise} \end{cases} ,$$

and G' is a $(k-1)$-tree with n vertices acquired by removing vertex $n+1$ with exactly $k-1$ neighbors, $i \in I$, from G. Note $dim(ker(T)) = dim(ker(S'(G'))) + 1 \le k$ by assumption. Now assume $x \in ker(S'(G))$. Then

$$0 = x^T S'(G)x = x^T T x + x^T v_{n+1}v_{n+1}^T x.$$

Since T and $v_{n+1}v_{n+1}^T$ are both PSD, this implies $x^T T x = 0$ and $x^T v_{n+1}v_{n+1}^T x = 0$. Therefore $ker(S'(G)) = ker(T) \cap ker(v_{n+1}v_{n+1}^T)$. However, note that $x = (0, \cdots, 0, 1) \in ker(T)$, but $x \notin ker(v_{n+1}v_{n+1}^T)$. Then

$$ker(S'(G)) = ker(T) \cap ker(v_{n+1}v_{n+1}^T) \subsetneq ker(T).$$

This implies $dim(ker(S'(G)) < dim(ker(T)) \le k$, so $rank(S'(G)) \ge (n+1) - k + 1$.

$\square$

Theorem 1. *Given a $(k-1)$-tree G with n vertices, there is an optimal dual solution S to (SVCN-D) with rank at least $n-k+1$, and thus any optimal primal solution X to (SVCN-P) has rank at most $k-1$.*

Proof. Theorem 1 follows as an immediate consequence of Lemma 4, Theorem 3, and Theorem 4.

We now turn to showing that the reference solution is indeed the optimal solution in the case of $(k-1)$-trees.

Theorem 2. *The reference solution is the unique optimal primal solution (up to rotation) for a $(k-1)$-tree $G = (V, E)$.*

Theorem 2 shows that we can partition the vertices of a $(k-1)$-tree into k sets with each set associated to a different vector assigned in the low rank primal solution. Since vertices u, v are only in the same set in the partition if they were assigned the same vector in the primal solution, it is not possible for neighbors to be in the same set. We can then produce a valid coloring of the vertices by associating one color to each set in the partition.

We now turn to characterizing cases in which we cannot find dual solutions of sufficiently high rank by looking at potential solutions of vector colorings for graphs without unique colorings. In particular, we restrict our attention to graphs that have multiple distinct k-colorings and contain a k-clique. These assumptions provide information about the optimal objective function values.

Theorem 5. *Let G be a graph with n vertices, multiple distinct k-colorings, and a k-clique. There exists a primal solution to the strict vector chromatic number program for G with rank greater than $k-1$, and thus by Fact 2 the rank of any optimal dual solution must be less than $n-k+1$.*

While we have shown that $(k-1)$-trees have sufficiently high dual rank for the standard vector chromatic number SDP, it would be nice if we could completely characterize which graphs have high dual rank. A reasonable guess would be that a k-colorable graph G containing a k-clique has high dual rank if and only if it is uniquely colorable. This assertion is true for the important special case of planar graphs.

Corollary 1. *A planar graph with n vertices has dual rank at least $n-3$ if and only if it is uniquely colorable.*

Proof. Fowler [6] shows that uniquely-colorable planar graphs are exactly the set of planar 3-trees. By Theorem 1 we know such graphs have dual rank at least $n-3$. Furthermore, Theorem 5 shows that graphs with multiple colorings have primal solutions with rank more than 3 and therefore do not have dual solutions with rank $n-3$. □

Unfortunately, the following example shows unique colorability is not sufficient in general for a sufficiently high dual rank. Modifying a uniquely 3-colorable example of Hillar and Windfeldt [8] and computing the primal and dual SDPs of this graph returns solutions with objective value -0.5, primal rank of 24, and dual rank of 1. If the claim were true, we would expect all dual solutions to have rank at least 23.

Thus it remains an interesting open question to characterize in general cases in which graphs have sufficiently high dual rank and have the reference solution as the optimal primal solution.

4   A Semidefinite Program with Costs

Unfortunately, Theorem 5 seems to indicate that this method of looking for graphs that have high dual rank with the standard vector chromatic number SDP cannot be generalized to graphs with multiple colorings. To extend this method, we consider a modified SDP described next. The new program utilizes a new objective function. Here, we introduce the notion of a cost matrix $C(G)$. The goal is to identify a $C(G)$ such that minimizing $C(G) \bullet X$ forces X to have our desired rank. In particular, we consider the SDP given by (CP) in the introduction, along with its dual (CD).

To demonstrate how this cost matrix influences the behavior of $rank(X)$, assume that $G = (V, E)$ is a k-colorable graph containing a k-clique, but is not a $(k-1)$-tree. We still know there is a solution to the strict vector chromatic number program with $\alpha = -1/(k-1)$, and thus it is possible to find an X feasible for (CP). Now fix $c : V \to [k]$ to be a valid k-coloring of G. With this coloring, we can define an associated matrix $C(G)$ in the following way:

$$
C(G)_{ij} = \begin{cases} -1 & i < j, c(i) = c(j), \forall \ell \text{ such that } i < \ell < j, c(i) \neq c(\ell) \\ -1 & i > j, c(i) = c(j), \forall \ell \text{ such that } i > \ell > j, c(i) \neq c(\ell) \\ 0 & \text{otherwise.} \end{cases}
$$

Intuitively, the reference solution corresponding to the coloring given by c is the solution that will minimize total cost since we'll look for a solution X with $X_{ij} = 1$ exactly when $C(G)_{ij} = -1$; for such entries, we'll have the same vectors corresponding to vertices i and j. But we can show additionally that there is a dual optimal solution for cost function $C(G)$ that has sufficiently high rank.

Theorem 6. *For G and $C(G)$ as described, there is an optimal dual solution with rank at least $n - k + 1$, so that any optimal primal solution has rank at most $k - 1$.*

Let K be a k-clique in our k-colorable graph G. Let s_i denote the sum of entries in column i of $C(G)$. Consider the assignment of dual variables given by $y_i = s_i$ for $i \notin K$, $y_i = s_i - 1$ for $i \in K$, $z_e = -1$ for $e = (i, j), i, j \in K, i \neq j$, and $z_e = 0$ otherwise. We denote this assignment by (y, z).

Lemma 5. *The dual matrix S constructed with (y, z) is positive semidefinite.*

Table 1. This table depicts the number of times the heuristic algorithms failed on maximally planar graphs with between 5 and 14 vertices.

# nodes	# maximally planar graphs with K_4	# heuristic 1 failures	# heuristic 2 failures
5	1	0	0
6	1	0	0
7	4	0	0
8	12	0	0
9	45	0	0
10	222	0	0
11	1219	0	0
12	7485	18 ($\sim$ .24%)	18 ($\sim$ .24%)
13	49149	108 ($\sim$ .22%)	116 ($\sim$ .24%)
14	337849	619 ($\sim$ .18%)	811 ($\sim$ .24%)

Theorem 7. *The assignment (y, z) is an optimal dual solution, and the reference solution is an optimal primal solution.*

Theorem 8. *For G and $C(G)$ as described, the reference solution is the unique optimal primal solution.*

5   Experimental Results

Two heuristics have been implemented and experimentally demonstrated success returning low-rank primal solutions for planar graphs. Neither algorithm assumes knowledge of a graph coloring. We tested these heuristics on all maximal planar graphs of up to 14 vertices that contain a K_4. These graphs were generated via the planar graph generator plantri due to Brinkmann and McKay [4] found at https://users.cecs.anu.edu.au/~bdm/plantri/. The '-a' switch was used to produce graphs written in ascii format. The code was implemented in Python using the MOSEK Optimizer as the SDP solver. Both the graph data files and algorithm implementation can be found at https://github.com/rmirka/four-coloring.git. Our results are shown in Table 1. The heuristics successfully colored all graphs with up to 11 vertices, and successfully colored 99.75% of the graphs of 12–14 vertices. We do not record the running time of the heuristics; because the heuristics involve repeatedly solving semidefinite programs, they are not competitive with other greedy or local search style heuristics. Our primary reason for studying these heuristics was to find whether we could reliably find a cost matrix C giving rise to a four-coloring for planar graphs.

For space reasons, we only describe the first heuristic. It is based on the coloring-dependent cost matrix discussed in Sect. 4. The algorithm first identifies a $K_4 = \{k_1, k_2, k_3, k_4\}$ and finds an initial solution with $C = 0$. If the primal solution does not have low enough rank, the returned solution is used to update the cost matrix. Let $S_i = \{v \in V : X_{vk_i} = 1\}$ for $i = 1, 2, 3, 4$. Let v be a vertex in $V \setminus (\cup_{i=1}^{4} S_i)$. Then there must exist $i^* \in \{1, 2, 3, 4\}$ such that $X_{vk_{i^*}} \neq 1$ and $X_{vk_{i^*}} \neq -1/3$; we update this S_{i^*} by adding v to it. Now, C is constructed based on the $S_j, j = 1, 2, 3, 4$. In particular, for $i = 1, 2, 3, 4$, if n_i denotes the number of vertices in S_i, then for $j = 1, \ldots, n_i - 1$, we set $C_{rs} = C_{sr} = -1$ where r and s are the jth and $j + 1$st vertices in S_i. This new cost matrix C is used to compute an updated solution $\hat{X}$. If $\hat{X}$ is of the desired rank, the algorithm terminates. If not, we first check to see if $\hat{X}_{vk_{i^*}} = 1$, i.e. if our selected vertex from the previous iteration was successfully colored. If yes, we repeat the process beginning with our solution $\hat{X}$ and selecting a currently uncolored vertex. If v was not successfully colored, we remove the entry in the cost matrix corresponding to this assignment from the previous iteration and resolve the SDP while adding k_{i^*} to a list of 'bad' colors for v. We now repeat the process by selecting a new feasible color class for v (following the same rules as previously in addition to requiring it not be in the list of 'bad' colors for v) and constructing S_i, $i = 1, 2, 3, 4$ and C accordingly.

In both heuristics, the termination condition is that the primal rank is equal to 3, but this doesn't necessarily guarantee that the dual rank is $n - 3$. If instead one wanted to guarantee high dual rank, one could run the algorithm one more time, i.e. once the low-rank primal solution is achieved, extract the coloring and construct the corresponding C matrix as previously described in Theorem 6.

The example in Fig. 1 causes both heuristics to fail without coloring the graph. First we note the $K_4 = \{2, 5, 6, 7\}$. In the first iteration of the heuristic, these are the only four vertices that are assigned colors. In the second iteration, both heuristics successfully color vertex 1 to match vertex 6. However, afterwards each heuristic is unable to color any more vertices (it tries and fails on all other possible colors for the remaining vertices).

6 Open Questions

We close with several open questions. We were unable to give a complete characterization of the k-colorable graphs with a K_k for which the strict vector chromatic number (SVCN-P) has a unique primal solution of the reference solution. Such graphs must be uniquely colorable, but clearly some further restriction is needed.

When we know the coloring, we can produce a cost matrix C for the semidefinite program (CP) such that the reference solution is the unique optimal solution and it must have rank $k-1$. We wondered whether one could use (CP) in a greedy coloring scheme, by incrementally constructing the matrix C; the graph in Fig. 1 shows that our desired scheme does not work in a straightforward manner. Possibly one could consider an algorithm with a limited amount of backtracking, as long as one could show that the algorithm continued to make progress against some metric.

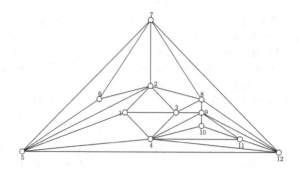

Fig. 1. Algorithm Obstacle: $K_4 = \{2, 5, 6, 7\}$

Another open question is whether one can somehow directly produce a cost matrix C leading to a dual solution of sufficiently high rank that does not need knowledge of the coloring. And we conclude with the open question that first motivated this work: is it possible to use the Colin de Verdière parameter to produce this matrix C?

References

1. Alizadeh, F.: Interior point methods in semidefinite programming with applications to combinatorial optimization. SIAM J. Optim. **5**, 13–51 (1995)
2. Appel, K., Haken, W.: Every planar map is four colorable. Part I: discharging. Ill. J. Math. **21**, 429–490 (1977)
3. Appel, K., Haken, W., Koch, J.: Every planar map is four colorable. Part II: reducibility. Ill. J. Math. **21**, 491–567 (1977)
4. Brinkmann, G., McKay, B.D.: Fast generation of planar graphs. Math.- Commun. Math. Comput. Chem. **58**, 323–357 (2007)
5. de Verdiére, Y.C.: Sur un nouvel invariant des graphes et un critére de planarité. J. Comb. Theory Ser. B **50**, 11–21 (1990)
6. Fowler, T.G.: Unique coloring of planar graphs. Ph.D. thesis, Georgia Institute of Technology, Department of Mathematics (1998)
7. Fritsch, R., Fritsch, G.: The Four-Color Theorem. Springer, Berlin, Germany (1994)
8. Hillar, C.J., Windfeldt, T.: Algebraic characterization of uniquely vertex colorable graphs. J. Comb. Theory Ser. B **98**, 400–414 (2008)
9. Jensen, T.R., Toft, B.: Graph Coloring Problems. John Wiley and Sons, New York, NY, USA (1995)
10. Karger, D., Motwani, R., Sudan, M.: Approximate graph coloring by semidefinite programming. J. ACM **45**(2), 246–265 (Mar 1998). https://doi.org/10.1145/274787.274791
11. Knuth, D.E.: The sandwich theorem. Electron. J. Comb. 1 (1994). https://www.emis.de/journals/EJC/Volume_1/PDFFiles/v1i1a1.pdf
12. Lovász, L.: On the Shannon capacity of a graph. IEEE Trans. Inf. Theory **IT–25**, 1–7 (1979)
13. Molloy, M., Reed, B.: Graph Colouring and the Probabilistic Method. No. 23 in Algorithms and Combinatorics. Springer, Berlin, Germany (2002)

14. Ore, O.: The Four-Color Theorem. Academic Press, New York, NY, USA (1967)
15. Robertson, N., Sanders, D., Seymour, P., Thomas, R.: The four-colour theorem. J. Comb. Theory Ser. B **70**, 2–44 (1997)
16. van der Holst, H., Lovász, L., Schrijver, A.: The Colin de Verdière graph parameter. Graph Theory Comb. Biol. **7**, 29–85 (1999)
17. Wilson, R.A.: Graphs, Colourings, and the Four-Colour Theorem. Oxford University Press, New York, NY, USA (2002)

A Competitive Algorithm for Throughput Maximization on Identical Machines

Benjamin Moseley[1], Kirk Pruhs[2], Clifford Stein[3]($\boxtimes$), and Rudy Zhou[1]

[1] Tepper School of Business, Carnegie Mellon University, Pittsburgh, USA
{moseleyb,rbz}@andrew.cmu.edu
[2] Computer Science Department, University of Pittsburgh, Pittsburgh, USA
kirk@cs.pitt.edu
[3] Industrial Engineering and Operations Research, Columbia University,
New York, USA
cliff@ieor.columbia.edu
https://www.andrew.cmu.edu/user/moseleyb/, https://rudyzhou.github.io/,
https://people.cs.pitt.edu/~kirk/

Abstract. This paper considers the basic problem of scheduling jobs online with preemption to maximize the number of jobs completed by their deadline on m identical machines. The main result is an $O(1)$ competitive deterministic algorithm for any number of machines $m > 1$.

Keywords: Scheduling · Competitive analysis · Online algorithm

1 Introduction

We consider the basic problem of preemptively scheduling jobs that arrive online with sizes and deadlines on m identical machines so as to maximize the number of jobs that complete by their deadline.

Definition 1 (Throughput Maximization). *Let J be a collection of jobs such that each $j \in J$ has a* release time r_j, *a* processing time *(or size) x_j, and a* deadline d_j. *The jobs arrive online at their release times, at which the scheduler becomes aware of job j and its x_j and d_j.*

At each moment of time, the scheduler can specify up to m released jobs to run, and the remaining processing time of the jobs that are run is decreased at a unit rate (so we assume that the online scheduler is allowed to produce a migratory schedule.) A job is completed if its remaining processing time drops to zero by the deadline of that job. The objective is to maximize the number of completed jobs.

B. Moseley—Benjamin Moseley and R. Zhou were supported in part by NSF grants CCF-1824303, CCF-1845146, CCF-1733873 and CMMI-1938909. Benjamin Moseley was additionally supported in part by a Google Research Award, an Infor Research Award, and a Carnegie Bosch Junior Faculty Chair.

K. Pruhs—Supported in part by NSF grants CCF-1907673, CCF-2036077 and an IBM Faculty Award.

C. Stein—Research partly supported by NSF Grants CCF-1714818 and CCF-1822809.

K. Aardal and L. Sanitá (Eds.): IPCO 2022, LNCS 13265, pp. 402–414, 2022.
https://doi.org/10.1007/978-3-031-06901-7_30

A key concept is the *laxity* of a job j, which is $\ell_j = (d_j - r_j) - x_j$, that is, the maximum amount of time we can not run job j and still possibly complete it.

We measure the performance of our algorithm by the *competitive ratio*, which is the maximum over all instances of the ratio of the objective value of our algorithm to the objective value of the optimal offline schedule (Opt) that is aware of all jobs in advance.

This problem is well understood for the $m = 1$ machine case. No $O(1)$-competitive deterministic algorithm is possible [2], but there is a randomized algorithm that is $O(1)$-competitive against an oblivious adversary [9], and there is a scalable ($O(1 + \epsilon)$-speed $O(1/\epsilon)$-competitive) deterministic algorithm [7]. The scalability result in [7] was extended to the case of $m > 1$ machines in [11].

Whether an $O(1)$-competitive algorithm exists for $m > 1$ machines has been open for twenty years. Previous results for the multiple machines setting require resource augmentation or assume that all jobs have high laxity [5,11].

The main issue issue in removing these assumptions is determining which machine to assign a job to. If an online algorithm could determine which machine each job was assigned to in Opt, we could obtain an $O(1)$-competitive algorithm for $m > 1$ machines by a relatively straight-forward adaptation of the results from [9]. However, if the online algorithm ends up assigning some jobs to different machines than Opt, then comparing the number of completed jobs is challenging. Further, if jobs have small laxity, then the algorithm can be severely penalized for small mistakes in this assignment. One way to view the speed augmentation (or high laxity assumption) analyses in [5,11] is that the speed augmentation assumption allows one to avoid having to address this issue in the analyses.

1.1   Our Results

Our main result is an $O(1)$-competitive deterministic algorithm for Throughput Maximization on $m > 1$ machines.

Theorem 1. *For all $m > 1$, there exists a deterministic $O(1)$-competitive algorithm for Throughput Maximization on m machines.*

We summarize our results and prior work in Table 1. Interestingly, on a single machine there is no constant competitive deterministic algorithm, yet a randomized algorithm exists with constant competitive ratio. Our work shows that once more than one machine is considered, then determinism is sufficient to get a $O(1)$-competitive online algorithm.

1.2   Scheduling Policies

We give some basic definitions and notations about scheduling policies.

A job j is *feasible* at time t (with respect to some schedule) if it can still be feasibly completed, so $x_j(t) > 0$ and $t + x_j(t) \le d_j$, where $x_j(t)$ is the remaining processing time of job j at time t (with respect to the same schedule.)

Table 1. Competitiveness results

	Deterministic	Randomized	Speed Augmentation
$m = 1$ [2]	$\omega(1)$ [9]	$O(1)$	$O(1 + \epsilon)$-speed $O(1/\epsilon)$-competitive [7]
$m > 1$	$O(1)$ [This paper]	$O(1)$ [This paper]	$O(1 + \epsilon)$-speed $O(1/\epsilon)$-competitive [11]

A schedule $\mathcal{S}$ of jobs J is defined by a map from time/machine pairs (t, i) to a feasible job j that is run on machine i at time t, with the constraint that no job can be run one two different machines at the same time. We conflate $\mathcal{S}$ with the scheduling policy as well as the set of jobs *completed* by the schedule. Thus, the objective value achieved by this schedule is $|\mathcal{S}|$.

A schedule is *non-migratory* if for every job j there exists a machine i such that if j is run at time t then j is run on machine i. Otherwise the schedule is *migratory*.

If $\mathcal{S}$ is a scheduling algorithm, then $\mathcal{S}(J, m)$ denotes the schedule that results from running $\mathcal{S}$ on instance J on m machines. Similarly, $\mathrm{Opt}(J, m)$ denotes the optimal schedule on instance J on m machines. We will sometimes omit the J and/or the m if they are clear from context. Sometimes we will abuse notation and let Opt denote a nearly-optimal schedule that additionally has some desirable structural property.

1.3 Algorithms and Technical Overview

A simple consequence of the results in [8] and [9] is an $O(1)$-competitive randomized algorithm in the case that $m = O(1)$. Thus we concentrate on the case that m is large. We also observe that since there is an $O(1)$-approximate non-migratory schedule [8], changing the number of machines by an $O(1)$ factor does not change the optimal objective value by more than an $O(1)$ factor. This is because we can always take an optimal non-migratory schedule on m machines and create a new schedule on m/c machines whose objective value decreases by at most a factor of c, by keeping the m/c machines that complete the most jobs.

These observations about the structure of near-optimal schedules allow us to design a $O(1)$-competitive algorithm that is a combination of various deterministic algorithms. In particular, on an instance J, our algorithm, FINALALG, will run a deterministic algorithm, LMNY, on $m/3$ machines on the subinstance $J_{hi} = \{j \in J \mid \ell_j > x_j\}$ of high laxity jobs, a deterministic algorithm SRPT on $m/3$ machines on the subinstance $J_{lo} = \{j \in J \mid \ell_j \leq x_j\}$ of low laxity jobs, and a deterministic algorithm MLAX on $m/3$ machines on the subinstance J_{lo} of low laxity jobs. Note that we run SRPT and MLAX on the same jobs. To achieve this, if both algorithms decide to run the same job j, then the algorithm in which j has shorter remaining processing time actually runs job j, and the other simulates running j.

We will eventually show that for all instances, at least one of these three algorithms is $O(1)$-competitive, from which our main result will follow. Roughly, each of the three algorithms is responsible for a different part of Opt.

Our main theorem about FINALALG is the following:

Theorem 2. *For any* $m \geq 48$, FINALALG *is a* $O(1)$-*competitive deterministic algorithm for Throughput Maximization on* m *machines.*

We now discuss these three component algorithms of FINALALG.

LMNY. The algorithm LMNY is the algorithm from [11] with the following guarantee.

Lemma 3. *[11] For any number of machines* m, *and any job instance* J, LMNY *is an* $O(1)$-*competitive deterministic algorithm on the instance* J_{hi}.

SRPT. The algorithm SRPT is the standard shortest remaining processing time algorithm, modified to only run jobs that are feasible.

Definition 2 (SRPT). *At each time, run the* m *feasible jobs with shortest remaining processing time. If there are less than* m *feasible jobs, then all feasible jobs are run.*

We will show that SRPT is competitive with the low laxity jobs completed in Opt that are not preempted in Opt.

MLax. The final, most challenging, component algorithm of FINALALG is MLAX, which intuitively we want to be competitive on low-laxity jobs in Opt that are preempted.

To better understand the challenge of achieving this goal, consider $m = 1$ and an instance of *disagreeable jobs*. A set of jobs is disagreeable if, for any two jobs j and k, if j has an earlier release date than k, it also has a later deadline than k. Further, suppose all but one job in Opt is preempted and completed at a later time.

To be competitive, MLAX must preempt almost all the jobs that it completes, but cannot afford to abandon too many jobs that it preempts. Because the jobs have low laxity, this can be challenging as it can only preempt each job for a small amount of time, and its hard to know which of the many options is the "right" job to preempt for. This issue was resolved in [9] for the case of $m = 1$ machine, but the issue gets more challenging when $m > 1$, because we also have to choose the "right" machine for each job.

We now describe the algorithm MLAX. Let α be a sufficiently large constant (chosen later.) MLAX maintains m stacks (last-in-first-out data structures) of jobs (one per machine), $H_1, \ldots, H_m$. The stacks are initially empty. At all times, MLAX runs the top job of stack H_i on machine i. We define the *frontier* F to be the set consisting of the top job of each stack (i.e. all currently running jobs.) It remains to describe how the H_i's are updated.

There are two types of events that cause MLAX to update the H_i's: reaching a job's pseudo-release time (defined below) or completing a job.

Definition 3 (Viable Jobs and Pseudo-Release Time). *The pseudo-release time (if it exists) $\tilde{r}_j$ of job j is the earliest time in $[r_j, r_j + \frac{\ell_j}{2}]$ such that there are at least $\frac{7}{8}m$ jobs j' on the frontier satisfying $\alpha x_{j'} \geq \ell_j$.*

We say a job j is viable *if $\tilde{r}_j$ exists and* non-viable *otherwise.*

At job j's pseudo-release time (note $\tilde{r}_j$ can be determined online by MLAX), MLAX does the following:

a) If there exists a stack whose top job j' satisfies $\alpha x_j \leq \ell_{j'}$, then *push* j onto any such stack.
b) Else if there exist at least $\frac{3}{4}m$ stacks whose second-top job j'' satisfies $\alpha x_j \leq \ell_{j''}$ and further some such stack has top job j' satisfying $\ell_j > \ell_{j'}$, then on such a stack with minimum $\ell_{j'}$, *replace* its top job j' by j.

While the replacement operation in step b can be implemented as a pop and then push, we view it as a separate operation for analysis purposes. To handle corner cases in these descriptions, one can assume that there is a job with infinite size/laxity on the bottom of each H_i.

When MLAX completes a job j that was on stack H_i, MLAX does the following:

c) *Pop* j off of stack H_i.
d) Keep *popping* H_i until the top job of H_i is feasible.

Analysis Sketch. There are three main steps in proving Theorem 2 to show FINALALG is $O(1)$-competitive:

– In Sect. 2, we show how to modify the optimal schedule to obtain certain structural properties that facilitate the comparison with SRPT and MLAX.
– In Sect. 3, we show that SRPT is competitive with the low-laxity, non-viable jobs. Intuitively, the jobs that MLAX is running that prevent a job j from becoming viable are so much smaller than job j, and they provide a witness that SRPT must also be working on jobs much smaller than j.
– In Sect. 4, we show that SRPT and MLAX together are competitive with the low-laxity, viable jobs. First, we show that SRPT is competitive with the number of non-preempted jobs in Opt. We then essentially show that MLAX is competitive with the number of preempted jobs in Opt. The key component in the design of MLAX is the condition that a job j won't replace a job on the frontier unless at there are at least $\frac{3}{4}m$ stacks whose second-top job j'' satisfies $\alpha x_j \leq \ell_{j''}$. This condition most differentiates MLAX from m copies of the LAX algorithm in [9]. This condition also allows us to surmount the issue of potentially assigning a job to a "wrong" processor, as jobs that satisfy this condition are highly flexible about where they can go on the frontier.

We combine these results in Sect. 5 to complete the analysis of FINALALG.

1.4   Related Work

There is a line of papers that consider a dual version of the problem, where there is a constraint that all jobs must be completed by their deadline, and the objective is to minimize the number of machines used [1,4,6,12]. The current best known bound on the competitive ratio for this version is $O(\log \log m)$ from [6].

The speed augmentation results in [7,11] for throughput can be generalized to weighted throughput, where there a profit for each job, and the objective is to maximize the aggregate profit of jobs completed by their deadline. But without speed augmentation, $O(1)$-approximation is not possible for weighted throughput for any m, even allowing randomization [10].

There is also a line of papers that consider variations on online throughput scheduling in which the online scheduler has to commit to completing jobs at some point in time, with there being different variations of when commitment is required [3,5,11]. For example, [5] showed that there is a scalable algorithm for online throughput maximization that commits to finishing every job that it begins executing.

2   Structure of Optimal Schedule

The goal of this section is to introduce the key properties of (near-)optimal scheduling policies that we will use in our analysis.

By losing a constant factor in the competitive ratio, we can use a constant factor fewer machines than Opt, which justifies FINALALG running each of three algorithms on $\frac{m}{3}$ machines. The proof, which is an extension of results in [8], is omitted in this extended abstract.

Lemma 4. *For any collection of jobs J, number of machines m, and $c > 1$, we have $|\mathrm{Opt}(J, \frac{m}{c})| = \Omega(\frac{1}{c}|\mathrm{Opt}(J, m)|)$.*

A non-migratory schedule on m machines can be expressed as m schedules, each on a single machine and on a separate set of jobs. To characterize these single machine schedules, we introduce the concept of forest schedules. Let $\mathcal{S}$ be any schedule. For any job j, we let $f_j(\mathcal{S})$ and $c_j(\mathcal{S})$ denote the first and last times that $\mathcal{S}$ runs the job j, respectively. Note that $\mathcal{S}$ does not necessarily complete j at time $c_j(\mathcal{S})$.

Definition 4 (Forest Schedule). *We say a single-machine schedule $\mathcal{S}$ is a forest schedule if for all jobs j, j' such that $f_j(\mathcal{S}) < f_{j'}(\mathcal{S})$, $\mathcal{S}$ does not run j during the time interval $(f_{j'}(\mathcal{S}), c_{j'}(\mathcal{S}))$ (so the $(f_j(\mathcal{S}), c_j(\mathcal{S}))$-intervals form a laminar family.) Then $\mathcal{S}$ naturally defines a forest (in the graph-theoretic sense), where the nodes are jobs run by $\mathcal{S}$ and the descendants of a job j are the the jobs that are first run in the time interval $(f_j(\mathcal{S}), c_j(\mathcal{S}))$.*

A non-migratory m-machine schedule is a forest schedule if all of its single-machine schedules are forest schedules.

With these definitions, we are ready to construct the near-optimal policies to which we will compare SRPT and MLAX. We omit the proof, which follows from results in [9], in this extended abstract.

Lemma 5. *Let J be a set of jobs satisfying $\ell_j \leq x_j$ for all $j \in J$. Then for any times $\hat{r}_j \in [r_j, r_j + \frac{\ell_j}{2}]$ and constant $\alpha \geq 1$, there exist non-migratory forest schedules S and S' on the jobs J such that:*

1. *Both S and S' complete every job they run.*
2. *Let J_i be the set of jobs that S runs on machine i. For every machine i and time, if there exists a feasible job in J_i, then S runs such a job.*
3. *For all jobs $j \in S$, we have $f_j(S) = \hat{r}_j$.*
4. *If job j' is a descendant of job j in S, then $\alpha x_{j'} \leq \ell_j$*
5. *$|\{leaves\ of\ S'\}| + |S| = \Omega(|\mathrm{Opt}(J)|)$.*

Intuitively, the schedule S captures the jobs in Opt that are preempted and S' captures the jobs in Opt that are not preempted (i.e. the leaves in the forest schedule.)

3   SRPT is Competitive with Non-viable Jobs

The main result of this section is that SRPT is competitive with the number of non-viable, low-laxity jobs of the optimal schedule (Theorem 6). We recall that a job j is non-viable if for *every time* in $[r_j, r_j + \frac{\ell_j}{2}]$, there are at least $\frac{1}{8}m$ jobs j' on the frontier of MLAX satisfying $\alpha x_{j'} < \ell_j$.

Theorem 6. *Let J be a set of jobs satisfying $\ell_j \leq x_j$ for all $j \in J$. Then for $\alpha = O(1)$ sufficiently large and number of machines $m \geq 16$, we have $|\mathrm{SRPT}(J)| = \Omega(|\mathrm{Opt}(J_{nv})|)$, where J_{nv} is the set of non-viable jobs with respect to $\mathrm{MLAX}(J)$.*

We omit the proof of Theorem 6 in this extended abstract. The main idea of the proof is that for any non-viable job j, MLAX is running many jobs that are much smaller than j (by at least an α-factor.) These jobs give a witness that SRPT must be working on these jobs or even smaller ones. The following technical lemma is needed in the proof as well as in Sect. 4.

Lemma 7. *Let J be any set of jobs and S be any forest schedule on m machines and jobs $J' \subset J$ that only runs feasible jobs. Let L be the set of leaves of S. Then $|\mathrm{SRPT}(J)| \geq \frac{1}{2}|L|$.*

4   SRPT and MLax Are Competitive with Viable Jobs

We have shown that SRPT is competitive with the non-viable, low-laxity jobs. Thus, it remains to account for the viable, low-laxity jobs. We recall that a job j is viable if there exists a time in $[r_j, r_j + \frac{\ell_j}{2}]$ such that there are at least $\frac{7}{8}m$ jobs j' on the frontier satisfying $\alpha x_{j'} \geq \ell_j$. The first such time is the pseudo-release time, $\tilde{r}_j$ of job j. For these jobs, we show that SRPT and MLAX together are competitive with the viable, low-laxity jobs of the optimal schedule.

Theorem 8. *Let J be a set of jobs satisfying $\ell_j \leq x_j$ for all $j \in J$. Then for $\alpha = O(1)$ sufficiently large and number of machines $m \geq 8$, we have $|\mathrm{SRPT}(J)| + |\mathrm{MLAX}(J)| = \Omega(|\mathrm{Opt}(J_v)|)$, where J_v is the set of viable jobs with respect to $\mathrm{MLAX}(J)$.*

Proof of Theorem 8. Let $\mathcal{S}, \mathcal{S}'$ be the schedules guaranteed by Lemma 5 on the set of jobs J_v with $\hat{r}_j = \tilde{r}_j$ for all $j \in J_v$. We re-state the properties of these schedules for convenience:

1. Both $\mathcal{S}$ and $\mathcal{S}'$ complete every job they run.
2. Let J_i be the set of jobs that $\mathcal{S}$ runs on machine i. For every machine i and time, if there exists a feasible job in J_i, then $\mathcal{S}$ runs such a job.
3. For all jobs $j \in \mathcal{S}$, we have $f_j(\mathcal{S}) = \tilde{r}_j$.
4. If job j' is a descendant of job j in $\mathcal{S}$, then $\alpha x_{j'} \leq \ell_j$
5. $|\{\text{leaves of } \mathcal{S}'\}| + |\mathcal{S}| = \Omega(|\mathrm{Opt}(J_v)|)$.

By Lemma 7, we have $|\mathrm{SRPT}(J)| = \Omega(|\{\text{leaves of } \mathcal{S}'\}|)$. Thus, it suffices to show that $|\mathrm{SRPT}(J)| + |\mathrm{MLAX}(J)| = \Omega(|\mathcal{S}|)$. We do this with two lemmas, whose proofs we defer until later. First, we show that MLAX pushes (not necessarily completes) many jobs. In particular, we show:

Lemma 9. $|\mathrm{SRPT}(J)| + \#(\text{pushes of } \mathrm{MLAX}(J)) = \Omega(|\mathcal{S}|)$

The main idea to prove Lemma 9 is to consider sequences of preemptions in Opt. In particular, suppose Opt preempts job a for b and then b for c. Roughly, we use viability to show that the only way MLAX doesn't push any of these jobs is if in between their pseudo-release times, MLAX pushes $\Omega(m)$ jobs.

Second, we show that the pushes of MLAX give a witness that SRPT and MLAX together actually complete many jobs.

Lemma 10. $|\mathrm{SRPT}(J)| + |\mathrm{MLAX}(J)| = \Omega(\#(\text{pushes of } \mathrm{MLAX}(J)))$.

The main idea to prove Lemma 10 is to upper-bound the number of jobs that MLAX pops because they are infeasible (all other pushes lead to completed jobs.) The reason MLAX pops a job j for being infeasible is because while j was on a stack, MLAX spent at least $\frac{\ell_j}{2}$ units of time running jobs higher than j on j's stack. Either those jobs are completed by MLAX, or MLAX must have have done many pushes or replacements instead. We show that the replacements give a witness that SRPT must complete many jobs.

Combining these two lemmas completes the proof of Theorem 8. □

Now we go back and prove Lemma 9 and Lemma 10.

4.1 Proof of Lemma 9

Recall that $\mathcal{S}$ is a forest schedule. We say the *first child* of a job j is the child j' of j with the earliest starting time $f_{j'}(\mathcal{S})$. In other words, if j is not a leaf, then its first child is the first job that pre-empts j. We first focus on a sequence of first children in $\mathcal{S}$.

Lemma 11. *Let $a, b, c \in S$ be jobs such that b is the first child of a and c is the first child of b. Then* $\mathrm{MLAX}(J)$ *does at least one of the following during the time interval* $[\tilde{r}_a, \tilde{r}_c]$:

– *Push at least $\frac{m}{8}$ jobs,*
– *Push job b,*
– *Push a job on top of b when b is on the frontier,*
– *Push c.*

Proof. Because S is a forest schedule, we have $\tilde{r}_a < \tilde{r}_b < \tilde{r}_c$. It suffices to show that if during $[\tilde{r}_a, \tilde{r}_c]$, $\mathrm{MLAX}(J)$ pushes strictly fewer than $\frac{m}{8}$ jobs, $\mathrm{MLAX}(J)$ does not push b, and if $\mathrm{MLAX}(J)$ does not push any job on top of b when b is on the frontier, then $\mathrm{MLAX}(J)$ pushes c.

First, because $\mathrm{MLAX}(J)$ pushes strictly fewer than $\frac{m}{8}$ jobs during $[\tilde{r}_a, \tilde{r}_c]$, there exists at least $\frac{7}{8}m$ stacks that receive no push during this interval. We call such stacks *stable*. The key property of stable stacks is that the laxities of their top- and second-top jobs never decrease during this interval, because these stacks are only changed by replacements and pops.

Now consider time $\tilde{r}_a$. By definition of pseudo-release time, at this time, there exist at least $\frac{7}{8}m$ stacks whose top job j' satisfies $\alpha x_{j'} \geq \ell_j$. Further, for any such stack, let j'' be its second-top job. Then because b is a descendant of a in S, we have:

$$\alpha x_b \leq \ell_a \leq \alpha x_{j'} \leq \ell_{j''}.$$

It follows that there exist at least $\frac{3}{4}m$ stable stacks whose second-top job j'' satisfies $\alpha x_b \leq \ell_{j''}$ for the entirety of $[\tilde{r}_a, \tilde{r}_c]$. We say such stacks are *b-stable*.

Now consider time $\tilde{r}_b$. We may assume b is not pushed at this time. However, there exist at least $\frac{3}{4}m$ that are b-stable. Thus, if we do not replace the top of some stack with b, it must be the case that the top job j' of every b-stable stack satisfies $\ell'_j \geq \ell_b$. Because these stacks are stable, their laxities only increase by time $\tilde{r}_c$, so $\mathrm{MLAX}(J)$ will push c on some stack at that time.

Otherwise, suppose we replace the top job of some stack with b. Then b is on the frontier at $\tilde{r}_b$. We may assume that no job is pushed directly on top of b. If b remains on the frontier by time $\tilde{r}_c$, then $\mathrm{MLAX}(J)$ will push c, because $\alpha x_c \leq \ell_b$. The remaining case is if b leaves the frontier in some time in $[\tilde{r}_b, \tilde{r}_c]$. We claim that it cannot be the case that b is popped, because by (2), S could not complete b by time $\tilde{r}_c$, so $\mathrm{MLAX}(J)$ cannot as well. Thus, it must be the case that b is replaced by some job, say d at time $\tilde{r}_d$. At this time, there exist at least $\frac{3}{4}m$ stacks whose second-top job j'' satisfies $\alpha x_d \leq \ell_{j''}$. It follows, there exist at least $\frac{m}{2}$ b-stable stacks whose second-top job j'' satisfies $\alpha x_d \leq \ell_{j''}$ at time $\tilde{r}_d$. Note that because $m \geq 8$, there exists at least one such stack, say i, that is not b's stack. In particular, because b's stack has minimum laxity, it must be the case that the top job j' of stack i satisfies $\ell_{j'} \geq \ell_b$. Finally, because stack i is stable, at time $\tilde{r}_c$ we will push c.   □

Now using the above lemma, we give a charging scheme to prove Lemma 9. First note that by Lemma 7, we have $|\mathrm{SRPT}(J)| = \Omega(\#(\text{leaves of } S))$. Thus, it suffices to give a charging scheme such that each job $a \in S$ begins with 1 credit, and charges it to leaves of S and completions of $\text{MLax}(J)$ so that each job is charged $O(1)$ credits. Each job $a \in S$ distributes its 1 credit as follows:

- (Leaf Transfer) If a is a leaf or parent of a leaf of S, say ℓ, then a charges ℓ for 1 credit.

Else let b be the first child of a and c the first child of b in S

- (Push Transfer) If $\text{MLax}(J)$ pushes b or c, then a charges 1 unit to b or c, respectively.
- (Interior Transfer) Else if job b is on the frontier, but another job, say d, is pushed on top of b, then a charges 1 unit to d.
- (m-Push Transfer) Otherwise, by Lemma 11, $\text{MLax}(J)$ must push at least $\frac{m}{8}$ jobs during $[\tilde{r}_a, \tilde{r}_c]$. In this case, a charges $\frac{8}{m}$ units to each of these $\frac{m}{8}$ such jobs.

This completes the description of the charging scheme. It remains to show that each job is charged $O(1)$ credits. Each job receives at most 2 credits due to Leaf Transfers and at most 2 credits due to Push Transfers and Interior Transfers. As each job is in at most $3m$ intervals of the form $[\tilde{r}_a, \tilde{r}_c]$, each job is charged $O(1)$ from m-Push Transfers.

4.2   Proof of Lemma 10

Recall in MLax, there are two types of pops: a job is popped if it is completed, and then we continue popping until the top job of that stack is feasible. We call the former *completion pops* and the later *infeasible pops*. Note that it suffices to prove the next lemma, which bounds the infeasible pops. This is because #(pushes of $\text{MLax}(J)$) = #(completions pops of $\text{MLax}(J)$) + #(infeasible pops of $\text{MLax}(J)$). To see this, note that every stack is empty at the beginning and end of the algorithm, and the stack size only changes due to pushes and pops.

Lemma 12. *For $\alpha = O(1)$ sufficiently large, we have:*

$$|\text{SRPT}(J)| + |\text{MLax}(J)| + \#(\text{pushes of } \text{MLax}(J)) \geq 2 \cdot \#(\text{infeasible pops of } \text{MLax}(J)).$$

Proof. We define a charging scheme such that the completions of $\text{SRPT}(J)$ and $\text{MLax}(J)$ and the pushes executed by $\text{MLax}(J)$ pay for the infeasible pops. Each completion of $\text{SRPT}(J)$ is given 2 credits, each completion of $\text{MLax}(J)$ is given 1 credit, and each job that $\text{MLax}(J)$ pushes is given 1 credit. Thus each job begins with at most 4 credits. For any $z \geq 0$, we say job j' is z-below j (at time t) if j' and j are on the same stack in $\text{MLax}(J)$ and j' is z positions below j on that stack at time t. We define z-above analogously. A job j distributes these initial credits as follows:

- (SRPT-transfer) If $\text{SRPT}(J)$ completes job j and MLAX also ran j at some point, then j gives $\frac{1}{2^{z+1}}$ credits to the job that is z-below j at time $f_j(\text{MLAX}(J))$ for all $z \geq 0$.
- (m-SRPT-transfer) If $\text{SRPT}(J)$ completes job j at time t, then j gives $\frac{1}{2^{z+1}}\frac{1}{m}$ credits to the job that is z-below the top of each stack in $\text{MLAX}(J)$ at time t for all $z \geq 0$.
- (MLAX-transfer) If $\text{MLAX}(J)$ completes a job j, then j gives $\frac{1}{2^{z+1}}$ credits to the job that is z-below j at the time j is completed for all $z \geq 0$.
- (Push-transfer) If $\text{MLAX}(J)$ pushes a job j, then j gives $\frac{1}{2^{z+1}}$ credits to the job that is z-below j at the time j is pushed for all $z \geq 0$.

It remains to show that for $\alpha = O(1)$ sufficiently large, every infeasible pop gets at least 4 credits. We consider any job j that is an infeasible pop of $\text{MLAX}(J)$. At time $\tilde{r}_j$ when j joins some stack in $\text{MLAX}(J)$, say H, j's remaining laxity was at least $\frac{\ell_j}{2}$. However, as j later became an infeasible pop, it must be the case that while j was on stack H, $\text{MLAX}(J)$ was running jobs that are higher than j on stack H for at least $\frac{\ell_j}{2}$ units of time.

Let I be the union of intervals of times that $\text{MLAX}(J)$ runs a job higher than j on stack H (so j is on the stack for the entirety of I.) Then we have $|I| \geq \frac{\ell_j}{2}$. Further, we partition I based on the height of the job on H that $\text{MLAX}(J)$ is currently running. In particular, we partition $I = \bigcup_{z \geq 1} I_z$, where I_z is the union of intervals of times that $\text{MLAX}(J)$ runs a job on H that is exactly z-above j.

By averaging, there exists a $z \geq 1$ such that $|I_z| \geq \frac{\ell_j}{2^{z+1}}$. Fix such a z. We can write I_z as the union of disjoint intervals, say $I_z = \bigcup_{u=1}^{s}[a_u, b_u]$. Because during each sub-interval, $\text{MLAX}(J)$ is running jobs on H that are much smaller than j itself, these jobs give a witness that $\text{SRPT}(J)$ completes many jobs as long as these sub-intervals are long enough. We formalize this in the following proposition, whose proof is omitted in this extended abstract.

Proposition 13. *In each sub-interval $[a_u, b_u]$ of length at least $4\frac{\ell_j}{\alpha^z}$, job j earns at least $\frac{1}{2^{z+3}}\frac{b_u - a_u}{\ell_j/\alpha^z}$ credits from SRPT-transfers and m-SRPT-transfers.*

On the other hand, even if the sub-intervals are too short, the job j still gets credits from MLAX-transfers and Push-transfers when the height of the stack changes. We formalize this statement in the following proposition, whose proof is omitted in this extended abstract.

Proposition 14. *For every sub-interval $[a_u, b_u]$, job j earns at least $\frac{1}{2^{z+2}}$ credits from MLAX-transfers and Push-transfers at time b_u.*

Now we combine the above two propositions to complete the proof of Lemma 12. We say a sub-interval $[a_u, b_u]$ is *long* if it has length at least $4\frac{\ell_j}{\alpha^z}$ (i.e. we can apply Proposition 13 to it) and *short* otherwise. First, suppose the aggregate length of all long intervals it at least $4 \cdot 2^{z+3}\frac{\ell_j}{\alpha^z}$. Then by Proposition 13, job j gets at least 4 credits from the long intervals. Otherwise, the aggregate length of all long intervals is less than $4 \cdot 2^{z+3}\frac{\ell_j}{\alpha^z}$. In this case, recall that the long and short intervals partition I_z, which has length at least $\frac{\ell_j}{2^z+1}$. It follows, the aggregate length of the short intervals is at least $\frac{\ell_j}{2^z+1} - 4 \cdot 2^{z+3} \frac{\ell_j}{\alpha^z}$. For $\alpha = O(1)$ large enough, we may assume the aggregate length of the short intervals is at least $4 \cdot 2^{z+2} \frac{4\ell_j}{\alpha^z}$. Because each short interval has length at most $4\frac{\ell_j}{\alpha^z}$, there are at least $4 \cdot 2^{z+2}$ short intervals. We conclude, by Proposition 14, job j gets at least 4 credits from the short intervals. We conclude, in either case job j gets at least 4 credits. □

5 Putting it all Together

In this section, we prove our main result, Theorem 1, which follows from the next meta-theorem:

Theorem 15. *Let J be any set of jobs. Then for number of machines $m \geq 16$, we have $|\mathrm{LMNY}(J_{hi})| + |\mathrm{SRPT}(J_{lo})| + |\mathrm{MLAX}(J_{lo})| = \Omega(|\mathrm{Opt}(J)|)$, where $J_{hi} = \{j \in J \mid \ell_j > x_j\}$ and $J_{lo} = \{j \in J \mid \ell_j \leq x_j\}$ partition J into high- and low-laxity jobs.*

Proof. We have $|\mathrm{LMNY}(J_{hi})| = \Omega(|\mathrm{Opt}(J_{hi})|)$ by Lemma 3. Also, we further partition $J_{lo} = J_v \cup J_{nv}$ into the viable and non-viable jobs with respect to $\mathrm{MLAX}(J_{lo})$. Then Theorem 6 and Theorem 8 together give $|\mathrm{SRPT}(J_{lo})| + |\mathrm{MLAX}(J_{lo})| = \Omega(|\mathrm{Opt}(J_v)| + |\mathrm{Opt}(J_{nv})|)$. To complete the proof, we observe that $J = J_{hi} \cup J_v \cup J_{nv}$ partitions J, so $|\mathrm{Opt}(J_{hi})| + |\mathrm{Opt}(J_v)| + |\mathrm{Opt}(J_{nv})| = \Omega(|\mathrm{Opt}(J)|)$. □

The proof of Theorem 2, which gives our performance guarantee for FINALALG is immediate:

Proof of Theorem 2. By combining Theorem 15 and Lemma 4, the objective value achieved by FINALALG is:

$$\Omega(|\mathrm{LMNY}(J_{hi}, \frac{m}{3})| + |\mathrm{SRPT}(J_{lo}, \frac{m}{3})| + |\mathrm{MLAX}(J_{lo}, \frac{m}{3})|) = \Omega(|\mathrm{Opt}(J, \frac{m}{3})|)$$
$$= \Omega(|\mathrm{Opt}(J, m)|).$$

□

Finally, we obtain our $O(1)$-competitive deterministic algorithm for all $m > 1$ (recall FINALALG is $O(1)$-competitive only when $m \geq 48$) by using a two-machine algorithm when m is too small:

Proof of Theorem 1. Our algorithm is the following: If $1 < m < 48$, then we run the deterministic two-machine algorithm from [9] which is $O(1)$-competitive with the optimal single-machine schedule. Thus by Lemma 4, this algorithm is also $O(m) = O(1)$-competitive for all $m < 48$. Otherwise, $m \geq 48$, so we run FINALALG. □

References

1. Azar, Y., Cohen, S.: An improved algorithm for online machine minimization. Oper. Res. Lett. **46**(1), 128–133 (2018)
2. Baruah, S.K., Koren, G., Mao, D., Mishra, B., Raghunathan, A., Rosier, L.E., Shasha, D.E., Wang, F.: On the competitiveness of on-line real-time task scheduling. Real Time Syst. **4**(2), 125–144 (1992)
3. Chen, L., Eberle, F., Megow, N., Schewior, K., Stein, C.: A general framework for handling commitment in online throughput maximization. Math. Program. **183**(1), 215–247 (2020)
4. Chen, L., Megow, N., Schewior, K.: An o(log m)-competitive algorithm for online machine minimization. SIAM J. Comput. **47**(6), 2057–2077 (2018)
5. Eberle, F., Megow, N., Schewior, K.: Optimally handling commitment issues in online throughput maximization. In: Grandoni, F., Herman, G., Sanders, P. (eds.) European Symposium on Algorithms, LIPIcs, vol. 173, pp. 41:1–41:15. Schloss Dagstuhl - Leibniz-Zentrum für Informatik (2020)
6. Im, S., Moseley, B., Pruhs, K., Stein, C.: An o(log log m)-competitive algorithm for online machine minimization. In: 2017 IEEE Real-Time Systems Symposium, pp. 343–350. IEEE Computer Society (2017)
7. Kalyanasundaram, B., Pruhs, K.: Speed is as powerful as clairvoyance. J. ACM **47**(4), 617–643 (2000). Also 1995 Symposium on Foundations of Computer Science
8. Kalyanasundaram, B., Pruhs, K.: Eliminating migration in multi-processor scheduling. J. Algorithms **38**(1), 2–24 (2001)
9. Kalyanasundaram, B., Pruhs, K.: Maximizing job completions online. J. Algorithms **49**(1), 63–85 (2003). Also 1998 European Symposium on Algorithms
10. Koren, G., Shasha, D.E.: MOCA: a multiprocessor on-line competitive algorithm for real-time system scheduling. Theor. Comput. Sci. **128**(1&2), 75–97 (1994)
11. Lucier, B., Menache, I., Naor, J., Yaniv, J.: Efficient online scheduling for deadline-sensitive jobs. In: Blelloch, G.E., Vöcking, B. (eds.) ACM Symposium on Parallelism in Algorithms and Architectures, pp. 305–314. ACM (2013)
12. Phillips, C.A., Stein, C., Torng, E., Wein, J.: Optimal time-critical scheduling via resource augmentation. Algorithmica **32**(2), 163–200 (2002)

The Limits of Local Search for Weighted k-Set Packing

Meike Neuwohner[(✉)] [iD]

Research Institute for Discrete Mathematics, University of Bonn, Bonn, Germany
neuwohner@or.uni-bonn.de

Abstract. We consider the weighted k-Set Packing Problem, where, given a collection $\mathcal{S}$ of sets, each of cardinality at most k, and a positive weight function $w : \mathcal{S} \to \mathbb{R}_{>0}$, the task is to find a sub-collection of $\mathcal{S}$ consisting of pairwise disjoint sets of maximum total weight. As this problem does not permit a polynomial-time $o(\frac{k}{\log k})$-approximation unless $P = NP$ [11], most previous approaches rely on local search. For twenty years, Berman's algorithm *SquareImp* [2], which yields a polynomial-time $\frac{k+1}{2} + \epsilon$-approximation for any fixed $\epsilon > 0$, has remained unchallenged. Only recently, it could be improved to $\frac{k+1}{2} - \frac{1}{63,700,992} + \epsilon$ by Neuwohner [16]. In her paper, she showed that instances for which the analysis of SquareImp is almost tight are "close to unweighted" in a certain sense. But for the unit weight variant, the best known approximation guarantee is $\frac{k+1}{3} + \epsilon$. Using this observation as a starting point, we conduct a more in-depth analysis of close-to-tight instances of SquareImp. This finally allows us to generalize techniques used in the unweighted case to the weighted setting. In doing so, we obtain approximation guarantees of $\frac{k+\epsilon_k}{2}$, where $\lim_{k\to\infty} \epsilon_k = 0$. On the other hand, we prove that this is asymptotically best possible in that searching for local improvements of logarithmically bounded size cannot produce an approximation ratio below $\frac{k}{2}$.

Keywords: Weighted k-Set Packing · Local search · d-Claw free graphs · Independent set

1 Introduction

For a positive integer k, the *weighted k-Set Packing Problem* is defined as follows: Given a family $\mathcal{S}$ of sets each of size at most k together with a positive weight function $w : \mathcal{S} \to \mathbb{R}_{>0}$, the task is to find a sub-collection A of $\mathcal{S}$ of maximum weight such that the sets in A are pairwise disjoint. For $k \leq 2$, the weighted k-Set Packing Problem reduces to the maximum weight matching problem, which can be solved in polynomial time [7]. However, as soon as $k \geq 3$, weighted k-Set Packing becomes NP-hard since it generalizes the optimization variant of the NP-complete 3-dimensional-matching problem [13]. Even more, Hazan, Safra and Schwartz [11] have shown that there cannot be a polynomial-time $o\left(\frac{k}{\log k}\right)$-approximation for weighted k-Set Packing unless $P = NP$.

© Springer Nature Switzerland AG 2022
K. Aardal and L. Sanitá (Eds.): IPCO 2022, LNCS 13265, pp. 415–428, 2022.
https://doi.org/10.1007/978-3-031-06901-7_31

On the positive side, a simple greedy algorithm yields an approximation guarantee of k. In order to improve on this, the technique that has proven most successful so far is *local search*. The basic idea is to start with an arbitrary solution (e.g. the empty one) and to iteratively improve the current solution by applying some sort of local modifications until no more of these exist. More precisely, given a feasible solution A to the weighted k-Set Packing problem, we call a collection $X \subseteq \mathcal{S} \setminus A$ consisting of pairwise disjoint sets a *local improvement of A of size $|X|$* if $w(X) > w(\{a \in A : \exists x \in X : a \cap x \neq \emptyset\})$, that is, if replacing the collection of sets in A that intersect sets in X by the sets in X increases the weight of the solution. Note that whenever A is sub-optimum and A^* is a solution of maximum weight, then $A^* \setminus A$ defines a local improvement of A. However, if one aims at designing a polynomial-time algorithm, it is of course infeasible to check subsets of $\mathcal{S}$ of arbitrarily large size.

1.1 The Unit Weight Case

For the special case of unit weights, Hurkens and Schrijver [12] showed that searching for local improvements of arbitrary large, but constant size results in approximation guarantees arbitrarily close to $\frac{k}{2}$. Their paper also provides matching lower bound examples proving their result to be tight. Since then, a lot of progress has been made regarding the special case where $w \equiv 1$, which we will also refer to as the *unweighted k-Set Packing Problem*. In 1995, at the cost of a quasi-polynomial running time, Halldórsson [10] achieved an approximation factor of $\frac{k+2}{3}$ by applying local improvements of size logarithmic in the total number of sets. Cygan, Grandoni and Mastrolilli [6] managed to get down to an approximation factor of $\frac{k+1}{3} + \epsilon$, still with a quasi-polynomial running time.

The first polynomial-time algorithm improving on the result by Hurkens and Schrijver [12] was obtained by Sviridenko and Ward [19] in 2013. By combining means of color coding with the algorithm presented in [10], they achieved an approximation ratio of $\frac{k+2}{3}$. This result was further improved to $\frac{k+1}{3} + \epsilon$ for any fixed $\epsilon > 0$ by Cygan [5], obtaining a polynomial running time doubly exponential in $\frac{1}{\epsilon}$. The best approximation algorithm for the unweighted k-Set Packing Problem in terms of performance ratio and running time is due to Fürer and Yu [9] from 2014. They achieve the same approximation guarantee as Cygan [5], but a running time only singly exponential in $\frac{1}{\epsilon^2}$. Moreover, they show that their result is best possible in that there exist arbitrarily large instances that feature solutions that do not permit any local improvement of size $o(|\mathcal{S}|^{\frac{1}{5}})$, but that are by a factor of $\frac{k+1}{3}$ smaller than the optimum.

1.2 General Weights and the MWIS in d-Claw Free Graphs

In the weighted setting, much less is known. Arkin and Hassin [1] have shown that unlike the unit weight case, searching for local improvements of constant size cannot produce an approximation ratio better than $k-1$ for general weights. Both papers improving on this deal with a more general problem, the *Maximum Weight Independent Set Problem (MWIS) in $k+1$-claw free graphs* [2,4]:

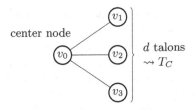

Fig. 1. A d-claw C for $d = 3$.

For $d \geq 1$, a d-*claw* C [2] is defined to be a star consisting of one *center node* and a set T_C of d *talons* connected to it (see Fig. 1). An undirected graph $G = (V, E)$ is said to be d-*claw free* if none of its induced sub-graphs forms a d-claw. For $d \leq 3$, the MWIS in d-claw free graphs can be solved in polynomial time (see [14,18] for the unweighted, [15] for the weighted variant), while for $d \geq 4$, again no $o(\frac{d}{\log d})$-approximation algorithm is possible unless $P = NP$ [11].

If we define an independent set $X \subseteq V \setminus A$ to be a local improvement of A if the weight of X exceeds the weight of its neighborhood in A, then most of the previous results for the (weighted or unweighted) k-Set Packing Problem also apply to the more general context of the MWIS in $k + 1$-claw free graphs. However, it is not known how to get down to a polynomial (instead of quasi-polynomial) running time for the algorithms in [5,19] and [9] since there is no obvious equivalent to coloring the underlying universe.

By considering the *conflict graph* G_S associated with an instance of weighted k-Set Packing, one obtains a weight preserving one-to-one correspondence between feasible solutions to the k-Set Packing Problem and independent sets in G_S. The vertices of G_S are given by the sets in S and the edges represent non-empty set intersections. It is not hard to see that G_S is $k + 1$-claw free.

The first significant improvement over the approximation guarantee of $d - 1$ achieved by the greedy approach for the MWIS in d-claw free graphs was made by Chandra and Halldórsson [4]. In each iteration, their algorithm *BestImp* picks a certain type of local improvement that maximizes the ratio between the total weight of the vertices added to and removed from the current solution. By further scaling and truncating the weight function to ensure a polynomial number of iterations, Chandra and Halldórsson [4] obtain a $\frac{2d}{3} + \epsilon$-approximation algorithm for the MWIS in d-claw free graphs.

1.3 Berman's Algorithm SquareImp

For 20 years, Berman's algorithm *SquareImp* [2] has been the state-of-the-art for both the MWIS in d-claw free graphs and weighted k-Set Packing. SquareImp proceeds by iteratively applying local improvements of the squared weight function that arise as sets of talons of claws in G, until no more exist. In doing so, SquareImp achieves an approximation ratio of $\frac{d}{2}$, leading to a polynomial-time $\frac{d}{2} + \epsilon$-approximation algorithm for the MWIS in d-claw free graphs for any fixed $\epsilon > 0$ (and a $\frac{k+1}{2} + \epsilon$-approximation for weighted k-Set Packing).

Berman [2] also provides an example for $w \equiv 1$ showing that his analysis is tight. As the example uses unit weights, he concludes that applying the same type of local improvement algorithm for a different power of the weight function does not provide further improvements. However, as also implied by the result in [12], while no small improvements *forming the set of talons of a claw* in the input graph exist in the tight example given by Berman [2], once this additional condition is dropped, improvements of small constant size can be found quite easily. This observation is the basis of a recent paper by Neuwohner [16], who managed to obtain an approximation guarantee slightly below $\frac{d}{2}$ by taking into account a broader class of local improvements, namely all improvements of the squared weight function of size at most $(d-1)^2 + (d-1)$.

2   Our Contribution

In this paper, we revisit the analysis of the algorithm SquareImp proposed by Berman [2]. Following [16], we show that whenever the analysis is close to being tight, the instance is locally unweighted in the sense that almost every time when a vertex from the solution chosen by SquareImp and a vertex from any optimum solution share an edge, their weights must be very similar. While [16] merely focuses on one of the two major steps in Berman's analysis, we consider both of them, allowing us to derive much stronger statements concerning the structure of instances where SquareImp does not do much better than a $\frac{d}{2}$-approximation. In particular, we are able to transfer techniques that are used in the state-of-the-art works on the unweighted k-Set Packing Problem [5,9] to a setting where vertex weights are locally similar. This is the main ingredient for our algorithm *LogImp*. In addition to the type of improvements considered by SquareImp, LogImp searches for a certain type of local improvement of logarithmic size. In doing so, it obtains an approximation guarantee of $\frac{d-1+\epsilon_d}{2}$ for the MWIS in d-claw free graphs for $d \geq 3$, where $0 \leq \epsilon_d \leq 1$ and $\lim_{d \to \infty} \epsilon_d = 0$. While we can only guarantee a quasi-polynomial running time for the MWIS, we manage to obtain a polynomial-time $\frac{k+\epsilon_{k+1}}{2}$-approximation for our main focus, the weighted k-Set Packing Problem, by means of color coding.

We further prove this result to be asymptotically tight by providing examples which show that any local improvement algorithm that, for an arbitrarily chosen, but fixed parameter $\alpha \in \mathbb{R}$, searches for local improvements of w^α of size $\mathcal{O}(\log(|\mathcal{S}|))$, cannot produce an approximation guarantee better than $\frac{k}{2}$ for the weighted k-Set Packing Problem with $k \geq 3$.

The latter significantly extends the state of knowledge in terms of lower bound examples. Even more importantly, we can finally (at least asymptotically) answer the long-standing question of how far one can get by using pure local search in the weighted setting. In doing so, we are also the first ones to port the idea of searching for local improvements of logarithmic size, which has proven very successful for unit weights [5,6,9,10,19], to the weighted setting.

The rest of this paper is organized as follows: Sect. 3 recaps some of the definitions and main results from [2] that we will employ in the analysis of our local improvement algorithm *LogImp* in Sect. 4. Section 5 then sketches our lower bound construction and Sect. 6 provides a brief conclusion.

3   Preliminaries

Definition 1 (Neighborhood [2]). *Given an undirected graph $G = (V, E)$ and subsets $U, W \subseteq V$ of its vertices, we define the* neighborhood $N(U, W)$ *of U in W as $N(U, W) := \{w \in W : \exists u \in U : \{u, w\} \in E \vee u = w\}$. For $u \in V$ and $W \subseteq V$, we write $N(u, W)$ instead of $N(\{u\}, W)$.*

Notation. Given $w : V \to \mathbb{R}$ and $U \subseteq V$, we write $w^2(U) := \sum_{u \in U} w^2(u)$.

Definition 2 ([2]). *Given an undirected graph $G = (V, E)$, a weight function $w : V \to \mathbb{R}_{\geq 0}$ and an independent set $A \subseteq V$, we say that a vertex set $B \subseteq V$* improves $w^2(A)$ *if B is independent in G and $w^2((A \setminus N(B, A)) \cup B) > w^2(A)$ holds. For a claw C in G, we say that C improves $w^2(A)$ if T_C does and call T_C a* claw-shaped *improvement in this case. We further define a 0-claw to consist of a single talon and an empty center.*

Note that in contrast to the introduction, we do not require a local improvement B to be disjoint from A anymore. Further observe that an independent set B improves A if and only if we have $w^2(B) > w^2(N(B, A))$.

Using the notation introduced above, Berman's algorithm SquareImp [2] can now be formulated as in Algorithm 1. As all weights are positive, every $v \notin A$ such that $A \cup \{v\}$ is independent constitutes the talon of a 0-claw improving $w^2(A)$, so the algorithm returns a maximal independent set.

The main idea of the analysis of SquareImp presented in [2] is to charge the vertices in A for preventing adjacent vertices in an optimum solution A^* from being included into A. The latter is done by spreading the weight of the vertices in A^* among their neighbors in the maximal independent set A in such a way that no vertex in A receives more than $\frac{d}{2}$ times its own weight. The suggested distribution of weights proceeds in two steps:

First, each vertex $u \in A^*$ invokes costs of $\frac{w(v)}{2}$ at each $v \in N(u, A)$. As G is d-claw free, no vertex $v \in A$ can have more than $d - 1$ neighbors in A^*. This implies that the total amount of weight v receives in this step is bounded by $|N(v, A^*)| \cdot \frac{w(v)}{2} \leq (d - 1) \cdot \frac{w(v)}{2}$.

In a second step, each vertex $u \in A^*$ sends an amount of $w(u) - \frac{w(N(u,A))}{2}$ to a heaviest neighbor it possesses in A, which is captured by the following definition of *charges*:

Definition 3 (Charges [2]). *For each $u \in A^*$, pick a vertex $v \in N(u, A)$ of maximum weight and call it $n(u)$ (recall that A is maximal).*
For $u \in A^*$ and $v \in A$, define

$$\text{charge}(u, v) := \begin{cases} w(u) - \frac{w(N(u,A))}{2} & , \text{ if } v = n(u) \\ 0 & , \text{ otherwise} \end{cases}.$$

Algorithm 1: SquareImp [2]

Input: an undirected d-claw free graph $G = (V, E)$, $w : V \rightarrow \mathbb{R}_{>0}$
Output: an independent set $A \subseteq V$

1   $A \leftarrow \emptyset$
2   **while** *there exists a claw C in G that improves $w^2(A)$* **do**
3     $\lfloor$ $A \leftarrow (A \setminus N(T_C, A)) \cup T_C$
4   **return** A

Algorithm 2: LogImp

Input: an undirected d-claw free graph $G = (V, E)$, $w : V \rightarrow \mathbb{R}_{>0}$
Output: an independent set $A \subseteq V$

1   $A \leftarrow \emptyset$
2   **while** *there exists a claw-shaped or circular local improvement X of $w^2(A)$* **do**
3     $\lfloor$ $A \leftarrow (A \setminus N(X, A)) \cup X$
4   **return** A

As we have already seen that each vertex $v \in A$ receives at most $\frac{d-1}{2} \cdot w(v)$ during the first step of the weight distribution, it suffices to show that the total amount of positive charges it has to pay is bounded by $\frac{w(v)}{2}$. In order to prove this, we want to exploit the fact that when SquareImp terminates, there is no improving claw centered at v. To this end, suppose that we want to construct an improving claw C centered at v and consider adding $u \in N(v, A^*)$ to its set T_C of talons. On the one hand, this increases $w^2(T_C)$ by $w^2(u)$. On the other hand, $w^2(N(T_C, A))$ may also increase by up to $w^2(N(u, A) \setminus \{v\})$ (if our claw should be centered at v, we have to pay for v anyways). In case $w^2(u) > w^2(N(u, A) \setminus \{v\})$, we surely want to add u to our claw, otherwise, we may choose not to. This is captured by the definition of the *contribution*:

Definition 4 (Contribution [16]). *Define a contribution map*

$$\text{contr} : A^* \times A \rightarrow \mathbb{R}_{\geq 0} \text{ by setting}$$

$$\text{contr}(u, v) := \begin{cases} \max\left\{0, \frac{w^2(u) - w^2(N(u,A) \setminus \{v\})}{w(v)}\right\} & , \text{ if } v \in N(u, A) \\ 0 & , \text{ else} \end{cases}.$$

The fact that there is no improving claw directly implies that the total contribution to $v \in A$ is bounded by $w(v)$. Moreover, a simple calculation shows that $2 \cdot \text{charge}(u, v) \leq \text{contr}(u, v)$ for all $u \in N(v, A^*)$ [16]. This finishes the analysis of SquareImp.

4   Our Algorithm LogImp

Our algorithm *LogImp* (Algorithm 2) starts with the empty solution, and then iteratively checks for the existence of improving claws as in Berman's algorithm

SquareImp and another type of local improvement, which we call *circular*. It corresponds to a cycle of logarithmically bounded size in a certain auxiliary graph. In particular, when LogImp terminates, there cannot be any further improving claw, so we can apply the analysis of SquareImp presented in Sect. 3 to our algorithm. Similar to [16], the key idea in analyzing LogImp is the following: Either the analysis of SquareImp is far enough from being tight to achieve the desired approximation ratio, or the instance at hand bears a certain structure that allows us to derive the existence of a circular improvement. But this contradicts the termination criterion of LogImp. Our main result is the following:

Theorem 5. *There is a sequence $(\epsilon_d)_{d\geq 3} \in [0,1]^{\mathbb{N}_{\geq 3}}$ with $\lim_{d\to\infty} \epsilon_d = 0$ such that LogImp yields a $\frac{d-1+\epsilon_d}{2}$-approximation for the MWIS in d-claw free graphs.*

As applying the analysis of SquareImp to LogImp proves that the approximation guarantee of LogImp is no larger than $\frac{d}{2}$, it suffices to show that for every $\delta > 0$, there is $d_0 \geq 3$ such that for any $d \geq d_0$, LogImp is a $\frac{d-1+\delta}{2}$-approximation for the MWIS in d-claw free graphs.

Fix $\delta > 0$, denote the solution returned by LogImp by A, and let A^* be an optimum solution. Moreover, fix two maps n and n_2 mapping each vertex in $V \setminus A$ to a heaviest neighbor in A, and each vertex in $V \setminus A$ with at least two neighbors in A to a second heaviest neighbor in A, respectively.

4.1 Classification of Vertices from A^*

We now provide a classification of the vertices in A^* that helps us to understand the structural properties of near-tight instances. For this purpose, we fix two constants $0 < \epsilon' \ll \sqrt{\epsilon'} \ll \tilde{\epsilon} < \delta < 1$ subject to certain inequalities (see [17]).

Lemma 6. *Each $u \in A^*$ with $\mathrm{charge}(u, n(u)) > 0$ is of one of the following three types:*
single: $\frac{w(u)}{w(n(u))} \in [1 - \sqrt{\epsilon'}, 1 + \sqrt{\epsilon'}]$ *and* $w(N(u, A)) \leq (1 + \sqrt{\epsilon'}) \cdot w(n(u))$
double: $|N(u, A)| \geq 2$, $\frac{w(u)}{w(n(u))} \in [1 - \sqrt{\epsilon'}, 1 + \sqrt{\epsilon'}]$, $\frac{w(n_2(u))}{w(n(u))} \in [1 - \sqrt{\epsilon'}, 1]$ *and* $(2 - \sqrt{\epsilon'}) \cdot w(n(u)) \leq w(N(u, A)) < 2 \cdot w(u)$
contributive: $\mathrm{contr}(u, n(u)) \geq 2 \cdot \mathrm{charge}(u, n(u)) + \frac{\epsilon'}{2} \cdot w(u)$

For a *single* vertex u, the heaviest neighbor $n(u)$ of u in A has almost the same weight as u and makes up almost all of $N(u, A)$ in terms of weight. Single vertices correspond to vertices of degree 1 to A in the unit weight case.

For a *double* vertex u, its two heaviest neighbors in A, $n(u)$ and $n_2(u)$, have roughly the same weight as u and make up most of $N(u, A)$. Double vertices can be thought of as having degree 2 to A.

For a *contributive* vertex u, we gain a constant fraction of $w(u)$ in the analysis.

Lemma 7. *Each $u \in A^*$ with $\mathrm{charge}(u, n(u)) \leq 0$ is of one of the following three types:*
payback: $w(N(u, A)) \geq (2 + \epsilon') \cdot w(u)$

good: $|N(u, A)| \geq 2$, $\frac{w(u)}{w(n(u))} \in \left[1 - \sqrt{2\epsilon'}, \frac{1}{1-\sqrt{2\epsilon'}}\right]$, $\frac{w(n_2(u))}{w(n(u))} \in [1 - \sqrt{2\epsilon'}, 1]$ *and*
$2 \cdot w(u) \leq w(N(u, A)) < (2 + \epsilon') \cdot w(u)$
contributive: $\mathrm{contr}(u, n(u)) \geq \frac{\epsilon'}{2} \cdot w(u) = \max\{0, 2 \cdot \mathrm{charge}(u, n(u))\} + \frac{\epsilon'}{2} \cdot w(u)$

The weight of a *payback* vertex u is overestimated in the first step of the weight distribution, so u can pay back $\frac{\epsilon'}{2} \cdot w(u)$, improving our bound on $w(A^*)$.

For a *good* vertex $u \in A^*$, $n(u)$ and $n_2(u)$ have almost the same weight as u and make up most of $N(u, A)$ in terms of weight. Like double vertices, good vertices can be thought of as having degree 2 to A.

4.2   Missing, Profitable and Helpful Vertices

We now discuss the role the different types of vertices in A^* play in our analysis. We start by recalling that in the first step of the weight distribution in the analysis of SquareImp, each $v \in A$ pays $|N(v, A^*)| \cdot \frac{w(v)}{2}$, which we bound by $(d-1) \cdot \frac{w(v)}{2}$. In particular, if the number of neighbors of v in A^* happens to be less than $d - 1$ (we say that v has $d - 1 - |N(v, A^*)|$ *missing neighbors* in this case), we gain $\frac{w(v)}{2}$ for each missing neighbor of v (cf. Lemma 8).

We partition the neighbors a vertex $v \in A$ has in A^* into those that are *helpful* for v, and those that are *profitable* for v. While helpful vertices are those vertices that would be considered neighbors of v in an unweighted approximation of our instance, and that *help* us to construct local improvements of logarithmic size, profitable vertices are the remaining neighbors that in some sense keep the instance from being close to unweighted and hence, tight. Therefore, they improve the analysis (i.e. our bound on $w(A^*)$ *profits* from these). Formally speaking, we say that $u \in N(v, A^*)$ is *helpful* for v if u is single and $v = n(u)$, or if u is double or good and $v \in \{n(u), n_2(u)\}$. Otherwise, we call u *profitable* for v.

One can show that for each profitable neighbor that a vertex $v \in A$ possesses in A^*, we gain a constant fraction of $w(v)$ in bounding $w(A^*)$. Intuitively, this is because for every profitable neighbor u of v, v makes the estimate $2 \cdot \mathrm{charge}(u, n(u)) \leq \mathrm{contr}(u, n(u))$ less tight. Lemma 8 formalizes the way missing and profitable neighbors improve our bound on $w(A^*)$.

Lemma 8.

$$w(A^*) \leq \frac{d}{2} \cdot w(A) - \sum_{v \in A} (d - 1 - |N(v, A^*)|) \cdot \frac{w(v)}{2}$$

$$- \sum_{v \in A} \frac{\epsilon'}{10} \cdot w(v) \cdot \# \textit{ profitable neighbors of } v.$$

This implies that if the total weight of vertices $v \in A$ with more than $\frac{d-1}{4}$ neighbors that are not helpful, and, hence, missing or profitable for v, is at least $\frac{20}{\epsilon' \cdot (d-1)} \cdot w(A)$, we obtain $w(A^*) \leq \frac{d-1}{2} \cdot w(A)$ and are therefore done. As a consequence, we can assume that in terms of weight, all but a fraction of $\frac{20}{\epsilon' \cdot (d-1)}$ of the vertices in A have at least $\frac{3}{4} \cdot (d-1)$ helpful neighbors.

We can further bound the number of these neighbors that are single: If we choose ϵ' small enough, each single helpful neighbor u of v contributes a large constant fraction of $w(u)$ and, hence, also $w(v)$, to $v = n(u)$. For d large enough, no $v \in A$ can therefore have more than $\frac{d-1}{4}$ neighbors that are single and helpful for v, as this would result in an improving claw. This allows us to conclude that the total weight of all vertices from A with at least $\frac{d-1}{2}$ neighbors in A^* that are helpful for them and either good or double (i.e. correspond to vertices of degree 2 to A in the unweighted setting) is at least $\left(1 - \frac{20}{\epsilon' \cdot (d-1)}\right) \cdot w(A)$.

4.3 Local Improvements of Logarithmic Size

Our goal is to use these neighbors towards a local improvement of logarithmic size. In order to get a better idea of what we are aiming at, we take a brief detour and recapitulate how these vertices are handled in the unit weight case. In [9], an auxiliary graph G_A is constructed, the vertices of which correspond to the vertices in the current solution A. Each vertex from an optimum solution A^* with exactly one neighbor in A creates a loop on that neighbor, while every $u \in A^*$ with exactly two neighbors in A results in an edge connecting these. Now, it is not hard to see that there is a one-to-one correspondence between local improvements only featuring vertices from A^* with degree one or two to A, and sub-graphs of G_A with more edges than vertices. A minimal such sub-graph is called a *binocular* [3]. Now, a result by Berman and Fürer [3] comes into play:

Lemma 9 ([3]). *For any $s \in \mathbb{N}_{>0}$, any graph $G = (V, E)$ with $|E| \geq \frac{s+1}{s} \cdot |V|$ contains a binocular of size at most $4 \cdot s \cdot \log(|V|)$.*

In particular, Lemma 9 implies the existence of a cycle of the given size. Moreover, if the number of vertices of degree 1 or 2 to A exceeds $(1 + \epsilon) \cdot |A|$, one can find local improvements of size $\mathcal{O}(\frac{1}{\epsilon} \cdot \log(|A|))$, which is one of the key ingredients of the result in [9].

We would like to port this approach to our weighted setting where vertex weights are locally similar. If we could ensure that for each good or double vertex u, its squared weight is at least as large as the average squared weight of its neighbors $n(u)$ and $n_2(u)$, plus the squared weight of neighbors of u in A other than $n(u)$ and $n_2(u)$, then we would be done since in a binocular, every vertex has degree at least two, and there is at least one vertex of degree more than two. However, this need not be the case in general, and even if we only lose an ϵ' fraction of the weight of each vertex involved, the total loss might be arbitrarily large if we consider improvements of logarithmic size. In order to overcome this issue, we have to make sure that locally, we have some additional vertices with a positive contribution to the endpoints of the edges that we can add to guarantee that we can make up for the slight inaccuracies caused by the weight differences.

To this end, for $v \in A$, we consider the set of vertices T_v sending positive charges to v (in particular, $v = n(u)$ for all $u \in T_v$), and define $\bar{B}$ to consist of all $v \in A$ for which the total contribution of T_v to v exceeds $\tilde{\epsilon} \cdot w(v)$. (Recall that

$0 < \epsilon' \ll \sqrt{\epsilon'} \ll \tilde{\epsilon} < \delta < 1$.) Then all vertices $v \in A \setminus \bar{B}$ receive charges of at most $\frac{\tilde{\epsilon}}{2} \cdot w(v)$. Hence, if $w(\bar{B}) \leq \frac{40}{\epsilon' \cdot (d-1)} \cdot w(A)$, we can bound the total weight the vertices in A receive in the second step of the weight distribution by

$$\frac{\tilde{\epsilon}}{2} \cdot w(A \setminus \bar{B}) + \frac{1}{2} \cdot w(\bar{B}) \leq \left(\tilde{\epsilon} + \frac{40}{\epsilon' \cdot (d-1)} \right) \cdot \frac{w(A)}{2}.$$

For d large enough, this is at most $\delta \cdot \frac{w(A)}{2}$, yielding the desired statement. This means that we can assume $w(\bar{B}) > \frac{40}{\epsilon' \cdot (d-1)} \cdot w(A)$ in the following. In particular, this means that the set B consisting of all vertices in $\bar{B}$ that have at least $\frac{d-1}{2}$ helpful neighbors that are good or double is of weight at least $w(B) \geq w(\bar{B}) - \frac{20}{\epsilon' \cdot (d-1)} \cdot w(A) > \frac{20}{\epsilon' \cdot (d-1)} \cdot w(A)$.

Our goal for the rest of the analysis is to lead this assumption to a contradiction, implying that we have to be in one of the previously handled cases where we obtain the desired approximation guarantee of $\frac{d-1+\delta}{2}$. To this end, we want to show that the current setting implies the existence of a circular improvement. We call a local improvement $X \subseteq V \setminus A$ *circular* if

- $\exists U \subseteq X$ s.t. $|U| \leq 4 \cdot \log(|V|)$, each $u \in U$ has at least two neighbors in A and $C := (\bigcup_{u \in U} \{n(u), n_2(u)\}, \{e_u = \{n(u), n_2(u)\}, u \in U\})$ is a cycle.
- If we let $Y_v := \{x \in X \setminus U : n(x) = v\}$, then $X = U \cup \bigcup_{v \in V(C)} Y_v$.
- $w^2(u) + \frac{1}{2} \cdot \left(\sum_{z \in Y_{n(u)}} \mathrm{contr}(z, n(u)) \cdot w(n(u)) + \sum_{z \in Y_{n_2(u)}} \mathrm{contr}(z, n_2(u)) \cdot w(n_2(u)) \right)$
  $> \frac{1}{2} \cdot (w^2(n(u)) + w^2(n_2(u))) + w^2(N(u, A) \setminus \{n(u), n_2(u)\})$ for all $u \in U$.

The intuition behind this definition is the following: Similar to the unweighted case, we build up an auxiliary graph H on the vertex set A, where each vertex $u \in V \setminus A$ with at least two neighbors in A induces an edge between its two heaviest ones $n(u)$ and $n_2(u)$. For the analysis, we will only consider edges induced by double or good vertices from A^*, i.e. those corresponding to vertices of degree two in the unweighted setting. The backbone of our circular improvement is given by a cycle C of logarithmic size in H. Additionally, for each $v \in V(C)$, we can add some additional vertices u with $n(u) = v$ that contribute a positive amount to v. Now, we want to cover for the weight of each $v \in V(C) \subseteq N(X, A)$ by using the vertices corresponding to the two incident edges in C as well as the contributions from the vertices in Y_v. This means that for each edge induced by $u \in U$, we would like $w^2(u)$, together with half of the contributions to $n(u)$ and $n_2(u)$, to be able to pay for all neighbors of u in A other than $n(u)$ and $n_2(u)$, as well as half of $n(u)$ and $n_2(u)$, which is precisely the last constraint.

Now, in order to find such circular improvements when $w(B) > \frac{20}{\epsilon' \cdot (d-1)} \cdot w(A)$, we consider the auxiliary graph H' with vertex set A, where each double or good vertex u induces an edge between $n(u)$ and $n_2(u)$, provided at least one of them is contained in B. For $v \in B$, the total contribution of T_v to v is at least $\tilde{\epsilon} \cdot w(v)$, whereas each good or double vertex contributes at most $\mathcal{O}(\sqrt{\epsilon'}) \cdot w(v)$ with

$\sqrt{\epsilon'} \ll \tilde{\epsilon}$. Hence, each cycle C in H' of logarithmic size gives rise to a circular improvement by choosing Y_v to contain all of T_v except for the at most two vertices inducing edges incident to v if $v \in V(C) \cap B$, and setting $Y_v := \emptyset$ otherwise.

To finally see that H' contains a cycle of logarithmic size, we use the facts that the weighted average degree $\frac{1}{w(A)} \cdot \sum_{v \in A} w(v) \cdot |\delta_{H'}(v)|$ is at least $\frac{d-1}{2} \cdot \frac{w(B)}{w(A)} > \frac{10}{\epsilon'}$, together with the fact that the weights of the endpoints of each edge only differ by a factor close to 1, to conclude that there has to be a sub-graph of H' that is dense enough to apply Lemma 9. (Note that if we had unit weights, this would follow immediately since the average degree would be at least $\frac{10}{\epsilon'} > 2$.) This implies the existence of a circular improvement, contradicting the termination criterion of LogImp and finishing the proof. See [17] for a more detailed analysis.

4.4   A Polynomial Running Time

To achieve a polynomial number of iterations of LogImp, we scale and truncate the weight function as in [4] and [2]. This only results in an arbitrarily small additive error in the approximation guarantee.

In order to search for circular improvements in polynomial time, we employ the color coding technique in a similar fashion as in [9]. Note that this is the only point where we need the additional structural properties of a k-Set Packing instance (as opposed to an instance of the MWIS in $k + 1$-claw free graphs). Detailed descriptions and proofs can be found in [17].

5   The Lower Bound

In this section, we show that our result is asymptotically best possible in the sense of Theorem 10.

Theorem 10. *Let* $k \geq 3, \alpha \in \mathbb{R}, 0 < \epsilon < 1$ *and* $C > 0$. *Then for each* $N_0 \in \mathbb{N}$, *there exist an instance* $(\mathcal{S}, w)$ *of the weighted* k-Set Packing Problem with $|\mathcal{S}| \geq N_0$ *and feasible solutions* $A, A^* \subseteq \mathcal{S}$, *such that for* A, *there is no local improvement of size at most* $C \cdot \log(|\mathcal{S}|)$ *with respect to* w^α, *but* $w(A^*) \geq \frac{k-\epsilon}{2} \cdot w(A)$.

Proof. For $k \leq 2$, there is nothing to show and for $\alpha \leq 0$, we can just choose the weights in A to be arbitrarily small. Hence, we can assume $k \geq 3$ and $\alpha > 0$. Now, there is a result by Erdős and Sachs [8] telling us that for every $N_0 \in \mathbb{N}$, there is a k-regular graph H on $|V(H)| \geq N_0$ vertices such that the girth of H, i.e. the minimum length of a cycle in H, is at least $\frac{\log(|V(H)|)}{\log(k-1)} - 1$. Consider the graph G with vertex set $V(G) := V(H) \cup E(H)$ and edge set $E(G) := \{\{v, e\} : v \in e \in E(H)\}$, i.e. each edge of H is connected via edges in G to both of its endpoints. We define $\mathcal{S} := \{\delta_G(x), x \in V(G)\}$, where $\delta_G(x)$ is the set of incident edges of x in G. By k-regularity of H, $|\delta_G(v)| = k \geq 3$ for $v \in V(H)$ and $|\delta_G(e)| = 2$ for $e \in E(H)$, so each element of $\mathcal{S}$ has cardinality at most k. By definition, G is simple, so no two vertices share more than one edge. As all degrees are at least two, the sets $\delta_G(x), x \in V(G)$ are pairwise distinct. Finally, $V(H)$ and $E(H)$ constitute independent sets in G, implying that $A := \{\delta_G(v), v \in V(H)\}$ and $A^* := \{\delta_G(e), e \in E(H)\}$ each consist of pairwise disjoint sets. We define positive weights on $\mathcal{S}$ by setting $w(\delta_G(v)) := 1$ for $v \in V(H)$ and $w(\delta_G(e)) := (1 - \bar{\epsilon})^{\frac{1}{\alpha}}$, where $\bar{\epsilon} > 0$ is chosen such that $\frac{1}{\bar{\epsilon}} \in \mathbb{N}$ and $(1 - \bar{\epsilon})^{\frac{1}{\alpha}} \geq 1 - \frac{\epsilon}{k}$. By k-regularity of H, $|A^*| = |E(H)| = \frac{k}{2} \cdot |V(H)| = \frac{k}{2} \cdot |A|$, so $w(A^*) \geq \frac{k-\epsilon}{2} \cdot w(A)$. To see that there is no local improvement X of $w^\alpha(A)$ with $|X| \leq C \cdot \log(|V(G)|) = C \cdot \log(\frac{k+2}{2} \cdot |V(H)|)$, first note that we can w.l.o.g. assume $X \subseteq \mathcal{S} \setminus A = A^*$ (otherwise, consider $X \setminus A$). Then, X being a local improvement implies that $|X| > \frac{1}{1-\bar{\epsilon}} \cdot |\{y \in A : \exists x \in X : y \cap x \neq \emptyset\}|$ by our choice of weights. But the sets from A that X intersects are precisely the sets $\delta_G(v)$ for those vertices $v \in V(H)$ that are endpoints of edges $e \in E(H)$ with $\delta_G(e) \in X$. Hence, we have found as sub-graph J of H with

$$C \cdot \log\left(\frac{k+2}{2} \cdot |V(H)|\right) \geq |X| = |E(J)| > \frac{1}{1-\bar{\epsilon}} \cdot |V(J)| \geq (1 + \bar{\epsilon}) \cdot |V(J)|,$$

which implies the existence of a cycle of length $\frac{4}{\bar{\epsilon}} \cdot \log(C \cdot \log(\frac{k+2}{2} \cdot |V(H)|))$ by Lemma 9. But as a function of $|V(H)|$, this grows asymptotically slower than our lower bound of $\frac{\log(|V(H)|)}{\log(k-1)} - 1$ on the girth, resulting in a contradiction for N_0 and, hence, $|V(H)|$ chosen large enough. For a more detailed proof, see [17].

6 Conclusion

In this paper, we have seen how to use local search to approximate the weighted k-Set Packing Problem with an approximation ratio that gets arbitrarily close to $\frac{k}{2}$ as k approaches infinity. At the cost of a quasi-polynomial running time, this result applies to the more general setting of the Maximum Weight Independent Set Problem in d-claw free graphs, yielding approximation ratios arbitrarily close to $\frac{d-1}{2}$. Moreover, we have seen that this result is asymptotically best possible in the sense that for no $\alpha \in \mathbb{R}$, a local improvement algorithm for the weighted k-Set Packing Problem that considers local improvements of w^α of logarithmically bounded size can produce an approximation guarantee better than $\frac{k}{2}$.

As a consequence, our paper seems to conclude the story of (pure) local improvement algorithms for both the MWIS in d-claw free graphs and the weighted k-Set Packing Problem. Hence, the search for new techniques beating the threshold of $\frac{d-1}{2}$, respectively $\frac{k}{2}$, is one of the next goals for research in this area.

References

1. Arkin, E.M., Hassin, R.: On local search for weighted k-set packing. Math. Oper. Res. **23**(3), 640–648 (1998). https://doi.org/10.1287/moor.23.3.640
2. Berman, P.: A d/2 approximation for maximum weight independent set in d-claw free graphs. In: SWAT 2000. LNCS, vol. 1851, pp. 214–219. Springer, Heidelberg (2000). https://doi.org/10.1007/3-540-44985-X_19
3. Berman, P., Fürer, M.: Approximating maximum independent set in bounded degree graphs. In: Proceedings of the Fifth Annual ACM-SIAM Symposium on Discrete Algorithms, pp. 365–371 (1994). https://dl.acm.org/doi/pdf/10.5555/314464.314570
4. Chandra, B., Halldórsson, M.M.: Greedy local improvement and weighted set packing approximation. J. Algorithms **39**(2), 223–240 (2001). https://doi.org/10.1006/jagm.2000.1155
5. Cygan, M.: Improved approximation for 3-dimensional matching via bounded pathwidth local search. In: 54th Annual IEEE Symposium on Foundations of Computer Science, FOCS 2013, Berkeley, CA, USA, 26–29 October 2013, pp. 509–518. IEEE Computer Society (2013). https://doi.org/10.1109/FOCS.2013.61
6. Cygan, M., Grandoni, F., Mastrolilli, M.: How to sell hyperedges: the hypermatching assignment problem. In: Proceedings of the 2013 Annual ACM-SIAM Symposium on Discrete Algorithms, pp. 342–351. SIAM (2013). https://doi.org/10.1137/1.9781611973105.25
7. Edmonds, J.: Maximum matching and a polyhedron with 0,1-vertices. J. Res. Natl. Bureau Stand. Sect. B Math. Math. Phys. **69B**, 125–130 (1965). https://doi.org/10.6028/jres.069b.013
8. Erdős, P., Sachs, H.: Reguläre Graphen gegebener Taillenweite mit minimaler Knotenzahl. Wiss. Z. Martin-Luther-Univ. Halle-Wittenberg Math.-Natur. Reihe **12**(3), 251–257 (1963)
9. Fürer, M., Yu, H.: Approximating the k-set packing problem by local improvements. In: Fouilhoux, P., Gouveia, L.E.N., Mahjoub, A.R., Paschos, V.T. (eds.) ISCO 2014. LNCS, vol. 8596, pp. 408–420. Springer, Cham (2014). https://doi.org/10.1007/978-3-319-09174-7_35
10. Halldórsson, M.M.: Approximating discrete collections via local improvements. In: Proceedings of the Sixth Annual ACM-SIAM Symposium on Discrete Algorithms, pp. 160–169. Society for Industrial and Applied Mathematics (1995). https://dl.acm.org/doi/10.5555/313651.313687
11. Hazan, E., Safra, S., Schwartz, O.: On the complexity of approximating k-set packing. Comput. Complex. **15**, 20–39 (2006). https://doi.org/10.1007/s00037-006-0205-6
12. Hurkens, C.A.J., Schrijver, A.: On the size of systems of sets every t of which have an SDR, with an application to the worst-case ratio of heuristics for packing problems. SIAM J. Disc. Math. **2**(1), 68–72 (1989). https://doi.org/10.1137/0402008
13. Karp, R.M.: Reducibility among Combinatorial Problems. In: Complexity of computer computations, pp. 85–103. Springer, Heidelberg (1972). https://doi.org/10.1007/978-1-4684-2001-2_9
14. Minty, G.J.: On maximal independent sets of vertices in claw-free graphs. J. Comb. Theory Ser. B **28**(3), 284–304 (1980). https://doi.org/10.1016/0095-8956(80)90074-X

15. Nakamura, D., Tamura, A.: A revision of Minty's algorithm for finding a maximum weight stable set of a claw-free graph. J. Oper. Res. Soc. Jpn. **44**(2), 194–204 (2001). https://doi.org/10.15807/jorsj.44.194
16. Neuwohner, M.: An improved approximation algorithm for the maximum weight independent set problem in d-claw free graphs. In: 38th International Symposium on Theoretical Aspects of Computer Science (STACS 2021), Leibniz International Proceedings in Informatics (LIPIcs), vol. 187, pp. 53:1–53:20 (2021). https://doi.org/10.4230/LIPIcs.STACS.2021.53
17. Neuwohner, M.: The limits of local search for the maximum weight independent set problem in d-claw free graphs (2021). https://arxiv.org/abs/2106.03555
18. Sbihi, N.: Algorithme de recherche d'un stable de cardinalité maximum dans un graphe sans étoile. Disc. Math. **29**(1), 53–76 (1980). https://doi.org/10.1016/0012-365X(90)90287-R
19. Sviridenko, M., Ward, J.: Large neighborhood local search for the maximum set packing problem. In: Fomin, F.V., Freivalds, R., Kwiatkowska, M., Peleg, D. (eds.) ICALP 2013. LNCS, vol. 7965, pp. 792–803. Springer, Heidelberg (2013). https://doi.org/10.1007/978-3-642-39206-1_67

The Secretary Problem with Distributions

Pranav Nuti[✉]

Stanford University, Stanford, CA 94305, USA
pranavn@stanford.edu

Abstract. In the secretary problem, we are presented with n numbers, adversarially chosen and then randomly ordered, in a sequential fashion. After each observation, we have to make an irrevocable decision about whether we would like to accept a number or not, with the goal of picking the largest number with the highest probability. The secretary problem is a fundamental online selection problem with a long history.

A natural variant of the problem is to assume that the n numbers come from n independent distributions known to us. Esfandiari et al. [AISTATS 2020] studied this problem, found the optimal probability under a restrictive "no-superstars" assumption on the distributions, and conjectured that this assumption could be dropped. In this paper, we prove that this is indeed the case while significantly simplifying both the optimal algorithm and its analysis. We then extend this result in two directions. First, we are able to relax the assumption of independence to a kind of negative dependence, demonstrating the robustness of our algorithm. Second, we are able to replace knowledge of the distributions with a (more realistic) knowledge of samples from the distributions.

Keywords: Online algorithms · Secretary problem · Negative dependence · Incomplete information

1 Introduction

In the classical secretary problem, we are presented with n numbers, adversarially chosen and then randomly ordered, in a sequential fashion. After each observation, we have to make an irrevocable decision about whether we would like to accept a number or not, with the goal of picking the largest number with the highest probability. It is well known that this goal can be achieved with a probability of $\frac{1}{e}$.

In this paper, we are interested in how access to more information about how the n numbers are chosen affects the optimal probability of success. Most relevant to our results is the work of Gilbert and Mosteller [1] who studied the case where the n numbers, instead of being adversarially chosen, are n independent draws from a *single* distribution known to us, and found the optimal probability of success to be $\gamma \approx 0.5801$.

A natural extension of the work of Gilbert and Mosteller is to assume that the n numbers come from n *different* independent distributions which are known

© Springer Nature Switzerland AG 2022
K. Aardal and L. Sanitá (Eds.): IPCO 2022, LNCS 13265, pp. 429–439, 2022.
https://doi.org/10.1007/978-3-031-06901-7_32

to us. As in the classical secretary problem, the numbers are randomly ordered before we observe them.

Esfandiari et al. [13] studied this problem (which they call the "best-choice prophet secretary problem") under a restrictive "no-superstars" assumption on the distributions and showed that the optimal probability of success is $\gamma \approx$ 0.5801, the same as in the work of Gilbert and Mosteller. Distributions are said to satisfy a no-superstars assumption if none of the distributions are (a priori) particularly likely to yield the largest draw. Esfandiari et al. conjectured that this assumption could be dropped, and asked for a simpler proof of their result.

1.1 Our Contributions

In this paper, we prove the conjecture of Esfandiari et al. [13] and significantly simplify the optimal algorithm and its analysis.

Our proof shows that this optimal probability γ can be attained by using "threshold-based" algorithms, i.e., algorithms of the following form: accept the ith number if it is the largest seen so far, and is greater than a threshold τ_i. The most important piece of our analysis is a lemma regarding the distribution of the maximum of n random variables which plays a vital role in allowing us to generalize the work of Gilbert and Mosteller.

We then investigate our result in two further directions. First, we are able to relax the assumption of independence to a kind of negative dependence, demonstrating the robustness of our algorithm. A crucial tool in this analysis is the use of Han's inequality for submodular set functions. Our method of applying Han's inequality together with this particular negative dependence criterion is flexible, and can also be used to generalize other results about secretary problems beyond the case of independent distributions. The number of balls in each bin when m balls are dropped randomly and independently into n bins satisfy our negative dependence criterion.

Second, we are able to replace knowledge of the distributions with a (more realistic) knowledge of samples from the distributions. Specifically, we are able to show that given enough samples from each distribution, we can approximate the thresholds in our optimal algorithm so well that we can succeed with a probability as close to $\gamma \approx 0.5801$ as we please. It is critical to note that the number of samples needed from each distribution *does not* depend on n.

1.2 Related Literature

Over the years, many variations on the secretary problem in which the numbers come from independent distributions have been explored. We discuss here some work different from our own along four particular dimensions.

The first dimension concerns the objective. In the secretary problem with distributions (or the "best-choice prophet secretary problem"), the goal is to maximize the probability of picking the largest number (the "secretary objective"). Alternatively, one's goal might be to maximize the expected value of the picked number. This problem was introduced in [6] by Esfandiari et al., and their results were improved in [7]. The optimal answer remains unknown, but the work of Correa et al. [9] provides the best known bounds.

The second dimension considers whether the distributions are identical or not. In this paper, we show that the distributions being non-identical does not decrease the probability of success; it remains the same as the IID (identical distribution) setting of Gilbert and Mosteller. However, the situation with the expected value objective is different, and the distributions being non-identical *does* decrease the optimal reward (see [5] and [9] for further discussion).

The third dimension considers the order in which the distributions are observed. In our paper, the numbers are observed in random order. Instead, the distributions may be adversarially ordered. When considered with the expected value objective, this is the the classical prophet inequality (see [8] for a recent survey). The same problem with the secretary objective is addressed in [13]. Recent work has also considered the possibility of other orderings (the so-called "constrained-order" prophet inequalities [10]).

The fourth dimension concerns how much information about the distributions is known. Some recent papers have focused on exploring which of the results from the full-knowledge-of-distributions setting discussed above can be replicated under access to samples from the distributions (see, for example, [11,12,14,15]). The most surprising of these results is perhaps the work of Rubinstein et al. [14], which demonstrates that the reward of the classical prophet inequality can be obtained with access to just a *single* sample from each distribution.

1.3   Organization

In Sect. 2, we present a statement and proof of the conjecture of Esfandiari et al. In Sect. 3, we discuss how the independence assumption can be relaxed to a kind of negative dependence. In Sect. 4, we describe how knowledge of samples from the distributions suffices to obtain a success probability arbitrarily close to $\gamma \approx 0.5801$.

2   The Secretary Problem with Distributions

An adversary chooses independent continuous distributions $\mathcal{D}_1, \ldots, \mathcal{D}_n$ and generates draws $X_1, \ldots, X_n$ from these distributions. We are told what the distributions $\mathcal{D}_1, \ldots, \mathcal{D}_n$ are. Next, the draws are randomly ordered and we are sequentially presented with them. We have to make an irrevocable decision about whether we would like to accept a draw or not, with the goal of selecting the largest number. What is the probability with which we can succeed?

Theorem 1. *The probability of success in the secretary problem with distributions is at least $\gamma \approx 0.5801$, the probability of success when the distributions are all identical.*

Proof. Fix parameters $\{d_i\}_{i=1}^n$ (to be picked later) and let us choose τ_i, which will be our thresholds, such that $\Pr[\max_{k=1}^n X_k \leq \tau_i] = d_i^n$ (all superscripts are exponents). Assume these parameters are monotonically decreasing in i. (If the X_i were IID, and uniform on $[0, 1]$, then $\tau_i = d_i$.)

We have the following simple, but vital lemma:

Lemma 1. *Suppose that* $X_1, X_2, \ldots, X_n$ *are independent. If r of the X_k are randomly chosen, say,* $X_{j_1}, X_{j_2}, \cdots, X_{j_r}$, *then,* $\Pr[\max_{k=1}^r X_{j_k} \leq \tau_i] \geq d_i^r$.

Proof. Fix a constant T, and suppose $\Pr[X_k \leq T] = a_k$. Then,

$$\Pr[\max_{k=1}^n X_k \leq T] = \prod_{k=1}^n a_k$$

If r of the X_k are randomly chosen, say, $X_{j_1}, X_{j_2}, \cdots, X_{j_r}$ then

$$\Pr[\max_{k=1}^r X_{j_k} \leq T] = \frac{\sum_{(i_1, i_2, \ldots, i_r)} \prod_{k=1}^r a_{i_k}}{\binom{n}{r}}$$

By the AM-GM inequality (which tells us that the arithmetic mean of a set of numbers is at least as large as its geometric mean), this is at least

$$\left(\prod_{k=1}^n a_k^{\binom{n-1}{r-1}} \right)^{\frac{1}{\binom{n}{r}}} = \left(\prod_{k=1}^n a_k \right)^{\frac{r}{n}}$$

In particular, we can let $T = \tau_i$, and the desired result follows.   □

Shortly, we shall be applying the lemma to the situation where $X_{j_1}, X_{j_2}, \cdots, X_{j_r}$ are the first r draws we are presented with (which indeed constitute a randomly chosen subset of the X_i of size r).

Consider a strategy in which we accept the ith number if and only if it is the largest seen so far, and is more than its threshold τ_i (Correa et al. [9] call such strategies *blind*). We analyze the probability of succeeding with this strategy, following the proof of Gilbert and Mosteller [1].

Note that for any $i \leq r$, the probability that the ith number is largest amongst the first r, but is less than its threshold is just

$$\frac{\Pr[\max_{k=1}^r X_{j_k} \leq \tau_i]}{r} \geq \frac{d_i^r}{r}$$

since this happens if (and only if!) all the first r numbers are less than τ_i and they're ordered so that the ith number is largest.

Next, note that the probability the ith number is less than its threshold, is largest amongst the first r, but is not largest amongst all numbers is at least

$$\frac{d_i^r}{r} - \frac{\Pr[\max_{k=1}^n X_k \leq \tau_i]}{n} = \frac{d_i^r}{r} - \frac{d_i^n}{n}$$

since the second term simply represents the probability that the ith number is largest amongst the first n and is less than its threshold.

Now, note that the probability that the ith number is less than its threshold, is largest amongst the first r, but the $r + 1^{\text{th}}$ number is largest amongst all numbers is at least

$$\frac{\frac{d_i^r}{r} - \frac{d_i^n}{n}}{n - r}$$

since it's equally likely that each of the last $n - r$ numbers turns out to be the largest amongst them.

Furthermore, note that the above is also actually the probability *no number before the $r + 1^{\text{th}}$ is chosen*, the ith number is largest amongst the first r, but the $r + 1^{\text{th}}$ number is largest amongst all numbers. This is because if the largest number amongst the first r is less than its threshold, it cannot be chosen; certainly, no number after it (amongst the first r) can be chosen; no number before it can be chosen either since it wouldn't have been able to meet its threshold (since the τ_i are decreasing).

Thus we conclude that the the probability no number before the $r + 1^{\text{th}}$ is chosen, and the $r + 1^{\text{th}}$ number is largest amongst all numbers is at least

$$\sum_{i=1}^{r} \frac{\frac{d_i^r}{r} - \frac{d_i^n}{n}}{n - r}$$

The probability that the $r + 1^{\text{th}}$ number is largest amongst all numbers but is less than its threshold is just

$$\frac{\Pr[\max_{k=1}^n X_k \leq \tau_{r+1}]}{n} = \frac{d_{r+1}^n}{n}$$

We conclude, that the probability of succeeding by picking the $r+1^{\text{th}}$ number is at least

$$\left(\sum_{i=1}^{r} \frac{\frac{d_i^r}{r} - \frac{d_i^n}{n}}{n - r} \right) - \frac{d_{r+1}^n}{n}$$

The probability of succeeding by picking the first number is

$$\frac{1 - d_1^n}{n}$$

The overall probability of succeeding is the sum of the probabilities of succeeding by picking any particular number, so our arguments thus far demonstrate that the probability of success for our threshold-based algorithm is at least

$$\frac{1 - d_1^n}{n} + \sum_{r=1}^{n-1} \left(\left(\sum_{i=1}^{r} \frac{\frac{d_i^r}{r} - \frac{d_i^n}{n}}{n - r} \right) - \frac{d_{r+1}^n}{n} \right)$$

In the IID setting, we actually have $\Pr[\max_{k=1}^r X_{j_k} \le \tau_i] = d_i^r$. Therefore, the above formula is in fact exact in the IID setting! This demonstrates that for any threshold-based algorithm in the IID setting, there is a threshold-based algorithm in our more general setting that does at least as well.

Since it is known that a threshold algorithm of this sort is in fact optimal when the distributions are identical (this claim is proved in Sect. 2 of [2]), we conclude that we can always do at least as well as we do in the IID setting and to achieve a probability of success of at least γ, we just need to choose the d_i which are optimal when the distributions are identical.

Note of course that it is impossible to do better than γ in general, because γ is the optimal answer when the distributions are identical.

We now briefly discuss how the optimal probability of success is calculated in the IID setting (see [2] for details of these calculations).

Our computation above immediately yields that the optimal probability of success in the IID setting is decreasing in n, since if Z is a random variable deterministically equal to 0 and $X_1, X_2, \ldots, X_{n+1}$ are IID uniform $[0, 1]$, then

$$
\begin{aligned}
&\Pr[\text{succeeding with draws } X_1, X_2, \ldots, X_n] \\
&= \Pr[\text{succeeding with draws } X_1, X_2, \ldots, X_n, Z] \\
&\ge \Pr[\text{succeeding with draws } X_1, X_2, \ldots, X_{n+1}]
\end{aligned}
$$

where the equality in the first line follows from the fact that the optimal algorithm with draws $X_1, X_2, \ldots, X_n, Z$ is to just act as if the Z never appeared, and the inequality from the second line is just our claim that we can always do at least as well in the IID setting.

In particular, this means that the worst case for the IID setting arises when we take the limit as $n \to \infty$. (This is proved slightly differently in Sect. 3 of [2].)

The optimal value of d_{n-i} turns out to not depend on n. Explicit expressions for the optimal values of d_{n-i} can be found ($d_{n-i} = 1 - \frac{c}{i} + O\left(\frac{1}{i^2}\right)$, $\int_0^c \frac{e^x - 1}{x} dx = 1$) [1].

We can then evaluate the success probability in the limit as $n \to \infty$ by plugging in the optimal value of d_i into our formula for the success probability of a threshold algorithm. The success probability turns out to equal ($e^c - c - 1) \int_c^\infty \frac{e^{-x}}{x} dx + e^{-c} \approx 0.5801$ (see Sect. 4 of [2]). □

We should note that the assumption that the distributions are continuous was not important. We can achieve the same result when the distributions are possibly discrete. To do this, we employ the same trick used in [14]. Replace $\mathcal{D}_i$ with a bivariate distribution, where the first coordinate is from $\mathcal{D}_i$ and the second coordinate is a value drawn independently and uniformly from $[0, 1]$. Impose the lexicographic order on $\mathbb{R}^2$, and then, there is no difficulty in determining the value of τ_i from d_i. While implementing the algorithm, we can generate a random value from $[0, 1]$ for every draw presented to us to decide whether it is large enough to accept, and the guarantee on the success probability remains the same.

3 Extension to Negative Dependence

In this section, we shall drop the assumption that the draws, X_i, are independent. Let us write $Y_i = \mathbb{1}_{X_i \leq T}$, and assume instead that the Y_i are *conditionally negatively associated* for every T. What does this mean?

We say that random variables $Y_1, Y_2, \ldots, Y_n$ are *negatively associated* if for any non-intersecting $I, J \subset [n]$, and any pair of non-decreasing functions f and g,

$$\mathbb{E}[f(Y_I)g(Y_J)] \leq \mathbb{E}[f(Y_I)]\mathbb{E}[g(Y_J)]$$

Furthermore, the random variables $Y_1, Y_2, \ldots, Y_n$ are said to be *conditionally negatively associated* if they remain negatively associated even after conditioning upon the values of some subset of the random variables $Y_1, Y_2, \ldots, Y_n$.

A full discussion of this particular negative dependence criterion is outside the scope of this paper (see [3] for an introduction to various negative dependence criteria). One reason to consider this criterion is that there is a very natural example of distributions that satisfy it: Suppose X_i is a draw from distribution $\mathcal{D}_i$, the number of balls in the ith bin when m balls are dropped randomly and independently into n bins. Then, the Y_i are conditionally negatively associated [4].

The following result recovers the lemma from Sect. 2 under the assumption that the Y_i are conditionally negatively associated:

Lemma 2. *Suppose that $Y_i = \mathbb{1}_{X_i \leq T}$, and the Y_i are conditionally negatively associated. If r of the X_k are randomly chosen, say, $X_{j_1}, X_{j_2}, \cdots, X_{j_r}$, then, $\Pr[\max_{k=1}^r X_{j_k} \leq T] \geq \Pr[\max_{k=1}^n X_k \leq T]^{\frac{r}{n}}$.*

Proof. Since the Y_i are conditionally negatively associated, we know that after conditioning on $S = \{\prod_{i \in A \cap B} Y_i = 1\}$, we have

$$\mathbb{E}\left[\prod_{i \in A \setminus B} Y_i \bigg| S\right] \cdot \mathbb{E}\left[\prod_{i \in B \setminus A} Y_i \bigg| S\right] \geq \mathbb{E}\left[\prod_{i \in A \setminus B \cup B \setminus A} Y_i \bigg| S\right]$$

which implies (with no conditioning!) that

$$\frac{\mathbb{E}\left[\prod_{i \in (A \setminus B) \cup (A \cap B)} Y_i\right]}{\mathbb{E}\left[\prod_{i \in A \cap B} Y_i\right]} \cdot \frac{\mathbb{E}\left[\prod_{i \in (B \setminus A) \cup (A \cap B)} Y_i\right]}{\mathbb{E}\left[\prod_{i \in A \cap B} Y_i\right]} \geq \frac{\mathbb{E}\left[\prod_{i \in A \setminus B \cup B \setminus A \cup (A \cap B)} Y_i\right]}{\mathbb{E}\left[\prod_{i \in A \cap B} Y_i\right]}$$

or in other words,

$$\mathbb{E}\left[\prod_{i \in A} Y_i\right] \cdot \mathbb{E}\left[\prod_{i \in B} Y_i\right] \geq \mathbb{E}\left[\prod_{i \in A \cup B} Y_i\right] \cdot \mathbb{E}\left[\prod_{i \in A \cap B} Y_i\right]$$

So it follows that if we define $g(A) = \log \mathbb{E}\left[\prod_{i \in A} Y_i\right] = \log \Pr[\max_{k \in A} X_k \leq T]$, then g is a submodular set function.

Therefore, it follows by Han's inequality for submodular set functions (see pages 16–18 in [16] for a statement and proof of Han's inequality),

$$\frac{r}{n} \cdot g([n]) \leq \frac{\sum_{|A|=r} g(A)}{\binom{n}{r}}$$

Hence, we must have

$$\frac{r}{n} \cdot \log \mathbb{E}\left[\prod_i Y_i\right] \leq \frac{\sum_{|A|=r} \log \mathbb{E}\left[\prod_{i \in A} Y_i\right]}{\binom{n}{r}}$$

This means that

$$\Pr[\max_{k=1}^n X_k \leq T]^{\frac{r}{n}} \leq \prod_{|A|=r} \mathbb{E}\left[\prod_{i \in A} Y_i\right]^{\frac{1}{\binom{n}{r}}}$$

By the AM-GM inequality, we know that

$$\prod_{|A|=r} \mathbb{E}\left[\prod_{i \in A} Y_i\right]^{\frac{1}{\binom{n}{r}}} \leq \frac{\sum_{|A|=r} \mathbb{E}\left[\prod_{i \in A} Y_i\right]}{\binom{n}{r}}$$

But, if r of the X_k are randomly chosen, say, $X_{j_1}, X_{j_2}, \cdots, X_{j_r}$ then

$$\Pr[\max_{k=1}^r X_{j_k} \leq T] = \frac{\sum_{|A|=r} \mathbb{E}\left[\prod_{i \in A} Y_i\right]}{\binom{n}{r}}$$

The desired result follows.   □

Now, the only place in which the results of Sect. 2 depended on the independence of the X_i was in the proof of the lemma. It follows that the results of Sect. 2 also apply to the X_i when the Y_i are conditionally negatively associated.

In particular, this means, in the secretary problem with balls and bins, in which we observe a sequence of n bins (into which m balls have been randomly and independently dropped) and we have to select the bin with the most balls, we can succeed with a probability of at least γ.

Our method of applying Han's inequality together with this particular negative dependence criterion is flexible, and can also be used to generalize other results about secretary problems beyond the case of independent distributions (for example, Theorem 1 in [13] which is similar to the question that this paper addresses, but with an adversarial ordering of distributions).

4   Limited Knowledge: Samples from Distributions

In this section, we aim to show that with sufficiently many samples, we can find thresholds T_i amongst numbers in the samples which estimate the "real" thresholds we would've used with full knowledge of the distributions. We have not attempted to optimize any of the arising constants.

Suppose we have a distribution $\mathcal{D}$, and independent of $\mathcal{D}$, probabilities p_i, with $p_i > \delta$. Fix a parameter $\varepsilon < \frac{1}{10}$. Let X be a random variable with distribution $\mathcal{D}$ and suppose we draw m samples independently from $\mathcal{D}$. Essential to the discussion in this section is the following lemma (quite similar to Lemma 8 in [14]):

Lemma 3. *If m is sufficiently large (as a function of only ε and δ), with probability arbitrarily close to 1 we can find numbers T_i amongst the m samples so that $\frac{p_i}{(1+\varepsilon)^2} \leq \Pr[X \leq T_i] \leq p_i(1+\varepsilon)$ for every i simultaneously.*

Proof. We need to find numbers T_i amongst the samples so that $\Pr[X \leq T_i] \approx p_i$. To do this, we only need to find numbers M_k such that $\Pr[X \leq M_k] \approx \frac{1}{(1+\varepsilon)^k}$ for k such that $\frac{1}{(1+\varepsilon)^k} > \frac{\delta}{1+\varepsilon}$, because then, if $p_i \in \left[\frac{1}{(1+\varepsilon)^k}, \frac{1}{(1+\varepsilon)^{k-1}} \right)$, $\Pr[X \leq M_k]$ is also approximately p_i, so we can let $T_i = M_k$.

Accordingly, let us define for $k \in \mathbb{N}$, $m_k = \frac{m}{(1+\varepsilon)^k}$, $M_k = m_k^{\text{th}}$ smallest sample, and A_k is such that $\Pr[X \leq A_k] = \frac{1}{(1+\varepsilon)^k}$. We know that the expected number of samples $\leq A_k$ is m_k, and the number of samples $\leq A_k$ is a sum of independent Bernoulli random variables, so by the multiplicative Chernoff bound, the probability that the number of samples is not between m_{k-1} and m_{k+1} is bounded by $2 \exp(\frac{-\varepsilon^2 m_k}{3})$.

By the union bound, we conclude that the number of samples $\leq A_k$ is between m_{k-1} and m_{k+1} simultaneously for every k such that $\frac{1}{(1+\varepsilon)^k} > \frac{\delta}{(1+\varepsilon)^2}$ with probability at least $1 - 2 \left(\frac{-\log \delta}{\log(1+\epsilon)} + 2 \right) \exp(\frac{-\varepsilon^2 \delta m}{3(1+\varepsilon)^2})$. Note that if m is sufficiently large, this probability is arbitrarily close to 1.

It follows that $M_{k-1} \geq A_k \geq M_{k+1}$ (for $k = 1, 2, \ldots, \frac{-\log \delta}{\log(1+\epsilon)} + 2$) with probability close to 1, and so, we have $A_{k+1} \leq M_k \leq A_{k-1}$ (for $k = 1, 2, \ldots, \frac{-\log \delta}{\log(1+\epsilon)} + 1$) with probability close to 1. Hence, we conclude that $\frac{1}{(1+\varepsilon)^{k+1}} \leq \Pr[X \leq M_k] \leq \frac{1}{(1+\varepsilon)^{k-1}}$ with probability close to 1.

So if $p_i \in \left[\frac{1}{(1+\varepsilon)^k}, \frac{1}{(1+\varepsilon)^{k-1}} \right)$, then $\frac{p_i}{(1+\varepsilon)^2} \leq \Pr[X \leq M_k] \leq p_i(1 + \varepsilon)$, as required.   □

We can now prove the main theorem of this section:

Theorem 2. *There is a function N so that with just $N(\epsilon)$ (independent of n) samples from each distribution $\mathcal{D}_i$ (but without knowledge of the distributions themselves), we can succeed with a probability of $\gamma - \epsilon$ in the secretary problem with distributions $\mathcal{D}_i$.*

Proof. In the following, $f_1(\epsilon), f_2(\epsilon)$, and $f_3(\epsilon)$ are functions which tend to 0 as $\epsilon \to 0$ which we have not attempted to optimize (but we can think for the purpose of the proof that $f_1(\epsilon) \approx f_2(\epsilon) \approx \frac{\epsilon}{10}, f_3(\epsilon) \approx -\frac{\epsilon}{100 \log \epsilon}$).

Note that we can safely ignore the last $f_1(\epsilon)$ fraction of draws, and never choose them, and this will effect our success probability by at most $f_1(\epsilon)$. Furthermore, as long as $i \leq (1 - f_1(\epsilon))n$, for $d_i = 1 - \frac{c}{n-i} + O(\frac{1}{(n-i)^2})$ (the optimal value), d_i^n is at least $\approx e^{\frac{-c}{f_1(\epsilon)}}$.

Applying the lemma while letting $\mathcal{D}$ be the distribution of $\max_{k=1}^n X_k$, $p_i = d_i^n$, $\delta \approx e^{\frac{-c}{f_1(\epsilon)}}$ and $\varepsilon \approx \frac{f_3(\epsilon)}{4}$, we conclude that we can find T_i such that with probability $1 - f_2(\epsilon)$, we have that $\Pr[\max_{k=1}^n X_k \leq T_i]$ is within a factor of $1 + f_3(\epsilon)$ of d_i^n for every i simultaneously (where X_i are the draws we must choose from).

Now, if $\Pr[\max_{k=1}^n X_k \leq T_i]$ is within a factor of $1 + f_3(\epsilon)$ of d_i^n, then we must have that the lower bound we have used for $\Pr[\max_{k=1}^r X_{j_k} \leq T_i]$, $\Pr[\max_{k=1}^n X_k \leq T_i]^{\frac{r}{n}}$ is within a factor of $1 + f_3(\epsilon)$ of d_i^r as well (where $\Pr[\max_{k=1}^r X_{j_k} \leq T_i]$ is the probability that r randomly chosen draws are all less than T_i).

From the expression for the success probability derived in Sect. 2 (remembering that we only need to consider this for $r \leq (1 - f_1(\epsilon)n)$), we see that using the estimated thresholds (which are good with a probability of $(1 - f_2(\epsilon))$), we succeed with probability $\gamma - \epsilon$, as long as $f_1(\epsilon)$ $f_2(\epsilon)$, and $f_3(\epsilon)$ are sufficiently small (which can all be achieved with sufficiently many samples).   □

References

1. Gilbert, J., Mosteller, F.: Recognizing the maximum of a sequence. J. Am. Stat. Assoc. **61**(313), 35–73 (1966)
2. Samuels, S.: Exact solutions for the full information best choice problem. Purdue Univ. Stat. Dept. Mimeo Ser., 82–17 (1982)
3. Pemantle, R.: Towards a theory of negative dependence, pp. 1371–1390 (2000)
4. Kahn, J., Neiman, M.: Conditional negative association for competing urns (2010). arXiv: 1001.0610 [math.PR]
5. Correa, J., Foncea, P., Hoeksma, R., Oosterwijk, T., Vredeveld, T.: Posted price mechanisms for a random stream of customers. In: Proceedings of the 2017 ACM Conference on Economics and Computation, EC 2017, pp. 169–186. Association for Computing Machinery, Cambridge (2017)
6. Esfandiari, H., Hajiaghayi, M., Liaghat, V., Monemizadeh, M.: Prophet secretary. SIAM J. Disc. Math. **31**(3), 1685–1701 (2017)
7. Azar, Y., Chiplunkar, A., Kaplan, H.: Prophet secretary: surpassing the 1–1/e barrier. In: Proceedings of the 2018 ACM Conference on Economics and Computation, EC 2018, pp. 303–318. Association for Computing Machinery, Ithaca (2018)
8. Correa, J., Foncea, P., Hoeksma, R., Oosterwijk, T., Vredeveld, T.: Recent developments in prophet inequalities. SIGecom Exch. **17**(1), 61–70 (2019)
9. Correa, J., Saona, R., Ziliotto, B.: Prophet secretary through blind strategies. In: Proceedings of the Thirtieth Annual ACM-SIAM Symposium on Discrete Algorithms, SODA 2019, pp. 1946–1961. Society for Industrial and Applied Mathematics, San Diego (2019)
10. Arsenis, M., Drosis, O., Kleinberg, R.: Constrained-Order Prophet Inequalities. CoRR abs/2010.09705 (2020)

11. Correa, J.R., Cristi, A., Epstein, B., Soto, J.A.: Sample-driven optimal stopping: from the secretary problem to the i.i.d. prophet inequality. CoRR abs/2011.06516 (2020)
12. Correa, J.R., Cristi, A., Epstein, B., Soto, J.A.: The two-sided game of googol and sample-based prophet inequalities. In: SODA, pp. 2066–2081 (2020)
13. Esfandiari, H., Hajiaghayi, M., Lucier, B., Mitzenmacher, M.: Prophets, secretaries, and maximizing the probability of choosing the best. In: Chiappa, S., Calandra, R. (eds.) Proceedings of the Twenty Third International Conference on Artificial Intelligence and Statistics. Proceedings of Machine Learning Research, pp. 3717–3727. PMLR (2020)
14. Rubinstein, A., Wang, J.Z., Weinberg, S.M.: Optimal single-choice prophet inequalities from samples. In: Vidick, T. (ed.) 11th Innovations in Theoretical Computer Science Conference (ITCS 2020), Leibniz International Proceedings in Informatics (LIPIcs), 60:1–60:10. Schloss Dagstuhl-Leibniz-Zentrum fuer Informatik, Dagstuhl (2020)
15. Correa, J., Cristi, A., Feuilloley, L., Oosterwijk, T., Tsigonias-Dimitriadis, A.: The secretary problem with independent sampling. In: Proceedings of the Thirty-Second Annual ACM-SIAM Symposium on Discrete Algorithms, pp. 2047–2058. Society for Industrial and Applied Mathematics, USA 2021
16. http://www.stat.yale.edu/~yw562/teaching/itlectures.pdf

Approximate CVP in Time $2^{0.802n}$ - Now in Any Norm!

Thomas Rothvoss[1] and Moritz Venzin[2(✉)]

[1] University of Washington, Seattle, USA
`rothvoss@uw.edu`
[2] Ecole polytechnique fédérale de Lausanne (EPFL), Lausanne, Switzerland
`moritz.venzin@epfl.ch`

Abstract. We show that a constant factor approximation of the shortest and closest lattice vector problem in any norm can be computed in time $2^{0.802\,n}$. This contrasts the corresponding 2^n time, (gap)-SETH based lower bounds for these problems that even apply for small constant approximation.

For both problems, SVP and CVP, we reduce to the case of the Euclidean norm. A key technical ingredient in that reduction is a twist of Milman's construction of an M-ellipsoid which approximates any symmetric convex body K with an ellipsoid $\mathcal{E}$ so that $2^{\varepsilon n}$ translates of a constant scaling of $\mathcal{E}$ can cover K and vice versa.

Keywords: Lattice algorithms · Sieving · Integer programming

1 Introduction

For some basis $B \in \mathbb{R}^{d \times n}$, the d dimensional lattice $\mathscr{L}$ of rank n is a discrete subgroup of $\mathbb{R}^d$ given by

$$\mathscr{L}(B) = \{Bx \ : \ x \in \mathbb{Z}^n\}.$$

In this work, we will consider the *shortest vector problem* (SVP) and the *closest vector problem* (CVP), the two most important computational problems on lattices. Given some lattice, the shortest vector problem is to compute a shortest *non-zero* lattice vector. When in addition some target $t \in \mathbb{R}^d$ is given, the closest vector problem is to compute a lattice vector closest to t.

Here, "short" and "close" are defined in terms of a given norm $\|\cdot\|_K$, induced by some symmetric convex body $K \subseteq \mathbb{R}^d$ with $\mathbf{0}$ in its interior. Specifically, $\|x\|_K = \min\{s \in \mathbb{R}_{\geq 0} \mid x \in s \cdot K\}$. When we care to specify what norm we are working with, we denote these problems by SVP_K and CVP_K respectively and by SVP_p and CVP_p respectively for the important case of ℓ_p norms.

Supported by NSF CAREER grant 1651861, a David & Lucile Packard Foundation Fellowship, SNF grant (Nr. 185030) and a SNF Doc.Mobility scholarship.

K. Aardal and L. Sanitá (Eds.): IPCO 2022, LNCS 13265, pp. 440–453, 2022.
https://doi.org/10.1007/978-3-031-06901-7_33

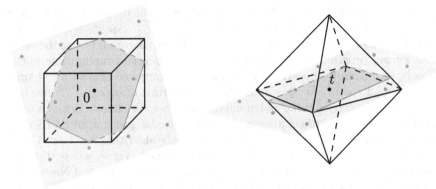

Fig. 1. The shortest and the closest vector problem.

It is important to note that the dimension, rank and span of the lattice have a considerable effect on the norm $\| \cdot \|_K$ that is *induced* on $\mathscr{L}$. For the shortest vector problem, the norm induced on $\mathscr{L}$ corresponds to K *intersected* with the span of $\mathscr{L}$, see Fig. 1 for an illustration. For different n and different rotations of the lattice, these resulting convex bodies vary considerably. However, when the convex body K is centered in the span of the lattice, these sections of K still satisfy the required properties to define a norm. In particular, up to changing the norm, any (α-approximation of the) shortest vector problem in dimension d can be directly reduced to (α-approximate) SVP with $d = n$ (from the definition of $\mathscr{L}$, we can assume that $d \geq n$ to begin with). The situation is slightly more delicate for CVP. Whenever $t \notin \mathrm{span}(\mathscr{L})$, the function measuring the distance to t, i.e. $\|t - \cdot\|_K$, can be asymmetrical on $\mathrm{span}(\mathscr{L})$, meaning it does not define a norm on $\mathrm{span}(\mathscr{L})$. This can be seen by lifting t and the cross-polytope with it in Fig. 1. However, up to a loss in the approximation guarantee, we can always take the target to lie in $\mathrm{span}(\mathscr{L})$ and consider the norm induced by K intersected with $\mathrm{span}(\mathscr{L})$. More precisely, we can always reduce $(2 \cdot \alpha + 1)$-approximate CVP to α-approximate CVP in any norm with $d = n$. This loss in the approximation factor is not surprising, seeing that exact CVP under general norms is extremely versatile. In fact, *Integer Programming* with n variables and m constraints reduces to CVP_∞ on a m-dimensional lattice of rank n.

Both SVP and CVP and their respective (approximation) algorithms have found considerable applications. These include Integer Programming [31,34], factoring polynomials over the rationals [33] and cryptanalysis [42]. On the other hand, the security of recent cryptographic schemes are based on the worst-case hardness of (approximations of) these problems [10,25,45]. In view of their importance, much attention has been devoted to understand the complexity of SVP and CVP. In [11,14,21,24,28,32,36,46], both SVP and CVP were shown to be hard to approximate to within almost polynomial factors under reasonable complexity assumptions. However, the best polynomial-time approximation algorithms only achieve exponential approximation factors [15,33,47]. This huge gap is further highlighted by the fact that these problems are in co-NP and co-AM for small polynomial factors, $\sqrt{n}$ and $\sqrt{n/\log(n)}$ respectively, [9,27,43].

The first algorithm to solve SVP and CVP in any norm and even the more general integer programming problem with an exponential running time in the rank only was given by Lenstra [34]. Kannan [31] improved this to $n^{O(n)}$ time and polynomial space[1]. To this date, the running time of order $n^{O(n)}$ remains best for algorithms only using polynomial space. It took almost 15 years until Ajtai, Kumar and Sivakumar presented a randomized algorithm for SVP_2 with time *and* space $2^{O(n)}$ and a $2^{O(1+1/\varepsilon)n}$ time and space algorithm for $(1+\varepsilon)$-CVP_2 [12,13]. Here, c-CVP is the problem of finding a lattice vector, whose distance to the target is at most c times the minimal distance. Blömer and Naewe [17] extended the randomized sieving algorithm of Ajtai et al. to solve SVP_p and CVP_p respectively in $2^{O(d)}$ time and space and $O(1+1/\varepsilon)^{2d}$ time and space respectively. For CVP_∞, using a geometric covering technique, Eisenbrand et al. [22] improved this to $O(\log(2+1/\varepsilon))^d$ time. This covering idea was adapted in [41] to all (sections of) ℓ_p norms. Their algorithm for $(1+\varepsilon)$-CVP_p requires time $2^{O(p \cdot n)}(1+1/\varepsilon)^{n/\min(2,p)}$ and is based on the current state-of-the-art, $2^{O(n)}(1+1/\varepsilon)^n$ deterministic time CVP solver for general (even asymmetric) norms from Dadush and Kun [19].

Currently, exact and singly-exponential time algorithms for CVP are only known for the ℓ_2 norm. The first such algorithm was developed by [37] and is *deterministic*. In fact, this algorithm was also the first to solve SVP_2 in deterministic time (as there is an efficient reduction from SVP to CVP, [26]) and has been instrumental to give deterministic algorithms for SVP and $(1+\varepsilon)$-CVP, [19,20]. Currently, the fastest exact algorithms for SVP_2 and CVP_2 run in time and space 2^n and are based on Discrete Gaussian Sampling [1,4,7].

Recently there has been exciting progress in understanding the *fine-grained complexity* of exact and constant approximation algorithms for SVP and CVP [2, 6,16]. Under the assumption of the *strong exponential time hypothesis (SETH)* and for $p \neq 0 \pmod 2$, exact CVP_p and SVP_∞ cannot be solved in time $2^{(1-\varepsilon)n}$. For a fixed $\varepsilon > 0$, the dimension of the lattice can be taken linear in n, i.e. $d = O_\varepsilon(n)$. Under the assumption of a *gap-version* of the strong exponential time hypothesis *(gap-SETH)* these lower bounds also hold for the approximate versions of CVP_p and SVP_∞. More precisely, in our setting these results read as follows. For each $\varepsilon > 0$ and for some norm $\|\cdot\|_K$ there exists a constant $\gamma_\varepsilon > 1$ such that there exists no $2^{(1-\varepsilon)n}$ algorithm that computes a γ_ε-approximation of SVP_K and CVP_K (where the corresponding target lies in the span of the lattice).

Until very recently, the fastest approximation algorithms for SVP_p and CVP_p did not match these lower bounds by a large margin, even for large approximation factors [5,18,39]. The only exception was SVP_2 (where no strong, fine-grained lower bound is known) where a constant factor approximation is possible in time $2^{0.802n}$ and space $2^{0.401n}$, see [8,35,38,44]. Last year, Eisenbrand and Venzin

---

[1] For the sake of readability we omit polynomials in the encoding length of the matrix B and the target vector t when stating running times and space requirements.

presented a $2^{0.802n+\varepsilon d}$ algorithm for SVP_p and CVP_p for all ℓ_p norms [23]. Their algorithm exploits a specific covering of the Euclidean (ℓ_2) norm-ball by ℓ_p norm-balls and uses the fastest (sieving) algorithm for SVP_2 as a subroutine. This approach was then further extended to yield generic, $2^{\varepsilon d}$ time reductions from SVP_q to SVP_p, $q \geq p \geq 2$ and CVP_p to CVP_q, $q \geq p$, see [3]. While this improved previous algorithms for SVP_p and CVP_p (and even constant factor approximate CVP_2), these techniques *only* apply to the very specific case of ℓ_p norms with the added restriction that the dimension d is small, further accentuating the issue of rank versus dimension. For any other norm and even for ℓ_p norms with, say, $d = \Omega(n \cdot \log(n))$, their approach yields no improvement.

In this work, we close this gap. Specifically, for any d-dimensional lattice of rank n and for any norm, we show how to solve constant factor approximate CVP and SVP in time $2^{0.802n}$.

Theorem 1. *For any lattice $\mathscr{L} \subseteq \mathbb{R}^d$ of rank n, any norm $\|\cdot\|_K$ on $\mathbb{R}^d$ and for each $\varepsilon > 0$, there exists a constant γ_ε such that a γ_ε-approximate CVP_K and SVP_K can be solved in (randomized) time $2^{(0.802+\varepsilon)n}$ and space $2^{(0.401+\varepsilon)n}$.*

We note that the constant 0.802 in the exponent can be replaced by a slightly smaller number, [30]. Thus, we indeed get the running time as advertised in the title. We use γ_ε throughout this paper. It can be assumed to be of order $1/\varepsilon^5$. For the shortest vector problem, we can significantly generalize this result.

Theorem 2. *For any lattice $\mathscr{L} \subseteq \mathbb{R}^d$ of rank n, any norm $\|\cdot\|_K$ on $\mathbb{R}^d$ and for each $\varepsilon > 0$, there exists a constant γ_ε such that there is a $2^{\varepsilon n}$ time, randomized reduction from $(\alpha \cdot \gamma_\varepsilon)$-approximate SVP_K to an oracle for α-approximate CVP_2 (or even α-approximate CVP in any norm).*

Our main idea is to cover K by a special class of ellipsoids to obtain the approximate closest vector by using an approximate closest vector algorithm with respect to ℓ_2. This covering idea draws from [23] and is also similar to the approach of Dadush et al. in [19,20]. Specifically, for any $\varepsilon > 0$, one can compute some ellipsoid $\mathcal{E}$, so that K can be covered by $2^{\varepsilon n}$ translates of $\mathcal{E}$, and, conversely, $\mathcal{E}$ can be covered by $2^{\varepsilon n}$ translates of $c_\varepsilon \cdot K$. Here, c_ε is a constant that only depends on ε. Such a covering will be sufficient to reduce approximate SVP_K to $2^{\varepsilon n}$ calls to an oracle for (approximate) CVP_2. Specifically, using an oracle for α-CVP_2, we will obtain a $O(\alpha \cdot c_\varepsilon)$-approximation to the shortest vector problem with respect to $\|\cdot\|_K$. Similarly, using these covering ideas twice, it is straight-forward to extend this approach to reduce $O(\alpha \cdot c_\varepsilon)$-approximate SVP_K to $2^{\varepsilon n}$ calls to an oracle for α-approximate CVP_Q, for any two norms $\|\cdot\|_K, \|\cdot\|_Q$. For space reasons, we defer this proof to the full version of the paper.[2] These reductions are randomized and use lattice sparsification. This covering idea can be used to solve constant factor approximate CVP_K as well. However, we can no longer assume to only have access to an oracle for the approximate closest vector problem. Instead, we will have to use one very specific property of the currently

---

[2] Available at https://arxiv.org/abs/2110.02387.

fastest, $2^{0.802n}$ time and randomized approximate algorithm for SVP_2 that is based on Ajtai et al.'s sieving approach [12]. In fact, this property is inherent in any sieving algorithm with respect to any norm. Our geometric ideas then also carry over to this general setting. Any sieving algorithm for (approximate) SVP_Q that runs in time $2^{\beta n}$ can be used to compute a γ_ε-approximate solution to CVP_K in $2^{(\beta+\varepsilon)n}$ time. This generalization is also discussed in the full version.

2   Approximate SVP in $2^{0.802n}$ Time

In this section, we describe our main geometric observation and how it leads to an algorithm for the approximate shortest and closest vector problem respectively. In a first part, we state the main geometric theorem and informally present how it leads to a reduction from approximate SVP_K to an oracle for approximate CVP_2. In the second part, we make this formal using lattice sparsification and some further geometric considerations.

2.1   Covering K with Few Ellipsoids and Vice Versa

The (approximate) shortest vector problem in the norm $\|\cdot\|_K$ can be rephrased as follows.

Does K contain a lattice point different from $\mathbf{0}$?

This follows by guessing the length ℓ of the shortest non-zero lattice vector, scaling the lattice by $1/\ell$ and then confirm the right guess by finding this lattice vector in K. By guessing we mean to try out all possibilities for ℓ of the form $(1+1/\text{poly}(n))^k$, this can be limited to a polynomial in the relevant parameters. Imagine now that one can cover K using a collection of (Euclidean) balls $\mathcal{B}_i$ of any radii such that

$$K \subseteq \bigcup_{i=1}^{N}(x_i + \mathcal{B}_i) \subseteq c \cdot K.$$

To then find a c-approximation to the shortest vector, one could use a solver for CVP_2 using targets $x_1 \cdots, x_N$. However, this naïve approach is already doomed for $K = B_\infty^n$. To achieve a constant factor approximation, $N = n^{O(n)}$ translates of (any scaling of) the Euclidean ball $B_2^n := \{x \in \mathbb{R}^n \mid \|x\|_2 \leq 1\}$ are required and can only be brought down to $N = 2^{O(n)}$ if one is willing to settle for $c = O(\sqrt{n})$ [29].

However, if we impose a second condition on these balls (and even relax the above condition), this idea will work. Specifically, if we can cover $K = -K \subseteq \mathbb{R}^n$ by N translates of $c_1 \cdot B_2^n$ and B_2^n by N translates of $c_2 \cdot K$,

$$K \subseteq \bigcup_{i=1}^{N}(x_i + c_1 \cdot B_2^n) \quad \text{and} \quad B_2^n \subseteq \bigcup_{i=1}^{N}(y_i + c_2 \cdot K),$$

we can solve $(\alpha \cdot c_1 \cdot c_2)$-approximate SVP_K using (essentially) N calls to a solver for $\alpha\text{-CVP}_2$. It turns out that such a covering is always possible for any symmetric and convex K, and, in particular, the number of translates N can be made smaller than $2^{\varepsilon n}$ for any $\varepsilon > 0$. To be precise, the covering is by ellipsoids (affine transformations of balls), but we can always apply a linear transformation to restrict to Euclidean unit norm-balls, i.e. we may take $c_1 = 1$. The precise properties of the covering are now stated in the following theorem, where we denote by $N(T, L)$ the *translative covering number* of T by L, i.e. the least number of translates of L required to cover T.

Theorem 3. *For any symmetric and convex body $K \subseteq \mathbb{R}^n$ (given by a weak separation oracle) and for any $\varepsilon > 0$, there exists an (invertible) linear transformation $T_\varepsilon : \mathbb{R}^n \to \mathbb{R}^n$ and a constant $c_\varepsilon \in \mathbb{R}_{>0}$ such that*

1. $N(T_\varepsilon(K), B_2^n) \leq 2^{\varepsilon n}$ *(even* $\mathrm{Vol}(T_\varepsilon(K) + B_2^n) \leq 2^{\varepsilon n} \cdot \mathrm{Vol}(B_2^n))$ *and*
2. $N(B_2^n, c_\varepsilon \cdot T_\varepsilon(K)) \leq 2^{\varepsilon n}$ *(even* $\mathrm{Vol}(B_2^n + c_\varepsilon \cdot T_\varepsilon(K)) \leq 2^{\varepsilon n} \cdot \mathrm{Vol}(c_\varepsilon \cdot T_\varepsilon(K)))$.

The linear transformation T_ε can be computed in (randomized) $n^{O(\log(n))}$ time.

The volume estimate $\mathrm{Vol}(T_\varepsilon(K) + B_2^n) \leq 2^{\varepsilon n} \cdot \mathrm{Vol}(B_2^n)$ in (1) is stronger than $N(T_\varepsilon(K), B_2^n) \leq 2^{\varepsilon n}$, similar in (2). It makes the covering of $T_\varepsilon(K)$ by translates of B_2^n constructive. Indeed, any point inside $T_\varepsilon(K)$ will be covered with probability at least $2^{-\varepsilon n}$ if we sample a random point within $T_\varepsilon(K) + B_2^n$ and place a copy of B_2^n around it. Repeating this for $O(n \cdot \log(n) \cdot 2^{\varepsilon n})$ iterations yields, with high probability, a full covering of $T_\varepsilon(K)$ by translates of B_2^n. See [40] for details.

We defer the proof of Theorem 3 to the full version of the paper. We now discuss how we intend to use it to solve the shortest vector problem in arbitrary norms. To do so, we first fix some notations and do some simplifications. We will denote by $\mathbf{s}$ a shortest, non-zero lattice vector of the given lattice $\mathscr{L} \subseteq \mathbb{R}^n$ with respect to $\|\cdot\|_K$, the norm under consideration. We will assume that $1 - 1/n \leq \|\mathbf{s}\|_K \leq 1$, i.e. $\mathbf{s} \in K \setminus (1 - 1/n) \cdot K$. We fix $\varepsilon > 0$, and denote by T_ε be the linear transformation that is guaranteed by Theorem 3. Up to replacing $\mathscr{L}$ by $T_\varepsilon^{-1}(\mathscr{L})$ and K by $T_\varepsilon^{-1}(K)$, we can also assume that $T_\varepsilon = \mathrm{Id}$ ($\|\cdot\|_K = \|T_\varepsilon^{-1}(\cdot)\|_{T_\varepsilon^{-1}(K)}$). We can now describe how we will use the covering guaranteed by Theorem 3. Since K is covered by translates of B_2^n, there is some translate that holds $\mathbf{s}$. We denote it by $t + B_2^n$. Now, suppose there is a procedure that either returns $\mathbf{s}$ or generates at least $2^{\varepsilon n} + 1$ *distinct* lattice vectors lying in $t + \alpha \cdot B_2^n$. For the latter case, while these vectors may all have very large norm with respect to $\|\cdot\|_K$ or may even equal the zero vector, by taking pairwise differences, we are still able to find a $O(\alpha \cdot c_\varepsilon)$-approximation to the shortest vector. Indeed, by property (2) of Theorem 3, $t + \alpha \cdot B_2^n$ can be covered by fewer than $2^{\varepsilon n}$ translates of $(\alpha \cdot c_\varepsilon) \cdot K$. Thus, one translate of $(\alpha \cdot c_\varepsilon) \cdot K$ must hold *two distinct* lattice vectors. Their pairwise difference is then a $(2 \cdot (1 - 1/n)^{-1} \cdot \alpha \cdot c_\varepsilon)$-approximation to the shortest vector $\mathbf{s}$. This is depicted in Fig. 2.

To finish the argument, it remains to argue that such a procedure can be simulated with an oracle for α-approximate CVP_2 at hand. This will be achieved through the use of *lattice sparsification*. This technique will allow us to delete lattice points in an almost uniform manner. Specifically, when the oracle has

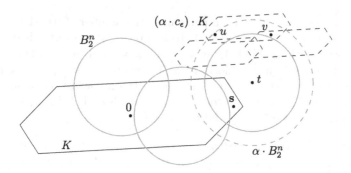

Fig. 2. The covering of K by translates of the Euclidean norm ball. One translate of $(\alpha \cdot c_\epsilon) \cdot K$ covering $t + \alpha \cdot B_2^n$ holds two lattice vectors u and v. Their difference $u - v$ is a $(2 \cdot (1 - 1/n)^{-1} \cdot \alpha \cdot c_\epsilon)$-approximation to the shortest vector problem.

already returned lattice vectors $v_1, \cdots, v_N$, sparsifying the lattice ensures that with sufficiently high probability, s is retained and $v_1, \cdots, v_N$ are deleted. This then *forces* the α-approximate CVP_2 oracle to return a lattice vector v_{N+1} distinct from $v_1, \cdots, v_N$ with $\|t - v_{N+1}\|_K \leq \alpha \cdot \|t - s\|_K$.

2.2   Approximate SVP_K Using an Oracle for Approximate CVP_2

In this subsection we are going to formalize the exponential time reduction from approximate SVP to an oracle for (approximate) CVP_2 as outlined in the previous section. We are going to make use of the following theorem from [48], slightly rephrased for our purpose.

Theorem 4. *For any prime p, any lattice $\mathscr{L} \subseteq \mathbb{R}^d$ of rank n and lattice vectors $w, v_1, v_2, \cdots, v_N \in \mathscr{L}$ with $v_i \notin w + p \cdot \mathscr{L} \setminus \{\mathbf{0}\}$, one can, in polynomial time, sample a shifted sub-lattice $u + \mathscr{L}'$ with $\mathscr{L}' \subseteq \mathscr{L}$ and $u \in \mathscr{L}$ such that:*

$$\Pr\left[w \in u + \mathscr{L}' \text{ and } v_1, \cdots, v_N \notin u + \mathscr{L}'\right] \geq \frac{1}{p} - \frac{N}{p^2} - \frac{N}{p^n}.$$

The condition $v_i \notin w + p \cdot \mathscr{L} \setminus \{\mathbf{0}\}$ is slightly inconvenient. We will deal with these type of vectors by showing that, for large enough p, they will be too large and will not be considered by our α-approximate CVP_2 oracle. This is done by the following lemma.

Lemma 1. *Let $\mathscr{L}$ be a lattice, let K be a symmetric and convex body containing no lattice vector other than $\mathbf{0}$ in its interior and suppose that $N(K, B_2^n), N(B_2^n, \beta \cdot K) \leq 2^{\varepsilon n}$. Then, the following three properties hold:*

1. $\forall v \in p \cdot \mathscr{L} \setminus \{\mathbf{0}\} \, (p \in \mathbb{N}) : \|v\|_2 \geq (2^{-\varepsilon n}/\beta) \cdot p$.
2. $K \subseteq 2^{\varepsilon n} \cdot B_2^n$
3. $(2^{-\varepsilon n}/\beta) \cdot B_2^n \subseteq K$

Proof. We first show the last two properties by using the translative covering numbers. Since $N(K, B_2^n) \le 2^{\varepsilon n}$, we must have $K \subseteq 2^{\varepsilon n} \cdot B_2^n$. To see this, we note that otherwise, the largest segment contained in K cannot be covered using $2^{\varepsilon n}$ translates of B_2^n. Conversely, since $N(B_2^n, \beta \cdot K) \le 2^{\varepsilon n}$ and the convexity of $K = -K$, the smallest inradius of K cannot be smaller than $2^{-\varepsilon n}/\beta$. It follows that $(2^{-\varepsilon n}/\beta) \cdot B_2^n \subseteq K$.

We can now show the first property. Let $v \in p \cdot \mathscr{L} \setminus \{0\}$. By assumption $\frac{v}{p} \notin \text{int}(K)$ and so $\frac{v}{p} \notin (2^{-\varepsilon n}/\beta) \cdot \text{int}(B_2^n)$, where $\text{int}(K)$ denotes the interior of the body K. This means that $\|v\|_2 \ge (2^{-\varepsilon n}/\beta) \cdot p$.

We now state our randomized reduction. We rotate the lattice and K so that $\text{span}(\mathscr{L}) = \mathbb{R}^n \times \{0\}^{d-n}$. Up to replacing K by $K \cap \text{span}(\mathscr{L})$ and deleting the last $d - n$ zeros, we may assume that $d = n$. For details, we refer to the second part of the proof of Lemma 2. We fix some $\varepsilon > 0$ and compute the invertible linear transformation T_ε guaranteed by Theorem 3. Up to applying the inverse of T_ε and scaling the lattice, we may assume that $T_\varepsilon = \text{Id}$ and $1 - 1/n \le \|\mathbf{s}\|_K \le 1$, where $\mathbf{s}$ is a shortest lattice vector with respect to the norm $\|\cdot\|_K$. One iteration of the reduction consists of the following steps and succeeds with probability $2^{-O(\varepsilon)n}$.

(1) Fix a prime number p between $2^{3\varepsilon n}$ and $2 \cdot 2^{3\varepsilon n}$.
(2) Sample a random point $t \in K + B_2^n$.
(3) Using the number p, sparsify the lattice as in Theorem 4. Denote the resulting lattice by $u + \mathscr{L}'$.
(4) Run the oracle for α-approximate CVP_2 for $\mathscr{L}'$ with target $t - u$, add u to the vector returned and store it.
(5) Repeat steps (3) and (4) $n^2 \cdot 2^{5\varepsilon n}$ times. Among the resulting lattice vectors and their pairwise differences, output the shortest (non-zero) with respect to $\|\cdot\|_K$.

Theorem 5. *Let $\mathscr{L} \subseteq \mathbb{R}^d$ be any lattice of rank n. For any $\varepsilon > 0$, there is a constant γ_ε, such that there is a randomized, $2^{\varepsilon n}$ time reduction from $(\alpha \cdot \gamma_\varepsilon)$-approximate SVP_K for $\mathscr{L}$ to an oracle for α-approximate CVP_2 for n-dimensional lattices.*

Proof. Let us already condition on the event that the sampled point t verifies $\mathbf{s} \in t + B_2^n$. This happens with probability at least $2^{-\varepsilon n}$ by property (1) of Theorem 3. This probability can be boosted to $1 - 2^{-n}$ by repeating steps (2) to (5) $O(n \cdot 2^{\varepsilon n})$ times and outputting the shortest non-zero vector with respect to $\|\cdot\|_K$.

We are now going to show that, with high probability, $\mathbf{s}$ is going to be retained in the shifted sub-lattice and the lattice vector that is returned in that iteration is different from the ones returned from a previous iteration (where also $\mathbf{s}$ was retained). To this end, denote by $L := \{v_1, \cdots, v_j\}$ the (possibly empty) list of lattice vectors that were obtained in an iteration where $\mathbf{s}$ belonged in the corresponding (shifted) sub-lattice. This implies that $\|t - v_i\|_2 \le \alpha \cdot \|t - \mathbf{s}\|_2$, and, by the triangle inequality and Lemma 1, $\|v_i\|_2 \le (2 \cdot \alpha) \cdot 2^{\varepsilon n}$ for all $i \in \{1, \cdots, j\}$. On the other hand, by slightly rescaling K in Lemma 1, the triangle inequality and our choice of p, any vector in $\mathbf{s} + p \cdot \mathscr{L} \setminus \{\mathbf{0}\}$ is larger than $2^{2\varepsilon n}/c_\varepsilon$ in the Euclidean norm. It follows that

$$v_1, \cdots, v_j \notin \mathbf{s} + p \cdot \mathscr{L}$$

(We are assuming that $\alpha = 2^{o(n)}$, this is without loss of generality. For $\alpha = 2^{n \log \log(n)/\log(n)}$, Babai's algorithm runs in polynomial time, [15].)

Thus, provided $j \leq 2^{\varepsilon n}$ and by Theorem 4, in any iteration of the algorithm and with probability at least $2^{-3\varepsilon n}/2$, we add a lattice vector to the list L that is either distinct from all other vectors in L or that equals $\mathbf{s}$. In other words, the number of distinct lattice vectors in our list follows a binomial distribution with parameter $2^{-3\varepsilon n}/2$. Since we repeat this $n^2 \cdot 2^{5\varepsilon n}$ times, by Chernov's inequality and with probability at least $1 - 2^{-n}$, after the final iteration the list L contains at least $2^{\varepsilon n} + 1$ distinct lattice vectors lying within $t + \alpha \cdot B_2^n$ or contains $\mathbf{s}$. In the latter case we are done. In the former case, since $N(\alpha \cdot B_2^n, (\alpha \cdot c_\varepsilon) \cdot K) \leq 2^{\varepsilon n}$, by checking all $2^{10\varepsilon n}$ pairwise differences, we will find a non-zero lattice vector with $\| \cdot \|_K$-norm at most $\alpha \cdot (2 \cdot c_\varepsilon)$. Since $\|\mathbf{s}\|_K \geq 1 - 1/n$ and setting $\gamma_\varepsilon := (1 - 1/n)^{-1} \cdot 2 \cdot c_\varepsilon$, this vector is a $(\alpha \cdot \gamma_\varepsilon)$-approximation to $\mathbf{s}$.

Remark 1. We can bring down the space requirement of this reduction to polynomial space. Secretly sample two numbers i, j from $\{1, 2, \cdots, n^2 \cdot 2^{5\varepsilon n}\}$ and only store the lattice vectors returned from the ith and j^{th} iteration of step (4). Since this is independent from the success of the reduction, the overall probability of success is still $2^{-\Omega(\varepsilon)n}$. The reason for this complication is that we must assume that the CVP_2 oracle is malicious and returns lattice vectors that, dependent on previous inquiries, help us the least.

3   Approximate CVP in Time $2^{0.802n}$

In this section we are going to describe a $2^{0.802n}$ time algorithm for a constant factor approximation to the closest vector problem in any norm. Specifically, for any d-dimensional lattice $\mathscr{L}$ of rank n, target t, any norm $\| \cdot \|_K$ and any $\varepsilon > 0$, we show how to approximate the closest vector problem to within a constant factor γ_ε in time $2^{(0.802+\varepsilon)n}$. The space requirement is of order $2^{(0.401+\varepsilon)n}$.

To achieve this, we will adapt the geometric ideas as outlined in the previous subsection to the setting of the closest vector problem. To do so, we are first going to describe how to restrict to the full-dimensional case.

Lemma 2. *Consider an instance of the closest vector problem, $\text{CVP}_K(\mathscr{L}, t)$, $\mathscr{L} \subseteq \mathbb{R}^d$ of rank n. In polynomial time, we can find a lattice $\tilde{\mathscr{L}} \subseteq \mathbb{R}^n$ of rank and dimension n, target $\tilde{t} \in \mathbb{R}^n$ and norm $\| \cdot \|_{\tilde{K}}$ so that an α-approximation to $\text{CVP}_{\tilde{K}}$ on $\tilde{\mathscr{L}}$ with target $\tilde{t}$ can be efficiently transformed in a $(2\alpha + 1)$-approximation to $\text{CVP}_K(\mathscr{L}, t)$. Whenever $t \in \text{span}(\mathscr{L})$, the latter is a α-approximation to $\text{CVP}_K(\mathscr{L}, t)$.*

Proof. Let us define $t' := \operatorname{argmin}\{\|t - x\|_K, x \in \operatorname{span}(\mathscr{L})\}$. Such a point may not be unique, for instance for $K = B_1^n$ or $K = B_\infty^n$, but it suffices to consider any point t' realizing this minimum or an approximation thereof. Given a (weak) separation oracle for K, this can be computed in polynomial time. We now show that an α-approximation to $\operatorname{CVP}_K(\mathscr{L}, t')$ yields a $(2 \cdot \alpha + 1)$ approximation to $\operatorname{CVP}_K(\mathscr{L}, t)$. Indeed, since $\|t - t'\|_K$ is smaller than $\operatorname{dist}_K(t, \mathscr{L})$, the distance of t to the closest lattice vector, we have that

$$\operatorname{dist}_K(t', \mathscr{L}) \le 2 \cdot \operatorname{dist}_K(t, \mathscr{L}).$$

Denote by $c_\alpha \in \mathscr{L}$ an α-approximation to the closest lattice vector to t'. By the triangle inequality,

$$\|t - c_\alpha\|_K \le \|t - t'\|_K + \|t' - c_\alpha\|_K \le \operatorname{dist}(t, \mathscr{L}) + \alpha \cdot \operatorname{dist}(t', \mathscr{L}) \le (2 \cdot \alpha + 1) \cdot \operatorname{dist}(t, \mathscr{L}).$$

This means that an α-approximation to the closest vector to t' is a $(2 \cdot \alpha + 1)$ approximation to the closest vector to t. Now that $t' \in \operatorname{span}(\mathscr{L})$, we can restrict to the case $d = n$.

Let $\mathcal{O}_n$ be a linear transformation that first applies a rotation sending $\operatorname{span}(\mathscr{L})$ to $\mathbb{R}^n \times \{0\}^{d-n}$ and then restricts onto its first n coordinates. The transformation $\mathcal{O}_n : \operatorname{span}(\mathscr{L}) \to \mathbb{R}^n$ is invertible. The n-dimensional instance of the closest vector problem is then obtained by setting $\tilde{\mathscr{L}} \leftarrow \mathcal{O}_n(\mathscr{L})$, $\tilde{t} \leftarrow \mathcal{O}_n(t')$ and $\| \cdot \|_{\tilde{K}}$ where $\tilde{K} \leftarrow \mathcal{O}_n(K)$. Whenever c_α is an α-approximation to $\operatorname{CVP}_{\tilde{K}}(\tilde{\mathscr{L}}, \tilde{t})$, $\mathcal{O}_n^{-1}(c_\alpha)$ is a $(2 \cdot \alpha + 1)$-approximation to $\operatorname{CVP}_K(\mathscr{L}, t)$ (or an α-approximation to $\operatorname{CVP}_K(\mathscr{L}, t)$, if $t \in \operatorname{span}(\mathscr{L})$).

In our algorithm, we are going to make use of the main subroutine of [23]. We note that their subroutine is implicit in all sieving algorithms for SVP_2 and was first described by [38,44]. For convenience, we slightly restate it in the following form.

Theorem 6. *Given $\varepsilon > 0$, $R > 0$, $N \in \mathbb{N}$ and a lattice $\mathscr{L} \subseteq \mathbb{R}^d$ of rank n, there is a randomized procedure that produces independent samples $v_1, \cdots, v_N \sim \mathcal{D}$, where the distribution $\mathcal{D}$ satisfies the following two properties:*

1. *Every sample $v \sim \mathcal{D}$ has $v \in \mathscr{L}$ and $\|v\|_2 \le a_\varepsilon \cdot R$, where a_ε is a constant only depending on ε.*
2. *For any $\mathbf{s} \in \mathscr{L}$ with $\|\mathbf{s}\|_2 \le R$, there are distributions $\mathcal{D}_0^{\mathbf{s}}$ and $\mathcal{D}_1^{\mathbf{s}}$ and some parameter $\rho_{\mathbf{s}}$ with $2^{-\varepsilon n} \le \rho_{\mathbf{s}} \le 1$ such that the distribution $\mathcal{D}$ is equivalent to the following process:*
   (a) With probability $\rho_{\mathbf{s}}$, sample $u \sim \mathcal{D}_0^{\mathbf{s}}$. Then, flip a fair coin and with probability $1/2$, return u, otherwise return $u + \mathbf{s}$.
   (b) With probability $1 - \rho_{\mathbf{s}}$, sample $u \sim \mathcal{D}_1^{\mathbf{s}}$.

This procedure takes time $2^{(0.802+\varepsilon)n} + N \cdot 2^{(0.401+\varepsilon)n}$ and requires $N + 2^{(0.401+\varepsilon)n}$ space and succeeds with probability at least $1/2$.

With this randomized procedure, we obtain our main result.

Theorem 7. *For any $\varepsilon > 0$ and lattice $\mathcal{L}$ of rank n, we can solve the γ_ε-approximate closest vector problem for any norm $\|\cdot\|_K$ in (randomized) time $2^{(0.802+\varepsilon)n}$ and space $2^{(0.401+\varepsilon)n}$.*

Proof. Using the reduction given by Lemma 2, we may assume that $\mathcal{L}$ is of full rank, i.e. $\mathcal{L} \subseteq \mathbb{R}^n$. We denote by $\mathbf{c}$ the closest lattice vector with respect to $\|\cdot\|_K$ to t. Up to scaling and applying the inverse of the linear transformation given by Theorem 3 to $\mathcal{L}$, K and t, we may assume that $1-1/n \leq \|\mathbf{c}-t\|_K \leq 1$, $\mathrm{Vol}(K + B_2^n) \leq 2^{\varepsilon n} \cdot \mathrm{Vol}(B_2^n)$ and $N(B_2^n, c_\varepsilon \cdot K) \leq 2^{\varepsilon n}$ for the constant $c_\varepsilon > 0$.

We now sample a uniformly random point $\tilde{t}$ within $t + K + B_2^n$. With probability at least $2^{-\varepsilon n}$, $\mathbf{c} \in \tilde{t} + B_2^n$. For the remainder of the proof, we condition on $\mathbf{c} \in \tilde{t} + B_2^n$.

We now use Kannan's embedding technique [31] and define a new lattice $\mathcal{L}' \subseteq \mathbb{R}^{n+1}$ of rank $n + 1$ with the following basis:

$$\tilde{B} = \begin{pmatrix} B & \tilde{t} \\ 0 & 1/n \end{pmatrix} \in \mathbb{Q}^{(n+1) \times (n+1)}.$$

Finding a (α-approximate) closest lattice vector to $\tilde{t}$ in $\mathcal{L}$ is equivalent to finding a (α-approximate) shortest lattice vector in $\mathcal{L}' \cap \{x \in \mathbb{R}^{n+1} \mid x_{n+1} = 1/n\}$. The vector $\mathbf{s} := (\tilde{t} - \mathbf{c}, 1/n)$ is such a vector (although not necessarily shortest), its Euclidean length is at most $1 + 1/n$.

Now, consider the $n + 1$-dimensional scaled Euclidean ball $((1 + 1/n) \cdot a_\varepsilon) \cdot B_2^{n+1}$. Here, a_ε is the constant from Theorem 6. Each of its n-dimensional layers of the form $((1 + 1/n) \cdot a_\varepsilon) \cdot B_2^{n+1} \cap \{x \in \mathbb{R}^{n+1} \mid x_{n+1} = k/n\}$ for $k \in \mathbb{Z}$ can be covered by at most $2^{\varepsilon n}$ translates of $((1+1/n) \cdot a_\varepsilon \cdot c_\varepsilon) \cdot K \times \{0\}$. It follows that all lattice vectors inside $a_\varepsilon \cdot (1 + 1/n) \cdot B_2^{n+1} \cap \mathcal{L}'$ can be covered by at most $(2 \cdot n \cdot a_\varepsilon + 1) \cdot 2^{\varepsilon n}$ translates of $((1 + 1/n) \cdot a_\varepsilon \cdot c_\varepsilon) \cdot K \times \{0\}$.

We now use the procedure from Theorem 6 with $R := 1 + 1/n$ and sample $N := 2$ lattice vectors from $a_\varepsilon \cdot R \cdot B_2^{n+1} \cap \mathcal{L}'$. With probability at least $\frac{1}{2} \cdot \rho_{\mathbf{s}}^2 \geq \frac{1}{2} \cdot 2^{-2\varepsilon n}$ this succeeds and both lattice vectors are generated according to (2a). Let us condition on this event. That means the sampled lattice vectors are of the form $v_1 + \sigma_1 \mathbf{s}$ and $v_2 + \sigma_2 \mathbf{s}$ where $v_1, v_2, \sigma_1, \sigma_2$ are independently distributed with $v_1, v_2 \sim \mathcal{D}_0^{\mathbf{s}}$ and $\sigma_1, \sigma_2 \sim \{0, 1\}$ uniformly. Since v_1 and v_2 are i.i.d., with probability at least $(2 \cdot n \cdot a_\varepsilon + 1)^{-2} \cdot 2^{-2\varepsilon n}$, there must be one translate of $((1 + 1/n) \cdot a_\varepsilon \cdot c_\varepsilon) \cdot K \times \{0\}$ that contains both v_1 and v_2. Put differently and using $K = -K$,

$$v_1 - v_2 \in (2 \cdot (1 + 1/n) \cdot a_\varepsilon \cdot c_\varepsilon) \cdot K \times \{0\}.$$

Next, we decide the independent coin flips and with probability of $1/4$ we have $\sigma_1 = 1$ and $\sigma_2 = 0$. We condition on this event. Then the difference of the sampled lattice vectors is

$$(v_1 + \mathbf{s}) - v_2 \in (2 \cdot (1 + 1/n) \cdot a_\varepsilon \cdot c_\varepsilon) \cdot K \times \{0\} + \mathbf{s}.$$

We can rewrite it as

$$(v_1 + \mathbf{s}) - v_2 = \begin{pmatrix} u \\ 0 \end{pmatrix} + \mathbf{s} = \begin{pmatrix} u - \mathbf{c} \\ 0 \end{pmatrix} + \begin{pmatrix} \tilde{t} \\ 1/n \end{pmatrix},$$

where $u \in \mathcal{L}$ and $\|u\|_K \leq (2 \cdot (1 + 1/n) \cdot a_\varepsilon \cdot c_\varepsilon)$. The vector $u - \mathbf{c}$ can be found by adding $-\tilde{t}$ to the first n coordinates of $(v_1 + \boldsymbol{s}) - v_2$. Then $\mathbf{c} - u$ will be our approximation to the closest lattice vector. Indeed, by the triangle inequality,

$$\|t - (\mathbf{c} - u)\|_K \leq \|t - \mathbf{c}\|_K + \|u\|_K \leq 1 + (2 \cdot (1 + 1/n) \cdot c_\varepsilon \cdot a_\varepsilon) := \beta_\varepsilon.$$

We set $\gamma_\varepsilon := (1 - 1/n)^{-1} \cdot (2 \cdot \beta_\varepsilon + 1)$; recall that we have used Lemma 2 and scaled the lattice so that $\|t - \mathbf{c}\|_K \geq 1 - 1/n$. The lattice vector $\mathbf{c} - u$ is a γ_ε-approximation to the closest lattice vector to t.

To boost the probability of success from $2^{-\Omega(\varepsilon)n}$ to $1 - 2^{-n}$, we can repeat the steps starting from where we defined $\mathcal{L}'$ $2^{-\Omega(\varepsilon)n}$ times and only store the currently closest lattice vector to t. Finally, to ensure that a $\tilde{t}$ with $\mathbf{c} \in \tilde{t} + B_2^n$ is found, we repeat the whole procedure starting from (2) $O(n \cdot 2^{\varepsilon n})$ many times. This boosts the overall success probability to $1 - 2^{-n}$ and yields a total running time of $2^{(0.802 + O(\varepsilon))n}$ and space $2^{(0.401 + O(\varepsilon))n}$.

References

1. Aggarwal, D., Dadush, D., Stephens-Davidowitz, N.: Solving the closest vector problem in 2^n time - the discrete gaussian strikes again! In: 2015 IEEE 56th Annual Symposium on Foundations of Computer Science, pp. 563–582 (2015)
2. Aggarwal, D., Bennett, H., Golovnev, A., Stephens-Davidowitz, N.: Fine-grained hardness of CVP(P)–everything that we can prove (and nothing else). In: Proceedings of the 2021 ACM-SIAM Symposium on Discrete Algorithms (SODA) (2021)
3. Aggarwal, D., Chen, Y., Kumar, R., Li, Z., Stephens-Davidowitz, N.: Dimension-preserving reductions between SVP and CVP in different p-norms. In: Proceedings of the 2021 ACM-SIAM Symposium on Discrete Algorithms (SODA), pp. 2444–2462 (2021)
4. Aggarwal, D., Dadush, D., Regev, O., Stephens-Davidowitz, N.: Solving the shortest vector problem in 2^n time using discrete gaussian sampling. In: Proceedings of the Forty-Seventh Annual ACM Symposium on Theory of Computing, pp. 733–742 (2015)
5. Aggarwal, D., Mukhopadhyay, P.: Improved algorithms for the shortest vector problem and the closest vector problem in the infinity norm. In: 29th International Symposium on Algorithms and Computation (ISAAC 2018), pp. 35:1–35:13 (2018)
6. Aggarwal, D., Stephens-Davidowitz, N.: (Gap/S)ETH hardness of SVP. In: Proceedings of the 50th Annual ACM SIGACT Symposium on Theory of Computing, STOC 2018, pp. 228–238. Association for Computing Machinery (2018)
7. Aggarwal, D., Stephens-Davidowitz, N.: Just take the average! an embarrassingly simple 2^n-time algorithm for SVP (and CVP). In: 1st Symposium on Simplicity in Algorithms, SOSA 2018, pp. 12:1–12:19 (2018)
8. Aggarwal, D., Ursu, B., Vaudenay, S.: Faster sieving algorithm for approximate SVP with constant approximation factors. Cryptology ePrint Archive, Report 2019/1028 (2019). https://eprint.iacr.org/2019/1028
9. Aharonov, D., Regev, O.: Lattice problems in NP ∩ coNP. J. ACM **52**(5), 749–765 (2005)

10. Ajtai, M.: Generating hard instances of lattice problems (extended abstract). In: Proceedings of the Twenty-Eighth Annual ACM Symposium on Theory of Computing, STOC 1996, pp. 99–108 (1996)

11. Ajtai, M.: The shortest vector problem in ℓ_2 is NP-hard for randomized reductions (extended abstract). In: Proceedings of the Thirtieth Annual ACM Symposium on Theory of Computing, STOC 1998, pp. 10–19 (1998)

12. Ajtai, M., Kumar, R., Sivakumar, D.: A sieve algorithm for the shortest lattice vector problem. In: Proceedings on 33rd Annual ACM Symposium on Theory of Computing, pp. 601–610 (2001)

13. Ajtai, M., Kumar, R., Sivakumar, D.: Sampling short lattice vectors and the closest lattice vector problem. In: Proceedings of the 17th Annual IEEE Conference on Computational Complexity 2002, pp. 53–57 (2002)

14. Arora, S.: Probabilistic Checking of Proofs and Hardness of Approximation Problems. Ph.D. thesis, University of California at Berkeley, Berkeley, CA, USA (1995)

15. Babai, L.: On lovász' lattice reduction and the nearest lattice point problem. Combinatorica **6**, 1–13 (1986)

16. Bennett, H., Golovnev, A., Stephens-Davidowitz, N.: On the quantitative hardness of CVP. In: 2017 IEEE 58th Annual Symposium on Foundations of Computer Science (FOCS), pp. 13–24 (2017)

17. Blömer, J., Naewe, S.: Sampling methods for shortest vectors, closest vectors and successive minima. Theor. Comput. Sci. **410**(18), 1648–1665 (2009)

18. Dadush, D.: A $O(1/\varepsilon^2)^n$ time sieving algorithm for approximate integer programming. In: LATIN 2012: Theoretical Informatics - 10th Latin American Symposium (2012), pp. 207–218 (2012)

19. Dadush, D., Kun, G.: Lattice sparsification and the approximate closest vector problem. Theory Comput. **12**(1), 1–34 (2016)

20. Dadush, D., Peikert, C., Vempala, S.S.: Enumerative lattice algorithms in any norm via m-ellipsoid coverings. In: IEEE 52nd Annual Symposium on Foundations of Computer Science, FOCS 2011, pp. 580–589. IEEE Computer Society (2011)

21. Dinur, I., Kindler, G., Raz, R., Safra, S.: Approximating CVP to within almost-polynomial factors is NP-hard. Combinatorica **23**(2), 205–243 (2003)

22. Eisenbrand, F., Hähnle, N., Niemeier, M.: Covering cubes and the closest vector problem. In: Proceedings of the 27th ACM Symposium on Computational Geometry, 2011, pp. 417–423 (2011)

23. Eisenbrand, F., Venzin, M.: Approximate CVP_p in time $2^{0.802n}$. J. Comput. Syst. Sci. **124**, 129–139 (2022)

24. van Emde Boas, P.: Another NP-complete problem and the complexity of computing short vectors in a lattice. Technical Report 81–04, Mathematische Instituut, University of Amsterdam (1981)

25. Gentry, C.: Fully homomorphic encryption using ideal lattices. In: Proceedings of the Forty-First Annual ACM Symposium on Theory of Computing, STOC 2009, pp. 169–178 (2009)

26. Goldreich, O., Micciancio, D., Safra, S., Seifert, J.P.: Approximating shortest lattice vectors is not harder than approximating closest lattice vectors. Inf. Process. Lett. **71**(2), 55–61 (1999)

27. Goldreich, O., Goldwasser, S.: On the limits of nonapproximability of lattice problems. J. Comput. Syst. Sci. **60**(3), 540–563 (2000)

28. Haviv, I., Regev, O.: Tensor-based hardness of the shortest vector problem to within almost polynomial factors. In: Proceedings of the Thirty-Ninth Annual ACM Symposium on Theory of Computing, pp. 469–477 (2007)

29. John, F.: Extremum problems with inequalities as subsidiary conditions. In: Studies and Essays Presented to R. Courant on his 60th Birthday, 8 January 1948, pp. 187–204. Interscience Publishers, Inc., New York (1948)

30. Kabatiansky, G.A., Levenshtein, V.I.: On bounds for packings on a sphere and in space. Problemy Peredachi Informatsii **14**(1), 3–25 (1978)

31. Kannan, R.: Minkowski's convex body theorem and integer programming. Math. Oper. Res. **12**(3), 415–440 (1987)

32. Khot, S.: Hardness of approximating the shortest vector problem in lattices. J. ACM **52**(5), 789–808 (2005)

33. Lenstra, A.K., Lenstra, H.W., Lovász, L.: Factoring polynomials with rational coefficients. Mathematische Annalen **261**(4), 515–534 (1982)

34. Lenstra, H.W.: Integer programming with a fixed number of variables. Math. Oper. Res. **8**(4), 538–548 (1983)

35. Liu, M., Wang, X., Xu, G., Zheng, X.: Shortest lattice vectors in the presence of gaps. IACR Cryptol. ePrint Arch. **2011**, 139 (2011)

36. Micciancio, D.: The shortest vector in a lattice is hard to approximate to within some constant. SIAM J. Comput. **30**(6), 2008–2035 (2001)

37. Micciancio, D., Voulgaris, P.: A deterministic single exponential time algorithm for most lattice problems based on voronoi cell computations. In: Proceedings of the 42nd ACM Symposium on Theory of Computing, STOC 2010, pp. 351–358 (2010)

38. Micciancio, D., Voulgaris, P.: Faster exponential time algorithms for the shortest vector problem. In: Proceedings of the Twenty-First Annual ACM-SIAM Symposium on Discrete Algorithms, SODA 2010, USA, pp. 1468–1480 (2010)

39. Mukhopadhyay, P.: Faster provable sieving algorithms for the shortest vector problem and the closest vector problem on lattices in p norm. Algorithms **14**(12), 362 (2021)

40. Naszódi, M.: On some covering problems in geometry. Proc. Am. Math. Soc. **144**, 3555–3562 (2014)

41. Naszódi, M., Venzin, M.: Covering convex bodies and the closest vector problem. arXiv preprint arXiv:1908.08384 (2019)

42. Odlyzko, A.M.: The rise and fall of knapsack cryptosystems. In: Cryptology and Computational Number Theory, pp. 75–88. A.M.S (1990)

43. Peikert, C.: Limits on the hardness of lattice problems in l_p norms. Comput. Complex. **17**(2), 300–351 (2008)

44. Pujol, X., Stehlé, D.: Solving the shortest lattice vector problem in time $2^{2.465n}$. IACR Cryptol. ePrint Arch. **2009**, 605 (2009)

45. Regev, O.: On lattices, learning with errors, random linear codes, and cryptography. J. ACM **56**(6), 1–40 (2009)

46. Regev, O., Rosen, R.: Lattice problems and norm embeddings. In: Proceedings of the Thirty-Eighth Annual ACM Symposium on Theory of Computing, STOC 2006, pp. 447–456 (2006)

47. Schnorr, C.P.: A hierarchy of polynomial time lattice basis reduction algorithms. Theor. Comput. Sci. **53**(2–3), 201–224 (1987)

48. Stephens-Davidowitz, N.: Discrete gaussian sampling reduces to CVP and SVP. In: Proceedings of the Twenty-Seventh Annual ACM-SIAM Symposium on Discrete Algorithms, SODA 2016, pp. 1748–1764. SIAM (2016)

Author Index

Printed in the United States
by Baker & Taylor Publisher Services